Switching Power Supply Design & Optimization

Sanjaya Maniktala

Second Edition

New York Chicago San Francisco
Athens London Madrid
Mexico City Milan New Delhi
Singapore Sydney Toronto

Copyright © 2014, 2005 by McGraw-Hill Education. All rights reserved. Printed in the United States of America. Except as permitted under the United States Copyright Act of 1976, no part of this publication may be reproduced or distributed in any form or by any means, or stored in a data base or retrieval system, without the prior written permission of the publisher.

1 2 3 4 5 6 7 8 9 0   DOW/DOW   1 2 0 9 8 7 6 5 4

ISBN 978-0-07-179814-3
MHID 0-07-179814-5

Sponsoring Editor
Michael McCabe

Editing Supervisor
Stephen M. Smith

Production Supervisor
Pamela A. Pelton

Acquisitions Coordinator
Bridget L. Thoreson

Project Manager
Yashmita Hota, Cenveo® Publisher Services

Copy Editors
Patti Scott and Lucy Mullins

Proofreader
Eina Malik

Indexer
Robert Swanson

Art Director, Cover
Jeff Weeks

Composition
Cenveo Publisher Services

Printed and bound by RR Donnelley.

McGraw-Hill Education books are available at special quantity discounts to use as premiums and sales promotions or for use in corporate training programs. To contact a representative, please visit the Contact Us page at www.mhprofessional.com.

This book is printed on acid-free paper.

Information contained in this work has been obtained by McGraw-Hill Education from sources believed to be reliable. However, neither McGraw-Hill Education nor its authors guarantee the accuracy or completeness of any information published herein, and neither McGraw-Hill Education nor its authors shall be responsible for any errors, omissions, or damages arising out of use of this information. This work is published with the understanding that McGraw-Hill Education and its authors are supplying information but are not attempting to render engineering or other professional services. If such services are required, the assistance of an appropriate professional should be sought.

Contents

Preface xiii
Acknowledgments xxi

1 References to Regulators 1

PART 1 OVERVIEW

Walking a Design Tightrope 1
A Creative Experience 2
Static and Dynamic Regulation 3

PART 2 UNDERSTANDING VOLTAGE REFERENCES

Voltage References in General 5
The Bandgap Reference 6
Understanding the BJT Better (the PTAT and CTAT) 7
The Principle behind the Bandgap Reference 13
    The Basic BJT Current Mirror and Widlar's Bandgap
        Reference Cell 13
    A Modified Current Mirror and the CMOS Bandgap
        Reference Cell 17
    Simulation or the Lab? 18
    Comparing Voltage References 19
    Specifying and Interpreting the Tempco of Voltage
        References 20
    Understanding Bit Accuracy 22

PART 3 DESIGNING THE VOLTAGE DIVIDER

Sources of Output Error: Voltage Divider Input Bias Current . 23
Sources of Output Error: Voltage Divider Resistor Tolerance . 25
Sources of Output Error: Commercial Resistor Values 26
Voltage Divider: Constraints Imposed by Type of Error Amplifier ... 30
Voltage Divider: Correct PCB Placement 31

2 DC-DC Converters: Topologies to Configurations 33

PART 1 INTRODUCTION TO THE PRINCIPLES OF SWITCHING POWER CONVERTERS

Watch Out for the Not So Obvious 33
What Is Ground? 33
The Three Basic Switching Topologies 35
Why Step-Up/Down? 35
Current Ripple Ratio 37
Average Input and Output Currents 42
Energy Relationships of the Three Topologies 42
    Buck (Assuming 100 Percent Efficiency) 44
    Boost (Assuming 100 Percent Efficiency) 44
    Buck-boost (Assuming 100 Percent Efficiency) 45
Loss Relationships in Converters 45
Nonideal Duty Cycle Equations 49
Power Scaling Guidelines in Power Converters 51

PART 2 BASIC WAVEFORM ANALYSIS AND STRESS COMPUTATIONS

General Method for Piecewise Linear Waveforms 52
Other RMS and Average Values 55

Shape of Capacitor Currents 58
Wide Input Voltage and the Design Entry Point 59
How r Varies with Changes in Line and Load 61
How Capacitor RMS Varies with Changes in Line and Load 61
Stress Spiders 62
Boundary (Critical) Conduction 62
Using Too High an Inductance (Small r) 67
The Flattop Approximation 68

PART 3 TOPOLOGY MORPHOLOGY

Introduction 68
The N-Switch and P-Switch 69
The LSD Cell 70
Configurations of Switching Regulator Topologies 71
Basic Types of Switcher ICs 74
Flyback, Buck-boost, and Boost ICs Compared 75
Inductor Selection Criteria 76
Other Possible Applications of Buck and Buck-boost (Type 1 and
Type 2) ICs 77
Some Practical Cases 79
Differential Voltage Sensing 82
Some Topology Nuances 83

3 Contemporary Converters, Composites, and Related Techniques . . . **87**

PART 1 FUNDAMENTAL TOPOLOGIES

Synchronous DC-DC Converters 87
Modes of Operation of a Synchronous Buck 88
    Deadtime 88
The FAN5340 Lesson 90
Another Shoot-through Possibility: $C\,dV/dt$ Induced Turn-on 91
The Boost within the Buck and the Buck within the Boost 91
    Popular Current-Sensing Scheme in Voltage Mode
        Controlled Synchronous Buck Converters 91
    Popular Current-Sensing Scheme in Current Mode
        Controlled Synchronous Buck Converters 92
Paralleling and Interleaving of Buck Converters 95

PART 2 COMPOSITE TOPOLOGIES

4 Understanding and Using Discontinuous Conduction Mode **107**
Introduction 107
How DCM Duty Cycle Equations Are Calculated 111
Treatment of DCM in the Related Literature 112
Simplified Treatment of DCM Based on Optimal Setting of r 114
Tabulating the Stress Equations in DCM 114
Plotting the Key Stresses in Going from CCM to DCM 115
Varying the Input Voltage in a Buck-boost 118
Studying the Universal Input Flyback 119
Overload Margin of the Universal Input Flyback 121
Closed-Form Equations for Buck CCM and DCM Compared 124
Closed-Form Equations for Buck, Boost, and Buck-Boost DCM 127

5 Comprehensive Front-End Design in AC-DC Power Conversion . . . **129**

PART 1 FRONT-END WITH NO POWER FACTOR CORRECTION

Flyback and the Search for Closed-Form Equations 129
Droop Curves for Different Capacitances per Watt 131
Different UVLO? 131

Entry Point into Iteration for Closed-Form Equations for Normal
Operation 133
Universal-Input Single-Ended Forward (without PFC) 135
Compiled Design Tables 137
Typical Minimum Capacitance Requirements for Holdup Time and
RMS Capacitor Currents (Numerical Examples) 137
Estimating RMS Capacitor Current (High-Frequency
Component) 137
Total Input Capacitor RMS Current for Input Capacitors with
Non-Frequency-Dependent ESR 141
Total Input Capacitor RMS Current for Input Capacitors with
Frequency-Dependent ESR 141
Efficacy of EMI Filters 143
Forward Converters for Low Power? 143

PART 2 FRONT-END WITH POWER FACTOR CORRECTION

Instantaneous Duty Cycle in a PFC Boost 146
Holdup Time Considerations with PFC Boost Stages 148
Synchronization and Anti-Synchronization Techniques 149
Anti-Synchronization over a Wide Input Range 151
Calculating the High- and Low-Frequency RMS Components 151
Quick Estimate of RMS of Low-Frequency (Line) Component in a
Bulk Capacitor 153
Quick Estimate of RMS of High-Frequency (Line) Component in a
Bulk Capacitor 154
Numerical Table for Capacitor Selection and Comparison 154
Design of a PFC Choke and PFC Design Tables 155
Practical Nuances of PFC Design 155

6 Topologies for AC-DC Applications: An Introduction 161
Introduction 161

PART 1 THE (STRAIGHT) FORWARD CONVERTER

Flux Balancing in the Push-Pull 167
Flux Staircasing in the Half-Bridge 167

PART 2 THE (TRICKY) FLYBACK CONVERTER

The Integrated Power (IP) Switch 169
The Equivalent Buck-Boost Models of a Flyback 169
    Dealing with Multi-output Flyback Converters 175
The Primary-Side Leakage Term 176
The Secondary-Side Leakage Term 178
Flyback Optimization and Deeper Analysis 180
    Step 1: V_Z/V_{OR} 180
    Step 2: Dissipation in Zener 180
    Step 3: Duty Cycle 181
    Step 4: Peak Current 181
The RCD Clamp 183
    Step 5: Holdup Time 184
    Step 6: Inductor/Transformer Energy 185
Dissipation Estimates (Graphical) 187
Careful with That Calculation: Correct Dissipation in Switch and
Sense Resistors 188
Practical Flyback Designs Using 600-V Switches 190
How Higher V_{OR} Impacts the Output Diode Rating 191
Pulse-Skipping and Required Preload 191
Overload Protection at High-Line (Feedforward) 193

Contents

7 The Holy Grail: An Overview of Magnetics in Power Conversion 195
- Basic Magnetic Concepts and Definitions (MKS units) 195
- The Electrical-Magnetic Analogy 197
- The Inductor Equation 198
- The Voltage-Independent Equation 198
- The Voltage-Dependent Equation 200
- Units in Magnetics 204
  - The Magnetomotive Force (mmf) Equation 205
- Effective Area and Effective Length in Toroids 206
- Effective Area and Effective Length in E-cores 207
- The Effect of the Air Gap 208
- The Gap Factor z 210
- The Origin and Significance of z 211
- Relating B to H 213
- Gapped E-Cores 213
- Energy Storage Considerations: How to Vary the Gap in Practice, Optimally 214
  - First Attempt (N Constant) 215
  - Second Attempt (N Proportional to z: B Kept Constant) 215
  - Third Attempt (N Proportional to $\sqrt{z}$: Energy Kept Constant) 216
- How Much Energy Is in the Gap and How Much Is in the Core? ... 217
- Optimum Design Target Values for z 218
- The BH Curve 218
- Careful Design Principles Help Decrease Core Size 219
- Understanding L Better 220
- Scaling: Difference between an Inductor and (Flyback) Transformer 221
- (Real) Transformers 222
- Fringing Flux Correction 227
- Worked Example Using Fringing Flux Correction 229

8 Tapped-Inductor (Autotransformer-Based) Converters 231
- Introduction 231
- The Tapped-Inductor Buck 232
- Other Tapped-Inductor Stages and Duty Cycle 236

9 Selecting Inductors for DC-DC Converters 239
- Introduction 239
- The Basics 239
- Specifying the Current Ripple Ratio r 240
- Mapping the Inductor 241
  - Voltseconds 241
- Choosing r and L 242
- B in Terms of Current 243
- Counterintuition in Magnetics 244
- Core-Loss Optimization by Tweaking Geometry at a Fixed Frequency 244
  - Case 1: Varying the Volume of the Core (N Fixed) 244
  - Case 2: Varying the Volume of the Core (L Fixed) 245
  - Case 3: Varying the Volume of the Core by Varying Area (Length and N Fixed) 245
  - Case 4: Varying the Volume of the Core by Varying Area (Length and L Fixed) 245
  - Case 5: Varying the Volume of the Core by Varying Length (Area and L Fixed) 246

Core-Loss Optimization by Tweaking Geometry as We Vary
    Frequency 246
        Case 6: Varying the Frequency (Core Volume and *L* Fixed) ... 246
        Case 7: Varying the Frequency (Core Volume and *r* Fixed) ... 246
        Case 8: Varying the Frequency (*r* Fixed, *N* Fixed, Each
            Dimension Changed) 247
        Case 9: Varying the Frequency (*r* Fixed, *N* Fixed, Only Area
            Changed) 247
        A Walk-Through Example 247
        Choosing an Inductor 248
Evaluating the Inductor in Our Application 249
    a) Current Ripple Ratio 250
    b) Peak Flux Density 251
    c) Peak Current 251
    d) Temperature Rise 252

10 Basics of Flyback Transformer Design 255
The Voltage-Dependent Equation: A Practical Form 255
Energy Stored, as Related to Core Volume in an Ungapped
    Core 256
General Energy Relationships for a Gapped Core 256
General Relationships for A_L and μ 257
Duty Cycle of Universal-Input Flyback with $V_{OR} = 100$ V 257
The Area × Turns Rule 257
Worked Example (Part 1) 257
Some Finer Points of Optimization 260
Rule of Thumb for Quick Selection of Flyback Transformer
    Cores 260
Worked Example (Part 2) 261
Impact of Current Limit on Core Size 262
Circular Mils (cmil) 264
Current Carrying Capacity of Wires 265
Skin Depth 267
Graphical Solutions and a Useful Nomogram 269
A Feel for Wire Gauges 271
Diameter of Coated Wire 271
SWG Comparison 272

11 Basics of Forward Converter Magnetics Design 275
Introduction 275
The Transformer and Choke (Inductor) Compared 275
Turns Ratio Calculation (at Minimum Input) 277
Summary of Observations 277
Introducing the Proximity Effect 279
More about Skin Depth 279
    Dowell's Equations 280
The Equivalent Foil Transformation 286
Some Useful Equations for Quick Selection of Forward
    Converter Cores 287
Stacking Wires and Bundles 288
Core-Loss Calculations 289

12 Forward and Flyback Converters: Step-by-Step Design and
    Comparison 293
Introduction 293
Finer Classes of Window Area and Area Product (Some New
    Terminology) 295
Power and Area Product Relation 296

Current Density and Conversions Based on D	298
Optimum Current Density	299
Industry-Recommended Equations for the Area Product of a Forward Converter	301
Fairchild Semi Recommendation	301
TI/Unitrode Recommendation	302
Basso/On-Semi Recommendation	303
ST Micro Recommendation	303
Keith Billings and Pressman Recommendation and Explanation	303
Plotting Industry Recommendations for Forward Converters	304
Area Product for Symmetric Converters	307
More Accurate Estimate of Power Throughput in Safety Transformers	308
Number of Primary Turns	310
Worked Example: Flyback and Forward Alternative Design Paths	314
Step-by-Step Forward Converter Design	315
Core Selection	315
Primary Turns	315
Magnetization Inductance and Peak Magnetization Current	315
Turns Ratio	315
Voltage Ratings	316
Secondary Turns	316
Sense Resistor	316
Minimum Duty Cycle	316
Choke Inductance and Rating	316
Overall Loss Estimation in Transformer	316
Core Loss and Total Estimated Loss	317
Copper Sizing and Transformer Windings	319
P-S Winding Arrangement	319
Input Capacitor Selection	325
Paralleling Ceramic and Electrolytic Capacitors at the Input	326
Output Capacitor Selection	326
Step-by-Step Flyback Converter Design	328
Choosing V_{OR}	329
Turns Ratio	330
Core Selection	330
Primary Turns	330
Secondary Turns	330
Primary Inductance	330
Zener Clamp	330
Voltage Ratings	330
Input Capacitor Selection	331
Output Capacitor Selection	331
Copper Windings	332
Industrywide Current Density Targets in Flyback Converters	332
Comparison of Energy Storage Requirements in a Forward and Flyback	333

13 PCBs and Thermal Management **335**

PART 1   PCBs AND LAYOUT

Introduction	335
Trace Analysis	335
Miscellaneous Points to Note	337
Sizing Copper Traces	340
Routing the Feedback Trace	341

Routing the Current-Sense Trace	341
The Ground Plane	341
Some Manufacturing Issues	343
PCB Vendors and Gerber Files	345

PART 2 THERMAL MANAGEMENT

Introduction	345
Thermal Measurements and Efficiency Estimates	346
The Equations of Natural Convection	347
Historical Definitions	348
Available Equations	348
Manipulating the Equations	349
Comparing the Two Standard Equations	350
h from Thermodynamic Theory	351
Working with the Tables of the Standard Equations	351
PCBs for Heatsinking	353
Natural Convection at an Altitude	355
Forced Air Cooling	356
Radiative Heat Transfer	357
Miscellaneous Issues	357

14 Closing the Loop: Feedback and Stability 359
Basic Terminology 359

PART 1 STABILIZING CURRENT-MODE CONVERTERS

Background	363
Input (Line) Feedforward in Voltage-Mode Control (Wannabe CMC)	366
How Much Slope Compensation?	366
Generalized Rule for Avoiding Subharmonic Instability	371

PART 2 RETROSPECTIVE: VOLTAGE MODE, CURRENT MODE TO HYSTERETIC

Introduction	372
Plant Transfer Functions	373
Conclusions on CMC versus VMC	378
Hysteretic Control: Energy on Demand	378
Summary of Pros and Cons of the Different Control Techniques	388

PART 3 DESIGN EXAMPLES FOR VMC AND CMC NONISOLATED, AND FOR TL431 + OPTO ISOLATED FLYBACK CASE

Design Examples for VMC and CMC, Using Type 2 and 3 Error Amps or an OTA	388
Design Examples for TL431 and Opto	391

15 Practical EMI Filter Design 401
The CISPR 22 Standard	401
The LISN	401
Fourier Series	402
The Trapezoid	403
Practical DM Filter Design	404
Practical CM Filter Design	405

16 Reset Techniques in Flyback and Forward Converters 409

PART 1 FLYBACK CONVERTER TRANSFORMER RESET (OF LEAKAGE INDUCTANCE)

Zener Clamp	409
RCD Clamp	411
Lossless Snubbers	412

PART 2 FORWARD CONVERTER TRANSFORMER RESET
(OF MAGNETIZATION INDUCTANCE)

Introduction 413
General Tertiary Winding 414
Active Clamp Reset 417
Analysis and Conclusions (Figs. 16.9 to 16.11) 418
Which Is Better: High-Side or Low-Side Active Clamp? 420

17 Reliability, Testing, and Safety Issues 425
Introduction 425
Reliability Definitions 425
Chi-Square Distribution 426
Chargeable Failures 427
Warranty Costs 428
Calculating Reliability 429
Testing and Qualifying Power Supplies 429
    HAST/HALT, HASS, and ESS 430
Safety Issues 431
Calculating Working Voltage 432
Estimating Capacitor Life 434
Safety Restrictions on the Total Y-Capacitance 440
Safety and the 5-cent Zener 440

18 Watts in It for Us: Unraveling Buck Efficiency 443

PART 1 BREAKUP OF LOSSES AND ANALYSIS

Introduction 443
Only One Loss Term at a Time: Understanding Each 444
    Crossover Losses 445
    Deadtime Losses 445
    Input Capacitor ESR Losses 445
    Conduction Losses (R_{DS} and DCR) 446
    Controller IC Losses 447
Cumulating Losses: Adding Them up One by One 447
The Underlying Buck Spreadsheet 451

PART 2 PREDICTING EFFICIENCY AND REVERSE-
ENGINEERING TRICKS

19 Soft-Switching and Designing *LLC* Converters 461

PART 1 OVERVIEW OF THE TRANSITION FROM CONVENTIONAL
PWM POWER CONVERSION TO RESONANT TOPOLOGIES

Introduction 461
Soft and Hard Switching 462
Two Key Concerns (Guiding Criteria) 464
Switching Losses in a Synchronous Buck and Lessons Learned ... 465
Building Basic Resonant Circuits from Components 468

PART 2 BUILDING RESONANT TANK CIRCUITS FOR
FUTURE CONVERTERS

Series Resonant Tank Circuit 473
Introducing the *LLC* Tank for Creating Future *LLC* Converters 474
Analyzing the Gain-Phase Relationships of the *LLC* Tank Circuit ... 479
Two Resonances in the *LLC* Tank 483

PART 3 GENERALIZING VIA SCALING STRATEGIES

Step 1 of Design Validation Process: An "AC-AC Converter" 486
Derating Maximum Power for Increased Input Range 488

Step 2 of Design Validation Process: An AC-DC Converter with
Diodes and Transformer (Still No Output Capacitor) 490
Step 3 of Design Validation Process: A DC-DC Converter
with Diodes and Transformer (and Also an Output
Capacitor) 491

PART 4 DESIGNING A PRACTICAL *LLC*-BASED PD

PART 5 VALIDATING OUR THEORETICAL PD DESIGN VIA SIMULATIONS

Simulation Run 1 (52 V, Square-Wave Driven, See Fig. 19.22) . . . 498
Simulation Run 2 (32 V, Sine-Wave Driven, See Fig. 19.23) . . . 498
Simulation Run 3 (32 V, Square-Wave Driven, See Fig. 19.24) . . . 498
Conclusion 498

PART 6 RATIO OF INDUCTANCES (PROS AND CONS)

PART 7 HALF-BRIDGE IMPLEMENTATION OF LLC CONVERTER AND SOLVED EXAMPLES

Efficiency Estimates 505
Solved Example for LLC Selection with Maximum 1:3.16 Frequency
Spread 507
Final Selection 509
Alternate *LLC* Seed 509
Solved Example for LLC Selection with Maximum 1:2 Frequency
Spread 510
Final Selection 511
Breakdown of First Harmonic Approximation and Other
Subtleties 511

20 Things to Try 513
Introduction 513
Synchronizing Two 3844 ICs 513
A Self-Oscillating Low-Cost Standby/Auxiliary Power Supply 513
An Adapter with Battery Charging Function 515
Paralleling Bridge Rectifiers 516
Self-Contained Inrush Protection Circuit 516
Cheap Power Good Signal 517
An Overcurrent Protection Circuit 517
Another Overcurrent Protection Circuit 517
Adding Overtemperature Protection to the 384x Series 518
Turn-On Snubber for PFC 518
A Unique Active Inrush Protection Circuit 519
Floating Drive from a 384x Controller 519
Floating Buck Topology 520
Symmetrical Boost Topology 520
A Slave Converter 521
A Boost Preregulator with a Regulated Auxiliary Output 522

A Design Tables and Aids, and Component FAQs 525

Index 537

Preface

It's been almost a decade since I held the previous edition of this book ever so lovingly in my hands. It was my very first time. How can I ever forget that moment? The thrill lingers on, probably destined to last a lifetime. But, like all special moments and relationships in our lives, it needed nurturing. So, though it seems a trifle late in coming, here it finally is—the second edition of my very first book. To me personally, the book still remains what it was on that thrilling day in 2004: *a labor of love*. That's one aspect which hasn't changed one bit over the years. Because I wouldn't even have embarked on the difficult journey of writing this lengthy book, if I hadn't felt I was going to enjoy the process one hundred percent. *And I did*. So, now I truly hope you will love it equally. Because in the final reckoning, that is what makes it worth all the effort. Nothing else matters.

You will notice that there is something which has definitely evolved, if not changed over the years—my writing style. You can perhaps already tell this isn't going to be your typical run-of-the-mill, dry, technical book. Sorry, I agree I get a bit wordy at times. But I just couldn't bring myself to write "normally" anymore. Perhaps I enjoy power electronics so much that it has finally started to show—almost uncontrollably by now.

Looking back over the years, our times have changed incredibly since the previous edition. Technology has made huge strides. Barely a few years ago, this book could have served as another dependable, UL-certified paperweight on your maple desk—guaranteed to keep your bank or mortgage statements firmly pinned down, even through recurrent storms. But today, it may just be a bunch of weightless pixels floating on a slim e-reader or tablet clutched gingerly in your hand. Perhaps beamed down from a virtual cloud somewhere up there. No rain though, yet! Yes, there is no escaping *change* anymore: we need to embrace it faster than it tries to envelop us. And this book is certainly no exception to that trend. You will realize it too has changed beyond imagination. Certainly no drought in these parts, I assure you!

While pitching the initial idea for this edition, I had insisted that I needed to do *complete* justice to my first book ever. A lot of colleagues and their acquaintances had said to me that of all my subsequent writing efforts, they had liked that particular work of mine the most. Providing me with no-holds-barred feedback, they had indicated that it was a "straight-to-the-point, no beating around the bush, professionals' handbook" (I have edited out the part invoking remnants of a rampaging male herbivore). Well, I didn't necessarily agree to all of that, but I was definite that I wouldn't be happy just adding a chapter or two, inserting footnotes here and there to "bring it up-to-date," and *pretending* the second edition was somehow "new." That's not my style. Check out the latest incarnation of my A-Z book too, to realize that I truly treat any second edition as a new book entirely, not just a second chance, for pinning the tail on a donkey, or a game of throwing darts, hoping one sticks. Nor do I subscribe to any preset dogma that only if you belabor yourself with a certain approach, say simulations and small-signal models, will you ever understand power and build successful power supplies. Quite the contrary, actually. It can be great fun too. I wanted to emphasize that and convey it somehow. That's always been my approach. My first book reflected that belief as much as this one does today. I guess that's what makes it "practical" to some of my readers. I just cut to the chase, respecting their valuable time. I realize you too perhaps want tangible results very quickly, not just reams of yellowing paper withering on your shelves.

Most people are not aware that in late 2004, I had been the one first offered to take the late Mr. Abraham Pressman's book to its third edition. Of course I had felt honored, but I had also felt strongly that "mix and match" wouldn't work too well, in my case in particular, because my writing style was already so different. I had therefore regretfully declined. But that's also why, when I say that this new book is "new," it *really* is. For one, it has been almost completely rewritten. I might even go as far as to claim that any residual resemblance to any previous edition, dead or alive, is purely coincidental!

There is another thing I had promised myself at the initial proposal stage: that this book will *complement* my other books, not replace them, or compete with them. You will discover there is a large amount of new material in my A-Z book, which is frankly *not* included here. Coupled inductors, for example. Buy that book if that is what you are looking to develop, not this one. Yet there is a lot this new book has to offer which you will likely find neither in A-Z nor in any other book out there. Such as: the world's first simplified LLC design methodology

(Chap. 19), active reset techniques with design equations and charts (Chap. 17), tapped inductors with design charts (Chap. 8), unraveling Buck efficiency layer by layer to maximize DC-DC converter performance in battery-powered equipment (Chap. 18), simplified yet thorough discontinuous conduction mode analysis (Chap. 4), the most thorough front-end design procedure ever available for AC-DC converters with and without power factor correction (Chap. 5), a complete side-by-side worked example of both Forward and Flyback converters for a sample telecom application using exhaustive proximity analysis charts and comprehensive E-core selection tables (Chap. 12), detailed worked examples and wall-charts for loop compensation, including a no-sweat TL431-opto-based feedback design procedure (Chap. 14). Even bandgap references (Chap. 1) and the Sepic-Cuk-Zeta composite-topology triad (Chap. 3). Not to forget an overdose of z-factor simplified magnetics in Chap. 7. Plus an updated component FAQ in the Appendix, with all design equations compiled in one easy look-up place, and also a medley of neat circuits to try out (Chap. 20). It's mostly all original material. However, if you are still looking for derivations, you may also need to refer to my new A-Z book. Almost all the derivations you can hope for are probably there! Incidentally, you may have noticed, the equations are not "coming out of thin air". But this particular Design and Optimization book is, as before, still meant more to be an on-the-go *professional's quick reference*, not an academician's classroom support tool, and probably not a good entry-level book. It's certainly not intended for your typical, high-tech senior VP, or a chemistry student trying to build a science project or flirting with a career change.

From time to time, you may discover buried layers of complexity in this book, which you hardly expected, or noticed at first sight. For example, in Chap. 12 you will see the simple procedure on how to parallel ceramic and aluminum electrolytic capacitors. In Chap. 18, you will find a unique technique which I had introduced years ago in my troubleshooting book actually: *How to extract a lot of hidden device information from published efficiency curves*, and also how you can take just three points to predict the efficiency at almost any other input-output condition. Is it an example of the well-known *Fermi problem* perhaps? Maybe not.[1] You will realize it's amazing what all we can do through astute deductions, as opposed to a brute-force armada of equations (yes, the square root of −1 is often misused for providing an imaginary sense of comfort). I had originally proposed my predictive procedure only for switchers with a BJT and catch diode. But this time, it is far more detailed, encompassing even synchronous Buck switcher ICs. You will definitely not find it in any other publication out there. I feel it can be very useful in unearthing what a competitor may be doing—whether he or she is "fudging" or "massaging" the curves, for example. I remember in one analysis, a very respectable and big vendor's switcher IC's datasheet efficiency curves revealed to me that the vendor had apparently, for the purpose of generating the curves in the lab, deliberately selected a part whose internal FETs were way better than the typical values declared in the part's datasheet. Because, even if I put crossover losses to zero, DC resistance losses and controller section losses to zero, and so on, I still couldn't get to their published efficiency numbers—unless I really "improved" the drain-to-source resistances (even assuming 25°C die temperature). No mistake in my calculations, I triple-verified on other parts too, *even from the same vendor*. As a result, chances were very high that no one who ever used that part would get the datasheet results in any real application scenario. Not even close. But for you, as a possible competitor to that vendor, it may already be too late: you may have just *lost the bid*. Whoever said honesty is the best policy? However, times may be starting to change now. Because, armed with the simple equations of Chap. 18, not a mere bunch of inscrutable simulations to impress a fellow paper tiger with, you have the means not only of getting your efficiency curves to (honestly) rise past the magic threshold of 90 percent, if at all possible, but also a way of catching others' lies (if any), *well in time*. It goes to show that commercial power conversion is really about *insight*, first and foremost. It's about *thinking power*. And that, incidentally, is the title of one of the chapters in my so-called *blue book*.

You will discover that each book of mine is unique and relatively complete in itself, despite differing scope and coverage. For example, when you finally start building your power supply, and run into seemingly intractable practical problems, or mysteriously blown-up transistors, which no oxymoron headlined as *practical simulations* appears to even warn you about, leave aside shed light on, there is my troubleshooting book to lean on. Recently, an "Amazon-verified purchase" reviewer hailed it as a "life-saver," and a "gritty, down-in-the-trenches battle manual for winning the struggle against a misbehaving design." I felt vindicated. The

---

[1] See http://en.wikipedia.org/wiki/Fermi_problem and several other websites dedicated to the Fermi problem.

book was useful after all! The effort had been worth it. I already knew the book was extremely popular in China (its Chinese translation), but this was heartening to hear nevertheless.

At this point, I have to pause—to sincerely thank unknown reviewers like the above gentleman. They are the real reason why most technical writers can summon up the grit to return with another huge body of work. And that ultimately helps the larger design community. The system seems to work—what goes around, comes around. We all benefit. But that does not happen always! It is imperative that authors and readers alike learn to recognize and perhaps ignore *carpet bombers*, as described by Forbes.[2] I've had to mentally bypass a few of these myself in the past. I knew they were fake, and the rest weren't, yet I had to make peace in my heart like a true Gandhian. Certainly, if I had taken these to heart, I would have stopped writing. And this book would not have been in your hands today. So, people with an axe to grind actually do no service to the larger design community. They are self-serving. You just don't know their hidden agendas. They give themselves away though, by (a) trying so hard to tell you that *all the other reviews* are fake, just not *theirs*, whereas the reverse is actually true. In my case they start off by complaining that the book is misleadingly called "A to Z" but does not include *everything* under the sun. I wondered: Is that *really* how you purchase things? So, can I now sell you the *Secrets of the Universe* too? Or perhaps the *World's Most Comfortable Pillow*? Or the *Most Exotic Vacation Ever Created*? Hey, those are but titles, and anyone designing power supplies should be smart enough to figure out that undercompensated authors do *not* control the marketing departments of major publishers. Finally, (b) these carpet bombers give themselves away, when right after sowing doubts about the other genuine reviews, they direct people to a competing product. Why on earth would they do that? Oh, for the wider good of the population? Well, in that case, how come they took the trouble of only posting one review in 5 years?

And now over to the flip side of the same coin: Asking friends to post positive fake reviews to help you out (perhaps on the basis of some hidden quid pro quo), or posting fake negative reviews against perceived rivals, are one and the same thing. It takes the *very same bent of mind* in my opinion: essentially unethical and dishonest. Certainly not a (good) engineer's mind. It makes me think: If I can't trust an *engineer*, what is his *data* worth? In fact Wikipedia goes as far as to say that "shills" break the law. See http://en.wikipedia.org/wiki/Shill. But the more practical question facing us is: How can we spot fake positive or fake negative reviews?[3] For me personally, only the genuine reviews (positive, or *constructively* negative) matter, and they are the ones that make my day. I still remember the lessons of Lance Armstrong and Milli Vanilli.

I have endless patience to wait for genuine reviews to come along. But I also recently realized that not everyone sees it my way. Some are in a tearing hurry to achieve "success" and glory at all costs in our modern times. So, a few months ago, I queried a certain writer who had gotten about a dozen reviews in one year alone—six times the industry average for good technical books. I had wondered how that was possible. Maybe he was really that good. In fact, I'd figured if that was the case, I would happily become his biggest fan. Because that's how I learned power electronics in the first place—not in any classroom, but in the lab and by paying humble obeisance to Gurus. But I had also seen some puzzling, or simply curious, signs, such as reviewers stating in rather suspiciously flowery language that the said author had for example: "… *worked very hard giving years of his life to produce the best Switch-Mode Power Supply book available, by far… working so hard and smart to produce the best Switch-Mode Power Supply book ever written.*" And so on. Lots of curious repetitions too. Why? That seemed odd to me. I even wanted to say: *Hey! Only I write that way!* Seriously, however, very few people still refer to these devices as Switch-Mode Power Supplies. Because, the word *Switchmode* was historically a registered trademark of Motorola (now On-Semi). See http://www.onsemi.com/pub_link/Collateral/SMPSRM-D.PDF. In most companies I have been to, and I have been all over the world, it's a term in increasing disuse, if not completely extinct by now. But I also wondered about the following: If I was just a reader or an unknown reviewer, how would I ever know (or care) whether a certain author had spent *years of his life*, or months, hours, or even nanoseconds in his book-writing process? In my case, even my wife doesn't know that for sure, certainly not to be able to state it as a fact on Amazon of all the places. Besides, all that would matter to *me* as a buyer/reviewer would be: What the book ultimately did for *me* for all the money I spent. It's business after all. It is also not logical for anyone to make any assumptions, because not *all* authors spend years writing each book. I happen to

---

[2] http://www.forbes.com/sites/suwcharmananderson/2012/08/28/fake-reviews-amazons-rotten-core/.

[3] Try this link if interested: http://consumerist.com/2010/04/14/how-you-spot-fake-online-reviews/. And this: http://newsfeed.time.com/2013/09/25/how-to-spot-a-fake-review-4-clues-somethings-fishy/.

know that very well by personal experience. Typically, I only spend a few (extremely intense) months writing any of my books. The one in your hands took about 6 months. Its first edition had actually taken less than 3. Yes months, not years! The reason is, most of us always have a body of work ongoing in our respective workplaces, and we are constantly trying out something or the other at any given moment, so *we never really start from scratch*. That's the truth. But granted, some authors do take years perhaps. But how would *you* or any unknown reviewer know about that? And mostly, that is probably because of *language issues*, not technical constraints. So finally, I was curious enough to query the author directly, not wanting to believe all these tell-tale signs. That would be a blow to me too. But on April 21, 2013, he admitted as much by Email (the italics are mine): "To *fuel the reviews* on Amazon, *I've asked a few* of them to *let others know* what they thought of the book. I did not drive anything… Regarding the last book, I did *not ask specifically* my friends to post reviews. One of them was in the review team. He disclosed that upfront and his review was *technically sound*. For the rest, *I can't be blamed* for *having friends or people that I know* writing comments on my work?" This is quoted verbatim. But I could still see, the author was *not being fully truthful*, because not one of his Amazon reviewers had disclosed any ties to the author upfront as he claimed. I checked. By definition, they were "shills." And worse, by then, I knew there were at least ten planted shills on that page. It's a small world after all. I also found it a bit amusing, if not alarming, that a certain 5-star review was being declared "technically sound" (and therefore well-deserved) by the recipient of the review himself. And that too, coming from someone in his own review team! That was to me, a biased review of a biased review, of a biased reviewer! Oh good for him I thought, and just moved on. It's our modern world. Times have obviously changed. I have been left a little behind, I guess.

So, let's return to the relatively primitive, but safe, 2003 for a brief moment. I remember I had nervously approached the well-known acquisitions editor of McGraw-Hill at the time, Steve Chapman, with my first writing proposal. By the way, he is also the one who got this new edition going. But way back in 2003, I only had to my credit a couple of App Notes at National Semiconductor (now Texas Instruments). No, no, in 2001 I had also published my very first article—in *Electronic Engineering*, a UK magazine long since defunct (for no obvious fault of mine). Soon there was a well-known cover story in 2002 too, in *Power Electronics* (formerly *PCIM*) magazine, titled "Reducing Converter Stresses." The previous year, *PCIM* had kindly submitted a slightly altered version of my UK article under the title "Current Ripple Ratio Simplifies Selection of Off-the-Shelf Inductors for Buck Converters". Also, one article on thermal management shortly thereafter.[4] Oh, in 2003, I also had a string of popular articles on EMI, on Planet Analog (*EE Times*). And yet another on slave converters in *EDN* print magazine. Perhaps a few more too, I really forget. I am pointing all this out only because I want to clarify where the now well-known term *current ripple ratio* came from. To be very precise, it was first mentioned in the application note AN-1197 at National Semiconductor.[5] I remember I had thought for days, not only what to call it, but what symbol I should assign to it. I had settled on the simplicity of r (for ripple). I had initially targeted the concept only for a Buck, but soon extended it to all topologies—in application note AN-1246.[6] Yes, I notice some companies still strip out all authors' names, denying them any credit. It is company property they insist. And of course it is. But the rules don't seem to apply evenly to some of their own. Sometimes all they often end up saying is that only their "visionary" CEO's name and face should appear everywhere. It is being implied that the rest are all minions, mere extensions of his great mind. And in doing so, they also make it clear they really don't care that they are removing any incentive for future authors to do any more than their minimum 9 to 5 jobs. Eventually, only the design community suffers. National Semiconductor was thankfully very different (when it was still around). I do owe them a lot in retrospect. I shudder to think what would have happened had I stayed on at some egoistic, three-terminal integrated offline switcher IC maker. Where would I be today? Completely unknown! A shadow of their CEO. I would certainly never have introduced r, or even the LLC topology to the world.

By introducing this unique parameter, r, I believe in one fell sweep I managed to bring all switching power converters under one roof, irrespective of their frequency, output voltage, load, input voltage, and even their topology. I gave all of them a common design entry

---

[4]See all the *PCIM* articles at http://powerelectronics.com/search/results/maniktala.

[5]See its modified version at http://www.ti.com/lit/an/snva038b/snva038b.pdf and its original Chinese version at http://www.bdtic.com/DownLoad/NSC/ApplicationNote/AN-1197.pdf.

[6]See its modified version at http://www.ti.com/lit/an/snoa425c/snoa425c.pdf, and its original at http://www.thierry-lequeu.fr/data/AN-1246.pdf.

point: Just set $r = 0.4$. Always! Can anyone ever make it simpler than that? Yet, not everyone got the point initially. One mildly disappointed Amazon reviewer even had this to say about my A to Z book (its first edition): "Unfortunately I also felt the author talked too much about ripple etc. and made some of the simpler concepts more arcane (for example the introductory chapters on inductors and capacitors and energy sloshing back and forth)." At this point my sense of conviction stepped in. This was certainly not a carpet-bomber, or a shill. It was definitely constructive, thank you. It made me think. But soon I became clear about one thing: that I had to do what I was convinced was in the best interests of my readers, even though *they may not realize it fully as of that moment*. I recognized these were all (honest) fellow practitioners—pushing the very same art that I love so deeply. In return, I had to do what I was convinced would likely most benefit them *ultimately*. So I stuck to the concept. And today, it is increasingly taught in schools everywhere.[7] Consultants are using it too (with my symbol r).[8] Even rival chip vendors refer to my very first app note.[9] Magnetics vendor, Wurth Elektronik lifted all the design tables I had initially presented in AN-1246, based on current ripple ratio (without acknowledgment, but they have promised to fix it in the next edition), in their handbook *Trilogy of Magnetics*.[10]

Recently, in a well-known IEEE article,[11] the authors write:

> The general way to design the output inductor is taking the critical discontinue current as the basis, then selecting the inductor current ripple according to the experimental equations to calculate the inductor value. This method neglects the effects of the inductor current ripple on the overall stresses of the converter, and can't provide ideal output inductor values, so it will affect the actual performance of the converter. In order to solve the shortcoming, the theoretical basis for selecting the current ripple ratio of Buck converter is analyzed deeply and an optimal design method of the output inductor is proposed in this paper. The overall stresses of the converter, the size of the filter inductor, and dynamic performance of the converter are all considered in the proposed method. The optimum value of current ripple ratio is set at about 0.4 finally. The filter inductor value, which is calculated in terms of the determined current ripple ratio, would not affect the dynamic performance of the converter, and the design of the converter is optimized. The correctness of theoretical analysis and the feasibility of the proposed method are verified by the simulation and experimental results.

In other words, even for a Buck, my design suggestion based on current ripple ratio, was proven by this research team to be correct by both simulations and experiments. We should try it on the other topologies too.

Nowadays, there is a certain expert amongst us, deservedly well-known for unraveling the intricacies of current-mode control. I am going to respectfully simply call him "Doctor" from this point on. He is also the person behind an Excel-based software design tool which I believe he gives out at the end of his very expensive 4- or 5- or 6-day seminar sessions (hence its name I guess!). I didn't think it stood for $4000 or $5000 or $6000! Early in 2013, he sent me a complimentary license for this software, asking if in return, I could evaluate the user interface, and so on. I am sure he was not expecting me to find errors, and that too in his loop compensation equations. Very simply put, he was trying to use me, and I knew from others he likes to do that. When you ask for return help, he is known to invariably proffer a quote or bill instead. No problem! I too had thought I would just use his tool to selfishly confirm the equations in my book. But finally, after some surprising twists, on Feb 2, he replied: "…25 years and no one ever noticed this. I think I just lost my weekend! Do you have your schematics for a type II calculation as well?" Thereafter, I had to send him the entire loop compensation design procedure from my books (free of charge), while also consoling him that *to err is human*. He had realized my equations had been right all along, and so finally, a renowned expert on loop compensation had to sheepishly release a corrected version of his design tool shortly thereafter. And it's *that* exact loop design procedure, the one that helped Doctor clean up his design tool after *25 years*, which is presented in Chap. 14 of this book, now in an even more easy-lookup form, with lots of new design charts. It is much more comprehensive than my other (red) book incidentally. In this chapter, I have also provided an effective, but concise procedure to deal with the TL431-optocoupler combination in isolated (e.g., AC-DC) converters, along with simple, numerical design examples to have you on your way, instead

---

[7]See http://classes.soe.ucsc.edu/ee174/Fall13/References/AppNotes/AN-1197.pdf.

[8]See http://www.daycounter.com/LabBook/BuckConverter/Buck-Converter-Equations.phtml.

[9]See http://www.onsemi.com/pub_link/Collateral/AND9135-D.PDF.

[10]See http://www.we-online.com/web/en/electronic_components/produkte_pb/fachbuecher/Anforderung_Trilogie.php.

[11]See http://ieeexplore.ieee.org/xpl/articleDetails.jsp?arnumber=5515878.

of asking you to meander through endless pages of simulations, models, and abstractions in the s-plane. To add to that, in Chap. 14, I have also given a lot of information on *hysteretic control*, something that is silently appearing all around us in a big way nowadays. It is likely to be a major thrust area in the years ahead. Look out for it. Oh, by the way, by then I had respectfully dropped the fleeting idea of asking Doctor to pen this preface for me.

The most striking part of this new book is probably the long chapter on resonant converters, chiefly the LLC topology. In my last company, my rather foresighted boss (a true diamond in a coal mine) had been urging me to try and shed light on this new, upcoming topology for several seemingly logical reasons: (a) It offers very high efficiency, so it seems perfect for our green-planet thrust (b) It offers very low EMI, which makes it an ideal candidate for helping ensure signal integrity—especially in applications where data and power sections fall adjacent to each other (which is almost *everywhere* without exception nowadays). So, how could we afford to ignore the LLC topology any longer? Good point! However, my mind was trained over decades to think only in terms of conventional, square/trapezoidal power conversion. That was as "modern" at it gets, wasn't it? Not really. I suddenly remembered in my previous job at a networking company, we had always run into seemingly insurmountable issues trying to get to higher data rates, with or without PoE (Power over Ethernet), simply because the on-board Buck converters were throwing out so much noise all across the board. Bit error rates would shoot up and we were stuck trying to get the data to move any faster. The LLC topology would likely make things much easier, since it dealt with sine waves, not square waves with their usual complement of tons of harmonics riding on them. Yes, that was probably the reason why so many companies seem to have started making LLC controller chips. But I noticed, all of them either hesitate to give a simple design procedure, or provide a procedure for a very limited input range. On blogs around me, a common opinion seemed to be that the "key drawback of the LLC topology is its limited input range." That is why it seems to have found a niche market in the form of LED backlighting circuitry in flat-panel LCD televisions, positioned after the front-end power factor correction stage, where a relatively steady 400 Vdc exists as the input of the LLC stage. In fact, another learned Doctor actually announced to a fellow-blogger: "Fantastic! You answered all the questions on this group in one posting! First—*no LLC needed, the phase shifted bridge wins.*" *Not so fast Doctor*—I'd warned (wait for *this* Optimization book). Maybe he should have stuck to loop compensation. Oh, maybe not that either![12]

That was the background, when in December 2012, my family left for a trip to India, leaving me alone. Having lots of time (and two constantly licking dogs) on my hands, I challenged myself. I told myself: with my two (hitherto useless) master's degrees in physics, if I can't understand resonances, who can? Thus literally shaming myself mentally, I set to work painfully on the LLC, armed with Mathcad and Simplis (PSpice). In twenty days I had got to the heart of the topology. It was truly breathtaking, I admit. I felt I had landed on the moon. But I'd also managed to come up with a startlingly simple design procedure, *by applying power and frequency scaling principles used in conventional power conversion to resonant power conversion*. Very simply put, I start with a low-frequency LLC seed, study it thoroughly, decide its best operating region, then scale it to any desired frequency and power, by scaling techniques that people always seem to know instinctively in conventional power conversion, but never seem to fully comprehend or use effectively! Applied to the LLC topology, scaling makes the latter devastatingly simple. And it's that technique which I am publishing here for the very first time. Note that in this chapter, I didn't just throw resonant power conversion in your face. Once again, I have started with perhaps unnecessarily "arcane" stuff like energy sloshing back and forth between inductors and capacitors, because I felt there is no way you will make the transition from conventional power conversion to resonant power conversion, without understanding that. I have carefully constructed a "bridge" into an alternate world of power conversion, using several successive simulations of increasing complexity. I presumed you wouldn't be happy if I just threw myriad equations at you, with some muscle-flexing and bravado—all gift-wrapped in feel-good statements like: "I single-handedly derived all of these equations from scratch working every single night for 3 years," or "look Ma, no hands," and blah-blah-blah... leaving you to your own devices, steeped in admiration for *me*! I didn't think *you* deserve that. And nor did the field we dedicate ourselves to every day. As I said, it should never be about the author, it's about you, the reader.

The LLC topology is likely to become mainstream, provided we understand it and use it well. Chap. 19 does give you that power I feel. And it is tried and tested too, by now.

---

[12]See the exchange at http://www.gozuk.com/forum/high-efficiency-converters-next-generation-449570.html.

Over the past year, I have used the same (hitherto secret) procedure to build several low- and high-voltage converters. These have included what I believe is the world's first 25-W PD (powered device) working from 32 to 57 Vdc input with an output of 12 V. I have also built a 25-W universal-input AC-DC LLC-based power supply working from 100 Vac to 270 Vac with efficiency almost flat at 90 percent over the entire input range. I have also built a 25-W wireless mat charger in my spare time, and so on. It's been practically tested, many times over. I can now handle a 100 to 400 percent input range variation, as compared to the usually accepted 100 to 114 percent. Maybe now I will get some fresh carpet bombers.

Finally, I must point out that I sincerely believe that despite all the changes around us, certain things should never change. Which takes me back to the basic question: Why do we, as authors, even do it? Being technical writers, there are no big bucks at stake for us in this. Nor do we expect a hefty ego boost (*usually* at least, I can't speak for others). After a decade, I still shudder at the mere thought of being counted among those with self-esteem apparently ten times the size of their entire body of knowledge. As usual, I still maintain we have to be humbler than the totality of our art. We need to be in the position of a supplicant, to learn anything at all. We need to absorb, well before we dare to preach. No, no, we should never *preach*, instead, we must *teach*. That is the true meaning of the word *Guru*. Unfortunately, I still see a lot of self-styled *Gurus* around me today. To them I will point to Wikipedia, and hope they understand the origin of the word *Guru*[13]:

> **Guru:** As a noun, the word means the imparter of knowledge (*jñāna*; also Pali: *ñāna*) .... The word has its roots in the Sanskrit *gri* (to invoke, or to praise), and may have a connection to the word *gur*, meaning "to raise, lift up, or to make an effort". The importance of finding a guru who can impart transcendental knowledge (vidyā) is emphasised in Hinduism. ... One of the main Hindu texts, the Bhagavad Gita is a dialogue between God in the form of Krishna and his friend Arjuna ... "Acquire the transcendental knowledge from a self-realized master by humble reverence, by sincere inquiry, and by service. The wise ones who have realized the truth will impart the knowledge to you".

In other words, the word *Guru* is not just a heady label to stick onto your LinkedIn profile. We must first *learn* very patiently from nature. Thereafter, any acquired knowledge needs to be *passed on*. Because knowledge alone is eternal, none of us are. To repeat a famous quote by Isaac Newton, one which I had mentioned in my first edition too, and one which I still try to abide by today:

> I do not know what I may appear to the world, but to myself I seem to have been only like a boy playing on the sea-shore, and diverting myself in now and then finding a smoother pebble or a prettier shell than ordinary, whilst the great ocean of truth lay all undiscovered before me.

In effect, we are all obviously far smaller than what we are trying so hard to decipher. But on the other hand, we can consciously choose to take brash Susan Powter's approach instead (at our own risk). Ms. Powter happens to be a fairly well-known Australian-born motivational speaker who apparently has this to say: "*I know the guru route, I know you go sit on a mountain. But screw India. I ain't going there.*"[14] I would encourage her in that case to just "Go for it!"

With those final thoughts, I will leave you to explore the book, and discover it for what it really is. Our work should always speak louder than anything we can say on its behalf. So, I sincerely hope you will like it once again! May it help you break new ground, in your personal life, and also in technology. Let's keep the passion alive.

Sanjaya Maniktala

---

[13]See http://en.wikipedia.org/wiki/Guru.
[14]See http://www.brainyquote.com/quotes/authors/s/susan_powter.html.

Acknowledgments

I will keep this very short this time. First, I need to thank my technical reviewers and key supporters. They are (in no particular order) from old times too: Daniel Feldman, Thomas Chiang, Stephen K. Lee, Anton Bakker, Edward Lam, Craig Lambert, Philip Dunning, Jose Rangel, Behzad Shahi, Jeff Rupp, Ajithkumar Jain, Arman Naghavi, Inder Dhingra, and Anand Kudari.

The first Optimization book, and this latest edition too, were possible only because of Steve Chapman at McGraw-Hill. The second edition was subsequently handled very professionally indeed by Joy Bramble (sadly, no longer at McGraw-Hill), and finally, the wonderful Michael McCabe, who I know made this book possible at the last moment, when the manuscript was discovered to be much bigger than ever anticipated, or planned for.

Once again, following close on the heels of my PoE book, I received outstanding support from the very same production team: Yashmita Hota and all her competent colleagues at Cenveo Publisher Services at Noida, India. Thank you once again, for making it so easy on me. You guys rock.

I had promised my wife this would be my last book. Hey where is she gone? Oh, here she is now (still with me)! I really need to thank Disha, and my daughter Aartika, tremendously once again, for convincingly pretending to all that I was still at home, both physically and mentally, when I wrote this book. And thanks for not abandoning me years ago. I am done now, I promise! Now can I stay?

And finally, eternal thanks to my lasting Guru in India, Doctor GT Murthy (retired). Sir, I wouldn't have done any of this if I hadn't met you on that life-changing day in Bombay, years ago. Thank you so much once again, for all your faith. I wish you good health and a very long life. This book too is inevitably, an extension of your life, unquestionable integrity, and unforgettable teachings.

About the Author

Sanjaya Maniktala is the author of *Power over Ethernet Interoperability* (McGraw-Hill, 2013) and several other books on power electronics. He has held lead engineering and managerial positions in India, Singapore, Germany, and the United States. Mr. Maniktala holds several patents in power conversion and Power over Ethernet, including the floating Buck regulator topology and Mode transitioning in a Buck-boost converter using a constant duty cycle difference technique.

CHAPTER 1

References to Regulators

PART 1 OVERVIEW

Walking a Design Tightrope

The underlying expectation of any voltage regulator—switching or nonswitching, inductor- or capacitor-based, mundane or exotic—is that it produces an output voltage rail that is *accurate*, referring to its nominal (set) value, and *well regulated*, referring to its spread around that value, over what can be widely varying input (line) voltage and load conditions.

Several questions arise out of this seemingly simple requirement. For example, what do we consider accurate? What is *well* regulated? Unfortunately, there are no boilerplate answers for any of that here—both the center (nominal) value of the set output voltage and its allowable spread depend on the *application* at hand.

The output rail of the regulator forms the input rail of a certain target device or circuit (its load). The basic requirement is that the latter's performance and reliability remain virtually unaffected under the stated or allowed variation. For example, our mobile phone doesn't show increasing difficulty connecting to the cellular network over the course of a day as its battery gradually depletes. Its performance remains virtually unaffected right until the moment the battery actually runs out. We owe that little feat, something we usually take for granted, to a bunch of tiny regulators ticking away inside it.

In many modern-day applications, mainly because of the shrinking process geometries involved in chip fabrication, the requirements imposed on a regulator's output level have started to resemble a design tightrope of sorts, with many, near-conflicting, concerns and compulsions. A clear elucidation of this is contained in a white paper titled "Power Delivery for Platforms with Embedded Intel® Atom™ *Processor*," supporting the Intel mobile voltage positioning (IMVP) specification, as applied to voltage regulator modules (VRMs). These are DC-DC converters placed right next to a modern central processing unit (CPU). The paper explains the very tight situation they are in, in more ways than one, as follows:

> Generally, higher core voltages (CPU core or graphics core) enable higher performance and faster logic circuits, but if the voltage is too high it may damage or degrade the silicon and limit the lifespan of the CPU. Generally, lower core voltages reduce the energy requirements, and reduce the power that must be dissipated as heat, but if the voltage is too low, you run the risk of causing a logic failure that may cause the system to hang, reboot or blue screen. If the core voltage is fixed it must be fixed at a voltage level high enough to enable the required performance, yet low enough that the part is not damaged, and low enough that the CPU maximum operating temperature is not exceeded.

We begin to realize it is entirely possible that while we were griping about our computer's recurring "blue screen of death," attributing it to a "buggy" operating system or sinister malware, the culprit was actually some tiny, errant voltage regulator hidden deep inside our notebook, immune to scrutiny and thus considered innocent (yes, because we couldn't prove it guilty!). This highlights the importance of trying to understand the intricacies of the underestimated voltage regulator in our electronic world of today. It's not just about endless streams of data going back and forth.

> **NOTE** *The late Bob Widlar (1937–1991), one of the foremost analog designers of our time, some of whose breathtaking achievements are covered later in this chapter, had apparently opined that "every idiot can count to 1." See http://en.wikipedia.org/wiki/Bob_Widlar. That "tradition" was maintained steadfastly by his former colleague, the late analog guru, Bob Pease (1940–2011), who became rather well known for ceremonially throwing computers off the roof of Building C at the Santa Clara headquarters of National Semiconductor (now Texas Instruments). The latter Bob had this to say about the former Bob: "Obviously, there will never be another engineer like Widlar. He led the linear IC industry in many amazing new directions." Incidentally, both these hall-of-famers eschewed, if not despised, simulation, for reasons that we will start to recognize as we get deeper into this chapter.*

A Creative Experience

Voltage regulators based on switching power conversion principles are referred to in many ways: switch-mode power supplies (SMPS), switching power supplies, power supply units (PSUs), switchers, converters, and so on. These terms are somewhat loosely defined and applied, depending on the application. However, there also seems to be some sort of unifying attempt at progress in the form of an emerging standard called the IPC-9592 (available from www.ipc.org), which refers to them as PCDs (power conversion devices). Now all we have to do is to remember that!

Despite the impressive plethora of names, and an equally "impressive" history in the form of veritable waves of unexpected field returns—early aluminum electrolytic capacitors, followed by the crumbling of powdered iron cores (reported at Artesyn Technologies, now Emerson Electric), followed by strange shorts due to tin whisker growth (reported at Power Integrations), and so on—switchers are still sometimes taken for granted, especially by those who instinctively tend to equate design complexity to the size of the printed circuit board, or the solution cost, or the number of components. In reality, the design tightrope extends deep into almost every aspect of a switching regulator's design. For example, in a *Boost converter* (in which the output *exceeds* the input), we quickly discover we can speed up the switch (usually a FET) to decrease its transition (switching) losses, but end up with worse overall efficiency because of shoot-through (*cross-conduction*) through the *catch (freewheeling) diode*. Or we can increase the voltage ratings of the FET and the catch diode in switching regulators in an effort to enhance *field reliability*, and end up with higher, not lower, field failures—a direct outcome of the significantly higher temperatures arising from the higher forward (conduction) losses of the newly selected FET and diode. We thus realize that switching power supplies are nothing if not *the art of careful design compromises* and *tradeoffs* or, in short, *optimization*. And lest we forget; ultimately, cost is the key factor against which we must eventually learn to weigh *all* our design choices and decisions. The ultimate thrill is to eke out maximum reliability and performance at minimum cost. That is the ballgame that makes switchers so challenging, yet also so satisfying in the long term.

In the short term, many engineers feel somewhat overwhelmed at the demands being constantly placed on their shoulders. Their expertise needs to eventually encompass several almost distinct areas—electrical, magnetics, electromagnetics, thermals, control-loop theory, Fourier/Laplace transforms, and so on. Add to that simulation, and expertise in various CAD tools such as Mathcad and MATLAB. Not to mention *physics*! Now, we may hate the very sound of that, coming all the way at us from high school, but the truth is that physics is totally vital to understanding the *inductor*, the key component behind modern switchers. If we don't take the trouble to develop a keen conceptual and intuitive understanding of inductors (with the help of physics, naturally), pretty soon we will be the ones feeling saturated to the core. But if we do use it and thereby develop a broader perspective, we will suddenly start noticing and appreciating the exciting *interplay* of several varied engineering disciplines in the design of any good commercial power supply.

One thing we learn very quickly while working in the trenches is *we just can't take anything for granted*, certainly not in *switching* power conversion. Just when we thought adding a few more turns to the transformer would increase its inductance and thereby help "hold off" the high applied input voltage better, we find to our dismay that doing so only causes all our remaining prototype boards to blow up in a steady stream, all because of transformer saturation. Quite inevitably, in this new world of hard knocks (of the switching knocks variety), we learn the following lesson very quickly: don't assume anything, *test it*. For example, perhaps we "thought" step A would lower the temperature of the FET. But did it really? *Let's check*. And no simulations please, let's get *real*—real, as in "on the bench real"! That may be the only way we will discover that *there is really almost nothing we can do at a given point in a switching converter that will fail to affect something else, somewhere else*, usually to our eternal dismay. So, maybe the FET really is really running cooler now. But what about the catch diode? Do we now need step B to get its temperature down now? And how is the zener clamp of the Flyback holding up (literally)? Do we need step C to steady that before its solder melts? And so on.

In brief, we can rarely consider any switcher problem solved unless we are completely sure (by *bench-testing* of course, not from an ivory tower staring at a Simplis model) that we haven't just *transferred* the problem elsewhere or, worse, created several new problems (some almost astutely hidden under our noses right until the moment we ship boards off to the customer!). We must learn to avoid going to a series of unforeseen "containment steps" (Band-Aids galore), in an almost endless spiral of futility. That's really not *good* engineering. Knowing or *anticipating* where *else* the impact of a contemplated change is likely to be felt, looking for it proactively, then minimizing or limiting it before it catches us—all that is what distinguishes an experienced designer from a rookie.

But sometimes we will also need to accept, or shrug off, what is best described as the "fear of the unknown," that we may find residing somewhere within the wide arena of modern switching power supply development. It seems that historically the all-pervading sense of false comfort about switchers, bordering on near-cockiness initially, has finally given way to something akin to nervous overreaction in certain quarters. The author remembers one of his experienced colleagues being instructed by his manager, as he handed him a power supply that a customer was having problems with, saying, "Fix it, *but don't change anything."* The concerned engineer was thereafter seen walking around for days shaking his head in disbelief, wondering how he could abide by *that* mandate. Would he get fired if he *did* or if he *didn't*? And incidentally, did *what*? Or didn't do *what*? Bystanders like us had surmised that perhaps the manager was having a recurring nightmare: featuring an overly creative engineer, creating a web of changes, getting everyone from design to production hopelessly entangled, and the spider (the customer in this case) finally swooping down with exemplary timing. However, insecurities aside, creativity is *absolutely necessary* and (should be) welcomed with open arms in power conversion. We just don't throw the baby out with the bathwater. But it should all be buttressed and/or tempered by experience.

Static and Dynamic Regulation

With that brief introduction, we return to the basic aim of all regulators—*regulation*. For many broad market applications, a typical regulator's datasheet may describe it as being accurate to within ±2 percent, or simply 2 percent (which is actually a spread of 4 percent). That means that if the output of the regulator is set to 3.3 V, we are expecting, or allowing, the output to vary between the values $[1 - (2/100)] \times 3.3 = 0.98 \times 3.3 = 3.234$ V and $[1 + (2/100)] = 1.02 \times 3.3 = 3.366$ V.

Note that the defined ±mV or ±% output variation is in effect an *onion*. As we peel, we will unearth several layers that have gone into creating it. The first obvious contribution is the initial-accuracy term. In other words, we may *think* we have set the output precisely to 3.3 V, but because of component tolerances and so on, we will have a certain output error right off the bat—usually specified with zero load, at nominal input and at room temperature. It is important to know what this initial accuracy is, and we will discuss some of the factors affecting it very shortly (as well as some ways to deal with it if necessary).

On top of that, there is a certain *drift with temperature*. This is the temperature coefficient, or "tempco" (sometimes just called TC). The output will change as the unit heats up or cools down, and we still need to stay within a certain acceptable window. We therefore need to know the tempco of the regulator.

The other, most obvious (but not necessarily largest) contribution comes from the (static) line and load regulation characteristics of the regulator. This expresses the change in the output of the regulator in response to slow changes in its input voltage (line) and load. For example, we can gradually reduce the load from maximum to zero (or to a specified minimum) and record the initial and final *steady* output readings. That will indicate to us the static regulation capability of the regulator (measured over the specified min-max load variation).

In other words, besides an initial-accuracy term, there are several other contributions to the final total observed output accuracy, measured over the full range of the application. This includes line and load regulation.

Line regulation quantifies the error in the output resulting from a (slow) change in the input voltage. It is considered a DC (steady) specification and does *not* include the effects of any input ripple voltage e.g., or a sudden input (line) transient.

Load regulation quantifies the error in the output produced by a (slow) change in load current. It is also usually considered a DC (steady) specification and does *not* include the effects of any (sudden) load transients.

We could also subject the regulator to sudden or dynamic changes in line and load. These sudden transients can cause a rapid but usually momentary change in the output rail, and the resultant behavior can be quantified under the category of dynamic line and load regulation of the regulator. This can become the most significant, or dominant, layer of our proverbial "regulation onion." And for that reason, a typical power supply product specification (or regulator chip datasheet) may specifically allow for a wider dynamic variation window, say ±5 percent, as compared to the declared static regulation window (including initial accuracy and temperature drift) of, say, ±2 percent. There may also be a specified maximum *settling time* for such dynamic excursions. That numbers can typically be 1 to 2 μs for converters intended for microprocessor applications (VRMs), 100 to 200 μs for typical off-the-shelf DC-DC converters, and typically 1 to 2 ms for general-purpose AC-DC power supplies.

NOTE *The load transient requirement of a modern microprocessor chip can be in the hundreds of amperes per microsecond. However, the regulator itself may not "see," or need to support, such a high dI/dt, because a good part of that transient requirement is typically met by a bunch of paralleled high-frequency decoupling capacitors (often called* decaps*), sitting very close to the microprocessor chip. But that also raises the philosophical question: do we consider this bank of decaps as part of the total* output *capacitance of the regulator or part of the* input *capacitance of its load? Alternatively, where does the regulator end and the load start? It can get really hard to tell in VRM or point-of-load (POL) converter applications. But nowadays, the view is increasingly "chip-centric." That means that we look* outward *from the microprocessor (or other similar) chip, and everything outside connecting to its supply pins, including the regulator, the bank of decoupling capacitors, and so on, becomes the* power delivery network (PDN) *of the chip. It is the PDN that needs to provide the dI/dt demanded by the chip. The chip doesn't care what the constituents of the PDN want to call themselves, or how they want to split up the task among themselves. Similarly, on our part, we need to test it all out as a* complete system*, with the regulator only being part of that.*

The left side of Fig. 1.1 represents a typical dynamic response "scope shot" (an oscilloscope screen capture). Note that settling time is usually measured from the start of the disturbance to the moment the output returns to within the static regulation window (also note that it steadies down fully *a short while later*). Also the "settling voltage" (for want of a better name here) is commensurate with the static regulation characteristics of the regulator. That means we will reach exactly *that* level when we perform a *slow* (or *quasi-static*) load variation, as shown on the right side of Fig. 1.1. We see that the only difference is that we no longer get the dynamic excursion portion of the curve on the left. And that in effect distinguishes a slow perturbation from a sudden one.

We thus realize that the scope shot of dynamic response, as shown on the left side of Fig. 1.1, gives us a lot of "buried" or hidden information about the functioning and basic design of a voltage regulator, including its static regulation characteristics. In fact, it can go further and answer one of the most fundamental questions that can be asked of any voltage regulator: *is the regulator feedback (output correction) loop stable?* For example, if we see a lot of output ringing in the scope shot, due to which the output never really seems to *decisively* return to a settling value (within a reasonable time), then very likely we have an instability problem. Even if it does settle down quickly enough, the *amount of ringing* before that (settling) happens, reveals a lot about the *phase margin* of the feedback loop (in effect, the safety margin *before* the thresholds of full-blown instability). We will discuss these aspects in greater detail later. Here we just note and recognize that the left side of Fig. 1.1 is one of the most important bench measurements we can perform on any voltage regulator, switching or otherwise, well before we get down to validating its efficiency, thermal performance, electromagnetic interference (EMI), and so on. None of that really matters if the output is hopelessly ringing. With an unstable output, there really is no "regulator" to speak of (or test any further), is there? Yet we may find engineers doing exactly that, though thankfully only in their early years on the job.

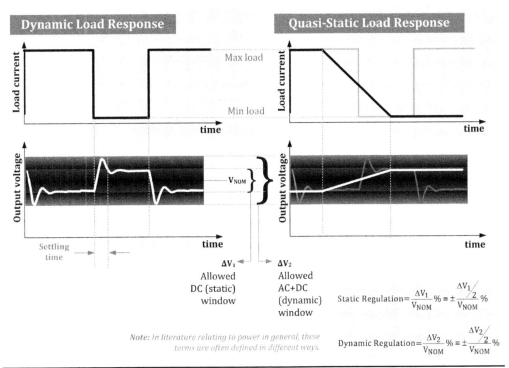

FIGURE 1.1 Static and dynamic load response of a typical regulator.

PART 2 UNDERSTANDING VOLTAGE REFERENCES

Voltage References in General

If we want to design a circuit that will somehow automatically set the output rail to the required level, say, 5 V, we should ask: how will our circuit ever "know" what is 5 V? Doesn't it need something to *compare* the output *against* (and correct accordingly)? In other words, we need a *voltage reference*. This vital component of any power supply/regulator design is often taken for granted, but especially in critical applications, such as the one mentioned in the preceding section "Walking the Design Tightrope." It becomes very important that *we understand our voltage reference well*. After all, how can we hope to create a regulator with, say, ±1 percent output accuracy if the reference voltage itself is only accurate to within ±2 percent? Note that just like the regulator, the reference itself also has line and load regulation characteristics, initial accuracy, tempco, and so on. It is the seed from which the full plant called the *regulator* grows. This is shown graphically as several layers of a *regulation onion* in Fig. 1.2. This diagram applies to the regulator, but has key components from the reference as indicated at the left.

Discrete zener diodes have been used as two-terminal voltage references for years. For example, a small zener diode can be connected to the base terminal of a hefty bipolar junction transistor (BJT), or to the gate terminal of a metal-oxide semiconductor field-effect transistor (MOSFET). This entire circuit becomes a *linear regulator* (also called a *series-pass regulator*, or simply *pass regulator* or *series regulator*). Under control of the zener diode, the transistor behaves as a *variable resistor* and adjusts its resistance to block any "excess voltage" across itself. Series regulators are typically three-terminal regulators, with an input, an output, and a shared ground. Incidentally, this behavior on the part of the transistor is the reason why its name evolved historically as a combination of two words "*trans*fer re*sistor*." See Fig. 1.3 for an overview of regulators.

Sometimes, hefty zener diodes are used as shunt regulators independently. They are then two-terminal regulators placed directly across the load in a brute-force attempt to clamp the voltage appearing across the load (they do need a limiting series resistor to control the current). Unfortunately, the common feature of all regulators involving discrete zener diodes, series or shunt types, is that the output regulation is usually rather poor (typically ±10 to 20 percent), and they can also display significant drift over temperature, besides noise.

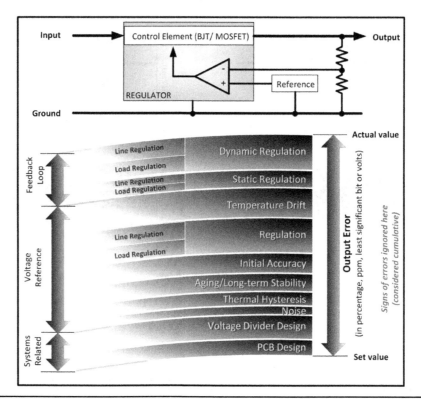

Figure 1.2 The regulation onion of a typical regulator and some key contributory errors in its output.

There are steps we can take to improve matters somewhat. For example, it is well known that low-voltage zener diodes have a *negative* temperature coefficient. So, a 3.3-V zener will typically have a tempco of about −3.5 mV/°C. On the other hand, high-voltage zeners have a *positive* tempco. A typical 15-V zener will exhibit around +10 mV/°C. Somewhere in the middle, around 5.1 to 5.6 V, the tempco changes sign and passes through zero. That is why it is often said that a 5.1- or 5.6-V discrete zener diode is a good choice for zener-based regulators. But zeners can also be fabricated on chips, and for such cases, a 6.3-V zener is considered the most stable value. Note that zeners can be fabricated on a chip in two ways—on its surface (a cheaper process) or as buried (subsurface) zeners. At the surface of a chip there are significantly more impurities, mechanical stresses, and defects (crystal-lattice dislocations) than within the chip. All these contribute to noise and long-term instability. In contrast, buried zener references are far more stable and accurate. Some of the most accurate monolithic chip references around today are therefore based on buried zeners. An example is the VRE3050A from Apex (Cirrus Logic). Its output (initial) accuracy is 5 V ± 0.5 mV (±0.01 percent) at 25°C. However, there is a potential showstopper here too: any monolithic chip reference using a zener reference needs a minimum supply voltage of at least 7 V. How else can you get a 6.3-V zener diode to exhibit zener breakdown? Also, usually the zener needs to itself draw several hundred microamperes for optimum operation (to come up to the "knee" of its current-voltage curve). Therefore, the VRE3050A uses a 6.3-V buried zener, but its minimum operating voltage is 10 V. Clearly, with today's shrinking supply voltages, such a high supply rail may not be available; and further, such a high idle current consumption ("quiescent current"), and resulting no-load dissipation is usually unacceptable, especially for battery-powered (portable) devices. Therefore, the bandgap reference has become the reference of increasing choice for most types of regulator applications today.

The Bandgap Reference

The most common bandgap reference value in use is "1.2 V." (The actual value ranges somewhat around this, but we will keep things simple for building up concepts.) From this basic reference value, by the use of scaling resistors and other circuitry such as op-amps, other common "fixed" reference values can also be generated. An example of a monolithic (standalone) bandgap reference is the LT1790A from Linear Technology. It is based on a 1.25-V bandgap reference. Its output (initial) accuracy is stated to be 5 V ± 2.5 mV (±0.05 percent) at 25°C. One of its key advantages is that it needs a minimum supply rail of only 5.5 V compared to the 10 V required by the VRE3050A. Its quiescent current is also just 35 μA compared to the 3.5 mA of the VRE3050A.

Since the 1.2-V bandgap reference is found in a majority of modern regulator ICs, we will focus on it in the next few sections. We start by trying to understand the BJT better, since that is the basic building block of bandgap references.

Understanding the BJT Better (the PTAT and CTAT)

The first silicon BJT was made by Texas Instruments in 1954. Thereafter, some interesting properties of the device were discovered, which eventually led to the bandgap reference. The first empirically discovered, reasonably stable voltage reference originated in 1964 from David Hilbiber of Fairchild. But its properties were not well understood until 6 years later when Bob Widlar (1937–1991) came along. He converted one of his basic current sources to a stable bandgap reference of nominal value 1.2 V, in what has been nicknamed "Widlar's Leap." This was released commercially in 1971 as a two-terminal "reference diode" IC, the LM113 (obsoleted just recently), from National Semiconductor (now Texas Instruments). Three years later Paul Brokaw came up with his (adjustable) version of the bandgap reference, at Analog Devices. The *Brokaw cell* was released in 1974 in an historic 2.5-V three-terminal (input, output, and ground) reference called the AD580, which is still in production. It has an initial accuracy of ±0.4 percent. All these events are not only part and parcel of but also the underlying reasons for the momentous period of growth that we now consider to be the birth of Silicon Valley.

Let's start the discussion with something everybody seems to intuitively understand: to get an NPN bipolar junction transistor to conduct, the voltage drop between the base and emitter of a BJT needs to be raised to around 0.6 V (at room temperature). So, basically we try to overcome this 0.6-V offset (it appears as a diode drop to us) and thereby force a small current into the base. In doing so, we observe a "lever effect" of sorts: the tiny base current opens the floodgates to a much larger current flowing from collector to emitter. We learn that this collector current is proportional to the base current, and the proportionality factor (= I_C/I_B) is called the β of the transistor (in a slightly different context it is called *hfe*).

The current-voltage (*I-V*) curves for the BJT are presented in Fig. 1.3. Note that *so far we are implicitly assuming temperature is constant*. The "ideal" (flat) curves assume that once the BJT is conducting, the collector current is fixed (and, as indicated, is in effect a fixed ratio of the base current). In other words, we are assuming that the collector current is *independent of the collector-to-emitter voltage* (it depends only on base current). This is a reasonable assumption to make for most first-order calculations, as we too have done in the ensuing discussions. But keep in mind that in reality, there is a slight divergence of the curves from ideal, and if extrapolated backward, the curves seem to be coming from a fixed voltage point $-V_A$ (typically around −75 to −100 V). This is called the *Early effect*, after its discoverer James Early (1922–2004). In the same figure, we have presented the *I-V* curves for the metal-oxide semiconductor field effect transistor (MOSFET). The curves are indeed similar. But note that a prevalent source of confusion lies in the way the saturation region of the BJT curves is called the *triode* or ohmic/resistive region in the MOSFET curves, whereas the forward active or linear region of the BJT is called the *saturation region* in the MOSFET curves. It is unfortunate terminology for sure.

We have mentioned in the same figure that the collector current varies exponentially as we increase the base-emitter voltage V_{BE}. The same is indicated at the top of Fig. 1.4, where we now also indicate that the collector current is β times the base current. We also clarify that the BJT has *four* quadrants of operation. One is of virtually no use (all its so-called diodes are reverse-biased). On the other hand, a switching regulator (one that uses a BJT switch) operates constantly between saturation and cutoff regions (very briefly passing through the active region as it changes state). However, current mirrors, amplifiers, oscillators, and voltage references operate continuously in the *forward active* (also called *active* or *linear*) region. Note that linear (series-pass) regulators also work in the linear region (or ohmic/resistive region) when using a BJT (or MOSFET) as the pass element.

We will focus on the forward active region of the BJT now. The guiding equation in this region is the exponential equation

$$I_C \approx I_S \times e^{qV_{BE}/kT}$$

This indicates that a small change in base-emitter voltage V_{BE} produces an *exponential* increase in collector current I_C (constant temperature still being assumed).

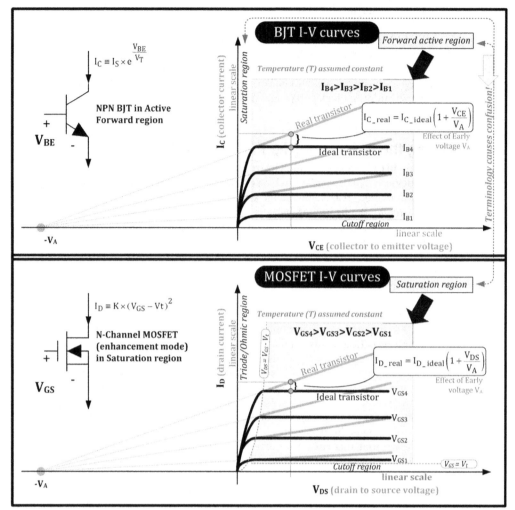

FIGURE 1.3 The I-V curves of a BJT and MOSFET compared, along with the Early effect.

NOTE *Though it is commonly stated that the BJT is a current-controlled device versus the MOSFET, which is a voltage-controlled device, the equations for a BJT indicate that it too is best visualized as a voltage-controlled device, since we can see from the above equation that a voltage (V_{BE}) determines current (I_C), quite similar to a MOSFET. No wonder their I-V equations in Fig. 1.2 are also so alike. One major difference, however, is that V_{BE} leads to a small, but constant base current in the BJT, whereas in a MOSFET the gate current stops when the voltage V_{GS} (between gate and source) reaches a steady value. So, the input of a BJT behaves as a diode, whereas the input of a MOSFET behaves as a small capacitor—but to pass drain current, you must first overcome an offset called the* **threshold voltage** *(V_t in Fig. 1.3).*

We can take the natural logarithm of both sides of the BJT equation to get the equation of a straight line

$$\ln(I_C) \approx \ln(I_S) + \frac{q}{kT} V_{BE}$$

We can also take the more common logarithm (to base 10) of both sides to get

$$\log(I_C) \approx \log(I_S) + \frac{q}{kT} V_{BE} \times \log(e) = \log(I_S) + \frac{q}{2.303\, kT} V_{BE}$$

which, as expected, is also a straight line.

References to Regulators 9

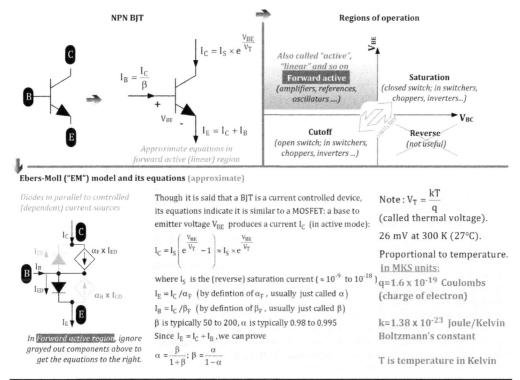

FIGURE 1.4 The Ebers-Moll model of an NPN BJT and its regions of operation.

***

NOTE *The scales "ln" (i.e., natural log) and "log" (i.e., to base 10) are related by a simple factor, because $\log(e) = 0.434$, $\ln(10) = 2.303$, and $\log(e) \times \ln(10) = 1$.*

Incidentally, the I_C versus V_{BE} exponential equation is very similar to the well-known *Shockley's equation* for a diode, and therefore it is no surprise that we discover the base-emitter junction of a BJT is best modeled as a diode. However, the formal BJT model, one that covers all its regions of operation, is the *Ebers-Moll model*, presented in simplified form within Fig. 1.4. But there is no need to get into that level of detail however, if you are feeling overwhelmed. Consider it as being there only for completeness sake.

In looking at various online forums, it is obvious there are several sources of deep confusion concerning the above I_C versus V_{BE} textbook equation. Yes, it seems to be a simple, exponential equation, but one that needs to be seriously clarified here. In the course of that endeavor, we will also develop the key ideas that go into making a band-gap reference.

1. *What is I_S?* This is often called the *saturation current*, a phrase borrowed from the similar-looking diode equation (from Shockley). But for a transistor the use of this phrase can get confusing. It has led some to erroneously think that I_S is the collector current in the *saturation region* of the BJT. Not so: it is a very small current of around 10^{-14} typically, so it can't possibly be that. Therefore, some prefer to call I_S the "reverse saturation current" and thereby implicitly assume it is the reverse leakage current of the base-emitter diode junction. And then, right after that, they usually compound their error by assuming that since I_S is the *reverse* current, it has no effect on the *forward* characteristics of the BJT. In fact none of these statements is true. Under reverse bias, there is a parasitic leakage current term that can dwarf I_S. Therefore I_S is best measured in the *active* region, not in the reverse region (same recommendation as for a diode). So, let's be clear on what I_S is *mathematically*, for therein lies our answer. Looking at the simplified Ebers-Moll equation above, we see that *if V_{BE} is set to zero, we get $I_C = I_S$*. So, that's what it is. Graphically, if we increase the voltage at the base, starting from zero (with respect to the emitter), and plot I_C *on a log scale* versus V_{BE} on a linear scale, we get a straight line, which *intersects* the y axis at the exact value I_S. It is therefore perhaps better to call I_S the *scaling current* to avoid confusion, because the *entire I_C* versus V_{BE} curve *scales* proportionately to I_S. Thus I_S is simply the *starting point* of the entire forward-biased I_C versus

V_{BE} curve, and therefore it does have a profound influence on the entire curve. For example, changing I_S by a factor of 10, say, from 10^{-14} to 10^{-13}, will cause *all* I_C values (for a given V_{BE}) to increase by the exact same factor (10 in this case). So it is very important to know what I_S really is: don't get fooled into thinking that since I_S is a small number (typically picoamperes), it is insignificant, or second-order. It is vital; hence the need to bring it up here.

In the SPICE (Simulation Program with Integrated Circuit Emphasis) model of the well-known discrete small-signal NPN transistor 2N2222 of a certain vendor, we come across the following line: "IS = 14.340000E-15." This means I_S is 14.34 × 10^{-15} A. The SPICE model of a well-known power transistor, 2N3055, from another vendor, states "IS=2.37426e-14," which means I_S is 2.37426 × 10^{-14}. And so on. However, this also points to a major limitation, if not error, in the commonly used BJT SPICE models, as we will see shortly.

2. *What is the slope of $log(I_C)$ versus V_{BE}?* Engineers may have heard this particular slope being expressed alternatively as 40 mV or 60 mV or 80 mV per decade of current. What exactly is it and *why*? [Note that actually we are talking about the slope of V_{BE} versus log (I_C) here.] From the above equations, the slope of ΔV_{BE} versus $\Delta[\log(I_C)]$ is $2.303\ kT/q$, as also indicated in Fig. 1.5. So, we see that the *slope $\Delta V_{BE}/\Delta[\log(I_C)]$ is proportional to temperature*. Or, equivalently, for a fixed $\Delta[\log(I_C)]$, or equivalently a fixed ΔI_C (i.e., a fixed *change* in current), ΔV_{BE} (the corresponding *change* in base-emitter voltage) *is proportional to temperature*. So, at 300 K (27°C), the slope of ΔV_{BE} versus $\log(I_C)$ is 60 mV per decade of current. It becomes 80 mV per decade at 400 K (127°C) and is 40 mV per decade at 200 K (–73°C). It goes from 40 to 80 mV/decade from 200 to 400 K. We can see the obvious proportionality with respect to T.

Note *One decade of current change (say, 0.1 to 1 A, or 1 to 10 A), corresponds to $\Delta[\log(I_C)] = 1$.*

Note *The Kelvin (absolute temperature) scale is just the Celsius scale translated by 273°C. So, 0 degrees K (or equivalently 0 degrees absolute temperature) equals –273°C.*

We are seeing a change of ΔV_{BE} from 40 to 80 mV for a temperature change of 400 K – 200 K = 200°C. In other words, the tempco of ΔV_{BE} is 40 mV/200°C =

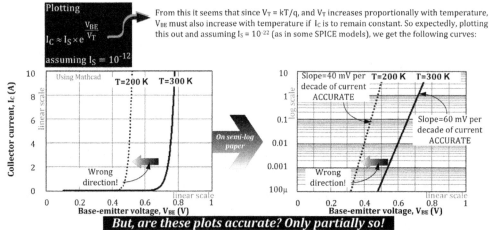

Figure 1.5 Attempting to plot I_C versus V_{BE} for different temperatures.

0.2 mV/°C (*for a decade of current variation*). This is referred to as a **p**roportional **t**o **a**bsolute **t**emperature (PTAT) voltage. We will shortly see that this very property of ΔV_{BE}, along with a complementary (*inversely proportional*) property of V_{BE}, is what goes into creating the well-known 1.2 V bandgap reference.

3. *How does V_{BE} vary with temperature?* This is the source of the biggest confusion. As mentioned, the base-emitter voltage for forward conduction is 0.6 V *at room temperature*. We ask: what is it, e.g., at exactly −273°C (0 K)? In fact, it is about 1.2 V *for any collector current* (though as an extrapolated value, it is not real, since at absolute zero all atomic motion is at a standstill and there can be no "current").

But is that what the exponential *I-V* equation predicts? In the *I-V* equation, to keep I_C fixed as we vary temperature, it seems we need to keep the following term fixed:

$$\frac{qV_{BE}}{kT} = \text{constant}$$

We can thus easily (but *erroneously*) conclude that if T (i.e., temperature, measured on the Kelvin scale) *increases* (or decreases), then V_{BE} must *increase* (or decrease). The tempco *seems* to be

$$\frac{V_{BE}}{T} = \frac{k}{q} = 86.25\ \mu V/°C$$

NOTE *kT/q is commonly called the* thermal voltage *as it has the units of voltage and is also proportional to temperature. At 300 K (27°C) it has the well-known value 86.25 $\mu V \times 300$ = 26 mV.*

In other words, the above equation seems to say that V_{BE} will increase with temperature at the rate of 86.25 μV/°C. But that doesn't sound right: most engineers have heard (and rightly so) that V_{BE} *falls* with increasing temperature—by about −2 mV/°C. Where is the error coming from?

The problem is what no one seems to clearly tell you: *I_S is actually a very strong function of temperature*. That fact is surprisingly overlooked by many engineers, especially those who rely blindly on SPICE and its commonly available BJT/diode models—in which I_S is usually set to some *fixed* (default) value. In reality, I_S roughly doubles for every 5°C rise in temperature. And *if that variation is factored into the I-V equation, then to keep I_C constant, V_{BE} actually needs to decrease substantially with increasing temperature, not increase*.

As indicated in Fig. 1.6, a fairly good empirical fit to data is obtained by setting

$$I_S = I_{S0} \times e^{\frac{-V_{G0}}{V_T}}$$

where $V_T = kT/q$, $V_{G0} \gg 1.2$ V, and I_{S0} = constant. With this model for $I_S(T)$, the *correct* curves are now plotted in Fig. 1.6. We finally see that because of the dependency of I_S on temperature, *all* the log(I_C) versus V_{BE} curves we talked about earlier (for different temperatures), if extrapolated, point to the voltage of 1.2 V (same as V_{G0}), though at some very high, hypothetical (and yes, *absurd*) value of collector current. More importantly, we see that as temperature increases, the curve shifts to the *left, not to the right* as wrongly indicated by Fig. 1.5 (in which we had *wrongly* assumed that I_S was a constant). So now to keep I_C constant as we vary temperature, we need to keep

$$\frac{q(-V_{G0} + V_{BE})}{kT} = \text{constant} \quad \text{or} \quad \frac{(-V_{G0} + V_{BE})}{T} = \text{constant}$$

Let's talk numbers here. For example, at $T = 300$ K and at the offset value $V_{BE} = 0.6$ V, the numerator of the above fraction is $-V_{G0} + V_{BE} = -1.2 + 0.6 = -0.6$. If the temperature decreases to 200 K, the numerator must decrease by the same factor, i.e., to $(2/3) \times (-0.6) = -0.4$. The only way that could happen is if V_{BE} became $-0.4 + V_{G0} = -0.4 + 1.2 = 0.8$ V. In other words, for a 100°C increase in temperature (200 K to 300 K), V_{BE} decreases from 0.8 V to 0.6 V, that is, by 0.2 V. The effective tempco is $-0.2\ V/100°C = -2.0\ mV/°C$, which is the often-mentioned value in literature.

Now, if the base-emitter voltage is 0.6 V at 300 K and its tempco is −2 mV/°C, we can imagine that at a theoretical temperature of 0 K (−273°C), the base-emitter

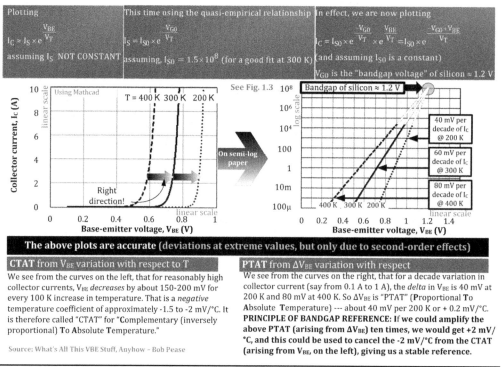

Figure 1.6 Correctly plotting I_C versus V_{BE} and understanding CTAT and PTAT generation.

voltage will be $0.6\,\text{V} + (300\,\text{K} \times 2\,\text{mV}) = 1.2\,\text{V}$ at 0 K (i.e., at absolute zero), which is V_{G0}, the *bandgap voltage* of 1.2 V (we won't bring in atomic physics to explain the name *bandgap* here). But clearly 1.2 V is the value we had set for V_{G0} while modeling I_S. And that's where it comes from.

We can plot several V_{BE} variations with respect to T in Fig. 1.7, and we see that *all* the V_{BE} curves (for different currents and biasing) intersect at $V_{BE} = 1.2\,\text{V}$ at

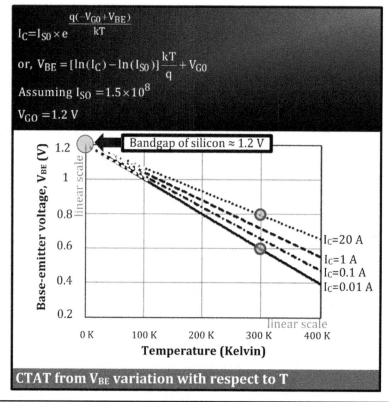

Figure 1.7 Plotting V_{BE} versus T and understanding CTAT generation.

absolute zero (0 K). This type of curve is a very useful curve for analog IC designers. It also tells us that contrary to popular myth, the tempco of V_{BE} is −2 mV/°C, but *only for the case of V_{BE} = 0.6 V at 300 K*. For example, if in Fig. 1.7 we look at the curve for 20 A, we see that it has a V_{BE} of 0.8 V at 300 K. That must have a different tempco, because we now know this curve too *must converge to 1.2 V at 0 K*. So obviously its tempco must be (0.8 V − 1.2 V)/300 K = −1.43 mV/°C, not −2 mV/°C.

We can clearly observe an *inverse* proportionality relationship here (with respect to temperature). We thus say that V_{BE} is a CTAT voltage.

***

NOTE *Maybe, a more contemporary or easier name would have been iPTAT (inversely proportional to absolute temperature).*

***

The Principle behind the Bandgap Reference

For example, the difference in the base-emitter voltages (ΔV_{BE}) between two *matched* transistors, one running at current I and the other at $10 \times I$ (one decade difference), is 60 mV at room temperature (300 K). Its tempco is +0.2 mV/°C. If we amplify this PTAT difference voltage 10 times, we will get a voltage value of 0.6 V with a tempco of +2 mV/°C. This is a PTAT voltage. On the other hand, the actual V_{BE} (of either of the two transistors above) is about 0.6 V at 300 K, with a tempco of −2 mV/°C. This is a CTAT voltage. If we add this CTAT V_{BE} voltage to the 10 times amplified ΔV_{BE} PTAT voltage mentioned above, we will get a net voltage of 0.6 + 0.6 = 1.2 V, *with a net tempco of +2 mV/°C − 2 mV/°C = 0* (ideally). This is the well-known bandgap reference voltage. And that's how we get the number 1.2 V here (0.6 V CTAT plus 0.6-V PTAT).

To understand how it was actually implemented by Widlar and Brokaw, we need to first understand the basic current mirror.

The Basic BJT Current Mirror and Widlar's Bandgap Reference Cell

In Fig. 1.8 we start by explaining the advantages of the *diode-connected BJT*. This forms the first element of the basic two-transistor current mirror. The input is considered to be the current passing through the collector of the first transistor, and we get a corresponding output—the base-emitter voltage. If this base-emitter voltage is then applied as an input to a second *matched* transistor, by virtue of the fact that both transistors have the same *I-V* curves and have the same V_{BE} values (externally forced) too, we necessarily get the same I_C (collector current) as an output from the second transistor. In other words, the current in the first transistor has been buffered and "mirrored" (1:1) onto another transistor, just by having them share their control voltages (base-emitter drops).

What if the "second transistor" is actually a set of, say, *four* identical/matched transistors in parallel? Now, since each constituent transistor of this set receives the same applied base-emitter voltage, each one will produce a collector current equal to the current through the first transistor. Therefore, all together, the four paralleled transistors produce $4I_C$. In effect we have created a 1:4 current mirror.

It is actually not necessary to have four separate base, collector, and emitters joined together. By using a second transistor with an emitter area 4 times that of the first transistor, we can get the same 1:4 current mirror. And similarly we can produce a 1:10 current mirror, as shown in the figure. Note that rather than draw so many emitters to one BJT, we could simply write "10" or "10E" (sometimes also written as "10A") next to the second transistor to indicate that its emitter area is 10 times that of the first transistor. The underlying assumption is these are *matched transistors* and therefore *monolithic* (on the same integrated chip).

Not shown in the figure is the fact that instead of a 1:10 mirror, we could create a 10:1 current mirror by using a first transistor with an emitter area 10E compared to the second transistor with an area 1E. Similarly we could produce a 3:2 current mirror, with emitter areas 3E and 2E, respectively. And so on.

Widlar gave an interesting twist to the current mirror circuit and created the bandgap reference. We will explain it through Figs. 1.9 and 1.10. In Fig. 1.9, we have something similar to a current mirror, *but it really isn't one anymore*: because we have a new resistor R_{DIFF} in place, that alters the V_{BE} values of the two transistors (or rather lets them be different). Also, we are externally *forcing equal* currents into the two transistors. However, as before, similar

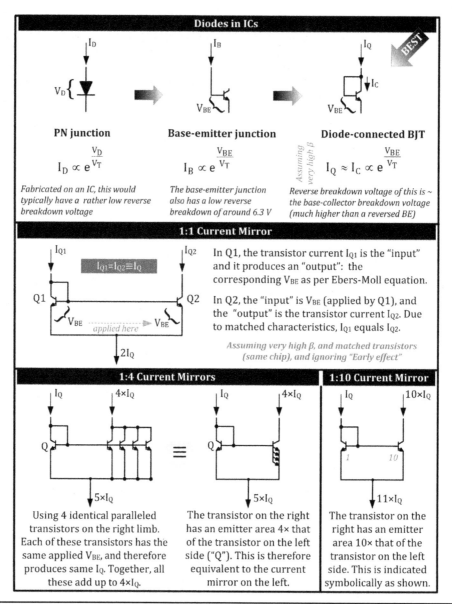

Figure 1.8 Understanding BJT current mirrors.

to the 1:10 current mirror case, the right-side transistor has been chosen to have 10 times the emitter area compared to the left-side transistor. We can mentally think of the larger emitter area as 10 paralleled transistors, each with a collector current of one-tenth of the transistor on the left side. So the corresponding V_{BE} is also less, per the Ebers-Moll equation (applied to each paralleled transistor). In effect we have a certain V_{BE} on the left-side transistor that corresponds to a collector current I_Q. We have a different (lesser) V_{BE} on the right side, corresponding to a lesser current of $I_Q/10$. Note that this is a difference of one decade of current, and so from Fig. 1.4, we realize the difference voltage ΔV_{BE} (a PTAT voltage) must be 60 mV at 300 K, with a tempco of 0.2 mV/°C.

Next step: Since the voltage drop between the shared base connection and the (common) ground must be the same whether we go down from the left transistor side or the right transistor side, we realize the difference voltage ΔV_{BE} *must appear across* R_{DIFF}. Further, if this is amplified 10 times and added to the –2 mV/°C CTAT voltage V_{BE}, then, as explained in Fig. 1.10, we get a 1.2 V reference as desired, and the net tempco is ideally zero.

Note that instead of *forcing equal currents* through *unequal* (1:10) transistors to create a certain ΔV_{BE}, we could equivalently force *unequal currents* (10:1 ratio) through *identical* transistors, and we would get the same ΔV_{BE}. Eventually, it is all about the emitter current *density*.

In the right limb: 10 transistors in parallel or one transistor with 10 times the emitter area.
*Note: **looks like a current mirror, but is not!***

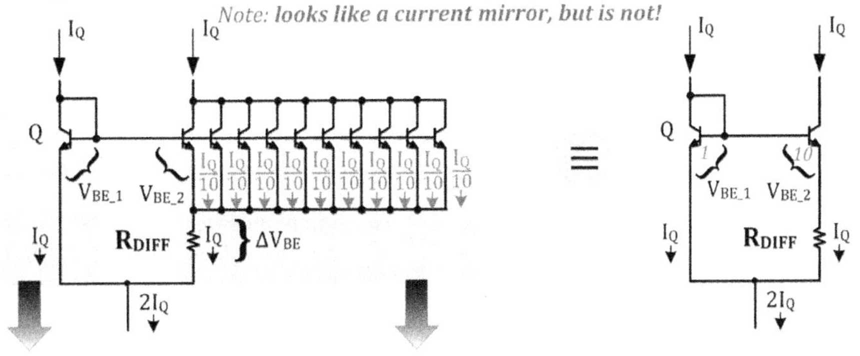

Using Ebers-Moll equation (approximate) and assuming matched transistors and very high β

For the transistor in the left limb

$$I_Q = I_S \times e^{\frac{V_{BE\_1}}{V_T}}$$

$$V_{BE\_1} = V_T [\ln(I_Q) - \ln(I_S)]$$

For each transistor in the right limb

$$\frac{I_Q}{10} = I_S \times e^{\frac{V_{BE\_2}}{V_T}}$$

$$V_{BE\_2} = V_T \left[\ln\left(\frac{I_Q}{10}\right) - \ln(I_S)\right]$$

Therefore, the delta in V_{BE}'s is

$$\Delta V_{BE} = V_{BE\_1} - V_{BE\_2} = V_T \left[\ln(I_Q) - \ln\left(\frac{I_Q}{10}\right)\right] = V_T [\ln(I_Q) - \ln(I_Q) + \ln(10)]$$

$$\Delta V_{BE} = V_T \times 2.303 = \frac{k}{q} \times 2.303 \times T \quad \Longleftarrow \text{Generating a PTAT voltage}$$

$$\Delta V_{BE} = 198.6 \times T \ \mu V$$

Example: At 27°C (300 K), we get $\Delta V_{BE} = 198.6 \times 300 \ \mu V = 60 \ mV$.

Also, $\Delta V_{BE}/T$ is about 200 μV or 0.2 mV/°C. ←Its temperature coefficient

Therefore we now have a PTAT voltage of 60 mV (at 300 K — room temperature) with a temperature coefficient of 0.2 mV/°C. If we amplify this ten times, and add it to the 0.6 V of V_{BE} coming from, say, another transistor, which we know is a CTAT voltage with a temperature coefficient of -2 mV/°C, we will get an absolute voltage value of 0.6 V + 10 × 0.06 = 1.2 V with a net temperature coefficient of -2 +10 × 0.02 = 0 mV/°C. We implement this in the next figure, and thus get a basic 1.2V stable bandgap reference.

NOTE: This requires two transistors, one with emitter area _ten_ times that of the other: ONLY then we get a difference of 60 mV with a temperature coefficient of +0.2 mV/°C (PTAT)

FIGURE 1.9 First step in understanding the basic Widlar bandgap reference cell.

In Fig. 1.11 we show Brokaw's version. In his own words, as quoted from his historic 1974 paper*: "A new configuration for realization of a stabilized bandgap voltage is described. The new two-transistor circuit uses collector current sensing to eliminate errors due to base current. Because the stabilized voltage appears at a high impedance point, the application to circuits with higher output voltage is simplified." So, now we can use the 1.2-V bandgap reference with scaling resistors, to easily derive reference voltages of say 3.3 V, 5 V, and so on.

Theoretically, we could take a 1.2-V bandgap reference and use an op-amp with gain less than 1 to generate stable references of less than 1.2 V. That is not the problem. The problem is that any circuit that is based on a 1.2-V reference would necessarily require a supply rail significantly *higher than 1.2 V* (at least 1.9 V to 2.3 V typically). That voltage could be applied through a separate pin in a three-terminal device, or connected to a biasing resistor in a two-terminal (shunt) device (e.g., LT1389). In any case, such a resulting voltage reference would, for that reason, never be suited for, say, a single 1.5-V alkaline cell application, because an alkaline cell can go down to below 1 V before being considered fully discharged. To combat

---

*http://www.cems.uvm.edu/~abonacci/ee222/Images/JSSC_Brokaw.pdf.

Widlar's first bandgap reference (simplified)

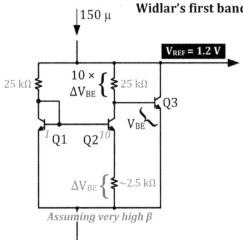

Q2 has an emitter area ten times that of Q1 and thus, for the same current, has a lower V_{BE}. The delta between the two V_{BE}'s of Q1 and Q2 shows up across the 2.5 kΩ resistor. It has a temperature coefficient of +0.2 mV/°C. That is voltage-amplified ten times by the ratio 25 kΩ/2.5 kΩ = 10 to give us 600 mV with a temperature coefficient of + 2 mV/°C. This PTAT voltage further gets added to the 0.6 V of V_{BE} from Q3, which we know has a temperature coefficient of -2 mV/°C (is a CTAT). Hence we get a net 1.2 V with a theoretical temperature coefficient of zero: our basic **bandgap reference:**

$$V_{REF} = 10 \times \Delta V_{BE} + V_{BE}$$

Rather than scaled (1:10) transistors, Q1 and Q2, operated with the identical collector currents, we could use identical/matched transistors, but run Q1 at 10× the current through Q2. We would get the same ΔV_{BE}

Source: Designing Analog Chips – Hans Camenzind

FIGURE 1.10 The basic Widlar bandgap reference cell.

Brokaw's bandgap reference (simplified)

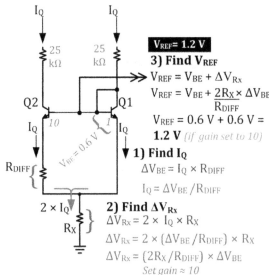

Once again, Q2 has an emitter area ten times that of Q1 and thus, for the same current, it has a lower V_{BE}. The difference is 60 mV as per previous figure. (Note: We need to externally force equal currents through both limbs, because now we can have a significant drop across the new resistor R_X, and therefore, we cannot depend on the fact the two equal 25 kΩ collector resistors will create equal currents). The voltage across R_{DIFF} (i.e. ΔV_{BE} = equal to 60 mV at 300 K) is voltage-amplified by the ratio $2R_X/R_{DIFF}$. So if R_X is, say, 5 times R_{DIFF}, we get a gain of 10 as required, and so, 60 mV becomes 600 mV. We do not need an extra transistor to add another V_{BE}. We use the V_{BE} across Q1. So we end up adding a PTAT voltage (600 mV from the amplified ΔV_{BE}, with a temperature coefficient of + 2 mV/°C) to a CTAT voltage (0.6 V from the V_{BE} of Q1, with a temperature coefficient of -2 mV/°C), to get a net 1.2 V with a temperature coefficient of near zero: a *1.2 V bandgap reference*.

Brokaw's bandgap reference cell for voltages > 1.2 V (simplified)

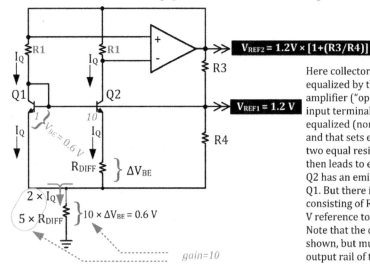

Here collector currents are proactively equalized by the use of an operational amplifier ("op-amp"). The voltages at input terminals of the op-amp get equalized (normal action of an op-amp), and that sets equal voltages across the two equal resistors marked "R1", which then leads to equal currents. As before, Q2 has an emitter area ten times that of Q1. But there is also a voltage divider consisting of R3 and R4 that scales the 1.2 V reference to a higher level *if required*. Note that the op-amp supply rails are not shown, but must obviously exceed the output rail of the op-amp.

FIGURE 1.11 The basic Brokaw bandgap reference cell.

that problem, a lot of effort is ongoing to produce "sub-bandgap references" (less than 1.2 V) and/or "sub-volt bandgap references" (less than 1 V).

But Widlar was ahead of everyone even in that day. In 1978, the LM10 was introduced by National Semiconductor, and it is still in production. It can work down to 1-V supply rail, and its reference output can go as low as 200 mV—because it is based on a 200-mV sub-bandgap reference from Widlar, one that remained unmatched by competitors for a decade after it was created. Basically, by using BJT transistors with a 50:1 emitter area ratio, Widlar produced a PTAT of 100 mV. Then, rather than amplify this to get the tempco up to the CTAT level (with opposite sign), he added to it a 100-mV CTAT derived by *scaling down* V_{BE}. Adding the two, he got a stable reference of 200 mV.

A Modified Current Mirror and the CMOS Bandgap Reference Cell

In Fig. 1.12, we modify the basic Brokaw bandgap reference cell by introducing an op-amp and moving the difference resistor (with the PTAT voltage) onto the collector side of the transistor, for higher gain and better *power supply (noise) rejection ratio* (PSRR). We then transition to a version using P-MOS transistors (the MOSFET equivalent of a PNP BJT) and low-side PNP BJTs. Without going into the details of IC fabrication processes, it suffices to say that there are parasitic bipolars available within a typical CMOS (complementary metal-oxide

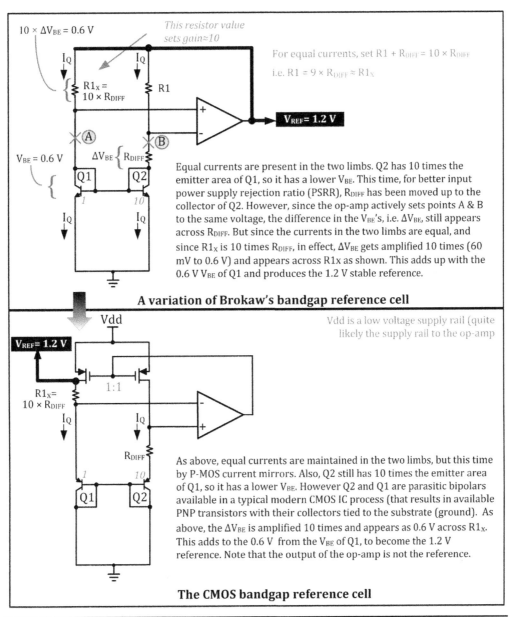

FIGURE 1.12 Brokaw's bandgap reference with an op-amp, and the transition to its CMOS version.

semiconductor) fabrication process, and the most useful ones for the purpose are usually those that happen to be PNP types with their *collectors tied together to the substrate* (ground). That is how we generate the basic CMOS bandgap reference cell shown in Fig. 1.12. This at least illustrates the principle commonly used for CMOS.

NOTE *The Brokaw cell usually requires a start-up circuit (not shown), because it can have two stable values for transistor currents: one that leads to the bandgap voltage and one that corresponds to zero current and voltage. It can get stuck in the latter without some "help." Also, CMOS op-amps have a larger offset and are therefore self-starting only if the applied offset is right.*

Simulation or the Lab?

In general, almost all bandgap references are based on the same basic principle: a PTAT and a CTAT canceling tempcos mutually to get a stable reference. But as we just found out, the only reason the bandgap reference exists is because I_S *is a function of temperature*. Now consider the fact that *most SPICE models set a default, fixed value for* I_S. And that leads us to some remarks about simulation in general.

Bob Pease (1940–2011) wrote: "... there are many ways to get false results in SPICE and other simulation schemes. Most transistor models do not accurately model the shape of the curve of V_{BE} vs. temperature. It is sometimes possible to tweak the characteristics of the model of a transistor until the tempco of the breadboard or analog model matches that of the computer model. However, after you have done this, it is not safe to assume that reasonable changes in the operating points will cause reasonable changes in the tempco. The actual changes may be different from the computed changes, even for a minor tweak."

Intusoft, a well-known vendor of a popular SPICE-type tool, warns: Designing circuits using computer simulation requires that models accurately reflect device behavior within a specific circuit context. Models with excessive detail will obscure the circuit designer's insight and will quickly reach both runtime and complexity limitations of the simulation program. Overly simple models will fail to predict key circuit performance parameters and may lead to costly design mistakes. Device modeling is one of the most difficult steps in the circuit simulation process. It requires not only an understanding of the device's physical and electrical properties, but also an intimate knowledge of the particular circuit application.

Therefore, we must ask: if the conventional device models of simulation do not even predict a simple bandgap voltage reference, how *trustworthy* do you expect them to be in replicating complete switching regulators (especially on paper)? We also ask: did Widlar, Brokaw, or Pease ever base any of their legendary commercial op-amp or voltage reference designs on a *simulation package*? We are quite sure they didn't. They collected data fast and furiously and first really understood what they were dealing with. For example, Wikipedia reports that Widlar locked himself in at Fairchild for 170 hours of continuous experiments (related to the BJT) and emerged with a certain robust design.

NOTE *It is interesting that the Ebers-Moll model (also called the coupled diode model because of the EB and EC diodes) was released in 1954, much earlier than Widlar's insights. Clearly, the correct variation of V_{BE} with temperature could not have been present in the Ebers-Moll model. And since traditional BJT SPICE models are largely based on the Ebers-Moll model, they too have the same limitations as regards temperature.*

In November 2004, Paul Tuinenga responded to an online article on the topic of SPICE: "Sanjaya . . . I have this recollection of a passage from an out of print book (I think it was) 'Paper Money' by Adam Smith, in which the author tells of a conversation with a Southern stock broker while accompanying the said broker on a morning bird hunt with the broker's dog. The topic is computers and their use in stock trading. The broker counsels (something along the lines of): The computer is like a bird (hunting) dog. Very useful. Wouldn't think of hunting without one. They spot birds and retrieve. But you don't give the gun to the dog!" Incidentally, Tuinenga is the co-creator of PSPICE and the cofounder of MicroSim Corp. (taken over by Orcad, followed by Cadence). He is also the author of the best-selling book on SPICE.

So perhaps we really shouldn't make the mistake of using a tool to learn a subject—that does sound a lot like giving the gun to the dog.

With that, it is time to go back to looking at voltage references as "black boxes" and understanding the specifications and performance of voltage references in general as applied to regulators.

Comparing Voltage References

One of the most basic comparisons we can make is based on the initial accuracy term discussed previously. The simplest way to express this is as a percentage of the *nominal value* (which is usually specified at 25°C, with no load applied to the reference). For example, a 2.5-V reference accurate to 0.5 percent initial accuracy has the following absolute voltage spread.

$$\frac{\pm(2.5 \text{ V} \times 0.5)}{100} = \pm\frac{1.25 \text{ V}}{100} = \pm 0.0125 \text{ V} \Rightarrow 12.5 \text{ mV (or 25 mV)}$$

In other words, the maximum is 2.5 V + 12.5 mV = 2.513 V. The minimum is 2.487 V.

Instead of percentage, *parts per million* (ppm) is often used: 1 PPM is 10^{-6}. The factor connecting ppm and percentage is 10^4. So *ppm is 10^4 times %*. A 100 ppm initial accuracy is $\pm 100 \times 10^{-4}\% = \pm 10^{-2}\%$, or 0.01 percent.

NOTE *For stating initial accuracy, the ± is often not written out explicitly, but is certainly always implied (whether expressed as a percentage or in terms of ppm). So an initial accuracy of 100 ppm or 0.01 percent always implies ±100 ppm and ±0.01 percent respectively.*

NOTE *If the reference is stated to have a certain initial accuracy expressed in, say, X ppm or Y percent, only when talking of initial accuracy does that mean ±X ppm or ±Y percent— because the nominal value is always picked to be the center of the initial accuracy spread. This ± prefix does* not *automatically apply to any other source of output error, such as that due to temperature.*

Accuracy can also be specified in terms of *bit accuracy*. We may come across this term in comparative analysis like the one below, and since this phrase is the source of some confusion, it will be discussed shortly. Meanwhile, here is a boilerplate summary of the pros and cons of the two main categories of monolithic voltage references.

Bandgap references are relatively inexpensive and typically used for data acquisition systems where 8- to 10-bit accuracy is required. They have a typical initial error of ±0.2 to ±1 percent, a tempco of 20 to 40 ppm/°C. The output noise is typically 15 to 30 μV peak to peak (measured over 0.1 to 10 Hz), and they have a long-term stability of 20 to 30 ppm/kilohour. The quiescent current is typically 40 to 200 μA. Supply voltages can be as low as 1 V, but typically are as low as 2.4 V.

Buried-zener references are relatively expensive and typically used for data acquisition systems where 12-bit or higher accuracy is required. They have a typical initial error of ±0.04 to ±0.06 percent, a tempco of 5 to 20 ppm/°C. The output noise is typically less than 10 μV peak to peak (measured over 0.1 to 10 Hz), and they have a long-term stability of 10 to 15 ppm/kilohour. The quiescent current can be typically around 2 to 3.5 mA. The supply voltage needs to be larger than about 8 to 10 V.

NOTE *For stating tempco, most references are characterized by the "box method" (see further below). So the tempco in ppm/°C is the total ΔV (occurring over the specified temperature range), divided by the voltage at 25°C (its nominal value), divided by the ΔT (the specified temperature range of the reference), and finally multiplied by 10^6 (for ppm). Note that for stating tempco, there is no implied ±, unlike for the initial-accuracy specification.*

NOTE *Long-term drift is usually expressed in ppm per kilohour. But this can be misleading. For example, there is about 8760 hours in 1 yr. So some engineers pessimistically multiply the 1000-h specification of the part by the factor 8760/1000 = 8.76 to find the annual drift. In doing so, they are implicitly assuming linearity of drift with time whereas, in reality, drift decreases logarithmically as time passes, because things tend to stabilize as the part ages. For a better estimate we should take the* square root of the *elapsed time. That is why long-term drift is written in some datasheets as ppm/$\sqrt{kilohour}$ instead of ppm/kilohour. For example, for the correct estimate of the annual drift, we should multiply the specified drift per 1000 h, not by 8.76, but by $\sqrt{8.76} \approx 3$. That is roughly 3 times better than the (wrong) pessimistic calculation.*

NOTE *Voltage reference datasheets typically specify both low- and high-frequency noise. The latter (broadband noise) is usually specified as an RMS (root-mean-square) value in microvolts, over the range 10 Hz to 10 kHz. The low-frequency noise component is usually expressed as a peak-to-peak value in microvolts (or ppm), over the range 0.1 Hz to 10 Hz. The broadband noise component can be more readily filtered out, by better decoupling capacitors and so on. But the low-frequency component is particularly troublesome since filtering below 10 Hz is impractical. Therefore, the low-frequency noise specification is considered more critical and important, insofar as it contributes directly to the total reference error.*

NOTE *There are modern proprietary voltage reference technologies offering significant improvements. We will not discuss them here, but the reader can investigate further if desired. One is the XFET voltage reference from Analog Devices. Another is the FGA technology from Xicor (now part of Intersil).*

Now, look at an interesting tabulated comparison in Fig. 1.13. In particular, notice point D where a state-of-the-art bandgap is at par in terms of accuracy and drift, with a buried zener device.

Specifying and Interpreting the Tempco of Voltage References

What really is tempco? We mentioned that a 15-V discrete zener for instance, will have a tempco of about +10 mV/°C. So at 125°C (a Δ of 100°C above nominal), we expect the zener voltage to be 15 + 100 × (10 mV) = 16 V. That agrees with the typical zener data collected and shown in Fig. 1.14 (based on discrete diodes from Vishay). Expressed in terms of percentage, this gives us a voltage shift (called *temperature drift*) of 1 V referenced to 15-V nominal, which is an error of 1/15 = 6.7 percent. In terms of ppm, that is 67,000 ppm (25 to 125°C). Indeed, that is very high, but it is a discrete zener after all.

Based on the above analysis, we see that the zener voltage shift is proportional to the temperature (for discrete zeners). This comes about from the fact that the tempco of a discrete zener is almost *constant* (see the straight solid lines in Fig. 1.14). And it is because of that property alone that we expect the following statement to be true: at 75°C (a Δ of only 50°C above nominal), the error will be roughly one-half—close to 15.5 V. And that is borne out by the curves in Fig. 1.14. In other words, *we get proportionally less error if the temperature range is reduced, provided the voltage reference has a constant tempco.*

A constant tempco is a greatly simplifying property that engineers inadvertently assume and apply even when they shouldn't! For example, from the point of view of overall system requirements, suppose there is a defined maximum acceptable error (expressed in ppm, or %, or LSB, as explained later). Then if we halve the operating temperature range, we may relax (double) the required tempco of the reference—but only *provided the voltage reference we*

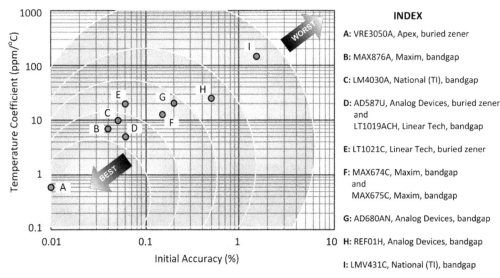

Note: The temperature range over which the tempco is guaranteed is very important in knowing the actual output error in any application.

FIGURE 1.13 Some well-known voltage references compared.

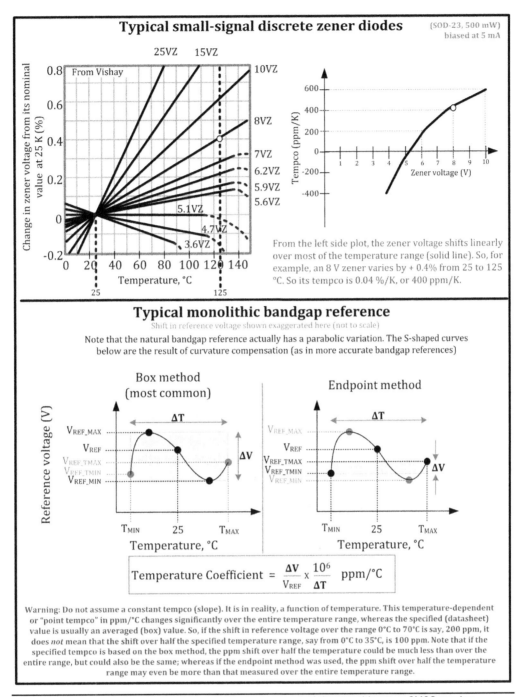

FIGURE 1.14 Brokaw's bandgap reference with an op-amp, and the transition to a CMOS version.

plan to use has constant tempco characteristics. Not otherwise! If we do not heed this little fact, we can seriously underestimate the required tempco (under reduced operating temperature ranges) and thus end up with a system with greater than acceptable error.

We said previously that in a bandgap reference, the PTAT and CTAT terms cancel, leaving net zero tempco. But that is an ideal assumption. In reality, because of several second-order effects and the behavior of associated circuitry placed to correct that, all monolithic voltage references (both bandgap and buried-zener types) have an output curve that is very rarely a straight line, and (with active curvature correction circuitry), is actually typically close to the horizontal S-shaped curve in the lower half of Fig. 1.14. That is obviously not a constant-tempco curve as in the upper half of the figure. And so, irrespective of the method used for evaluating the tempco of the reference, *we can no longer just assume that reducing the operating temperature range will allow us to pick a worse voltage reference (with higher ppm/°C).* But surprisingly, that is a common error and assumption in some related literature.

What is the solution? We have to recognize that the tempco (usually expressed in ppm/°C) *for most monolithic references* is some sort of *averaged* value over the full temperature range, and it is actually of a rather complex shape. The old-world, traditional approach is intuitively based on assuming constant tempco, perhaps because of the predictable behavior of passive references such as discrete zeners in the past. Once we are clear about this difference, we then realize that perhaps the best and safest option is to just stop talking in terms of ppm/°C entirely, *and stick to ppm—the total drift over the full specified temperature range of the voltage reference* (do not use the temperature range of the specific application, since this can cause underestimation of tempco requirements as explained above). Note that ppm could also be expressed as a percentage, or in terms of LSB, or in more absolute terms as millivolts or microvolts, e.g.—just don't use any *ratios to temperature*, because that implies a certain constancy (linear drift), which is misleading.

We finally understand the word *tempco* is a bit of a misnomer when applied to monolithic references. In Fig. 1.14, we took the liberty of coining a new term: *point tempco*. That sounds strange, but goes a long way in explaining and understanding the above issues in selecting a monolithic reference.

Finally note that the temperature characteristics of voltage references may be expressed over the *commercial* temperature range (0 to 70°C, but sometimes −10 to 70°C, or 0 to 85°C), or the *industrial* temperature range (−40 to 85°C, but sometimes −25 to 85°C, or −40 to 100°C), or the *extended* temperature range (−40 to 125°C, but sometimes −55 to 125°C, or even −55 to 150°C), or the *military* temperature range (−55 to 125°C, but sometimes −65 to 175°C). Clearly people don't fully agrees on the descriptions of the temperature ranges!

Understanding Bit Accuracy

With the above explanation of tempco behind us, we are now in a better position to understand what was really meant by *bit accuracy* previously. But note carefully that in the relevant sentence, the phrase referred to the *system accuracy* and was not stated to be the accuracy of the voltage reference itself. There are subtle differences in that as we will explain.

First, the use of voltage references is multifold. Examples follow.

1. In a data converter (e.g., data logger), a reference is used for providing a voltage level to the *analog-to-digital converter* (ADC), to compare the input voltage against so as to determine the proper digital code.

2. In a voltage regulator, the reference provides a voltage level against which the output is compared so as to develop a feedback signal that eventually regulates (corrects) the output voltage.

3. In a voltage detector circuit, a relatively crude reference is used to set up some sort of internal threshold for operation. For example, it can be used to set a safe start-up voltage level (undervoltage lockout, also called UVLO), or to set some safety trip point (overvoltage protection, also called OVP). But since we can usually accept much larger errors here, we will ignore this last item from now on.

In item 1, the term *bit accuracy* would usually refer to the entire system. In item 2, it would refer to the reference as a stand-alone device. Note that modern monolithic precision references are often set (trimmed) in production. Various means such as laser trimming and so on are used to achieve a certain trim target. To do any of that, the reference voltage needs to be first read by a digital test system. So system accuracy becomes an important aspect to understand even when talking about references as stand-alone devices in voltage regulator applications.

Let us start by understanding, say, 3-bit accuracy first. For 3 bits, the digital (binary) numbers are 000, 001, 010, 011, 100, 101, 110, 111. We see that we have $2^3 = 8$ levels (increments) leading up to a certain "full-scale" value. For n bits, we will have 2^n levels. Each tiny step is called the *least significant bit* (LSB), and it can be expressed either in terms of an absolute value (say, in millivolts or microvolts), or as a *fraction* of the full-scale value, which is just the nominal reference voltage in an ADC. Expressed as a basic fraction, the LSB is therefore $1/(2^n)$. Since this number is usually too small to write out as it is, it is more conveniently expressed in terms of ppm. This is what we should remember:

3-bit system:	125,000 ppm/LSB
8-bit system:	3,906 ppm/LSB
10-bit system:	977 ppm/LSB
12-bit system:	244 ppm/LSB
16-bit system:	15 ppm/LSB

In all cases, ½ LSB is one-half the above-stated ppm. When we talk of voltage references as stand-alone devices, 1-LSB accuracy is implied. For data loggers and ADC applications, however, since the drift can be either plus or minus, to have no observable error on the (digital) output (attributable to the tempco of the device), ½-LSB accuracy is required. But if not, the output error can be calculated.

For example, the VRE3050A in Fig. 1.13 has a 5-V output, with an initial accuracy of 0.01 percent ($0.01 \times 10^4 = 100$ ppm) and a tempco of 0.06 ppm/°C over the commercial temperature range of 0 to 70°C. So the temperature drift is $70 \times 0.06 = 4.2$ ppm. Assuming these two are the dominant sources of error, the worst-case ppm is $100 + 4.2 = 104.2$ ppm. Since a 12-bit ADC with ½-LSB requirement can accept $244/2 = 122$ ppm, the VRE3050A is suitable for 12-bit systems. If the system can calibrate out the initial error, the net error (drift) is only 4.2 ppm, which makes it suitable even for an ADC in a 16-bit system.

Note *Among the typical full-scale (reference voltage) values specified for an ADC input, 2.048 and 4.096 V are most commonly used because they provide an integral or half-integral number of millivolts for any bit value (they are multiples of 2). For example, a 12-bit ADC with 4.096-V full-scale input (based on a 4.096-V reference) gives $4.096/2^{12} = 1$ mV/LSB. In terms of ppm this is $(1 \text{ mV}/4.096 \text{ V}) \times 10^6 = 244$ ppm/LSB, which agrees with the values of the numbered items above.*

PART 3 DESIGNING THE VOLTAGE DIVIDER

Sources of Output Error: Voltage Divider Input Bias Current

In the lower half of Fig. 1.11, R_3 and R_4 constituted a resistive "voltage divider" (inside the monolithic voltage reference). We can also have an external voltage divider to set the output of a complete regulator: see Figs. 1.2 and 1.15.

If the resistors of the voltage divider of a regulator are inside the chip, the systems designer can ignore their exact contribution to the output error, because the regulator/controller is a "black box" with a known total output tolerance. For example, there are monolithic (chip) regulators available in fixed-output versions, such as Texas Instruments' LM2676T-3.3 and LM2676T-5.0, which provide 3.3- and 5-V outputs, respectively. However, LM2676T-ADJ is one where you need to add an *external* voltage divider. And with external dividers, we have to be very careful, since they can contribute a thick new layer of error in the regulation onion of Fig. 1.2 (see the layer called *voltage divider design*).

Going forward, we are adopting the convention that the "lower resistor" of the divider is R_1 and R_2 is the "upper resistor." See Fig. 1.15. First, we ignore the small current flowing out of node A, called I_{IN}. We then have a perfect voltage divider. We can derive the relationship that essentially sets the *ratio R_2/R_1* for a required output. Once the ratio is known, the absolute values can be determined by first selecting either R_1 or R_2, based on several other concerns that we will discuss shortly. For now, assume that we are interested only in the divider *ratio*.

Note that, by definition, a voltage divider is a voltage divider only when absolutely zero current enters or leaves the shared node (between the two resistors, called A in Fig. 1.15). Node A goes to the *feedback pin* of the controller, where it is compared against the reference voltage V_{REF}. Note that an op-amp (error amplifier) always swings its output in such a way as to equalize the voltages at its two inputs. So, in steady state, we can assume that the voltage at A (V_A) is equal to V_{REF}. Now let us account for the fact that the input impedance of the feedback pin appears as some arbitrary R_{IN}, which is not an infinite value, and therefore pulls in a nonzero current I_{IN} from node A. So, what is the effect of this current on the output voltage? And how do we minimize its effect?

If the feedback pin "steals" a tiny current I_{IN} away from the divider, usually referred to as the *input bias current* of the feedback pin, we are left with $I_1 = I_2 - I_{IN}$ flowing through R_1. This causes a shift in the output to a new value, which we call V'_O here. We get

$$\frac{V'_O - V_{REF}}{V_{REF} - 0} = \frac{I_2 \times R_2}{I_1 \times R_1}$$

Comparing it with the "perfect divider" equation in Fig. 1.15, and solving for the error ΔV, we get alternative forms for computing ΔV (also stated in the figure):

$$\Delta V = (V_O - V_{REF}) \times \frac{I_{IN}}{I_1} \approx (V_O - V_{REF}) \times \frac{I_{IN}}{I} \approx I_{IN} \times R_2$$

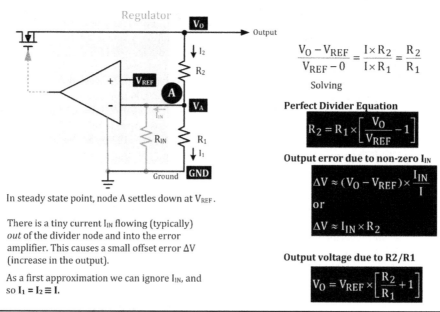

FIGURE 1.15 Basic voltage divider and some useful equations.

where I is the current flowing *through* the divider ($\approx I_1 \approx I_2$), and I_{IN} is the current into the feedback pin. Note that if this current flows not into but *out of* the feedback pin (i.e., *into* node A), the sign of ΔV will become negative, indicating that the output decreases as a result, instead of increasing.

For a given output and feedback voltage, *the way to minimize any error due to the current in/out of the feedback pin is to increase I, the current in the divider* (compared to the current I_{IN}). So small values for R_2 (and therefore R_1) are preferred for this particular reason (though that might not be always practical, for reasons mainly related to the feedback loop, as explained shortly).

Some quick-lookup charts are available in Fig. 1.16, for the most common cases of $V_{REF} = 2.5$ V (as is the case when we use the LM431 or TL431 voltage reference) and $V_{REF} = 1.23$ V (a typical bandgap reference). Observe that for the *same bias current–divider current ratio, we get higher output errors for high output voltages (expressed as a percentage, or otherwise)*. The small feedback pin bias current clearly has a crowbar/lever effect on the output voltage accuracy. What is important is that it is the *ratio* of this bias current to the main divider current,

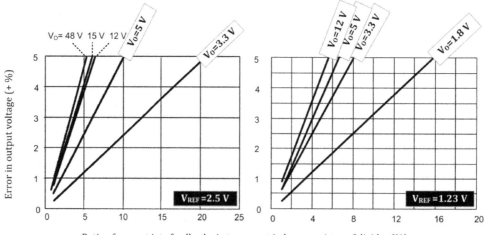

FIGURE 1.16 Minimizing the output error due to the feedback pin bias current.

that determines the percentage output error. So we can try to either decrease the bias current or increase the divider current (by lowering R_1 and R_2).

NOTE *If I_{IN} were fixed, we could compensate for it by suitably tweaking the resistors of the divider. The reason that we can't use this approach in production has to do with the variability of this input bias current. It can vary anywhere between the MIN and MAX limits in a typical regulator datasheet. To know the worst-case error, we must take the maximum value of the input bias current spread, not its typical value.*

Example 1.1 We have a 2.5-V reference, and the lower resistor of the divider (R_1) is set to 62.5 kΩ. We set up a divider with the ideal divider equation to get 12 V on the output. What is the upper resistor value? What is the output voltage we will see if the pin flowing into the feedback pin is 2 µA?

With 62.5 kΩ at the lower position, the divider current is

$$I_1 = \frac{2.5 \text{ V}}{62.5 \text{ k}\Omega} = 0.04 \text{ mA}$$

So the ratio of the bias current to the divider current is

$$\frac{I_{IN}}{I_1} = \frac{2 \times 10^{-6}}{4 \times 10^{-5}} = 0.05$$

This is a current ratio of 5 percent. The corresponding output error is

$$\Delta V = (12 \text{ V} - 2.5 \text{ V}) \times 0.05 = 0.475 \text{ V}$$

We can find the upper resistance too.

$$R_2 = R_1 \times \left[\frac{V_O}{V_{REF}} - 1\right] = 62.5 \text{ k}\Omega \times \left[\frac{12}{2.5} - 1\right] = 237.5 \text{ k}\Omega$$

Based on R_2, we have an alternative calculation for error (Fig. 1.15): $I_{IN} \times R_2 = 2$ µA $\times$ 237.5 kΩ = 0.475 V which agrees with the previous method of calculation.

We have confirmed that the output voltage will be 12.475 V, which is an error of +0.475/12 = +4 percent. This is a ppm of 40,000. It is way too high. We really need to reduce the input bias current I_{IN} (chip design goal), and/or we need to significantly increase the current in the voltage divider (which will certainly add to the dissipation, and that can become significant in battery-powered applications).

Graphically, we could have seen this directly from Fig. 1.16. Note also that under the same conditions (same lower resistor and reference), for a 3.3-V output, we would have gotten only about +1.2 percent error for a 2.5-V reference. *So lower outputs are less affected by input bias currents.*

Therefore for setting high output voltages, we must pass more current in the divider (decrease R_1 and R_2). This unfortunately works against minimizing the power dissipated in the divider, when we need it most. In particular the dissipation in the upper resistor R_2 will increase significantly.

Sources of Output Error: Voltage Divider Resistor Tolerance

If we want an accurate output, it is obvious that we need resistors of tight tolerance. What is not so obvious is how tight the tolerance of each resistor needs to be and whether the error contribution from the tolerance of R_1 is the same as that from R_2, or different, and why.

We start from the equation

$$V_O = V_{REF} \times \left[\frac{R_2}{R_1} + 1\right]$$

If R_1 increases, V_O will fall. If R_2 increases, V_O will rise.

Let us look at Example 1.1 to understand what happens.

Example 1.2 We have a 2.5-V reference. The lower resistor of the divider (R_1) is set to 62.5 kΩ and the upper resistor R_2 is 237.5 kΩ, for a nominal output of 12 V. Ignore the feedback pin bias current here. What is the highest voltage due to resistor tolerances? Assume the tolerance of the resistors is ±5 percent.

To achieve the highest output, we need to use the lowest value of R_1 and the highest value of R_2. We therefore get

$$V_O = V_{REF} \times \left[\frac{R_2}{R_1} + 1\right] = 2.5 \times \left[\frac{237.5 \times 1.05}{62.5 \times 0.95} + 1\right] = 13.0 \text{ V}$$

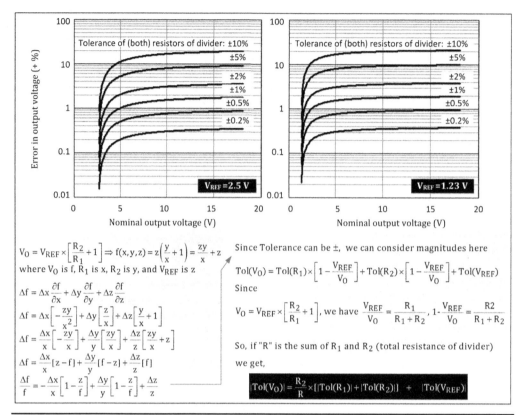

FIGURE 1.17 Effect of resistor tolerance on output error (sensitivity analysis).

The error is therefore 1 V/12 V = 8.3 percent. Note that it is not the sum of the tolerances of two resistors. So, what is the closed-form equation? We perform a sensitivity analysis in Fig. 1.17, where the formula is provided and also plotted out for easy graphical reference. The formula is based on the assumption of small increments (partial differentiation), so it gives slightly different results, but well within the ballpark. For example, for the above example we get (assuming a reference tolerance of zero)

$$\left|\text{Tol}(V_O)\right| = \frac{R_2}{R} \times \left[\left|\text{Tol}(R_1)\right| + \left|\text{Tol}(R_2)\right|\right] + \left|\text{Tol}(V_{REF})\right|$$

$$\left|\text{Tol}(V_O)\right| = \frac{237.5}{237.5 + 62.5} \times [5\% + 5\%] + 0 = 7.9\%$$

Note the following statements applicable to a voltage divider:

1. Two resistors, each of x percent tolerance, do not produce an error of $\pm 2x$ percent as some believe.

2. The upper resistor's tolerance is as important as the tolerance of the lower resistor (some believe the upper resistor can have worse tolerance because it usually has the larger resistance of the two). The sum of the tolerances of the resistors is what matters.

3. If the reference has $\pm x$ percent accuracy, it will produce $\pm x$ percent (additional) output error.

4. The chances of getting two resistors, one with exactly $+x$ percent error and the other with exactly $-x$ percent, so as to produce the worst-case output error computed above, are very slim. Monte Carlo analysis is often carried out to get a more *realistic* (rather than worst-case) estimate of the output error.

Sources of Output Error: Commercial Resistor Values

In the past, circuit designers were able to precisely specify the value of resistor required at a particular location based on calculations. But this created difficulties for the resistor manufacturers as there was little or no standardization, and hence all resistors were custom-made.

Even today, in the author's experience, it is bewildering to find that even some very large manufacturers routinely ask for whatever value they calculate—and usually get it. Perhaps resistor manufacturers don't think it prudent to risk losing such a big account by "educating" the customer about preferred values. But most engineers around the world are not so lucky. Especially in Europe, the *E series*" is fully known and widely used. The chosen resistor values of the divider need to conform to this standard set of values, and therein arises another major contributor to the output error of a regulator.

Note that overall costs can go up for several reasons, e.g., if we use custom parts or nonpreferred values and if we have a large number of possibly *unnecessary* part values. For example, if we look carefully at some power supply circuits, we may find a 10-kΩ resistor at some point, and for some odd reason (probably associated with blindly cutting and pasting a "nice" circuit block from a previous product), we find 10.5 kΩ at another point. On closer analysis we will often see no reason why the 10.5 kΩ could not have been made into a 10 kΩ. So we forget that we want to reduce not only the number of parts in our power supply, but also the *number of types* of different parts. This would serve us logistically, especially because each new part's type and reel need to be separately loaded onto a pick-and-place machine.

One of our design goals is not to *cater* to a circuit's sensitivity to tolerances, but to *diminish* that sensitivity. With some thought we can usually *make* our circuit robust enough to accept a wider range of values, thus also increasing manufacturing yield. Sometimes we may be willing to compromise the component count to keep the number of different *types* of components down. So, to get 10.5 kΩ, we could consider using 10 kΩ in series with a 470-Ω resistor, if these two values in any case were being used elsewhere in the circuit. In doing so, not only do we reduce ordering costs, but also we may get parts cheaper because we are using larger quantities of each. In the worst-case scenario we may end up adding a couple of cents to our Bill of Materials (BOM), but it could save a lot of other related costs and headaches in the long run. As component manufacturer Vishay advises, "use of standard values is encouraged, because stocking programs (of vendors) are designed around them."

But what are *preferred (standard) values*? We all remember using values such as 220, 470, and so on in the past. Today those familiar numbers are being replaced by 221, 464, and so on. There are reasons for this. First look at Fig. 1.18, which gives the tolerances of

Series	Tolerance (±%)	Values/dec	Typical	Thick-Film (e.g. SMD)	Thin-Film (e.g. SMD)	Metal-Film (e.g. MELF)	Metal-Foil
E3 (obs.)	50	3	Resistance (Ω)	1-100M	10-100k	0.22-22M	2m-1M
E6	20	6	Tolerance + and − (%)	0.5-5	0.1-5	0.1-2	0.005-5
E12	10	12	Tempco (TCR) + or − (ppm/K)	50-400	10-50	5-50	2-50
E24	5, (2), (1)	24	Stability + or − (%/k-hr)	1-3	0.05-0.1	0.15-0.5	0.05
E48	2, (1), (0.5), (0.25), (0.1)	48	Rated P_{DISS} (W at 70°C)	1/16-1/2	1/16-1/4	1/16-1	1/4-10
E96	**1**, (0.5), (0.25), (0.1)	96	Oper. Voltage (V)	50-200	50-100	50-500	200-500
E192	(1), **0.5**, (0.25), (0.1)	192	Thick-film is the most popular SMD (rectangular profile), but thin-film is preferred for voltage dividers (better stability and tempco)				

FIGURE **1.18** Standard resistor ranges (E series) and comparison of resistor technologies.

E6	E12	E24	E48	E96	E192	E6	E12	E24	E48	E96	E192	E6	E12	E24	E48	E96	E192
100	100	100	100	100	100	220	220	220	215	215	215	470	470	470	464	464	464
					101						218						470
			102	102	102					221	221					475	475
					104						223						481
		105	105	105	105				226	226	226				487	487	487
					106						229						493
			107	107	107					232	232					499	499
					109						234						505
	110	110	110	110	110		240	237	237	237	237		510	511	511	511	511
					111						240						517
			113	113	113					243	243					523	523
					114						246						530
		115	115	115	115			249	249	249	249				536	536	536
					117						252						542
			118	118	118					255	255					549	549
					120						258						556
	120	120	121	121	121		270	270	261	261	261		560	560	562	562	562
					123						264						569
			124	124	124					267	267					576	576
					126						271						583
		127	127	127	127			274	274	274	274				590	590	590
					129						277						597
			130	130	130					280	280					604	604
					132						284						612
		130	133	133	133			300	287	287	287			620	619	619	619
					135						291						626
			137	137	137					294	294					634	634
					138						298						642
			140	140	140				301	301	301				649	649	649
					142						305						657
			143	143	143					309	309					665	665
					145						312						673
150	150	150	147	147	147	330	330	330	316	316	316	680	680	680	681	681	681
					149						320						690
			150	150	150					324	324					698	698
					152						328						706
			154	154	154				332	332	332				715	715	715
					156						336						723
			158	158	158					340	340					732	732
					160						344						741
		160	162	162	162			360	348	348	348			750	750	750	750
					164						352						759
			165	165	165					357	357					768	768
					167						361						777
			169	169	169				365	365	365				787	787	787
					172						370						796
			174	174	174					374	374					806	806
					176						379						816
	180	180	178	178	178		390	390	383	383	383		820	820	825	825	825
					180						388						835
			182	182	182					392	392					845	845
					184						397						856
			187	187	187				402	402	402				866	866	866
					189						407						876
			191	191	191					412	412					887	887
					193						417						898
		200	196	196	196			430	422	422	422			910	909	909	909
					198						427						920
			200	200	200					432	432					931	931
					203						437						942
			205	205	205				442	442	442				953	953	953
					208						448						965
				210	210					453	453					976	976
					213						459						988

Figure 1.19 Standard (preferred) resistor values.

the standard E series (based on IEC 60063). The values we were used to are the 5 percent tolerance values, or the E24 series. Its values are all listed in Fig. 1.19.

In general, preferred (standard) values (E series) are a modern system for selecting nominal values within a given decade of resistance, *based on the accuracy with which they are able to be manufactured*. For example, if we had the capability to produce only 10 percent tolerance resistors, and we arbitrarily picked the first preferred value to be 100 Ω, then it would make little sense to produce a 105-Ω resistor since that value falls well within the 10 percent tolerance range of the 100-Ω resistor (90 to 110 ohms). The next *reasonable* value is 120 Ω (about 20 percent higher), because its lower (−10 percent tolerance) value is 120 × 0.9 = 108 (only a slight overlap with the previous value's upper spread). We thus get the well-known E12 series in Fig. 1.19. As manufacturing capability improved, it made sense to introduce more values. Now we have progressed up to 0.1 percent tolerances, and several new series have been added. All the preferred series start with the letter E followed by a number denoting how many nominal values there are within a *decade* of resistors. A *decade* is 10 to 100 Ω, or 100 Ω to 1 kΩ, and so on. So between 100 Ω and 1 kΩ, we will have 48 resistor values in the E48 (2 percent tolerance) series. It is obvious that all resistor values of a given E series in a given decade can be generated by multiplying the values of the previous decade by 10.

NOTE *Per common practice, when we talk about 1 percent resistors, what we really mean is ±1 percent resistors. We omit the signs, but they are understood, since the nominal value is always the center of the spread.*

Can we describe a closed-form equation for the resistor values in any E series? The resistors must be selected in a geometric progression, since by doing so we keep the same *ratio* from one value to the next—just as the tolerance behaves. So, if we divide a decade (factor of 10) by 96 values, to generate the E96 series, the ratio of the geometric progression must be

$$\text{Ratio} = 10^{\frac{1}{96}} = 1.024$$

This means that the next value is 2.4 percent away. Clearly, based on our previous analysis of 10 percent tolerance, the E96 series is appropriate for 1 percent tolerance. Similarly, E192 is most appropriate for 0.5 percent tolerance. However, nothing stops vendors from, say, producing E192 resistors in 0.1 percent, or 1 percent or even 5 percent, if they can sell them! And we may even find them if we look around. But they are not preferred.

Returning to the E96 series, suppose the first value R_1 is 100 Ω; then the next value R_2 must be $100 \times 1.024 = 102.4$ Ω. The next value R_3 would be $102.4 \times 1.024 = 104.86$ Ω. In general, the mth resistor value in an "Ex" series would be

$$R_m = 100 \times [10^{1/x}]^{m-1}$$

For example, the fifth value after 100 Ω in the E48 series would be

$$R_5 = 100 \times [10^{1/48}]^{5-1} = 121.15$$

In all cases, the values are actually rounded up to the nearest integer. So 121.15 Ω becomes 121 Ω. We can confirm this value from Fig. 1.19. Similarly, for E96 series we get the following rounded values: 100, 102, 105, 107, 110, and so on. We can confirm these values from Fig. 1.19 too.

Since every calculated value of resistor is not available, we will end up using something "close." So what is the error produced on the initial accuracy of the output? And what "golden" combination of, say, standard 1 percent (E96) resistors will produce the lowest possible output error? We are assuming we have full flexibility in choosing the upper and lower resistors, and we are also ignoring all the other sources of output error, such as input bias current. In addition, we want to avoid parallel or series combinations of resistors for achieving the required accuracy.

We will need to do a lot of number crunching, using a Mathcad file if possible, to answer the above questions. In Fig. 1.20 we have presented some useful results based on such an exercise. Note that the recommended resistor solutions provided here can be multiplied by 10 or 100, and so on, to go to the next decade of values, if we want. *The error reported in this figure has nothing to do with the tolerance of the resistors used; it is just based on nominal values, and it represents the error solely due to the use of discrete (preferred) values.* In the lower part of the figure we have allowed a combination of E96 and E24 values, because we recognize that E24 series is also available nowadays in tolerances similar to E96 series (1 percent or better).

NOTE *Chip designers struggle hard to produce the right S-shaped curve corresponding to the lowest tempco in monolithic references. Depending on their semiconductor process and so on, they end up with values typically ranging anywhere from 1.2 to 1.25 V. We have therefore used a center value of 1.23 V in our Mathcad-based number-crunching exercise, where we tried to find the lowest error that the use of preferred resistor values has on the initial accuracy. That is clearly a systems designer's exercise. In fact, the best results will arise if we combine the endeavors of the two types of designers, chip and systems, and calculate what exact value of the bandgap reference produces the lowest possible output error, in terms of initial accuracy combined with temperature effects.*

NOTE *Can we always use two resistors per divider? A commercial AC-DC power supply systems designer may need to think twice before using any single resistor larger than about 0.5 MΩ. Some extremely quality-conscious power supply companies have internal rules prohibiting any value greater than 100 kΩ. Contamination on the PCB, or moisture and humidity, can cause a large change in the resistance. So they ask their engineers to put several 100-kΩ resistors in series rather than use a single resistor.*

$V_{REF} = 2.5$ V					
Output Rail (V)	3.3	12	15	18	24
R2 (upper)	115	523	590	806	1.18k
R1 (lower)	357	137	118	130	137
Error (%)	0.161	+0.364	0	0	+0.137
$V_{REF} = 1.23$ V					
Output Rail (V)	1.5	1.8	2.5	3.3	5
R2 (upper)	301	133	137	1.07k	1.02k
R1 (lower)	1.37k	287	133	634	332
Error (%)	+0.016	0	−0.12	+0.177	+0.178
$V_{REF} = 1.23$ V (continued)					
Output Rail (V)	12	15	18	24	48
R2 (upper)	931	10.2k	1.5k	10.7k	5.23k
R1 (lower)	107	909	110	576	137
Error (%)	−0.569	+0.213	+0.015	+0.328	+0.385

(Only E96 series available — error does not include tolerance)

$V_{REF} = 1.23$ V					
Output Rail (V)	1.5	1.8	2.5	3.3	5
R2 (upper)	180	133	124	180	1.2k
R1 (lower)	820	287	120	107	392
Error (%)	0	0	+0.04	−0.025	−0.094
$V_{REF} = 1.23$ V (continued)			**$V_{REF} = 0.6$ V**	**$V_{REF} = 0.3$ V**	
Output Rail (V)	12	15	18	1.05 *(IMVP voltage rail)*	
R2 (upper)	1.2k	1.6k	1.5k	105	255
R1 (lower)	137	143	110	140	102
Error (%)	+0.031	−0.052	+0.015	0	0

(Both E96 + E24 series available — error does not include tolerance)

FIGURE 1.20 Best combinations of E96, and E96 + E24, resistors for lowest errors.

Voltage Divider: Constraints Imposed by Type of Error Amplifier

From Fig. 1.20 we see that for an output of 3.3 V with 1.23-V reference, using only E96 values, we get 1.07 kΩ and 634 Ω for minimum error (using only two resistors in the divider). The question is: Can we use, say, 10.7 and 6.34 kΩ instead? Yes, certainly it will reduce dissipation in the divider by a factor of 10, but it will also increase the error due to any feedback pin bias current. But are there any other restrictions? *Not* if we are using a transconductance type of error (operational) amplifier. As we will see in the Chap. 14 on loop stability, only the ratio of the resistors of the divider enters the feedback loop equations. So if we keep the ratio unchanged, the loop is unchanged. However, if we use a regular, voltage-based operational amplifier, only the upper resistor enters the feedback loop equations. The lower resistor is just a DC-biasing resistor, and it disappears from the AC loop response. So, if we change the upper resistor, such as when we change the ratio, we can significantly alter the loop response. So we have to pick the right decade at least, provided we can keep to the ratios given in Fig. 1.20.

> **HINT** *For a regular voltage-based op-amp, if we have a converter with satisfactory loop response, we should try to keep the* upper *resistor (and all other compensation components) unchanged, as we tweak the output for a somewhat different output voltage, say, from 3.3 to 5 V. In other words, we should always change the lower resistor,* not the upper one, *despite the error that may create in the initial accuracy by the use of preferred values.*

Voltage Divider: Correct PCB Placement

In the regulation onion of Fig. 1.2, we drew a layer called *PCB design*. What does that mean? In Fig. 1.21, we show a typical regulator (could be series-pass or switching). We exaggerate the PCB trace resistances to make a point about the correct location of the resistors of the

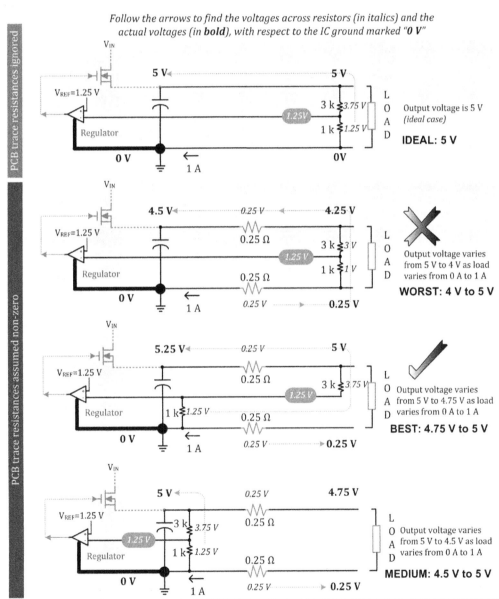

FIGURE 1.21 How to correctly position the voltage divider to minimize impact of PCB traces and output leads on load regulation.

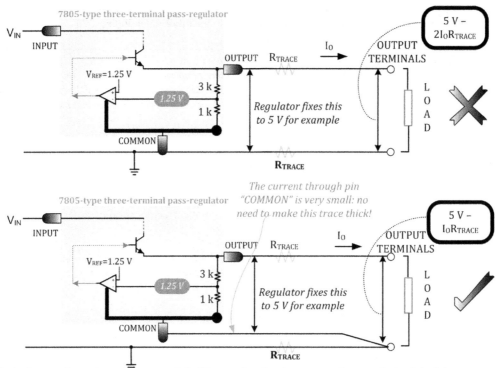

In the lower schematic, we have routed the IC ground so that it connects closer to the load. In doing so, we find that load regulation improves. In an adjustable-output linear regulator such as the LM317, which has an external voltage divider connected to its COMMON (or "ADJ") pin, by the same logic, the upper resistor of the divider must be connected closest to the OUTPUT pin (NOT TO THE OUTPUT TERMINAL), whereas the lower end of the lower resistor of the divider, along with the ADJ pin, are connected as close to the load as possible (to the lower OUTPUT terminal). This is not the same as shown in Fig. 1.21 for a switcher. See also *http://www.ti.com/lit/an/snva558/snva558.pdf* by Chester Simpson

FIGURE 1.22 How to correctly position the IC ground of a monolithic regulator (with internal divider) to minimize the impact of PCB traces on load regulation.

divider. We see that the lower ends of the resistor must connect as close to the IC ground as possible, whereas the upper end of the upper resistor must connect as close to the load as possible. This helps improve static load regulation. Note that some inexperienced engineers think that both resistors must connect as close to the load as possible. They imagine some sort of "kelvin sensing" but forget to do the math as shown. In Fig. 1.22 we apply the same principle to a monolithic regulator with an internal divider.

However, there is something called droop positioning, in which resistances toward the load that clearly degrade static regulation can help significantly to improve the dynamic regulation. In fact, the total load regulation window (static + dynamic) may actually improve by droop positioning. See Chap. 9 of this author's *Switching Power Supplies A–Z*, Second Edition, for further details on this aspect.

CHAPTER 2

DC-DC Converters: Topologies to Configurations

PART 1   INTRODUCTION TO THE PRINCIPLES OF SWITCHING POWER CONVERTERS

Watch Out for the Not So Obvious

As we enter the area of switching power conversion, we must always expect the unexpected. We need to keep the focus, at least initially until we garner experience, on what was not so obvious to us. We should in fact proactively look out for it, almost welcome it, because that is what helps us learn faster. The biggest "gotcha" in switching power conversion is the *switching* aspect of it ! But there are many more pitfalls, not related to switching, that no amount of computer bravado will prepare us for. A well-known pioneer behind current mode control wrote this to the author in a personal e-mail exchange in January 2013: "I see so many engineers spending months on simulation and Mathcad, not realizing the real world will find totally different modes of failure than they will ever find with a computer."

For example, having studied voltage dividers in seemingly excruciating detail in Chap. 1, we think we know it all by now. But do we really? Before we delve into power conversion in greater detail in this chapter, we point out something seemingly innocuous: The feedback pin of any regulator, like all pins, has absolute maximum voltage ratings that should not be exceeded. On the low side of the allowed range in particular, we learn that most controller ICs' datasheets state that the feedback pin is not allowed to go lower than about 0.3 to 0.4 V below the IC ground (as for most other pins too). The feedback pin also seems reassuringly self-stabilizing, and it settles down in steady state to the internal reference level. But we need to watch out for what stresses it sees during transients, such as during a hard start-up or shutdown. This is of greater concern when the IC is used "unconventionally," that is, in a configuration other than its basic intended application (e.g., a Buck IC being used as a Buck-boost, as we will discuss later in this chapter). But what we hardly expected is that even simply shorting the output of the converter in normal operation, in a simple Buck with a feed-forward capacitor present, can also lead to exceeding the published absolute maximum rating of the feedback pin significantly! See Fig. 2.1. This effect usually does not cause instantaneous or direct damage, but does induce "substrate currents"—currents in the reverse direction through the die. These currents have been known to cause unpredictable behavior of the controller IC (or switcher). In one case, that even included the obviously temporary, but complete disappearance of the usual current limit—and that led to the destruction of the FET/switcher IC eventually. The author actually captured this entire sequence of events right up to the subsequent destruction, on two oscilloscopes with deliberately different time scales (one zoom in, the other zoom out). A warning was therefore proactively added by this author, in several datasheets of the same family, e.g., in the LM2593HV datasheet at http://www.ti.com/lit/ds/symlink/lm2593hv.pdf. See under Pin Functions (Pin 6). That's one gotcha that no book of power supply simulation will probably ever put you on guard against, to emphasize the point a bit.

What Is Ground?

We need to start with some basics. In a regulator, there are two input rails (connected to the DC source) and two output rails (connected to the load). That is actually two voltage rails, input and output, and their respective returns. Of these, only those cases in which one rail is

34 Chapter Two

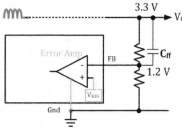
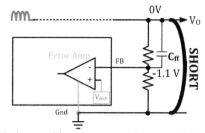

C_{ff} is charged up in steady state to 3.3 − 1.2 = 1.1 V

Under a sudden short, since C_{ff} cannot discharge immediately through the resistors of the divider, its lower end gets dragged down toward −1.1 V. In the process, C_{ff} actually gets discharged through the IC ground pin, in a reverse direction, and this causes substrate currents, which can impair functionality of the IC in an unpredictable manner, even causing its destruction (if for example, its current limit block is affected momentarily).

FIGURE 2.1 How we may be able to destroy a switcher IC by simply shorting its output.

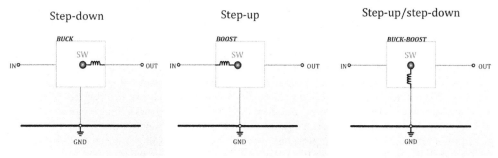

There are two input rails and two output rails, including one rail called the system ground (GND), which is shared by both the input and the output. In effect, there are only three rails as shown.

In all cases, one end of the inductor is tied to one of the three available DC rails (IN, OUT or GND). That end is its steady, or DC end. It determines the three basic topologies possible. The other end of the inductor in all cases, gets alternately connected, via the switch or the catch diode, either to pull in energy into the inductor from the input voltage source during the ON-time, or release it to the output during the OFF-time. Therefore the voltage on this alternating end of the inductor, keeps switching between two voltage levels, and is called the switching node (SW above). The inductor voltage reversal phenomena (flipping of the SW node during the OFF-time) occurs with respect to the steady (DC) end of the inductor, to equalize volt-seconds across the inductor in steady state. And this voltage reversal, based on the end it is being referred to (the DC end), is what leads to the observed input voltage step-down (Buck), step-up (Boost) or step-up/step-down (Buck-boost) behavior of the said topology.

FIGURE 2.2 Why are there three basic topologies?

shared between the input and the output are considered practical in system designs. By convention, this common rail is designated the (system) *ground*. See Fig. 2.2. Historically, this was often the upper rail, but nowadays we always work with lower (negative) rail grounds. The former is akin to a clothesline, the latter is a city skyline, rising from the ground up.

This shared ground, passing across the PCB, is in effect the ground for the entire power stage. Of course there is another ground too: the IC/controller ground. This may or may not always be connected to the system ground, especially if the IC is being used in a manner other than its primary intended application. We will come to those permutations later, but it is enough to point out here that if we do connect the system and IC grounds differently, there could be problems. For example, it may not be possible to have a simple direct connection between the voltage divider and the feedback pin. The divider node voltage may need to be translated, or level-shifted, because, feedback voltage almost always needs to be referenced to the IC ground, unless the IC itself has a *differential voltage amplifier* just before the feedback block to do exactly that.

The Three Basic Switching Topologies

Why are there three basic topologies only? The underlying reason is that they all use only one inductor. Topologies with more than one inductor are best considered to be composite, not basic or fundamental, topologies. In a basic topology, the single inductor carries out the basic energy storage and transfer function. One end of it is held fixed always; the other end is switched alternately between two paths, one in which energy is drawn from the input and the other in which energy is delivered to the output. If there are only three rails, as explained above, there can be only three topologies, based on which rail the fixed-voltage end of the inductor rests on, while it is alternately switched between the other two rails. Hence there are three of them only: three basic topologies corresponding to the three available power rails.

To complete the connection to the rail which causes energy to flow into the system from the input DC source requires one switch—the control switch, or control FET. The other switch connects it to the remaining rail which causes energy to flow into the output. That is often a diode (a Schottky most often), but nowadays in synchronous topologies this may be another FET too.

We see all this in Fig. 2.2. The node that is being switched constantly between the two remaining rails is called the *switching node*, or *swinging node*, often designated SW; but we should be clear that V_{SW} is sometimes used for the forward voltage drop across the switch (i.e., when it is fully conducting). We should watch out for this possible source of confusion in this book and in related literature, but with a little thought it is always obvious what is intended.

Why Step-Up/Down?

Let's introduce the most basic idea in switching power conversion.

In normal "square-wave" (nonresonant) power conversion we always apply a certain *constant* voltage (denoted here by V_{ON}) during the switch ON-time (T_{ON}) and then a constant voltage (of opposite sign, whose magnitude is denoted here as V_{OFF}) during the OFF-time (T_{OFF}). This leads to piece wise linear current segments. So we can write (in terms of magnitudes)

$$V_{ON} = L\frac{\Delta I_{ON}}{T_{ON}} \qquad V_{OFF} = L\frac{\Delta I_{OFF}}{T_{OFF}}$$

A *steady state* in power conversion can be defined as

$$\Delta I_{ON} = \Delta I_{OFF}$$

(again in terms of magnitudes). This equality in effect implies that the *current at the end of a given switching cycle returns to the exact instantaneous value it had at the start of the same cycle, every cycle*. Thus the entire current (and voltage) pattern becomes repetitive, and the operation is in that sense steady. If it were not, energy and current would keep "staircasing." Sooner or later, if the ΔI's did not equalize (with opposite signs), this would no longer be considered a viable topology. A topology must be able to self-stabilize, from a power-flow viewpoint, *even with no control loop present*. A control loop only ensures regulation to a *desired set point* under line and load variations, *provided a set point exists*—it cannot ensure there is a possible set point, because it cannot coax physics. Physics is what determines power and energy flow and balance conditions—we cannot hope to tame that with any clever algorithm driving tiny op-amps.

This basically means that in steady state

$$V_{ON} \times T_{ON} = V_{OFF} \times T_{OFF}$$

The product of the applied voltage and the duration for which it is applied is called *volt-seconds*. The above equation therefore forms the *volt-seconds law*.

The OFF-time, called T_{OFF} above, is not necessarily equal to the entire available OFF-time, which is $T - T_{ON}$ [i.e., $(1 - D)/f$], where $T = 1/f$ is the time period of the switching cycle. In discontinuous conduction mode (DCM), e.g., the voltage reversal across the inductor lasts for a duration less than $T - T_{ON}$. During the remaining part of the cycle, the voltage across it remains zero, and so do the current and its slope with respect to time.

NOTE *It is easy to understand and confirm this volt-seconds law for an inductor, but what about a multiwinding magnetic element such as a transformer or coupled inductor? In fact, this law applies to any chosen winding on a magnetic structure. We can check it out, but in doing so, each winding should be considered individually, without regard to any other winding that may or may not be present, or passing current, or not. We cannot, e.g., use the voltage applied during the ON-time across one winding and draw a simple and direct volt-seconds relationship with the voltage present across another winding during the OFF-time. Yes, if we converted to volts per turn, we could do it! That would be a volt-seconds per turn rule, and that would be valid. In general, two parameters are key to a transformer (or multiwinding structure): (1) the volts divided by the turns of any winding and (2) the current multiplied by the number of turns of any winding. So we focus on volts per turn or/and ampere-turns.*

Returning to the basic volt-seconds law for converters, any topology that *exists* (discovered or to be) tends to automatically move toward a steady state, in which during every cycle, the *net* volt-seconds across the inductor is zero. Because volt-seconds is energy, and we must end each cycle with no increase or decrease in that quantity, to ensure continuous repetition. The volt-seconds law, being directly related to energy flow, is therefore fundamental. Despite its seeming simplicity, it is the basic tool for confirming the viability or existence of any topology.

The input-to-output transfer function V_O/V_{IN} for any topology thus follows simply from the volt-seconds law. Alternatively, we can express the duty cycle t_{ON}/T (where $T = 1/f$) in terms of the input and output voltages (their magnitudes). See Table 2.1 for derivations for a Buck, a Boost, and a Buck-boost [all assumed to be in continuous conduction mode (CCM) and ignoring parasitic voltage drops].

Having understood this, we realize that to achieve steady state, since the current ramps up during the ON-time, and we want to ensure that it ramps down during the OFF-time (by the same amount), *the voltage must change sign*. It is similar to the gas pedal being pressed during the ON-time and the brake pedal pressed during the OFF-time. The car still moves forward,

	Buck	Boost	Buck-boost
V_{ON}	$V_{IN} - V_O$	V_{IN}	V_{IN}
V_{OFF}	V_O	$V_O - V_{IN}$	V_O
T_{ON}	D/f	D/f	D/f
T_{OFF}	$(1-D)/f$	$(1-D)/f$	$(1-D)/f$
Volt-seconds Method 1	$V_{ON} \times T_{ON} = V_{OFF} \times T_{OFF}$ $\dfrac{T_{ON}}{T_{OFF}} = \dfrac{V_{OFF}}{V_{ON}}$ $\dfrac{T_{ON}}{T_{OFF} + T_{OFF}} = \dfrac{V_{OFF}}{V_{ON} + V_{OFF}}$		
	$D = \dfrac{V_{OFF}}{V_{ON} + V_{OFF}}$		
Volt-seconds Method 2 $AB = CD$	$(V_{IN} - V_O)D = V_O(1-D)$	$V_{IN}D = (V_O - V_{IN})(1-D)$	$V_{IN}D = V_O(1-D)$
$\dfrac{A}{C} = \dfrac{D}{B}$	$\dfrac{V_{IN} - V_O}{V_O} = \dfrac{1-D}{D}$	$\dfrac{V_{IN}}{V_O - V_{IN}} = \dfrac{1-D}{D}$	$\dfrac{V_{IN}}{V_O} = \dfrac{1-D}{D}$
$\dfrac{A+C}{C} = \dfrac{D+B}{B}$	$\dfrac{V_{IN} - V_O + V_O}{V_O} = \dfrac{1-D+D}{D}$ $\dfrac{V_{IN}}{V_O} = \dfrac{1}{D}$	$\dfrac{V_{IN} + V_O - V_{IN}}{V_O - V_{IN}} = \dfrac{1-D+D}{D}$ $\dfrac{V_O}{V_O - V_{IN}} = \dfrac{1}{D}$	$\dfrac{V_{IN} + V_O}{V_O} = \dfrac{1-D+D}{D}$ $\dfrac{V_{IN} + V_O}{V_O} = \dfrac{1}{D}$
	$D = \dfrac{V_O}{V_{IN}}$	$D = \dfrac{V_O - V_{IN}}{V_O}$	$D = \dfrac{V_O}{V_{IN} + V_O}$

TABLE 2.1 Derivations of Input-Output Transfer Functions from Volt-Seconds Law

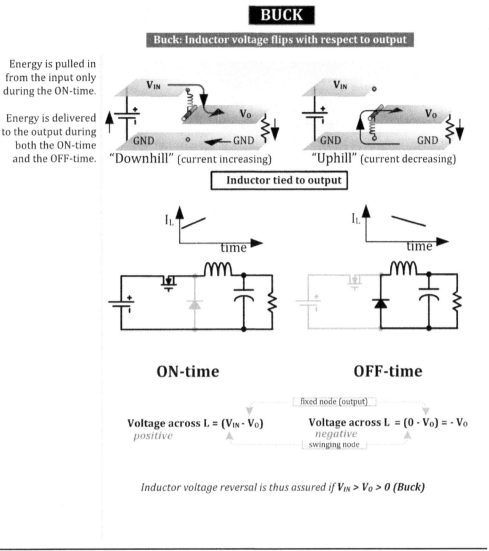

FIGURE 2.3 Analogy and explanation of why this is a step-down topology.

on average, as we alternate between the two pedals. In a real converter, this "lurching" is smoothed by the output capacitor of course, so a steady DC current flows into the output.

Now we combine this with the fact that the inductor is constantly "swiveling" around one fixed rail. We can see that, e.g., in the case of the inductor swiveling around the *output rail*, the only way to ensure the occurrence of *voltage reversal* (in sign) is for the output (fixed) rail to be *between* the other two (alternating) rails, in this case between the input rail and ground (in terms of potential, or voltage). This is illustrated both in a circuit and in a rather intuitive (gravitational analogy) way, in Fig. 2.3. That is for a Buck. It is followed by Figs. 2.4 and 2.5 for the Boost and Buck-boost, respectively. In the last case we see that the output changes sign itself, and so its magnitude becomes irrelevant: that can be greater or less than the incoming rail and still produce voltage reversal purely based on the flipped sign of the output itself. So this is a Buck-boost (an inverting topology).

Current Ripple Ratio

In 2001, the author introduced what he declared was the most simplifying, yet fundamental, concept in designing switching power converters: the *current ripple ratio*, which after much thought, he christened r, to indicate its inherent simplicity.

This concept was first published as the author's Application Note AN-1197 at National Semiconductor, now Texas Instruments (see http://www.ti.com/lit/an/snva038a/snva038a.pdf), and shortly thereafter by Editor Sam Davis in *Power Electronics* magazine (see http://powerelectronics.com/mag/power_current_ripple_ratio/). Since then, this concept has

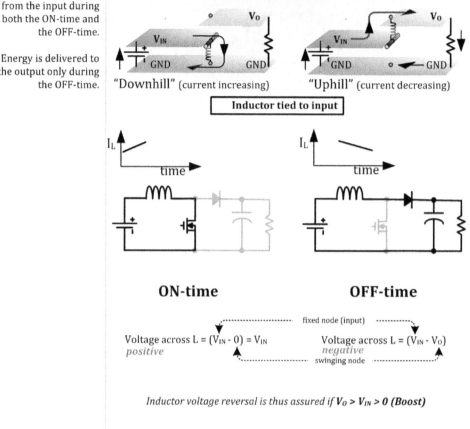

FIGURE 2.4 Analogy and explanation of why this is a step-up topology.

become very widely accepted, even by well-known persons such as Dr. Ray Ridley (see http://www.powersystemsdesign.com/library/resources/documents/europe/2009/psde_nov09.pdf), Dennis Feucht (see http://www.en-genius.net/includes/files/col_080607.pdf), and various professors writing in IEEE publications and teaching crucial courses (see http://en.cnki.com.cn/Article_en/CJFDTOTAL-DLDZ201005028.htm and http://www.usna.edu/EE/ee320/Supplements/dcdc 3_inductors.pdf). Even the entire design table based on r, created by this author originally for AN-1197, was recently lifted (without acknowledgment) by notable magnetic vendors such as Wurth Electronics (see http://www.we-online.com/web/en/electronic_components/download_center_pbs/Trilogie.php). The concept has by now vindicated itself, since imitation (and even piracy) is the best form of flattery.

The key advantage is that r provides a clean design entry point into the heart of any switching power converter (any topology, any power level, any frequency), by specifying an optimum of sorts: *set $r = 0.4$ at maximum load (at maximum line for a Buck, and at minimum line for Boost and Buck-boost)*. This value provides the best compromise between stresses in the converter and the size of magnetic and associated power components (consult AN-1197). But it also enables power scaling. It does that by breaking up most design equations into one part that connects only to the *average power* flowing through the converter (based on the center-of-ramp, or COR, value, which does not change if we change the inductance or frequency) and another part that corresponds to only to the geometric shape factor related to the delta of the waveform (the up and down current ramp, which does not change if we change the load current, but is frequency-dependent). We thus enable the concept of power scaling easily. Note that the concept of power scaling has in turn been recently applied by the author with great success to resonant LLC power conversion too, as discussed in Chap. 19 of this book.

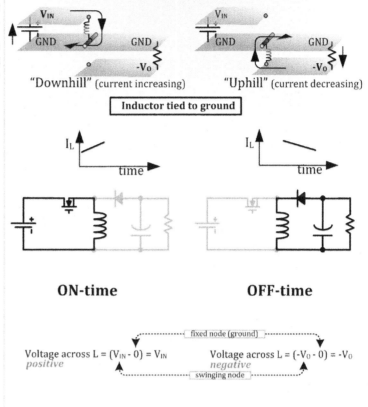

Figure 2.5 Analogy and explanation of why this is a step-up/step-down topology.

For example, per the design tables in the Appendix of this book, all based on r, we have the following equation for L for a Boost converter:

$$L \approx \frac{V_O}{I_O \times r \times f} \times D(1-D)^2$$

where I_O is the maximum load current. Compare this with the usual equations in literature, as in http://www.ti.com/lit/an/slva 372b/slva 372b.pdf.

$$L \approx \frac{V_{IN} \times (V_O - V_{IN})}{\Delta I_L \times f \times V_O}$$

This looks simple enough, except that it doesn't highlight the simple scaling law of inductance, which we almost intuitively use every day and have learned to recognize: *if we double the load current in a fixed (say, 5- to 3.3-V) application, we simply need to halve the inductance!*

After that, the above reference almost arbitrarily states that we need to set

$$\Delta I_L = (0.2 \text{ to } 0.4) \times I_{O_{MAX}} \times \frac{V_O}{V_{IN}}$$

Why is this an optimum? Does the optimum ΔI_L really depend on the input and output voltage *ratio*? And why so? It is all nonintuitive. Now, if we plug this ΔI_L into the preceding equation, we get something close to our equation

$$L \approx \frac{V_{IN}^2 \times (V_O - V_{IN})}{I_{O_{MAX}} \times (0.2 \text{ to } 0.4) \times f \times V_O^2}$$

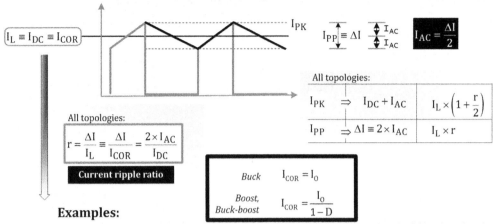

Figure 2.6 Current ripple ratio and basic terms used in this book.

It turns out that the equation in the Appendix of this book, and the derived equation based on the above-referenced application note, both give the same ballpark value for L, as they should, but are cast very differently.

In Fig. 2.6 we show more clearly how we define r, and we also provide a numerical example. The important point to note is that in a Buck, the COR value is equal to the load current. In a Boost and Buck-boost, since energy flows into the output only during the OFF-time [i.e., in $T_{OFF} = (1-D)/f$], and since the average diode current is $I_{COR} \times (1-D)$, and must equal the desired load current I_O, we get

$$I_{COR} = \frac{I_O}{1-D} \quad \text{for Boost and Buck-boost;} \quad I_{COR} = I_O \quad \text{for Buck}$$

In Fig. 2.7, we compare with other similar geometric factors, akin to r, seen in literature sometimes, in particular what we call r_{alt} and K_{RP} (the latter introduced and widely used by Power Integrations as an for example). We also provide their conversions. Note that all the equations using r appear much more elegant, more intuitive, and simpler than all the other ways to express the same, such as the equations from power integrations (which use K_{RP}). For example, as in the Appendix, for a Buck-boost we ask for

$$L \approx \frac{V_O}{I_O \times r \times f} \times (1-D)^2$$

Compare this with the Power Integrations equation in AN-17 (see http://www.thierry-lequeu.fr/data/TOPSWITCH/an17.pdf):

$$L \approx \frac{P_O}{I_P^2 \times f \times K_{RP}(1-K_{RP}/2)}$$

We have simplified the above equation for the ideal case of 100 percent efficiency. But this is still very nonintuitive, because it seems to imply at first sight that if we double power, we need to double the inductance. In reality, for a given input-output voltage, if we double power, we halve the inductance always. In this equation I_P^2 actually dominates from the denominator, but it is not so obvious at first sight.

Also, all the RMS equations using K_{RP} are generally complicated and nonintuitive. For example, the equation for RMS current in the primary winding of a Flyback is, per the Power Integrations equations,

$$I_{RMS} = I_{PK} \cdot \sqrt{D \times (K_{RP}^2/3 - K_{RP} + 1)}$$

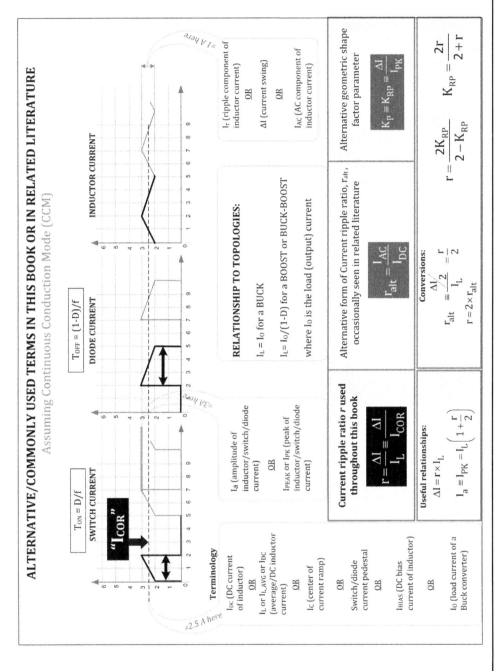

FIGURE 2.7 Current ripple ratio and basic terms used in this book, compared with industry.

In the Appendix, we provide this:

$$= \frac{I_{OR}}{1-D} \times \sqrt{D \times \left[1 + \frac{r^2}{12}\right]}$$

The difference is, by using r, we can clearly separate the RMS into three distinct parts (and dependencies), using n for the turns ratio

$$= \left[\frac{I_O}{n}\right] \times \left[\frac{\sqrt{D}}{1-D}\right] \times \left[1 + \frac{r^2}{12}\right] \Rightarrow \text{[load-dependent][voltage-dependent][inductance-dependent]}$$

In contrast, there is no such way to examine the equations using K_{RP}. Which is why almost all the Power Integrations equations on Flyback design are numerically accurate perhaps, but intuitively misleading. The author therefore does not recommend them for that reason alone.

Average Input and Output Currents

In Figs. 2.8, 2.9, and 2.10, we have presented the basic relations for a Buck, Boost, and Buck-boost, respectively. Note the embedded statements on duty cycle, DC transfer function (V_O/V_{IN}), and the average input/output currents. These figures can be used as a quick reference for picking the ratings of the power components (first pass or barely acceptable limiting values).

Energy Relationships of the Three Topologies

What is the energy per cycle out of any converter? Assuming 100 percent efficiency, it is $\varepsilon = P_{IN}/f = P_O/f$. What is the energy stored in the inductor every cycle (corresponding to the current ramp up of ΔI) and released to the output? If we call that quantity $\Delta\varepsilon$, it must be equal to the work done $V \times I$ to drive the (average) current I_{COR} through the inductor.

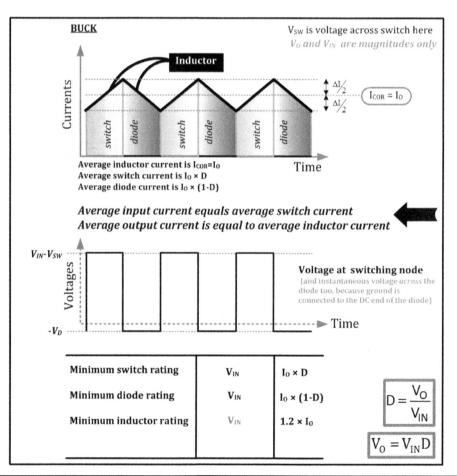

FIGURE 2.8 Waveforms, ratings, and basic relations in a Buck.

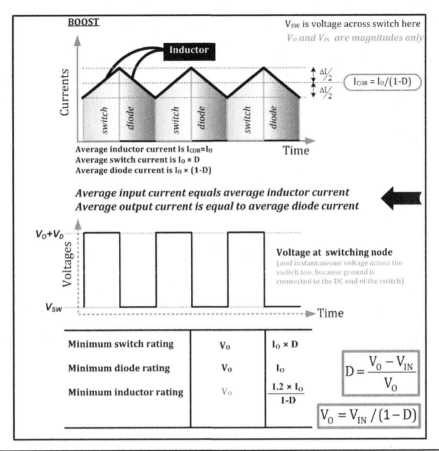

FIGURE 2.9 Waveforms, ratings, and basic relations in a Boost.

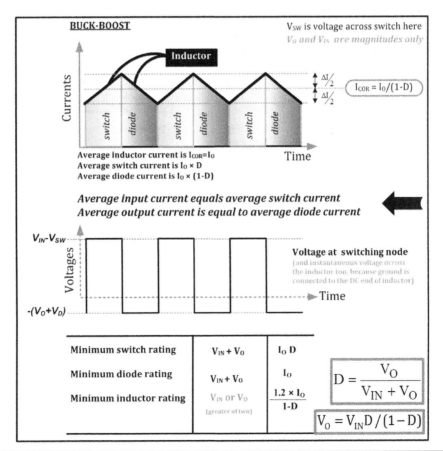

FIGURE 2.10 Waveforms, ratings, and basic relations in a Buck-boost.

44 Chapter Two

Buck (Assuming 100 Percent Efficiency)

We get, using the OFF-time (Vsec is the volt-seconds)

$$\Delta \varepsilon = \text{Vsec} \times I_{COR} = V_O \times I_O \times \left(\frac{1-D}{f}\right)$$

Using

$$I_O \times D = I_{IN} \qquad P_{IN} = P_O$$

we get

$$\Delta \varepsilon = \frac{P_{IN}}{f} \times (1-D) = \frac{P_O}{f} \times (1-D) = \varepsilon \times (1-D)$$

In other words, not all the energy that makes it to the output gets stored in the inductor, only $1 - D$ times the total energy. This is made clear graphically, in the table embedded on the right side of Fig. 2.11.

Note that above we have used the fact that the average switch current is equal to the DC (input current). The switch current is a pedestal (ignoring the ramp portion) of height I_O, with duty cycle D. So the average switch current, that is, the input current, is equal to $I_O \times D$. See Fig. 2.8.

For other topologies, we now proceed similarly.

Boost (Assuming 100 Percent Efficiency)

Using volt-seconds during the ON-time gives

$$\Delta \varepsilon = \text{Vsec} \times I_{COR} = V_{IN} \times \frac{I_O}{1-D} \times \frac{D}{f}$$

Using

$$\frac{I_O}{1-D} = I_{IN} \qquad P_{IN} = P_O$$

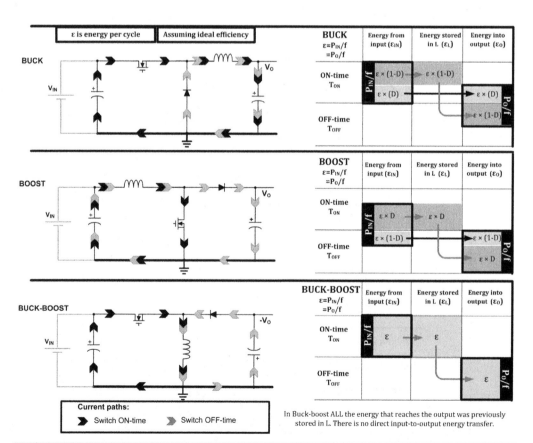

Figure 2.11 Current flow and energy transfer timings.

gives

$$\Delta\varepsilon = \frac{P_{IN}}{f} \times D = \frac{P_O}{f} \times D = \varepsilon \times D$$

In other words, not all the energy making it to the output gets stored in the inductor, only D times the total energy. This is made clear graphically in the table embedded on the right side of Fig. 2.11.

Note that above we have used the fact that the average inductor current is equal to the DC (input current). The inductor current is a constant (ignoring the ramp portion) current of height $I_O/(1 - D)$, which is the I_{COR} value indicated in Fig. 2.25. So the input current is equal to $I_O/(1 - D)$. See Fig. 2.9.

Buck-boost (Assuming 100 Percent Efficiency)

Using volt-seconds during the ON-time gives

$$\Delta\varepsilon = \text{Vsec} \times I_{COR} = V_{IN} \times \frac{I_O}{1-D} \times \frac{D}{f}$$

Using

$$\frac{I_O}{1-D} \times D = I_{IN} \qquad P_{IN} = P_O$$

gives

$$\Delta\varepsilon = \frac{P_{IN}}{f} = \frac{P_O}{f} = \varepsilon$$

In other words, *all* the energy making it to the output *does* get stored in the inductor. This is made clear graphically, in the table embedded on the right side of Fig. 2.11.

Note that above we have used the fact that the average switch current is equal to the DC (input current). The switch current is a pedestal of (ignoring the ramp portion) current of height $I_O/(1 - D)$, with duty cycle D. So the input current is equal to $D \times I_O/(1 - D)$. See Fig. 2.10.

These relationships tell us three things in particular about the Buck-boost (and its transformer-based version, the Flyback):

1. Since all the output power must cycle through the inductor, its size tends to be bigger than the inductors of the remaining two topologies, *for the same power throughput*.

2. It really doesn't matter what the input or output voltages are: The size of the Buck-boost inductor (or Flyback transformer) need not vary according to the voltages at all, just according to the power. *For example, a properly designed transformer for a universal input Flyback is not any larger than one just designed for European voltages!* We will present some examples for this in Chap. 7.

3. This does say that if we double the output power, we have to store twice the energy, so the core size will double too (even though, as we learned, its inductance must halve).

Loss Relationships in Converters

In Fig. 2.12 we summarize the relationships in nonideal converters, that is, for efficiency (η) not equal to 1. We can calculate the loss from efficiency and vice versa in many different ways, depending on what we know. Some numerical examples will make that clearer. Refer to the equations in the Appendix.

Example 2.1 We have a nonsynchronous Buck converter operating in continuous conduction mode (CCM). Its input is 12 V and its output is 5 V. It uses a BJT (bipolar junction transistor) switch with a forward drop $V_{CE}(\text{sat}) \equiv V_{SW} = 0.2$ V. The catch diode is a Schottky device with forward drop $V_D = 0.4$ V. The load current is 1.5 A. What is the duty cycle? What is the dissipation in the BJT and in the diode? What is the estimated efficiency?

We set $V_O = 5$ V, $V_{IN} = 12$ V, $V_D = 0.4$ V, $V_{SW} = 0.2$ V, and $I_O = 1.5$ A. Using the full equation for D, we find

$$D = \frac{V_O + V_D}{V_{IN} + V_D - V_{SW}} = \frac{5 + 0.4}{12 + 0.4 - 0.2} = 0.4426$$

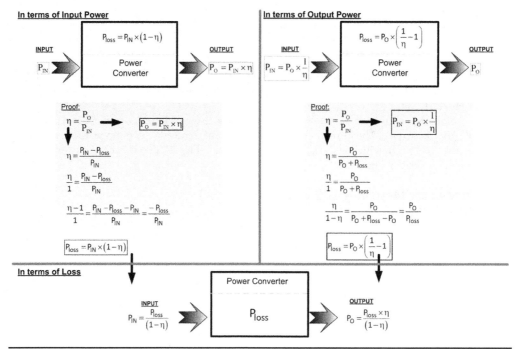

FIGURE 2.12 Loss and efficiency relationships.

Dissipation in both the BJT and diode and the total loss are

$$P_{BJT} = I_O \times D \times V_{SW} = 0.1328 \text{ W} \qquad P_D = I_O \times (1-D) \times V_D = 0.3344 \text{ W}$$
$$P_{LOSS} = P_{BJT} + P_D = 0.4672 \text{ W}$$

Note that we averaged the switch dissipation over a complete cycle by multiplying it by D similarly, we averaged the diode dissipation by multiplying it by $1 - D$.

The output power, input power, and efficiency are

$$P_O = V_O \times I_O = 7.5 \text{ W} \qquad P_{IN} = P_O + P_{LOSS} = 7.9672 \text{ W}$$
$$\eta = \frac{P_O}{P_{IN}} = 0.9414 \qquad (94.14\%)$$

Example 2.2 We have a nonsynchronous Buck converter operating in continuous conduction mode. Its input is 12 V and its output is 5 V. It uses a FET switch with $R_{DS} = 0.1\ \Omega$. The catch diode is a Schottky device with forward drop $V_D = 0.4$ V. The load current is 1.5 A. What is the duty cycle?

We set $V_O = 5$ V, $V_{IN} = 12$ V, $V_D = 0.4$ V, $I_O = 1.5$ A, and $R_{DS} = 0.1\ \Omega$.

For a BJT, to a first approximation, we usually assume that its forward voltage drop is almost constant with respect to the current through it, which is the main reason why the BJT (along with its FET-driven cousin, the IGBT) is still often used in high-power applications. For a FET, the forward drop varies significantly, being considered virtually proportional to the current through it. In the simple duty cycle equation, however, we need to plug in a certain *fixed* number V_{SW}. So for a FET, we need to average the forward switch drop *over the ON-time* (note: here we do not average over the *entire* switching cycle). This is equivalent to taking the voltage drop associated with the *average* current through the switch during the ON-time, which is simply the center of ramp (of the inductor current). Further, in a Buck topology, the center of ramp is equal to the load current I_O. Hence, denoting I_{SW} as the average current in the switch during the ON-time (corresponding to the average drop V_{SW}), we get

$$I_{SW} = I_O \qquad V_{SW} = I_O \times R_{DS} = 1.5 \times 0.1 = 0.15 \text{ V}$$
$$D = \frac{V_O + V_D}{V_{IN} + V_D - V_{SW}} = \frac{5 + 0.4}{12 + 0.4 - 0.15} = 0.4408$$

Example 2.3 What is the efficiency of the Buck converter in Example 2.2 if we disregard both the switch and diode drops?

Now we set $V_O = 5$ V, $V_{IN} = 12$ V, $V_D = 0$ V, $V_{SW} = 0$ V, and $I_O = 1.5$ A. This leads to the "ideal" duty cycle equation for a Buck. We will also confirm that, in effect, it assumes 100 percent efficiency.

DC-DC Converters: Topologies to Configurations 47

$$D_{IDEAL} = \frac{V_O + \cancel{V_D}}{V_{IN} + \cancel{V_D} - \cancel{V_{SW}}} = \frac{V_O}{V_{IN}} = \frac{5}{12} = 0.4167$$

The input current of a buck is the switch current averaged over the entire ON-time. So the input current corresponding to this duty cycle is

$$I_{IN\_IDEAL} = I_{SW} \times D_{IDEAL} = I_O \times D_{IDEAL} = 0.625 \text{ A}$$

The corresponding input power is thus

$$P_{IN\_IDEAL} = V_{IN} \times I_{IN\_IDEAL} = 12 \times 0.625 = 7.5 \text{ W}$$

The output power is

$$P_O = I_O \times V_O = 7.5 \text{ W}$$

Therefore, the efficiency is 7.5 W/7.5 W = 1 (or 100 percent) as expected, validating our statement that if the switch and diode drops are set to zero, we get an "ideal" situation, with no losses.

NOTE *Of course, the only losses that were allowed in the first place, by the duty cycle equation currently in use in previous examples, are those losses related to the forward drops in the switch and diode, that is, the conduction losses in the semiconductors, no more. This indicates that since quite obviously not all switcher losses have been accounted for, the duty cycle equation in use so far is itself limited and is clearly just an approximation.*

Example 2.4 What is the efficiency of the Buck converter in Example 2.2 if we disregard (only) the switch drop (i.e., we assume only a diode drop is present)? Also what is the loss in this diode?

We set $V_O = 5$ V, $V_{IN} = 12$ V, $V_D = 0.4$ V, $V_{SW} = 0$ V, and $I_O = 1.5$ A.

$$D_{IDEAL} = \frac{V_O + V_D}{V_{IN} + V_D} = \frac{5 + 0.4}{12 + 0.4} = 0.4355$$

The input current of a Buck is the switch current averaged over the entire ON-time. So the input current corresponding to this duty cycle is

$$I_{IN} = I_{SW} \times D = I_O \times D = 1.5 \times 0.4355 = 0.6532 \text{ A}$$

The corresponding input power is thus

$$P_{IN} = V_{IN} \times I_{IN} = 12 \times 0.6532 = 7.8387 \text{ W}$$

The output power is clearly

$$P_O = I_O \times V_O = 7.5 \text{ W}$$

Therefore, the efficiency is 7.5 W/7.8387 W = 0.9568.

Now, the average diode current in a Buck is $I_O \times (1 - D)$. So we get $I_{D\_AVG} = 1.5 \times (1 - 0.4355) = 0.8468$ A. The loss in the diode is therefore

$$P_D = I_{D\_AVG} \times V_D = 0.8468 \times 0.4 = 0.3387 \text{ W}$$

We can see that this is exactly equal to the difference between the input power and output power: $P_{IN} - P_O = 7.8387 - 7.5 = 0.3387$ W, as expected. So the balance sheet of losses is complete and accurate.

Example 2.5 In a Buck converter we assume, as above, that the switch is "ideal" (very low R_{DS}), and the catch diode has a voltage drop of 0.4 V. If the efficiency of the converter is 95.679 percent and the diode loss is 0.3387 W, what is the input power? What is the output power?

Here we are just working backward. Further, we are not assuming any specific input and output voltages, or even a certain load current. We are just talking in terms of power. Looking at Fig. 2.12, we see all the possible relationships between input and output power versus loss and efficiency. Keep in mind these are valid equations for any power converter in general, not necessarily just switchers. We focus our attention on the lowermost diagram in the figure (under "In terms of loss"). To use the relationship here, we need to know the loss, which in this example is the loss in the diode.

We set $P_{LOSS} = 0.3387$ W and $\eta = 0.95679$.

So

$$P_{IN} = \frac{P_{LOSS}}{1-\eta} = \frac{0.3387}{1-0.95679} = 7.8387 \text{ W}$$

$$P_O = \frac{P_{LOSS} \times \eta}{1-\eta} = 7.5 \text{ W}$$

This agrees with Example 2.4. We have thus validated the relevant equations in Fig. 2.12 and also our previous calculations.

Example 2.6 In Example 2.4, correlate the diode dissipation to the additional energy drawn from the input and the increase in input current, as compared to the ideal case.

The diode loss was $P_D = 0.3387$ W. This must correspond to the additional energy per unit time drawn from the input. We recall from Example 2.3, that the ideal duty cycle was $D_{IDEAL} = 0.4167$. Now, with diode loss included, the duty cycle is $D = 0.4355$. The general equation for the (average) input current of a Buck is $I_O \times D$. Note that in a Buck, the input current is the switch current first averaged over the ON-time, that is, $I_{SW} \equiv I_O$ and then further averaged over the entire cycle (by multiplying it by D). So for the ideal case, we get

$$I_{IN\_IDEAL} = I_O \times D_{IDEAL} = 1.5 \times 0.4167 = 0.625 \text{ A}$$

whereas for the nonideal case (using the value of D calculated in Example 2.4)

$$I_{IN} = I_O \times D = 1.5 \times 0.4355 = 0.6532 \text{ A}$$

The additional energy per unit time inputted when the duty cycle stretches out from its ideal value (in turn leading to the observed increase in input current) is

$$V_{IN} \times (I_{IN} - I_{IN\_IDEAL}) = 12 \times (0.6532 - 0.625) = 0.3387 \text{ W}$$

This is equal to the diode loss in Example 2.4. It thus validates the following general statement: I_{IN_IDEAL} is the baseline current level for a given P_O and input/output, corresponding to all the incoming energy being fully converted to useful energy (i.e., no losses). Any increase above and beyond this baseline level coincides exactly with the losses in the converter.

Example 2.7 Suppose we have a 12- to 5-V synchronous buck, using a control FET with R_{DS} of 1 Ω, and a synchronous FET with R_{DS} of 0.8 Ω. The output current is 1.5 A. The inductor has a DC resistance (DCR) of 0.1 Ω. What are the duty cycle, the breakup of the losses, and the efficiency? Continue to ignore switching losses, as we have been doing so far.

We set $V_O = 5$ V, $V_{IN} = 12$ V, $R_{DS\_1} = 1$ Ω, $R_{DS\_2} = 0.8$ Ω, and DCR = 0.1 Ω.

Let us call the average current in the two FETs *during the ON-time* (not averaged over the whole cycle) as I_{SW_1} and I_{SW_2}. So, since in a Buck that is equal to the center-of-ramp I_O, we get

$$I_{SW\_1} = I_{SW\_2} = I_O = 1.5 \text{ A}$$

The corresponding switch drops (i.e., their average over the ON-time) are

$$V_{SW\_1} = I_O \times R_{DS\_1} = 1.5 \times 1 = 1.5 \text{ V} \qquad V_{SW\_2} = I_O \times R_{DS\_2} = 1.5 \times 0.8 = 1.2 \text{ V}$$

So, from the general duty cycle equation, with $V_D = V_{SW\_2'}$

$$D = \frac{V_O + V_{SW\_2} + I_O \times \text{DCR}}{V_{IN} + V_{SW\_2} - V_{SW\_1}} = 0.5427$$

The remaining calculations are

$$I_{IN} = I_O \times D = 0.8141 \text{ A} \qquad P_{IN} = I_{IN} \times V_{IN} = 9.7692 \text{ W}$$

$$P_O = I_O \times V_O = 7.5 \text{ W}$$

The computed efficiency is therefore $\eta = P_O/P_{IN} = 7.5/9.7692 = 0.7677$ (or 76.8 percent)

The losses are $P_{IN} - P_O = 2.2692$ W. Let us confirm where they went.

We get the FET and inductor losses as

$$P_{FET\_1} = (I_O^2 \times R_{DS}) \times D = 1.2212 \text{ W}$$

$$P_{FET\_2} = (I_O^2 \times R_{DS}) \times (1-D) = 0.8231 \text{ W}$$

Summing up all loss terms gives $P_{LOSS} = 1.2212 + 0.8231 + 0.225 = 2.2692$ W. This agrees with the difference between the input and output power $P_{IN} - P_O$, thus validating our equations above.

Nonideal Duty Cycle Equations

The ideal equations for duty cycle, provided in Figs. 2.8, 2.9, and 2.10, are the most inaccurate and lead to the smallest duty cycle possible for a given input/output condition. Real-world equations, which we will discuss here, are the most accurate and lead to the highest duty cycle (lowest-efficiency estimate). Between these two sets of equations lie many other forms of duty cycle equations found in literature, all with varying degrees of accuracy. For example, by using the fundamental principle of volt-seconds balance in steady state, but not ignoring switch and diode drops, we can derive the following duty cycle equations:

$$D_{\text{BUCK}} \approx \frac{V_O + V_D}{V_{\text{IN}} - V_{\text{SW}} + V_D} \qquad D_{\text{BOOST}} \approx \frac{V_O - V_{\text{IN}} + V_D}{V_O - V_{\text{SW}} + V_D}$$

$$D_{\text{BUCK-BOOST}} \approx \frac{V_O + V_D}{V_O + V_{\text{IN}} - V_{\text{SW}} + V_D}$$

We realize that though these equations explicitly include the drop across the diode and switch, and therefore factor in the *conduction losses* inside those *two* components, *they continue to ignore several other smaller loss terms*, such as the I^2R conduction loss in the DC resistance (DCR) of the inductor, or the various switching losses, or the AC resistance losses in the inductor, or the capacitor ESR losses, and so on—*all of which, if factored in somehow, will cause the duty cycle to increase further*.

How do we derive accurate equations? Suppose we pick the Buck-boost. We have

$$I_{\text{COR}} = \frac{I_O}{1-D}$$

On the input side,

$$I_{\text{COR}} \times D = I_{\text{IN}} \qquad I_{\text{COR}} = \frac{I_{\text{IN}}}{D}$$

Equating gives

$$\frac{I_O}{1-D} = \frac{I_{\text{IN}}}{D}$$

$$\frac{I_O}{I_{\text{IN}}} = \frac{1-D}{D}$$

But we also know that efficiency is, by definition,

$$\eta = \frac{P_O}{P_{\text{IN}}} = \frac{V_O \times I_O}{V_{\text{IN}} \times I_{\text{IN}}}$$

So

$$\frac{I_O}{I_{\text{IN}}} = \frac{\eta V_{\text{IN}}}{V_O}$$

Equating the above two equations for I_O/I_{IN}, we get

$$\frac{\eta V_{\text{IN}}}{V_O} = \frac{1-D}{D}$$

which simplifies to

$$D = \frac{V_O}{\eta V_{\text{IN}} + V_O}$$

If η is the actual (measured) efficiency this is the most accurate equation to use for predicting duty cycle as it is not constrained by only R_{DS} conduction losses. The general efficiency number η, in effect, allows us to include all losses.

Note that mathematically *we can think of this as having increased the input from V_{IN} to $\eta \times V_{IN}$*. So if we start with our "ideal" (most inaccurate) duty cycle equations and carry out the substitution $V_{IN} \to \eta V_{IN}$, we will arrive at the most accurate duty cycle equation for all topologies. We get

$$\text{Buck:} \quad D = \frac{V_O}{V_{IN}} \to \frac{V_O}{\eta V_{IN}}$$

$$\text{Boost:} \quad D = \frac{V_O - V_{IN}}{V_O} \to \frac{V_O - \eta V_{IN}}{V_O}$$

$$\text{Buck-boost:} \quad D = \frac{V_O}{V_{IN} + V_O} \to \frac{V_O}{\eta V_{IN} + V_O}$$

In Fig. 2.13, however, we reveal that mathematically, for a Buck as shown, we can start off applying the volt-seconds law to a decreased input of $\eta \times V_{IN}$ or to an increased output of V_O/η and still end up with the same duty cycle. For all other topologies, we can equivalently write

$$D_{BUCK} = \frac{V_O/\eta}{V_{IN}} \qquad D_{BOOST} = \frac{V_O/\eta - V_{IN}}{V_O/\eta} \qquad D_{BUCK\text{-}BOOST} = \frac{V_O/\eta}{V_O/\eta + V_{IN}}$$

Despite leading to the same duty cycles, the two possibilities (decreased input versus increased output) lead to different ON-time/OFF-time *volt-seconds* and therefore to magnetic

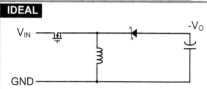

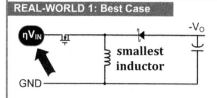
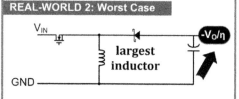

Figure 2.13 Same duty cycle, different inductor size requirements.

components of different sizes. The former interpretation (i.e., an effective decrease in V_{IN}) leads to an optimistic (and possibly undersized) core, whereas the latter (an effective increase in V_O) leads to a relatively larger core. In general, the *latter* model is a safer bet in design, especially if we don't know where exactly the losses corresponding to the less-than-unity estimated/measured efficiency are occurring inside the converter. What really happens in a practical converter lies somewhere in between the two real-world models of Fig. 2.13.

Note *The Power Integrations method of transformer design gets closest to recognizing and modeling this subtlety by creating an* allocation factor, *which they refer to as* Z: *the ratio of secondary-side losses (presumably in the diodes and output capacitors) to the total losses, such as those including losses of the "primary side" (the switch, input capacitors, EMI filter). However, they typically (arbitrarily) recommend you set* Z *to around 50 percent. This implies something exactly between the two real-world models mentioned above. But that ratio is not necessarily true—it is more like a fudge factor. The best way from an engineering viewpoint is to assume worst-case transformer sizing, based on an imagined "increased output" of* V_O/η.

Power Scaling Guidelines in Power Converters

Previously we talked about how defining *r* helps to achieve an easy intuitive understanding of power scaling. Let us understand this more clearly and how power supplies "scale" with load. Let us take one of the equations from the Appendix to illustrate something quite interesting and useful here. Let us take the RMS of the Buck converter switch current for an example. It is

$$I_{SW\_RMS} = I_{COR}\sqrt{D\left(1+\frac{r^2}{12}\right)} \quad \text{or equivalently} \quad I_{SW\_RMS} = I_{COR} \times \sqrt{D} \times \sqrt{1+\frac{r^2}{12}}$$

We do remember that, by definition, $r = \Delta I / I_{COR}$. In the case of a buck, $I_{COR} = I_O$. So we can also write the above equations as

$$I_{SW\_RMS} = I_O\sqrt{D\left(1+\frac{\Delta I^2}{12 \times I_O^2}\right)} \quad \text{or equivalently} \quad I_{SW\_RMS} = \sqrt{D\left(I_O^2+\frac{\Delta I^2}{12}\right)}$$

The latter equation is seen more commonly in literature. Notice that it looks "messier" than our simpler-looking equations expressed in terms of *r*. But cosmetics aside, the usual way of writing out the RMS currents also misses out on a potentially huge simplification. In contrast, using *r*, we express the switch RMS current in a more intuitive manner—as a product of three relatively *orthogonal* terms: an AC term, that is, a term involving *only r*; multiplied by a DC term, that is, involving *only* load current I_O, and a term related to duty cycle *D* (input/output voltages) as expected. We can thus separate the terms and reveal the underlying concept of *power scaling in DC-DC converters*, something which is very hard to see from the usual way of writing out the RMS equations, as indicated above.

Using our unique method of writing out the RMS current stress equations, we now recognize the fact that current stresses (AVG and RMS) are all *proportional to load current* (for a given *r* and fixed *D*). We thus start to realize what scaling implies. For example, this can mean several things:

1. In terms of ability to handle stresses, a 100-W power supply will require an output capacitor roughly *twice* the value, in terms of capacitance and size, of a 50-W power supply (for the same input and output voltages). Here we assume that if we are using only one output capacitor, its ripple (RMS) current rating is almost proportional to its capacitance. That is not strictly true, however. More correctly, we can say that if a 50-W power supply has a single output capacitor of value *C* with a certain ripple rating I_{RIPP}, then a 100-W power supply will require two such identical capacitors—each of value *C* and ripple rating I_{RIPP}—*paralleled together*. That doubles the capacitance and the ripple rating (ensuring the PCB layout is conducive to good sharing too). We could then justifiably assert that *output capacitance (and its size) is roughly proportional to* I_O. Note that we are implicitly assuming the switching frequency is the same for the 50-W and 100-W converters. Changing the frequency can impact capacitor selection too.

2. Similarly, rather generally speaking, a 100-W power supply will require an input capacitor twice that of a 50-W power supply. So *input capacitance (and its size) will also be roughly proportional to* I_O.

52 Chapter Two

3. Since heating in a FET is $I_{RMS}^2 \times R_{DS}$, and I_{RMS} is proportional to I_O, then for the same dissipation, we might initially think we would want a 100-W power supply to use a FET with one-fourth of the R_{DS} of a 50-W supply. However, we are actually not interested in the *absolute* dissipation (unless thermally limited), only its *percentage*. In other words, if we double the output wattage of a converter, say, from 50 to 100 W, we typically expect/allow twice the dissipation too (i.e., the same efficiency). Therefore, it is good enough if the R_{DS} of the FET of the 100-W power supply is only one-half (not one-quarter) of the R_{DS} of the FET used in the 50-W power supply. So in effect, the *FET R_{DS} is inversely proportional to I_O*.

4. We also know that for any power supply, we usually always like to set $r \approx 0.4$ for any output power. So from the equations for L, we see that for a given r, L is inversely proportional to I_O. This means the inductance of a 100-W power supply choke will be one-half that of a 50-W power supply choke. *Therefore L is inversely proportional to I_O*. Note that we are implicitly assuming the switching frequency is unchanged.

5. Energy of an inductor is $½ \times L \times I^2$. If L halves (for twice the wattage) and I doubles, then the required energy-handling capability of a 100-W choke must be twice that of a 50-W choke. In effect, the *size of an inductor is proportional to I_O*. Note that since L is dependent on frequency, we are again implicitly assuming that the switching frequency is unchanged here.

These scaling relationships can help us easily generate a BOM of the components for a 200-W supply, if we just know, e.g., all the correct components for a 50-W supply. Using r was the key to building up this powerful intuitive method of visualizing converters.

PART 2   BASIC WAVEFORM ANALYSIS AND STRESS COMPUTATIONS

In this part we gain mastery over how to calculate the RMS and average of DC-DC converter waveforms (can be current and/or voltage, but is usually intended for calculating current stresses).

General Method for Piecewise Linear Waveforms

In Fig. 2.14 we present a shortcut method as an alternative to memorizing or integrating. We can arrive at the average and the RMS squared of any applicable waveform. What we basically do here is the following:

- We pick any part of the waveform that is repetitive. It does not matter whether we start at the moment the switch turns ON or some other point, provided that at the end of the cycle we return to exactly the same point.

- We break the cycle into segments of *constant* slope. So the ends of a segment are usually break points of slope, though it is more important to ensure that no designated segment *encloses* a break point.

- We calculate the average value of each segment (considered independently), and then we sum over all segments to get the average for the entire waveform. Thus

$$I_{AVG} = I_{AVG\_1} + I_{AVG\_2} + I_{AVG\_3} + \cdots$$

where I_{AVG} of each segment is

$$I_{AVG\_n} = \frac{I_n + I_{n+1}}{2} \times \delta_n$$

- Similarly we calculate the RMS values of each segment and then sum the *squares* of the RMS of each segment to get the square of the RMS for the entire waveform. So,

$$I_{RMS}^2 = I_{RMS\_1}^2 + I_{RMS\_2}^2 + I_{RMS\_3}^2 + \cdots$$

where I_{RMS} of each segment can be found from

$$I_{RMS\_n}^2 = \frac{I_n^2 + I_{n+1}^2 + I_n I_{n+1}}{3} \delta_n$$

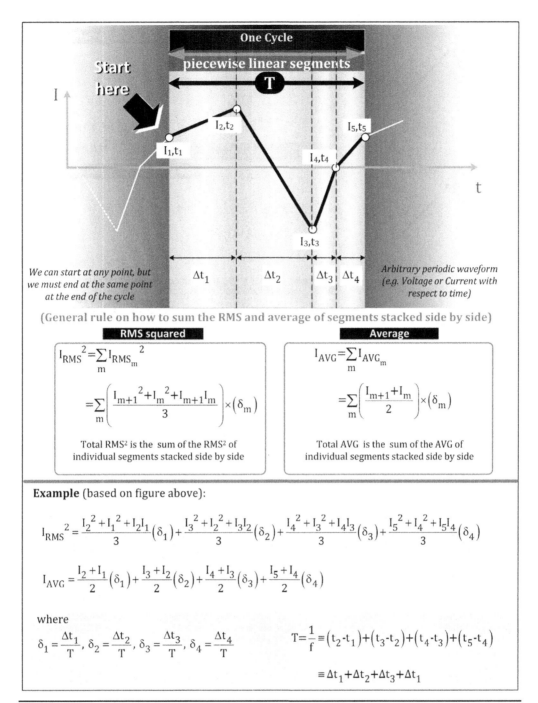

FIGURE 2.14 Way to calculate the RMS and average of most DC-DC converter waveforms.

Here δ is the *geometric* duty cycle of that segment, that is, the ratio of its duration to the time period of the entire waveform.

- The RMS of a waveform does not care if the waveform goes below zero (ground). We could therefore take the entire part of the waveform that is below ground and "fold" it to be above ground. This is equivalent to rectifying it. See Fig. 1.3 for a worked example. The average value, however, does change in the process.
- Though it may seem obvious, we should note that reflecting a waveform horizontally does not change its RMS or average. So a switch waveform can change its familiar shape and become a diode waveform, and the same equations for RMS and average would still apply. However, the duty cycle δ of each segment in the equations above is D for the switch, but is $1-D$ for the diode (converter operated in CCM). We can do the same for, say, a transformer isolated Flyback, but we will need to first "reflect" the currents to the same side of the transformer for a comparison.

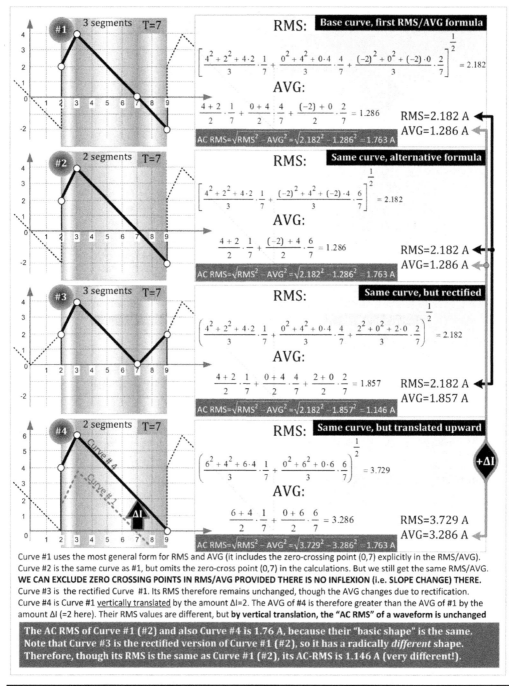

Figure 2.15 Exercises on using the piecewise linear method for RMS and average.

In Fig. 2.15 we present examples, using this formula. These reveal a few interesting facts.

1. Curve 1: This uses the formula in Fig. 2.14 with a slight difference: the zero crossing point is also used to divide up segments. That is the way the formula is presented in E.J. Bloom's course. However, provided there is no change of slope at the zero crossing point, we need not segment the waveform at that point, as is evident from the next curve.

2. Curve 2: This strictly follows the formula in Fig. 2.14 and gives the same results as the method used above. This is obviously less work, and so this is preferred.

3. Curve 3: Here we "rectify" the waveform. Since RMS depends on the square of the waveform, indeed it is not affected. But the average is, as expected.

4. Curve 4: Here we translate the entire waveform vertically. Yes, both the RMS and average values are affected, but the surprise is that there is a quantity called AC

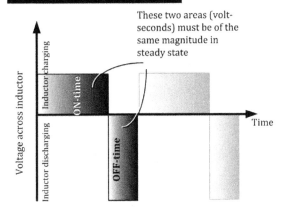

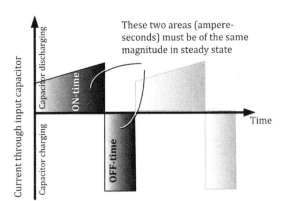

FIGURE 2.16 Volt-seconds law for inductors is analogous to ampere-seconds law for capacitors.

RMS that remains invariant in all cases (except for the rectified waveform case, of course, since that changes the basic shape). AC RMS is pure RMS, without any DC; that is, it is defined as

$$(AC\ RMS)^2 = RMS^2 - AVG^2$$

In switching power conversion, in steady state, just as the inductor voltage has equal and opposite volt-seconds (area under curve segments, see Fig. 2.15), similarly, the capacitor current has equal and opposite ampere-seconds. See Fig. 2.16. As explained, this prevents current (and magnetic energy) runaway in inductors and voltage (and electric energy) runaway in capacitors. Both situations are analogous.

Keep in mind that current is equal to charge per unit time, so ampere-seconds is charge, and we know that no capacitor can accumulate charge in steady state.

In effect this means that the average current through a capacitor in steady state is zero. So the computed RMS of the current waveform in the input/output capacitors of any converter in steady state must be the AC RMS in effect. Therefore, we can start with the (parent) waveform from which the capacitor waveform is derived and "reset" that waveform to have a zero average value, in effect calculating the AC RMS of the associated parent waveform. That process is shown in Fig. 2.17. We thus arrive at the RMS capacitor currents, and for convenience, we have also tabulated the current stresses for the other components (derived from the formula in Fig. 2.14).

Other RMS and Average Values

We may encounter other types of waveforms in power conversion. For convenience we have listed them in Fig. 2.18. These are the most common.

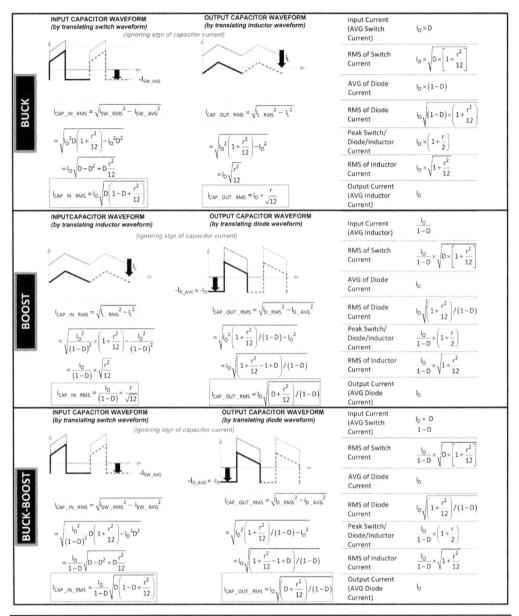

FIGURE 2.17 Calculating capacitor RMS currents (and other current stresses tabulated).

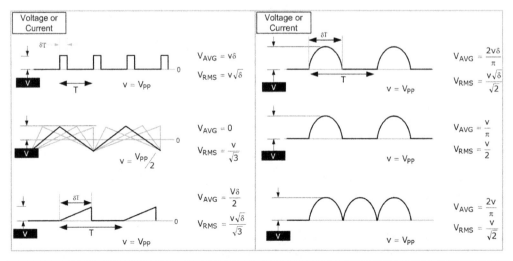

FIGURE 2.18 RMS and average of some commonly encountered waveforms.

Note *Note that some of the equations provided for arbitrary waveforms involve δ, which refers to the duty cycle of the particular segment of the waveform. This duty cycle is only in a geometric sense, and it is not necessarily the duty cycle as we know it in a converter. Therefore, we when apply the equations to the switch and diode waveforms, e.g., δ is the same as D when it comes to the switch, but is 1 – D for the diode. This is an easy way to second-guess diode RMS equations from switch RMS equations. We can also estimate capacitor stresses. Keep in mind that horizontally reflecting a waveform (or translating it horizontally) has no effect on its RMS or average. Translating it vertically has no effect on its AC RMS.*

In Fig. 2.19 we have provided a more detailed lookup table for sine wave–related waveforms. These are useful to know when you are dealing with power factor correction applications or resonant topologies. Note the difference between the average value, mean value, and median value. We also show how to quickly convert between them (for the last two full-cycle cases).

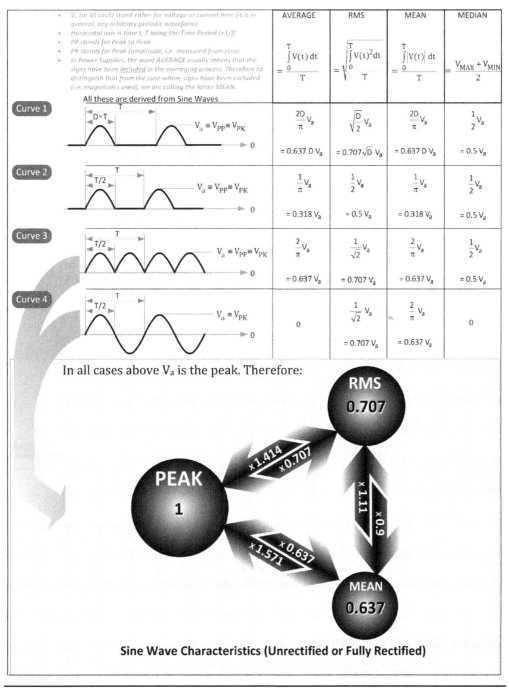

FIGURE 2.19 RMS, average, mean, and median of some commonly encountered sine wave–related waveforms.

58 Chapter Two

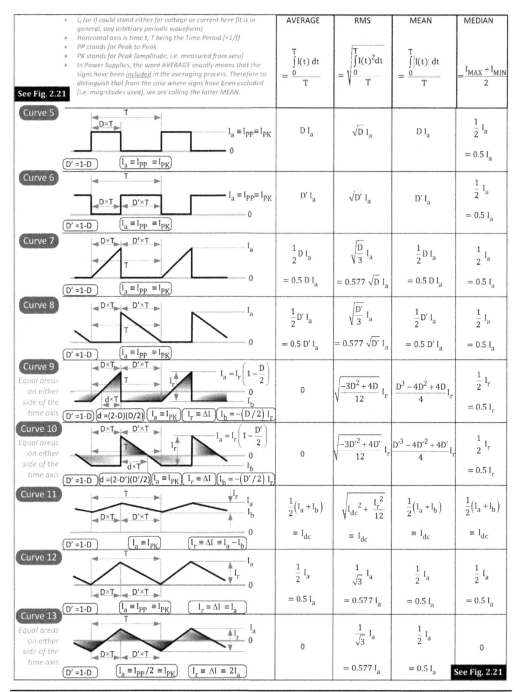

FIGURE 2.20 Part 1 of RMS, average, mean, and median of some commonly encountered square wave–related waveforms.

In Figs. 2.20 and 2.21 we provide a detailed lookup table to help us evaluate stresses for square and triangular-shaped waveforms, and in Fig. 2.21 we provide where we will encounter this waveform.

Finally, in Fig. 2.22, we go through the RMS calculations of a switch, and we show how other, more complicated forms appear in literature, but give the same results. Compare these to our simple equations found in the Appendix, based on r.

Shape of Capacitor Currents

A fundamental difference between the topologies concerns the basic shape of the input and output waveforms. For a Buck (or Forward) converter, e.g., the output current into the capacitor is relatively smooth as it comes through an inductor. However, the input current is "chopped" (pulsating). For a Boost the situation reverses, and it is the output current

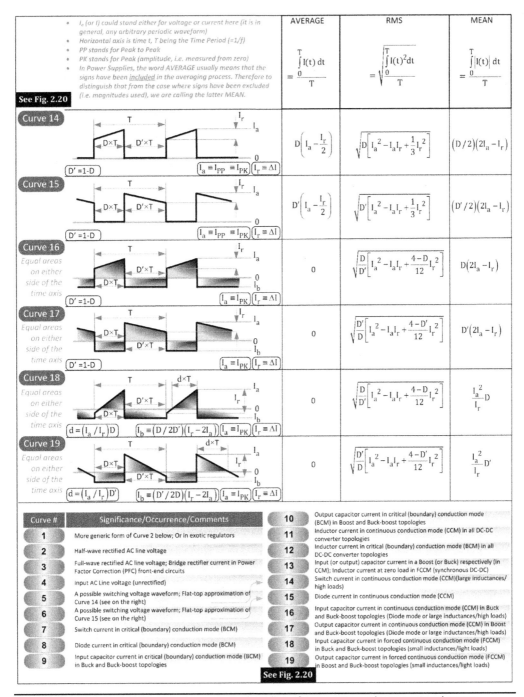

Figure 2.21 Part 2 of RMS, average, mean, and median of some commonly encountered square wave–related waveforms.

that is pulsating. For a Buck-boost or a Flyback, both the input and output currents are pulsating. This contributes to the inability of the Flyback to handle larger power (another factor is leakage inductance). For the Cuk converter, discussed in Chap. 3, which is essentially a composite of a Boost stage input and a Buck stage output, we get the best of two worlds. So *both* the input and output currents are smooth, and this topology is therefore sometimes called an *ideal DC-DC converter*.

Wide Input Voltage and the Design Entry Point

When we introduced r, we did not point out what happens when we have a *wide* input range. Do we set r to the suggested 0.4 at V_{INMIN} or at V_{INMAX}? For that we have to understand what constitutes worst case in a general sense for the entire power supply, especially

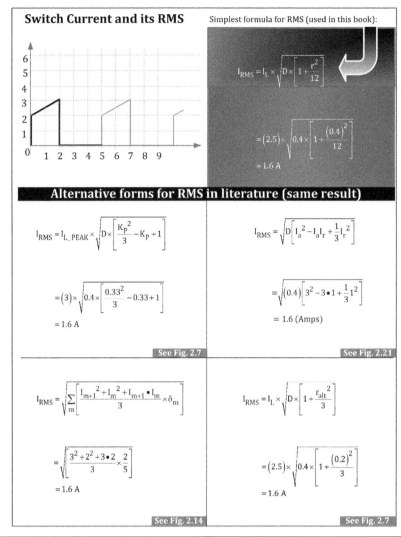

Figure 2.22 Switch RMS equations and calculations in literature, compared.

for the inductor design. A key to converter design is ensuring that the magnetic components do not saturate. So the worst case is the point where we should "ensure" our magnetic design. This forms our design entry point. And it is topology-dependent, as we learn.

In general, the ramp portion of the inductor waveform will add 20 percent to the COR (center of ramp) value, when we set r to 0.4. So the primary criterion for knowing what is the worst-case (design entry) point for a given topology is knowing at what input voltage I_{COR} (the average inductor current) is the highest.

We know from Fig. 2.6 that the COR has the following value for the Boost and the Buck-boost (also Flyback):

$$I_{L\_AVG} \equiv I_{COR} \propto \frac{1}{1-D}$$

For a Buck

$$I_{L\_AVG} \equiv I_{COR} = \text{constant}$$

Since, for all topologies, a high D corresponds to a low V_{IN}, for the Boost and the Buck-boost, the worst case corresponds to high D, that is, low input. For the Buck, the average inductor current is always equal to the load current I_O, so its worst case cannot be determined by the average inductor current alone. We turn to its peak value, and we see that it has a maximum at high input voltage. *So for a Buck (and for the Forward converter choke), the design must start at high line. For a Boost and Buck-boost (and for a Flyback transformer), the design must start at low line.* Note also that since duty cycle increases as input falls, and because a single-ended Forward converter has a duty cycle maximum of 50 percent, just to ensure we do not hit the

DC-DC Converters: Topologies to Configurations

maximum duty cycle "brick wall" and can continue to deliver full power, *the transformer of a Forward converter must also be designed at lowest input voltage, that is*, this time for an entirely different reason than peak current and energy storage capability.

How *r* Varies with Changes in Line and Load

We first note that *for all topologies, a low D implies a high V_{IN} and a high D implies a low V_{IN}*. We are of course only referring to the magnitudes of the voltages involved.

In Table 2.1 we provided the equations for *r*. We thus know how it will vary from any selected set point. We have also included the capacitor RMS equations from Fig. 2.17, for convenience and for subsequent discussion.

Assume we have set *r* to a value r_{SET} (typically 0.4) at the designated design entry point. Now as we change the voltage, the actual current ripple ratio *r* will change. The way it varies is determined by the equations in Table 2.2. In words:

- For a Buck, having set $r = 0.4$ at D_{MIN}, as we increase *D*, the current ripple ratio will decrease because it has the form $1 - D$.
- For the Buck-boost, having set $r = 0.4$ at D_{MAX}, as we decrease *D*, the current ripple ratio will increase, as it has the form $(1 - D)^2$. This is attributable to the fact that $r = \Delta I/I_{COR}$, and I_{COR} starts falling.
- For the Boost, having set $r = 0.4$ at D_{MAX}, as we decrease *D*, the current ripple ratio will increase at first; but as *D* becomes even smaller, it will eventually decrease. This is so because it has the form $D(1 - D)^2$, and this function *has a maximum at D = 0.33*.

The input voltage at which $D = 0.5$ is generically designated as $V_{IN\_D=0.5}$ in Table 2.2. For the Boost, this corresponds to an input voltage equal to one-half the set output.

NOTE *For completeness we have included the forward drops across the switch and diode, V_{SW} and V_D respectively, in the table. But we can ignore them for now.*

How Capacitor RMS Varies with Changes in Line and Load

From Table 2.2, we see that the equations for capacitor RMS include both *r* and *D*. And both these do change if we vary the line (input voltage) because *r* is also a function of *D*. So, in general, the actual (net) variation of the capacitor RMS with changes in input voltage will need to include the variations of both terms, as indicated in the following example.

	Buck	Boost	Buck-boost
RMS Current in Input Capacitor (A)	$I_O \sqrt{D\left[1 - D + \dfrac{r^2}{12}\right]}$ $\approx \dfrac{I_O}{2}$ (small *r*, $D = 0.5$)	$\dfrac{I_O}{1-D} \times \dfrac{r}{\sqrt{12}}$ ≈ 0 (small *r*)	$\dfrac{I_O}{1-D}\sqrt{D[1 - D + r^2/12]}$ $\approx I_O$ (small *r*, $D = 0.5$)
RMS Current in Output Capacitor (A)	$I_O \times \dfrac{r}{\sqrt{12}}$ ≈ 0 (small *r*)	$I_O \times \sqrt{\dfrac{D + r^2/12}{1-D}}$ $\approx I_O$ (small *r*, $D = 0.5$)	$I_O \times \sqrt{\dfrac{D + r^2/12}{1-D}}$ $\approx I_O$ (small *r*, $D = 0.5$)
r (L in H, f in Hz)	$\dfrac{V_O + V_D}{I_O \times L \times f} \times (1-D)$	$\dfrac{V_O - V_{SW} + V_D}{I_O \times L \times f} \times D \times (1-D)^2$	$\dfrac{V_O + V_D}{I_O \times L \times f} \times (1-D)^2$
$V_{IN\_D=0.5}$ (V)	$2V_O + V_{SW} + V_D$ $\approx 2 \cdot V_O$	$\dfrac{V_O + V_{SW} + V_D}{2}$ $\approx V_O/2$	$V_O + V_{SW} + V_D$ $\approx V_O$

TABLE 2.2 RMS Capacitor Currents for the Three Main Topologies

Relating the variation of r to the input capacitor current of a Boost (see Table 2.2), we see that

$$I_{\text{CIN\_RMS\_BOOST}} \propto \frac{r}{1-D} \propto \frac{D(1-D)^2}{1-D} = D(1-D)$$

But we already know that the function $D(1-D)$ has a maximum at $D = 0.5$. So we conclude that for a Boost we must design and test the input capacitor at $D = 0.5$ (or the closest point to it in the valid input range).

This was actually almost the only exception, because in this particular case r happened to be a (prominent) multiplicative factor in the capacitor RMS equation. Only one more such case exists: the output capacitor RMS of a Buck, as we can see from Table 2.2. For all other cases, the term in r is just a small adder ($r^2/12$ is typically $0.4^2/12 = 0.013$). It can therefore be neglected for the purpose of this discussion. We can simplify our equations to decipher the dependency with respect to D. We therefore say that for small r, the input capacitor of a Buck has the following dependency on D:

$$I_{\text{CIN\_RMS\_BUCK}} \propto \sqrt{D(1-D)}$$

Since $D = 1 - D$ at $D = 0.5$, the worst case input capacitor current occurs at $D = 0.5$ for a Buck. We have to be mindful of this when we design and test the capacitor. If we do not have $D = 0.5$ within the input range of our application, we have to *pick the closest input voltage to this point*.

Example 2.8 A Buck with a 12-V output has an input of 15 to 20 V. What input voltage represents the worst case for the input capacitor current?

At $D = 0.5$ we require an input of 24 V for a 12-V output since $D = V_O/V_{IN}$. The point closest to 24 V in our input range is V_{INMAX}, that is, 20 V, and we must design and test the input capacitor at this voltage.

Example 2.9 A Buck with a 12-V output has an input of 28 to 50 V. What input voltage represents the worst case for the input capacitor current?

At $D = 0.5$ we require an input of 24 V for a 12-V output since $D = V_O/V_{IN}$. The point closest to 24 V in our input range is V_{INMIN}, that is, 28 V, and we must design and test the input capacitor at this voltage.

Stress Spiders

Based on the type of analysis of dependencies, as carried out for the RMS of a Boost input capacitor above, we can, in general, generate what we call a *stress spider*. We have presented three such spiders, in Figs. 2.23, 2.24, and 2.25, for the Buck, Boost, and Buck-boost, respectively. Keep in mind that for all topologies, a low D implies a high V_{IN} and a high D implies a low V_{IN}.

So after having set the inductance and size of magnetic components at the designated worst-case design entry point, we can step back and figure out from these stress spiders which is the worst-case input voltage at which other stresses maximize. This is essential for proper commercial design and test methodologies. And in most cases, it is hardly obvious! Yet, if we do not have this information upfront while selecting components, we will end up with either an overdesigned or an underdesigned converter. The stress spiders enable reliable wide-input designs. There is no alternative approach in literature. And even this necessity has not been recognized in related literature. It was first published by this author in 2002, as Application Note AN-1246 at National Semiconductor (now Texas Instruments, see http://www.ti.com/lit/an/snoa425b/snoa425b.pdf), and later as a cover story in *Power Electronics* magazine titled "Reducing Converter Stresses" (see http://powerelectronics.com/mag/power_reducing_converter_stresses/).

Boundary (Critical) Conduction

It can be shown that any converter reaches critical (boundary) conduction mode (current just returning to zero) as we reduce the load. This happens at a load current equal to $r_{\text{SET}}/2 \times$ maximum load. See Fig. 2.26 for an explanation.

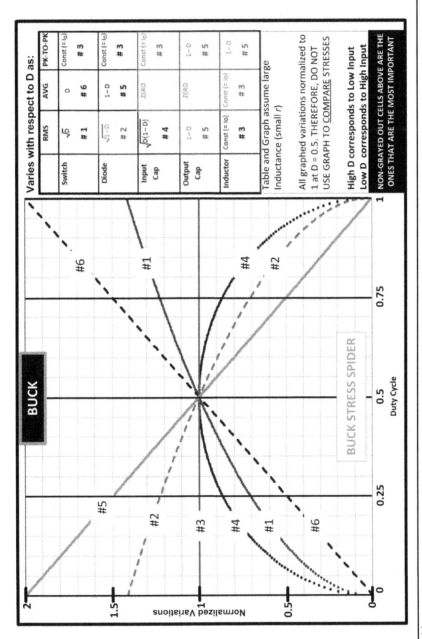

FIGURE 2.23 Buck stress spider.

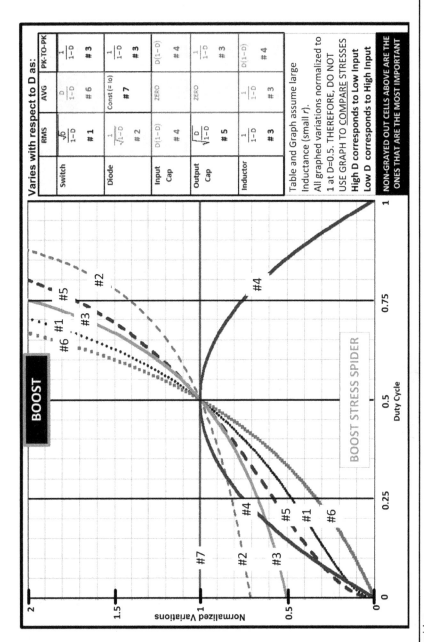

FIGURE 2.24 Boost stress spider.

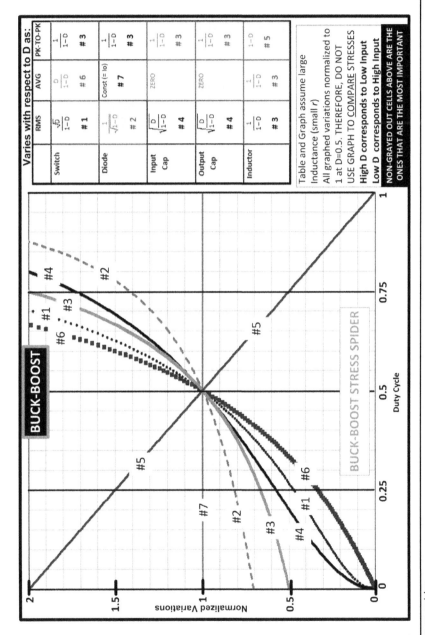

FIGURE 2.25 Buck-boost stress spider.

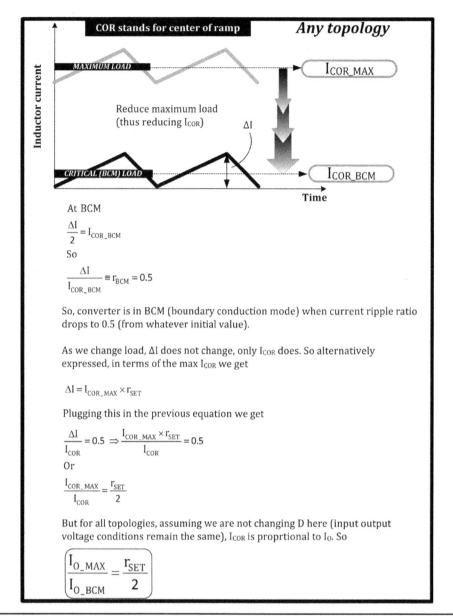

FIGURE 2.26 Relationship of load current to set r, at which BCM is reached.

Example 2.10 A Buck converter is designed for r_{SET} equal to 0.4 at maximum load of 5 A. If we keep the input and output voltages unchanged, as we reduce load, at what load will the converter try to enter DCM (discontinuous conduction mode)?

This happens at $(r_{SET}/2) \times I_{O\_MAX} = (0.4/2) \times 5 = 1$ A.

In words, *a converter enters DCM at $(r_{SET}/2 \times 100)$ percent of maximum load (keeping the same input-output voltages as used for the design entry point r_{SET}).*

So a converter with r_{SET} of 0.4 will try to enter DCM at exactly 20 percent of maximum load.

As we reduce the load past BCM, what really happens depends on whether we are operating synchronously (forced CCM) or in DCM. This is the subject of Chaps. 3 and 4.

Keep in mind that it was always intuitively felt that in DCM the efficiency gets worse. So every attempt was made to stay in CCM (continuous conduction mode) down to very light loads. Historically, it was often suggested, in effect, to set r to 0.1 (despite the fact that the vital role of r in the entire converter design was never clearly noticed or pointed out). But if we do that, yes indeed, we will enter DCM at 5 percent of maximum load. However that leads to oversized magnetic components. In additions, the lessons of efficiency, as unraveled in Chap. 18, confirm that it is not so easy to intuitively conclude that DCM is really the worst option! In fact if we continue in forced CCM (as we often do in synchronous topologies), the efficiency falls off faster than if we had stuck with DCM. All that is described in Chap. 18.

Dr. Ray Ridley did an industrywide analysis in 2009 of the recommended current shape factors, and he concluded that this author's recommendation in particular of $r_{SET} = 0.4$ is indeed the best entry point into the design of a converter. See http://www.catagle.com/41-10/psde_nov09.htm and http://www.powersystemsdesign.com/library/resources/documents/europe/2009/psde_nov09.pdf. The same was also validated in this IEEE reference: http://ieeexplore.ieee.org/xpl/login.jsp?tp=&arnumber=5515878&url=http%3A%2F%2Fieeexplore.ieee.org%2Fxpls%2Fabs_all.jsp%3Farnumber%3D5515878. In addition, Dennis Feucht, noted analog author, also agreed with the author's shape factor (he called it the *form factor*) at this link: http://www.en-genius.net/includes/files/col_080309.pdf.

Using Too High an Inductance (Small *r*)

We learn that in an effort to reduce the transition to DCM to very light loads, engineers sometimes overdesign their magnetics. We know now that that a very large inductor (small *r*) is not a cost-effective or the optimum choice. But what are some other issues associated with it?

One major concern is the leading-edge spike as shown in Fig. 2.27. This can cause jitter and, in severe cases, a consequent inability to deliver full power. If we increase the inductance, we could cause premature termination of the switching pulses in current-mode controllers, because by increasing inductance (I_{COR} remains the same) we are perhaps inadvertently raising the pedestal on which the spike is riding. But since less than the required energy will be delivered for that (prematurely terminated) cycle, in the next cycle the converter will try to compensate by a larger duty cycle. In this process it gets some unexpected help, because after the early termination of the previous pulse, the inductor current had a longer time to slew down, and thus the pedestal on which the leading-edge spike is riding comes down, probably enough to help it evade early pulse limiting in the next cycle. Finally, what we may see on the oscilloscope are *alternate wide and narrow pulses*, which mimic what we get under genuine subharmonic instability (see more on this in Chap. 14).

We also remember that an inductor takes time for current through it to change. So it cannot react fast enough to sudden changes in the load demand. The loop response may thus be poor with a large inductance.

In addition, high inductances shift the ominous right half-plane zero closer to operating frequencies (see Chap. 14).

Another problem is that an inductor working in steady state never fully parts with all energy. A certain minimum of residual energy is stored every cycle (in CCM). But under a fault condition, even if the controller reacts by ceasing all switching action, the energy stored in the inductor will get dumped into the output capacitor. If the output capacitance

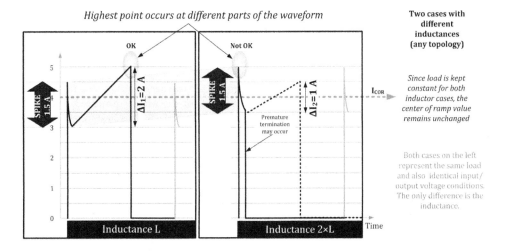

Conclusion: If L is too large, it pushes up the peak of the spike relative to the rest of the waveform. Beyond a certain point, the spike may become the highest point of the waveform causing premature (spurious) triggering of the ON-pulse in current-mode control (unless the spike is filtered out or blanked), and/or premature triggering of the current limit comparator in all types of converters, thereby preventing full calculated maximum load from being achieved.

FIGURE 2.27 Abnormal pulse patterns can result if inductance is too large (small *r*).

is small and inductance is large, we may get a rather high output voltage spike, despite ceasing switching. So we have to be careful when using a large inductance or/and low output capacitances.

Some mistakenly think that by using a large inductance, we are somehow reducing the heating in the FET substantially. Actually that doesn't bear out either. To see this, we take up the flattop approximation next.

Too small an inductance can, however, cause subharmonic instability in converters using (peak) current-mode control (for $D > 50$ percent and for CCM only). See Chap. 14.

The Flattop Approximation

For a practicing engineer, there is always a constant struggle to work with less imposing equations and yet compromise little in terms of accuracy. One of these is the *flattop approximation*. This approximation is in fact used extensively by engineers, because it is easier to handle and because of its sheer availability. We will discuss various concerns and pitfalls when using this popular approximation, and we provide some correction curves to get more-accurate results, despite them. But we will also show why at some stage *we do need the more exact form of the equations*.

The flattop approximation is equivalent to not actually using, but *assuming* an extremely large inductance. Let us see what are the possible issues with this.

- If we need to set a current limit, we cannot afford to forget that *the peak is at least 20 percent higher than the flattop assumption (for $r_{SET} = 0.4$)*. If we set the current limit based on the flattop approximation, we will have a power supply that can't deliver its expected power.

- By assuming that the current swing is zero, we are in effect assuming that the inductor has no ripple. So it has a pure DC current. Therefore, we cannot estimate the core losses either with this approximation.

- In peak current-mode control, we encounter subharmonic oscillations at about $D > 0.5$ unless we set slope compensation. The slope that we need to set is typically between 1 and 2 times larger than the slope of the down ramp of the inductor current. By assuming no down ramp (zero slope), we are solving the subharmonic instability problem trivially—by not requiring any slope compensation at all. Clearly, the flattop approximation is incompatible with the design of a practical slope compensation scheme.

- The flattop approximation does underestimate the switch RMS current somewhat, but as we can see from the error curves in Fig. 2.28, the loss in accuracy is fairly insignificant under the normal range of selection of r. We must, however, keep in mind that heating is proportional to RMS squared, and the error in that may be a bit higher.

- The RMS of the capacitor current is also underestimated by using the flattop approximation, *but much more so*. In Fig. 2.28 we have the error for a Buck-boost/Flyback. *We see that we get a large flattop approximation error in the input capacitor current estimate for large D and a large error in the output capacitor for low D.*

As a shortcut, in most cases, we can suggest that the designer take the flattop equations, work with them, but then apply the corrections per the errors indicated in Fig. 2.30. But keep in mind that in string calculations, *the errors cumulate*. So it is best to use abundant caution in using the flat top approximation in the first place.

PART 3 TOPOLOGY MORPHOLOGY

Introduction

One of the first things we must realize is that a switcher IC for a DC-DC converter application can actually be used across other topologies quite freely. For example, a Buck IC can be used in a Buck-boost application and vice versa. This is not so puzzling once we understand that a switcher is just that: It basically switches a transistor between ON and OFF states. How we use the transistor and configure our circuit to switch a voltage across an associated inductor and how we finally route the energy to the output are not necessarily preordained. Indeed there are limitations on what we can do, but that has more to do with how the transistor is actually driven, how the internal control of the IC is referenced, etc. It is certainly important to

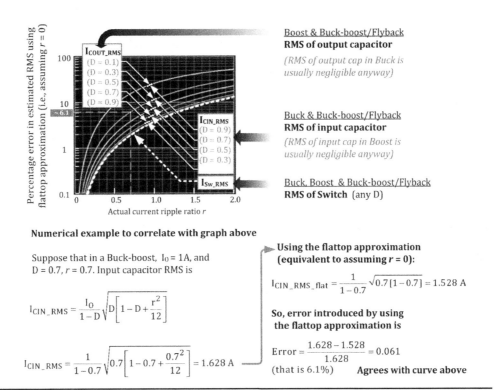

Figure 2.28 Error in RMS estimates from using the flattop approximation.

understand the internal construction of the IC itself, but to finally see the other "hidden" applications needs a better (and more abstract) understanding of switcher topologies themselves.

Related literature occasionally refers to some of these hidden applications, but in a rather scattered and unintelligible way. One major problem we noticed was the way a schematic is often drawn. There are schematics that may require several right-angle turns and/or several horizontal and vertical mirror reflections to make any sense of then. To compound the confusion, there is rarely any attempt to even explicitly state the fact that the underlying topology had changed. In the accompanying text, it may just have been mentioned in passing that the Buck IC was being used in an "inverting" configuration (or a positive-to-negative converter). Go figure! What is the maximum load current in this new configuration, and what is the safe input operating voltage range? Why are these different now? Is that the feedback? And how does it really work? Is the output regulation as good as it was for the original Buck? Questions like these just add to the general mystery surrounding switcher ICs and the host of other possible applications possible for them.

The N-Switch and P-Switch

For bringing the myriad possibilities under smaller umbrellas, we need to make some rather unconventional definitions in this chapter. The reader should bear with us, as she or he will see that it really does help in isolating the common threads among the various hidden applications.

In Fig. 2.29 we have indicated that a voltage of magnitude v needs to be applied with respect to the source terminal of both N-channel FETs and P-channel FETs to turn ON the FET (assuming enhancement-type FETs only). The dotted triangles (alongside the label v) indicate the direction of *increasing* voltage (in terms of magnitude). We have also indicated that a voltage of magnitude v needs to be applied with respect to the emitter terminals of both NPN and PNP BJTs to turn ON the BJT.

Throughout this chapter *we will usually attempt to keep the lower-voltage input rail on the bottom side of the figure and the higher-voltage input rail on the top side. Further, the input is on the left side, and the output on the right*. These steps will help in keeping all the schematics visually appealing, easy to follow, and mutually consistent.

In Fig. 2.29 we have also shown the easiest way to turn OFF the FET or BJT is to connect the Gate/Base to the Source/Emitter.

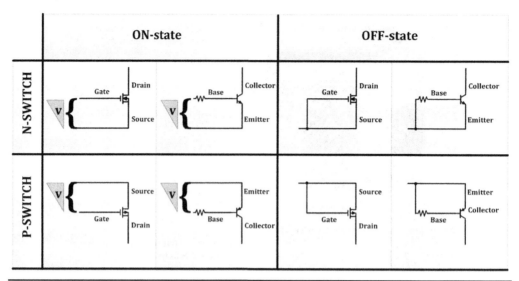

FIGURE 2.29 N-switches and P-switches.

Because of the drive similarities, in this chapter we will generally talk in terms of an N-switch (being either an N-channel FET or an NPN BJT) or a P-switch (either a P-channel FET or a PNP BJT). There are differences between BJTs and FETs, but we will discuss these later.

The LSD Cell

The LSD cell as introduced here goes a long way in understanding which hidden applications are a natural possibility, for a given IC, and which are not. It will be shown that if we identify the LSD cell type occurring in the original intended function of a switcher IC, we can easily apply it to any other topology in which the *same LSD cell type occurs, irrespective of the topology*. We will also see that, in some cases, by a special technique we can even make an IC perform in an LSD cell type other than its originally intended (familial) LSD cell.

The first recommendation is that the designer start thinking in terms of HI and LO rails rather than positive or negative rails or even ground. The reason is that the designations can change as we change topologies and configurations, but HI and LO will remain unaltered. Coming to the basic structure of power conversion circuits, in all cases we have an inductor L, a switch S, and a diode D connected to one another. This is hereafter generically nicknamed an *LSD cell*. From it we can generate the "– cell" which is so called because, as we can see from Fig. 2.30, this requires the switch to be physically *below* the switching node. The "+ cell" has the switch positioned physically above the switching node. That is understandable because by putting HI on top and LO below, current always flows from the top down. And in the + cell, the cathode connects to the switching node, whereas in the – cell the anode connects to the switching node. See how the current bifurcates at the switching node as the FET turns ON and OFF, and why the diodes need to be pointed in the manner shown.

Each + or – type leads to two more types, one with a P-channel FET and the other with an N-channel FET. So we get the N+ and, P+ and the N– and P– cells. In the figure, note the body diode for each FET as shown in gray. It cannot be the other way round clearly, or the supply will short.

We also see that *two of the four cases require a Gate-drive voltage outside the input rails* whereas two do not. This has a great bearing on the usefulness and applicability of the device. If a drive signal is to be outside the rails, we need a *bootstrap circuit*. In this the bootstrap capacitor needs to be "refreshed" every cycle. But this can happen only when the switch turns OFF because then the internal circuitry releases a current source for charging the capacitor up close to the HI of the input rails. So any IC requiring a bootstrap is not allowed to have a duty cycle of 100 percent.

We note that a duty cycle of 100 percent is often useful for a Buck because as the input falls below the set output level, the output tracks the input very closely (less a switch forward voltage drop). Then the device is just functioning as a *linear dropout* regulator (LDO) of sorts. In portable battery-operated devices, this greatly extends the operating time of the unit as the batteries slowly discharge. So for the Buck, the P+ cell is popular for portable devices such as MP3 players. However, we also note that *100 percent duty cycle is not acceptable for either the Boost or the Buck-boost*. These topologies are different in that energy is delivered to the output only when the switch turns OFF. So we can end up with a situation at start-up where the

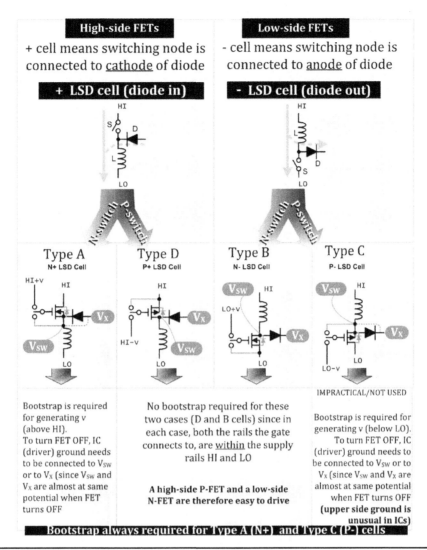

FIGURE 2.30 Types of LSD cells and their drive signal level requirements.

control maximizes the duty cycle because the output voltage is low. But since the duty cycle is 100 percent, the output cannot rise anyway.

Note also that when the diode conducts, that is, the switch turns OFF, V_X and V_{SW} in Fig. 2.30 become almost the same node (within a forward diode drop). To avoid confusion, note that V_{SW} now *refers to the voltage at the switching node* (not the forward drop across the switch, which is also called V_{SW} elsewhere in this book). Also, V_X is the voltage on the other side of the diode (the fixed voltage end of the diode in any topology). So in these two cases, we could have turned the switch OFF by connecting to V_X instead of the more obvious node V_{SW}. In some applications we actually do need to do this (implementation).

Summarizing, we see that we can have all four LSD cell types. Two correspond to N-switches and two to P-switches. We have called them A, B, C, D cells in the order shown in the figure.

Configurations of Switching Regulator Topologies

We note that the term *Boost*, or *Buck* or *Buck-boost* always refer to only the magnitudes of the input and output voltages. So now we see the need for qualifiers such as *negative to negative*, *negative to positive*, etc. to fully describe the actual configurations. The negative (or inverted) form of the popular positive-to-positive Boost converter is in full a negative-to-negative Boost converter. It would convert say −12 to −48 V relative to the (common) ground rail.

We have now four possible configurations for each of the standard topologies, corresponding to type A, type B, type C, and type D cells. In Figs. 2.31, 2.32 and 2.33 we show the possibilities for the Buck, the Boost, and the Buck-boost converters, respectively. The Gate drive levels are also shown.

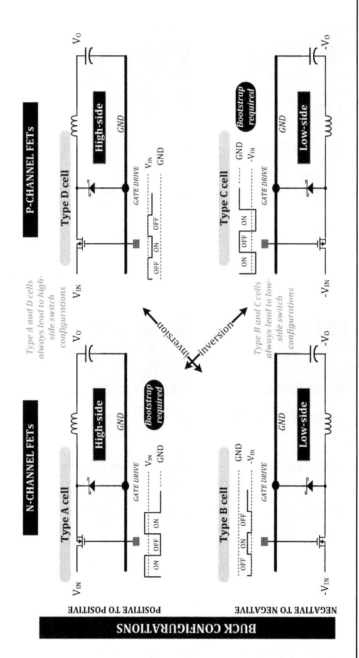

Figure 2.31 Buck configurations.

DC-DC Converters: Topologies to Configurations

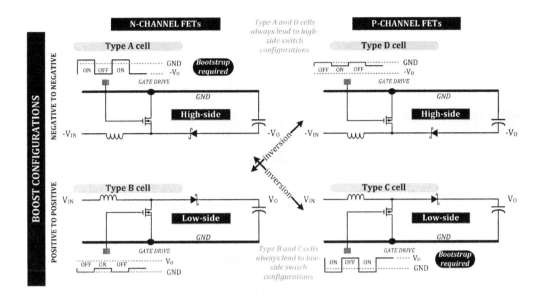

FIGURE 2.32 Boost configurations.

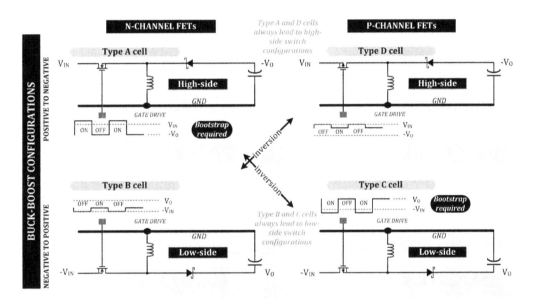

FIGURE 2.33 Buck-boost configurations.

Note the following:

1. By *inversion* we can convert an entire schematic (including switch drive signals) involving an N+ cell to a P− cell, and vice versa (i.e., type A ⇔ type C). See Figs. 2.31, 2.32, and 2.33 (N+ goes to P−).

2. Similarly we can invert a schematic with a N− cell to a P+ cell, that is, type B ⇔ type D (N− goes to P+).

3. *Inversion* is the following procedure: (*a*) Change the polarity of all voltages (for example, V to $-V$). (*b*) Change the polarities of the capacitors. (*c*) Reverse all diode directions. (*d*) Change an N-channel FET to a P-channel FET (and vice versa), or a NPN BJT to a PNP BJT etc. (*e*) Remember, however, that for the transistors we do *not* change which pin is which; e. g., Source still remains Source, Drain too remains the same. (*f*) Note that in the process HI changes and becomes a LO, and therefore we will need to flip the schematic vertically to comply with our convention of keeping HI on top and LO below.

The reader can try this out and see how it works. Configurations with a type A cell will lead to type C, and configurations with a type B cell will lead to type D.

Example 2.11 If we need to create a schematic for a negative-to-positive Buck-boost using a P-channel FET (type C from Fig. 2.33), we can start by generating a positive-to-negative Buck-boost schematic using an N-channel FET (type A from Fig. 2.33) and then *invert* it.

In discussing configurations, note that e.g., a positive-to-positive Buck is simply called a positive Buck (or just Buck usually).

Basic Types of Switcher ICs

In the previous figures, the details of the IC and control were not shown. Let us now study typical ICs first to see how they are internally configured.

Focusing on FET switches, we see the two (or three) most common types of IC constructions in Fig. 2.34. We say they fall into two basic categories, hereby designated type 1 and type 2 (for control ICs). Type 1 ICs are generally considered to be Flyback/Buck-boost/Boost ICs, and type 2 ICs are considered Buck ICs. We will see that type 1 ICs are generally the most versatile.

Here are some related comments:

- *Type 1 ICs are generically considered Buck-boost/Boost ICs, whereas type 2 ICs are generically considered to be Buck ICs.* We have generalized this based on their inherent internal architecture, since we now understand that Buck or Buck-boost is just their primary intended application. There are more.

- Type 1 hard-connects the Source (lower-voltage switch pin) to the lower-voltage pin of the control block (IC ground). This is often called an IC with *low-side FET/drive*.

- Type 2 connects the Drain (or the higher-voltage switch pin) to the higher-voltage pin of the control block. This is called an IC with a *floating drive* or *high-side FET/drive*.

- A type 2 IC in its normal intended application certainly needs a bootstrap circuit if an N-FET is used.

- The primary intended application for type 1 IC is the positive Boost, and we can see from Fig. 2.30 that this uses a type B cell.

- The type 2 IC is intended for a positive Buck, and we can see from Fig. 2.30 that it corresponds to type A cell (for an N-FET).

- In all configurations shown in Figs. 2.31, 2.32, and 2.33, we see that for a type A (N+) cell, the drive signal is outside the available rails, and therefore bootstrap is required

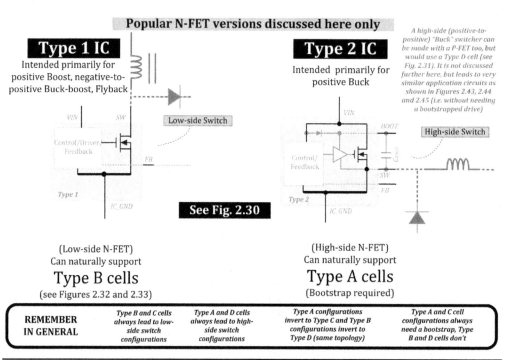

FIGURE 2.34 The two common IC types.

(since type A is an N-FET high-side case). This is what we always expect in a positive Buck using an N-FET.

- Note that NPN BJTs are generally easier to drive than FETs since the Base of the BJT has to be taken only slightly higher than the Emitter to turn ON the switch (even the small existing collector-Base drop can be used for this purpose, as in a Darlington/ β-multiplier drive arrangement). So we do not need a bootstrap (or floating driver) since we can "force" its base drive signals to be within the available rails. On the other hand, we know the limitations of BJTs in terms of higher conduction losses typically, as well as slower speed with higher switching losses.

- Note that the actual SW pin labeling, shown in Fig. 2.34, depends on the *perceived* application for the part, not necessarily on how it is *actually* used. *It cannot always be assumed that the SW pin is the switching node of the regulator power stage.* Look upon it as very simply, the *uncommitted pin* of the transistor. As we will see later, this pin can be connected to a fixed rail. And in that case, either the so-called V_{DD} pin or the IC GND (IC ground pin) of the IC may become the actual swinging/switching node! Therefore in all cases, the systems designer exploring topology morphology needs to take the labeling of the IC pins with a pinch of salt, keeping in mind what they really are in terms of the *internal construction* of the IC. Forget names!

- In all cases the feedback signal node is referred by the control to the lower (IC GND) rail of the IC. So, in unconventional configurations, we may need to translate the sensed output rail to get it referenced correctly with respect to the IC control (and its internal reference). This will be discussed later.

- Type 1 ICs are more versatile mainly because they usually have two pins with two voltage ratings: one for the control (V_{DD} pin in Fig. 2.34) and another, higher rating for the switch pin (Drain), *both of course referenced to the IC ground* (the IC GND pin in Fig. 2.34).

- Type 2 ICs, though having a supposedly uncommitted SW pin, do not offer much freedom, because, almost invariably, by internal design of driver etc., the SW pin cannot be taken more than about 1 V below the lower rail (IC ground). One reason could be that an ESD structure is present between the SW pin and the IC GND pin, in effect behaving as a diode between the two. In addition, it is possible that, in general, if we take any pin of an IC below IC GND, we get substrate currents that can cause very strange behavior (as discussed in the first section of this chapter). This therefore limits some possible applications with type 2 ICs, particularly involving tapped-inductors. In fact, the author knows of only one commercial Buck IC that allows the SW pin to go far below IC GND. That is a type 2 IC, but is BJT-based: the LT1074 from Linear Technology. In its typical published applications, you will see the tapped-inductor Buck for that reason. This is also why several more theoretical topological configurations are enabled by this uncommitted SW pin—only for this IC, however.

Now we will take up type 1 and type 2 ICs in greater detail.

Flyback, Buck-boost, and Boost ICs Compared

We mentioned that type 1 ICs are intended for all these applications. There is no essential difference between an IC intended primarily for a Boost application and one, say, for a Flyback/ Buck-boost application. We should first be aware of the basic topological difference between a Boost and a Buck-boost power stage.

In Fig. 2.35 we can see that the change from a positive-to-positive Boost to a negative-to- positive Buck-boost is actually very simple: it involves just redirecting the connection of the negative terminal of the output capacitor from the lower rail to the upper rail. Therefore the two topologies are not all that different. In fact as *far as the drive of the switch is concerned, it sees absolutely no difference between these topologies, because basically only the designation (or labeling) of the rails has changed. The output voltage rail is exactly 30 V above the IC ground (not system ground) in both cases (for the same duty cycle).* So the IC "doesn't know better."

But there is one difference: the *feedback*. Since for a Boost the IC control is typically always connected to the lower rail, a simple resistive divider across the output capacitor can be used to connect directly to the feedback pin of the IC control. But for the Buck-boost, the output voltage is with respect to the system ground (the upper rail), whereas the IC control is still referenced to the lower rail. Therefore a more elaborate solution is required. This usually takes

76 Chapter Two

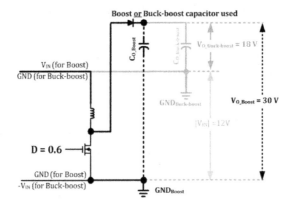

$$V_{O\_Boost} = \frac{V_{IN}}{1-D} = \frac{12}{1-0.6} = 30 \text{ V}$$
$$V_{O\_Buck-boost} = \frac{D}{1-D} \times V_{IN} = \frac{0.6}{1-0.6} \times 12 = 12 \text{ V}$$

Figure 2.35 Comparing the Boost and the Buck-boost.

the form of a differential amplifier stage to sense the output voltage of the Buck-boost, or to "translate" it (re-reference it) to the lower rail. But the requirements, specifications, and ratings of such a differential stage are so diverse depending on the input-output levels that this extra stage is rarely (if ever) integrated into the switcher IC. This means that a true negative-to-negative-to-positive Buck-boost integrated switcher (with integrated feedback) may be nearly impossible to find. So, since the feedback implementation is generally external to the IC, there is no remaining architectural difference left between a Boost IC and a Buck-boost IC. They are one and the same. With that in mind, it comes as no surprise that any switcher meant for a Flyback/Buck-boost application can always successfully be used for a Boost application, and vice versa.

Note that if we are dealing with a positive-to-negative Buck-boost application, direct feedback with a simple divider is possible by simply *floating the IC on the switching node* (for a type 1 IC) or on the negative output rail (for a type 2 IC). This is a powerful technique to use in morphology. We will discuss this further.

Generally, the word *Flyback* refers exclusively to a Buck-boost stage with inherent primary-to-secondary isolation (transformer-based). but we could also have a transformer-based Buck-boost with no isolation present: then the primary and secondary windings are connected for correcting the polarity inversion (by re-referencing the output ground in effect), and this leads to an easier implementation of feedback. In both cases, with and without isolation, the feedback method involves using two resistors in a divider network positioned at the output, and no differential sense stage is required. In the isolated case we just go through an opto-coupler thereafter.

Coming back to the main focus of this section, we now look for the other possible applications of a Flyback/Boost IC.

Inductor Selection Criteria

We know that for the Buck-boost and Boost topologies the average inductor current goes as $I_O/(1-D)$. So with a current ripple ratio of around 0.4, we can add 20 percent to the average value to get the peak current. The current limit I_{CLIM} must therefore be at least that high. This clearly depends on the duty cycle, that is, input and output voltages. Therefore for a type 1 (Buck-boost) IC used in a Buck-boost/Boost configuration, when we say it is a 3-A device, we are talking about its current limit, not its maximum load current, which is a function of D and therefore depends on the input-output voltages.

For a type 2 (Buck) IC, the average inductor current is I_O. We can add about 20 percent to get the peak, but we can see that the peak is almost independent of the input voltage. Therefore when we talk of a type 2 IC as being a 3-A device, we are talking about the load current possible irrespective of application, *provided we are using it for a Buck application*. Then its current limit will typically be set 20 to 40 percent higher than its declared load current capability. So if this IC is used, say, as a Boost topology, *then it cannot deliver 3 A of load current*—just as a 3-A type 1 IC can't necessarily deliver 3-A load current, but in a Buck topology, it would. In general, we have to calculate how much load current we can get for each case based on the current limit and the topology.

Note that unlike off-line power supplies where we size the inductor/transformer according to the upper limit (maximum) of the current limit tolerance, for low-voltage applications (up to about 40 V) we usually never need to size the inductor according to the current limit, but according to the load current. In such cases, the inductor is expected to saturate momentarily under start-up or step load conditions, but since especially integrated switchers (with built-in FETs) have such fast-acting current limits and protection, the switch (particularly a FET) easily survives. But when the voltages are "high" (empirically determined to be greater than 40 V), we usually have to change our inductor selection criteria for DC-DC converters too, from one being based on load current to one being sized per the (maximum) current limit.

Other Possible Applications of Buck and Buck-boost (Type 1 and Type 2) ICs

In Table 2.3 we have provided all the equations to use in qualifying devices depending upon the application. The reader should check them out for each schematic to follow for the possible applications of type 1 and type 2 ICs.

Some of the configuration conditions and equations may depend on the minimum and/or maximum input voltages, V_{INMIN} and V_{INMAX}, respectively. In addition, every controller is designed with a certain maximum possible duty cycle limit D_{MAX}. Clearly, if the input and output voltages demand more than D_{MAX}, the circuit cannot work. Therefore the equation to check this possible limitation is also provided. The feedback scheme is shown, and the equations to set the resistor values are provided. The voltage on the feedback pin of the IC under regulation is V'_{FB} (e.g., it is the reference voltage to the internal error amplifier for an adjustable output part, or just connected to the output for a fixed-voltage option part). Keep in mind that many integrated switcher ICs come as either "adjustable," requiring an external resistive divider to set output, or "fixed" voltage parts, in which the divider is internal to the IC.

In all the equations to follow, the switch and diode forward drops are generally assumed to be negligible. So a little additional guard-banding may be necessary to take these into account.

Now we come to the chain of logic behind the "hidden" applications of ICs:

- The primary intended application for a type 1 IC is the positive-to-positive Boost. We discover that this involves an N– cell (type B cell, see Fig. 2.30). Therefore we conclude that this IC is "comfortable" *with any topology or configuration, provided it involves a (similar) type B cell*. This cell becomes a natural choice for this IC. Let us make this analogous to the caste system of some older societies, by calling this *natural morphology*.

    Note that we finally start seeing the advantage in talking in terms of LSD cells rather than directly in terms of topologies and configurations. This common thread would have been missed otherwise.

- The primary intended application for a type 2 IC is the positive-to-positive Buck. We discover that this involves a type A cell. Therefore we conclude that this IC is most "comfortable" with any topology or configuration, provided it involves a (similar) type A cell. This cell is a natural choice for this IC. We call this too the *natural morphology*. Note that with a P-FET, a type 2 IC can also support type C cases. See Fig. 2.34.

- However, because the type 1 IC has an *additional degree of freedom*, because the rail going to the switch is not connected to the supply rail of the control (Fig. 2.34), *there is a certain technique (let us call it unnatural morphology, to distinguish it from natural), by which we can use a type 1 IC in an application that uses the "opposite" cell*, that is, a type A cell.

    Yes, there are limitations when we implement these *unnatural* configurations and topologies. First, regulation will likely suffer by this trick. Second, some ICs just may not 'like' to be operated in this manner, with their IC GND being the switching node in effect (because that is how these *unnatural* applications are created. So these "possibilities" have to be actually tried out on the bench to confirm. There is no shortcut to that. There are also subtle nuances of topologies (besides layout), such as loop stability, that might get in the way. But let us say, in principle, that all these

(All voltages and currents are in terms of magnitudes only.)				
Topology	Configuration	IC Type	Figure	Equations
Buck	Positive to Positive	1	2.42	$V_{SWMAX} \geq V_{INMAX}$, $V_{ICMAX} \geq V_{INMAX}$, $V_{ICMIN} \leq V_{INMIN}$, $I_{O\_MAX} \leq 0.8 \times I_{CLIM}$, $R_2 \approx R_1 \times \left[\dfrac{V_O}{V_{FB}} - 1\right]$, $D_{MAX} \geq \dfrac{V_O}{V_{INMIN}}$
		2	2.43	$V_{ICMAX} \geq V_{INMAX}$, $V_{ICMIN} \leq V_{INMIN}$, $I_{O\_MAX} \leq 0.8 \times I_{CLIM}$, $R_2 = R_1 \times \left[\dfrac{V_O}{V_{FB}} - 1\right]$, $D_{MAX} \geq \dfrac{V_O}{V_{INMIN}}$
	Negative to Negative	1	2.39	$V_{SWMAX} \geq V_{INMAX}$, $V_{ICMAX} \geq V_{INMAX}$, $V_{ICMIN} \leq V_{INMIN}$, $I_{O\_MAX} \leq 0.8 \times I_{CLIM}$, $R_2 \approx R_1 \times \left[\dfrac{V_O - 0.6}{V_{FB}}\right]$, $D_{MAX} \geq \dfrac{V_O}{V_{INMIN}}$
		2	Does not exist (opposite cell)	
Boost	Positive to Positive	1	2.37	$V_{SWMAX} \geq V_O$, $V_{ICMAX} \geq V_{INMAX}$, $V_{ICMIN} \leq V_{INMIN}$, $I_{O\_MAX} \leq 0.8 \times I_{CLIM} \times \dfrac{V_{INMIN}}{V_O}$, $R_2 = R_1 \times \left[\dfrac{V_O}{V_{FB}} - 1\right]$, $D_{MAX} \geq \dfrac{V_O - V_{INMIN}}{V_O}$
		2	Does not exist (opposite cell)	
	Negative to Negative	1	2.40	$V_{SWMAX} \geq V_O$, $V_{ICMAX} \geq V_O$, $V_{ICMIN} \leq V_{INMIN}$, $I_{O\_MAX} \leq 0.8 \times I_{CLIM} \times \dfrac{V_{INMIN}}{V_O}$, $R_2 \approx R_1 \times \left[\dfrac{V_O}{V_{FB}} - 1\right]$, $D_{MAX} \geq \dfrac{V_O - V_{INMIN}}{V_O}$
		2	2.45	$V_{ICMAX} \geq V_O$, $V_{ICMIN} \leq V_{INMIN}$, $I_{O\_MAX} \leq 0.8 \times I_{CLIM} \times \dfrac{V_{INMIN}}{V_O}$, $R_2 = R_1 \times \left[\dfrac{V_O}{V_{FB}} - 1\right]$, $D_{MAX} \geq \dfrac{V_O - V_{INMIN}}{V_O}$

Note: By convention, R_2 is always connected to the higher-voltage rail of output and R_1 to the lower.

TABLE 2.3 IC Selection Table for the Various Topologies (Part 1)

unnatural applications are possible by using a type 1 IC with its IC GND connected to the switching node of the topology.

- If we use a transformer-based Buck-boost converter, we can connect the primary and secondary windings in any manner, to get any combination of positive or negative input and outputs. We can also leave the two windings isolated, in which case we need an error amplifier on the secondary side and an opto-coupler. In this case V_{FB} would refer to the reference voltage of the secondary-side error amplifier. The required equations are provided in Table 2.4.

(All voltages and currents are in terms of magnitudes only)				
Topology	Configuration	IC Type	Figure	Equations
Buck-boost	Positive to Negative	1	2.41	$V_{SWMAX} \geq V_{INMAX} + V_O$ $V_{ICMAX} \geq V_{INMAX} + V_O$ $V_{ICMIN} \leq V_{INMIN}$ $I_{O\_MAX} \leq 0.8 \times I_{CLIM} \times \dfrac{V_{INMIN}}{V_{INMIN} + V_O}$ $R_2 = R_1 \times \left[\dfrac{V_O}{V_{FB}} - 1 \right]$ $D_{MAX} \geq \dfrac{V_O}{V_{INMIN} + V_O}$
		2	2.44	$V_{ICMAX} \geq V_{INMAX} + V_O$ $V_{ICMIN} \leq V_{INMIN}$ $I_{O\_MAX} \leq 0.8 \times I_{CLIM} \times \dfrac{V_{INMIN}}{V_{INMIN} + V_O}$ $R_2 = R_1 \times \left[\dfrac{V_O}{V_{FB}} - 1 \right]$ $D_{MAX} \geq \dfrac{V_O}{V_{INMIN} + V_O}$
	Negative to Positive	1	2.38	$V_{SWMAX} \geq V_{INMAX} + V_O$ $V_{ICMAX} \geq V_{INMAX}$ $V_{ICMIN} \leq V_{INMIN}$ $I_{O\_MAX} \leq 0.8 \times I_{CLIM} \times \dfrac{V_{INMIN}}{V_{INMIN} + V_O}$ $R_2 \approx R_1 \bullet \left[\dfrac{V_O - 0.6}{V_{FB}} \right]$ $D_{MAX} \geq \dfrac{V_O}{V_{INMIN} + V_O}$
		2	Does not exist (opposite cell)	

Note: By convention, R_2 is always connected to the higher-voltage rail of output and R_1 to the lower.

TABLE 2.4 IC Selection Table for the Various Topologies (Part 2)

In Fig. 2.36 we present a quick lookup chart on what exactly is possible (though with small add-ons for feedback or IC supply), using commonly available, type 1 (Flyback) ICs and type 2 (Buck) ICs.

In Tables 2.3 and 2.4 we have provided the equations used to verify that a chosen IC suits the application at hand, mainly in terms of voltage and current stresses. See some examples below. In Figs. 2.37 through 2.39, we present the configurations for a Type B cell (commonly found in a commercial Buck IC), and in Figs. 2.40 through 2.45, we present the possible configurations of a Type A cell (commonly found in a commercial Boost IC).

Some Practical Cases

We now present some typical examples to clarify the selection procedure further.

Example 2.12 The LM2585 is a 3-A Flyback regulator. The minimum value of its internal current limit (see its table of electrical characteristics) is 3 A. Its input operating voltage range is 4 to 40 V. Its switch can withstand 65 V. Can it be used in a Boost topology? For what applications?

The following steps are required in this analysis.

1. We identify that the LM2585 is a type 1 IC by our nomenclature.
2. Referring to Fig. 2.36 and Tables 2.3 and 2.4, we see that it can be used as, say, a positive-to-positive Boost.
3. From the equations we see that the input voltage must be below 40 V and the output voltage must be below 65 V (since $V_{SWMAX} > V_O$ and $V_{ICMAX} > V_{INMAX}$). These define the input-output voltage conditions for any suitable application.
4. The maximum load current is (with allowance for 20 percent peak over the COR value)

$$I_{O\_MAX} = 0.8 \times I_{CLIM} \times \left[\dfrac{V_{INMIN}}{V_O} \right]$$

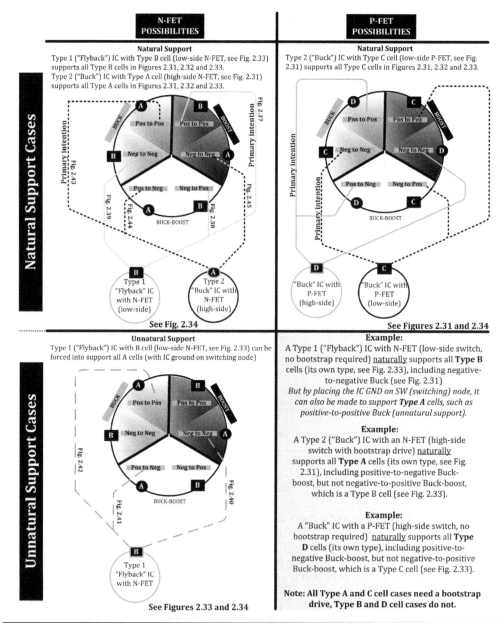

Figure 2.36 Chart showing all possible applications, configurations, and topologies possible with common existing control ICs.

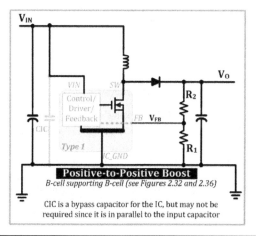

Figure 2.37 Positive-to-positive Boost using a Boost/Flyback IC (type B cell: natural and intended choice).

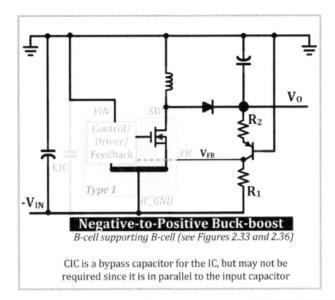

FIGURE 2.38 Negative-to-positive Buck-boost using a Boost/Flyback IC (type B cell: natural and intended choice).

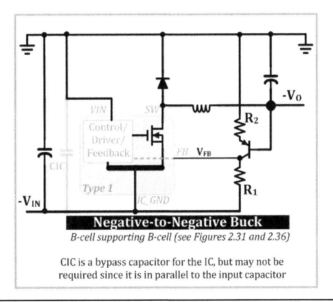

FIGURE 2.39 Negative-to-negative Buck using a Boost/Flyback IC (type B cell: natural choice).

So if the output is set to 60 V and the input ranges from, say, 20 to 40 V, the maximum load (with a suitably designed practical inductor) is

$$I_{O\_MAX} = 0.8 \times 3 \times \frac{20}{60} = 0.8 \text{ A}$$

Example 2.13 The required application conditions are V_{IN} ranging from 4.5 to 5.5 V. The output requirement is –5 V at 0.5 A. Can the LM2651 be used?

The LM2651 is a 1.5-A Buck regulator. Note first that this IC can deliver 1.5 A in a Buck configuration, but not so in any other configuration or topology. The load rating must then be recalculated. The following steps are performed.

1. We identify the LM2651 as a type 2 IC according to our nomenclature.
2. We refer to Fig. 2.36 and Tables 2.3 and 2.4, and we see that a positive-to-negative Buck-boost is possible with this IC.
3. Referring to the datasheet of this device, we get

$$V_{ICMIN} = 4 \text{ V} \quad V_{ICMAX} = 14 \text{ V}$$
$$I_{CLIM} = 1.55 \text{ A} \quad \text{(minimum of tolerance band)}$$
$$D_{max} = 92 \text{ percent} \quad \text{(minimum of tolerance band)}$$

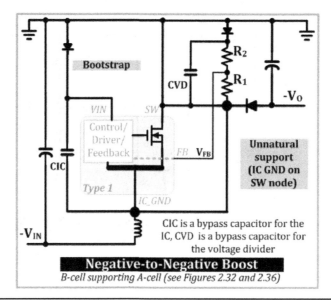

FIGURE 2.40 Negative-to-negative Boost using a Boost/Flyback IC (type A cell: unnatural choice).

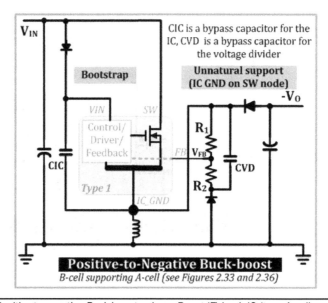

FIGURE 2.41 Positive-to-negative Buck-boost using a Boost/Flyback IC (type A cell: unnatural choice).

4. Therefore we now check sequentially for these conditions:

$V_{ICMAX} > V_{INMAX} + V_O$
$14\text{ V} > 5.5\text{ V} + 5\text{ V} = 10.5\text{ V}$ (OK)

$V_{ICMIN} < V_{INMIN}$
$4\text{ V} < 4.5\text{ V}$ (OK)

$I_{O\_MAX} < 0.8 \times I_{CLIM} \times (V_{INMIN}/V_{INMIN} + V_O)$
$0.5 < 0.8 \times 1.55 \times \{4.5/(4.5 + 5)\} = 0.587$ (OK)

$D_{MAX} > V_O/(V_O + V_{INMIN})$
$0.92 > 5/(5 + 4.5) = 0.53$ (OK)

Therefore the LM2651 is acceptable for the intended application.

Differential Voltage Sensing

In Figs. 2.38 and 2.39, we have used a crude differential sense stage to reference the feedback to the IC ground. A more accurate sensing scheme can be implemented by using an op-amp (such as the LM324), as shown in Fig. 2.46. There are two ways of setting up such a differential

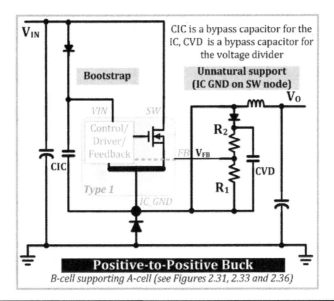

FIGURE 2.42 Positive-to-positive Buck using a Boost/Flyback IC (type A cell: unnatural choice).

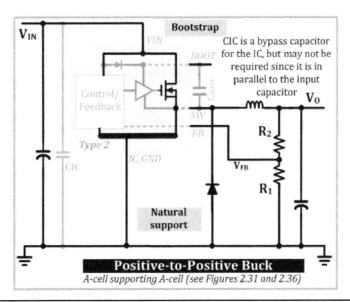

FIGURE 2.43 Positive-to-positive Buck using a Buck IC (type A cell: natural and intended choice).

amplifier. The lower schematic block has a higher gain. Note that the inputs to the op-amp are labeled V_{O-hi} and V_{O-lo}. This means that irrespective of how the schematic actually labels the output rails, the inputs to the op-amp connect to the HI rail or the LO rail, respectively. Some of the relevant aspects of op-amps must be kept in mind. For example, note that an op-amp has a specified input voltage common-mode range. For the LM324 series this number is specified to be 1.5 V below the upper supply rail, and this parameter is hereby called v' (v prime). We require that the voltage on both the input pins of the op-amp stay within this allowed range, or the op-amp cannot be considered fully functional. Since the voltages on these pins are fixed by virtue of the resistors, if the resistors are also considered fixed, the only way is to ensure that the common-mode condition is met is to set the op-amp supply rail $+V_{aux}$ sufficiently higher. This limit equation is therefore also provided in Table 2.5. Note that if the required minimum $+V_{aux}$ value is still low enough, it may be possible to connect it to an available DC rail. If not, an additional external rail will need to be created to run the op-amp stage.

Some Topology Nuances

Some of the concerns when we traverse topologies have to do with the nuances of the topologies themselves. In particular, a Buck topology has no right half-plane (RHP) zero, but the Boost and the Flyback/Buck-boost do. Therefore when we try to take a Buck IC

84 Chapter Two

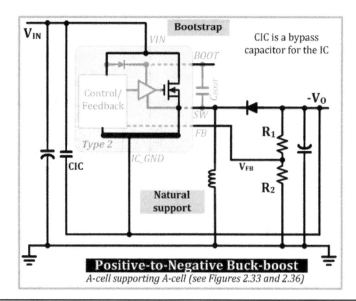

FIGURE 2.44 Positive-to-negative Buck-boost using a Buck IC (type A cell: natural choice).

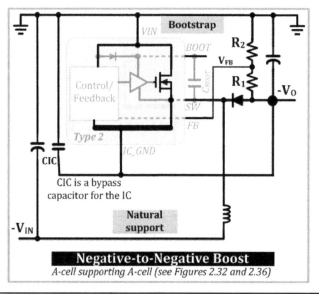

FIGURE 2.45 Negative-to-negative Boost using a Buck IC (type A cell: natural choice).

(with internal fixed compensation), we may not have the ability to tailor the crossover frequency to less than one-fourth of the RHP zero frequency, as is generally recommended for avoiding this particular mode of instability. So how do we successfully take a type 2 IC and apply it to other topologies?

To answer that, we first must remember the intuitive explanation behind the RHP zero. This is said to occur as follows. If we suddenly increase the load on the output of a Boost or Buck-boost-regulator, the output dips momentarily. The voltage on the feedback pin therefore falls slightly, and this commands the duty cycle to increase to try to correct for this. But both the Boost and the Buck-boost are different from the Buck in that during the switch ON-time, no energy flows into the output—we are basically just building up energy in the inductor during that time. So if the duty cycle increases in response to the load increase, in fact there is a smaller OFF-time, and therefore less, rather than more, current will flow into the output. This causes the output to decrease further. Eventually, after a few cycles, the average inductor current does ramp up progressively and the output dip will get corrected. But before that happens, we see a situation where the load disturbance is reinforcing itself. In severe cases this may lead to sustained oscillations. In the Bode plot, the RHP zero shows up as a normal (LHP) zero in the gain plot, causing the slope to

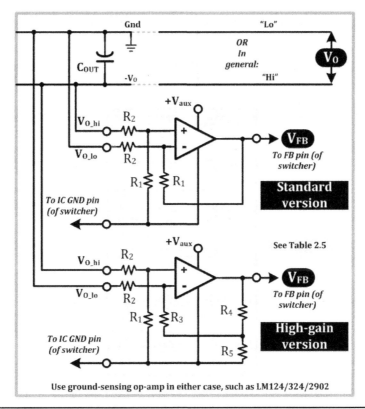

FIGURE 2.46 Two differential sensing techniques for improved output regulation.

Op-amp	Equation Set
Standard differential amplifier	$R_2 = R_1 \times V_O / V_{FB}$ $V_{aux} \geq v' + \left[\dfrac{R_1}{R_1 + R_2} \times (V_{INMAX} + V_O) \right]$
High-gain differential amplifier	$R_2 = R_X \times V_O / V_{FB}$ $R_X \equiv R_3 + R_4 + \dfrac{R_3 \times R_4}{R_5}$ $R_2 = R_3 + \dfrac{R_4 \times R_5}{R_4 + R_5}$ $V_{aux} \geq v' + \left[\dfrac{R_1}{R_1 + R_2} \times (V_{INMAX} + V_O) \right]$

TABLE 2.5 Design Table for Differential Sense Stages

increase (upward) by 20 dB per decade, but it behaves as a pole in the phase plot, causing a −90° shift (downward). The location of the RHP zero is

$$f_{RHPZ\_BOOST} = \frac{R_L \times (1-D)^2}{2\pi L}$$

$$f_{RHPZ\_BUCK\text{-}BOOST} = \frac{R_L \times (1-D)^2}{2\pi L \times D}$$

where R_L is the load resistance. As mentioned, we should generally ensure that we roll off the gain fast enough to be well clear of this frequency. What this implies is that we just don't let the control react too fast to the change in load requirement.

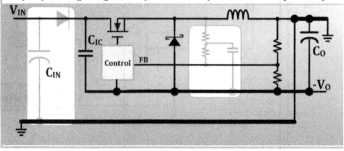

FIGURE 2.47 Practical techniques to alleviate RHP zero.

Two well-known practical RHP zero suppression techniques are used when a Buck IC is used to generate other topologies. We show them as applied to a positive-to-negative (Buck-boost) configuration. Both are shown together in Fig. 2.47, but one or both can be used. The one on the right senses when the duty cycle increases suddenly and pushes up the feedback pin slightly so that it doesn't dip too low in response to the sudden load demand. On the left we see that a diode has been inserted, and the IC bypass capacitor 'CIC' needs to be sized much bigger now. It is no wonder that the schematic then looks much closer to a Buck than to a Buck-boost. And in fact during a load transient it does behave temporarily as a Buck, because now some energy can be transferred to the output even during the switch ON-time. Note that the input diode is for reverse protection and can often be omitted if sudden input steps are not expected. Also, if we follow the current on initial application of power, we will see that a reverse current will flow through the output capacitor. So we may need to put in a reverse diode at that point too, whenever the second method is used.

CHAPTER 3

Contemporary Converters, Composites, and Related Techniques

PART 1 FUNDAMENTAL TOPOLOGIES

Synchronous DC-DC Converters

We know from Chap. 2 that as we lower the load current, we lower I_{COR}, where COR stands for *center of ramp* (average inductor current in CCM). That is true for all three topologies, because $I_{COR} = I_O$ for a Buck and $I_{COR} = I_O/(1-D)$ for the other two topologies. At some point the current returns to zero every cycle, and that is the *boundary conduction mode* (BCM). This corresponds to $r = 2$, as we discussed. If we lower the load current further, then for all diode-based topologies, since it cannot go the other way, the current stops at zero, and the core stays deenergized until the start of the next cycle. That is the *discontinuous conduction mode* (DCM).

In modern topologies, in an effort to reduce losses, the forward drop across the diode was considered excessive, and a FET was placed across it to reduce losses, since the drop across the FET could be made much smaller. Let us designate the main FET, also called the control FET, as Q and the synchronous FET as Qs. This is more consistent (and enduring) than calling them the upper FET or lower FET, since the position (high or low) changes with topology and configuration, and so the control FET will not necessarily be the upper FET.

The simplest way to implement this is to have "complementary" drives. If, e.g., both the FETs are N-FETs, then whenever one is ON, the other is OFF, and vice versa. If this is implemented, then as we lower the load current, just after $r = 2$ is reached, the current goes negative through the synchronous FET. This is often called *forced continuous conduction mode* (FCCM). It has several pros and and cons, the main pro being that the duty cycle remains (almost) constant, irrespective of load current. However, since this mode is based on circulating current, it is not very efficient. So, modern ICs usually offer the option of *diode emulation mode*, in which Qs is turned OFF as soon as the inductor current tries to become negative. So, in effect, Qs appears as a diode, but one with a very low forward drop. The converter enters DCM. These options are shown in Fig. 3.1.

We indicated that the duty cycle is constant in FCCM and is the same as in CCM. So can we say that FCCM is CCM in *all* respects? Almost all respects for sure. In particular, all the CCM equations provided in the Appendix apply in FCCM too, that is, for $r > 2$. We do, however, run into a *potential* problem in using our simplified equations involving r, at extremely light loads, when r is almost infinity (because r is $\Delta I / I_{COR}$, with ΔI remaining unchanged, but I_{COR} heading to zero). How do we deal with that? One way is to put a little mental preload, never going down to exactly zero load current. The other way is to revert to the full form of the equations. For example, in a Buck, the input capacitor RMS equation can be changed as follows.

With $I_O = 0$, we get

$$I_{CIN\_RMS\_NO\_LOAD} = I_O\sqrt{D \times \left[1 - D + \frac{r^2}{12}\right]} = \sqrt{D \times \left[I_O^2(1-D) + \frac{\Delta I^2}{12}\right]} = \sqrt{D \times \left[\frac{\Delta I^2}{12}\right]} = \frac{\Delta I}{2}\sqrt{\frac{D}{3}}$$

There is no singularity anymore.

The bottom line is that in FCCM, we can use the CCM equations provided in the Appendix. For DCM, we should turn to Chap. 4.

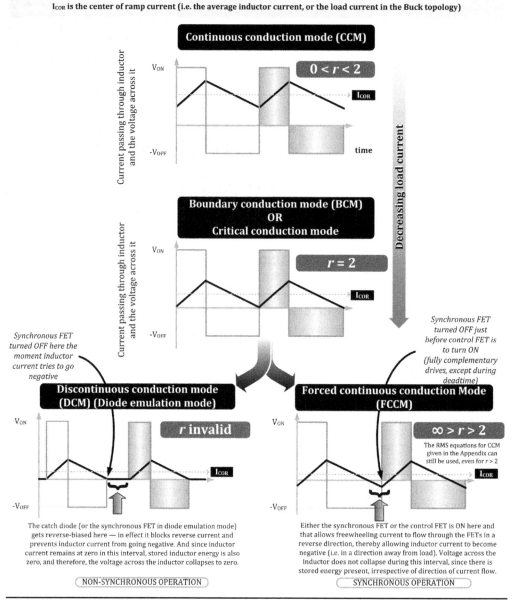

Figure 3.1 Fully synchronous operation or diode emulation mode.

Modes of Operation of a Synchronous Buck

In Fig. 3.2 we show how the currents get routed in a synchronous Buck. We have assumed "near complementary" drives; that is, they are complementary, with the exception of a small deadtime, which we discuss next. But also note that in part of the negative current region, phase D in particular, current goes the "wrong" way—from output to input. We can call this a *boost* phase within the Buck operation, and soon we will discuss this in greater detail too.

Deadtime

What is not represented very clearly in the inductor current waveform in Fig. 3.2 is the deadtime. But we can see from the Gate drives at the top of the figure that there is a certain deadtime—a break before make operation. The main reason that this tiny interval, typically about 100 ns, was introduced was the fear that Q_s may *turn ON a little before Q has (fully) turned OFF*. Even though we may command a FET to turn ON or OFF at a certain time, because of its internal parasitic Gate resistance and capacitance, there is a small delay. That is potentially disastrous, since both the FETs will be ON momentarily (overlapping ON-times), and the supply could get shorted through them, although there are parasitics, such as

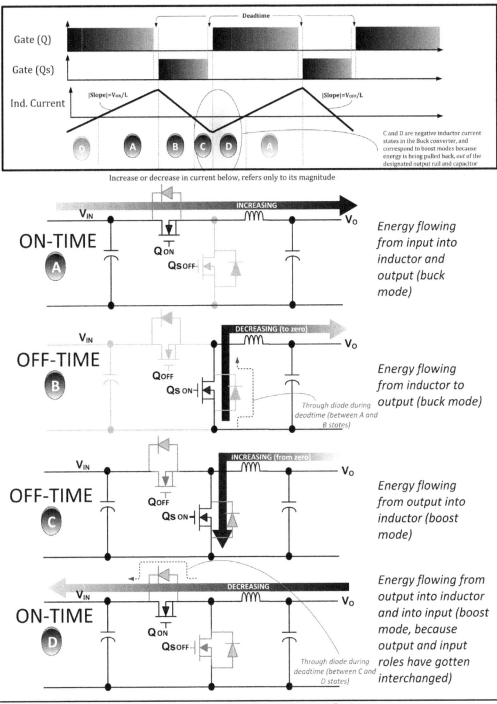

FIGURE 3.2 Explanation of the modes of operation of a synchronous Buck.

trace inductances, that will very likely not allow a complete dead short, but certainly an "inexplicable" fall in efficiency, particularly at light loads, and a higher than expected quiescent current (switching). This potentially dangerous *cross-conduction* spike of current, also called *shoot-through*, will flow from the input capacitor through both the FETs and return via ground. And it is very hard to see on the bench. *If we even insert a piece of wire for a current probe, the small inductance of the wire loop is usually enough to quell the cross-conduction spike, and we may not see anything unusual.*

A properly designed deadtime is a safety buffer against any accidental or unintentional overlap over manufacturing processes, temperature variations, diverse PCB layouts (in the case of switchers using external FETs), and so on. But it is a double-edged sword: It causes a significant loss in efficiency by itself too, since, during this deadtime, we may have momentary conduction through the (bad) body diode, as indicated in Fig. 2.3. This is harmful in two ways. First, the forward drop across the body diode can be 1 to 2.5 V, and this leads to obvious conduction-loss type of dissipation. Engineers try to offset these losses by reducing the deadtime as much as possible and *by retaining the Schottky diode*, the one across the synchronous FET in particular. The latter step actually can lead to another significant improvement in efficiency, as explained under the FAN5340 section below. Second, this is related to the fact that the body diode is a "bad" diode, because of both its high forward drop and its poor reverse recovery characteristics. This phenomenon is now explained.

The FAN5340 Lesson

Why does the Schottky diode help so much? In an actual test conducted in early 2007 at a major semiconductor manufacturer, adding a Schottky diode across the internal synchronous FET of a 2.7- to 16-V synchronous LED Boost IC, on the author's suggestion, led to an almost 10 percent increase in efficiency at maximum load. The IC was eventually, after seemingly years of internal debate as to the root cause of the demonstrated efficiency improvement, brought back to the drawing board and redesigned with an integrated Schottky diode across the synchronous FET (on the same die, since the vendor was one of the very few who possessed that technology) and was finally released as the FAN5340 (see http://www.fairchildsemi.com/ds/FA/FAN5340.pdf). One casualty was perhaps the time to market unfortunately, but this was a valuable lesson learned on the overwhelming effect of body diode characteristics in synchronous Boost regulators.

The root cause of the efficiency advantage posed by the paralleled Schottky diode is that the body diode of a FET has another unfortunate quality besides its high forward drop, one that we also wish to avoid. It is a "bad" diode in the sense that its PN junction absorbs a lot of minority carriers as it starts to conduct in the forward direction. Thereafter, to get it to turn OFF, all those minority carriers *need to be extracted*. Until that process is over, the body diode continues to conduct, and it does not reverse-bias and block voltage as a good diode (with only majority carriers, such as a Schottky diode) is expected to.

This sequence occurs if the Schottky diode is not present. During the deadtime interval between Q turning OFF and Q_s turning ON (i.e., the $Q \rightarrow Q_s$ crossover), to get the body diode of Q_s to conduct, we inject plenty of minority carriers into it (via the diode forward current). The full deleterious effects of that stored charge actually show up during the *next* deadtime interval—when Q_s needs to turn OFF just before Q starts to conduct (i.e., the $Q_s \rightarrow Q$ crossover). Now we discover that Q_s doesn't turn OFF quickly enough, and so a *shoot-through current spike* flows through Q and Q_s. This unwanted reverse current ultimately does extract all the minority carriers, and the diode finally does reverse-bias. But during the rather large duration of this shoot-through, we have a significant $V \times I$ product occurring inside the FETs, and therefore there is high instantaneous dissipation. The current is coming from the output capacitor in this case. The average dissipation (over the entire cycle) is almost proportional to the duration of the deadtime, since the diode is just not recovering fast enough.

The only way to avoid the above reverse recovery behavior and the resulting shoot-through is *to avoid forward-biasing the body-diode altogether*, which basically means we need to bypass it completely. The way to do that is by connecting a Schottky diode across the synchronous FET, which then provides an alternative and preferred path for the current even during the seemingly insignificant deadtime.

Note *However, to ensure that the Schottky diode is "preferred" in a* dynamic *(switching) scenario, we need to ensure the external Schottky diode is connected with very thick, short traces to the drain and source of the FET across it; otherwise the inductance of the traces will be high enough to prevent current from getting diverted from the body diode and into the Schottky diode as desired. The best solution is to have the Schottky diode integrated into the package of the FET itself,* preferably on the same die, *to minimize inductances as much as possible.*

Some modern ICs use *adaptive deadtime*. The control constantly monitors the situation and reduces the deadtime just to the point where cross-conduction is virtually ruled out. (But what about the reverse recovery effects, in a synchronous Boost, for example?) Keep in mind that a little deadtime does have some great benefits too: If we leave enough deadtime for the voltage to swing across the FETs, we can induce zero-voltage switching. This aspect is discussed in greater detail in Chap. 19.

Another Shoot-through Possibility: *C dV/dt* Induced Turn-on

Note that there is a phenomenon called *C dV/dt induced turn-on* that is known to cause cross-conduction, especially in low-voltage VRM-type applications—despite sufficient deadtime apparently being present. For example, in a synchronous Buck, if the control N-FET turns ON very suddenly, it will produce a high *dV/dt* on the Drain of the synchronous FET. That can cause enough current to flow through the Drain-to-Gate capacitance (C_{GD}) of the synchronous FET, which can produce a noticeable voltage bump on its Gate, perhaps enough to cause it to turn ON momentarily (but it may be only a partial turn-on). This will therefore produce an unexpected FET overlap, one apparent perhaps only through an inexplicably low efficiency reading at light loads. To avoid this scenario, we may need to do one or more of the following: (1) slow down the top FET a bit; (2) have good PCB layout (in the case of controllers driving external FETs) to ensure that the Gate drive of the lower FET is held firmly down; (3) design the Gate driver of the lower FET to be "stiff"; (4) choose a bottom (synchronous) FET with slightly higher Gate threshold, if possible; (5) choose a bottom FET that has a low C_{GD}; (6) choose a bottom FET with a very small internal series Gate resistance; (7) choose a bottom FET with high Gate-to-Source capacitance (C_{GS}); and (8) perhaps even try to position the decoupling capacitor on the input rail slightly far away from the FETs (even a few millimeters of trace inductance can help), so that despite slight overlap, *at least the cross-conduction current flowing during that overlap time gets limited by the intervening PCB trace inductances.*

The Boost within the Buck and the Buck within the Boost

In FCCM, as the average inductor current dips below the critical conduction limit of $I_{COR} = \Delta I/2$, two phases emerge marked C and D in Fig. 3.2. In C, the inductor current is going in a negative direction through Q_s. But in D, Q_s has turned OFF and the current freewheels into the input capacitor. This latter phase represents energy being returned from the output to the input. Keeping in mind that the output is a lower-voltage rail than the input, this represents, in effect, a boost operation, albeit for part of the cycle. It is, however, still considered a Buck because on average energy still flows from left to right, that is, from input to output, and the proof of that is that the average inductor current I_{COR} is still positive. Note that positive or negative is simply arbitrary in synchronous converters—it merely reflects our expectation of the "correct" direction of current flow.

Therefore, in a Buck, if we reduce I_{COR} (load) to zero, then all the energy we put into the inductor is returned to the source, so we have 100 percent circulating current. In fact, there is no way to distinguish this from a Boost topology. It is like saying, What is the difference, if any, between +0 and −0? The Boost topology emerges when we (somehow) lower the current I_{COR} to below zero. That means we now have average current (and energy) flow from right to left. Indeed the current is "negative," but that impression can be quickly corrected by rotating the entire schematic by 180°. Now current is going from left to right again, except that this is now very clearly a synchronous Boost converter. This is all explained in Fig. 3.3.

Combining this with the fact that if there is an output short/overload, the energy in the inductor could find its way into the input or/and output capacitors, *we may need to oversize the input/output capacitors just to avoid damage to a synchronous Buck or Boost IC under overloads.*

Popular Current-Sensing Scheme in Voltage Mode Controlled Synchronous Buck Converters

In voltage mode control, a fixed ramp is injected onto the PWM comparator. So the only reason to sense the current is for protection during overloads. The most common way to detect current in synchronous Buck ICs nowadays is explained in Fig. 3.4. Starting from the ground voltage (zero) at the source of Q_s, going backward toward the pin marked CLIM_SENSE, we find two drops of opposite sign, one created by the current through Q_s (interval B of Fig. 3.2) and the other created by the tiny (example) 50-µA current source on the CLIM_SENSE pin, passing through the current limit programming pin RLIM. In this way, by setting RLIM, we can set ILIM.

If we are monitoring the CLIM_SENSE voltage on a scope, we will see it normally several volts above ground. But in current limit it falls toward zero, as indicated in the overload waveforms captured in Fig. 3.5. At that point, a typical control IC will just omit the next few ON pulses on Q, waiting for the voltage on CLIM_SENSE to rise a little above zero or just counting a certain number of skipped pulses before resuming.

92 Chapter Three

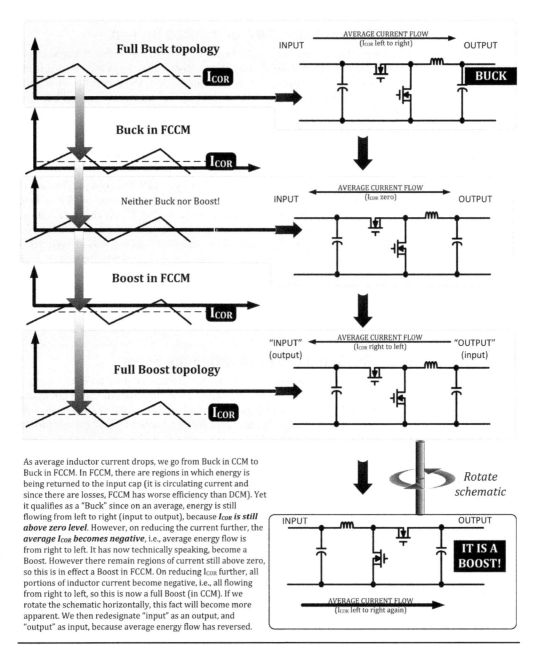

Figure 3.3 Buck to Boost.

Note that if we have an N-FET in the upper position, by skipping several ON pulses, we may exhaust the bootstrap capacitor, and the Gate drive of Q could collapse. To avoid that, some IC designers prefer to use a P-FET for Q. Some might force the top FET ON for a narrow minimum pulse width, just to drive the current into a typical bootstrap capacitor; or they may force an ON pulse every X number of missed pulses, and so on.

Also this method is very temperature-dependent, because temperature is not being compensated, whereas the R_{DS} of the FET is highly temperature-dependent. That can induce a large error. It is best to assume a "hot FET" to start with and to calculate the RLIM based on that, to avoid premature power limiting at elevated temperatures under normal operation.

Popular Current-Sensing Scheme in Current Mode Controlled Synchronous Buck Converters

In this case, we need information about the entire inductor current waveform, to apply on the PWM comparator pin. The most popular method ("lossless") for doing this is called *DCR sensing*. DCR is the *DC resistance* of the inductor winding.

Contemporary Converters, Composites, and Related Techniques

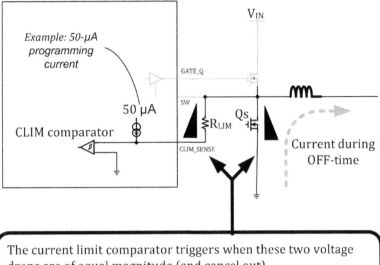

Drop across RLIM is: $V_{RLIM} = 50\ \mu A \times R_{LIM}$

Drop across Qs: $V_{Qs} \approx I_O \times R_{DS\_Qs}$

Equating at CLIM threshold: $50\ \mu A \times R_{LIM} = I_{CLIM} \times R_{DS\_Qs}$

So set CLIM is $I_{CLIM} = \dfrac{50\ \mu A \times R_{LIM}}{R_{DS\_Qs}}$

Actual load current is about 20% less than this because this is actually (1+r/2) times I_{O_MAX}.

FIGURE 3.4 Typical low-side current sense for voltage mode control ICs.

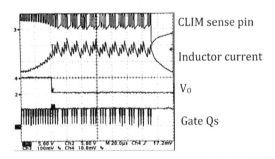

FIGURE 3.5 Waveforms of overload based on low-side sensing.

Perhaps it all started with this simple thought: instead of putting a sense resistor in series with the FET, why not put the sense resistor in series *with the inductor*? We would then obtain full information about the inductor current, rather than just the ON-time or OFF-time slice of it. This could conceivably help us in implementing various new control and protection techniques too. However, that thought process then went a step further to ask, since every inductor has a resistance built in already, called its *DCR*, could we somehow use that resistance in place of a separate sense resistor? We would then have "lossless current sensing," since we are at least not introducing any *additional* loss term into the circuit. The obvious problem here is that the DCR is itself *inaccessible*: We just can't put an error amplifier or multimeter "directly across DCR," because the DCR lies buried deep within the object we call an *inductor*. However, we want to persist. We ask: Is there some way to *extract* the voltage

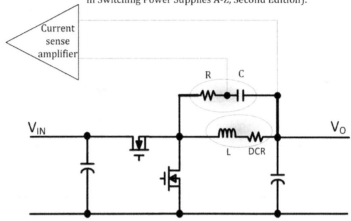

FIGURE 3.6 DCR sensing for current mode controlled converters.

drop across this DCR from the total inductor voltage waveform? The answer is yes, we can—and that technique is called *DCR sensing*.

In Fig. 3.6, we show how this is done rather commonly nowadays. In this case, the RC time constant is not large (as is acceptable for slightly delayed peak current limiting in voltage mode control), but is in fact *matched exactly to the time constant of the L-DCR combination*, which means mathematically $RC = L/DCR$. It is interesting that, in fact, both the DCR voltage and the capacitor voltage will increase or decrease *by exactly the same amount under transients too*. This indicates that both the voltages will change identically (i.e., track each other), and finally settle down identically too, at their common shared steady-state initial voltage condition. Note that the settling/initial value is $I_O \times DCR$.

The advantage of this DCR sensing technique is that the entire AC and DC information of the inductor current waveform is now available within the sensed capacitor voltage (assuming, however, that the inductor does not saturate somewhere in the process).

Is there any simple intuitive math to explain this time-constant matching and the resulting tracking? Yes, but it is not a rigorous proof, as we will see. Let us start by applying the duality principle. We know that the voltage on a capacitor, when charged by a current source, is similar to the current through an inductor when "charged" by a voltage source. Assuming steady-state conditions are in effect, and that the maximum voltage on the capacitor is low compared to the applied ON-time and OFF-time voltages (V_{ON} and V_{OFF}, roughly $V_{IN} - V_O$ and V_O, respectively), we get near *constant-current sources* (V_{ON}/R and V_{OFF}/R) charging and discharging the capacitor during the ON-time and OFF-time, respectively. The question is: What is the capacitor voltage swing during the ON-time and during the OFF-time? Further, are the swings identical in magnitude, or is there a leftover delta remaining at the end of the cycle, indicative of a nonsteady state?

Using $I = C\Delta V/\Delta t$, we get

$$\Delta V_{ON} = \frac{V_{ON}}{RC} \times \frac{D}{f} \qquad \Delta V_{OFF} = \frac{V_{OFF}}{RC} \times \frac{1-D}{f}$$

We know from the volt-seconds law that

$$V_{ON}D = V_{OFF}(1-D)$$

Therefore

$$\Delta V_{ON} = \Delta V_{OFF}$$

We can similarly calculate the inductor current increments as shown later.

Using $V = L\Delta I/\Delta t$, we get

$$\Delta I_{ON} = \frac{V_{ON}}{L} \times \frac{D}{f} \qquad \Delta I_{OFF} = \frac{V_{OFF}}{L} \times \frac{1-D}{f}$$

We know from the volt-seconds law that

$$V_{ON}D = V_{OFF}(1-D)$$

Therefore,

$$\Delta I_{ON} = \Delta I_{OFF}$$

We conclude that there is *no* leftover delta remaining at the end of the cycle for either the inductor current swing or the capacitor voltage swing. Therefore, both the inductor current and the capacitor voltage are in steady state. That is duality at work.

Let us now compare the capacitor voltage swing and the DCR voltage swing and try to *set them equal* so we can get them to be exact copies, rather than merely scaled copies, of each other. We set

$$\Delta V_{ON} = \frac{V_{ON}}{RC} \times \frac{D}{f}$$

equal to

$$(\Delta I_{ON} \times DCR) = \frac{V_{ON}}{L} \times \frac{D}{f} \times DCR$$

Therefore,

$$\frac{1}{RC} = \frac{DCR}{L}$$

or

$$RC = \frac{L}{DCR}$$

That is the basic matched time-constant condition for DCR sensing, and we see it follows rather naturally from the simplified intuitive discussion above.

This simple analysis does indicate that the capacitor voltage is mimicking the inductor current. It tells us quite clearly that the AC portion of the inductor current is certainly being copied by the capacitor voltage. But what can we say about the DC level? Is the DC capacitor voltage copying the DC inductor current too? It actually does, but that does *not* follow from the simple intuitive analysis presented above (or from any complicated *s*-plane AC analysis you might see in literature). A Mathcad file written for the purpose proves that (it is not presented here). But it can be found in Fig. 9.6 of *Switching Power Supplies A-Z*, 2d ed.

With good time-constant matching, the DCR voltage becomes a carbon copy of the capacitor voltage, and vice versa—with both AC and DC values replicated. Further, the replication is true not only in steady-state conditions, but under transient conditions too.

DCR sensing is not considered very accurate and should not be relied upon for critical applications. For example, it is well known that the nominal DCR value has a wide spread in production, and the DCR itself varies significantly with temperature. For that reason, many commercial ICs allow the user to either choose DCR sensing for cost and efficiency reasons or use an external sense resistor in series with the inductor for higher accuracy.

Paralleling and Interleaving of Buck Converters

Here we try to split a Buck power train into two "out of phase" (Gate drives being 180° apart) paralleled power trains. We need some load share circuitry to implement this sharing, but that is not the scope of the discussion here, since we assume that feature is already built into the chip that supports this. We are interested, more as systems designers, how that helps us and by how much.

96 Chapter Three

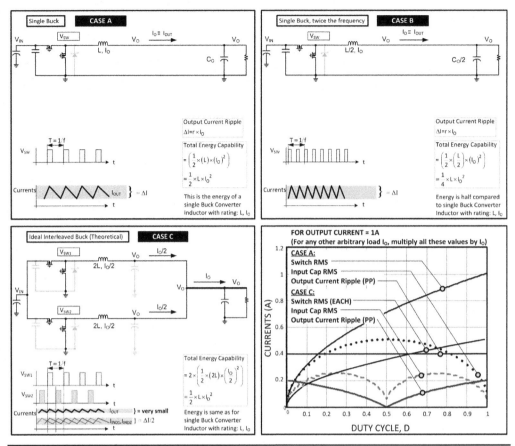

Figure 3.7 Interleaving and paralleling explained.

Hypothetically, at an abstract level, suppose we somehow implement sharing *exactly* (however we do it), that is, suppose we have somehow managed to implement two paralleled, identical converters, each delivering a load current $I_O/2$. This is case C in Fig. 3.7. Both power trains (individual converters) are connected to the same input V_{IN} and the same output V_O. They are driven at the same frequency (although the effect of synchronization, if any, between these two "phases" is only on the input/output capacitors, as discussed later). First we show that, in terms of inductor volume, paralleling may even make matters worse.

We know (or assume for now) from optimization principles that, for each inductor, r should be set to about 0.4. Since the two paralleled converters carry only $I_O/2$, we need to double the inductance of each, to achieve the same current ripple ratio for each inductor, and that is commensurate with the inductance scaling rule explained on page 52 of this book.

Now, let us look at the energy-handling capability of each of the two inductors above. That is proportional to LI^2, and that gets halved—because I halves and L doubles. We also have two such inductors. So the total core volume (both inductors combined) is still *unchanged* from the volume of the single inductor of the original single power train (the latter is case A in Fig. 3.7).

Note that in literature it is often stated rather simplistically that "interleaving helps reduce the total volume of the inductors." The "logic" presented is as follows:

Single-converter inductor with current I:

$$\varepsilon = \frac{1}{2}LI^2$$

Two phases, with each inductor carrying one-half of the current:

$$\varepsilon = \frac{1}{2}L\left(\frac{I}{2}\right)^2 + \frac{1}{2}L\left(\frac{I}{2}\right)^2 = \frac{1}{4}LI^2$$

So it is said that the total inductor volume halves due to interleaving. Yes, it does, but only for two inductors *each with an inductance equal to that of the single converter*. However, now each phase is carrying one-half of the current, and if we do not change L, the ΔI_L remains the same, but I_L (center of ramp) is one half, so the current ripple ratio $\Delta I_L/I_L$ is doubled! If we were willing to accept the same amount of current ripple ratio in a single-inductor converter, we would get the same reduction in volume. That has actually nothing to do with the concept of interleaving. It is a basic misunderstanding of energy storage concepts.

What exactly do we gain in using case C (paralleled converters) instead of case A (single converter)? In terms of the inductor volume required to store a certain amount of energy, there seems to be no way to "cheat" physics. We learned in Chap. 2 that for a given output power requirement and a given time interval $T (= 1/f)$, the energy transferred in a certain time interval t is $\varepsilon = P_O t$. Yes, we can split this energy packet into two (or more) energy packets, each handled by a separate inductor. But the total energy transferred in interval t must eventually remain unchanged, because ε/t (energy/time) has to equal P_O, and we have kept the output power fixed in our current analysis. Therefore, the total inductor volume (of two inductors) in case C is still ε (because $2 \times \varepsilon/2 = \varepsilon$). Note that, just for simplicity, we are assuming the logic of a Flyback above since, as we learned in Chap. 2, its inductor has to store *all* the energy that flows out of it (or we need to include D in the above estimates too; but the conclusions remain unchanged). See Fig. 2.11.

Yes, we could *double the frequency* and *spread* the energy packets. We thus get case B and it remains a single converter. Only this time, instead of two $\varepsilon/2$ packets every interval T, we need to store and transfer one packet of size $\varepsilon/2$ every $T/2$. In effect, rather intuitively speaking, we are using the *same* inductor to first deliver $\varepsilon/2$ in $T/2$ and then *reusing* it to deliver the next $\varepsilon/2$ in $T/2$. This is time-division multiplexing (TDM) in computer jargon. So eventually, we still get a total of ε joules (J) in T seconds (s) as required (equivalently, P_O J in 1 s, to make up the required output power). However (only) one inductor is present to handle $\varepsilon/2$ at any given moment. So the total inductor volume halves in case B, as indicated in Fig. 3.7.

However, going back and comparing situations belonging to the *same switching frequency* (i.e., cases A and C), we now point out that *in practice*, as opposed to theory, paralleling two converters may end up requiring *higher* total energy storage capability (and higher total inductor volume)—simply because there is no such thing as perfect sharing, however smart our implementation may be. For example, to deliver 50-W output, two paralleled 25-W rated converters just won't do the job. We will have to plan for a situation where due to inherent differences, *and despite our best efforts*, one power train (called a *phase*) may end up delivering more output watts, say, 30 W, and the other correspondingly only 50 W − 30 W = 20 W. But we also don't know in advance which of the two power trains will end up carrying more current. So we will need to plan ahead for *two* 30-W rated converters, just to guarantee a 50-W combined output. In effect, we need a total inductor volume sufficient to store *60 W*, as compared to a single-converter inductor that needs to store only *50 W*. Hence we actually get an increase in inductor size due to paralleling (for the same current ripple ratio r).

So why don't we just stick to doubling the frequency of a *single* converter? Why even bother to consider *paralleling* converters? Nothing seems to really impress about paralleling so far. Well, we certainly want to *distribute* current stresses and the resulting dissipation across the PCB to avoid "hot spots," especially in high-current *point-of-load* (POL) applications. But there is another good reason too. Looking closely at case C in Fig. 3.7, we note that the output current waveform is sketched with an AC swing that is described as "very small." No equation or number was provided here, and for good reason. We remember the output current is the *sum of the two inductor currents*. Suppose we visualize a situation where one waveform is falling at the rate of X A/μs, and the other simultaneously rising at the rate of X A/μs. If that happens, clearly *the sum of the two will remain unchanged*— we will get pure DC with no AC swing at all. How exactly can that happen? Consider the fact that in steady state, for a buck, the only way the falling slope $(-V_O/L)$ of the inductor current can be numerically equal to the rising slope $[(V_{IN} - V_O)/L]$ is if $V_{IN} - V_O = V_O$, or $V_{IN} = 2V_O$, that is, $D = 50$ percent. In other words, we expect that at $D = 50$ percent the output current ripple will be zero! Since for fairly large C, the voltage ripple is simply the inductor current ripple multiplied by the ESR of the output capacitor, we expect a low output *voltage* ripple too. That is great news. See Fig. 3.7 for the graphed output *current ripple* (peak to peak) on the lower right side (Mathcad-generated plot). It has a minimum of zero at $D = 50$ percent, as intuitively explained above.

We see that the output current ripple, and therefore the output voltage ripple, can be significantly reduced by *interleaving*—this means running the two converters with a phase shift of 180° (360° corresponds to one full clock cycle). But it also turns out that the *input RMS current is also almost zero at D = 50 percent* (see graph in Fig. 3.7). The reason is that as soon as one converter stops drawing current, the other starts drawing current, so the net input current appears closer and closer to DC as the duty cycle approaches 50 percent (except for the small component related to *r*). In other words, instead of the sharp edges of switch current waveforms that usually affect the input capacitor current waveforms of a buck, we now start approaching something more similar to the smoother undulating waveforms typically found on the output capacitor. All these improvements are graphed in Fig. 3.7 in comparing case A to case C (i.e., single converter of frequency *f* versus two paralleled ones, each with frequency *f* but phase-shifted, as indicated by the Gate drive waveforms).

Note that we have made the assumption above that the two converters switch exactly out of phase (180° apart, or $T/2$ apart). As mentioned, this is called *interleaving*, and in that case each power train is more commonly referred to in literature as one *phase* of the (combined or composite) converter. So in case C, the converter has two phases. We could also have more phases, and that would be generically a *multiphase (N-phase) converter*. We have to divide T by the number of phases we desire (T/N) and start each successive converter's ON-time exactly after that subinterval. If we run all the power trains (i.e., phases) *in phase* (all ON-times commencing at the same moment), the only resulting advantage in this case is that we distribute the heat around. But interleaving *reduces* overall stresses and improves performance. From the output capacitor's viewpoint, the frequency effectively doubles, so the output ripple not only is much smaller, but also can even be zero under the right duty cycle conditions.

At the input side, it is not the peak-to-peak ripple, but the RMS current that is very important. In general, the RMS of a waveform is independent of frequency. However, interleaving does reduce the input RMS stress. That happens not due to any frequency-doubling effect but because interleaving ends up changing the very shape of the input current waveform—to something closer to that of a steady stream of current (a gradual removal of the AC component depending on the duty cycle).

One drawback of interleaving, as described so far, is that the inductors each still "see" a switching frequency of $1/f$, as is obvious from their current ripple (their upslope and downslope durations). But by using coupled inductors (as discussed in *Switching Power Supplies A–Z*, 2d ed.), we can "fool" the inductors too into "thinking" they are at a higher switching frequency.

Now for some simple math to validate the key curves in Fig. 3.7, and to derive *closed-form equations*. The equations that follow can be rigorously derived over several rather intimidating pages, if so desired. Such derivations are readily available in related literature. Here we do the same, but intuitively, and hopefully more elegantly, if not very rigorously.

First, let us look at the output capacitor ripple. We know that in single-phase converters, since $T_{OFF} = (1 - D)/f$, the current ripple is $\Delta I = (V_O/L) \times [(1 - D)/f]$. So the peak-to-peak ripple depends on $1 - D$. Now if we look closely at Fig. 3.7, we will realize by looking at the *combined* output current, that it certainly has a repetition rate of $2f$, just as we expected, but its duty cycle is not D, *but $2D$*. That is so because the ON-time of each converter has remained the same, but the effective time period has been cut in half. So the effective duty cycle for the combined output current is $T_{ON}/(0.5T) = 2T_{ON}/T = 2D$. Since the peak-to-peak ripple (ΔI) is proportional to $1 - D$, for the combined output it becomes

$$I_{O\_RIPPLE\_TOTAL} = \frac{1-2D}{1-D} \times I_{O\_RIPPLE\_PHASE} \quad \text{if } D \leq 50\%$$

Alternatively stated,

$$\frac{I_{O\_RIPPLE\_TOTAL}}{I_{O\_RIPPLE\_PHASE}} = \frac{1-2D}{1-D} \quad \text{if } D \leq 50\% \quad \Leftarrow$$

This is true when the switch waveforms of both phases *do not overlap*, that is, $D < 50$ percent. That means the ON durations of the two phases are separated in time. Of course, in that case, the OFF durations are the ones that overlap. Recognizing that symmetry, we can actually quickly figure out what happens when the *reverse happens*; that is, when the switch waveforms overlap, that is, when $D > 50$ percent, because then the OFF-times do *not* overlap. We keep in mind that these are just geometric waveforms. There is no significance to what we call the ON-time and what we call the OFF-time as far as the waveforms go. Therefore, we can easily reverse the roles of D and D'. So now we can easily guess what the peak-to-peak ripple relationship is for the case $D > 50$ percent, that is, for the case $D' < 50$ percent! We thus get

$$\frac{I_{O\_RIPPLE\_TOTAL}}{I_{O\_RIPPLE\_PHASE}} = \frac{1-2D'}{1-D'} = \frac{2D-1}{D} \quad \text{if } D \geq 50\% \Leftarrow$$

Note that this represents a current ripple reduction for the *combined output of the interleaved buck*, compared to the current ripple *of each of its two phases* (we are not comparing this with the single-converter case anymore).

Second, now let us look at the input capacitor RMS current. From the viewpoint of the input capacitor, the switching frequency is again $2f$, and the duty cycle is $2D$. The current is drawn in spikes of height I_O, which is 0.5 A for every 1 A (combined) of output current. We are using the flattop approximation here (ignoring the small term involving r). For a Buck, we can approximate the input capacitor RMS as

$$I_{CAP\_IN\_RMS} = I_O \sqrt{D(1-D)} \quad \text{(single-phase converter)}$$

So for the interleaved converter, although changing the effective frequency has no effect on the RMS stress of the input capacitor, *the effective doubling of the duty cycle profoundly affects the waveshape and the computed RMS*. We get the following closed-form equation

$$I_{CAP\_IN\_RMS} = I_O \sqrt{2D(1-2D)} \quad \text{(2-phase converter)} \quad \text{if } D \leq 50\% \Leftarrow$$

where I_O is the output current of *each* phase (one-half of the combined current output). When the waveforms overlap, using the same logic as above, we can easily guess the input capacitor RMS is

$$I_{CAP\_IN\_RMS} = I_O \sqrt{2D'(1-2D')}$$
$$= I_O \sqrt{2(1-D)(2D-1)} \quad \text{(2-phase converter)} \quad \text{if } D \geq 50\% \Leftarrow$$

To cement all this, a quick numerical example is called for.

Example 3.1 We have an interleaved buck converter with $D = 60$ percent, rated for 5 V at 4 A. Compare the output ripple and input capacitor RMS to a single-phase converter delivering 5 V at 4 A.

Single-Phase Case: We typically set $r = 0.4$. So for a 4-A load, the inductance is chosen for a swing of 0.4×4 A $= 1.6$ A. That is just the peak-to-peak output current ripple. This agrees with Fig. 3.7, which is for 1-A load. So, scaling that 4 times for a 4-A load, we get 0.4 A $\times$ 4 $=$ 1.6 A.

The input capacitor RMS with flattop approximation is from our equation

$$I_{CAP\_IN\_RMS\_SINGLE} = I_O \sqrt{D(1-D)} = 4\sqrt{0.6(0.4)} = 1.96 \text{ A}$$

This also agrees with the Mathcad-based plot in Fig. 3.7 in which we get about 0.5 A at $D = 0.6$ for 1-A load. Scaling that for 4-A load current, we get 0.5 $\times$ 4 = 2 A RMS; slightly higher than the 1.96-A result from our closed-form equation.

Interleaved Case: This time we split the 4-A load into 2 A per phase. The reduction in the ripple equation for $D > 50$ percent gives us a "ripple advantage" of

$$\frac{I_{O\_RIPPLE\_TOTAL}}{I_{O\_RIPPLE\_PHASE}} = \frac{2D-1}{D} = \frac{2(0.6)-1}{0.6} = 0.333$$

The peak-to-peak ripple of each phase is 0.4×2 A $= 0.8$ A. Therefore the ripple of the combined output current must be 0.333×0.8 A $= 0.27$ A. Comparing this with the plot in Fig. 3.7, at $D = 0.6$, we have a peak-to-peak ripple of 0.067 A. But that is for 1-A load. So for a 4-A load we get 4×0.067 A $= 0.268$ A, which agrees closely with the 0.27-A result from the closed-form equation. Now calculating the input capacitor RMS from our equations, we get

$$I_{CAP\_IN\_RMS} = I_O \sqrt{2(1-D)(2D-1)} = 2\sqrt{2(1-0.6)[2(0.6)-1]} = 0.8 \text{ A}$$

Comparing this with the plot in Fig. 3.7, at $D = 0.6$ we get the input capacitor RMS to be 0.2 A. But that is for a 1-A load. So for a 4-A load we get 4×0.2 A $= 0.8$ A, which agrees exactly with the 0.8-A result from the closed-form equation.

Summarizing, we see that for a single-phase converter the output current ripple (peak to peak) was 1.6 A, and the input capacitor RMS was 2 A. For the interleaved solution, output ripple fell to 0.27 A, and the input capacitor RMS fell to 0.8 A. This represents a significant improvement and shows the beneficial effects of interleaving (paralleling out of phase).

We remember that in a Buck, the dominant concern at the input capacitor is mainly the RMS stress it sees, whereas on the output, it is the voltage ripple that determines the capacitance. So interleaving can greatly help in decreasing the sizes of *both* input and output capacitors. The latter reduction will *help improve loop response in the bargain*—smaller L and C components charge and discharge faster, and they can therefore respond to sudden changes in load much faster too. But in addition we can now *revisit our entire rationale for trying to keep inductors at an "optimum"* of $r = 0.4$. We recall that was considered an optimum for the entire (single-phase) converter (see Fig. 5-7). But now we can argue that we *don't* have a single converter anymore. And further, if we are able to reduce RMS stresses and the output ripple by interleaving, why not consciously *increase r* (judiciously though)? That could dramatically reduce the size of the inductor. In other words, *for a given output voltage ripple* (not inductor current ripple), we really can go ahead and increase the r (reduce inductance) of each phase. Reducing inductance typically *reduces* the size of the inductor for a given application. Certainly the amount of energy we need to cycle through the total inductor volume is fixed by physics, as explained in Chap. 2. However, the *peak* energy storage requirement significantly goes down if we decrease inductance (increase r).

One limitation of increasing r is that as we reduce load, we will approach the *discontinuous conduction mode* (DCM) for higher and higher minimum load currents. That is why we used synchronous Buck converters in Fig. 3.7 to illustrate the principle of interleaving. Another advantage of intelligent multiphase operation is that at light loads we can start to "shed" phases. For example, we can change over from, say, a six-phase multiphase converter to a four-phase converter at medium loads, then go to a two-phase converter at lighter loads, and so on. That way *we reduce switching losses*, because, in effect, we are reducing the overall switching frequency, thereby greatly improving the light-load efficiency.

PART 2   COMPOSITE TOPOLOGIES

Here we focus our attention on three main topologies in existence. They are not fundamental: They consist of two inductors (sometimes wound on one core), and they can be shown to consist of a Boost/Buck-boost cell followed by a Buck cell.

In Chap. 2 we saw that the Boost and Buck-boost are pretty much the same topology, except for the way we draw the output. So they contain "hidden" outputs when combined into a composite with a Buck cell. These hidden rails are shown grayed out in Fig. 3.8, but they can be revived if desired, to provide low-power, regulated auxiliary rails.

In Fig. 3.8, we have also referred to type A, B, C, and D cells that we defined under the topology morphology section in Chap. 2. It is instructive to continue to identify them in that manner because it tells us two key things:

1. Is it an N-FET or P-FET based topology? (We may need bootstrapping of course, but the important difference is whether it is a high-side configuration or a low-side one.)

2. Do we have the diode "pointing" toward or away from the switching node?

However, we also have an additional degree of freedom, as we can take the Buck cell and connect its lower rail to either the upper or the lower rail of the Boost/Buck-boost cell.

In fact, historically something similar seems to have happened. After Cuk had discovered his "ideal DC-DC converter," with nonpulsating current waveforms into both the input and output capacitors (see Fig. 3.9), the remaining issue was that it was an inverting topology. We certainly can't redefine system ground on either side of a converter stage on a board. Yes, it would work either if it were either completely at the front-end or if it were the very last stage. But in the middle of the board, polarity inversion is unacceptable. That is one reason why it continues to have a low adoption rate. The SEPIC (single-ended primary inductance converter) simply redefined the outputs by calling the inverted rail *ground*, whereas the ground became the previously inverted output rail. See Fig. 3.10. The polarity inversion did get "corrected," but it introduced a new issue: All the output current now flowed out of the catch diode, much as in a Boost or Buck-boost topology. So once again, we

Contemporary Converters, Composites, and Related Techniques

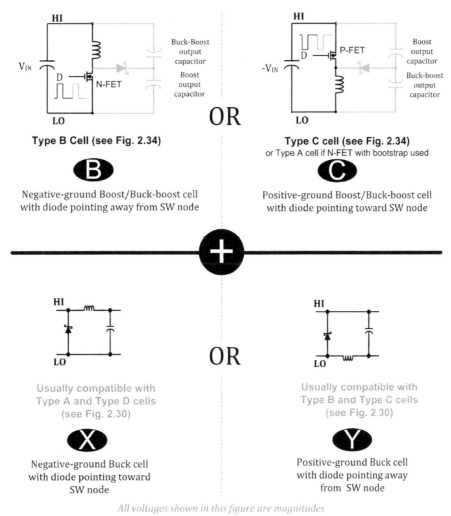

FIGURE 3.8 Building blocks of composites.

had resurrected the sharp pulsating output capacitor current waveforms, having high RMS content and thus leading to lower overall efficiency. The zeta topology in effect flipped the rails at the input. Another way to look at it is by comparing the positive-to-positive SEPIC with the positive-to-positive zeta. We realize that the zeta takes the low-side FET configuration of the SEPIC and converts it to a high-side FET configuration. It does correct the pulsating output current problem of the SEPIC, but lands up with a new one: that of a pulsating input capacitor current, with correspondingly high input RMS and thereby lower efficiency compared to a Cuk diode again. There is no ideal, noninverting DC-DC converter yet!

So how do we go about computing the stresses in these three composite topologies?

1. In all cases the duty cycle depends on volts, volt-seconds, and not on current flow. All appear to be a cascaded Boost and a Buck with the same duty cycle. So we get the usual Buck-boost DC transfer function for three topologies: $V_O = V_{IN} \times D/(1-D)$.

2. In all cases, the switch and diode RMS (and average) currents are the same as the switch and diode RMS (and average) currents in any Boost/Buck-boost topology, so we can use the same equations found in the Appendix.

3. In all cases, the switch and diode voltage stress is $V_{IN} + V_O$. That would be the minimum voltage rating for the components we select.

4. For the Cuk, both the input and output capacitor RMS stresses are negligible.

5. For the SEPIC, the input capacitor RMS is negligible, but the output capacitor RMS uses the same equation as the output capacitor of a Buck-Boost topology (use those equations from the Appendix).

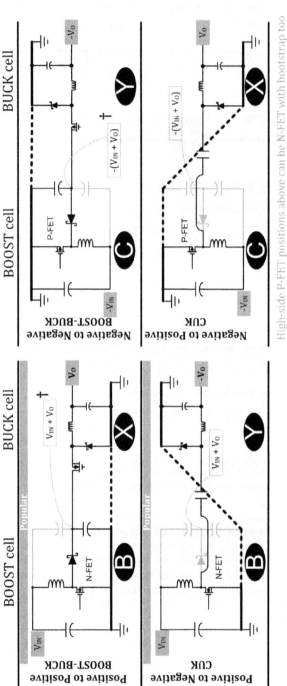

FIGURE 3.9 Boost-buck and Cuk topologies and configurations.

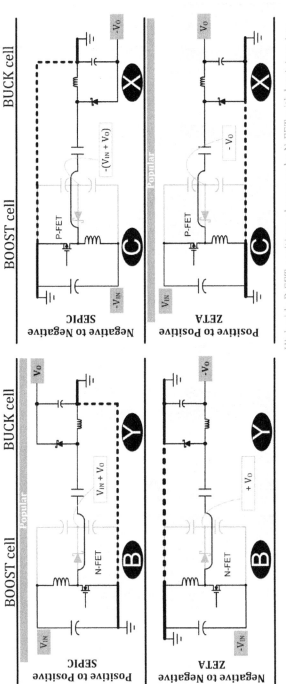

FIGURE 3.10 SEPIC and zeta topologies and configurations.

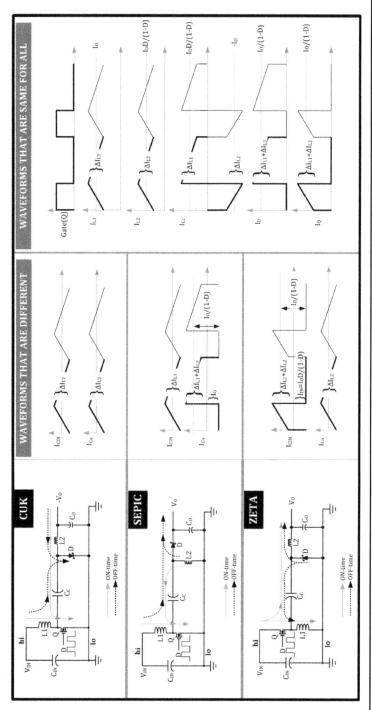

Figure 3.11 Cuk, SEPIC, and zeta current waveforms.

6. For the zeta topology, the output capacitor RMS is negligible, but the input capacitor RMS uses the same equation as the input capacitor of a Buck-boost topology (use those equations from the Appendix).

7. The last remaining component is the coupling capacitor. The voltage ratings of this capacitor are $V_O + V_{IN}$ or V_{IN} or V_O, respectively, in the Cuk, SEPIC, and zeta topologies.

8. The RMS rating of the coupling capacitor is (for all three topologies)

$$I_{Cc\_RMS} = I_O \sqrt{\frac{V_O}{V_{IN}}}$$

All the current waveforms are provided in Fig. 3.11 for easy reference. They have been derived in *Switching Power Supplies A–Z*, 2d ed.

Last, note that the simplest option for a step-up, step-down function is to literally cascade a complete Boost stage and a complete Buck stage, as shown for comparison in Fig. 3.9. It is not cost-effective since it uses two FETs and an intermediate filter capacitor. But it does have low RMS in both the input and output capacitors. So the efficiency is good. We can also run the two stages with staggered frequencies for low EMI.

CHAPTER 4

Understanding and Using Discontinuous Conduction Mode

Introduction

Let us recapitulate a little first. In previous chapters we identified the geometric center of the inductor current ramp in continuous conduction mode (CCM) as the *average inductor current*. This is alternatively, and equivalently, referred to in related literature (including this book) as I_{DC}, I_L, I_{AVG}, and so on. This is essentially the DC value of the inductor current waveform. Superimposed on that DC level, we have a certain current swing. The measured swing (trough/valley to peak) is called ΔI here and is also equal to $2 \times I_{AC}$ by the definition of I_{AC}. We thus have a symmetric swing equal to $\Delta I/2 = I_{AC}$ above the DC level (I_L) and exactly the same amount of swing I_{AC} below it. Clearly, I_{AC} is usually less than I_{DC} ($I_{DC} > I_{AC}$), and that is CCM because the entire inductor current waveform is "high enough" and does not "touch" the zero current axis.

Keep the following in mind too: the DC value of the inductor current is equal to I_O (load current) for the Buck and $I_O/(1 - D)$ for the Boost and Buck-boost. In all cases, I_{DC} *is thus proportional to* I_O for a given input-output combination. In the Buck they are in fact equal, but not in the other two basic topologies.

As we reduce I_O (and/or increase the input voltage, as discussed later), we lower I_{DC} *proportionally*, lowering the entire inductor current waveform. See Fig. 4.1 (Buck) for an example. In the process, I_{AC} is unchanged, since that depends on the inductance, switching frequency, the applied voltages across the inductor (which in turn depend on the input/output voltages), the durations involved (T_{ON} and T_{OFF}), *all of which remain fixed* if we are just lowering I_O. In such a situation, although I_{AC} is fixed, I_{DC} falls, and sooner or later reach we will transit from $I_{DC} > I_{AC}$ (CCM) to $I_{DC} < I_{AC}$. The latter is discontinuous conduction mode (DCM). On the way we will move through the *critical boundary* $I_{AC} = I_{DC}$. This is appropriately called BCM, for boundary conduction mode, but is also often called *critical conduction mode* (though it is clearly dangerous to abbreviate that as CCM).

We defined the *current ripple ratio* as $r = \Delta I/I_{DC} = 2 \times I_{AC}/I_{DC}$. So if I_{AC} equals I_{DC}, we will get $r = 2$. That is the maximum (limiting) value of r in normal CCM operation. In other words, in CCM we always have $r \leq 2$. Later we will see that in forced CCM (FCCM or FPWM, as occurring in synchronous CCM converters with *complementary* Gate drives), in fact r can exceed 2, and the system will still be considered CCM, with all our *usual CCM equations remaining valid*. But for now we ignore this special $r > 2$ CCM mode, and we stick to conventional catch-diode-based (*nonsynchronous*) DC-DC converters. In such conventional converters, if we reduce the load current just a little after reaching BCM, we will enter DCM because the inductor current has reached zero during the switch OFF-time, but cannot go negative. After reaching zero, the current is forced to *stay* at zero until the ON-time starts again. We are calling this the *idle time* interval of DCM.

Note that we actually defined r as a useful parameter in *CCM only*, and it really has no meaning in DCM. But in nonsynchronous converters, we can *mentally* think of DCM as corresponding to $r > 2$. And *that can thus become a simple mathematical criterion for judging whether to apply CCM or DCM equations to a nonsynchronous converter*. We have used that in some of the analysis that follows later.

In the above paragraph, we have said a little something *between the lines* that is worth spelling out clearly now: DCM is entered whenever the ramp-down current reaches zero

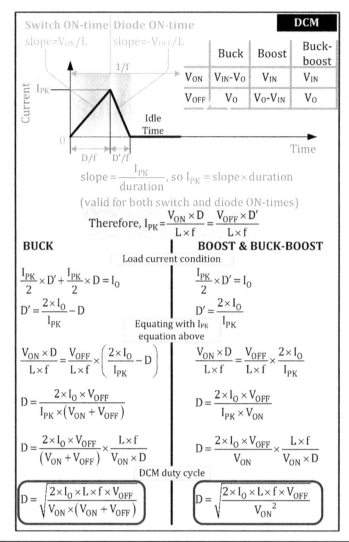

FIGURE 4.1 Calculating duty cycle for the three basic topologies.

before the ON-time starts, that is, before the OFF-time is fully over. As a corollary, any converter is *always* in CCM provided the inductor current has *not had enough time* to ramp down to zero during the OFF-time. So it is basically a *timing issue*: a race condition of sorts. We can ask: Does the inductor current reach zero first, or does the OFF-time time-out first? The answer to that determines whether we are in DCM or in CCM. Clearly, lots of factors come into play in answering that question. Frequency vis-à-vis inductor value would be one of them, since timing is the key here. For example, we know that the ramp-down slope is V_{OFF}/L (in magnitude), where V_{OFF} is the voltage across the inductor during the OFF-time. So, e.g., the slope can be made very steep if L is made very small (or if V_{OFF} is very large). But why would you *vary L* anyway? Many textbooks try to define a "critical conductance." In most cases, the optimum for any converter is characterized by about $r = 0.4$, and based on the switching frequency, that then determines L. In other words, L gets fixed anyway as a design entry point in almost all cases.

Once DCM is entered, the behavior of the converter changes a great deal, and we may need to treat it very differently. The starkest and most obvious change is that in DCM the duty cycle starts to pinch off as we reduce load. In contrast, in CCM, the duty cycle is almost independent of load. What else changes? For example, in this chapter we will come up with a completely fresh bunch of RMS, AVG, and peak current stress equations for DCM. We are not going to derive all these equations here for lack of space, but most are obvious by pure geometry once we have derived the basic equations for D (duty cycle) in DCM. Further, it is easy to tell if the stated DCM equations are valid or not, by just plotting them out along with the CCM equations. The general idea is this: for a given nonsynchronous topology we will always have a set of DCM equations, valid in the DCM region, and a completely different set of CCM equations, valid in the CCM region. *But logically, both sets of equations must give exactly the same numerical results at the critical conduction boundary*, because BCM is not just an extreme case of DCM, but also an extreme case of CCM. BCM happens to be the border/boundary between CCM and DCM. It is the critical boundary. So, we cannot have any *discontinuity* where they meet—they must merge *smoothly*. One obvious test of the validity of the CCM and DCM equations as provided in this book is that they must converge to the same numerical values at BCM. If not, one or both equations are definitely erroneous. However, if we find they *do* converge, as in fact we discover in the plots that we have presented later in this chapter, we can ask: What are the chances that the CCM and/or DCM equations are wrong *by exactly the same error amount* so that they end up yielding exactly the same numerical results at the critical conduction boundary (despite one or both being wrong)? There is a very slim chance indeed for that sort of coincidence. In other words, if the CCM equations provided here are assumed *valid* (and that is easily verified since CCM equations are much easier to derive, as well as readily available in related literature), the chances are then *extremely high* that the DCM equations provided here are *valid too*, since the two sets of equations so obviously merge smoothly together at the critical boundary.

Why are we taking all this trouble to delve into DCM anyway? From Chap. 7, we realize that increasing the current ripple ratio (more ripple for a given load), by decreasing inductance, leads to *smaller* magnetic core volumes (not larger, as erroneously stated in some related literature). Nevertheless, BCM/DCM is admittedly considered an oddity by a good number of engineers, and some even pretend that DCM does not exist! However, keep in mind that *every* nonsynchronous converter *will* go into DCM at light enough loads, and so we should at least, as a bare minimum, want to *document its behavior* in this not-so-infrequent region of operation. One possible reason why engineers may be studiously avoiding studying DCM is some sort of fear of the rather more complicated expressions of DCM as compared to CCM, including that of duty cycle. But keeping that aside, we must recognize that DCM has *disadvantages and advantages*, just as CCM does. A lot of known advantages of DCM (and BCM) relate to feedback loop stability, such as the absence of subharmonic instability in current-mode control. But there are other reasons why DCM is not necessarily such a bad idea after all, as discussed below.

First, let's be clear we are *not* recommending that converters be designed in DCM under *all* operating conditions. We should in fact almost invariably try to design *for CCM* at maximum load *as a starting point*. Otherwise the resulting currents will be too "peaky" and their RMS values very high, causing serious degradation of efficiency, which we know is paramount in today's product environment. Yes, we could opt for a notable compromise and force the converter to be in BCM *always*. That would simplify loop stability responses and perhaps provide acceptable efficiency. However, to enforce BCM under all conditions, we need to actively sense when the inductor current reaches zero and *at that very moment, turn the switch back ON*. We realize that we would thereby avoid the rather inefficient "idle time" spent at at zero current, which is so characteristic of DCM. But the price to pay for that would be a *variable switching frequency* with consequences like *unpredictable EMI*. Note that DCM also exhibits high EMI because of the natural, almost unrestrained/undamped ringing of voltage waveforms at the moment that idle time is entered (as the inductor gets suddenly deenergized). So compared to DCM, BCM may in fact have lower EMI, but BCM is perhaps far more *unpredictable in terms of EMI* because of the varying switching frequency. So though BCM is still quite popular for some rather low- to medium-power applications, it is generally avoided by a lot of engineers. However as mentioned, since a converter designed for CCM *will* go through BCM into DCM sooner or later, we do need to understand DCM better. With that aim in mind, we can now ask: *at what load current will the CCM-DCM transition occur, and what would be the effect of that transition on the various current/voltage stresses in the converter?* This is the key question we seek to answer further below.

Before we get there finally, here are some points to keep in mind in the eternal CCM versus DCM debate.

1. The efficiency may not be as bad in DCM as instinctively expected on the basis of "peaky" currents. This is attributable mainly to the fact that in DCM the switch turns ON when the freewheeling diode has already *recovered*. So there is no reverse recovery current spike through the diode. This spike can have a rather severe effect on efficiency, particularly in CCM-based Boost converters. That is why, e.g., in high-voltage Boost stages such as in PFC (power factor correction) AC-DC front-end preregulators, extremely fast-recovery diodes are used (or complicated low-loss inductive snubbers).

2. The shoot-through current spike mentioned above not only is bad for efficiency, but also can affect the overall behavior, especially in a current-mode controlled converter. In a typical current-mode control implementation, the switch current is constantly

monitored so as to provide a ramp for the pulse-width modulator (PWM) comparator to act on (instead of a conventional clock). The reverse recovery spike passes through the switch and can inadvertently trigger the PWM comparator. For that reason, current-mode controllers usually incorporate *leading-edge blanking*, which basically means the PWM comparator in effect "looks away" for a certain predetermined amount of time (typically 50 to 200 ns) at the start of the ON-time so as to avoid reacting to this spurious current spike (premature termination of ON pulse). Unfortunately, this also means that there is no protective *current limiting* present during the blanking time. And that can lower system reliability under various fault conditions. DCM/BCM helps in this regard because of the virtually absent shoot-through spike and the consequent ability to significantly reduce the blanking time.

3. The significant blanking time in CCM converters, in general, also means that there is a rather large *minimum T_{ON} pulse width* of 50 to 200 ns (typically the same as the enforced blanking time). Under *extreme down-conversions at high switching frequencies*, this can pose a problem, even when operating in CCM, due to a very low duty cycle requirement, which demands a T_{ON} pulse *smaller* than the minimum pulse width allowed by the controller. And even if there is no extreme down-conversion involved, when this converter enters DCM, *at very light loads*, once again the required duty cycle may be practically unsupportable. We may be able to avoid the situation in DCM by *actively monitoring the transition* from CCM to DCM and removing or significantly reducing the minimum pulse width when the converter enters DCM.

   In general, we should remember that whenever the *naturally demanded* duty cycle of the system, whether in CCM or DCM, is less than what the converter control can support *architecturally*, then we will necessarily see *chaotic pulsing*. This is so because the energy we put into the system every T_{ON} pulse interval is a *little higher* than the energy demanded naturally. So, after a few cycles, the feedback loop will suddenly sense the output climbing and will try to correct it by omitting several T_{ON} pulses in succession (provided "omitting pulses" is architecturally possible of course). The advantage of DCM, as opposed to CCM, is that we *can* either remove or at least significantly reduce the minimum T_{ON} pulse width in DCM, and this leads to smoother pulsing in such "corner cases."

   Note that in DCM, one other way to accept a larger minimum pulse width without chaotic pulsing is to simply fix the minimum natural duty cycle higher by applying a "preload" on the converter (usually with some resistors of high value placed at its output). But that technique is obviously not conducive for high efficiency at light loads.

4. As mentioned earlier, the magnetics are smaller in DCM, contrary to what is often mentioned in related literature. A larger r (closer to 2 instead of the typically recommended value of 0.3 to 0.4) actually leads to much *smaller* magnetics (though at a price in terms of larger neighboring power components)—because of the *smaller required core volume* to handle the *lower stored magnetic energy*. Why is the energy requirement lower as we decrease L (increase r)? The core volume is related to stored energy, which is $\varepsilon = \frac{1}{2} \times L \times I_{PEAK}^2$. If we reduce the number of turns on the inductor in a given application, inductance falls and so I_{PEAK} does increase; but at least in CCM, I_{PEAK} is equal to $I_{DC} + I_{AC'}$ and the AC portion (which is the component that changes as we lower the inductance) is usually a *smaller* contributor to the peak than the DC value (which does not change as we change L). In contrast, L, which depends directly on turns *squared*, falls off much faster in the equation for ε, leading to a lowered ε. So, in DCM, we get smaller magnetic components in general *except if the copper wire used is required to be so thick* just to handle the higher RMS currents that it demands a larger core simply to accommodate the extra copper on it. But that situation is rare.

5. DCM converters are usually easier to stabilize, though a little sloppy in their response to load and line transients. For example, subharmonic instability, an artifact of current-mode control, is absent if we use DCM (see Chap. 14 in which deals with loop stability). There is some debate on the significance of the right-half-plane (RHP) zero in DCM. Most, however, agree that though the RHP zero is still present in DCM as in CCM (for the Boost and Buck-boost topologies), but because of the relatively smaller inductor in DCM-based or BCM-based converters, the RHP zero frequency extends to well beyond the switching frequency, with negligible impact at typical loop-crossover frequencies. For all practical purposes the RHP zero does not exist in DCM.

6. In integrated switchers the current limit is usually internal and unfortunately is also usually fixed. If we can't tailor the current limit to the specific application at hand,

and if our maximum load current is much lower than the maximum load current rating of the switcher, we can end up with excessive *overload margin*. This implies too much "headroom" between the peak operating current of our application I_{PEAK} and the set current limit of the switcher I_{LIM}. Under fault conditions (e.g., shorted output), too much energy may be outputted from the converter, causing a possible safety hazard and destruction of switch. This is of serious concern in a universal input Flyback in particular. In such cases, DCM helps a lot in reducing this overload margin. We will discuss this in greater detail below.

How DCM Duty Cycle Equations Are Calculated

We know that in CCM the duty cycle does not depend on the load current or inductance. In DCM the picture looks comparatively complicated at first sight. But the duty cycle can still be quite easily calculated from basics, as we show below in Fig. 4.1. In Table 4.1, we present the generalized forms of the key stress components. This table suffices completely, provided we carry out the calculation from top to bottom.

To calculate the duty cycle from first principles, we must consider the following: in DCM the diode conducts for a period designated as D'/f. So, just as in CCM, we designate the diode duty cycle as D' in DCM too. The difference is that because of the idle time when neither the switch nor the diode is conducting, $D' \neq 1 - D$, in DCM as is true in CCM. We have to work out the duty cycle from the volt-seconds law and from the fact that we need a certain load current I_O at the output. See the derivation in Fig. 4.1. Here we need to remember that the Buck topology is different in that to get the required load current, we have to average over the current flowing during both the switch times *and* the diode conduction times

RMS/AVG/duty cycle equations for DCM are valid only if $r > 2$. (Calculate from top to bottom.)	Buck	Boost	Buck-boost
V_{ON} (Voltage across inductor during switch conduction)	$V_{IN} - V_O - V_{SW}$	$V_{IN} - V_{SW}$	$V_{IN} - V_{SW}$
	V_{SW} is forward drop across switch, V_D is forward drop across diode		
	Same as in CCM		
V_{OFF} (Voltage across inductor during diode conduction)	$V_O + V_D$	$V_O - V_{IN} + V_D$	$V_O + V_D$
	Same as in CCM		
I_{O_CRIT} (Load current at critical conduction)	$r_{SET} \times I_{O\_MAX}/2$		
	r_{SET} is the set value of r at I_{O_MAX} (in CCM, see Appendix)		
D (Duty cycle in DCM)	$\sqrt{\dfrac{2 \times I_O \times L \times f \times V_{OFF}}{V_{ON} \times (V_{ON} + V_{OFF})}}$	$\sqrt{\dfrac{2 \times I_O \times L \times f \times V_{OFF}}{V_{ON}^2}}$	
D' (Diode duty cycle in DCM)	$\dfrac{2 \times I_O}{I_{PK}} - D$	$\dfrac{2 \times I_O}{I_{PK}}$	
M (DC transfer function V_O/V_{IN})	D	$\dfrac{1}{1-D}$	$\dfrac{D}{1-D}$
	where the value of D plugged in above is the CCM duty cycle		
	$\dfrac{2}{1+\sqrt{1+(4K/D^2)}}$	$\dfrac{1+\sqrt{1+(4D^2/K)}}{2}$	$\dfrac{D}{\sqrt{K}}$
	where $K = 2Lf/R$ and R is the load resistor ($= V_O/I_O$), and the value of D plugged in above is the DCM duty cycle calculated at load current I_O (where I_O is less than I_{O_CRIT})		

TABLE 4.1 DCM Stress Equations Summarized in Top-to-Bottom Calculation Order

RMS/AVG/duty cycle equations for DCM are valid only if r > 2. (Calculate from top to bottom.)	Buck	Boost	Buck-boost
I_{PEAK} (Peak current)	$\dfrac{V_{ON} \times D}{L \times f}$		
I_{SW_AVG} (Average switch current)	$\dfrac{I_{PEAK}}{2} D$		
I_{SW_RMS} (RMS of switch current)	$I_{PEAK}\sqrt{\dfrac{D}{3}}$		
I_{L_AVG} (Average inductor current)	$\dfrac{I_{PEAK}}{2}D' + \dfrac{I_{PEAK}}{2}D$		
	Also referred to as I_{DC}, I_L, I_{AVG} and so on		
I_{L_RMS} (RMS of inductor current)	$I_{PEAK}\sqrt{\dfrac{D'}{3} + \dfrac{D}{3}}$		
I_{D_AVG} (Average diode current)	$\dfrac{I_{PEAK}}{2} D'$		
I_{D_RMS} (RMS of diode current)	$I_{PEAK}\sqrt{\dfrac{D'}{3}}$		
I_{Cin_RMS} (RMS of input capacitor current)	$\sqrt{I_{SW\_RMS}^2 - I_{SW\_AVG}^2}$	$\sqrt{I_{L\_RMS}^2 - I_{L\_AVG}^2}$	$\sqrt{I_{SW\_RMS}^2 - I_{SW\_AVG}^2}$
	Same as in CCM		
I_{Co_RMS} (RMS of output capacitor current)	$\sqrt{I_{L\_RMS}^2 - I_{L\_AVG}^2}$	$\sqrt{I_{D\_RMS}^2 - I_{D\_AVG}^2}$	
	Same as in CCM		
I_{IN} (Input DC current)	I_{SW_AVG}	I_{L_AVG}	I_{SW_AVG}
	Same as in CCM		

TABLE 4.1 DCM Stress Equations Summarized in Top-to-Bottom Calculation Order (*Continued*)

(since energy flows to the output during both these intervals). But for the Boost and Buck-boost, we have to average *only* over the diode conduction time.

As mentioned in Chap. 2 on DC-DC converters and configurations, the Boost and Buck-boost are very similar topologies. Thus we see they share the same general form of the DCM duty cycle equation too. The only difference is that the applied voltages across the inductor during the ON-time (V_{ON}) and the OFF-time (V_{OFF}) are different.

Treatment of DCM in the Related Literature

In related literature (e.g., Erickson's book, *Fundamentals of Power Electronics*, 2d ed.), there are differences (and similarities) in terminology compared to ours. So we need to get that out of the way first.

1. What we call the full swing ΔI or $I_{AC}/2$, they call Δi_L.
2. What we call I_{DC} or I_L, they just call I.
3. What we call V_{IN} (input voltage), they call V_g.
4. What we call V_O, they just call V.
5. What we call f (switching frequency), they too call f.
6. What we call T (time period = $1/f$), they call T_s.
7. In CCM, what we call D or D_{CCM} here (switch duty cycle), they just call D.
8. In CCM, what we call D' (diode duty cycle), they too call D'.
9. In DCM, we call the switch duty cycle D or D_{DCM}; they call it D.

10. In DCM, we call the diode duty cycle D'; they call it D_2.
11. In DCM, we call the idle duty cycle D''; they call it D_3.

Their condition for critical (boundary) mode is $\Delta i_L = I$. Ours is $\Delta I/2$ ($= I_{AC}) = I_{DC}$ (more on our particular treatment follows later). They further "simplify" this (for a Buck) as follows

$$I = \frac{V}{R} = \frac{D \times V_g}{R} \quad \text{(since } D = \frac{V}{V_g} \text{ by definition and } R \text{ is load resistor)}$$

$$\Delta I = \frac{V \sec}{2 \times L} = \frac{V_O \times D'}{2 \times L \times f} = \frac{V_g \times D \times D'}{2 \times L \times f} \quad \left(\text{since } T_{OFF} = \frac{D'}{f} \text{ and } D = \frac{V}{V_g}\right)$$

So their mathematical condition for critical conduction is obtained by setting the two equal:

$$\frac{\cancel{D} \times \cancel{V_g}}{R} = \frac{\cancel{V_g} \times \cancel{D} \times D'}{2 \times L \times f} \Rightarrow D' = \frac{2 \times L}{R \times T_s}$$

They also then define $2L/RT_s$ as a *general* parameter called K.

$$K = \frac{2 \times L}{R \times T_s} \quad \left(= \frac{2 \times L}{V_O/I_O \times 1/f} = \frac{2 \times I_O \times L \times f}{V_O}\right)$$

This way of representation is meant to imply that the load resistor R, is *varying*, and that the exact value it has at critical conduction is R_{crit} leading to a value of K at critical conduction equal to K_{crit}. They also assume that since the duty cycle D varies as a function of load (in DCM), K_{crit} is actually a function of D. So eventually, they write

$$K \leq K_{crit}(D) \quad \text{for DCM}$$
$$K \geq K_{crit}(D) \quad \text{for CCM}$$
$$K_{crit}(D) = D'$$

After that, they also define $M(D, K)$ as the *conversion ratio* (which is just output voltage divided by input voltage). They calculate that (for a Buck) as

$$M = \begin{cases} D & \text{for } K > K_{crit}(D) \quad \text{Buck in CCM} \\ \dfrac{2}{1+\sqrt{1+4K/D^2}} & \text{for } K < K_{crit}(D) \quad \text{Buck in DCM} \end{cases}$$

Similarly, working through the equations of a Boost, they get

$$M = \begin{cases} \dfrac{1}{1-D} & \text{for } K > K_{crit}(D) \quad \text{Boost in CCM} \\ \dfrac{1+\sqrt{1+4D^2/K}}{2} & \text{for } K < K_{crit}(D) \quad \text{Boost in DCM} \end{cases}$$

And similarly, working through the equations of a Buck-boost, they get (ignoring the sign change)

$$M = \begin{cases} \dfrac{D}{1-D} & \text{for } K > K_{crit}(D) \quad \text{Buck-boost in CCM} \\ \dfrac{D}{\sqrt{K}} & \text{for } K < K_{crit}(D) \quad \text{Buck-boost in DCM} \end{cases}$$

They also then present some rather complicated-looking plots of $M(K, D)$ versus D. But do these plots mean much *intuitively*?

There is a better answer. Note that we know that the time constant of any inductor is L/R, where L is the series resistance. The units of this are seconds. So somewhat more intuitively, there are others (e.g., Scott Dearborn at Microchip) who define a dimensionless quantity called the *normalized inductor time constant*, symbolized by τ_L, as

$$\tau_L = \frac{L}{R \times T}$$

where, as usual, $T = 1/f$, with f being the switching frequency. This differs just a little from K because this new parameter is one-half of K (since $K = 2 \times \tau_L$). It turns out that τ_L is actually a rather nice, physically valid parameter to use in switching power supplies, because it indicates how things scale per the switching frequency, in a manner quite similar to the current ripple ratio r. For example, if, expressing everything in terms of τ_L, we double the frequency, that is, we halve the time period T, we intuitively realize we should halve the inductance—to keep the normalized time constant fixed.

> **NOTE** *Readers will remember that we defined r in CCM only. And it has physical meaning in the CCM region alone. As opposed to using Δ's, use of r does lead to very elegant-looking equations in CCM. But we realize there are no corresponding r-based equations for DCM. On the other hand, τ_L certainly has physical significance in both CCM and DCM regions. We can cast all CCM and DCM equations in terms of τ_L as we have done later in this chapter. The disadvantage of that approach is that even the CCM equations look far more complicated in the CCM region than in the corresponding r-based ones. The advantage, however, is that the τ_L-based equations can be written in both CCM and DCM (whereas the r-based equations are for CCM only).*

In this author's opinion, K is a still quite nonintuitive. It may also be part of the reason that people are so scared of DCM, not recognizing its underlying elegance and simplicity, and preferring to shun it completely. We try out a simple "top-to-bottom" calculation approach next.

Simplified Treatment of DCM Based on Optimal Setting of r

Unlike what the "K treatment" above suggests, the load current at which critical conduction occurs (I_{O_CRIT}) in an actual *practical* converter is in fact *independent of D* (and therefore independent of both input and output voltages), and it is also independent of the *topology*. We do ourselves no favor by plotting anything as a *function of D*, or by compounding the issue by arbitrarily creating K.

What is the alternative? By now we have come to recognize that converters of *all* topologies, used in *all* applications, and at *any* switching frequency, are typically designed for a typical r of 0.4 at maximum load. That is a current swing of ±20 percent. This also means that *a practical converter with a set r of, say, 0.4 at maximum load will go critical at exactly 20 percent of maximum load*. Similarly, if we had set r to 0.6, the converter would go critical at exactly 30 percent of maximum load. This is just plain waveform geometry. We see that critical conduction boundary does not depend on either the CCM duty cycle or the topology—for a *practical* converter.

Summing up, we *will always have a target current ripple ratio in any real converter as a starting design point*. It is therefore best to *fix that* first, get it out of the way, and then proceed into further CCM-DCM analysis, *starting from that reference point*. If we fail to do that, the *theoretical* possibilities are almost endless, and mostly immaterial too. They do not conform to any *practical* converter design that we will ever carry out or even study. So why study them?

Our simple condition of critical conduction is this:

$$I_{O\_CRIT} = \frac{I_{O\_MAX} \times r_{SET}}{2} \quad \text{(any topology, any application)}$$

Here r_{SET} is the set current ripple ratio at *maximum load* I_{O_MAX}. In the Appendix, we have all the equations connecting r to L (in CCM). We can plug those into the above condition to find the inductance. See Figs. 4.2 through 4.4 for the summarized duty cycle and critical conduction equations for all three topologies.

Tabulating the Stress Equations in DCM

What are we really interested in here? As mentioned, we are not planning to design a converter that works in DCM under *all* line and load conditions—no one does that because that implies r > 2, and that will *not* lead to good efficiency because of the excessively "peaky" currents and high corresponding RMS. So yes, we will design the converter to be in *CCM at maximum load* and nominal input voltage (design entry point). But having done that, we have two key reasons for wanting to know more about DCM, and we intend to discuss those shortly:

1. We want to know the current stresses (peak, average, and RMS) in the power components at light loads corresponding to DCM, so we have a *better understanding of the efficiency at light loads*. Then perhaps we can make a conscious decision if and when to introduce pulse-skipping and so on by suitable control circuitry.

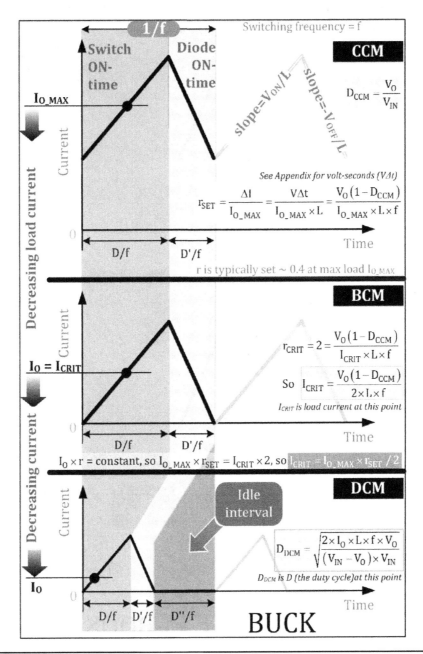

FIGURE 4.2 Duty cycle and critical conduction equations for Buck.

2. In universal input AC-DC Flybacks, we usually design the converter to be in CCM at low line (90 Vac). We ask: what happens at high line? We want to know whether DCM is entered, and if so, at what input voltage (and load current). In other words, we know that DCM is reached by reducing the load current, but we also want to understand how *raising the input voltage* may affect the critical conduction boundary. That can affect reliability too. We will see that, indeed, there is an indirect dependence on D in such *wide-input* converters.

Meanwhile, in Table 4.1 we present all the key DCM stress equations (all topologies) as a reference. These are in simple top-to-bottom calculation order, suitable for entering into Mathcad and plotting.

Plotting the Key Stresses in Going from CCM to DCM

In Figs. 4.5, 4.6, and 4.7 we have plotted the key stresses for a Buck converter (numerical example). In Fig. 4.8 we have compiled the key stresses for a Boost, and in Fig. 4.9 we have similarly plotted the key stresses for a Buck-boost. All are with a maximum load of 10 A.

116 Chapter Four

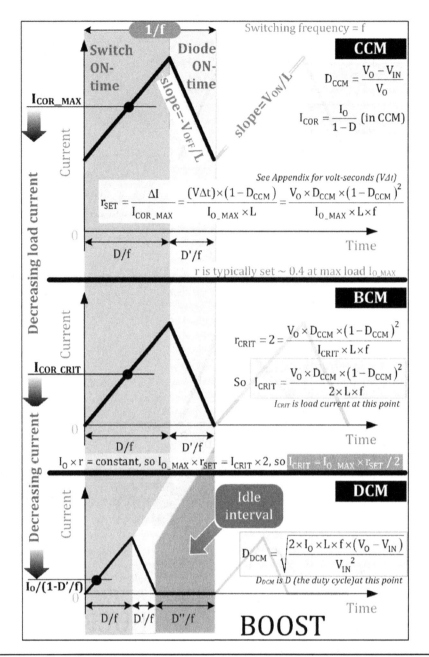

Figure 4.3 Duty cycle and critical conduction equations for Boost.

We have set r as 0.4 at maximum load in all cases, and we used Mathcad files to plot these. The equations used for the worksheet are the DCM ones found in Table 4.1 and the standard CCM ones found in the Appendix.

As expected, since the maximum load is 10 A in all cases, in all cases the converters go critical at exactly $10 \times 0.4/2 = 2$ A. Note that we have not bothered to mention the switching frequency f for any of these plots, because we realize f affects L (and vice versa), *but has no effect on anything else*, which highlights the beauty of talking in terms of r instead of constantly bringing up f and L separately. Here are a few observations based on the plots.

1. In Fig. 4.5, not only have we plotted the CCM equations in the CCM region and the DCM equations in the DCM region, but also we can see what the CCM equations would yield if *they were valid* in the DCM region ("CCM equations projected"). Similarly, we can see what the DCM equations would yield in the CCM region, if they were valid. We have also drawn the final valid stress curves (in black and bold lines), consisting of the relevant CCM equation in the CCM region and the DCM equation in the DCM region. We note there is *no discontinuity* as we traverse the critical boundary, which gives us confidence in the DCM equations presented in Table 4.1.

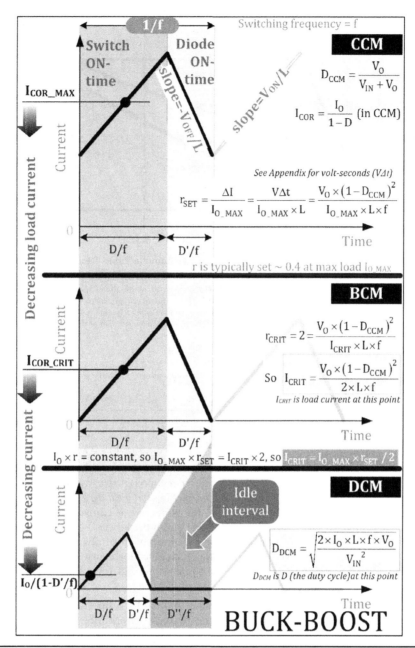

FIGURE 4.4 Duty cycle and critical conduction equations for Buck-boost.

2. We see that the CCM equations would have yielded *higher* RMS and peak currents *had they been valid*. This is a surprising result to many engineers who instinctively say, "DCM makes waveforms more peaky, leading to higher RMS stresses." Yes, at *a given load* that is true, but it is entirely untrue if we vary the load. The peak current actually decreases in the DCM region because the duty cycle *pinches off*, leaving less time for the current to ramp up and peak further. So in all cases, the switch RMS current (based on DCM equations) is similarly less in the DCM region.

3. In Fig. 4.6 we compare how the switch and diode duty cycles pinch off in DCM. Note that on a log versus log scale, the variation is linear. The DCM duty cycle always *varies 10-fold for every 100-fold variation in load* (all topologies). That can be traced back to the fact that both the DCM duty cycles are proportional to $\sqrt{I_O}$. We have some sample calculations embedded in the figure, to find out the duty cycle at any current, based on these rules.

4. In Fig. 4.7 we see how the input capacitor RMS varies (the output capacitor RMS is insignificant in a Buck and has been omitted). Once again we see that the DCM equations yield a lower stress (in the DCM region) than the CCM equations projected into the DCM region.

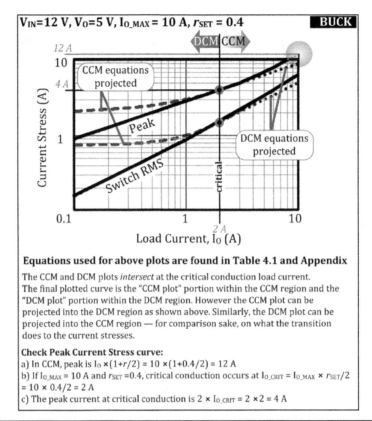

FIGURE 4.5 Peak and switch RMS currents plotted for a Buck.

5. We also see what the average diode current looks like as we vary the load. We realize that for the Buck and the other topologies, as we will see shortly, *the CCM and DCM equations for the diode average current yield the same result on either side of the critical boundary*. So that makes the *average diode current stress unique*. With a little thought we realize that for a Boost and a Buck-boost, this fact is relatively easy to visualize because the diode average current must always equal the load current, whether in CCM or DCM. And so, for a given load current, both the CCM and DCM equations must "point" to the same value. But for a Buck it is a little harder to visualize but nonetheless true, as we can see.

6. In Fig. 4.8 we have the key stresses plotted for a Boost. The output capacitor RMS current is of concern here, not the input capacitor RMS, so the latter is omitted here.

7. In Fig. 4.9 we have the key stresses for a Buck-boost. Both the input and output capacitor RMS values are important for this topology, and *both* are therefore plotted.

Varying the Input Voltage in a Buck-boost

We know that lowering I_O causes the system to enter DCM. What about raising the voltage? We are most interested in the Buck-boost here, because it relates to the behavior of a universal input Flyback discussed next.

In Fig. 4.10 we have used the same Buck-boost Mathcad file that was used for Fig. 4.9 to plot how r, the instantaneous current ripple ratio, varies as a function of input voltage. Note that we have fixed r as a starting point to $r_{SET} = 0.4$ at maximum load and at minimum voltage, as we always do for this topology. Then we increase the input voltage for different constant load currents. The intersection of these curves with the $r = 2$ boundary gives us V_{IN_CRIT} for that particular constant load current.

In Fig. 4.10 we have also compared the graphical results based on the Mathcad worksheet against a closed-form equation derived after some serious mathematical manipulation. That equation is

$$V_{IN\_CRIT} = V_O \times \left[\sqrt{\frac{1}{\gamma^2} + \frac{1}{\gamma}} + \frac{1}{\gamma}\right] \quad \text{where} \quad \gamma = \left(\frac{V_O}{2 \times I_O \times L \times f}\right) - 1$$

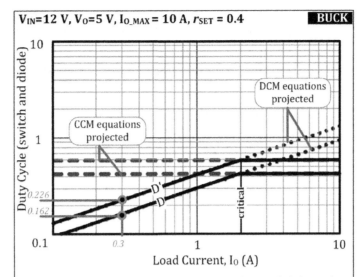

FIGURE 4.6 Diode and switch duty cycle equations plotted for a Buck.

Alternatively, using K defined earlier (from related literature)

$$K = \frac{2 \times L}{R \times T_s} = \frac{2 \times I_O \times L \times f}{V_O}$$

we get

$$V_{IN\_CRIT} = V_O \times \left(\frac{K + \sqrt{K}}{1 - K} \right)$$

Studying the Universal Input Flyback

The Flyback is essentially a transformer-based Buck-boost. However, there are some fine differences too. A key difference is that the input voltage of a typical universal input Flyback is not a steady DC, but has a fair amount of input voltage ripple, especially at low line, due to the peak charging and discharging of the input bulk capacitor. At 85 Vac, using 5 μF/W of input capacitance, it can be shown that the lowest point of this voltage ripple is close to 105 Vdc. So that is the V_{INMIN} selected for setting r (to r_{SET}) as a starting point in the curves presented in Fig. 4.11.

We plot results for different set values of r (r is the horizontal axis in Fig. 4.11).

We also plot results for two converters: a 70-W maximum-rated power supply and a 35-W maximum-rated power supply.

120   Chapter Four

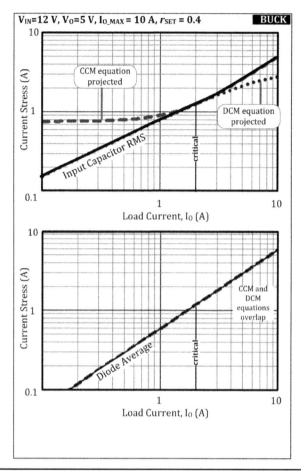

Figure 4.7 Input capacitor RMS and average diode currents plotted for a Buck.

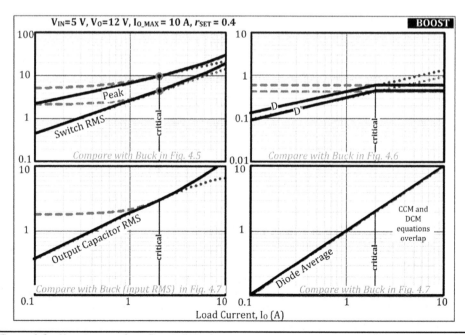

Figure 4.8 Key stresses for a Boost plotted.

Understanding and Using Discontinuous Conduction Mode

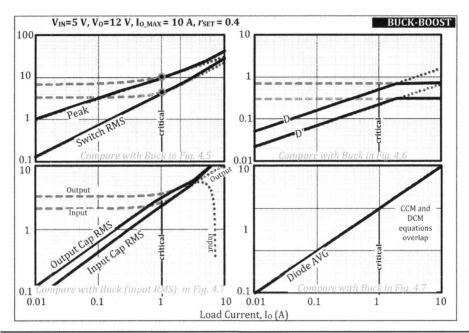

FIGURE 4.9 Key stresses for a Buck-boost plotted.

We need to fix the output voltage of the converter. We remember that for all practical purposes, using the primary-side equivalent Buck-boost model, the output voltage of the Flyback is V_{OR}, the reflected output voltage. We therefore do the analysis for two cases: V_{OR} = 105 V (typically chosen for Flybacks with a 600-V FET) and V_{OR} = 130 V (typically chosen for Flybacks with a 700-V FET).

While setting r at V_{INMIN}, we also need to know the maximum load current at which to set r. The effective output load current of the primary-side equivalent Buck-boost is P_O/V_{OR}. So, e.g., if V_{OR} = 105 V, then a 70-W Flyback power supply has an I_{O_MAX} of 70/105 = 0.67 A. The same 70-W Flyback with a selected V_{OR} of 130 V has an I_{O_MAX} of 70/130 = 0.54 A. These are the values used for generating Fig. 4.11. In other words, we are fixing the power supply unit (PSU) *wattage* for these curves, not the maximum load current per se as we do for a simple Buck-boost.

After setting r, we raise the input voltage and plot the *critical conduction voltage* curves, using the closed-form equation given in the previous section. We have plotted these for several r values (on the horizontal axis). We also change the actual power output of the PSU in discrete steps, and we use that as a parameter for the plots. For example, using the 70-W maximum-rated PSU, we plotted the results for several possible output wattages: 70, 60, 50, ..., 10 W. Similarly, using the 35-W PSU, we plotted the results for 35, 30, 25, ..., 5 W. By varying the power we want to see by how much we need to lower the load current/wattage to bring the converter into DCM.

For example, we see that the 70-W PSU, with V_{OR} = 105 V, when operated at 70 W, designed with r = 0.4, has a critical conduction voltage that is very high: the projected intersection of the two dashed arrowed curves. In other words, for all practical purposes, at 70-W loading, we will *never* encounter DCM *if we set r to 0.4*. However, if we lower the load on this particular 70-W PSU (with r_{SET} = 0.4) to about 35 W, we *will* start seeing DCM appear within the shaded region of Fig. 4.10 (which is the rectified operating European voltage range).

Note that if initially we had set r to 0.6 (at V_{INMIN} = 105 V and I_{O_MAX} = 0.67 A, as before) and lowered the load to about 52 W, then we would start to see DCM appear within the normal European operating range. If we had set r to about 0.82, we would start to see DCM at the high line even at the full maximum-rated load (70 W).

We realize the curves presented in the figure are indeed powerful and meaningful. We also notice that *the plots barely depend on V_{OR}, or on the maximum-rated wattage*. So, with some intelligence, looking at the curves in terms of the *ratio* of actual load to the maximum-rated load, we can apply these to almost any maximum-rated universal input Flyback PSU.

Overload Margin of the Universal Input Flyback

One of the most effective ways of destroying a poorly designed universal input Flyback is to go at high line and do a sudden and severe overload (perhaps just shorting the output).

122 Chapter Four

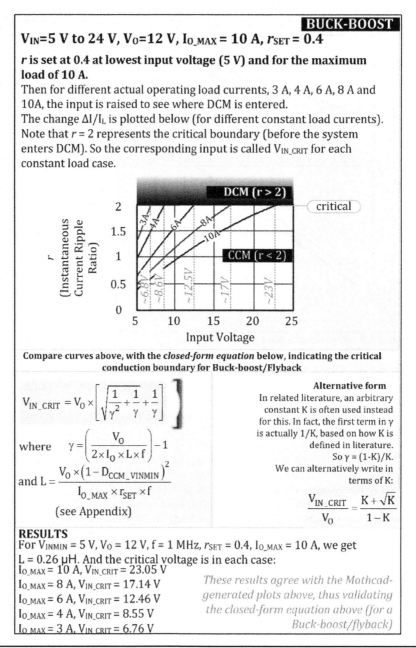

FIGURE 4.10 Critical conduction boundary and input voltage.

Consider a practical universal input AC-DC Flyback converter as an example. Its core size is dictated by the conditions at low line (maximum D), and that is the point at which its current limit is set per the normal design strategy. But a problem occurs *at high line*. Under normal operation, as the input increases, D decreases, and so the center of the inductor current waveform falls, since it goes as $I_O/(1-D)$ when operated in CCM. Basically, we end up with excessive *overload margin* at high line (see Fig. 4.12). Under abnormal load conditions, the energy ($V_{IN} \times I_{IN}$) that can accidentally flow in from the mains (and get delivered to the load) is potentially enormous. This can cause a safety hazard. Further, when using an RCD clamp with the Flyback, we are actually *counting on a lower primary-side current to lower the clamping voltage* and thus stay within the maximum voltage rating of the switch at high line. So, if under a fault condition the current goes to the full value (based on current limit set at low line), then the clamp voltage will become suddenly very large, applying a much higher voltage on the Drain of the switching FET. At high line, this can be disastrous. Therefore, many engineers have come up with unique *shaped current limits* that basically progressively reduce the switch current limit as the input increases so as to virtually "track" the natural reduction in the peak current of the Flyback, staying just a little above it. However, as Fig. 4.12 reveals, we can dramatically reduce the overload

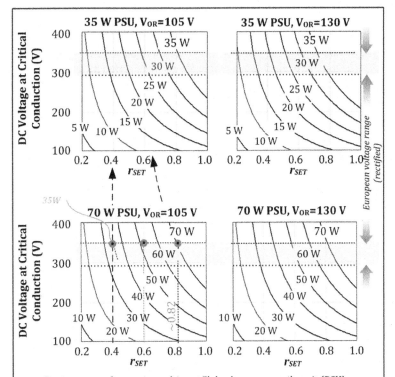

Design curves for a universal-input Flyback power supply unit (PSU). We set the r at V_{INMIN} of 105 V (minimum rectified DC input) *at maximum load* (based on the PSU rated wattage of 35 W or 70 W). The shaded area is the range of the rectified spread of European line voltage (assumed 292 VDC to 360 VDC).

How to interpret these curves

Example: A 70 W PSU has a typical V_{OR} set to 105 V, and is designed with a typical r of 0.4, set at maximum load and V_{INMIN} (105 VDC assumed here). To reach critical conduction (DCM), we have to raise the input voltage to a very high (unachievable) voltage (the intersection of the two dashed *arrowed* curves above). We conclude that *at maximum load* we will be in CCM throughout the normal operating input range. But if we reduce output wattage of the 70 W PSU to ~ 35 W we will start to get DCM (see above). We can also conclude that we can set r (at maximum load) right *up to about 0.82* without seeing CCM over entire input range (see above). If we set r any higher, we will start seeing DCM mode within the input range (even at maximum load).

FIGURE 4.11 Useful design curves for universal input Flyback power supplies.

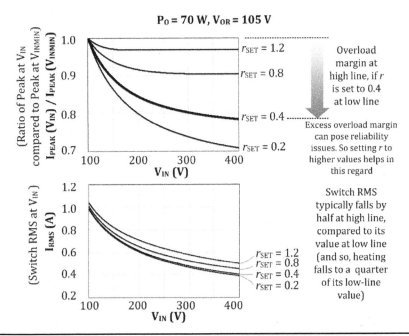

FIGURE 4.12 Overload margin and RMS variation at high line for universal input Flyback.

margin by just setting the *r* at low line to a much higher value than the optimum 0.4 (say, to 0.8 to 1.2). However, this has surprisingly little impact on the switch dissipation (the switch RMS is also plotted in the figure). So DCM is not such a bad idea in this application. As mentioned, it also helps reduce the core size. But watch out for high input and output capacitor RMS currents.

Closed-Form Equations for Buck CCM and DCM Compared

As mentioned earlier, there are engineers who use a parameter called τ_L, and this has the advantage that we can, at some cost of elegance, cast all the stress equations in both CCM and DCM in a closed-form equation format, rather than the top-to-bottom format presented earlier in Table 4.1. So this approach leads us to Table 4.2, where we have provided the (rather horrendous) equations for a Buck.

RMS/AVG/duty cycle equations for DCM are valid only if $r > 2$, or equivalently if $\tau_L < (1-M)/2$	Buck (CCM)	Buck (DCM)
I_{O_CRIT} (Load current at critical conduction)	$r_{SET} \times I_{O\_MAX}/2$	
	r_{SET} is the set value of r at I_{O_MAX} (in CCM, see Appendix)	
r (Current ripple ratio)	$\dfrac{\Delta I_L}{I_O}$	
	r is physically meaningful only in CCM, but we can calculate it in DCM too	
τ_L (Normalized inductor time constant)	$\dfrac{L}{R \times T}$	
	where $T = 1/f$ (Note: $\tau_L = K/2$; it is the normalized inductor time constant.)	
$\dfrac{V_O}{R} I_{PEAK}$	$I_O + \dfrac{\Delta I_L}{2}$	ΔI_L
	$I_O \left(1 + \dfrac{r}{2}\right)$	N.A.
	$\dfrac{V_O}{R} \times \left[1 + \dfrac{1-M}{2 \times \tau_L}\right]$	$\dfrac{V_O}{R} \times \sqrt{\dfrac{2 \times (1-M)}{\tau_L}}$
	where $M =$ output/input $= V_O/V_{IN}$ and $\tau_L = L/(RT)$; R is the load resistor $= V_O/I_O$	
ΔI_L (Inductor current swing)	$\dfrac{V_O \times D'}{L \times f}$	
	D' is the diode conduction duty cycle in CCM and DCM	
	$r \times I_O$	N.A.
	$\dfrac{V_O}{R} \times \left[\dfrac{1-M}{\tau_L}\right]$	$\dfrac{V_O}{R} \times \sqrt{\left[\dfrac{2 \times (1-M)}{\tau_L}\right]}$
D (Switch duty cycle)	M	$M \times \sqrt{\dfrac{2 \times \tau_L}{1-M}}$
D' (Diode duty cycle)	$1 - M$	$\sqrt{2 \times \tau_L \times (1-M)}$
M (DC transfer function V_O/V_{IN})	D	$\dfrac{2}{1 + \sqrt{1 + (8 \times \tau_L / D^2)}}$
	where $\tau_L = Lf/R$ and $R = V_O/I_O$ (load resistor)	

TABLE 4.2 Buck CCM and DCM Stress Equations Tabulated in Closed Form

Understanding and Using Discontinuous Conduction Mode

RMS/AVG/duty cycle equations for DCM are valid only if $r > 2$, or equivalently if $\tau_L < (1-M)/2$	Buck (CCM)	Buck (DCM)
I_{SW_AVG} (Average switch current)	$I_0 \times D$	$\dfrac{I_{PEAK}}{2} D$
	$\dfrac{M \times V_O}{R}$	
I_{SW_RMS} (RMS of switch current)	$I_0 \times \sqrt{D \times \left(1 + \dfrac{r^2}{12}\right)}$	$I_{PEAK}\sqrt{\dfrac{D}{3}}$
	$\dfrac{V_O}{R} \times \sqrt{M \times \left[1 + \dfrac{1}{12} \times \left(\dfrac{1-M}{\tau_L}\right)^2\right]}$	$\dfrac{V_O}{R} \times \sqrt{\dfrac{8 \times M^2 \times (1-M)}{9 \times \tau_L}}$
I_{L_AVG} (Average inductor current)	I_0	$\dfrac{I_{PEAK}}{2}D' + \dfrac{I_{PEAK}}{2}D$
	$\dfrac{V_O}{R}$	
I_{L_RMS} (RMS of inductor current)	$I_0 \times \sqrt{1 + \dfrac{r^2}{12}}$	$I_{PEAK}\sqrt{\dfrac{D'}{3} + \dfrac{D}{3}}$
	$\dfrac{V_O}{R}\sqrt{1 + \dfrac{1}{12} \times \left(\dfrac{1-M}{\tau_L}\right)^2}$	$\dfrac{V_O}{R} \times \sqrt{\dfrac{8 \times (1-M)}{9 \times \tau_L}}$
I_{D_AVG} (Average diode current)	$I_0 \times (1-D)$	$\dfrac{I_{PEAK}}{2}D'$
	$\dfrac{V_O}{R} \times (1-M)$	
I_{D_RMS} (RMS of diode current)	$I_0 \times \sqrt{(1-D) \times \left(1 + \dfrac{r^2}{12}\right)}$	$I_{PEAK}\sqrt{\dfrac{D'}{3}}$
	$\dfrac{V_O}{R} \times \sqrt{(1-M) \times \left[1 + \dfrac{1}{12} \times \left(\dfrac{1-M}{\tau_L}\right)^2\right]}$	$\dfrac{V_O}{R} \times \sqrt{\dfrac{8 \times (1-M)^3}{9 \times \tau_L}}$
I_{Cin_RMS} (RMS of input capacitor current)	$\sqrt{I_{SW\_RMS}^2 - I_{SW\_AVG}^2}$	
	$I_0 \times \sqrt{D \times \left(1 - D + \dfrac{r^2}{12}\right)}$	N.A.
	$\dfrac{V_O}{R} \times \sqrt{M \times \left[(1-M) + \dfrac{1}{12} \times \left(\dfrac{1-M}{\tau_L}\right)^2\right]}$	$\dfrac{V_O}{R} \times \sqrt{M \times \left(\dfrac{8 \times (1-M)}{9 \times \tau_L}\right)^{1/2} - M^2}$
I_{Co_RMS} (RMS of output capacitor current)	$\sqrt{I_{L\_RMS}^2 - I_{L\_AVG}^2}$	
	$I_0 \times \dfrac{r}{\sqrt{12}}$	N.A.
	$\dfrac{V_O}{R} \times \left[\dfrac{1-M}{\sqrt{12} \times \tau_L}\right]$	$\dfrac{V_O}{R} \times \sqrt{\dfrac{8 \times (1-M)}{9 \times \tau_L} - 1}$
I_{IN} (Input DC current)	I_{SW_AVG}	
	$\dfrac{M \times V_O}{R}$	

TABLE 4.2 Buck CCM and DCM Stress Equations Tabulated in Closed Form (*Continued*)

RMS/AVG/duty cycle equations for DCM are valid only if $r > 2$. (Calculate from top to bottom)	Buck	Boost	Buck-boost
V_{ON} (Voltage across inductor during switch conduction)	$V_{IN} - V_O - V_{SW}$	$V_{IN} - V_{SW}$	$V_{IN} - V_{SW}$
	V_{SW} is forward drop across switch, V_D is forward drop across diode		
	Same as in CCM		
V_{OFF} (Voltage across inductor during diode conduction)	$V_O + V_D$	$V_O - V_{IN} + V_D$	$V_O + V_D$
	Same as in CCM		
I_{O_CRIT} (Load current at critical conduction)	$r_{SET} \times I_{O\_MAX} / 2$		
	r_{SET} is the set value of r at I_{O_MAX} (in CCM, see Appendix)		
D (Duty cycle in DCM)	$\sqrt{\dfrac{2 \times I_O \times L \times f \times V_{OFF}}{V_{ON} \times (V_{ON} + V_{OFF})}}$	$\sqrt{\dfrac{2 \times I_O \times L \times f \times V_{OFF}}{V_{ON}^2}}$	
α (Ratio of load to peak currents)	$\dfrac{I_O}{I_{PK}}$		
D' (Diode duty cycle in DCM)	$2\alpha - D$	2α	
M (DC transfer function V_O/V_{IN})	D	$\dfrac{1}{1-D}$	$\dfrac{D}{1-D}$
	where the value of D plugged in above is the CCM duty cycle		
	$\dfrac{2}{1+\sqrt{1+(4K/D^2)}}$	$\dfrac{1+\sqrt{1+(4D^2/K)}}{2}$	$\dfrac{D}{\sqrt{K}}$
	where $K = 2Lf/R$ and R is the load resistor ($= V_O/I_O$), and the value of D plugged in above is the DCM duty cycle calculated at load current I_O (where I_O is less than I_{O_CRIT})		
I_{PEAK} (Peak current)	$\dfrac{I_O}{\alpha}$		
I_{SW_AVG} (Average switch current)	$\dfrac{I_O D}{2\alpha}$		
	where the value of D plugged in above is the DCM duty cycle		
I_{SW_RMS} (RMS of switch current)	$\dfrac{I_O}{\alpha}\sqrt{\dfrac{D}{3}}$		
I_{L_AVG} (Average inductor current)	I_O	$\dfrac{I_O(2\alpha + D)}{2\alpha}$	
	Also referred to as I_{DC}, I_L, I_{AVG} and so on		
I_{L_RMS} (RMS of inductor current)	$\dfrac{I_O}{\sqrt{1.5 \times \alpha}}$	$\dfrac{I_O\sqrt{2\alpha + D}}{\alpha\sqrt{3}}$	
I_{D_AVG} (Average diode current)	$I_O\left(1 - \dfrac{D}{2\alpha}\right)$	I_O	
I_{D_RMS} (RMS of diode current)	$\dfrac{I_O\sqrt{2\alpha - D}}{\alpha\sqrt{3}}$	$\dfrac{I_O\sqrt{2}}{\sqrt{3\alpha}}$	
I_{Cin_RMS} (RMS of input capacitor current)	$\dfrac{I_O\sqrt{D(4-3D)}}{2\alpha\sqrt{3}}$	$\dfrac{I_O\sqrt{4\alpha(2-3\alpha-3D)+D(4-3D)}}{2\alpha\sqrt{3}}$	$\dfrac{I_O\sqrt{D(4-3D)}}{2\alpha\sqrt{3}}$
I_{Co_RMS} (RMS of output capacitor current)	$\dfrac{I_O\sqrt{1-1.5\alpha}}{\sqrt{1.5 \times \alpha}}$		
I_{IN} (Input DC current)	$\dfrac{I_O D}{2\alpha}$	$\dfrac{I_O(2\alpha + D)}{2\alpha}$	$\dfrac{I_O D}{2\alpha}$

TABLE 4.3 DCM Stress Equations Summarized in Closed-Form Format

Closed-Form Equations for Buck, Boost, and Buck-Boost DCM

Summarizing, in this book we have defined α, which we believe to be a more elegant parameter as the ratio of the load current to the peak current in all three topologies. We cast all the DCM equations in closed-form format, using this new parameter, in Table 4.3. This is a quick and easy lookup chart for evaluating losses and computing efficiency in DCM.

Summarizing, we have three design tables to refer to in this chapter. They have been thoroughly cross-checked against one another using detailed Mathcad files and are all self-consistent and accurate.

CHAPTER 5

Comprehensive Front-End Design in AC-DC Power Conversion

PART 1 FRONT-END WITH NO POWER FACTOR CORRECTION

Flyback and the Search for Closed-Form Equations

In Fig 5.1 we start with a Flyback at 85 Vac. Under normal operation, the peak rectified voltage is 85 Vac $\times \sqrt{2}$ = 120 Vdc. We are ignoring the couple of diode drops across the bridge rectifier. Between successive rectified AC peaks, for the majority of the time, there is no current through the bridge and no charging current into the input capacitor. Because, though the voltage on the capacitor keeps falling as it struggles to "single-handedly" provide energy to the converter, the AC (shown as the rectified sine wave in Fig. 5.1) falls off even faster (usually, for reasonably large capacitances). So the bridge is reverse-biased. However, at some point, the AC voltage rises again, and in the process, the bridge is once again forward-biased. The AC tries to "overtake" the capacitor voltage but, in reality, lifts it up, along with itself. It achieves this by charging up the capacitor with a *very high surge of input current*. We must understand and carefully model this occurence because it is *the key to limiting line harmonics, ensuring efficacy of the EMI (electromagnetic interference) filter, and also guaranteeing the expected lifetime of the input capacitor*.

To know the exact current waveform, we first need to know the voltages accurately. Under normal operation, the capacitor voltage follows the rising AC voltage right up to the peak AC voltage, getting charged up in the process. During the following time interval when the capacitor is again discharging, its voltage follows a *constant-power* path, which is characterized by an increasingly steeper downward slope as time elapses. This characteristic nose-dive of sorts is related to the fact that the converter stage demands an almost constant input power (because its output power is assumed constant too). So as the capacitor voltage falls, it demands an increasing current to keep the instantaneous $V \times I$ product constant. But the more current it demands, the faster the capacitor voltage falls and, then the more current it draws and so on. It is in effect, quite close to a runaway situation. If the capacitance is too small, the extreme possibility is the capacitor voltage will fall so fast that it falls off even faster than the AC voltage waveform: Which is the point at which we should be getting ready to consign our Flyback to the dustbin? At a more benign level, the capacitor voltage may just reach the set (undervoltage lockout) UVLO threshold (typically 60 Vdc by design, but sometimes set to 80 Vdc) before the AC can overtake it and keep the switching fully sustained. Because if UVLO is reached first, the converter will stop working momentarily and will thereafter likely need a fresh start-up sequence. The power supply may sputter dangerously, and that is not normal operation.

By maintaining a minimum capacitance, we ensure this sputtering doesn't happen under normal operation. In Fig. 5.1 we have presented Mathcad-generated curves for 3 µF/W and 5 µF/W in particular. We will see that these values are enough to keep the capacitor voltage above UVLO, for the case of typical missing AC half-cycles, as discussed shortly. Note that by writing capacitance in this way (per watt), we can *scale the results and conclusions for any power level*. So, for example, for a 100-W power supply, 3 µF/W means 300 µF of input capacitance. But keep in mind that the *input* capacitor is only aware of the *input* power and needs to store that energy. In reality, 3 µF/W is more accurately 3 µF per *input* power, not output power, as is sometimes wrongly assumed.

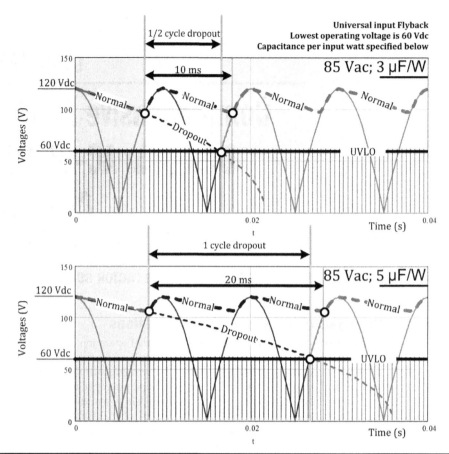

Figure 5.1 Universal input Flyback at 85 Vac with different input capacitances per watt.

We need to calculate the required input capacitance as it relates to the input surge current under normal operation, but there are also *power quality issues* to contend with in AC applications. For example, especially if we are in an industrial area, some heavy electrical equipment may suddenly activate, causing a huge draw of current and a severe dip in the local AC voltage. In effect, we say, that one or two half-cycles of AC may go entirely "missing" (or too low to count anyway). But we do not want our nearby office equipment, say a printer or desktop computer (a laptop will have a battery for holdup), to get reset mysteriously. We therefore need to guarantee a minimum *holdup time*.

Most power supply specifications ask for a fixed 10-ms or 20-ms holdup time. What they *should* be asking for, based on their obvious intent and the behavior of the incoming AC after a missing AC half-cycle, is really support for one missing half-cycle or two missing half-cycles. Yes, at a 50-Hz AC line frequency, two half-cycles really does seem to mean 20 ms, but as we see in Fig. 5.1, designing for one missing half-cycle will likely be noticeably cheaper (lower capacitance) than designing for a fixed 10-ms dropout. These are not the same intervals because as we notice from the figure, the AC overtakes the falling capacitor voltage a little *before* 10 ms (for the case of one half-cycle missing) and a little *before* 20 ms (for one cycle missing). High-voltage input capacitors (for AC) are expensive, so it makes economic sense to design the front-end without any fat (flab). No expensive overdesign here please.

Notice carefully that we are counting the 10 ms or 20 ms as starting from the worst-case point of the normal operating ripple, i.e., just before the AC comes up and starts charging up the capacitor again every half-cycle. Some people, however, start their dropout (holdup) timer from the highest (peak position) of normal operating voltage ripple. So, in effect, when they say "20-ms holdup," they mean only one missing half-cycle! That makes it doubly important to clarify the requirement in terms of missing half-cycles, not some arbitrary time interval (measured from where to where?). In our discussions in most of this chapter, whenever we say "~10 ms" we mean one missing half-cycle, and "~20 ms" means two missing half-cycles. In the case of power factor correction (PFC) we may have stated and defined it somewhat differently, but there is barely any difference in the end results for that case. Also, as we will see, the holdup time requirement for PFC front-end stages does not eventually affect the bulk capacitor selection in practical power supplies, so the point is moot.

In the case of no PFC, we want to *accurately* compute the capacitace, as it could be dollars saved. However, it is increasingly evident that the solution is likely to be rather approximate and perhaps iterative too. Yes, we could write a Mathcad file that would carefully compute the locus of the constant power curve accurately, and thus determine very accurately the shape of the input current waveform under normal operation, and hopefully also the required minimum capacitance for a certain holdup time. *But is there a closed-form equation that is accurate enough*, at least for computing the current stresses during normal operation, if not for dropouts and holdup time considerations?

Unfortunately, the equations abounding in related literature usually give results that are not even in the ±20 percent ballpark of the results from the (undeniably accurate) Mathcad file. In other words, the front-end design is usually not done with a great deal of accuracy. This needs to be corrected. For holdup, we can simply present accurate rules of thumb for capacitance based on the Mathcad file, as we will soon see.

This author found fairly accurate equations in one recent source at least: Christopher Basso's new book on simulation from McGraw-Hill titled *Switch-Mode Power Supplies Spice Simulations and Practical Designs*. But the author also discovered that those typical industry-wide results could be significantly improved by reiterating the equations in a certain way, as will be subsequently explained further. The final results show a very good match with the Mathcad file for the case of normal operation—excluding holdup computations for which we are not seeking closed-form equations here.

To aid in our subsequent discussions here, we have introduced some related terminology in Fig. 5.2. We also show the input current waveshape as derived in this author's book *Switching Power Supplies A-Z*, 2d ed. We have so far just shown the current waveshape through the bridge. That is actually very similar to the current through the input capacitor, but not quite, since the capacitor has zero direct current through it, so in effect, it DC-offsets the bridge current slightly. Here, V_{SAG0} (or just V_{SAG}), is the lowest voltage the capacitor voltage sags to before it rises, under normal operation (0 half-cycles dropout). V_{SAG1} is the lowest voltage for one half-cycle missing, V_{SAG2} is that for two half-cycles, and so on.

Droop Curves for Different Capacitances per Watt

In Fig. 5.3 we show Mathcad-computed graphs for a Flyback at 85 Vac, assuming Flyback UVLO is set to 60 Vdc. We conclude that if we want the converter to support one missing AC half-cycle (~ 10-ms dropout), we need to have at least 3 µF/W. For two missing AC half-cycles (~ 20-ms dropout), we need to have 5 µF/W.

In Fig. 5.4 we present the numbers behind these curves and computations. Besides telling us the minimum capacitor voltage V_{SAGx}, we also learn what exactly happens for normal operation. We have concluded that since the input voltage always has some ripple, for the purpose of, say, doing efficiency calculations, we should take the average value of the input ripple. So, for the case of 3 µF/W we should take 108 Vdc as the input of our DC-DC converter stage, and 113 Vdc for 5 µF/W.

Different UVLO?

We are obviously dealing with a design compromise here. To reduce input capacitance, we need to set the UVLO lower, and this may have implications like larger core size (unless we consciously design the transformer for $r = 0.4$ at 60 Vdc, and in the process face more

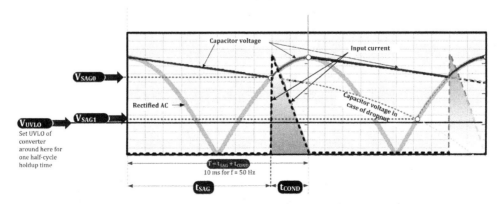

FIGURE 5.2 Input current shape and some terminology used here.

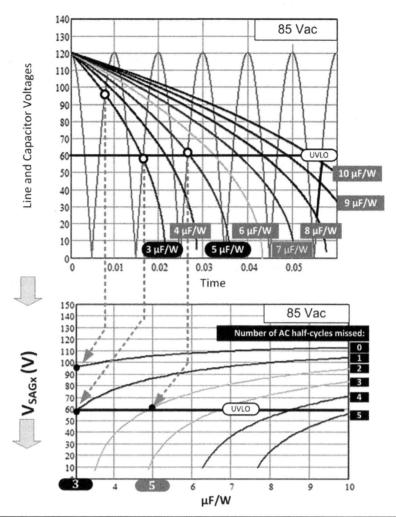

FIGURE 5.3 Droop curves for different capacitances per watt.

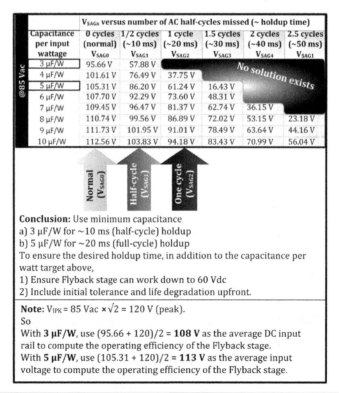

Capacitance per input wattage	V_{SAGx} versus number of AC half-cycles missed (~ holdup time)					
	0 cycles (normal) V_{SAG0}	1/2 cycles (~10 ms) V_{SAG1}	1 cycle (~20 ms) V_{SAG2}	1.5 cycles (~30 ms) V_{SAG3}	2 cycles (~40 ms) V_{SAG4}	2.5 cycles (~50 ms) V_{SAG5}
3 µF/W	95.66 V	57.88 V				
4 µF/W	101.61 V	76.49 V	37.75 V			
5 µF/W	105.31 V	86.20 V	61.24 V	16.43 V	No solution exists	
6 µF/W	107.70 V	92.29 V	73.60 V	48.31 V		
7 µF/W	109.45 V	96.47 V	81.37 V	62.74 V	36.15 V	
8 µF/W	110.74 V	99.56 V	86.89 V	72.02 V	53.15 V	23.18 V
9 µF/W	111.73 V	101.95 V	91.01 V	78.49 V	63.64 V	44.16 V
10 µF/W	112.56 V	103.83 V	94.18 V	83.43 V	70.99 V	56.04 V

Conclusion: Use minimum capacitance
a) 3 µF/W for ~10 ms (half-cycle) holdup
b) 5 µF/W for ~20 ms (full-cycle) holdup
To ensure the desired holdup time, in addition to the capacitance per watt target above,
1) Ensure Flyback stage can work down to 60 Vdc
2) Include initial tolerance and life degradation upfront.

Note: V_{IPK} = 85 Vac × √2 = 120 V (peak).
So
With **3 µF/W**, use (95.66 + 120)/2 = **108 V** as the average DC input rail to compute the operating efficiency of the Flyback stage.
With **5 µF/W**, use (105.31 + 120)/2 = **113 V** as the average input voltage to compute the operating efficiency of the Flyback stage.

FIGURE 5.4 Mathcad-generated numbers for normal operation and missing half-cycles.

"peaky," higher-RMS currents, under normal operation). What happens if we decide to set the UVLO to 80 Vdc instead of 60 Vdc? By accurate Mathcad estimates we will learn that

1. For the case of one missing half-cycle, we need a minimum capacitance of 4.3 µF/W (roughly 4.5 µF/W), instead of 3 µF/W for the 60 Vdc case.
2. For the case of two missing half-cycles (i.e., one full cycle), we need a minimum capacitance of 6.9 µF/W (roughly 7 µF/W), instead of 5 µF/W for 60 Vdc case.

This is around a 40 percent increase in capacitance for a 33 percent increase in UVLO.

Entry Point into Iteration for Closed-Form Equations for Normal Operation

This is only for normal operation (no missing half-cycles). Also the key parameter is bridge conduction time t_{COND}. Once we know that, we can calculate all other things. Here is a quick derivation of the basic parameters. Others are derived in *Switching Power Supplies A-Z*, 2d ed.

During the period just past the AC peak, the capacitor is the only source of input power to the switching converter. So the discharge is

$$\frac{1}{2} C \times \left(V_{AC\_PEAK}^2 - V_{CAP}(t)^2\right) = P_{IN} \times t \qquad \text{(since power is energy/time)}$$

Solving, we get

$$V_{CAP}(t) = \sqrt{2} \times \sqrt{V_{AC}^2 - \frac{P_{IN}}{C} t}$$

where C/P_{IN} is in farads per watt. This is the trajectory of discharge into a constant power load.

At the intersection point with the rising AC voltage, two equations (waveforms) combine to give

$$V_{AC} \times \sqrt{2} \times \cos(2\pi f t) = \sqrt{2} \times \sqrt{V_{AC}^2 - \frac{10^6}{\mu F \text{ per } W} t}$$

Solving,

$$\cos(2\pi f t) = \sqrt{1 - \frac{10^6}{\mu F \text{ per } W \times V_{AC}^2} t}$$

This occurs very close to $t = 1/2f$ (the next AC half-cycle peak). So approximating,

$$\cos(2\pi f t) \approx \sqrt{1 - \frac{10^6}{2 \times f \times (\mu F \text{ per } W) \times V_{AC}^2}} \equiv \pm A$$

$$A = \sqrt{1 - \frac{10^6}{2 \times f \times (\mu F \text{ per } W) \times V_{AC}^2}} \qquad \text{(two solutions)}$$

We finally get the first estimate for conduction time (found in literature) as

$$t_{COND\_EST} = \frac{\cos^{-1} A}{2 \times \pi \times f} \qquad \text{(bridge conduction time)}$$

This occurs at time

$$t_{SAG\_EST} = \frac{\cos^{-1}(-A)}{2 \times \pi \times f} \qquad \text{(lowest point of voltage)}$$

But we also have

$$t_{SAG\_EST} = \frac{1}{2f} - t_{COND\_EST} \qquad \text{(relationship between the two)}$$

This can be plugged in the following equation for the first estimate of V_{SAG}.

$$V_{SAG\_EST}(t) = \sqrt{2} \times \sqrt{V_{AC}^2 - \frac{10^6}{\mu F \text{ per } W} t_{SAG\_EST}}$$

But now, for much greater accuracy, we *iterate*: We know that if in a cosine curve, we shift the time axis origin by a quarter cycle, we will get a sine wave. So we also have

$$V_{SAG\_EST} = (\sqrt{2} \times V_{AC}) \times \sin\left[(2 \times \pi \times f)\left(\frac{1}{4f} - t_{COND}\right)\right]$$

Solving

$$t_{COND} = \frac{1}{4f} - \frac{\sin^{-1}[V_{SAG\_EST}/(\sqrt{2} \times V_{AC})]}{2 \times \pi \times f}$$

This estimate is indeed very accurate. In Fig. 5.5 we show how we calculate V_{SAG} from this too and how good the results are as compared to the Mathcad file.

The reason we get accurate predictions of RMS currents only if we estimate t_{COND} accurately is because t_{COND} is the interval where the input source will provide all the energy the converter will draw for a full AC half-cycle. So if t_{COND} is very small, or inaccurately estimated, it will have a dramatic effect on the height of the current pedestal drawn during t_{COND}. Yes, the

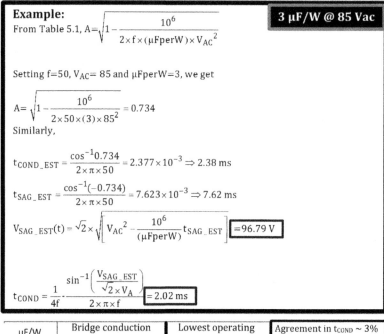

$\mu F/W$ @85 Vac (peak = 120.2 V)	Bridge conduction time (ms), t_{COND}		Lowest operating voltage (V), V_{SAG}		Agreement in t_{COND} ~ 3% Agreement in V_{SAG} ~ 1.5%
	Mathcad loop (exact)	Closed-form eqn. (estimated)	Mathcad loop (exact)	Closed-form eqn. (estimated)	**Conclusion:** The closed-form equations in Tables 5.1 and 5.2 are therefore almost exact, since they match the results of the accurate Mathcad file closely. Therefore, we can dispense with the Mathcad file altogether, and just use the closed-form equations instead.
3.0	2.05	2.021	95.66	96.79	
3.5	1.9	1.879	99.1	99.86	
4.0	1.75	1.763	101.61	102.23	
4.5	1.65	1.667	103.63	104.10	
5.0	1.6	1.585	105.31	105.62	
6.0	1.45	1.451	107.7	107.93	
7.0	1.35	1.347	109.45	109.60	
8.0	1.25	1.263	110.74	110.87	
9.0	1.15	1.193	111.73	111.86	
10.0	1.1	1.134	112.56	112.66	

Figure 5.5 Normal operation at 85 Vac: Mathcad-generated numbers compared to iterated closed-form equations.

average value of current does not change if we halve t_{COND}. (Check: $I_{AVG} = I \times D = 2I \times (D/2)$, where D is the duty cycle of the input AC pulse.) But the RMS value of the current goes up by a factor ~ 1.4. (Check: $I \times \sqrt{D} \rightarrow 2I \times \sqrt{D/2} = \sqrt{2} \times I \times \sqrt{D} = 1.4 \times I \times \sqrt{D}$.) Generalizing, if we reduce t_{COND} by the factor x, the RMS of the input current will increase by approximately the factor $\sqrt{x}$. Alternatively stated, the RMS of the input current must vary as $1/\sqrt{t_{COND}}$. We can check this relationship out: If t_{COND} halves, the RMS input current will increase by the factor $\sqrt{2}$ as expected. Further, we can say that the peak current is roughly proportional to $1/t_{COND}$. We can check this out too: If t_{COND} halves, the peak input current will increase by the factor 2 as expected, based on the fact that the input power and input voltage are unchanged.

Therefore, it is imperative we estimate t_{COND} accurately if we want to design our input stage well. We also see that being "ripple-phobic" is not helpful. If we increase the input capacitance too much, the RMS currents in the entire input stage will be very high.

Universal-Input Single-Ended Forward (without PFC)

Such a converter is rarely seen, because at Forward converter power levels, we usually need PFC by law. However, we may also design a low-power Forward converter. So for completeness sake we will also discuss this here.

This will include a *voltage-doubler circuit* at the input, as shown in Fig. 5.6. Then 110 Vac will be (almost) 220 Vac as far as the input capacitors are concerned. The reason we need a doubler is that the Flyback can handle a very wide input voltage range with the duty cycle [in continuous conduction mode (CCM)] swinging typically between 0.25 to 0.68, from 385 Vdc (the rectified AC input at 270 Vac) down to the UVLO level of 60 Vdc. This is because reason is it is based on the DC transfer function of the Buck-boost. But a Forward converter is based on a Buck transfer function, which doesn't accommodate such wide input swings to start with, and, further, if we limit the maximum duty cycle to 50 percent (as is required for typical Forward converters), then we just cannot make a practical universal-input converter. We definitely need a voltage doubler.

NOTE *With a voltage doubler we actually have two capacitors in series, each charged during alternate AC half-cycles. So, for example, if we need 1372 μF by a certain calculation (530 W, 180 Vdc case below), we need to use two 200 V (preferably 250 V) capacitors, each of $1372 \times 2 = 2744$ μF. But all capacitor RMS calculations now need to account for the fact that the computed RMS input current passes through each capacitor every alternate cycle. Since RMS depends on $\sqrt{D}$, if we halve the duty cycle (alternate half-cycle), the actual RMS falls to 0.707 of its calculated value based on the assumption that the current surge repeats every half-cycle. But also, now there are two capacitors and two RMS currents to account for. And that is how we calculate the voltage ratings, capacitances, and RMS currents for capacitors in doubler circuits (at low line).*

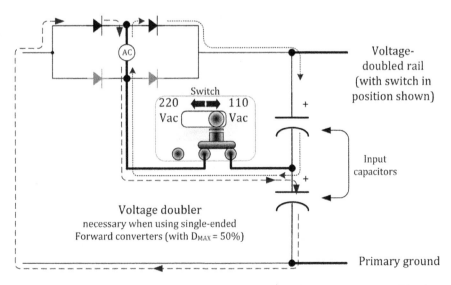

When switch is moved to 220 Vac position, this circuit behaves as a normal bridge rectifier input stage, except that the input capacitor is in effect, two capacitors in series.

FIGURE 5.6 Input doubler stage for single-ended Forward converters.

We typically want the nominal duty cycle of a Forward converter that is set at 110 Vac/220 Vac (about 311 Vdc rectified) to be somewhere between 0.26 to 0.29. The maximum duty cycle is fixed at 0.45 (to allow for tolerances). So we estimate that the input voltage can fall down to 311 × (0.26/0.45) = 180 Vdc, or alternatively, to 311 × (0.29/0.45) = 200 Vdc. The two duty cycle choices affect transformer design somewhat, but let us assume these are the two distinct design possibilities for sizing the input capacitor. We conclude:

1. The lowest operating voltage is 180 Vdc, corresponding to $D_{NOM} = 0.26$.
2. The lowest operating voltage is 200 Vdc, corresponding to $D_{NOM} = 0.29$.

However, to guarantee correct dropout performance, we wish to set the nominal duty cycle based on a minimum operating voltage of 85 Vac, which to the doubler is (almost) 170 Vac: This corresponds to a rectified peak of $170 \times \sqrt{2} = 240$ Vdc. Using the new possible UVLO levels just given, in our Mathcad spreadsheet, we summarize and conclude (see Fig. 5.1 again, and Fig. 5.7):

1. For Flyback, ~10-ms holdup: use 3 µF/W.
2. For Flyback, ~20-ms holdup: use 5 µF/W.
3. For Forward (with voltage doubler, UVLO 180 Vdc, $D_{NOM} = 0.26$), ~10-ms holdup: use 1.4 µF/W.
4. For Forward (with voltage doubler, UVLO 180 Vdc, $D_{NOM} = 0.26$), ~20-ms holdup: use 2.2 µF/W.

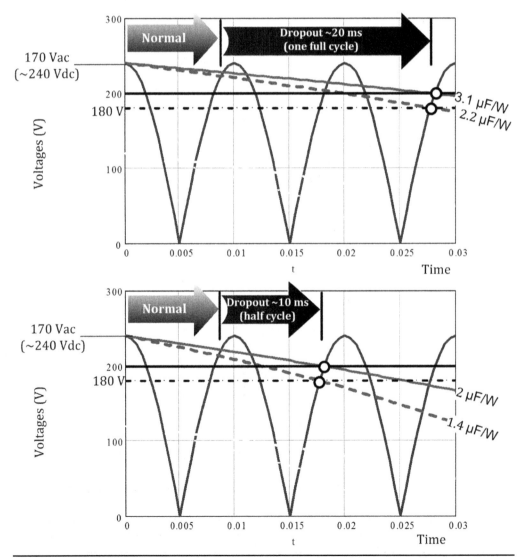

Figure 5.7 Droop curves for a Forward converter and computed capacitances per watt.

5. For Forward (with voltage doubler, UVLO 200 Vdc, $D_{NOM} = 0.29$), ~10-ms holdup: use 2 µF/W.

6. For Forward (with voltage doubler, UVLO 200 Vdc, $D_{NOM} = 0.29$), ~20-ms holdup: use 3.1 µF/W.

Keep in mind that in a Forward converter, there is no need for a formal UVLO circuit usually, since this input voltage point corresponds to the D_{MAX} limit being reached, and so, as the input falls further, the output just droops gracefully. In a Forward there are none of the associated high primary-side currents that occur in a Flyback, which as we know, arise from the rather troublesome equation $I_{COR} = I_{OR}/(1-D)$, where I_{COR} is the current at the center of the ramp (on the primary side) and I_{OR} is the reflected output current, that is, I_O/n, where $n = N_P/N_S$ is the turns ratio.

Compiled Design Tables

We thus have two tables, Tables 5.1 and 5.2, one for 85 Vac, intended for a universal-input Flyback, and the second at 170 Vac (which is actually a Forward converter at 85 Vac, with a doubler input). These present the closed-form equations for normal operation. We need to go from top to bottom, plugging in values in each equation, thereby iterating too. We have carried out the calculations for all the recommended capacitance-per-watt values, which we discovered using our Mathcad spreadsheet. The quoted numerical values are also per watt, and we can very easily use them to calculate all the input-stage current stresses for any power supply (without PFC). Note that these are all computed RMS values based on the same equation list. So, in the case of the doubler (Table 5.2), we still have to account for the fact that the RMS input current passes through two capacitors alternately, as discussed in the preceding section. These two tables do not account for that split.

Note that in Table 5.1, we also provide the numbers for 7 µF/W, which, if we refer back to Fig. 5.5, we will see correspond to three missing AC half-cycles (for a UVLO of 60 Vdc).

Typical Minimum Capacitance Requirements for Holdup Time and RMS Capacitor Currents (Numerical Examples)

In Table 5.3, we take several power supplies of different output wattages, all with an assumed efficiency of 85 percent. We show how we calculate the minimum capacitances, for two cases of holdup time (one half-cycle missing and two half-cycles missing). These are simply based on the recommended minimum µF/W. We have summarized the capacitor RMS requirements (line, i.e., low-frequency component) in Table 5.4 for easy reference.

Estimating RMS Capacitor Current (High-Frequency Component)

The low-frequency surge current is not the only contribution to heating in the input capacitor. The converter is also pulling out high-frequency pulses from the capacitor. Let us estimate that.

We have the following equations (using the Appendix):

$$I_{RMS\_CAP\_HF} = I_{OR} \times \sqrt{D\left(1 - D + \frac{r^2}{12}\right)} \quad \text{(for Forward converter, similar to Buck)}$$

$$I_{RMS\_CAP\_HF} = \frac{I_{OR}}{1-D} \times \sqrt{D\left(1 - D + \frac{r^2}{12}\right)} \quad \text{(for Flyback converter, similar to Buck-boost)}$$

where I_{OR} is the reflected output voltage, I_O/n, (where I_O is the load current and $n = N_P/N_S$ is the turns ratio) and r is the current ripple ratio $\Delta I/I_{COR}$, where ΔI is the total swing in inductor current per cycle, and I_{COR} is the center of ramp, i.e., average inductor current.

1. *Forward converter.* Assume nominal $D = 0.25$. Note that 230 Vac is 325 Vdc. Since $D = V_O/V_{INR}$, then $V_{INR} = 4 \times V_O$. Assume $V_O = 3.3$ V. Then $V_{INR} = 13.2$ V. So, the turns ratio is $n = 325 \text{ V}/13.2 \text{ V} = 24.6$. To account for the diode drop, assume the turns ratio is slightly less, say 24. So $I_{OR} = P_O/(nV_O) = P_O/79.2 \approx P_O/80$.

$$I_{RMS\_CAP\_HF} \approx I_{OR}\sqrt{D \times \left(1 - D + \frac{r^2}{12}\right)} = \frac{P_O}{80}\sqrt{0.25 \times \left(1 - 0.25 + \frac{0.4^2}{12}\right)} \approx 0.005 \times P_O$$

$$= 0.005 \times P_{IN}$$

Chapter Five

Description	Parameter	Equation	3 μF/W	5 μF/W	7 μF/W
1. Required to calculate parameters below	A	$= \sqrt{1 - \dfrac{10^6}{2 \times f \times (\mu F \text{ per } W) \times V_{AC}^2}}$	0.734	0.85	0.896
2a. Bridge conduction duration (*first estimate*)	t_{COND} (s)	$\approx \dfrac{\cos^{-1} A}{2 \times \pi \times f}$ (all calculations based on this approximate value will be approximate too)	2.38 (ms)	1.76 (ms)	1.47 (ms)
3. Time coordinate of minima of bulk capacitor voltage (*first estimate*)	t_{SAG} (s)	$= \dfrac{1}{2f} - t_{COND}$	7.62 (ms)	8.24 (ms)	8.53 (ms)
4. Lowest input operating voltage of converter	$V_{SAG}(t)$ (V)	$= \sqrt{2} \times \sqrt{V_{AC}^2 - \dfrac{10^6}{\mu F \text{ per } W} t_{SAG}}$	96.79 (V)	105.62 (V)	109.60 (V)
2b. Bridge conduction duration (*second estimate*)	t_{COND} (s)	$= \dfrac{1}{4f} - \dfrac{\sin^{-1}\left[V_{SAG}/(\sqrt{2} \times V_{AC})\right]}{2 \times \pi \times f}$ (this is an iterated value, far more accurate)	2.02 (ms)	1.59 (ms)	1.35 (ms)
5. Average bulk capacitor voltage and average input voltage of converter	V_{IN_AVG} (V)	$= \dfrac{(\sqrt{2} \times V_{AC}) + V_{SAG}}{2}$	108.50 (V)	112.91 (V)	114.90 (V)
6. Average input current (*in bridge and filter*)	I_{IN} (A) (for P_{IN} = 1 W)	$= \dfrac{1}{V_{IN\_AVG}}$ (this is an exact equation)	9.22 (mA/W)	8.86 (mA/W)	8.70 (mA/W)
7. Peak capacitor charging current	I_{PEAK} (A) (for P_{IN} = 1 W)	$= \dfrac{8\pi^2 \times (\mu F \text{ per } W)}{10^6 \times \sqrt{2}} \times (V_{AC} f^2) \times t_{COND}$	0.072 (A/W)	0.094 (A/W)	0.112 (A/W)
8. Required ripple (RMS) current rating of bulk capacitor	I_{CAP_RMS} (A)	$= \left[I_{IN}^2 + 2ft_{COND}\left(\dfrac{I_{PEAK}^2}{3} - I_{IN}^2\right)\right]^{1/2}$	0.02 (A/W)	0.023 (A/W)	0.025 (A/W)
9. Average input current (x $2V_D$ for bridge dissipation)	I_{BRIDGE_AVG} (A)	$= f \times t_{COND} \times (I_{PEAK} + 2I_{IN})$ (would be the same as I_{IN} above if t_{COND} value used here were 100 percent exact)	9.13 (mA/W)	8.85 (mA/W)	8.711 (mA/W)
10. RMS of line current (*in bridge and filter*)	I_{FILTER_RMS} (A)	$= \left[2ft_{COND}\left(\dfrac{I_{PEAK}^2}{3} + I_{IN}(I_{PEAK} + I_{IN})\right)\right]^{1/2}$	0.022 (A/W)	0.025 (A/W)	0.027 (A/W)
11. Peak of line current (*in bridge and filter*)	I_{FILTER_PEAK} (A)	$= I_{PEAK} + I_{IN}$	0.081 (A/W)	0.103 (A/W)	0.121 (A/W)

Note Calculations are meant to be performed from top to bottom, in that order.
All A/W and μF/W numbers are normalized to P_{IN} = 1 W.
All numbers are for 85 Vac and line frequency f = 50 Hz.

TABLE 5.1 Design Table for 85 Vac and Three Desired Holdup Time Cases (UVLO set to 60 Vdc in all cases)

Description	Parameter	Equation	1.4 µF/W	2 µF/W	2.2 µF/W	3.1 µF/W
1. Required to calculate parameters below	A	$= \sqrt{1 - \dfrac{10^6}{2 \times f \times (\mu F \text{ per } W) \times V_{AC}^2}}$	0.868	0.909	0.918	0.943
2a. Bridge conduction duration (*first estimate*)	t_{COND} (s)	$\approx \dfrac{\cos^{-1} A}{2 \times \pi \times f}$	1.66 (ms)	1.36 (ms)	1.30 (ms)	1.08 (ms)
3. Time coordinate of minima of bulk capacitor voltage (*first estimate*)	t_{SAG} (s)	$= \dfrac{1}{2f} - t_{COND}$	8.34 (ms)	8.64 (ms)	8.70 (ms)	8.92 (ms)
4. Lowest input operating voltage of converter	$V_{SAG}(t)$ (V)	$= \sqrt{2} \times \sqrt{V_{AC}^2 - \dfrac{10^6}{\mu F \text{ per } W} t_{SAG}}$	214.2 (V)	221.7 (V)	231.9 (V)	234.3 (V)
2b. Bridge conduction duration (*second estimate*)	t_{COND} (s)	$= \dfrac{1}{4f} - \dfrac{\sin^{-1}[V_{SAG}/(\sqrt{2} \times V_{AC})]}{2 \times \pi \times f}$	1.5 (ms)	1.26 (ms)	1.21 (ms)	1.02 (ms)
5. Average bulk capacitor voltage and average input voltage of converter	V_{IN_AVG} (V)	$= \dfrac{(\sqrt{2} \times V_{AC}) + V_{SAG}}{2}$	227.3 (V)	231.1 (V)	231.9 (V)	234.3 (V)
6. Average input current (*in bridge and filter*)	I_{IN} (A) (for P_{IN} = 1 W)	$= \dfrac{1}{V_{IN\_AVG}}$	4.4 (mA/W)	4.33 (mA/W)	4.31 (mA/W)	4.27 (mA/W)
7. Peak capacitor charging current	I_{PEAK} (A) (for P_{IN} = 1 W)	$= \dfrac{8\pi^2 \times (\mu F \text{ per } W)}{10^6 \times \sqrt{2}} \times (V_{AC} f^2) \times t_{COND}$	0.05 (A/W)	0.06 (A/W)	0.063 (A/W)	0.075 (A/W)
8. Required ripple (RMS) current rating of bulk capacitor	I_{CAP_RMS} (A)	$= \left[I_{IN}^2 + 2f t_{COND} \left(\dfrac{I_{PEAK}^2}{3} - I_{IN}^2 \right) \right]^{1/2}$	0.012 (A/W)	0.013 (A/W)	0.013 (A/W)	0.014 (A/W)
9. Average input current (× 2 V_D for bridge dissipation)	I_{BRIDGE_AVG} (A)	$= f \times t_{COND} \times (I_{PEAK} + 2 I_{IN})$	4.4 (m A/W)	4.33 (m A/W)	4.32 (m A/W)	4.27 (m A/W)
10. RMS of line current (*in bridge and filter*)	I_{FILTER_RMS} (A)	$= \left[2f t_{COND} \left(\dfrac{I_{PEAK}^2}{3} + I_{IN}(I_{PEAK} + I_{IN}) \right) \right]^{1/2}$	0.013 (A/W)	0.014 (A/W)	0.014 (A/W)	0.015 (A/W)
11. Peak of line current (*in bridge and filter*)	I_{FILTER_PEAK} (A)	$= I_{PEAK} + I_{IN}$	0.054 (A/W)	0.064 (A/W)	0.067 (A/W)	0.079 (A/W)

Note Calculations are meant to be performed from top to bottom, in that order.
All A/W and µF/W numbers are normalized to P_{IN} = 1 W.
All numbers are for **170 V**ac and Line Frequency f = 50 Hz.

TABLE 5.2 Design Table for 170 Vac, and Two Desired Holdup Time Cases (UVLO set to 180 Vdc or 200 Vdc)

| No PFC; Holdup Time Limited Starting at 85 Vac, ~ 10-ms (1/2 cycle) ||||
Example: $\eta = 0.85$		Flyback, 3 µF/W (60 Vdc limit)	Forward, 1.4 µF/W (180 Vdc limit)	Forward, 2 µF/W (200 Vdc limit)
P_O (W)	P_{IN} (W)	Capacitance (µF)	Capacitance (µF)	Capacitance (µF)
6	6/0.85 = 7.06	7.06 × 3 = 21.18	7.06 × 1.4 = 9.88	7.06 × 2.0 = 14.12
20	20/0.85 = 23.53	23.53 × 3 = 70.59	23.53 × 1.4 = 32.94	23.53 × 2.0 = 47.1
50	58.82	176.46	82.35	117.64
75	88.23	264.69	123.52	176.46
100	117.65	352.95	164.71	235.3
150	176.47	529.41	247.06	352.94
200	235.29	705.87	329.41	470.58
530	623.53	1870.59	872.94	1247.06
No PFC; Holdup Time Limited Starting at 85 Vac, ~ 20 ms (1 cycle)			
Example: $\eta = 0.85$		Flyback, 5 µF/W (60 Vdc limit)	Forward, 2.2 µF/W (180 Vdc limit)	Forward, 3.1 µF/W (200 Vdc limit)
P_O (W)	P_{IN} (W)	Capacitance (µF)	Capacitance (µF)	Capacitance (µF)
6	6/0.85 = 7.06	7.06 × 5 = 35.3	7.06 × 2.2 = 15.53	7.06 × 3.1 = 21.89
20	20/0.85 = 23.53	23.53 × 5 = 117.64	23.53 × 2.2 = 51.77	23.53 × 3.1 = 72.94
50	58.82	294.12	129.40	182.34
75	88.23	441.18	194.10	273.51
100	117.65	588.23	258.90	364.72
150	176.47	882.35	388.234	547.06
200	235.29	1176.47	517.64	729.40
530	623.53	3117.65	1371.77	1932.94

TABLE 5.3 Calculating Minimum Capacitances for Meeting Holdup Time Requirements (with No Power Factor Correction Front-End)

Minimum µF/W and Low-Frequency RMS component of Capacitor Current (no PFC)			
At 85 Vac (Flyback)		At 170 Vac (Forward)	
Capacitor per (Input) Watt	Low-Freq Input Capacitor RMS	Capacitor per (Input) Watt	Low-Freq Input Capacitor RMS
		1.4 µF/W	0.012 A/W
		(UVLO 180 Vdc, Dropout ~10 ms)	
		2 µF/W	0.013 A/W
		(UVLO 200 Vdc, Dropout ~10 ms)	
		2.2 µF/W	0.013 A/W
		(UVLO 180 Vdc, Dropout ~20 ms)	
3 µF/W	0.020 A/W		
(UVLO 60 Vdc, Dropout ~10 ms)			
		3.1 µF/W	0.014 A/W
		(UVLO 200 Vdc, Dropout ~20 ms)	
5 µF/W	0.023 A/W		
(UVLO 60 Vdc, Dropout ~ 20 ms)			

↓ Capacitance Increasing

TABLE 5.4 Minimum Capacitances and Corresponding RMS Capacitor Currents (Low-Frequency Component)

Alternative estimate. Suppose input power is 1 W. Then the average input current is $1\,W/V_{IN}$. Suppose duty cycle is $D = 0.25$. Then peak input current must be $1\,W/(D \times V_{IN}) = 4\,W/V_{IN}$. At 325 Vdc, this becomes $4/325 = 1/83$. So in general, $I_{OR} = P_{IN}/83$. This is close to what we originally found.

2. *Flyback.* Assume the nominal duty cycle is 0.5. Typical $V_{OR} = 100$ V. The turns ratio is 100 V/3.3 V = 30. So $P_O = V_{OR} \times I_{OR}$. Therefore, I_{OR} is $\mathbf{P_O/100}$. So RMS of the input capacitor is

$$I_{RMS\_HF} \approx \frac{I_{OR}}{1-D}\sqrt{D \times \left(1 - D + \frac{r^2}{12}\right)} = \frac{P_O}{100 \times 0.5}\sqrt{0.5 \times \left(1 - 0.5 + \frac{0.4^2}{12}\right)} \approx 0.01 \times P_O$$

$$= 0.01 \times P_{IN}$$

In Table 5.5 we summarize these values for the range of power supplies we have arbitrarily selected. We have used the same scaling factors.

Total Input Capacitor RMS Current for Input Capacitors with Non-Frequency-Dependent ESR

At this point we reach a crucial juncture. We have calculated the high-and low-frequency components of the RMS current. If the ESR of the input capacitor is relatively *frequency-independent*, we can simply combine the components orthogonally as follows to give us the total RMS current (and heating) in the capacitor.

$$I_{CIN\_RMS\_TOTAL} = \sqrt{I_{CIN\_RMS\_LF}^2 + I_{CIN\_RMS\_HF}^2}$$

where LF stands for low frequency and HF stands for high frequency. In this way we arrive at the numbers presented in Table 5.6, which includes a summary of the LF and HF components too. We will need to use them in a different manner shortly.

The total RMS can be used directly for, say, ceramic capacitors (rarely used). For aluminum electrolytic capacitors we need to follow a different procedure.

Total Input Capacitor RMS Current for Input Capacitors with Frequency-Dependent ESR

In aluminum electrolytic capacitors, ESR typically falls by a factor of 2 in going from low frequencies (AC line) to high (switching) frequencies (in the range of 100 kHz). So for the same heating, we can increase the RMS current by a factor of $\sqrt{2} = 1.4$ at high frequencies. For example, we can either pass 1 A of low-frequency (typically 120 Hz) RMS current, or 1.4 A of 100-kHz RMS current. Each component "sees" a different ESR. It is like two capacitors in one.

Ratings of capacitors are for the same reason published in two ways: low-frequency and high-frequency ripple (RMS) ratings. What is meant is not immediately obvious when we look at datasheets sometimes, and certainly not obvious when we go to web sites like www.digikey.com and try to use a master section table that lumps all vendors together. Invariably, we need to, and should, go back to the datasheet from the specific vendor and

Assuming 85% Efficiency		Flyback	Forward
Output Power (W)	Input Power (W)	Hi-Freq RMS Current (A)	Hi-Freq RMS Current (A)
6	6/0.85 = 7.06	7.06 × 0.01 = 0.07	7.06 × 0.005 = 0.035
20	20/0.85 = 23.53	23.53 × 0.01 = 0.24	23.53 × 0.005 = 0.12
50	58.82	0.59	0.29
75	88.23	0.88	0.44
100	117.65	1.18	0.59
150	176.47	1.76	0.88
200	235.29	2.35	1.18
530	623.53	6.23	3.12

TABLE 5.5 High-Frequency RMS Currents in the Input Capacitors of the Two Types of Converters

142 Chapter Five

Assuming 85 percent efficiency			~10-ms (1/2 Cycle) Dropout			~20-ms (1 Cycle) Dropout		
			3 µF/W	1.4 µF/W	2 µF/W	5 µF/W	2.2 µF/W	3.1 µF/W
			Flyback (60 Vdc UVLO)	Forward (180 Vdc UVLO)	Forward (200 Vdc UVLO)	Flyback (60 Vdc UVLO)	Forward (180 Vdc UVLO)	Forward (200 Vdc UVLO)
Output power (W)	Input power (W)	Capacitor RMS (A)	Capacitor RMS (A)	Capacitor RMS (A)	Capacitor RMS (A)	Capacitor RMS (A)	Capacitor RMS (A)	Capacitor RMS (A)
6	7.06	LF	7.06 × 0.02 = 0.14	7.06 × 0.012 = 0.085	7.06 × 0.013 = 0.092	7.06 × 0.023 = 0.16	7.06 × 0.013 = 0.092	7.06 × 0.014 = 0.099
		HF	7.06 × 0.01 = 0.07	7.06 × 0.005 = 0.035	7.06 × 0.005 = 0.035	7.06 × 0.01 = 0.07	7.06 × 0.005 = 0.035	7.06 × 0.005 = 0.035
		Total	0.16	0.092	0.098	0.177	0.098	0.105
20	23.53	LF	0.47	0.282	0.306	0.54	0.306	0.329
		HF	0.24	0.12	0.12	0.24	0.12	0.12
		Total	0.53	0.31	0.33	0.59	0.33	0.35
50	58.82	LF	1.176	0.71	0.765	1.353	0.765	0.824
		HF	0.59	0.29	0.29	0.59	0.29	0.29
		Total	1.315	0.765	0.82	1.475	0.82	0.874
75	88.23	LF	1.765	1.06	1.15	2.03	1.147	1.235
		HF	0.88	0.44	0.44	0.88	0.44	0.44
		Total	1.97	1.15	1.23	2.21	1.23	1.31
100	117.65	LF	2.35	1.41	1.53	2.71	1.529	1.647
		HF	1.18	0.59	0.59	1.18	0.59	0.59
		Total	2.63	1.53	1.64	2.95	1.64	1.75
150	176.47	LF	3.53	2.12	2.294	4.06	2.29	2.47
		HF	1.77	0.88	0.88	1.77	0.88	0.88
		Total	3.95	2.294	2.46	4.43	2.46	2.62
200	235.29	LF	4.71	2.824	3.06	5.41	3.06	3.29
		HF	2.35	1.18	1.18	2.35	1.18	1.18
		Total	5.26	3.06	3.277	5.90	3.277	3.50
530	235.29	LF	12.47	7.48	8.11	14.34	8.11	8.73
		HF	6.24	3.12	3.12	6.24	3.12	3.12
		Total	13.94	8.106	8.69	15.64	8.69	9.27

TABLE 5.6 Summarizing the Low-and High-Frequency Components of Capacitor Current for Non-PFC Cases

check whether the published RMS rating is expressed at 100 kHz or 120 Hz. The difference is 40 percent! It is the difference between 1 W of expected heat versus an actual 2 W. That could lower the expected life of the capacitor by at least half. Also, earlier we in effect assumed a *frequency multiplier* f_{MULT} of 1.4, but to be precise, we should see if the vendor has given a more exact value applicable to its capacitors.

We next demonstrate numerically the correct way to handle capacitor RMS components in aluminum capacitors.

Example 5.1 What capacitor should we pick for a 75-W Flyback (with a set UVLO of 60 Vdc)?
From Table 5.6, for the case of one missing half-cycle (~10 ms), we have the following LF and HF components respectively: 1.765 A and 0.88 A. The equivalent total low-frequency (120 Hz) RMS rating required from the capacitor is

$$I_{CIN\_RMS\_TOTAL\_LF} = \sqrt{I_{CIN\_RMS\_LF}^2 + \left(\frac{I_{CIN\_RMS\_HF}}{f_{MULT}}\right)^2} = \sqrt{1.765^2 + \left(\frac{0.88}{1.4}\right)^2} = 1.87 \text{ A RMS}$$

Or we could look for a capacitor with an equivalent total high-frequency (100 kHz) RMS rating of

$$I_{CIN\_RMS\_TOTAL\_HF} = \sqrt{(I_{CIN\_RMS\_LF} \times f_{MULT})^2 + I_{CIN\_RMS\_HF}^2} = \sqrt{(1.765 \times 1.4)^2 + 0.88^2} = 2.62 \text{ A RMS}$$

As expected, 2.62/1.4 = 1.87 A.

Now we carry out a practical selection:

1. For 10 ms with P_O = 75 W: We need a capacitor with about the following specifications: 265 µF/400 V/1.87 A (LF) (see Table 5.3 for µF/W). Picking a 270 µF/400 V capacitor (EETED2G271CA Panasonic, 3000 hours/105°C), we get an LF-RMS rating of 1.56 A typical. And the cost per unit is around $4.60. But to get the required RMS rating we need a 390 µF/400 V capacitor (EETED2G391DA). The cost per unit is around $6.70.

2. For 20 ms with P_O = 75 W: We need a capacitor with about the following specifications: 441 µF/400 V/2.125 A (LF). Picking a 470 µF/400 V capacitor (EETED2G471EA Panasonic, 3000 hours/105°C), we get an LF-RMS rating of 2.01 A typical. The cost per unit is around $7.50. But to get the required RMS rating we need a 560 µF/400 V capacitor (EETED2G561EA). The cost per unit is around $8.64.

CONCLUSION *Except for much larger holdup times, we are limited by RMS ratings of available capacitors, not by holdup and capacitance. So by somehow reducing RMS requirements, we could use smaller capacitances and save cost and energy, while still meeting the required holdup time.*

Things get better for smaller power. In fact, *until about 20 to 30 W, RMS current is not a limitation for Flyback converters*, and we can indeed go by our rule of 3 µF/W for one half-cycle holdup time. But as we see, even by the time we get to 75 W, we are in effect having to pick, *not 3 µF/W*, but *4.5 µF/W* just to meet the RMS rating.

What if we use a voltage doubler along with a single-ended Forward converter?

Efficacy of EMI Filters

In Tables 5.1 and 5.2, in row 11, we have also provided the peak current. For example, a 75-W Flyback power supply at 90 percent efficiency has an input power of 75 W/0.9 = 83.3 W. Assuming the power supply it is designed for an ~10-ms dropout case, using 3 µF/W (i.e., 250 µF), row 11 of Table 5.1 tells us that the peak current is 81 mA/W, that is, 0.081 × 83.3 = 6.75 A. We must ensure that any differential-mode inductor can handle this peak current without saturating, or it will be barely efficacious. As far as the common-mode choke goes, the forward and return currents' flux contributions cancel out, so core saturation is not of much concern here. But of great importance is its RMS handling capability (heat and temperature concerns). From row 10 of the same table, we see that the RMS is 22 mA/W. So in our case, we have 0.022 × 83.3 = 1.83 A (RMS). So we have to ensure the common mode choke has wire gauge thick enough to handle this RMS current.

Forward Converters for Low Power?

Suppose we use a voltage doubler at the input. We benefit by the higher voltage, to start with. The holdup time requirement is also easier to meet with a smaller capacitance, and the input current surge has a lower RMS because of the lower average current drawn at higher input voltages. The benefit of using a Forward converter (with a doubler at the front end), versus a Flyback, is only related to the lower high-frequency current drawn by a Forward converter (it is about half that for a Flyback, for the same duty cycle). Assuming a Forward converter, here is a practical selection.

For 10 ms with P_O = 75 W: The input power is 75/0.85 = 88.2 W. Using the recommendation of 1.4 µF/W (for a 180 Vdc UVLO, see Table 5.4), we need about 88.2 × 1.4 = 123.5 µF. In a doubler, that would be two capacitors in series, each of at least 2 × 123.5 = 247 µF. The RMS LF and HF components are respectively, 1.06 A and 0.44 A (see Table 5.6). As previously discussed, the low-frequency component goes through the two capacitors of the doubler on alternate half-cycles. And that reduces the RMS by a factor of 0.707. So, through each capacitor, the LF component is only 1.06 × 0.707 = 0.75 A. This gives us a total of $[0.75^2 + (0.44/1.4)^2]^{1/2} = 0.81$ (LF equivalent). Picking two 270 µF/250 V capacitors (LGJ2E271MELB Nichicon, 3000 hours/105°C), we get an LF RMS rating of 1 A typical. The cost per unit is around $2.50. We need two such capacitors for $5.00 in all. The capacitors are adequate. Compare this to the universal-input Flyback we looked at previously, where we chose a single 270 µF/400 V capacitor (EETED2G391DA), at a cost per unit of around $6.70. Our cost has improved somewhat by using a Forward with a doubler, but not as much, since we need two capacitors instead of one. Let us say, roughly, both cost and volume are comparable. But the RMS has fallen from 1.87 to 0.81 A, though through two capacitors in series (with lesser ESR). On account of this, efficiency will improve.

For higher powers, we realize we need PFC by law. But ignoring that for a moment, even if we use a Forward converter at say 530 W, we will discover we are by now *severely limited by RMS currents*, not by holdup time. That forces us into higher and higher μF/W. PFC improves (lowers) the input RMS significantly, as we discuss next.

In Fig. 5.8 we finally present very easy lookup curves, based on all our preceding number crunching, for estimating capacitance and RMS ratings for converters with different flavors, all with no PFC. An embedded example is also provided. Note that we have only presented the total RMS, but since that is mainly low-frequency content in these cases, in effect we can safely use the numbers from these curves for picking a capacitor with the stated 120-Hz ripple (RMS) rating.

PART 2 FRONT-END WITH POWER FACTOR CORRECTION

Apparent power is defined as

$$\text{Apparent power} = V_{\text{RMS}} \times I_{\text{RMS}}$$

When dealing with AC power distribution, in which we use sine-wave AC voltages, this is often equivalently written as

$$\text{Apparent power} = V_{\text{AC}} \times I_{\text{AC}}$$

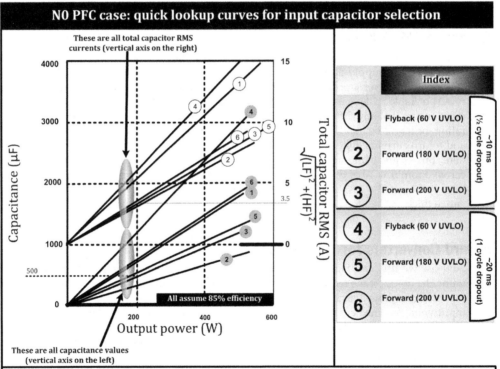

Figure 5.8 Quick selection curves for no PFC.

For example, when we refer to the U.S. household mains input as 120 Vac, this is a sine wave with an RMS value of 120 V. Its peak value is 120 V × $\sqrt{2}$ = 170 V.

The ratio of *real power* to apparent power is called the *power factor*. In the best case, the power factor (PF) is 1, and then apparent power equals real power. To achieve a power factor of 1, we try to achieve *resistor-like* characteristics somehow. This process is called power factor correction (PFC). If we succeed in doing that, the incoming AC sees a pure resistor across it, and if the input voltage is a sine (or rectified sine) wave, as it is in this case, the current drawn is also a sine wave—because that is just the way any resistance behaves: if we double the voltage across it, we double the current, and so on. So, current through a resistor literally tracks the exact shape of the applied voltage. We know that the instantaneous ratio of the two, V/I, is resistance by definition.

With near-perfect power factor correction, the real power, which is $V_{AC} \times I_{AC} \times PF$, becomes almost equal to the apparent power, which is $V_{AC} \times I_{AC}$. This means that the power factor is PF = $\cos\phi \times \cos\Theta \approx 1$. Here ϕ is the phase angle and Θ is the distortion angle. In non-PFC power supplies, $\phi \approx 0$, but $\cos\Theta$ is typically around 0.6 to 0.7. Too high a bulk capacitor (without PFC) will cause the power factor to decrease even lower than 0.6. It will also significantly increase the EMI and associated filtering costs. Hence the need for PFC.

Nowadays, equipment (typically drawing more than 75 W from the mains) needs to comply with mandatory line harmonics standards in particular those based on the international standard IEC61000-3-2. The most popular way to comply with the requirements unambiguously is to use a standard Boost PFC stage at the front-end. In particular, this buffers the bulk capacitor of the PWM stage from the mains, so there is no huge input surge current once every half-cycle, as we had in the case of non-PFC front-ends. That capacitor is the main culprit of line harmonics, and utilities just hate to have to deliver that odd-shaped current, which is not really contributing to the actual useful power anyway.

So, a typical power factor stage, based on a Boost topology, will also use various control techniques to sense the actual instantaneous line current and tailor it into a sine wave, thereby achieving power factor correction. That is because a simple Boost converter does not draw a sine wave current under a sine wave input. In fact, it does just the opposite: We know that if the input falls, to maintain constant output power from the converter, any typical DC-DC converter will attempt to draw a very high input current, in an effort to keep the $V \times I$ product constant. But what we really want out of a PFC stage is that if the input is low, the current drawn is also low! That is clearly not a typical DC-DC converter!

How do we accomplish that? We do that by *not* demanding a constant output power (or current) from the Boost PFC stage.

Note that the PWM stage which follows will eventually demand an almost constant direct current from the regulated output rail of the PFC stage. We will discuss how this happens a little later. For now remember that the intermediate rail (output of the Boost PFC and input of the PWM), is usually called HVDC, for high-voltage DC, or just V_{BUS}. The fully averaged direct current from it, flowing into the PWM stage, is I_{BUS}.

Let us first ask the reverse question: What should be the load profile at the output of the PFC stage for it to ask for a sine wave input current? Here is the simple math.

We first set the requirement equations

$$I_{IN}(t) = K|\sin(2\pi ft)|$$

where f is the AC (line) frequency and K is an arbitrary constant. The input voltage is a rectified sine wave of peak value V_{IN_PK}, with the same phase. So

$$V_{IN}(t) = V_{IN\_PK}|\sin(2\pi ft)|$$

The ratio of the input current and input voltage is thus independent of time and is called the *emulated resistance* R_E.

$$R_E = \frac{V_{IN\_PK}|\sin(2\pi ft)|}{K|\sin(2\pi ft)|} = \frac{V_{IN\_PK}}{K} \Rightarrow K = \frac{V_{IN\_PK}}{R_E}$$

We also have the power requirement equations

$$\frac{V_{AC}^2}{R_E} = \frac{V_{BUS} \times I_{BUS}}{\eta_{PFC}} = \frac{P_O}{\eta_{PFC}\eta_{PWM}}$$

V_{AC} is the RMS input voltage, V_{BUS} is the HVDC rail, I_{BUS} is the average direct current into the PWM stage from V_{BUS}, and P_O is the PWM output power.

Since $V_{AC} = V_{IN\_PK}/\sqrt{2}$, we get

$$R_E = \frac{\eta_{PFC}\eta_{PWM}V_{IN\_PK}^2}{2P_O} \equiv \frac{\eta V_{AC}^2}{P_O} \quad \Leftarrow \text{Remember! } (\eta = \eta_{PFC}\eta_{PWM})$$

We thus get the proportionality factor K as

$$K = \frac{V_{IN\_PK}}{R_E} = \frac{V_{IN\_PK}}{\eta V_{IN\_PK}^2/2P_O} = \frac{2P_O}{\eta V_{IN\_PK}}$$

To emphasize, P_O is the output power of the Boost + PWM stage.

Plugging this into our initial requirement

$$I_{IN}(t) = K|\sin(2\pi ft)| = \frac{2P_O}{\eta V_{IN\_PK}}|\sin(2\pi ft)|$$

Where f is the AC (line) frequency. Now, for any Boost stage, which this still is, we have a certain relationship between currents and voltages. There are little changes from a regular converter. For one, we have a time-varying duty cycle here, because the input voltage is varying. Also, as indicated, the load current cannot be assumed to be a constant, so let us call it $I_{OE}(t)$ (effective output current) instead of just I_O. Using the slightly modified standard Boost equations, we get

$$D(t) = \frac{V_O - \eta_{PFC}V_{IN}(t)}{V_O} \qquad I_{IN}(t) = \frac{I_{OE}(t)}{1-D(t)}$$

Solving, we get

$$I_{IN}(t) = \frac{I_{OE}(t) \times V_{BUS}}{\eta_{PFC}V_{IN}(t)}$$

Plugging this into our previous equation, and simplifying, we get

$$I_{OE}(t) = 2 \times I_{BUS} \times |\sin(2\pi ft)|^2$$

This simply means that the high-frequency averaged PFC diode current must be a low-frequency $\sin^2$ waveform with an average (center value) equal to the direct current drawn by the PWM stage (for ensuring power balance). This is how the PFC stage ensures that the input current is a sine waveform—and this is true for *any Boost PFC stage out there*, irrespective of its actual control strategy or implementation, whether clearly recognized or not. This must be the final result of all methods based on the Boost topology (like all roads leading to Rome).

The bulk capacitor between the PFC and PWM stages (called C_{BULK} usually) further averages the low-frequency component out, delivering a constant direct current I_{BUS} into the PWM stage. But the effective output load current of the Boost PFC stage is a $\sin^2$ waveform—called $I_{OE}(t)$. See Figs. 5.9 and 5.10. The former figure also has all the equations of the stresses for easy reference. Note that in literature, the equations usually found are inaccurate for the RMS current in the bulk capacitor, because they assume a resistive load across it. Our equations, verified against a Mathcad simulation, involve D_{PWM}, and we can see that if we put $D_{PWM} = 1$, implying a resistive load, we do get the familiar equations seen in literature.

NOTE *In Fig. 5.10, just for visual clarity, the PWM is running at half the frequency of the PFC stage. The normal implementation however, for maximum benefit in capacitor RMS, uses the same switching frequencies for both stages as discussed later.*

Instantaneous Duty Cycle in a PFC Boost

Since the input is varying, what exactly is the duty cycle? Keep in mind that assuming CCM, the duty cycle does not depend on load current. So though we are tailoring the load current to a $\sin^2$ profile, as explained in the previous section, this has little to do with the duty cycle. So load profile and duty cycle are two independent issues. Which takes us back to the question: What is the duty-cycle equation?

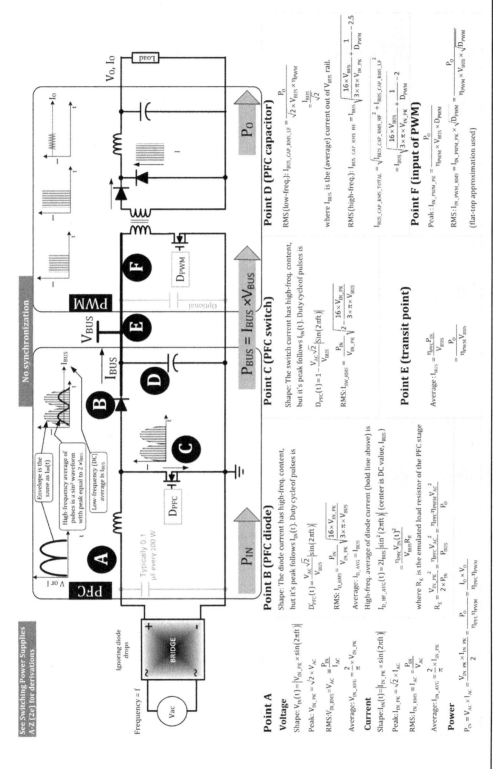

FIGURE 5.9 All currents and stress equations in the PFC Boost compiled.

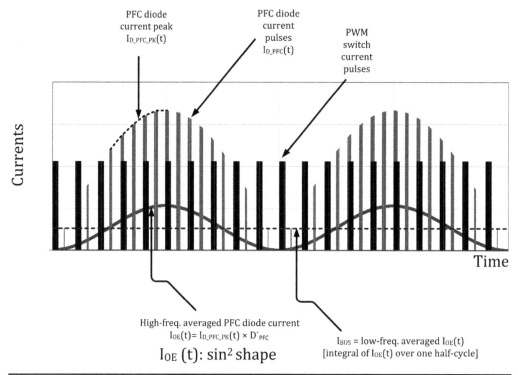

Figure 5.10 Currents in the Boost PFC and PWM stage analyzed.

NOTE *Yes, when $I_{OE}(t)$ is very low (as at low input voltages), the inductor will go into DCM and the duty-cycle equations will certainly change from the CCM value and will become load dependent. As a result, the input current at very low input voltages will certainly deviate from the expected sine wave. But usually that does not affect the Power Factor or the line harmonics much, on account of the low voltage where it happens, so we ignore that.*

The switching frequency is usually high enough that, relative to it, the slowly varying AC input can be considered as direct current at any given moment. So, in effect, we are constantly in quasi-steady-state, and so the normal CCM equations of the Boost topology will apply. We have

$$\frac{V_O}{\eta V_{IN}} = \frac{1}{1-D}$$

$$D = \frac{V_O - \eta V_{IN}}{V_O}$$

For our PFC Boost stage

$$D_{PFC}(t) = \frac{V_{BUS} - \eta_{PFC} \times |V_{IN\_PK} \times \sin(2\pi ft)|}{V_{BUS}} = 1 - \eta_{PFC} \times \frac{|V_{IN\_PK} \times \sin(2\pi ft)|}{V_{BUS}}$$

This is the equation we provided for point C in Fig. 5.9 (we assumed 100 percent efficiency, however).

Holdup Time Considerations with PFC Boost Stages

If we use our familiar Mathcad spreadsheet to plot the droop of capacitor voltage, then starting with a bus voltage of 385 Vdc, we see that we can design a typical single-ended Forward converter to hit its D_{MAX} at about 250 Vdc, starting with a nominal $D = 0.3$. Effectively, we have a UVLO of 250 Vdc. But we could also connect a Flyback after the PFC stage. This is not such a bad idea at low to medium power levels, because the Flyback can still be designed to hit UVLO at 60 Vdc, allowing a significant reduction in required capacitance per watt to meet a target holdup time. See a summary of this in Fig. 5.11 and the corresponding quick-lookup table in Table 5.7 (compare with to Table 5.3).

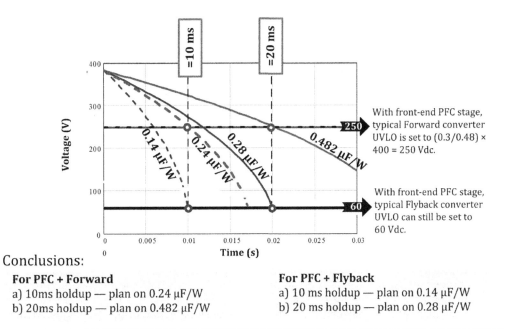

Conclusions:

For PFC + Forward
a) 10ms holdup — plan on 0.24 μF/W
b) 20ms holdup — plan on 0.482 μF/W

For PFC + Flyback
a) 10 ms holdup — plan on 0.14 μF/W
b) 20 ms holdup — plan on 0.28 μF/W

FIGURE 5.11 Capacitance per watt requirements for holdup time with PFC front-ends.

		Forward or Flyback Converter + PFC			
		10-ms Holdup		20-ms Holdup	
$\eta = 0.85$		Capacitor (μF) (0.14 μF/W)	Capacitor (μF) (0.24 μF/W)	Capacitor (μF) (0.28 μF/W)	Capacitor (μF) (0.482 μF/W)
Output Power (W)	Input Power (W)	Flyback (60 Vdc UVLO)	Forward (250 Vdc UVLO)	Flyback (60 Vdc UVLO)	Forward (250 Vdc UVLO)
6	6/0.85 = 7.06	7.06 × 0.14 = 0.99	7.06 × 0.24 = 1.69	7.06 × 0.28 = 1.98	7.06 × 0.482 = 3.40
20	20/0.85 = 23.53	23.53 × 0.14 = 3.29	23.53 × 0.24 = 5.65	23.53 × 0.28 = 6.59	23.53 × 0.482 = 11.34
50	58.82	8.234	14.11	16.47	28.35
75	88.23	12.35	21.175	24.7	42.53
100	117.65	16.47	28.24	32.94	56.71
150	176.47	24.706	42.35	49.41	86.06
200	235.29	32.94	56.47	65.88	113.41
530	623.53	87.29	149.65	174.59	300.54

TABLE 5.7 Calculating Minimum Capacitances for Meeting Holdup Time Requirements (with Power Factor Correction Front-end)

But in either case, we see very small capacitances are required with PFC, primarily because, whether we are operating at 110 Vac or 220 Vac, the bus voltage is very high (385 Vdc). And since, energy in a capacitor depends on V^2, at this voltage, even a small capacitor can hold enough energy to keep the PWM stage functioning for 10 or 20 ms as desired.

We also suspect that RMS through the capacitors becomes the dominant selection criterion for converters with Boost PFC-front-ends, not holdup time. And any method by which we can decrease the RMS current will lead to a massive reduction bulk capacitance (and overall size and cost of the power supply).

Synchronization and Anti-Synchronization Techniques

Some power factor control ICs offer synchronization capability, but that phrase is usually meant to mean that the PFC is *in phase* with the PWM. So whenever the PFC switch turns ON, so does the PWM switch. But in general, this synchronization scheme is not necessarily conducive to lowering EMI and certainly not cost. Admittedly, it may make the EMI spectrum more *predictable*, but in fact it can severely *increase* the required EMI filtering cost. So many engineers feel that it is actually better to run the PFC and PWM independently (no synchronization), just taking care to keep their respective switching frequencies spaced somewhat apart to avoid beat frequencies.

The anti-synchronization scheme was introduced as a combination IC, the ML4826, a rather long time ago, around 1995 by Microlinear, and billed as the "industry's first leading edge/trailing edge modulation scheme" PFC/PWM controller IC. What this essentially tried to accomplish was based on the following intuitive thought process: We know that the PWM draws current out from the bulk capacitor, whereas the PFC dumps current into it. What if we could make these opposing currents cancel? Maybe what we really want to do is to turn ON the PWM switch at the very same moment as the PFC starts to turn OFF. This is *out of phase* or simply, *anti-synchronization*. So we would then expect that the freewheeling current (coming out from the PFC diode) would head straight into the PWM stage (most of the time) *without having to recycle through the bulk capacitor*. Of course, the PFC duty cycle varies from very low to very high values, and the PWM duty cycle is fixed. So this cancellation would work but by different amounts over the AC line cycle (see Fig. 5.12). It is therefore hard to provide any easy closed-form equation for calculating the net reduction in the RMS current through the capacitor. Not surprisingly, there was in fact no such detailed application information provided even for the combination IC mentioned above. The lack of closed-form equations seems to have hindered more widespread adoption of this technique. Mathcad spreadsheets were created by this author which can very accurately estimate the reduction in RMS capacitor current in particular. The results are provided here.

NOTE *In Fig. 5.12, just for visual clarity, the PWM is running at half the frequency of the PFC stage. The normal implementation, however, for maximum benefit in capacitor RMS, uses the same switching frequencies for both stages.*

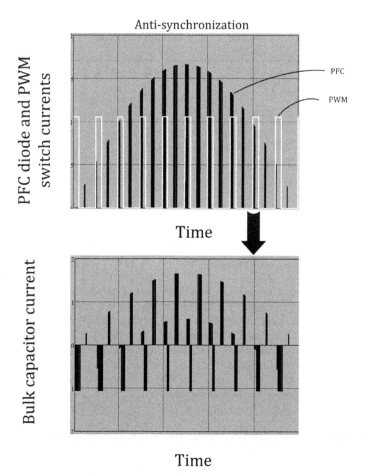

FIGURE 5.12 How anti-synchronization reduces capacitor current.

The ML4826 combo IC had a single clock of fixed frequency, and whereas the PWM switch would turn ON at the clock edge, the PFC would switch OFF at the same edge. Now, in general, in most DC-DC converters, regulation is carried out by varying the moment at which the switch turns OFF. This is called *trailing edge* modulation. In this IC, the PWM worked the same way, but for the PFC section, to create anti-synchronization, Microlinear opted for *leading edge* modulation in the ML4826. The problem is that we have to regulate by knowing *beforehand* that the switch will turn OFF at the clock edge and then determining the moment at which we need to turn the switch ON to regulate the output. Compensation is not straightforward. But it turns out that this is in reality an unnecessarily elaborate way of implementing out-of-phase synchronization. An easier way uses more standard parts: the common PFC IC, the UC3854, and a common UC3844 PWM IC. We can synchronize the two ICs much the way we normally do for such ICs (with a small resistor in series with the timing capacitor), except that now we use the OFF edge of the UC3844 output (Gate) signal to reset the clock of the UC3854. In principle we could also reverse the order and use the UC3854 as the *master*. We may then intuitively expect a slight inherent line frequency modulation in this process, but in practice it was found to be transparent at the output of the PWM. So we can end up with the more familiar trailing-edge modulation for both PFC and PWM, with the same reduction in RMS (as borne out by the author's Mathcad spreadsheet). This implementation method (both PFC and PWM trailing-edge modulated for anti-synchronization) is considered proprietary in the name of this author.

In Fig. 5.13 we have plotted the ripple currents for 90 and 270 Vac and also provided exact numbers from the Mathcad spreadsheet. We see that using anti-synchronization, we get the highest percentage improvement at low-line. We also see that at high-line, the RMS capacitor currents go up steeply again, because the much narrower PWM pulse widths can subtract much less from the diode current pulses. Note that the total RMS current through the capacitor is by definition

$$I_{RMS} = \sqrt{I_{RMS\_LF}^2 + I_{RMS\_HF}^2}$$

Low-frequency and high-frequency components have also been provided if we want to do a calculation for aluminum capacitors using the frequency multiplier, just as we did for non-PFC stages earlier in this chapter.

Anti-Synchronization over a Wide Input Range

We must not forget that a capacitor is chosen not on the basis of a perceived ripple current improvement (a factor), but on the *actual ripple current*. We may see a big improvement as compared to what it would have been *at that particular input voltage* without synchronization, but what is its absolute value? We should actually look at *both extremes* of input voltage, and pick the *higher* of the RMS currents so reported. And that would be a worst-case number to check the capacitor rating against. We can glean that information line by line from Fig. 5.13, but that is cumbersome. What we need to do is look at both the input voltage extremes (with synchronization present) and pick the higher of the two net RMS readings (note that in general, with synchronization, this could be occurring *at high-line or at low-line*). We then compare this higher reading with the maximum RMS current without synchronization (that we know from Fig. 5.13 always occurs at low-line). Thus we calculate the improvement over the *entire input range*. This is in the form $\Delta I/I_{NO\_SYNC}$. We have displayed these numbers too, for quick reference.

The conclusions from Fig. 5.13 are

1. The maximum improvement (*and* the lowest absolute value) for the RMS current is at around $D_{PWM} = 0.325$. For a conventional Forward converter with a duty cycle set to about 0.3 to 0.35, the improvement due to anti-synchronization is around 40 percent.

2. In fact over the duty-cycle range of 0.23 to 0.4, *we can expect more than a 32 percent improvement*.

Calculating the High-and Low-Frequency RMS Components

For the purpose of estimating the life of the aluminum capacitor, we need to know the breakup as explained earlier. Here is a sample calculation.

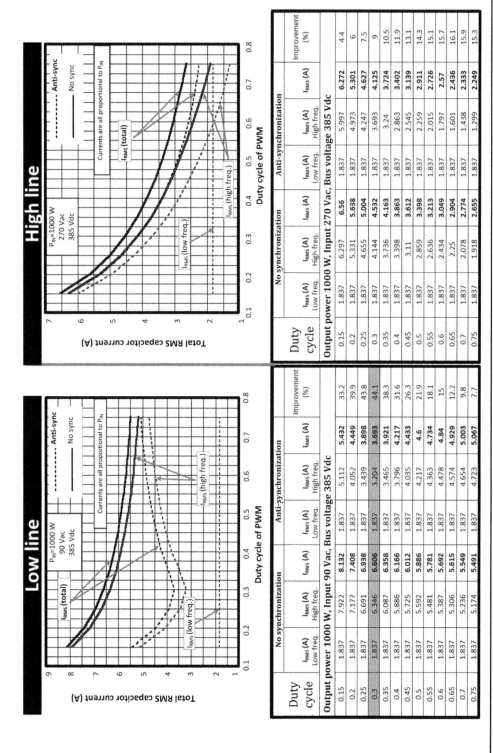

FIGURE 5.13 The improvement in RMS capacitor current by anti-synchronization.

Example 5.2 We have a worldwide input 70-W Flyback running off the PFC stage. Its efficiency is 70 percent. What are the components of the capacitor RMS current at a set duty cycle of 50 percent at 90 Vac?

Let us first calculate the results assuming an input power of 1000 W. At an HVDC of 385 V, the PFC stage load current is

$$I_{BUS} = \frac{1000}{385} = 2.597 \text{ A}$$

We know from Fig. 5.9 (point D) that the total unsynchronized capacitor RMS current at 90 Vac is

$$I_{BUS}\sqrt{\frac{16 \times V_{BUS}}{3 \times \pi \times V_{IN\_PK}} + \frac{1}{D_{PWM}} - 2} \text{ A}$$

So

$$2.597 \times \sqrt{\frac{16 \times 385}{3 \times \pi \times 127} + \frac{1}{0.50} - 2} = 5.891 \text{ A}$$

Note that this agrees very closely with the numerical results provided in the table within Fig. 5.13 (for $D = 0.5$, which we could have used directly here). This would be the value on whose basis we would have picked the capacitor had there been no synchronization.

The improvement from the table in Fig. 5.13 is 21.9 percent. So using anti-synchronization, the capacitor RMS current rating requirement would be

$$I_{RMS\_ASYNC} = 5.891 \times \left(1 - \frac{21.9}{100}\right) = 4.60 \text{ A}$$

We also know that the low-frequency component of this is (for any duty cycle)

$$I_{RMS\_ASYNC\_LF} = \frac{I_{BUS}}{\sqrt{2}} = \frac{2.597}{\sqrt{2}} = 1.836 \text{ A}$$

Therefore, the high-frequency component must be

$$I_{RMS\_SYNC\_HI} = \sqrt{I_{RMS\_SYNC}^2 - I_{RMS\_SYNC\_LO}^2}$$

$$I_{RMS\_ASYNC\_HF} = \sqrt{4.60^2 - 1.836^2} = 4.22 \text{ A}$$

But these numbers are based on an assumption of a 1000-W (input) converter. In our case we have 70 W/0.7 = 100 W. Therefore, all we need to do here is to reduce the numbers calculated above by a *factor of 10*. The high-frequency and low-frequency components of the RMS current are therefore 0.42 A and 0.18 A, respectively. Knowing the frequency multiplier for the chosen capacitor family, we can now normalize these values to be an equivalent low-frequency current as explained earlier, and thus select our capacitor.

NOTE *We should remember that after the preceding calculation which is based on a wide input comparison, we don't know (and don't really need to know) whether the (worst case) synchronized RMS we calculated occurs at high line or at the low line. For selection of the capacitor, the information is sufficient. But as a matter of fact, if we want to know, this worst case is at low line. This can be easily figured out by looking at the table in Fig. 5.13 for $D_{PWM} = 0.5$.*

Quick Estimate of RMS of Low-Frequency (Line) Component in a Bulk Capacitor

We have used this quick estimate earlier. The low-frequency component (valid for both synchronization and anti-synchronization cases) is

$$I_{CAP\_RMS\_LF} = \frac{I_{BUS}}{\sqrt{2}}$$

where I_{BUS} is the fully averaged *load current* of the PFC stage.

$$I_{CAP\_RMS\_LF} \approx \frac{P_{IN}}{V_{BUS} \times \sqrt{2}} = 1.77 \text{ mA} \times P_{IN} \approx I_{CAP\_RMS\_LF} \approx 1.77 \text{ mA / W}$$

We have an easy rule here: 1.77 mA/Watt. This is for a bus voltage of 400 V. It is slightly higher if the bus voltage is lowered to 385 Vdc.

Quick Estimate of RMS of High-Frequency (Line) Component in a Bulk Capacitor

For a Flyback operated at $D = 0.2$, the high-frequency capacitor RMS (computed for the unsynchronized case) is

$$I_{RMS\_HF} \approx \frac{I_{OR}}{1-D}\sqrt{D \times \left(1-D+\frac{r^2}{12}\right)} = \frac{P_O}{100 \times 0.8}\sqrt{0.2 \times \left(1-0.2+\frac{0.4^2}{12}\right)} \approx 0.005 \times P_O$$
$$= 0.005 \times P_{IN}$$

Similarly, for a Forward converter running at $D = 0.25$:

$$I_{RMS\_CAP\_HF} \approx I_{OR}\sqrt{D \times \left(1-D+\frac{r^2}{12}\right)}$$

Since

$$I_{OR} \times D = \frac{P_{IN}}{V_{bus}}$$

$$I_{RMS\_CAP\_HF} \approx \frac{P_{IN}}{V_{bus} \times D}\sqrt{D \times \left(1-D+\frac{r^2}{12}\right)}$$

It follows that

$$I_{RMS\_CAP\_HF} = \frac{P_{IN}}{400 \times 0.25}\sqrt{0.25 \times \left(1-0.25+\frac{0.4^2}{12}\right)} = 0.00437 \times P_{IN} \approx 4.4 \text{ mA / W}$$

Numerical Table for Capacitor Selection and Comparison

In Table 5.8 we have compiled numerical results for three typical power converters using PFC and single-ended Forward converters. We have provided the results for low-line and high-line, and we have also provided the results for the following cases: no-synchronization, anti-synchronized (out of phase) PFC + PWM, and the wrong (in-phase) PFC + PWM synchronization. In particular, note that for the 530-W converter we would normally look for a 400 V/450 V aluminum electrolytic Bulk capacitor with a LF equivalent RMS rating of at least

PFC (V_{BUS} = 400 V) + PWM (D_{NOM} = 0.25) (Forward Converter, UVLO = 250 Vdc)								
Assuming 85% Efficiency		Bulk Capacitor RMS						
Output /Input Power (W)		Low-Freq RMS		High-Freq RMS		Total RMS		Effective Low-Freq RMS (Capacitor Rating @ 120 Hz)	
		85 Vac	270 Vac	85 Vac	270 Vac	85 Vac	270 Vac	85 Vac	270 Vac
6/7.06	UNSYCH	0.012	0.012	0.047	0.032	0.049	0.034	0.036	0.026
	ASYNCH	0.012	0.012	0.024	0.029	0.027	0.031	0.021	0.024
	SYNCH	0.012	0.012	0.053	0.037	0.055	0.039	0.04	0.029
75/88.23	UNSYCH	0.156	0.156	0.59	0.399	0.61	0.429	0.449	0.325
	ASYNCH	0.156	0.156	0.297	0.359	0.335	0.391	0.263	0.3
	SYNCH	0.156	0.156	0.667	0.46	0.685	0.486	0.501	0.364
530/623.53	UNSYCH	1.102	1.102	4.168	2.822	4.311	3.03	**3.175**	2.298
	ASYNCH	1.102	1.102	2.098	2.534	2.37	2.764	1.86	**2.119**
	SYNCH	1.102	1.102	4.713	3.253	4.84	3.434	3.542	2.572

TABLE 5.8 Numerical Capacitor Selection Numbers for Three Power Converters for all Types of Synch/No-synch

3.18 A (for the popular non-synchronized case). But using anti-synchronization (ASYNCH), this rating drops to 2.12 A. That is a 33 percent reduction. On the basis of holdup, from Fig. 5.7, we see we need 0.24 µF/W for 10 ms and 0.482 µF/W for 20 ms. Using the input wattage of 623.53 W, we get respectively, 150 µF and 300 µF. Looking on digikey.com, we find very few choices for minimum 105°C capacitors. We finally decide to use two 250-V capacitors in series, each of 680 µF (LP681M250H5P3 from Cornel Dubilier). Each capacitor has a unit cost of $5.5. In addition, each has an RMS rating of 2.07 A, so two capacitors in series will also have the same rating and will just meet our RMS requirement for the anti-synchronization case. But the net capacitance is 340 µF (two in series). So this will meet our 10-ms and 20-ms holdup time target. (However, note that we should typically leave at least 10 percent margin for tolerances of nominal value and another 20 percent for the steady loss in capacitance till end of life is actually declared.)

However, if we do not use anti-synchronization, we would pick LP122M250H9P3 for its 3.1-A rating. But this is a 1200-µF capacitor. Its unit cost is $8.4 (over 50 percent increase compared to the previous choice). Two capacitors in series would give us a net capacitance of 600 µF, twice what we need purely based on a 20-ms holdup time requirement.

We see that except for much smaller holdup times, and very low powers, we are severely limited by RMS ratings, not by capacitance or holdup considerations. So reducing RMS requirements would allow us to use smaller capacitances and save cost and energy. The anti-synchronization technique is very promising. It can help significantly lower RMS ratings, and also cost and space.

Design of a PFC Choke and PFC Design Tables

In Figs. 5.14 and 5.15 we have the full design of a PFC choke, separated into two quick-lookup figures. In Table 5.9 we have consolidated all the PFC design equations for easy reference.

Practical Nuances of PFC Design

1. A sequence of external logic is necessary to maintain the integrity of sequencing during fault conditions. This should not be underestimated. A recommended sequence is as follows:
   a. A small low-power auxiliary Flyback switcher should be present. This should work all the way from very low line voltages (~30 Vac) to maximum line voltage.
   b. This switcher provides power from its primary side to the PFC and PWM ICs, as also for the primary-side sequencing and protection circuitry. From its secondary

---

Example: 500 W, 85-265 Vac, Universal PFC, η=0.9, 100 kHz (Part 1)

Inductance (set $r = 0.2$ at peak voltage of minimum Vac):

The maximum instantaneous inductor current occurs at low line (85 Vac), at the moment the input AC voltage peaks. This is the design point at which we set $r = 0.2$. And that in turn determines the inductance.

The rectified AC input peaks at 85 Vac × √2 = 120 Vdc. The duty cycle at that moment is

$$D_{INSTANT} = \frac{V_{BUS} - \eta V_{IN\_PK}}{V_{BUS}} = \frac{385 - (0.9 \times 120)}{385} = 0.72$$

Note: The instantaneous load current at this moment is 2 × I_{BUS} (peak of sine-squared curve) (see Fig. 5.10)

We can work out inductance based on first principles. At the peak voltage, in terms of voltseconds

$$V_{ON} = L \frac{\Delta I}{\Delta t_{ON}}, \text{ so } L = \frac{V_{ON} \times D}{\Delta I \times f_{SW}}$$

where D is 0.72 at the peak input voltage

(call it $D_{AT\_VIN\_PK}$, f_{SW} is the PFC switching frequency in Hz.

The peak current (center of inductor current at peak Vac) is related to the peak voltage by $P_{IN} = \frac{P_{O\_PFC}}{\eta_{PFC}} = V_{AC} \times I_{AC}$

So, $P_{IN} = \frac{V_{IN\_PK} \times I_{IN\_PK}}{2}$, and $I_{IN\_PK} = \frac{2 \times P_{O\_PFC}}{\eta_{PFC} V_{IN\_PK}}$

But, since I_{IN_PK} is the center of ramp (I_{COR}) at peak input AC voltage, $r = \frac{\Delta I}{I_{IN\_PK}}$. We write $\Delta I = r \times I_{IN\_PK}$, and since $V_{ON} = V_{IN\_PK}$ at the peak, $L = \frac{V_{ON} \times D}{r \times I_{IN\_PK} \times f}$

$$L = \frac{V_{IN\_PK} \times D_{AT\_VIN\_PK}}{r \times f_{SW} \times \frac{2 \times P_{O\_PFC}}{\eta_{PFC} V_{IN\_PK}}}$$

$$L_{\mu H} = \frac{\eta_{PFC} \times V_{IN\_PK}^2 \times D_{AT\_VIN\_PK} \times 10^6}{2 \times r \times P_{O\_PFC} \times f_{SW}}$$

where $D_{AT\_VIN\_PK} = \frac{V_{BUS} - \eta_{PFC} V_{IN\_PK}}{V_{O\_PFC}}$

(duty cycle at peak of minimum line input)

Using numerical values we get

$$L_{\mu H} = \frac{0.9 \times 120.2^2 \times 0.72 \times 10^6}{2 \times 0.2 \times 500 \times 100 \times 10^3} \Rightarrow 0.47 \text{ mH} \leftarrow$$

Figure 5.14 Part 1 of PFC design choke procedure.

Example: 500 W, 85-265 Vac Universal PFC, η=0.9, 100 kHz (Part 2)

Core, Air Gap, N:

For a DC-DC Boost converter, we worked out that the energy per cycle that goes in and out of the inductor is

$$\Delta\varepsilon = \frac{P_{IN}}{f} \times D = \frac{P_O}{\eta \times f} \times D \approx \frac{P_O}{f} \times D \quad \text{(see Chapter 3)}$$

Rule

The inductor for a normal 500 W, 90% efficiency Boost converter, running at a duty cycle of 0.72, would technically need to cycle only 500 x 0.72/0.9 = 400W. But a PFC Boost stage is different, because its effective load current is I_{OE}, which at its peak, is <u>twice</u> the time-averaged value I_{BUS}. In effect, the instantaneous power a 500 W PFC Boost converter is delivering at the moment the AC voltage peaks, is not 500 W, but 1000 W. So the PFC Boost inductor has to cycle twice the energy of a conventional DC-DC Boost stage. The equations and numerical results are as follows.

$$\Delta\varepsilon_{PEAK} = \frac{V_{BUS} \times (2 \times I_{BUS})}{\eta_{PFC} f_{SW}} \times D_{AT\_VIN\_PK} = \frac{2P_{O\_PFC}}{\eta_{PFC} f_{SW}}$$

$$\Delta\varepsilon_{PEAK} = \frac{385 \times (2 \times 1.3)}{0.9 \times 100 \times 10^3} \times 0.72 \Rightarrow 11\,mJ$$

From Fig. 5.6 in Switching Power Supplies A-Z, 2e, we get for $r = 0.2$, the following peak energy relationship

$$\varepsilon_{PEAK} = \frac{\Delta\varepsilon}{8} \times \left[r \times \left(\frac{2}{r} + 1\right)^2\right] = \frac{\Delta\varepsilon}{8} \times 24.2 = 3.025 \times \Delta\varepsilon$$

So the peak energy handling capability of the PFC choke must be $3.025 \times 11 = 34\,mJ$

For a ferrite choke, z (air gap factor) can be set to <u>40</u> (for a transformer we pick z = 10 usually), and for a PFC choke we should set <u>r = 0.2</u>. Therefore, starting from our basic equation for core volume (as per Switching Power Supplies A-Z, 2e), we get:

$$V_{e\_cm3} = \frac{31.4 \times P_{IN} \times \mu}{z \times f_{MHz} \times B_{SAT\_Gauss}^2} \times \left[r \times \left(\frac{2}{r} + 1\right)^2\right]$$

$$V_{e\_cm3} = \frac{0.00422 \times 500}{0.1 \times 0.9} = 23.44\,cm^3$$

We can therefore pick ETD-49 core ($V_e = 24.0\,cm^3$) ←

Using equation for air gap vs. z we get total center gap

$$l_{g\_mm} = (z-1) \times \frac{l_{e\_mm}}{\mu} = (40-1) \times \frac{114}{2000} = 2.22\,mm \leftarrow$$

$$(N \times A_{e\_cm2}) = \left(1 + \frac{2}{r}\right) \times \frac{V_{IN} \times D}{200 \times B_{PEAK} \times f_{MHz}}$$

$$N = \left(1 + \frac{2}{r}\right) \times \frac{V_{IN} \times D}{200 \times B_{PEAK} \times f_{MHz} \times A_{e\_cm2}}$$

$$= \left(1 + \frac{2}{0.2}\right) \times \frac{120.2 \times 0.72}{200 \times 0.3 \times 0.1 \times 2.11} = 75\,turns \leftarrow$$

FIGURE 5.15 Part 2 of PFC design choke procedure.

Input voltage before rectification	$\sqrt{2} \times V_{AC} \times \sin(2\pi f_{AC} t)$
Input voltage after rectification	$\left|\sqrt{2} \times V_{AC} \times \sin(2\pi f_{AC} t)\right|$
Input current before rectification	$\sqrt{2} \times \frac{P_{IN}}{V_{AC}} \times \sin(2\pi f_{AC} t)$
Input current after rectification	$\left|\sqrt{2} \times \frac{P_{IN}}{V_{AC}} \times \sin(2\pi f_{AC} t)\right|$
Input power	$V_{AC} \times I_{AC} = \frac{V_{IN\_PK} \times I_{IN\_PK}}{2}$
Average input current	$\frac{2}{\pi} \times I_{IN\_PK} \equiv 0.637 \times I_{IN\_PK}$
Average input voltage	$\frac{2}{\pi} \times V_{IN\_PK} \equiv 0.637 \times V_{IN\_PK}$
Duty cycle of PFC	$1 - \left|\sqrt{2} \times \frac{V_{AC}}{V_{BUS}} \times \sin(2\pi f_{AC} t)\right|$
	This is (almost) identical to the large signal duty cycle equation for a conventional Boost topology, i.e., $D = (V_O - V_{IN})/V_O$. A slight imbalance is required to cause the current to progressively ramp up or down with the AC cycle.
Conduction duty cycle of diode	$\left|\sqrt{2} \times \frac{V_{AC}}{V_{BUS}} \times \sin(2\pi f_{AC} t)\right|$
Switch RMS current	$\frac{P_{IN}}{V_{IN\_PK}} \sqrt{2 - \frac{16 \times V_{IN\_PK}}{3 \times \pi \times V_{BUS}}}$
	Proportional to P_{IN}
Inductor RMS current	$\sqrt{2} \times \frac{P_{IN}}{V_{IN\_PK}}$
	Proportional to P_{IN}, inversely proportional to input voltage

TABLE 5.9 Design Table for PFC

Diode RMS current	$\dfrac{P_{IN}}{V_{IN\_PK}}\sqrt{\dfrac{16\times V_{IN\_PK}}{3\times\pi\times V_{BUS}}}$	
	Proportional to P_{IN}	
Peak current (diode, switch and inductor)	$2\times\dfrac{P_{IN}}{V_{IN\_PK}}$	
	Proportional to P_{IN}, inversely proportional to input voltage	
Diode average current	I_{BUS}	
	Proportional to P_{IN}, does *not* depend on input voltage or Duty Cycle	
Bulk Capacitor Currents		
Low-Frequency Component	High-Frequency Component	Total RMS [$\sqrt{(HF^2 + LF^2)}$]
$\dfrac{I_{O\_PFC}}{\sqrt{2}}$	$I_{O\_PFC}\sqrt{\dfrac{16\times V_{BUS}}{3\times\pi\times V_{IPK}} - 1.5}$ This assumes resistive load on PFC	$I_{O\_PFC}\sqrt{\dfrac{16\times V_{BUS}}{3\times\pi\times V_{IPK}} - 1}$ This assumes resistive load on PFC
$\dfrac{I_{O\_PFC}}{\sqrt{2}}$	$I_{O\_PFC}\sqrt{\dfrac{16\times V_{BUS}}{3\times\pi\times V_{IPK}} + \dfrac{1}{D_{PWM}} - 2.5}$ Assumes switching converter	$I_{O\_PFC}\sqrt{\dfrac{16\times V_{BUS}}{3\times\pi\times V_{IPK}} + \dfrac{1}{D_{PWM}} - 2}$ Assumes switching converter
where $P_{IN} = V_{AC}\times I_{AC} = \dfrac{V_{IPK}\times I_{IPK}}{2}$ $V_{IPK} = \sqrt{2}\times V_{AC}$ $I_o = P_{IN}/V_o$ Flat top approximation used (high inductance)		

TABLE 5.9 (*Continued*)

    side it provides an auxiliary rail that powers the overcurrent protection (OCP) and the output overvoltage protection (OVP) circuitry.

  c. Two diodes *before* the bridge rectifier from the L and N lines should be OR-ed to sense the presence of the unrectified AC line. After the bridge, the bulk capacitor keeps the voltage up, and we therefore cannot immediately know from that point whether a dropout is occurring or not. If there is a momentary line dropout (say 10 to 20 ms), then this should be sensed rapidly and the PFC IC should be disabled (no switching, but stays IC alive) for the duration of the dropout to avoid *overworking* the PFC choke. The IC should also recover promptly to restore the output voltage as soon as the dropout is over. Note that the PWM must continue to function normally during this time.

  d. Another technique to simplify the sequencing logic and also to save on the size of the Boost choke is to *simply reprogram the HVDC rail to say 250 V (rather than 385 V) under severe line sags*. That way the PFC struggles much less to maintain regulation. In effect this shifts the burden from the PFC stage to the PWM stage for a short duration (assuming D_{LIMIT} is not reached), while keeping the PFC section alive.

2. The UC3854A/B featured an improvement over the previous generation UC3854 in that by limiting its maximum multiplier output current, it provides foldback under line brownout and extreme line conditions. This may help reduce the complexity of the external logic.

3. The UC3854B has a lower start-up threshold voltage of 10.5 V As compared to 16 V for the UC3854A and is usually considered a better choice for this reason. Also note that the drive signal from the PFC controller to the FET is almost equal to the upper supply rail of the IC, and for reliability reasons we want to keep the Gate only high enough to ensure the FET turns ON fully.

4. A reference voltage for the comparators (typically 2.5 V) is therefore required on the primary side, and this may, for example, come from a TL431 that derives its power from the auxiliary rail.

5. We should also use this 2.5 V to set up an accurate OVP for the HVDC rail. Normally this should only act to drag down the soft-start pin of the PFC controller to pinch off the duty cycle, not to reset the PFC IC completely. This helps avoid nuisance tripping under power-up or line/load transients. But in more severe cases we should be able to cause complete shutdown of the entire power supply and PFC.

NOTE *If a 400-or 450-V electrolytic capacitor is overcharged, there is danger of a catastrophic and hazardous failure. But we can remember that the failure in such a condition is purely a heat buildup issue. So typically we may actually be allowed to exceed the voltage rating of an electrolytic capacitor 1.2 to 1.4 times, provided the duration is less than about 1 second (consult the vendor). However, a sustained overvoltage just cannot be tolerated. There is also a safety issue involved. Some control ICs may simply sense the voltage on the feedback pin to initiate overvoltage protection. While this is OK in low-voltage applications, IEC does not allow any single-point failure to cause a hazardous condition. So effectively this means that for either the PFC IC or the PWM IC in any off-line power supply, the pin used for normal regulation cannot be used for fault protection (either internal to the IC or via external logic). Because, in doing so we are actually trying to use an object to sense its own malfunctioning! For example, what if that pin just had a bad solder joint? This single-pin issue was actually overlooked inadvertently by safety test agencies themselves until some years ago.*

---

6. We should include fault detection for the auxiliary switcher and disable the PFC if, for example, the auxiliary rail is not getting up close to its regulation point.

7. We can also have proper sequencing under normal power-up or if the user uses an ON/OFF logic switch to turn the unit ON. The inrush current protection circuit must first start working at the same time the auxiliary switcher turns ON. Then provided the voltage on the bulk capacitor has increased beyond a certain level, the PFC IC gets enabled. When the reference voltage of the PFC IC comes up, it enables the PWM IC.

8. Many companies do not wish to use a single-source part. While PFC ICs from different vendors are certainly not pin compatible, the UC3854 is manufactured under license by Toko at www.toko.com. It is then called the TK83854D.

9. For practical examples of inrush protection circuits, turn-on snubbers, and an interesting technique to get two UC3844s to mutually synchronize, see Chap. 20.

10. There is an "automatic" PFC control scheme used sometimes, which requires a Flyback PFC stage operating in critical conduction mode (boundary between continuous conduction and discontinuous conduction modes). Since $V = L dI/dt$, in critical conduction we get $V = LI_p/t_{ON}$. So if we keep a constant ON-time, I_p becomes proportional to V. But the average of the ON pulse is $I_p/2$, so the average is also proportional to V. If the input is a pure sine wave, then so is the average input current. This scheme, however, requires differential sensing of the output voltage for regulation, since the ground of the control IC is no longer the same as the power ground. We can use the differential sensing techniques in the Chap. 2 on DC-DC converter topologies and configurations.

11. The biggest hit in efficiency in a Boost stage may come from the severe shoot-through current from the bulk capacitor that passes through the yet unrecovered 600-V PFC diode when the switch turns ON. Since this current passes through the FET while there is still a high voltage present across it, we can get large $V \times I$ crossover losses. Besides the problem of the fall in efficiency, the required heatsinking of the FET is also affected significantly. In this position it is critical to use an extremely fast diode. Any diode with more than 20 to 30-ns recovery time is unacceptable except in very low power and noncritical applications.

HINT *Check not just the typical value of the recovery time in the vendor's datasheet but its maximum value.*

---

12. Therefore, for low-power applications engineers often prefer to use Boost PFC ICs that operate in critical conduction mode. In this case, since the current through the diode has fallen to zero when the switch turns ON, the diode has fully recovered and there are no reverse recovery issues and no shoot-through. But such ICs necessarily operate with variable frequency and their EMI filtering often poses a bigger problem.

13. In low-to-medium-power PFC stages, the lossless turn-on snubber which is virtually indispensable in high-power stages, may seem like a luxury (for a practical turn-on snubber circuit, see Chap. 20). In that case some engineers like to reduce the reverse recovery current spike by replacing the single 600-V PFC diode with two 300-V diodes in series. Here they are relying on the fact that low-voltage diodes recover much faster than high-voltage diodes. So, despite their higher combined forward drop and consequent increased conduction losses, we actually improve efficiency by reducing the $V \times I$ crossover losses. But note, we cannot allow the full voltage to appear across either one of the diodes at any moment, however brief. So they must be well matched, especially in terms of their dynamic characteristics. That is not easy. Some engineers try achieve this by placing ballasting resistors across each diode, much as we do for capacitors in series. But under dynamic conditions (during transitions), this cannot help even in principle, because the resistor lead inductances will prevent them from responding immediately to any sudden changes. The other option is to use two series diodes *in one package*, on the assumption that since they pass through exactly the same fabrication steps, they will automatically be well matched. Well they don't always. Some manufacturers take separate chips and combine them in a package. Part of the reason is that this diode combination requires an odd internal connection: the anode of one connected to the cathode of the other.

14. Recently some vendors (e.g., ST Microelectronics at www.st.com) have come up with *tandem diodes* based on this principle. They are rated (and look) like a single 600-V diode and have two leads, but they offer typical recovery times of about 12 ns. This is labeled "hyperfast".

15. A modern alternative is to use silicon carbide PFC diodes, as from Cree Inc.

16. A critical component of a commercial PFC implementation is a surge diode placed from the positive terminal of the bridge rectifier to the cathode of the PFC diode. This allows the bulk capacitor to charge up while keeping the huge inrush current away from the PFC section's choke and diode. But this component is also a key reliability issue for the entire power supply. *It is the most likely to fail under repeated application of AC input.* It need not be a fast diode as it goes out of the picture as soon as the PFC switch turns ON. It does not get hot and can be an axial component. But its nonrepetitive surge rating must be very high. Some designers still prefer to make this diode a fast (expensive) diode possibly because of EMI concerns. They use, for example, a freestanding 16 A/600 V diode (with *inrush current protection circuitry present*). But a slow diode like the 1N5408 (from a quality vendor) should also be tried out since it has higher surge ratings, on paper at least. Note that in this position, two diodes in series can also be tried out if we want to shift some of the surge current back into the PFC choke and diode. But the relative sharing then depends on various parasitics like the DC resistance of the choke. Therefore, we should do this only with great care.

NOTE *One general problem of using cheap multiple-source parts is that not all vendors have the same fabrication process or quality even for the "same part". So if we specify the 1N5408 for the surge diode and are careful enough to pick a quality source initially, there is no guarantee that tomorrow a "smart", cost-reducing purchase officer won't start buying the "same part" from an alternative source, with consequent reliability issues.*

17. For the Boost inductor of a DC-DC converter stage we normally set the inductance so that we get an inductor current ripple ratio $\Delta I/I_{AVG}$ of about 0.4, on the grounds that this represents a good compromise between the size of the inductor and the capacitors. But in a PFC Boost stage this holds little meaning, since the size of the capacitor is being determined by several other criteria, like the holdup time. Besides, the RMS current through the capacitor is also being determined by several factors, not just the current ripple ratio of the inductor. Therefore, for a Boost PFC, the inductance of the choke is typically chosen such that at low instantaneous line voltages the inductor is operating in discontinuous mode, and getting close to the peak of 90 Vac it starts operating in continuous conduction mode. This minimizes the size of the inductor. *In practice, we should set r = 0.2 at the peak of the rectified low-line alternating current,* as in the solved example given in Figs. 5.14 and 5.15.

CHAPTER 6
Topologies for AC-DC Applications: An Introduction

Introduction

In this chapter we will rather briefly introduce the Forward and Flyback converters. In Chaps. 10 and 11 we will discuss more about their magnetic design procedures. In Chap.12 we will consolidate by presenting a detailed step-by-step design, with a rather revealing topology comparison. In Chap. 16 we will describe modern techniques like active clamps and reset techniques related to these topologies.

PART 1 THE (STRAIGHT) FORWARD CONVERTER

This converter has two magnetic elements: the transformer and the output choke (inductor). Much of the skill involved in designing a good Forward converter is in understanding how to design its transformer optimally. That involves, among other things, understanding the rather difficult concept of *proximity effect*, explained in more detail in Chaps. 11 and 12.

In Fig. 6.1 we have a simple single-ended (one-switch) Forward converter. Besides the primary and secondary windings we also have an energy recovery winding (or *tertiary winding*, subscripted T). This tertiary winding is, in effect, the freewheeling path for the *magnetization current* I_M, shown as a gray area in the current waveforms.

Keep in mind some basic magnetics here: When the FET is ON, because of the polarities (dots next to the windings) and the direction of the output diode, the secondary winding also conducts. In doing so, the (changing) current through it cancels out the flux produced by the (changing) primary-winding current—all *except for the magnetization current component of the current in the primary winding*. So, if there was no load across the secondary winding, the flux in the transformer would be due to the magnetization current only. If there is a load connected, *additional* current flows into the primary winding to cancel out the flux from the current in the secondary winding. That is just the basic law of induced voltage playing out in the core. In other words, for any load condition, we are left only with the flux due to the magnetization current—same as in a no-load condition.

This also means that the Forward converter transformer is needed only for a very small energy-storage function: related to its magnetization (no-load primary-side) current. The transformer core does not store any of the energy associated with the output power of the converter. Which is why we talk of a Forward converter transformer not in terms of its energy-handling or energy-storage capability but in terms of its energy (or power) *throughput*, or sometimes its *power*-handling capability.

Since the flux in the Forward transformer remains unchanged for all loads, the question is: What does the power-handling (or throughput) capability of a Forward transformer depend on? We intuitively realize that we can't use just any transformer size for an output wattage! There must be some limit. But what is it? We learn that it is determined simply by *how much copper we can squeeze into the available window area of the core* (and more importantly, *how well we can utilize this available area*) without getting the transformer *too hot*. We will discuss this further in Chap. 12.

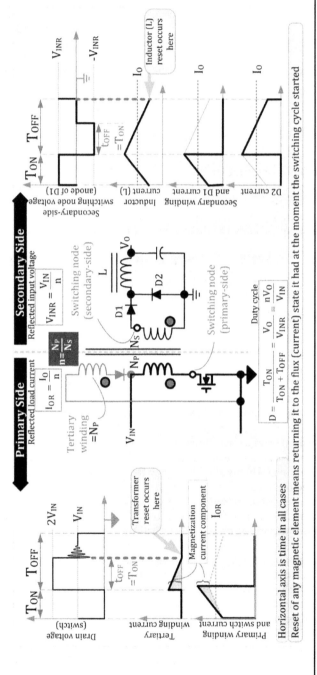

Figure 6.1 The single-ended Forward converter.

NOTE *The preceding is true for a Forward converter transformer, but not for a Flyback transformer as discussed later in this chapter. Engineers who try to apply the selection criteria of Forward converter transformers, such as area product (discussed in Chap. 12), to Flyback transformers (common mistake), or who try to apply energy storage concepts to Forward transformers (uncommon mistake), are both on the wrong path.*

We have learned very early with DC-DC converters that an inductor cannot tolerate discontinuities in its current. Thus catch diode is required to allow freewheeling current to continue to flow in the inductor after the main FET turns OFF. In a similar manner, a transformer cannot tolerate any discontinuity in the quantity $\Sigma N_a I_a$ (ampere-turns of all its windings summed up algebraically). Which also implies that if one winding stops, so long as another winding continues (with the appropriate current ratio), the ampere-turns are maintained at the switch crossover and the transformer does not "complain". This is very similar to the concept of current continuity in an inductor, only now it is the net ampere-turns through all the windings of a transformer. In other words, the magnetization energy that was stored in the transformer during the switch ON-time, freewheels into the *input bulk capacitor* through the tertiary winding (and tertiary winding diode) when the switch turns OFF. And this creates *transformer reset*.

By nature (i.e., the law of induced voltage), every inductor naturally tries to reset itself in an effort to achieve a steady state (provided we allow it to do so via an appropriate circuit configuration). It attempts to do so by reversing the voltage polarity across itself (which is the volt-seconds law in effect). Similarly, in a transformer in steady state, the volt-seconds during the ON-time must be equal and opposite (in sign) to the volt-seconds during the OFF-time. If we look at the voltages (and their related time intervals) across any selected winding, then, irrespective of whether that winding is actually conducting current or not, we will get volt-seconds balance for that winding, as for any other winding on the same core.

In a Forward converter, the ON- and OFF-time voltages across the primary winding have the same magnitude because the tertiary to primary turns ratio is 1:1 (by design). In Chap. 16 we will discuss the case of arbitrary turns ratios between primary and tertiary windings. But this is how it happens in the simple case: The voltage across the primary winding when the switch is ON is V_{IN}. The energy recovery (tertiary) winding has one of its ends connected to primary ground, while the other end gets clamped to the input bulk capacitor. So the voltage across the tertiary winding is V_{IN} as soon as the switching FET turns OFF. Since the turns ratio between the primary winding and the tertiary winding is 1:1, the tertiary winding reflects exactly the same voltage *across* the Primary, but its sign is reversed on account of the need for maintaining volt-seconds balance. Keep in mind that every winding of a transformer has exactly the same volts per turn at any given instant (except for leakage inductances, which constitute unlinked inductances in effect). So, when the switch turns OFF, the voltage *across* the primary winding is again V_{IN}, though now in the opposite direction.

So far we have been talking about the voltages *across* the primary winding. But now, looking at it from the viewpoint of the voltage across the FET, there is almost no across the FET when it is ON, and $2 \times V_{IN}$ across it during the OFF-time. And that determines the minimum voltage rating of the FET for this topology (we need to set it at the maximum input, V_{INMAX} of course).

With all this in mind, for a worldwide input power supply we realize we can usually make do with 800-V switches, despite the somewhat reduced voltage derating margins. However, if a front-end Boost PFC preregulator is present (HVDC rail of 385 Vdc or 400 Vdc; see Chap. 5), we would certainly need at least a 850- or 900-V FET for the Forward converter. Compare this to the typical 600- and 700-V FET minimum rating required for a Flyback in the same application without PFC, and with PFC respectively.

Ensuring transformer "reset" creates another restriction: *The duty cycle of such a Forward converter can under no circumstances be allowed to exceed 50 percent.* The reason for that is we have to unconditionally ensure that the transformer will reset, that is, *return to the same net flux/current value it had at the start of the switching cycle* (zero in this case). Otherwise every cycle there may be a net increment, causing *flux/current staircasing*, leading to transformer core saturation and eventual switch destruction. In a Forward converter, except for peak-current limiting on the primary side, which can be too slow or even ineffective *if the transformer is already saturating*, we really have no direct control on the transformer current to ensure that *on a cycle-to-cycle basis*, there is no staircasing. So, the practical option is to simply *leave enough time, every cycle*, for the current in the tertiary winding to ramp down to zero. This ensures volt-seconds balance by simply leaving enough seconds. Note that the slope of the rising (magnetization) current ramp during the ON-time is V_{IN}/L_{PRI}, where L_{PRI} is the primary-winding inductance (measured with all other windings open), and during the OFF-time, this current is continued by the current in the tertiary winding, and ramps down at $-V_{IN}/L_{PRI}$ since the tertiary winding has the same number of turns and also inductance, as the primary winding. So, it is obvious that we allow *volt-seconds balance* to occur naturally in the transformer, by simply leaving more time for the ramp-down to zero. Reset will occur as soon as t_{OFF} (the time the tertiary winding current takes to ramp down to zero) becomes equal to t_{ON} (the time for which the magnetization current ramped up). See Fig. 6.1. However, if the duty cycle exceeds 50 percent, t_{ON} (same as T_{ON} in this book) would certainly always exceed t_{OFF}, and therefore transformer reset can never occur, geometrically speaking either. Conversely, by ensuring the duty cycle never exceeds 50 percent (in practice this is set to about 45 percent maximum, to account for delays and tolerances), we can guarantee transformer reset every cycle.

Note *A practical reason why we cannot control flux staircasing by monitoring the current through the FET, is that the primary-side current is a combination of the magnetization current (which can cause flux staircasing) and a component related to the load current (which can be much larger, yet has nothing to do with the flux). So there is no way we can monitor one and not the other. In effect, we have no control over the magnetization component.*

We realize that the Forward converter transformer is always in DCM: Its current returns to zero every cycle. This is a special case of reset. In contrast, its choke (inductor) also resets, but is usually in CCM and with a designed r of 0.4 as for any Buck converter. See Fig. 6.1.

We can look upon the Forward converter, as a Buck stage (or cell), around the Forward converter's choke (inductor). The output of this Buck cell is V_O of course. The input, however, is in effect, the *reflected input voltage* $V_{INR} = V_{IN}/n$. This is the voltage that we can mentally imagine is the DC input rail of the Buck converter (or cell). Its duty cycle is (as for a Buck)

$$D = \frac{V_O}{V_{INR}} = \frac{V_O}{V_{IN}/n} = n\frac{V_O}{V_{IN}}$$

Alternatively

$$V_O = n \times D \times V_{IN}$$

There are two slightly different ways to intuitively look at these equations.

1. The input rail is reduced by a factor n and applied to the input of the secondary-side Buck cell.
2. If we want to go from a rectified AC input of, say, 230 Vac $\times \sqrt{2}$ = 325 Vdc, straight down to 5 V with a simple Buck stage, the required duty cycle would be 5/325 = 1.5 percent. That is completely impractical. Such narrow pulses are barely even possible. So, with the help of a brute step-down function from the turns ratio, say by setting n = 20, we manage to multiply this sliver of a duty cycle to something feasible: 20×1.5 percent = 30 percent. That is indeed what we do in a typical single-ended Forward converter, because we are also limited to the maximum duty controller cycle D_{MAX} of 50 percent to avoid staircasing. The turns ratio is in fact selected in this manner.

We see that the turns ratio is just one of the advantages of using a transformer that we need in high-voltage applications. It so happens, that for safety reasons, we also need *isolation* between the high AC input voltage and the output DC rail. The transformer ends up providing both these functions: step-down and isolation.

To design the primary side (pick a FET current rating, for example), we can mentally imagine that the (choke) inductor current I_L is reflected on to the primary side, with an average (DC) level of $I_{LR} = I_L/n$, where $n = N_P/N_S$. As for any Buck-derived topology, which the Forward converter is, the average (reflected) inductor current is equal to the (reflected) load current. Thus we get

$$I_L = I_O \quad \text{and} \quad I_{LR} = I_{OR}$$

where we have defined the *reflected output current* as $I_{OR} = I_O/n$. This I_{OR} is in effect the load as far as the switch is concerned. It is the *center of ramp* of the primary-side current, just as I_O

(load current) is the center of ramp for the secondary-side diode/inductor current waveforms. See Fig. 6.1. Of course, we need to add a small magnetization component on the primary side too, as per Fig. 6.1.

> **NOTE** *We suggested previously that the flux in the transformer is independent of load current. But the truth is that the Forward converter duty cycle, which is dictated completely by the Buck cell on the secondary side, pinches off under very light loads. This happens because the output choke enters DCM at light loads, and its duty cycle then suddenly shrinks (see Chap. 4). This reduced ON-time changes the shape of the magnetization current waveform. So now, I_M does actually depend on load current, despite what we had initially implied! Nothing is completely obvious in power conversion.*

As mentioned, in the simplest case, the tertiary winding has the same number of turns as the primary winding. It is usually wound *bifilar* with the primary winding: both are wound twisted together, or at least laid parallel alongside each other. This is usually considered necessary, since it is helpful if the tertiary winding couples well to the magnetization (primary) winding—with no leakage between the two; otherwise the remaining primary winding leakage can produce high-voltage spikes, which may destroy the FET. Note that one thing is unavoidable: There is always a large *voltage difference* between the two adjacent bifilar windings in this configuration because of the opposite polarities of the primary and tertiary windings. There may also be pinholes in the insulation of the windings due to manufacturing variations, or from accidental scratches incurred during production. Therefore, to avoid flashovers from occurring and destroying the converter, especially in off-line or similar applications, we may sometimes prefer *not* to wind the primary and tertiary windings bifilar—instead separating their copper layers by a thin layer of polyester tape. But then because of the poorer coupling, we will almost certainly need a small (dissipative) snubber or clamp to protect the switch from the resulting leakage inductance spikes. Note that the *leakage inductance* referred to here is the one between the primary and tertiary windings. Keep in mind that, since in a Forward converter, the secondary winding does not conduct when the primary winding is not conducting, in effect, the magnetization inductance behaves as an uncoupled leakage inductance term. And it is its energy we are trying to recover using the (well-coupled) tertiary winding.

There are also subtleties on where to position the diode of the energy recovery winding, as explained in Fig. 6.2. This is almost always missed in related literature.

A variant, the two-transistor (or asymmetric) Forward converter has become a workhorse of the medium-to high-power segment. See Fig. 6.3. Though still restricted to a D_{MAX} of 50 percent, it has no tertiary winding. It relies on two 400- to 500-V switches (driven in-phase), and two primary-side ultrafast diodes to route the magnetization current back into the bulk capacitor (see Fig. 6.2). This configuration, however, requires a floating (high-side) driver for the upper switch. This can be created either from a gate-drive transformer or by using a solid-state driver IC like the IRF2110 or IRF2112 from International Rectifier. Since there is no tertiary winding, and the primary winding is itself going to freewheel the magnetization current into the bulk capacitor, there is no question of leakage as discussed previously for the single-ended Forward. And therefore, there is no need for snubbers or clamps here. Further it can be shown that even if one FET is a little *sluggish* to turn ON or OFF compared to the other, this timing mismatch produces no major problems: certainly no attempted shoot-through or cross-conduction related effects. All this makes the asymmetric Forward a very robust topology. Think of it as a half-bridge with a buffer inductance between the switches (see Fig. 6.3 also, for other Buck derivatives).

The other possible design complexity in any Buck-derived Forward-converter-type design involves the coupled inductor for multiple regulated outputs. Here we replace individual output chokes of a multi-secondary-winding transformer with just one coupled component. This improves the cross-regulation and also helps in keeping the choke operating in continuous conduction mode under corner conditions: for example, if the main output is only lightly unloaded, but one or more of the other outputs are still loaded. The main design-related aspect we need to remember here is that the turns ratio we use between the windings in the choke must be as close as possible to their (mutual) ratios in the transformer, which in turn is simply based on the voltage requirements (ratio of voltages). So for example, ignoring diode drops, if the main (regulated) output is of 5 V, has 5 turns on the transformer and, say, 10 turns on the choke, then a desired 12-V output rail will require 12 turns on the transformer and 24 turns on the choke. So, the turns are always in the ratio of the output voltages, in both the transformer and choke.

166 Chapter Six

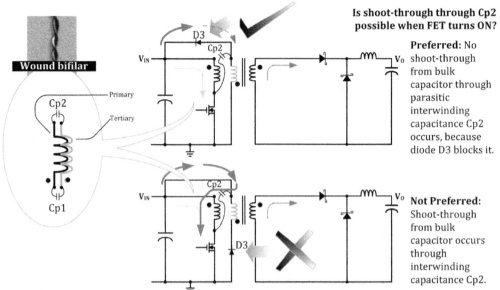

For minimizing leakage inductance, the energy recovery (tertiary) winding is usually wound bifilar to the primary winding. Therefore, the dotted polarity ends of the two windings fall adjacent to each other — with a certain lumped interwinding capacitance Cp1 between them. Similarly, the non-dotted ends lie adjacent too, with interwinding capacitance Cp2 between them. But in terms of voltages, the adjacent ends can have huge voltage differentials across their interwinding capacitors, and if permitted, a huge shoot-through current can then flow through the switch, causing a significant increase in its switching losses.
Note: The worst-case voltage differential between the two bifilar windings at any point along their length is V_{INMAX}, and the insulation of the enameled (magnet) wire must be adequate to handle that. So, the voltage rating of magnet wire insulation is very important from a reliability viewpoint. But safety requirements for AC-DC converters prohibit any primary-side and secondary-side windings to be placed physically next to each other (or wound bifilar), unless approved triple-insulated magnet wire is used for the purpose. So, in general, from a safety viewpoint, the voltage rating of (non-triple) insulated magnet wire is irrelevant.

FIGURE 6.2 Correct position for tertiary winding diode.

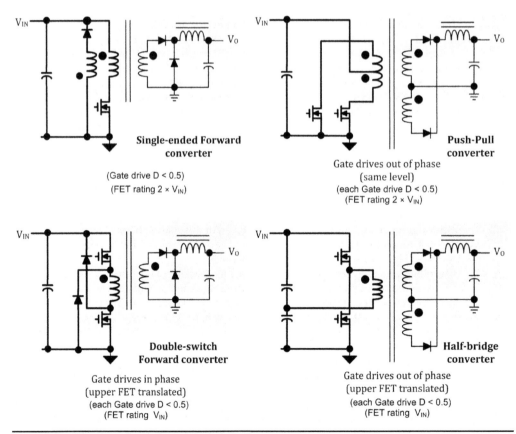

FIGURE 6.3 Other transformer-based Buck derivatives.

The coupled inductor technique also gives us an opportunity for *ripple steering*, as was promoted in early application notes from Unitrode (now Texas Instruments). But we should not count on always being able to implement this subtlety successfully in a *commercial* design, because ripple steering, though forming the basis of many excellent articles, relies on parameters that are not, or cannot, be guaranteed or tested easily in production. We should therefore assume the worst case for calculating output ripple in particular.

Note that in a coupled inductor of a Forward converter, the polarities of all the coupled windings are the same, unlike modern interleaved Buck converters using coupled inductors, in which the polarities are deliberately placed opposite for various reasons. See the book *Switching Power Supplies A-Z*, 2d ed, for a detailed study of the latter type of coupling.

Note that sometimes a separate auxiliary winding is thrown across the output choke, but is not from the transformer. Its polarity (and diode) is, however, such that it will conduct only when the current in the choke is ramping down; i.e., this low-energy winding operates essentially in Flyback mode, off the Forward converter choke. By coupling this auxiliary winding tightly with the main winding of the choke, and choosing the turns ratio according to the desired auxiliary output voltage, we can get fairly good regulation (about ±5 to ±10 percent) on this rail. This technique works well for auxiliary power outputs of up to about 5 to 10 percent of the power of the main winding. If our production department accepts, we can try twisting this auxiliary winding tightly with the main winding. In that case we may be able to double (up to 20 percent) the auxiliary power output without sacrificing regulation.

The Forward converter itself can actually be configured in several ways, depending on where the choke is placed and also on whether we are using the two secondary diodes in common-cathode or in common-anode formation. See Fig. 17.5.

Flux Balancing in the Push-Pull

In a Push-Pull, small differences in resistances and so on, in the two halves, result in a small net DC voltage across the Primary every cycle. Alternatively stated, even for the same applied voltage, slight differences in winding asymmetries cause the magnetizing current and core flux to slowly move in one direction until, finally, the transformer saturates. This is called *flux staircasing*. As indicated previously, the switch current is a sum of reflected output current and magnetization current. So relying on a fixed current-limiting device, for any load, as in voltage- mode control, is not enough to save the show. However, if we use the voltage drops across the two sense resistors (inserted in the source of the two FETs of a Push-Pull), on a *cycle-to-cycle basis*, as in current-mode control, we can make the control circuit perceive any differences in current between the two halves every cycle and correct for it automatically. Early push-pull converters used voltage-mode control and a hefty transformer to perhaps inadvertently avoid flux staircasing—because if the current in one half goes up steeply, it does tend to correct itself too somewhat, by creating larger drops across the switches and winding resistances, which ultimately reduce the voltage on the *runaway winding*—that is, if the transformer has not already saturated and blown up. However, the best solution in a Push-Pull is current-mode control. When the magnetizing current drifts off in one direction, it adds to the inductor current in one FET and subtracts in the other transistor. With voltage-mode control, current pulse inequality will only get worse and worse until the transformer core saturates. With current-mode control, the controller will terminate each pulse *when the same peak current is reached*. So, the current pulses will always have the same amplitude, although the pulse widths will be slightly different, on account of the original asymmetry. But since flux depends on current, not directly on pulse width, the flux staircasing gets corrected on a cycle-by-cycle basis. So, finally, we can make Push-Pull transformers small, as originally claimed. But how small? That topic is touched upon in Chap. 12.

We will also discuss the above-mentioned control schemes in greater detail in Chap. 14, including something else required to make current-mode control work well (in a stable fashion): slope compensation, which in effect mixes a little voltage-mode control (a small fixed-voltage ramp) to the larger sensed current ramp of current-mode control.

Flux Staircasing in the Half-Bridge

In the half-bridge, one side of the transformer Primary is connected to the midpoint of a capacitive voltage divider, although it could have just one capacitor connected to ground, since a capacitor blocks all direct current anyway. In the latter case we will find the single capacitor will eventually charge up very close to half the supply voltage. The reason is that the switches are driven opposite in phase, so when the top FET turns ON, the bottom FET is

OFF and vice versa. Further, each transistor turns ON, theoretically at least, for exactly the same time duration (same pulse width as the other). So the current goes in opposite directions in the two switching phases (or halves) of the switching cycle, ensuring transformer reset and also capacitor charge balance (the latter settling at half the supply voltage by symmetry in its charge and discharge cycles). It sounds promising enough, at least in principle.

There is one issue. In practice, we can use only one sense resistor, because only one FET is ground-referenced. So in any case we are in trouble already in correcting for any potential buildup in one direction by using current-mode control, because the only way we can correct for that is by comparing currents in the two halves of the cycle and fixing them to the same peak current value. So current-mode control would be ineffective (in fact impossible) with this simple implementation—because now we have information only about one-half of the cycle. But suppose we circumvent that practical issue and introduce a current-sense transformer to sense the current in both halves of the cycle in both FETs, or through the inductor. Now current-mode control seems a possible solution to the problem of flux staircasing. But was voltage-mode control really such a bad idea here?

Let us closely reexamine what would happen with pure voltage-mode control and with just one sense resistor. Suppose there were differences in the two switching halves, due to which the actual voltages applied across the winding in the two halves of the cycle were slightly different in magnitude. Suppose the top transistor was ON for a little longer. That would initially cause more current to flow from top to bottom in the primary winding, as compared to the opposite interval when the bottom FET turned ON. That would really not be so bad because this particular direction of current would end up charging up the capacitor more, which would then eventually decrease the voltage applied across the primary winding during the longer pulse width of the upper transistor. To create this self-balancing act, the capacitor needs to charge up quite quickly because that is how the voltage applied by the top FET will get ultimately reduced, which in turn will, despite the longer applied time, reduce the volt-seconds applied, thus reinstating volt-seconds balance. Putting very large blocking capacitors in a half-bridge would obviously tend to bring back to the fore the problem of staircasing and core saturation when using voltage-mode control.

Suppose we decide to use current-mode control instead. Of course, we need a way to sense current in both directions (both halves of the full switching cycle). However we do that, now when the circuit sees that the upper transistor is driving more current, it will try to limit it by reducing the pulse width. That will prevent the coupling capacitor from charging up. It could even end up making it discharge further, thereby increasing the voltage applied by the upper transistor. So current-mode control is really not suited for the half-bridge. Refer to Fig. 2.16 (Chap. 2) where we had talked about a certain duality principle at work in power supplies: capacitor ampere-seconds balance versus inductor volt-seconds balance. Viewed with that perspective we can see the basic problem with the half-bridge using current-mode control: We had indicated for the Push-Pull, that to correct for the current staircasing arising from reasons of asymmetry two halves, we would need to adjust pulse widths slightly. That could potentially correct for the difference in currents and thus ensure volt-seconds balance once again (for the inductor involved). Unfortunately, in the case of the half-bridge, by varying pulse widths proactively, we change the height of the *voltage ramp applied across the capacitor* (or capacitors), and this disturbs its ampere-seconds (charge) balance. In other words, in an effort to correct volt-seconds (flux) balance, the capacitor can end up getting charged up in one way—unfortunately, the wrong way, accentuating the problem we started with. So it is commonly said that the half-bridge is only compatible with voltage-mode control. However, Runo Nielsen tries to show that with slope compensation of a certain value, even the half-bridge can be used with current-mode control. See www.runonielsen.dk/Half_bridge_control.pdf. Understandable, because slope compensation is a little voltage-mode too.

In contrast, the single-ended and two-switch Forward can use either form of control in principle. This also explains their wide popularity.

PART 2 THE (TRICKY) FLYBACK CONVERTER

Nothing illustrates the finer twists and nuances of a power converter design process than the arguments put forth to either support or refute a certain design choice. In that spirit, we will delve into this rather tricky topology, also known as the Flyback. The Flyback forms the basis of most low-power off-line converters but is often underestimated in its complexity. It therefore deserves much more than a passing interest here. We all think we know all about it, but understanding it well is arguably one of the most challenging tasks a power supply designer will face. Among other things we will therefore also take up subtle issues that go into making a good commercial Flyback design, like minimum pulse width, feedforward, and protection.

To make it more interesting, we will present this topology from the angle of an ongoing comparison with the specifications and recommendations of a hypothetical product. We christen this off-line integrated switcher the *IP switch* (for integrated power switch). We will try to see what seems right in this product and what could have possibly been done better the next time around.

The Integrated Power (IP) Switch

Delving into the history of our fictitious benchmark product, we make a curious observation. The older generation of this product family has a maximum duty cycle of 67 percent. In the new generation, this number has mysteriously become 78 percent. The improved duty-cycle feature has even become an enduring marketing tool, with the front page of the datasheet bulleting "wider duty cycle for more power, and a *smaller input capacitor*".

As any engineer should, we take the trouble of going through the collateral, but are a little surprised to find that the evaluation boards themselves often don't seem to comply with the printed recommendations. We try the accompanying expert system software and ask it to auto-design for a given application, first using the older-generation and then the newer-generation device family. But we are surprised to see that it always suggests virtually the same input capacitor. The promised reduction in the size of the input capacitor, or in fact any lower associated cost, doesn't seem to be bearing out. Our curiosity is piqued, and we decide to check out this potential marketing hype further.

The Equivalent Buck-Boost Models of a Flyback

But first, we need to brush up on our basic understanding. Let us examine the Flyback shown in Fig. 6.4. The pertinent points are

- The gray and hatched areas of the voltage waveforms shown correspond to the switch being ON.
- The voltage *across* the primary winding scales according to the ratio N_S/N_P to become the voltage *across* the secondary winding.
- Keep in mind that the term *turns ratio* in literature, which we call n here, sometimes stands for N_P/N_S, sometimes N_S/N_P. We are using the former convention.
- The voltage across the secondary winding gets multiplied by n as it is scaled on to the primary side.
- Therefore, knowing that the voltage across the Secondary during switch turn off is V_O, we get a *reflected output voltage*, designated V_{OR} on the primary side. We have $V_{OR} = V_O \times n$, which is the transformer voltage scaling rule as shown in Fig. 6.4.
- We know that during switch turn-on, the voltage across the primary winding is V_{IN}, so the corresponding voltage level during that time, appearing across the secondary winding, must be V_{IN}/n. This is the *reflected input voltage*, i.e., the input reflected to the output side, though it is not popularly known by any particular name. Here we will just call it V_{INR}.

$$V_{INR} = \frac{V_{IN}}{n}$$

- The voltage across the secondary winding when the switch turns ON is the reflected input voltage V_{INR} because when the switch turns ON, it applies V_{IN} across the primary winding. When the switch turns OFF, the secondary winding voltage flips and forward-biases the output diode, so the winding gets clamped to the output voltage. Therefore, the *peak-to-peak* voltage appearing across the secondary winding (over the full cycle) is equal to $V_{INR} + V_O$.
- This also happens to be the amount of reverse voltage seen by the reverse-biased output diode, and so, the diode must be rated to handle this voltage at a bare minimum. Note that this diode voltage rating must be established at the maximum input voltage when V_{INR} is also at its highest.

NOTE *In literature, the term V_O is often neglected above. It is generally, but incorrectly, stated that the diode needs to withstand only V_{IN}/n. That could constitute a significant error in the component selection procedure, especially for high output voltage applications.*

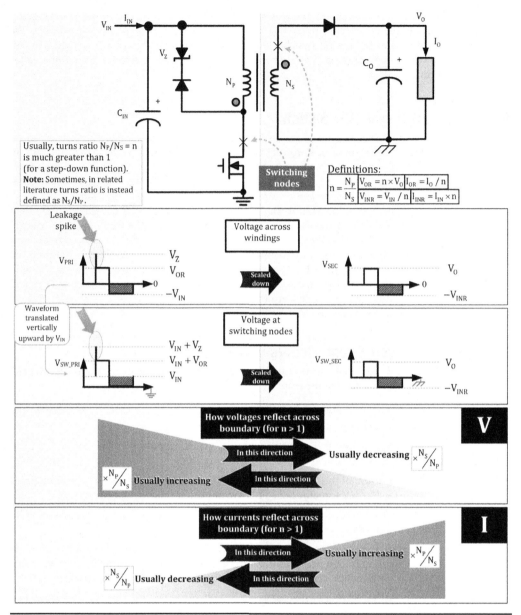

FIGURE 6.4 How voltages and currents reflect in a Flyback.

- Having applied scaling to the voltage levels *across* the windings and thereby having deciphered what they are, we then deduce the voltage at any point (with respect to ground). For this we apply any required amount of voltage *translation* (i.e., level shifting) to either the primary or secondary voltage waveforms, so as to express them with respect to the new desired reference level. But in doing so, we must keep the overall shape and peak-to-peak value unaltered as we shift the waveform vertically up or down. We have to be cautious of the fact that transformer voltage scaling, applies only to the voltage levels appearing *across* the windings, not necessarily to the absolute voltage levels at various points in the primary and secondary sections. Yes, if scaling does seem to apply elsewhere, it is just a coincidence. In general, transformer current and voltage reflection and the associated scaling rules apply only to the voltages and currents directly related to the transformer windings themselves.
- As indicated in Fig. 6.4, the currents in the windings scale in inverse manner to the voltage; i.e., in going from Primary to Secondary, the current *through* the windings goes as $\times n$. In the reverse direction we must use the factor $\times 1/n$. So we can, in effect, apply a form of transformer scaling to the (voltages and) currents involved in the Flyback transformer too, even though technically speaking there is really no

transformer action—at least not in the classic sense of a Forward converter transformer. *The Flyback transformer is actually an inductor with multiple windings.* Unlike a Forward converter transformer, in the Flyback, when current flows in the primary winding, it does *not* flow in the secondary, winding, and vice versa. However the two current waveforms do connect at the *moment of every transition* (turn ON and turn OFF). See Fig. 6.5, which we will discuss in more detail shortly. But it is at this precise moment, that a certain fundamental requirement applies, one which eventually produces "transformer action" for a Flyback too. The reason is as we discussed earlier: Since the total ampere-turns of the transformer ΣNI are directly related to the energy residing in the core, the net ampere-turns cannot change *abruptly*. This argument was previously used to discuss the energy-recovery (tertiary) winding of the Forward converter transformer. It so happens that the energy-recovery winding of the Forward converter, by virtue of its polarity, operates in Flyback mode too (it conducts only when the main winding turns OFF).

In our present case, this means that the peak of the primary-side current I_{PK} must be related to the peak of the secondary current I_{PKS} by

$$I_{PK} = \frac{I_{PKS}}{n}$$

The troughs (valleys) of the two waveforms are also likewise related.

- In Fig. 6.6 we reveal a useful trick for analyzing the primary and the secondary sides of an isolated Flyback. We achieve great simplification by drawing out the *equivalent primary-side Buck-boost* converter, *and* the *equivalent secondary-side Buck-boost* converter. Since the equations for a standard DC-DC Buck-boost topology are simpler and well known (as in the Appendix), these equivalent models are much easier to use and understand. The basic idea is that if we want to work out the stresses on the primary

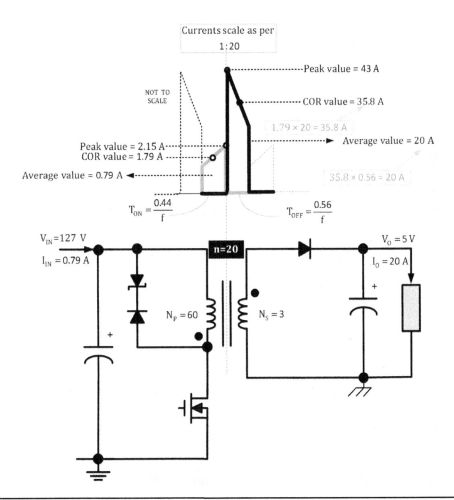

FIGURE 6.5 The calculated current components (see numerical example).

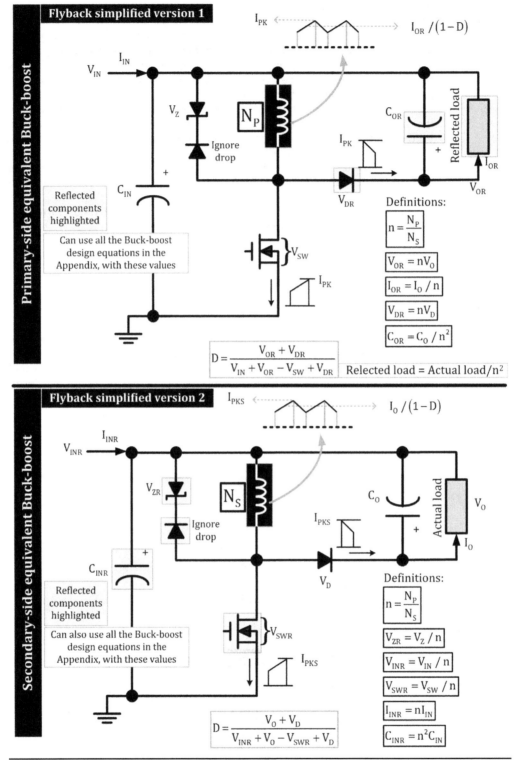

FIGURE 6.6 Equivalent Buck-boost models for the Flyback.

side, we convert the Flyback into an equivalent primary-side Buck-boost. If we want to figure out the stresses on the components on the secondary side, we convert the Flyback into an equivalent secondary-side Buck-boost.

- Note that the subscript R stands for *reflected*. So, for example, if any voltage is being reflected from the Primary to Secondary, we divide it by n, where n is the turns ratio N_P/N_S. If current is being reflected from the Primary to Secondary, it gets multiplied by n.

- The switch forward drop, represented by V_{SW}, subtracts from the applied input voltage when the switch is ON, and thus is effectively a component of the voltage *across* the primary winding. We can therefore apply scaling to it too, and so it gets reflected to the secondary side as V_{SW}/n as shown in the figure. On similar grounds, the output diode voltage drop appears on the primary side as nV_D. Notice also how the zener clamp voltage gets reflected.

- We see that voltages and currents scale in a simple linear manner. But energy is invariant and cannot change merely on the basis of any "equivalent circuit" abstraction we come up with. Therefore, we expect that all reactances, being by nature essentially energy storage elements, will reflect in a different manner compared to voltages and currents. For example, the output capacitor of the Flyback C_O has an energy $(1/2)(C_O V_O^2)$, and V_O^2 reflects as $n^2 V_O^2$ to the Primary, so C_O must reflect as C_O/n^2. This is shown as C_{OR} in Fig. 6.6. A table of how components reflect is presented in Fig. 6.7.

- We can also similarly show that *any leakage inductance on the secondary side will reflect to the primary side multiplied by n^2* (and that happens because $(1/2) \times LI^2$ must be invariant). *This reflected leakage has severe ramifications on the entire performance and power-handling capability of a Flyback,* as we will see.

- The equivalent primary Buck-boost "thinks" that its output voltage rail is

$$V_{OR} = nV_O$$

For that reason, V_{OR} *actually turns out to be the most fundamental design choice available to the Flyback designer.*

Note that simplistically speaking, a high V_{OR} leads to a high duty cycle whereas a low V_{OR} produces a low duty cycle. This is analogous to a simple Buck-boost,

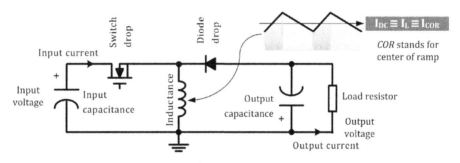

	Primary-side equivalent Buck-boost	Secondary-side equivalent Buck-boost
Input voltage	V_{IN}	$V_{INR} = V_{IN}/n$
Input current	I_{IN}	$I_{INR} = I_{IN} \times n$
Input capacitance	C_{IN}	$C_{INR} = n^2 \times C_{IN}$
Inductance	L_P	$L_S = L_P/n^2$
Switch foward drop	V_{SW}	$V_{SWR} = V_{SW}/n$
Output voltage	$V_{OR} = V_O \times n$	V_O
Output current	$I_{OR} = I_O/n$	I_O
Inductor current (I_{DC})	$I_{OR}/(1-D)$	$I_O/(1-D)$
Output capacitance	C_O/n^2	C_O
Diode drop	$V_{DR} = V_D \times n$	V_D
Duty cycle	D	D
Current ripple ratio	r	r
Load resistor	$R \times n^2$	R
Secondary-side leakage	$L_{LKS} \times n^2$	L_{LKS}

FIGURE 6.7 Component reflections in both the equivalent Buck-boost models of the Flyback.

where to raise the output, we have to increase the duty cycle. But in fact there are other effects of changing V_{OR}, that we need to study carefully.

We note that *as far as the primary side is concerned*, ignoring parasitics for now, it will see no difference between say a 5-V output with a turns ratio of 20 (i.e., a V_{OR} of $5 \times 20 = 100$ V), or a 10-V output with a turns ratio of 10 (i.e., a V_{OR} of $10 \times 10 = 100$ V). *So several different applications, each having the same V_{OR} (turns ratio adjusted according to the output voltage) are actually the same converter as far as the switch (primary) side is concerned.* All primary-side currents and voltages are identical. The only difference arises from the leakage inductance, which is an issue we will take up a little later.

- The equivalent load current at V_{OR} flowing out of the output terminals of the equivalent primary Buck-boost is the *reflected output current*

$$I_{OR} = \frac{I_O}{n}$$

Clearly, if losses are ignored, we get

$$P_{IN} = V_{IN}I_{IN} = P_O = V_O I_O = V_{OR} I_{OR}$$

- So, as far as the primary side is concerned, it 'thinks' that *it is delivering a load current of I_O/n at a voltage nV_O*.
- What is the load power? As expected it is $V_O \times I_O$. What is the load resistance? The load resistor on the secondary side is $R = V_O/I_O$. The primary side, however, thinks that the load resistor is $nV_O/(I_O/n) = n^2 V_O/I_O = n^2 R$. So we see that not only reactances, but even resistances, reflect from secondary to primary according to the square of the turns ratio. This is also indicated in Fig. 6.7.
- But we see that in going from Secondary to Primary, both L and R reflect according to n^2, but C reflects as C/n^2. Thus it is often stated that *impedances (in general) reflect from Primary to Secondary as n^2*. That is consistent with what happens to L, C, and R.
- Knowing that for a standard DC-DC Buck-boost converter, the average inductor current (in continuous conduction mode) is $I_O/(1-D)$, we can determine that the center of the secondary-side current ramp in an actual Flyback converter must be

$$I_{CORS} = \frac{I_{OR}}{1-D}$$

where I_{CORS} is the center of ramp on the secondary side.

- The center of the primary-side current must be by the scaling rule

$$I_{CORP} = \frac{I_O}{1-D} \times 1/n \equiv \frac{I_{OR}}{1-D}$$

where I_{CORP} is the center of ramp on the primary side.

- For a DC-DC Buck-boost converter, the duty cycle is (ignoring forward drops across switch and diode)

$$D = \frac{V_O}{V_{IN} + V_O}$$

For the Flyback (or its equivalent Buck-boost models)

$$D = \frac{V_{OR}}{V_{IN} + V_{OR}} \equiv \frac{V_O}{V_{INR} + V_O}$$

If we include the switch and diode drops, we get for the standard DC-DC Buck-boost converter

$$D = \frac{V_O + V_D}{V_{IN} + V_O - V_{SW} + V_D}$$

So for the *equivalent primary-side Buck-boost* we get

$$D = \frac{V_{OR} + V_{DR}}{V_{IN} + V_{OR} - V_{SW} + V_{DR}}$$

and for the *equivalent secondary-side Buck-boost* we get

$$D = \frac{V_O + V_D}{V_{INR} + V_O - V_{SWR} + V_D}$$

both of which are identical as we would expect if the models were truly equivalent.

Example 6.1 A Flyback has 60 turns on the Primary and 3 turns on the Secondary. The output voltage is set to 5 V. The load is 20 A. What are the currents in the inductor, switch, diode, windings, etc., at 90 Vac (i.e., V_{IN} of $90 \times \sqrt{2} = 127$ V)?

The V_{OR} is $n = 60/3 = 20$ times the output. Therefore, $V_{OR} = 5 \times 20 = 100$ V. We know the average load current is 20 A. Reflected to the Primary it appears as a load current of $I_{OR} = 20$ A$/20 = 1$ A. The duty cycle is $100/(100 + 127) = 0.44$. So the average inductor current in the equivalent primary Buck-boost model is 1 A$/(1 − 0.44) = 1.79$ A. The average value reflects in to the Secondary and is therefore 1.79 A $\times 20 = 35.8$ A in the equivalent secondary Buck-boost model.

Now, usually, for optimum results, we like to set the transformer inductance so that the current ripple ratio $r = \Delta I / I_{COR}$ in the (equivalent) inductor is 0.4 (±20 percent). In that case we will get I_P as 1.79 A $\times 1.2 = 2.15$ A. By the scaling rule, the corresponding secondary-side peak current is 2.15 A $\times 20 = 43$ A. We can check that, as expected, on the secondary side too, the peak of 43 A is 20 percent higher than the reflected average value of 35.8 A.

The average current into the switch is 1.79 A $\times 0.44 = 0.79$ A (multiplied by D). Let us double-check the average current through the diode. This is 35.8 A $\times (1 − 0.44) = 20$ A (multiplied by $1 − D$). This checks out because *we know that for a Flyback (or Buck-boost or Boost), the average diode current must equal the load current*. We have shown the calculated waveforms in Fig. 6.5.

The *average input current is equal to the average switch current* for this topology. So the input watts are 0.79 A $\times 127$ V $= 100$ W. We see that this implies 100 percent efficiency. But that is what we expected would be the result of ignoring parasitics like the diode and switch forward drops. So our calculations are clearly correct, since they are based on an ideal D equation (initial assumption of no losses).

Dealing with Multi-output Flyback Converters

Here the best way to proceed is to draw out the equivalent primary Buck-boost as before. But first we must lump all the output power into an equivalent single output (carrying full converter power). So if we have the outputs $V_1 @ I_{O1}$, $V_2 @ I_{O2}$, $V_3 @ I_{O3}$, $V_4 @ I_{O4}$, etc., we can lump them to behave as a single output with the following description:

$$V_O = V_1 \qquad I_O = I_{O1} + \frac{V_2 I_{O2}}{V_O} + \frac{V_3 I_{O3}}{V_O} + \frac{V_4 I_{O4}}{V_O} + \cdots$$

Then for the load current in the equivalent primary-side model, we have as before

$$I_{OR} = \frac{I_O}{n}$$

and

$$V_{OR} = n V_O$$

This will tell us the primary-side currents, zener dissipation, input capacitor requirements, etc. For calculating secondary-side parameters, we would need to consider each output on an individual basis.

We are warned that certain subtleties exist like how much cross-regulation we will get on the outputs. These issues are very hard to predict theoretically, and/or accurately, and usually a build-and-tweak approach is the fastest way to go.

For multi-output converters, the total power is shared between several windings. From the viewpoint of the primary side, they all appear as a lumped load. Therefore, reducing the power output of the converter on the primary side, though necessary, is not sufficient to handle an overload on *specific* outputs. For example, the entire power output can be shifted to one of the several outputs by overloading it, and the primary side would not know better. There are also safety considerations in how much energy is allowed from any output for it to qualify as an SELV-EL (safety extra-low-voltage energy limited) output. Therefore, for multi-output converters *it is usually practically necessary to add discrete current limiting on each output*. Safety agencies make an exception in cases where the output is coming from an integrated regulator chip such as the uA780x series. In that case, the output is considered inherently safe (i.e., current and energy limited). There is no need to add additional current limiting to that output.

The Primary-Side Leakage Term

In the top half of Fig. 6.8, we have included a leakage inductance term, which we assume is present on the primary side of the original Flyback. At the instant the switch turns OFF we get a voltage spike across the switch, and this is clamped by the zener as indicated, to a voltage $V_{IN} + V_Z$ (measured with respect to Primary ground). The voltage across the winding is momentarily V_Z. During this brief interval during which the zener conducts (called t_Z here), the current through the leakage current loop ramps down from I_{PK} to zero at a rate determined by the voltage across the leakage inductance. Let us calculate the energy dissipated in the zener as a result of that current. Since the voltage across the leakage inductance during the time the zener conducts is $V_Z - V_{OR}$, applying $V = L dI/dt$ to the leakage inductance we get

$$t_Z = \frac{L_{LK} \times I_{PK}}{V_Z - V_{OR}} \quad \text{s}$$

During this interval, the energy dissipated in the zener is

$$E_Z = V \times I \times t = V_Z \times \frac{I_{PK}}{2} \times t_Z \quad \text{J}$$

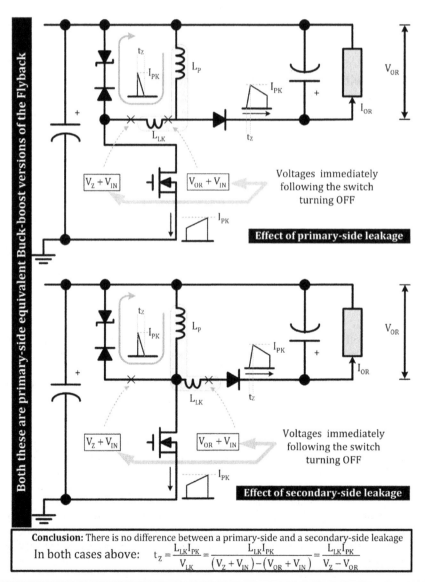

FIGURE 6.8 (Same) primary-side or secondary-side leakage inductances are equivalent.

where we have taken the average current through the zener during this interval as $I_{PK}/2$. Simplifying, we get

$$E_Z = \frac{1}{2}L_{LK}I_{PK}^2 \frac{V_Z}{V_Z - V_{OR}} \quad J$$

So the power dissipation in the zener is

$$P_Z = \frac{1}{2}L_{LK}I_{PK}^2 \frac{V_Z}{V_Z - V_{OR}} \times f \quad W$$

where f is the switching frequency in hertz.

The energy dissipated in the zener can be expanded as follows:

$$E_Z = \frac{1}{2}L_{LK}I_{PK}^2 \frac{V_Z}{V_Z - V_{OR}} = \frac{1}{2}L_{LK}I_{PK}^2 \times \left[1 + \frac{V_{OR}}{V_Z - V_{OR}}\right] \quad J$$

So assuming that $(1/2) \times L_{LK} \times I_{PK}^2$ is the energy present in the leakage, we seem to have an *unaccounted-for term*:

$$E = \frac{1}{2}L_{LK}I_{PK}^2 \frac{V_{OR}}{V_Z - V_{OR}} \quad J$$

The physical explanation for the presence of this term is that during the time t_Z, besides the energy residing in the leakage L_{LK}, some energy from the primary (magnetizing) inductance also gets dumped into the zener. This happens because the current flowing in the leakage inductance loop also has to pass through the *voltage gradient present across the magnetization inductance*. Let us see how much work is done in the process. Since the voltage gradient across this inductance is $(V_{OR} + V_{IN}) - V_{IN} = V_{OR}$ (see Fig. 6.8), the additional energy related to that is

$$V \times I \times t = V_{OR} \times \frac{I_{PK}}{2} \times t_Z \quad J$$

Simplifying by plugging in t_Z, this term is

$$\frac{1}{2}L_{LK}I_{PK}^2 \frac{V_{OR}}{V_Z - V_{OR}} \quad J$$

Clearly, this is the unaccounted-for term—now fully accounted for!

We can also make the following related conclusions:

- We must always set V_Z greater than V_{OR}; otherwise the zener will become the preferred freewheeling path for the magnetizing current (in preference to the output path). If V_Z is set greater than V_{OR}, the zener is the preferred (or the only available) path during t_Z only.

- To minimize dissipation we also need to set V_Z significantly higher than V_{OR}; otherwise the zener dissipation will climb almost exponentially on account of the $V_Z - V_{OR}$ term in the denominator.

- However, the required minimum voltage rating of the switch is $V_{IN} + V_Z$, so too high a zener voltage will have an effect on the overall cost and performance, due to the higher-rated FET required. Higher-rated FETs will also tend to have higher switching and conduction losses.

- We must also remember that the voltage across the zener is a function of the current passing through it. A zener is not a perfect device (like any other device we use), so we may get higher clamping levels across the zener than we theoretically expected. A safety margin must be left for this while selecting the FET voltage rating.

- V_Z is typically chosen to be about 40 to 100 percent higher than V_{OR}. For most worldwide input off-line applications, zeners will be found to be rated 150 to 200 V. The turns ratio is usually selected so that V_{OR} is set from about 70 to 140 V, and the switch is then required to be typically rated 600 to 700 V.

- Note that the best-designed commercial off-line Flybacks, those that are based on controller ICs and meant for worldwide input voltages, use standard low-cost 600-V external FETs. Anything more may represent overdesign.

The Secondary-Side Leakage Term

In the bottom half of Fig. 6.8 we have placed a secondary-side leakage inductance instead of a primary-side one. This could, for example, simply represent the secondary-winding lead lengths and trace inductances. We note that the voltages across the leakage inductance and at the drain of the FET are the same as for the top half of Fig. 6.8. *Therefore, for the equivalent primary-side Buck-boost model, there is actually no observable difference whether a certain leakage is primary-side or secondary-side.* The dissipation in the zener is still represented by the same equation calculated earlier. In effect, for all practical purposes, after we reflect the secondary-side leakage, it appears as a primary-side leakage (its reflected value).

But intuitively, how does a secondary-side leakage play the same role as a primary-side leakage? The truth is that by either introducing a secondary-side or a primary-side leakage inductance, we actually arrive at the very same result. The primary-side leakage current/energy, for example, has nowhere to go, as it is not linked magnetically to the secondary windings. So it insists on flowing until it manages to slew down and force a reset of the leakage inductance (in time t_Z). During this time, only the remainder of the primary-side current can slew into the secondary side. Similarly, if the leakage is purely secondary-side based, it acts by opposing immediate and full current buildup into the secondary side, for the same duration t_Z. In both cases, the magnetization current has no freewheeling path available until *that time has elapsed*. So once again, the magnetization current just courses through the zener for that duration, depositing energy in the process. In both cases, the slew-down condition and the slew-up condition of the leakage current, the governing equation is the same: $V = LdI/dt$. Note that as soon as the zener stops conducting, the voltage *across* the primary-side winding collapses from its clamped value V_Z to its natural operating value V_{OR}. See Fig. 6.1 once again.

Note *During the interval t_z (zener conduction time), what really happens is that as the primary current slews down in the primary winding, the secondary winding current ramps up simultaneously, thus maintaining the net ampere-turns across the switch crossover duration. In other words, the secondary current is not completely blocked until the leakage inductance resets, but provides an average output current equal to half the normal operating secondary current [which we know is $I_O/(1 - D)$]. An attempt to show this smooth transition has been made in Fig. 6.8 (see the slowly rising edge of the secondary current waveform).*

In summary, it can be shown that the current in the secondary loop of a Flyback is n times the reflected primary-side current. Since $(1/2) \times LI^2$ is an energy term that should remain the same for either method of representation, *any secondary-side-leakage inductance present in a Flyback (denoted by L_{LKS}) must get reflected to the primary side as an equivalent primary-side leakage inductance equal to $L_{LKS} \times (n)^2$.* This maintains the same energy. It is this reflected secondary inductance that behaves exactly as a genuine primary-side leakage. The zener dissipation equation given previously therefore applies to this reflected leakage too.

Example 6.2 A Flyback has 60 turns on the Primary and 3 turns on the Secondary. The output voltage is set to 5 V. An estimated 40 nH of secondary-side inductance is present, attributable to the winding terminations and PCB traces (the *rule of thumb is 20 nH/in*). What is the effective leakage inductance as seen by the switch? And what happens if the output was set to 12 V instead (keeping the same V_{OR})?

Note that

$$V_{OR} = V_O \frac{N_P}{N_S} = 5 \times \frac{60}{3} = 100 \text{ V}$$

The turns ratio is 60/3 = 20. So the reflected leakage is

$$L_{LK} = 20^2 \times 40 = 16{,}000 \text{ nH}$$

This is 16 µH of leakage inductance and is certainly not insignificant. Note that if V_{OR} is always set to 100 V irrespective of the set output voltage (and this is a normal design target for universal input Flybacks), then for a 12 V output we would need to set the following turns ratio:

$$n = \frac{N_P}{N_S} = \frac{V_{OR}}{V_O} = \frac{100}{12} = 8.33$$

In this case, if the secondary turns are increased to 5, then the calculated number of primary turns is $8.33 \times 5 = 41.67$. Clearly that cannot be, since turns are integral (sometimes half-integral) numbers. So suppose we choose 42 turns for the Primary.

Now

$$V_{OR} = V_O \frac{n_P}{n_S} = 12 \times \frac{42}{5} = 100.8 \text{ V}$$

The turns ratio is $42/5 = 8.4$. So the reflected leakage for a 12-V output is

$$L_{LK} = 8.4^2 \times 40 = 2822 \text{ nH}$$

We can see that the reflected leakage is only 2.8 μH here, i.e., about six times less than what we got for a 5-V output (ignoring any primary-side leakage). The dissipation in the zener being proportional to leakage inductance will also drop by about six times in going from an output of 5 V to 12 V.

Note *We said earlier that turns are typically integral numbers. We need to qualify this here. A half-turn is often used by power supply designers to get the required voltage in multi-output converters so as to get good centering of multiple outputs. They do this by simply not completing a full turn but passing the winding out from the opposite side of the transformer. First, there are safety issues involved here in terms of the clearance distances required between primary and secondary sides in off-line power supplies. Second, current will always flow in a complete loop one way or another. So by creating a half-turn what we are actually doing is creating some additional leakage inductance in series with that winding. This is an acceptable technique, provided a good freewheeling path is still available for the magnetization current. Therefore, half-turns are not normally placed in the* main *regulated winding. There are known textbook techniques on how to create a genuine half-turn from the point of view of the flux inside the core, and that is done by winding turns in a certain way on the side limbs of a standard E-core. But the author has never seen that implemented in high-volume commercial production.*

- In general we can say that effectively, *the secondary-side leakage inductance as seen by the switch is inversely proportional to V_O^2*. This assumes a standard (fixed) V_{OR} irrespective of output voltage (turns ratio adjusted accordingly).
- A *good* off-line Flyback transformer has a typical primary-side measured effective leakage inductance of less than 1 to 2 percent of its primary inductance. Anything more probably needs a major redesign of the winding arrangement (like using a split primary winding with a sandwiched secondary arrangement, etc).
- Small lead lengths and PCB traces on the secondary side are crucial for maintaining low effective leakage on a practical board assembly, *especially for low output voltages*.

Note *Leakage inductance is what ultimately restricts the power that a Flyback can deliver efficiently. Theoretically, even 600 W is feasible with a Flyback, provided the output voltage is set high. The author did, years ago, build a universal-input AC-DC Flyback with 60 V @ 10 A, using several paralleled BJTs on the primary side. The efficiency was about 70 percent. So, admittedly, a Flyback may not be the smallest or most cost-effective method for high-power applications, but it certainly can work in such cases.*

- Note that when a transformer's leakage is measured in production by putting a short across the transformer's secondary winding pins, we are actually not accounting for the trace inductances that will get added on to the complete secondary loop in real operation. So it may be OK to use this method to do a quality check on an incoming transformer but not for estimating dissipation in the clamp or efficiency.
- The overall efficiency of a Flyback can actually fall 5 to 10 percent simply on account of an inch or two of trace length on the secondary side, especially for low output voltages (high *n*).
- *Flying leads* are occasionally used when we run out of available transformer pins. But these should generally be avoided for the main output, especially if it is of low voltage. Because these will add to the leakage inductance considerably (and affect cross-regulation too).

- However, in flying lead cases, we can get quite a very significant reduction in the lead inductance by closely paralleling (or even twisting if possible) the forward and return leads of the winding.
- Secondary-side trace lengths present *after* the first output capacitor (or paralleled capacitors) don't count in this leakage reflection analysis. These post-filter traces essentially have near-DC flowing in them, and so their inductance does not pose any issue, from a high-frequency perspective at least.

***

HINT *A correct measurement strategy would be to take an actual converter board with the transformer in place. Then we should place thick wires across the diode and the capacitor. We should then cut any one trace leading to the primary winding and measure the inductance across the primary winding by an LCR meter. This will give us the effective L_{LK} present on the primary side, with all PCB trace inductances included. We may be very surprised to learn how much the traces can cause the leakage inductance to increase, especially for high turns ratios (as are typical for low- output-voltage applications).*

***

Flyback Optimization and Deeper Analysis

Here we will provide some curves to help in quick estimates. The idea is that if we don't see the larger picture, and instead get bogged down by detailed calculations, we will mistake the forest for the trees and never manage our ultimate goal of *optimization*. In the curves that follow, we may see proportionality statements at the top. This allows us to generate fairly accurate estimates for any specific application, using scaling techniques.

The steps toward understanding optimization are as follows.

Step 1: V_Z/V_{OR}

If the FET is rated 700 V (as in the IP switch), then for an input voltage of 270 Vac, the peak rectified value that forms the DC input to the converter is 270 V × $\sqrt{2}$ = 382 V. The voltage across the FET will be $V_{IN} + V_Z$. Keeping a safety margin of say 50 V, we are trying not to go above 650 V on the switch. Therefore, V_Z must be less than 650 − 382 = 268 V. A typical zener has a certain basic tolerance, and also its clamping voltage is a function of the current through it. Therefore, suppose we choose a 200-V zener. If we choose a V_{OR} of 100 V, the ratio V_Z/V_{OR} is 2. For a V_{OR} of 140 V, the V_Z/V_{OR} will be 1.4. *The ratio V_Z/V_{OR} is very important because it determines the zener clamp dissipation.*

Step 2: Dissipation in Zener

Looking now at Fig. 6.9, we see that if the switching frequency is 100 kHz, and the peak current in the switch just prior to turnoff is 1 A, then for a V_Z/V_{OR} of 2, as compared to a V_Z/V_{OR}

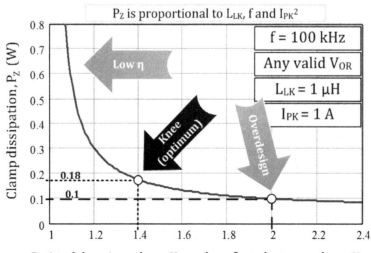

FIGURE 6.9 Zener dissipation as a function of V_Z/V_{OR}.

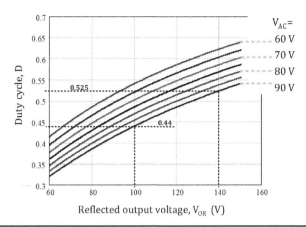

FIGURE 6.10 Duty cycle as a function of V_{OR}.

of 1.4, the dissipation goes up almost 80 percent. Realizing that at low-line conditions, the zener dissipation may account for 20 to 50 percent of the total energy dissipated inside the switcher, we can expect a steep fall in overall converter efficiency whenever we try to increase V_{OR} without being able to maintain the ratio V_Z/V_{OR} high enough.

Note that the zener dissipation formula can be rewritten as

$$P_Z = \frac{1}{2} L_{LK} I_{PK}^2 \frac{V\text{ratio}_{CLAMP}}{V\text{ratio}_{CLAMP} - 1} \times f \quad W$$

where

$$V\text{ratio}_{CLAMP} = \frac{V_{CLAMP}}{V_{OR}} \equiv \frac{V_Z}{V_{OR}}$$

This has been written in a more general form to indicate that the clamp can be a zener or an RCD type, and the *same dissipation equation applies*. Clearly, for ensuring even basic operation, $V\text{ratio}_{CLAMP}$ must always be *at least greater than unity* (otherwise the clamp becomes the preferred output!).

We learn that the next-generation IP switch has a frequency of 130 kHz as compared to 100 kHz used earlier. It also has an officially recommended V_Z/V_{OR} of about 180 V/135 V = 1.33. We see that the higher V_Z/V_{OR} will significantly *increase dissipation*. In addition, the higher switching frequency will cause a further 30 percent increase in the zener dissipation, because the zener will now take that many more "hits" every second as compared to the lower 100 kHz.

Step 3: Duty Cycle

In Fig. 6.10 we can read off the duty cycle for different input AC line voltages. Therefore, at 90 Vac, the duty cycle is 0.44 for a V_{OR} of 100 V, but it increases to 0.525 for a V_{OR} of 140 V.

Step 4: Peak Current

We use the fact that the center of ramp on the primary side, I_{COR} (the average inductor current in the primary-side equivalent Buck-boost), is (ignoring losses)

$$I_{COR} = \frac{I_{OR}}{1 - D}$$

The ramp typically adds about 20 percent to give the peak current (in the switch, inductor, and diode), with an "optimally" designed transformer ($r = 0.4$). So

$$I_{PK} = \frac{1.2 \times I_{OR}}{1 - D}$$

To account for losses inside the converter, we may want to typically add another 20 percent to this ideal peak estimate. The best option is to use the nonideal duty-cycle equation:

$$D = \frac{V_{OR}}{V_{OR} + \eta V_{IN}}$$

where η is the actual efficiency. This equation was discussed in reference to the Buck-boost in Chap. 2.

In Fig. 6.11, we can read off the peak current at 90 Vac for a load of 100 W (assuming full efficiency). We see that we get a peak of 2.15 A for a set V_{OR} of 100 V, but the peak *decreases* to 1.8 A for a V_{OR} of 140 V. The other curves in the figure tell us the *momentary* peak current that will be reached if, for example, the input collapses (assuming no current limit acted before the D_{MAX} limit of the controller was reached).

This reduction in peak current is real and was probably the intuitive reason why both the recommended operating duty cycle (at 90 Vac) and the D_{MAX} were set higher for the next-generation IP switch.

The physical reasoning behind the lower steady-state peak current is that by setting a high V_{OR} we set a higher steady-state D. Also, input power is virtually constant,

$$P_{IN} = I_{IN} \times V_{IN} = (I_{COR} \times D) \times V_{IN}$$

where I_{COR} is the primary-side center of ramp current here. We see that the *term in brackets must remain constant for a given input* V_{IN}. So if D increases, the center of the primary-side (reflected) current ramp I_{COR} must come down, along with its peak value.

However, looking at the equation

$$I_{PK} = \frac{1.2 \times I_{OR}}{1-D}$$

it may have seemed that increasing D should have caused I_{PK} to increase, not decrease. So, mathematically speaking, why is the peak lower for a higher V_{OR} (corresponding to a higher D)? First, understand that if we increase V_{OR}, this being the intermediate primary-side output rail, to lower it down to the (same) final output, we need to increase the turns ratio $n = N_P/N_S$ (lower N_S). What does that do to I_{OR}? I_{OR} is I_O/n. If n increases, for the same I_O, we see that I_{OR} decreases. So, even though D increases if we change V_{OR}, the decrease in I_{OR} overwhelms the decrease in the preceding denominator. So I_{PK} decreases, not increases, by setting a higher V_{OR}.

However, having set V_{OR}, and a fixed n too, if the duty cycle now momentarily increases, as for a sudden decrease in input, then I_{PK} will increase, as can also be seen from Fig. 6.11.

These are some of the subtle ways in which our visualization, or intuition, of a Flyback must differ from that of a straightforward Buck-boost. Because the turns ratio is an additional degree of freedom in a Flyback, not present in a Buck-boost. It can change our analysis significantly.

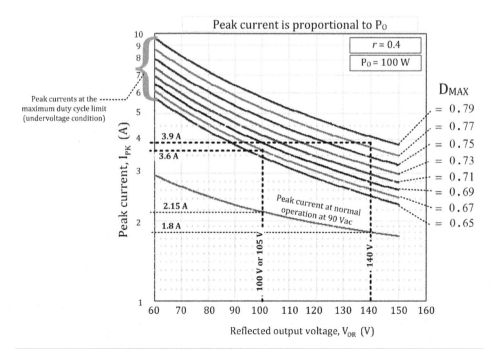

Figure 6.11 Peak current as a function of V_{OR}.

So, increasing V_{OR} sure helps reduce I_{PK}, and that was likely the whole intention of allowing both higher operating and maximum duty cycles, and also for recommending a higher V_{OR} (on paper at least if not implemented on evaluation boards). Unfortunately, the *zener dissipation goes up far more steeply, despite the peak current coming down*. This is on account of the higher D corresponding to a much higher V_{OR} (as obvious from the Flyback duty-cycle equation). As a result, zener dissipation, which depends on $V_Z - V_{OR}$, increased significantly, completely offsetting any mild efficiency improvement by the lower peak currents from setting a higher V_{OR}. Yes, if we had been able to increase V_Z simultaneously while increasing V_{OR}, we could have prevented $V_Z - V_{OR}$ from becoming smaller. But for the same maximum switch rating, we just cannot afford to increase V_{OR} blindly. That was the final lesson of the "improved" IP switch.

Now we illustrate how to use the curves presented so far for finding the zener dissipation for a more general case.

Example 6.3 Consider a worldwide input Flyback that has 60 turns on the Primary and 3 turns on the Secondary. The output is 5 V at 2 A, at an efficiency of 69 percent. The switching frequency is 130 kHz. What is the zener dissipation if the effective (lumped) leakage inductance as seen by the switch is 40 µH?

The worst-case dissipation occurs at 90 Vac (maximum peak current), which we take as the minimum line input. Let us set V_{OR} to 100 V (e.g., a turns ratio N_P/N_S of 20 for a 5-V output). In the earlier discussions we had assumed 100 percent efficiency. Now we account for it in the following manner. The output power here is 10 W. We increase it as per P_O/η to $10/0.7 = 14.5$ W and this becomes the equivalent load power at 100 percent efficiency. We need to first calculate the peak current from Fig. 6.11 before we can use Fig. 6.9, since the latter is in terms of I_{PK} and we need to know that. For 100 W, the peak current is 2.15 A. So scaling power and currents to our case, we obtain

$$I_{PK} = 2.15 \times \frac{14.5}{100} = 0.31 \text{ A}$$

Assuming we have a zener of 200 V, we see from Fig. 6.9 that we get a zener dissipation of 0.1 W with the following conditions: V_{OR} of 100 V, 1-µH leakage, 1-A peak current, and frequency of 100 kHz. So for this example, the zener dissipation in our case must be (using scaling techniques)

$$P_Z = 0.1 \text{ W} \times \frac{130 \text{ kHz}}{100 \text{ kHz}} \times \left(\frac{0.31 \text{ A}}{1 \text{ A}}\right)^2 \times \frac{40 \text{ µH}}{1 \text{ µH}} = 0.5 \text{ W}$$

We see that this accounts for $0.5/4.5 = 11$ percent of the total 4.5-W dissipation. Setting a higher V_{OR} (and lower zener clamp) could have increased the dissipation zener to over 1 W (> 22 percent).

NOTE *In actual bench measurements we will usually see that the peak current flowing into the zener is only about 70 to 80 percent of I_{PK}, which is the calculated peak current through the primary winding. The fact is that the model for the transformer is not so trivial. Some of the inductor current flowing just prior to turnoff freewheels into the parasitic capacitances of the transformer and the switch. This could typically end up reducing the dissipation in the zener to almost half of the preceding theoretical prediction above. But there is no closed-form equation for this. Therefore, the engineer is advised to actually place a current probe in the zener path and see the peak current flowing into the zener. Otherwise we may significantly overestimate the size (and cost) of the zener (besides underestimating efficiency). This is one of those rare cases where a mistake is actually over-conservative. See also Fig. 16.2.*

The RCD Clamp

For higher output currents, the zener dissipation will become almost intolerable. Then a resistor-capacitor-diode (RCD) clamp should be used to improve efficiency. The advantage of an RCD clamp is that *it automatically tends to increase the clamp voltage level at low-line* (aiding efficiency), but at high-line the clamp level subsides on account of the smaller energy residing in the leakage inductance (lower I_{PK}), and so the required FET voltage rating is not compromised. See the RCD clamp design chart in Fig. 16.2 (in Chap. 16).

During overloads or other transient conditions, the capacitor (C) of the RCD can get overcharged momentarily. So either we have to increase C substantially (to reduce the bump), or we need to retain a zener clamp parallel to the RCD clamp. The zener clamp level must be set higher than the steady-state clamping afforded by the RCD clamp over the *entire input range*, and we should ensure that the RCD is the only effective working clamp under normal operating conditions. The zener is present, just for reliability now.

Any RCD clamp design is, *under steady state, virtually independent of the value of C.* The average clamping level is determined strictly by R, and C only affects the voltage ripple around that average level. So R is critical, C much less. The best way to fix R is on the bench. For that we should operate the power supply at maximum load at the highest input voltage (e.g., 270 Vac), and ensure that the R is high enough to *just maintain the drain-to-source voltage to a little less than the FET rating*. This gives the most optimum and efficient RCD clamp.

C is typically 33 nF for a typical 75-W universal-input power supply, simply because under overloads we don't want it to get charged up faster than our protection circuitry can act. Note that some engineers "save" money by undersizing this key reliability component. That is not recommended. Also check the FET derating margins carefully, both at shorts and overloads at high-line and at low-line. That is the way to pick C, once R is fixed.

Step 5: Holdup Time

At 90 Vac, the duty cycle is 0.44 for a V_{OR} of 100 V, and it is 0.525 for a V_{OR} of 140 V. So why do we need to set a maximum duty cycle of the control IC, much higher than this in the first place? The main reason is that we need some *holdup time* as explained in Chap. 5. As the input voltage on the bulk capacitor terminals slews down, the duty cycle increases. The converter must continue to function for the specified holdup time, without loss of output regulation. Clearly the amount of holdup time available depends on the size of the input capacitance. If C_{IN} is very large, it will hold the voltage up much longer and we won't hit very high duty cycles (or a duty cycle limit). It also depends inversely on the output power since a high output demand will drain the input capacitor faster. A common "economical" choice for the input capacitance of a worldwide input power supply is 3 µF/W.

See Chap. 5 for a detailed analysis and calculations in this regard. Here, we should just remember that

1. Holdup time should be measured starting from the point of worst-case instantaneous voltage, i.e., just prior to the AC wave intersecting the capacitor droop curve in normal operation.

2. Of a stated "10-ms holdup time requirement," the bridge conduction time is around 2 ms, so the actual holdup we need is not 10 ms, but 10 − 2 = 8 ms.

3. We need to account for capacitor tolerances and steady degradation of capacitance before end-of-life is declared, so we may need to select a nominal value 20 to 40 percent higher than calculated based on holdup.

4. Also, while ensuring the maximum duty cycle brickwall is not prematurely encountered, looking at Fig. 2.13, keep in mind that the most accurate estimate of duty cycle is actually

$$D = \frac{V_{OR}}{V_{OR} + \eta V_{IN}}$$

Some other related thoughts:

- The high maximum duty cycle of the next-generation IP-switch was introduced not just to allow a maximum duty cycle, but to *enable* a higher *nominal* (steady-state) duty cycle (and thus a higher V_{OR}). That was supposedly the key to lowering the peak input current (but at great expense, as we just learned).

- Actually, opposite to what was claimed, in practice, a *low V_{OR}, not a high V_{OR}, helps in reducing the size of the input capacitor too*. We can understand this because, for example, at 90 Vac, the steady-state duty cycle is 0.44 for a V_{OR} of 100 V, and is 0.525 for a V_{OR} of 140 V. So a lower V_{OR} will allow for a larger *change* in duty cycle, and thus a larger input swing during a lost AC half-cycle. In other words, the "headroom" is more important than the absolute values of D and D_{MAX}. It can enable a reduction in input capacitance based on a target holdup time.

- We can reverse the question: What if we *don't want to incur any added cost* by increasing the input capacitance (beyond the value dictated by V_{OR} = 100 V and D_{MAX} = 0.67), nor compromise the holdup time, but we still want to increase the V_{OR} to 130 V (for arbitrary reasons)? What is the best way to do that, if we could? From Fig. 6.12 we see that we can do this, but we now need a D_{MAX} *of about 72.5 percent*. Unfortunately, we note that the IP switch D_{MAX} is set not to around this seemingly

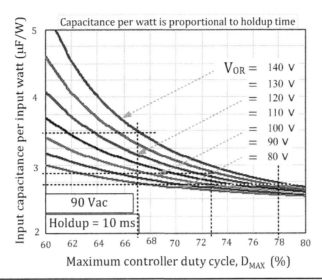

Figure 6.12 Holdup time design quick estimates.

more optimal 0.725, but arbitrarily to 0.78. We ask: Could that at least help in reducing the input capacitance further? It actually does, because we do have *more headroom* to swing the input. But not as much as we had hoped. Because looking at Fig. 6.12, we see that *at high D_{MAX} all the V_{OR} curves start bunching up together*.

- We can look upon Fig. 6.12 as intended for a 10-ms holdup time approximately (measured from the *lowest point of normal operating bulk-capacitor voltage ripple*, as in Chap. 5).
- We can see from Fig. 6.12 that for a V_{OR} of 140 V, with a D_{MAX} of 0.78, we get a requirement of 2.75 μF/W, which is actually *the same as the μF/W for a V_{OR} of 100 V using the older-generation IP switch*. So what did we really gain (or lose) in moving to the next-generation IP switch? Nothing! This is what happened: for a fixed D_{MAX}, decreasing the steady-state (nominal) D at 90 Vac (by setting a smaller V_{OR}) helps reduce the input capacitance. Also, increasing D_{MAX} *for a given steady-state D*, i.e., for a fixed V_{OR}, helps decrease the input capacitor too. In both cases the headroom increases, allowing for a bigger input swing. But *if we increase both V_{OR} (i.e., increase D) and D_{MAX}, we actually don't get any reduction in the input capacitance*, because the *headroom remains almost the same*. This is one reason, even after using the expert system software of the IP switch, we don't get any obvious improvement in the size of the input capacitor for the next-generation IP switch compared to the older generation.

Step 6: Inductor/Transformer Energy

We see from Fig. 6.12, that, for example, $D_{MAX}=0.67$ with a V_{OR} of 80 V gives us the same input capacitance (based on holdup time consideration) as $D_{MAX}=0.78$ with a V_{OR} of 140 V. So which of the two choices, if any, is better? What exactly is the last remaining design consideration?

Returning our attention to Fig. 6.11, we look at the D_{MAX} line. In the figure, we have shown the momentary *peak* currents expected in the converter for a 100 W load. Let us compare the specific cases based on the following specific official recommendations for the IP switch, i.e., a V_{OR} of 100 V for the first generation ($D_{MAX}=0.67$), and a V_{OR} of 140 V for the second generation ($D_{MAX}=0.78$). As we go from the normal operating duty cycle at 90 Vac to D_{MAX} during a line disturbance, we see the following:

- For the older-generation IP switch we get a peak current varying from 2.15 to 3.6 A.
- For the next-generation device we will get a variation from 1.8 A (that sounds good) to 3.9 A (*what?*).

We know that the transformer's physical size is virtually proportional to the peak energy stored in it, which is $(1/2) \times L \times I_{PK}^2$. So for every doubling of peak current we would need to increase the core four times. Therefore, in comparing a V_{OR} of 100 V to a V_{OR} of 140 V, the size of the inductor, calculated *only on the basis of steady operation* at 90 Vac (ignoring holdup for now),

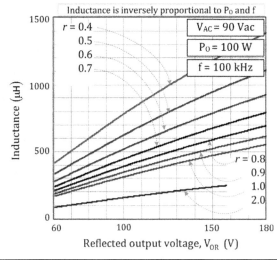

Figure 6.13 Inductance selector.

does go *down* by $(2.15^2 - 1.8^2)/2.15^2 = 30$ percent, on account of the higher V_{OR}. We also, however, note that since

$$L = \frac{V_{OR}}{I_{OR} \times r \times f}(1-D)^2 \times 10^6 \quad \mu H$$

and $P_O = V_{OR} \times I_{OR}$, and $D = V_{OR}/(V_{IN} + V_{OR})$, plotting this out we get Fig. 6.13. We see that *for a high V_{OR}, though the size of the inductor has come down, the required inductance goes up.* This would usually require more turns, so the copper losses will typically be higher, not lower.

But what happens as we slew down to higher duty cycles as the input falls (as during a holdup event)? For a low V_{OR} we go up to only 3.6 A (while meeting the holdup time requirement) whereas for the higher V_{OR} we go up to 3.9 A (albeit with a slightly smaller capacitance on 2.75 µF/W as compared to 2.9 µF/W). So, in fact, for the higher V_{OR}, we need a core size greater by $(3.9^2 - 3.6^2)/3.6^2 = 17$ percent. Yes, we may have got a 5 percent reduction in input capacitance from 2.9 to 2.75 µF/W in the process of increasing V_{OR} (and that is *assuming efficiency didn't change!*), but we now need to increase the core size by 17 percent. Though we can also see *that had we kept to a D_{MAX} of 72.5 percent, then with a V_{OR} of 130 V, we could have actually reduced the peak current compared to the case of $V_{OR} = 100$ V and $D_{MAX} = 67$ percent.* (Such are the intricacies of optimal Flyback design.)

Summarizing, the maximum duty cycle was excessive in the next-generation IP switch. It would have been more effective had the maximum duty cycle been set at 72.5 percent.

We have plotted this out in Fig. 6.14. We see that the *change* in the energy stored in the core (*in going from D to D_{MAX}*) is 4.3 µJ for a V_{OR} of 100 V ($D_{MAX} = 0.67$), but is 6 µJ for a V_{OR} of

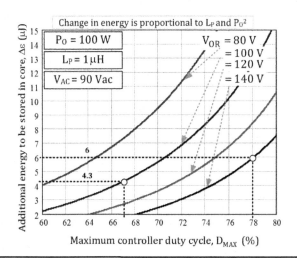

Figure 6.14 Energy of core.

140 V ($D_{MAX} = 0.78$), for every 1 µH of primary inductance and 100-W output power. So though the normal steady-state operating energy in the core may have been less with a higher V_{OR} (higher D) on account of the lower peak current, at D_{MAX} the peak energy requirement under transients, may be large enough to completely swamp out any perceived "advantage".

We are now clear how our reasoning must proceed as we try to optimize the Flyback. The ultimate choice will also depend on *break points* in the standard component values available. So, for example, if we choose the input capacitance to be 2.75 µF/W instead of 2.9 µF/W, it may mean nothing at a practical level if in either case we will eventually pick the same closest standard value. In that case we might as well try to optimize the other parameters and component values.

Dissipation Estimates (Graphical)

We show a first-order iterative calculation process, because otherwise we would actually require a complex math simulation file.

Example 6.4 A Flyback has 60 turns on the Primary and 3 turns on the Secondary. The output voltage is set to 5 V. The load is 20 A ($P_O = 100$ W, $V_{OR} = 100$ V). What are the losses at 90 Vac ($V_{IN} = 90 \times \sqrt{2} = 127$ V)?

We start with an assumption of 100 percent efficiency. Then, we know from Fig. 6.11 that for this condition the peak current is 2.15 A. This assumes $r = 0.4$. So the center of the switch ramp is $2.15/1.2 = 1.79$ A. Since $D = 0.44$ from Fig. 6.10, the average input current $1.79 \times 0.44 = 0.79$ A. The 1.79-A center reflects to the Secondary as per the turns ratio of 20 to become 1.79 A × 20 = 35.8 A. So the average current through the diode is $35.8 \times (1 - 0.44) = 20$ A. *Assuming that the switch is a FET, and has a specified forward drop of 10 V at 1 A*, we get its R_{DS} to be

$$R_{DS} = \frac{10\,V}{1\,A} = 10\,\Omega$$

The drop across the switch during the ON-time is

$$V_{SW} = 10\,\Omega \times 1.79\,A = 17.9\,V$$

The dissipation in the FET is about

$$P_{SW} = 1.79\,A^2 \times 10\,\Omega \times 0.44 = 14.1\,W$$

The diode dissipation is

$$P_D = V_D \times I_{D\_AVG} = V_D \times I_O = 0.6 \times 20 = 12\,W$$

Therefore, the total dissipation due to the switch and diode drops is about 26 W. The input power is not 100 W but is 126 W. This is a 26 percent increase. We have ignored zener dissipation and other parasitics. But we can quickly *recalculate some of the key parameters based on our new assumptions*. For example, the peak switch current is actually

$$I_{PK} = 126\% \times 1.79 = 2.25\,A$$

Using simple math applied to the equivalent primary-side Buck-boost model when the switch is ON, and the equivalent secondary-side Buck-boost model when the switch is OFF, we can include the resistance of the primary and secondary windings to get a more accurate estimate of the losses in a Flyback. We thus get Table 6.1.

Note that at a high duty cycle (as at low V_{IN} or/and high V_{OR}), the dissipation in the output capacitor will go up steeply. In fact, from a V_{OR} of 100 V to a V_{OR} of 130 V, the heating in the output capacitors (and its temperature) will increase by over 25 percent.

Also note that the dissipation in the input capacitor too has a $D/(1-D)$ dependency, so the remarks about the output capacitor generally apply here too. Except that if V_{OR} is increased, it helps to lower the input RMS current more than the $D/(1-D)$ tries to increase it. So a high V_{OR} helps in this case.

The equations provided in Table 6.1 assume a small current ripple ratio r. See Chap. 2 for more on the flat top approximation, and where this approximation creates a larger error and where it doesn't. We may want to use mor accurate equations, especially for the capacitors, as explained in Chap. 2.

The input capacitor RMS current is

$$I_{CIN\_RMS} = \frac{P_O}{V_{OR}} \times \sqrt{\frac{D}{1-D}}$$

and the output capacitor RMS current is

$$I_{COUT\_RMS} = \frac{P_O}{V_O} \times \sqrt{\frac{D}{1-D}}$$

Flyback Loss Terms	
$$D = \frac{V_{OR} + \left(\frac{V_{OR}}{V_O}\right)V_D + \left(\frac{V_{OR}}{V_O}\right)\left(1 + \frac{V_{OR}}{V_{IN}}\right)\frac{P_O}{V_O}R_{SEC}}{V_{IN} + V_{OR} + \left(\frac{V_{OR}}{V_O}\right)V_D - R_{DS}\left(1 + \frac{V_{OR}}{V_{IN}}\right)\left(\frac{P_O}{V_{OR}}\right) + \left(\frac{V_{OR}}{V_O}\right)\left(1 + \frac{V_{OR}}{V_{IN}}\right)\frac{P_O}{V_O}R_{SEC} - \left(\frac{V_O}{V_{OR}}\right)\left(1 + \frac{V_{OR}}{V_{IN}}\right)\frac{P_O}{V_O}R_{PRI}}$$	
Primary/switch peak current	$I_{PK} = \dfrac{1.2 \times (P_O/V_{OR})}{1-D}$
Dissipation in primary winding	$P_{PRI} = \left(\dfrac{P_O/V_{OR}}{1-D}\right)^2 \times R_{PRI} \times D$
Dissipation in secondary winding	$P_{SEC} = \dfrac{(P_O/V_O)^2 \times R_{SEC}}{1-D}$
Dissipation in zener clamp	$P_Z = \dfrac{1}{2} \times I_{PK}^2 \times \dfrac{V_Z}{V_Z - V_{OR}} \times f \times L_{LK} \times 10^{-6}$
Dissipation in switch (crossover loss)	$P_{CROSS} = \left(\dfrac{P_O}{V_{OR}}\right) \times V_{IN} \times t_{CROSS} \times f$
Dissipation in switch (conduction loss)	$P_{SW\_COND} = \left[\dfrac{P_O}{V_{OR}(1-D)}\right]^2 \times R_{DS} \times D$
Dissipation in diode	$P_D = \left(\dfrac{P_O}{V_O}\right) \times V_D$
Dissipation in the output capacitor	$P_{COUT} = \left(\dfrac{P_O}{V_O}\right)^2 \times \dfrac{D}{1-D} \times ESR_{COUT}$
Dissipation in the input capacitor	$P_{CIN} = \left(\dfrac{P_O}{V_{OR}}\right)^2 \times \dfrac{D}{1-D} \times ESR_{CIN}$
R_{PRI} is the resistance of the primary winding. R_{SEC} is the resistance of the secondary winding. ESR_{CIN} is the ESR of the input capacitor. ESR_{COUT} is the ESR of the output capacitor. L_{LK} is the lumped effective leakage in µH, as seen by the switch.	

TABLE 6.1 Estimating Dissipation Inside the Flyback.

So clearly, if the turns ratio is 1, or if this is a DC-DC Buck-boost converter (with a simple inductor), the input and output RMS currents are the same. This is an interesting coincidence.

In Fig. 6.15 we have provided the results of a sample calculation based on the IP switch, with converter parameters chosen as indicated. The inductance is assumed large (flat top approximation). We see that the zener dissipation is the main loss component (besides the switch and diode). Further, as we expected, it rises up steeply as V_{OR} increases. In Fig. 6.16 we have plotted the efficiency for various load conditions. We see that because of the zener dissipation term, *an ideal V_{OR} would be around 110 to 115 V purely on the grounds of efficiency*. The concave "bell" in the zener dissipation curve of Fig. 6.15 is clearly responsible for the convex bell in the efficiency curves of Fig. 6.16. The reason for that is *as we increase V_{OR}, up to some point we actually benefit, because the peak current falls in the process. But after that, the zener dissipation climbs again due to the $V_Z/(V_Z - V_{OR})$ term* in the corresponding equation given earlier in this chapter.

Careful with That Calculation: Correct Dissipation in Switch and Sense Resistors

In Example 6.4 we took a simplified equation for the dissipation in the switch. We used the flat top approximation. But other than that, there is a subtle assumption here. To recall, our steps were

Topologies for AC-DC Applications: An Introduction

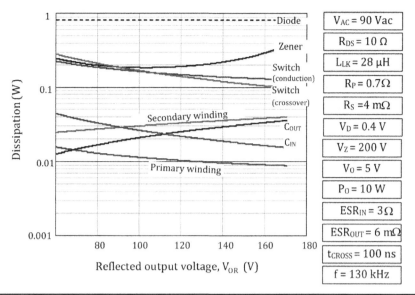

FIGURE 6.15 A sample calculation of parameters for various V_{OR}.

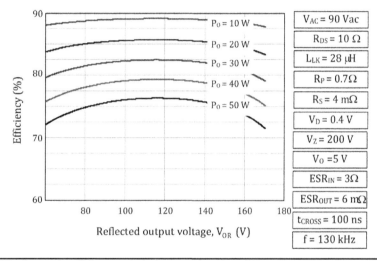

FIGURE 6.16 A sample efficiency calculation as a function of V_{OR}.

$$R_{DS} = \frac{10\text{ V}}{1\text{ A}} = 10\text{ }\Omega$$

$$V_{SW} = 10\text{ }\Omega \times 1.79\text{ A} = 17.9\text{ V}$$

$$P_{SW} = 1.79\text{ A}^2 \times 10\text{ }\Omega \times 0.44 = 14.1\text{ W}$$

So we found the I_{COR} (center of ramp) value of the primary-side current (1.79 A). Then we found out the (average) drop across the FET during the ON-time (17.9 V), but didn't use it. We multiplied I_{COR}^2 by the R_{DS} to get the dissipation during the ON-time. Then we averaged that dissipation by multiplying it by the duty cycle (0.44). In effect we used

$$P_{SW} = \overline{I_{COR}^2 \times R_{DS}} = I_{COR}^2 \times R_{DS} \times D \equiv P_{SW\_ON} \times D$$

We could have alternatively found the RMS value of the current and multiplied it by R_{DS}. In the flat top approximation, RMS of a flat pedestal of current of height I_{COR} is $I_{COR} \times \sqrt{D}$. So we could have used

$$P_{SW} = I_{RMS}^2 \times R_{DS} = \left(I_{COR} \times \sqrt{D}\right)^2 \times R_{DS} = I_{COR}^2 \times R_{DS} \times D$$

This gives us the same result as we got previously. So we can think of this in two equivalent ways:

1. Calculate the RMS of the current *during the ON-time* (this is equal to I_{COR} in the flat top approximation). Then, using R_{DS}, we get the dissipation *during the ON-time*. We can then average that ON-time dissipation over the entire cycle to get the final dissipation.

2. We can also first calculate the RMS of the current over the entire cycle ($I_{COR} \times \sqrt{D}$ in the flat top approximation). Then we multiply this full-cycle RMS current with the R_{DS} to get the dissipation over the whole cycle.

Very similarly, for the sense resistor (if any) we would have gotten:

$$P_{RSENSE} = I_{COR}^2 \times R_{SENSE} \times D$$

In Example 6.4, we did not include this because the IP switch uses R_{DS} sensing (no external sense resistor).

Another "method" of doing the preceding calculation (actually a pitfall) is to

3. Calculate the drop across the FET (or sense resistor) during the ON-time:

$$V_{SW\_ON} = I_{COR} \times R_{DS}$$

Then find the *average drop* across the FET during the entire cycle

$$V_{SW\_AVG} = I_{COR} \times R_{DS} \times D$$

Then calculate the average current in the switch during the ON-time:

$$I_{SW\_AVG} = I_{COR} \times D$$

Then find the average dissipation by multiplying the two averages to get

$$P_{SW} = V_{SW\_AVG} \times I_{SW\_AVG} = (I_{COR} \times R_{DS} \times D) \times (I_{COR} \times D)$$
$$= I_{COR}^2 \times R_{DS} \times D^2 \quad \text{(Wrong)}$$

This underestimates the switch and sense dissipation (by almost half typically). *Do not use this average voltage drop method for FET switches in particular.!*

Practical Flyback Designs Using 600-V Switches

Everything so far, points to lower performance and no significant reduction in cost, if we increase V_{OR} without due thought. As mentioned previously, most well-designed commercial Flybacks use only 600-V FETs. That is achieved by setting even a lower V_{OR} (around 70 V), and having effective current limiting for the FET and for each individual output, to avoid overvoltages during faults. Properly implemented input feedforward is also necessary (which we will discuss later). The efficiency is then maximized with an RCD clamp in place of a zener. The typical operating drain-to-source voltage is about 450 V at 270 Vac input, with a leakage spike of 50 V. This represents an 85 percent derating margin. Under overloads and startup into short-circuited outputs, the drain-to-source voltage doesn't exceed about 585 V.

Such power supplies typically fix the turns ratio at around 12–14 for a 5-V output. This translates to a V_{OR} of about 70 V. Actual turns are typically 85T:6T or 60T:5T (T stands for turns here). The choice of whether to have 5T for the secondary versus say 6T depends largely on the "other" secondary output. For example, if it is a 12-V output with a series-pass stage (linear postregulator), we will use 6T for the Primary. Assuming about 1-V drop from the 5-V output Schottky diode and the winding resistance, we are in effect getting 5 V per turn for the transformer. For the 12-V winding we then use 15½ turns to get 15.5 V at the winding. If we use an ultrafast low-drop diode for this output (e.g., the FEP6AT through FEP6DT series from www.vishay.com), we can assume a 1 V drop here too, so we get 14.5 V at the input of the 12 V postregulator. Then using a series-pass stage built around a cost-effective and popular FET like the MTP3055, we can drop the voltage down to 12 V (this works well up to about 2 to 3 A on the 12 V output). A typical headroom of 2.5 V across the series-pass regulator is required because of cross-regulation limitations. For example, if the 5 V output is at minimum load, the converter is going to try to move to a very low duty cycle because usually only the 5 V is being PWM-regulated. This is called a *starve* condition because that is exactly what it does: The other windings on the same transformer get starved due to inadequate volt-seconds and tend to droop. So a certain nominal headroom must be planned for, across the 12 V series-pass transistor. Sometimes, we may need to go up to 16T for the 12 V winding. Further, if the minimum specified load on the 5-V output is zero, we will also need to put in a preload on the 5-V output, inside the power supply itself, to maintain a certain minimum volt-seconds across the windings.

> **NOTE** *Sometimes, the series pass FET can be eliminated by a technique called double-point sensing in which both the outputs are sensed through upper resistors going to the LM431, but one is weighted much more than the other. This compromises the main regulation slightly in favor of the other one.*

How Higher V_{OR} Impacts the Output Diode Rating

The minimum voltage rating of the output diode is (alternative forms)

$$V_{D\_rating} = V_O + V_{INR} = V_O + \frac{V_{IN}}{n} = V_O + \frac{V_{IN}}{V_{OR}/V_O} = V_O \times \left(1 + \frac{V_{IN}}{V_{OR}}\right)$$

We must check this out at V_{INMAX}. We see that if V_{OR} is 100 V, we can use a 60-V Schottky diode for a 12-V output. But, to be able to use a 60-V diode for outputs as high as 15 V, we need to increase V_{OR} to about 140 V.

A corollary is that with 600-V switches, which require a lower V_{OR} setting, we will not be able to use any commonly available 60-V Schottky diode for 12-V or –12-V outputs.

Pulse-Skipping and Required Preload

At some lighter load, we may reach the minimum duty cycle capability of the controller. Then the controller will typically respond by randomly trying to omit cycles (unless designed specifically to handle this situation in a predefined manner such as a formal pulse-skip mode). What happens is that the energy being pushed into the converter, in even one (minimum) pulse width, is in excess of the output power requirement. The error amplifier then basically behaves in an almost random hysteretic (bang-bang) mode, causing increased output voltage ripple and poor transient response too. Suppose we want to avoid this behavior. Unfortunately, especially with current-mode control, we always need a certain *blanking time* of around 50 to 150 ns to avoid noise-induced jitter. That may in effect be too large a minimum duty cycle, especially at higher switching frequencies. A sure way to avoid this random-skipping/pseudo-hysteretic mode is to *maintain a certain minimum external load on the converter output, that is, a preload*.

The first question that arises is: As we increase input voltage, at what point does the converter enter discontinuous mode? The duty cycle in discontinuous conduction mode is (see Chap. 4):

$$D = \frac{\sqrt{2 \times P_O \times L \times f}}{V_{IN} \times 10^3}$$

where L is in microhenries, f in hertz, and P_O in watts. We can also show by equating duty cycle equations for continuous and discontinuous modes, that the transition occurs at the following input voltage:

$$V_{IN\_CRIT} = V_{OR} \times \left[\sqrt{\frac{1}{\gamma^2} + \frac{1}{\gamma}} + \frac{1}{\gamma}\right]$$

where

$$\gamma = \frac{V_{OR}^2 \times 10^6}{2 \times P_O \times f \times L} - 1$$

Note that L is in microhenries here. Quick design curves are also provided in Fig. 6.17.

> **Example 6.5** We have a Flyback switching at 100 kHz meant for a 10-W maximum load over the range 90 to 275 Vac. If the V_{OR} is 140 V, and we set $r = 0.9$ at minimum input, at what input voltage will it turn discontinuous? What is the minimum duty cycle we need to be able to handle if the load is reduced

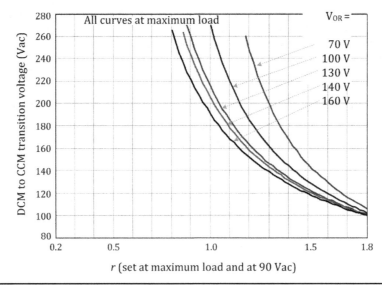

FIGURE 6.17 Voltage at which a universal-input Flyback will enter DCM.

to 1 W? Another design choice is $V_{OR} = 100$ V, implemented with a suitably designed transformer, everything else remaining the same. What is the minimum duty cycle now?

Since $V_{OR} = 140$ V, and $V_{OR} \times I_{OR} = 10$ W, we get $I_{OR} = 10/140 = 0.071$ A. At 90 Vac (ignoring diode and switch drops)

$$D = \frac{V_{OR}}{V_{OR} + V_{IN}} = \frac{140}{140 + 127} = 0.524$$

L for a Buck-boost (see the Appendix) is

$$L = \frac{V_O + V_D}{I_O \times r \times f} \times (1-D)^2 \times 10^6 \ \mu H$$

So since we can set V_O to V_{OR} and I_O to I_{OR}

$$L = \frac{140}{0.071 \times 0.9 \times 10^5}(1 - 0.524)^2 \times 10^6 = 4964 \ \mu H$$

We calculate γ as

$$\gamma = \frac{V_{OR}^2 \times 10^6}{2 P_O f L} - 1 = \frac{140^2 \times 10^6}{2 \times 10 \times 10^5 \times 4964} - 1 = 0.974$$

Note that if γ was negative, no solution would exist (i.e., the system could not go discontinuous by simply increasing the input voltage. We would need to also decrease the load current as we will see in a later example). So

$$V_{IN\_CRIT} = V_{OR} \times \left[\sqrt{\frac{1}{\gamma^2} + \frac{1}{\gamma}} + \frac{1}{\gamma} \right] = 140 \times \left[\sqrt{\frac{1}{0.974^2} + \frac{1}{0.974}} + \frac{1}{0.974} \right] = 346 \text{ V}$$

This is equivalent to $346/(2)^{1/2} = 245$ Vac.

Let us first go at maximum load from V_{IN_MIN} to V_{IN_CRIT}. Then we reduce the load and increase the voltage further. The duty cycle at the critical boundary (at maximum load) is

$$D_{CRIT} = \frac{V_{OR}}{V_{OR} + V_{IN\_CRIT}} = \frac{140}{140 + 346} = 0.29$$

Now if we decrease the load and increase the input voltage, we can see from

$$D = \frac{\sqrt{2 \times P_O \times L \times f}}{V_{IN} \times 10^3}$$

that D will vary as

$$\frac{D_{MIN}}{D_{CRIT}} = \frac{\sqrt{P_{O\_MIN}}}{\sqrt{P_{O\_CRIT}}} \times \frac{V_{IN\_CRIT}}{V_{IN\_MAX}}$$

So

$$D_{MIN} = \frac{\sqrt{1}}{\sqrt{10}} \times \frac{346}{389} \times 0.29 = 0.082$$

The D_{MIN} requirement is thus 8.2 percent.

Now let us change V_{OR} to 100 V. This time, $I_{OR} = 10/100 = 0.1$ A. We also know that $D = 0.44$ at 90 Vac with $V_{OR} = 100$ V. Thus L is recalculated for the required r, and we get $L = 3484$ µH. Then

$$\gamma = \frac{V_{OR}^2 \times 10^6}{2P_O fL} - 1 = \frac{100^2 \times 10^6}{2 \times 10 \times 10^5 \times 3484} - 1 = 0.435$$

So

$$V_{IN\_CRIT} = V_{OR} \times \left[\sqrt{\frac{1}{\gamma^2} + \frac{1}{\gamma}} + \frac{1}{\gamma}\right] = 100 \times \left[\sqrt{\frac{1}{0.435^2} + \frac{1}{0.435}} + \frac{1}{0.435}\right] = 505 \text{ V}$$

This is equivalent to 357 Vac, well outside our range. *Therefore, at 275 Vac, with maximum load, we are still running continuous.* The duty cycle at this point is

$$D_{VIN\_MAX} = \frac{100}{100 + 389} = 0.2$$

The current ripple ratio at this point is therefore

$$r_{VIN\_MAX} = \frac{V_{OR}}{I_{OR}Lf} \times (1 - D_{VIN\_MAX})^2 \times 10^6 = \frac{100}{0.1 \times 3484 \times 10^5} \times (1-0.2)^2 \times 10^6 = 1.84$$

Clearly, if we now decrease the load, we will go critical at

$$I_{CRIT} = \frac{r}{2} \times I_{OR} = \frac{1.84}{2} \times 0.1 = 0.092 \text{ A}$$

This is equivalent to an output power of $V_{OR} \times I_{OR} = 100 \times 0.092 = 9.2$ W. Thereafter, we are in discontinuous mode. Now, since the input voltage does not change, the duty cycle will shrink in going from 9.2 to 1 W as per

$$\frac{D_{MIN}}{D_{CRIT}} = \frac{\sqrt{P_{O\_MIN}}}{\sqrt{P_{O\_CRIT}}} = \frac{\sqrt{1}}{\sqrt{9.2}}$$

So finally

$$D_{MIN} = 0.2 \times \frac{1}{3.03} = 0.066$$

The D_{MIN} requirement is thus 6.6 percent. At 100-kHz switching frequency this is equivalent to a minimum pulse width requirement of 660 ns. With a minimum load of 0.2 W we would require less than 300 ns. We can see that a high V_{OR} does help somewhat in avoiding pulse-skipping at very light loads.

Finally we can also conversely calculate that given a minimum pulse width of the controller IC, what is the minimum load we need to have, so as to avoid the random pulse-skipping mode.

Overload Protection at High-Line (Feedforward)

One of the tricks we do with the popular 3842/3844 series is to place a resistor from V_{IN} to the I_{SENSE} pin. The purpose of this is to reduce the primary-side current limit and thus effectively lower the maximum possible duty cycle at high-line. Because otherwise, though at high-line the steady operating duty cycle is naturally lower, and so is the peak current, under overloads, the converter will try to hit the maximum duty cycle limit of the controller IC, until the slower secondary-side current limit can start to work and fold back the output power. But even during this small transition interval, damage can occur.

This situation cannot be understood in terms of any steady-state scenario. For example, if the output voltage is zero (perfect short), then the downslope of the primary-side inductor current V_{OR}/L_P (or equivalently the downslope of the secondary-side current V_O/L_S) is almost zero (actually it is V_D/L here). Therefore, the current can never reach the state it started the cycle off with, and a steady state by definition does not exist. So we will ultimately get, current-staircasing condition until we hit the current limit. But now, although the primary-side peak current is supposedly being limited precisely by the sense resistor, and should therefore protect the switch and the transformer from saturation, this does not happen in practice. It can be shown that the blanking time requirement of all current-mode control ICs translates to a minimum pulse width. And it can be shown that this can *effectively override any supposed current limit* in this particular condition. Because, the current limit, when reached, can only respond by commanding the controller to limit the duty cycle further. But even during this minimum pulse width of 150 ns or so, the slope of the up-ramp which is V_{IN}/L, is very high. And there is virtually no down-ramp. So the current will actually continue to staircase beyond the set current limit. Many modern current-mode DC-DC controllers/switchers respond by initiating a *frequency foldback* whenever the voltage on the feedback pin falls below a certain threshold. In doing so they effectively reduce the duty cycle under current limit, and this extends the OFF-time by a typical factor of four, giving enough time for the current to ramp down to a value less than what it started the cycle with, thereby quashing staircasing. But note that the effectiveness of this technique depends largely on the diode drop! So "good" diodes (with lower forward drops) actually make the fault currents even more severe, as they do not provide enough downslope under short-circuit conditions.

In the case of the 3842/3844, we need to protect ourselves from this situation by *reducing the primary-side current limit at high-line, so we have some enforced headroom available before the transformer can saturate*. We now give the equations to implement this.

Basically, by introducing a *feedforward* resistor R_{FF}, the current sense signal is DC shifted a little higher by an amount $R_{FF} \times I_{FF}$, so it will hit the current limit a little earlier. Note that we do not want to affect the current limit at low input voltages, because we still need to comply with holdup time. If R_{BL} is the (blanking) resistor normally connected between the sense resistor and the sense pin of the IC (typically 1 kΩ or so), the current limit at high-line V_{IN_MAX} as a ratio of the current limit at V_{IN_LO} (e.g., 60 Vac, i.e., 85 Vdc) is

$$\frac{CLIM_{VIN\_MAX}}{CLIM_{VIN\_LO}} = \frac{V_{CLIM} - [(V_{IN\_MAX}/R_{FF}) \times R_{BL}]}{V_{CLIM} - [(V_{IN\_LO}/R_{FF}) \times R_{BL}]}$$

where V_{CLIM} is the voltage on the current sense pin corresponding to current limit. So if, for example, $V_{IN\_MAX} = 389$ Vdc, $V_{IN\_LO} = 85$ Vdc, $R_{BL} = 1$ k, and $V_{CLIM} = 1$ V, we get the required value of the feedforward resistor for reducing the current limit threshold from 1 to 0.75 V as

$$0.75 = \frac{1 - [(389/R_{FF}) \times 1000]}{1 - [(85/R_{FF}) \times 1000]}$$

Solving we get

$$R_{FF} = 1.3 \text{ Meg}$$

At 60 Vac this will also lower the current limit threshold slightly below 1 V. The current I_{FF} is 85/1.3 M = 65 µA. Passing through the blanking resistor (1 k), it causes a drop of 0.065 V. So the current limit threshold is now 1 − 0.065 = 0.935 V. Note that the current limit threshold actually has a typical tolerance of ±10 percent. So we must assume that the worst-case threshold is actually 0.9 − 0.065 = 0.835 V. Knowing this we can correctly tweak R_{SENSE} too.

CHAPTER 7

The Holy Grail: An Overview of Magnetics in Power Conversion

It is rather unkindly said: Those who can, do; those who can't, teach. In power conversion this possibly translates to: Those who understand magnetics, make switching power supplies; those who don't, simulate using PSpice.

Magnetics, and we are not just talking about vanilla Buck inductors here, form the holy grail of power conversion. A lot of engineers are seemingly in denial about that fact though. Not unsurprisingly, there are therefore abundant misconceptions and myths prevailing, some even basic and fundamental. For example, even senior engineers are known to complain that an inductor/transformer used in a BCM/DCM design (such as a ringing choke Flyback) is much larger than in CCM. Their intuitive process is perhaps that "peak currents are much higher in BCM." But in fact BCM magnetics can be half the size of their CCM versions typically. Engineers also stumble in actually explaining "simple" transformer action; i.e., voltage and current scaling. It is certainly easy to write the equations, but rather hard to explain them. This lack of understanding is not conducive to building robust and optimized-power supplies. Yes, we can escape into the world of control-loop theory. But at the most basic level, power conversion is all about using *reactive elements* for energy storage. And the inductor (or transformer) is king. One possible reason for our inherent fear of magnetics is that the transformer/inductor is a constant-*current* element, whereas in our world we are more comfortable with constant-*voltage* elements: such as DC sources, lab power supplies, AC outlets, batteries, and capacitors. Constant current is anathema to our natural, intuitive way of thinking and perceiving. It is therefore going to take us the next five or six chapters to fully put our arms around this difficult topic. Once done, control-loop theory will certainly follow for the magnetic skeptics, in rather gory detail in Chap. 14.

In this chapter we will only present the most basic magnetic concepts, but in a rather unconventional and hopefully straightforward way. That should take away some of the unnecessary fear or mystique behind magnetics. The reader can safely assume we are using MKS (i.e., SI) units, unless otherwise stated.

Basic Magnetic Concepts and Definitions (MKS units)

A quick summary of terms follows first, as a refresher. We will go into more details after that.

1. *H-field*. Also called *field strength, field intensity, magnetizing force, applied field*, etc. Its units are ampere per meter (A/m).

2. When the magnetic intensity, or *H*-field, is integrated over a *closed loop*, we get the current enclosed by the loop

$$\int_{CL} H\, dl = I \quad A$$

where CL stands for *closed loop*. This is called *Ampere's circuital law*.

3. *B-field*. Also called *flux density* or *magnetic induction*. Units of B are tesla (T), or webers per square meter (Wb/m^2).

4. Flux is the integral of B over a surface area: $\Phi = \int_S B\, dS$. It is expressed as webers (Wb). If B is constant over the surface, we get the more common form $\Phi = BA$. The

integral of B (its "normal," i.e., perpendicular component actually), over a closed surface, is zero since *flux lines* (used for representing flux visually) do not start or end at any given point, but are continuous.

5. B and H are related *at any given point* by the equation $B = \mu H$, where μ is the permeability of the material.

6. In air, the permeability μ is denoted by μ_0, which is the symbol for permeability of free space/air. Numerically, $\mu_0 = 4\pi \times 10^{-7}$ henry per meter (H/m) (in MKS units). (In CGS units it equals 1.)

7. Faraday's law of induction (also called Lenz's law) relates the induced voltage V developed across the ends of a coil of N turns, resulting from a (time-varying) B-field passing through the coil:

$$V = NA\frac{dB}{dt} = N\frac{d\Phi}{dt}$$

The changing magnetic field could be due to changing current through another coil (induction) or through the same coil (self-induction).

8. Flux linkages λ, expressed as Wb-turns, is $N \times \Phi$. Here ϕ is the flux impinging on N turns of a coil. The flux could be from the same coil, and in that case, the effect (i.e., opposing itself) leads to the concept of "inductance" (self-inductance actually). But if the flux is from another coil, its effect is described in terms of the "mutual inductance" between the coils.

9. We have

$$V = N\frac{d\Phi}{dt} = \frac{d\lambda}{dt}$$

So, a rate of change in flux linkages of 1 Wb-turn/s will produce 1-V induced voltage (irrespective of inductance). More generally, if the flux linkages change at the rate of x Wb-turn/s, we will generate x V across the coil.

10. Alternatively stated, a uniform rate of change in flux, at the rate of 1 Wb/s, in the flux linking a single-turn coil ($N = 1$), or, alternatively, a rate of change of 1 flux-linkage per second (1 Wb-turn/s), will generate 1 V across the coil (which is the induced voltage). This too follows from

$$V = N\frac{d\Phi}{dt} = \frac{d\lambda}{dt}$$

11. Note that we are often going to ignore the sign of the induced voltage for convenience (where we can afford to). But we should always remember that the induced voltage is in such a direction that it opposes its cause (and thereby tries to reduce its effect). Expressed in somewhat more detail: Lenz's law states that the polarity of the induced voltage is such that the voltage attempts to produce a current (through an external resistance) that opposes the original change in flux linkages.

12. (Self) Inductance L, expressed in henry (H), provides the last term included in the following equation:

$$V = NA\frac{dB}{dt} = N\frac{d\Phi}{dt} = L\frac{dI}{dt}$$

13. We can also say from the last equation that an inductance of 1 H develops an induced voltage of 1 V across it, if the current through it changes at the rate of 1 A/s.

14. Equating the last two terms in the previous equation we get

$$\frac{d\lambda}{d\cancel{t}} = L\frac{dI}{d\cancel{t}} \Rightarrow \Delta\lambda = L\,\Delta I$$

So

$$L = \frac{\Delta\lambda}{\Delta I} \quad \text{(flux linkages per ampere)}$$

We can also say that an *inductance of xH means that the coil produces x flux linkages per ampere of current in the coil.*

15. Keep in mind that: *A steady current produces a steady flux. But voltage is induced only when the flux changes.*

16. Alternatively, the (electromagnetic) *inertia* of a coil, to a change in flux through it arising from its own time-varying current, is its *self-inductance*, or just *inductance*, designated L, and defined as

$$L = \frac{N\Phi}{I} \quad \text{H}$$

L (i.e., inductance) is alternatively defined as the number of *flux linkages* $N\phi$ divided by the current causing the flux Φ. Be warned, this expression is deceptive. L is actually proportional to N^2, not N. The reason is the flux produced is proportional to N, and it also gets linked to itself (N turns). So the final effect is proportional to N^2.

The unit henry is usually written as H. Strictly speaking, the plural is Henrys not Henries.

17. The proportionality constant, connecting L and N^2, is called the *inductance index*, or simply A_L. This value is usually expressed as nH/turns^2 (though sometimes it is also considered to be mH/1000 turns^2, but both are *numerically the same*). So

$$L = A_L \times N^2 \times 10^{-9} \quad \text{H}$$

The Electrical-Magnetic Analogy

Engineers are often confused about the difference between B and H, for example. We can also ask: If $B = \mu H$, and they are proportional, why even bother to talk of *two* fields? Indeed, in ferrites that is true: We can write all the equations in terms of either B or H, and it wouldn't matter. But that doesn't always happen. Which is why we need two magnetic fields and two electric fields.

We try to resolve the confusion through Fig. 7.1. Analogies always seem to help. In effect what we are implying here is that E and H are "natural," whereas their derivatives, D and B, are fields that depend on the material. So we see that D has a component called *polarization* just as B has a component called *magnetization*. So in effect we are looking at the effect of an underlying field (the cause, i.e., E or H) summed with its result (D or B, both being material-dependent).

In general, we could create an H-field as per Ampere's circuital law (H being proportional to current) and apply it to a material, create magnetization in it, and then reduce H to zero and still be left with some magnetization. That is called *remanence*. It is the principle behind magnetic disk drives, for example. In that case we could *not* say that B is proportional to H. In Fig. 7.1 we used the substitution $M = \chi_m H$. We implicitly assumed that χ_m is a

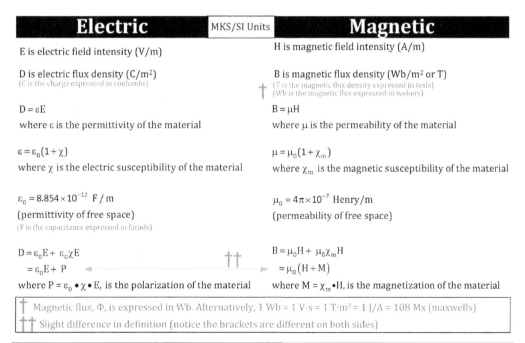

FIGURE 7.1 The electric-magnetic analogy (in MKS units).

constant. That is not generally true. Materials for which magnetization is really proportional to the magnetic (magnetizing) field intensity (H) are called linear materials. They also have *no remanence* since their response (magnetization) is proportional to the cause (I). If you remove the current, you remove the magnetization field.

Ferrites are, to an approximation, linear (until they start saturating, of course). So in such a case we can say that B and H are truly proportional to each other, and then, we can substitute either one for the other, in all our equations, without feeling mentally challenged as to their intuitive or real "difference." It really doesn't matter.

However, B and H, even in a ferrite, do have an important difference in *behavior* as we will soon see. The fact is that in the typical magnetic configurations we deal with in power, B does not change as we go from ferrite to air and back to ferrite. So it is *continuous*, with no sudden jumps at material interfaces. But in contrast, H is not! It jumps at every interface, because μ changes. So the fact is *all our computations become much easier if we focus on calculating the B-field only*—it would apply everywhere and just make life (*mathematically*) much easier for us. Other than that, it is a matter of preference whether we want to deal with B, or H, or both, in linear materials.

The Inductor Equation

Now we delve into some details. Combining Faraday's law with the definition of L, we get the most common equation of power conversion (we simply call it the *inductor equation*)

$$V = L\frac{dI}{dt}$$

Comparing this with the *law of induction* as discussed earlier and combining all the equations, we get the following *basic reference set of design equations (all expressing the induced voltage, V)*:

$$V = N\frac{d\Phi}{dt}$$

$$V = NA\frac{dB}{dt}$$

$$V = L\frac{dI}{dt}$$

In power conversion, we have mostly piecewise linear waveforms. So these become (using $V \Delta t$ equal to volt-seconds)

$$\text{volt-seconds} = N \times \Delta\Phi$$
$$\text{volt-seconds} = NA \times \Delta B$$
$$\text{volt-seconds} = L \times \Delta I$$

In Chap 2, we talked about volt-seconds balance and how by having equal and opposite volt-seconds in steady state during the ON-time and OFF-time, we force the net change in current in the inductor every cycle to zero—a process called *reset*. If we do not ensure reset, we get a runaway condition—called *flux-staircasing*. Now we know why. We can see from the preceding three equations that I, ϕ, and B are all related to each other.

Conversely, if the current in the inductor returns to the same value it started the switching cycle off with, then so will the flux and so will the B-field (flux density). As mentioned, they are all related, but in fact are *proportional to each other* through their common left-hand-side term (volt-seconds).

The Voltage-Independent Equation

Eliminating V from Faraday's law (by equating two of the three preceding equations), we get a key equation used in transformer/inductor design

$$NA\frac{dB}{dt} = L\frac{dI}{dt}$$

or

$$\Delta B = \frac{L\Delta I}{NA}$$

The Holy Grail: An Overview of Magnetics in Power Conversion

or simply

$$\Delta B \propto \Delta I \quad \text{(proportionality constant being } L/NA\text{)}$$

It so happens that for soft ferrites and most other magnetic materials used in power conversion, there is almost no field remaining in the core if the current in the windings goes to zero (the *remanence* is almost zero). Therefore, *since B is 0 when I is zero*, we can also write the above equation in terms of absolute quantities, such as

$$B_{PK} = \frac{LI_{PK}}{NA}$$

From Fig. 7.2 we can see that in fact this holds for any *instantaneous* values of B and I. So generically,

$$B = \frac{LI}{NA}$$

Keep in mind that the inductance can be considered to be a constant provided we are not close to saturation. Because only then is L, and therefore L/NA, really a proportionality *constant*! But if L changes, B is no longer proportional to I.

Therefore, in the linear region, the current ripple ratio r applies equally to current and also to the field. If, for example r is 0.4, then $B_{AC} \cong 0.2 \times B_{DC}$. So, in general,

$$B_{DC} = \frac{2}{r+2} \times B_{PK} \quad \text{or} \quad I_{DC} = \frac{2}{r+2} \times I_{PK}$$

$$B_{AC} = \frac{r}{r+2} \times B_{PK} \quad \text{or} \quad I_{AC} = \frac{r}{r+2} \times I_{PK}$$

$$\Delta B = \frac{2r}{r+2} \times B_{PK} \quad \text{or} \quad \Delta I = \frac{2r}{r+2} \times I_{PK}$$

Example 7.1 If the current swing is doubled (by halving inductance, since $V \Delta t = L \Delta I =$ constant), by what percentage does the peak current change?

$$\Delta I = \frac{2r}{r+2} \times I_{PK} \quad \text{or} \quad I_{PK} = \Delta I \times \frac{r+2}{2r}$$

Assuming an initial optimum value of $r = 0.4$, if the current swing doubles, r goes to 0.8. Therefore, in the first case

$$I_{PK1} = \Delta I_1 \times \frac{0.4 + 2}{2 \times 0.4} = 3 \times \Delta I_1$$

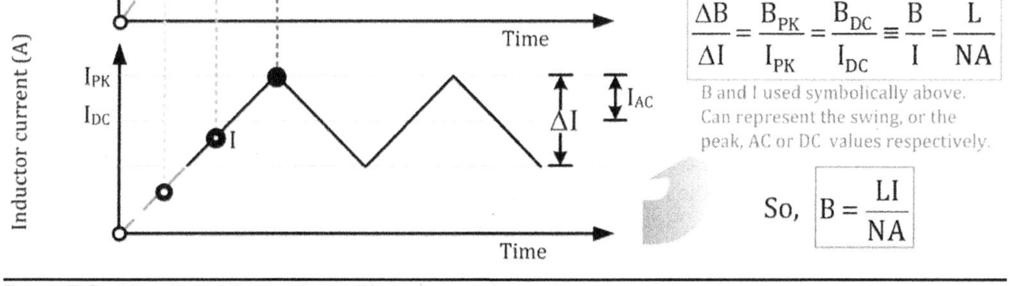

FIGURE 7.2 How *B* and *I* are related (if remanence is zero).

In the second case

$$I_{PK2} = \Delta I_2 \times \frac{0.8+2}{2 \times 0.8} = \Delta I_1 \times \frac{0.8+2}{0.8} = 3.5 \times \Delta I_1$$

Therefore, the ratio by which peak current changes is 3.5/3 = 1.17. In other words, *by changing inductance by 50 percent, the change in peak current and peak field, is only 17 percent.*

Example 7.2 The voltage-independent equation is useful if, for example, we want to do a quick check on saturation of a core. If we have wound 40 turns on a core with a datasheet specified value of A = 2 cm² (effective area, also called A_e), its measured inductance being 200 μH, and the peak current (measured or estimated) being 10 A, then the peak flux density is

$$B_{PK} = \frac{LI_{PK}}{NA} = \frac{200 \times 10^{-6} \times 10}{40 \times 2 \times 10^{-4}} = 0.25 \text{ T}$$

For a typical ferrite, this is quite close to its saturation flux density, typically 0.3 T. Therefore, peak currents greater than 10 A should not be passed through this inductor. *Nor should the turns be increased any further.* This latter observation actually seems contradictory to the observation that in B = LI/NA, N is in the denominator, which seems to indicate that increasing N should help in reducing the peak field. However, that is not so. The reason is that L in the numerator increases according to N^2. So the field is effectively *proportional* (not inversely proportional) to the number of turns (and obviously also proportional to the current passing through it). We must keep in mind that this implies that for a given core there is a *maximum ampere-turns it can handle before it saturates* (and becomes less effective, even ineffective).

NOTE *In Example 7.2, we may have thought that since increasing N causes I_{PK} to decrease, that should have lowered B_{PK}. But that effect was totally swamped out by the increase in L, which caused B_{PK} to increase instead. The underlying reason is that L only changes the ramp (swing) portion of the inductor current. Yes, if we double L, we halve ΔI, but ΔI itself is only a small fraction of the average and peak current values. So increasing N does not have that much of an impact on I_{PK}. Example 7.1 revealed that changing inductance by 50 percent, changed I_{PK} by only 17 percent.*

CONCLUSION *Never increase the number of turns on a core injudiciously: That may help lower the peak current slightly, but the peak B-field will go up dramatically, and that can only help saturate the core. In fact* decreasing *the number of turns is the preferred direction to go in if we suspect core saturation is causing failure.*

The Voltage-Dependent Equation

There are so many forms of this equation in the industry, that this is the point where the average designer loses all interest in magnetics. Let us, therefore, follow this through, if for nothing else but to demystify the topic and point us in the right direction (which is actually much simpler than we may have thought).

The basic design equation set is

$$V = N\frac{d\Phi}{dt} = NA\frac{dB}{dt} = L\frac{dI}{dt}$$

or

$$\Delta B = \frac{V_{AVG} \Delta t}{NA}$$

where in the latter form, we have, generically speaking, taken the *average value* of the applied voltage (V_{AVG}) just in case it is varying over the evaluation interval. Typically, in nonresonant switching converters, the applied voltage is a constant. It is either V_{ON} or V_{OFF}. V_{ON} is the voltage applied across the inductor during the switch ON-time, and V_{OFF} is the voltage during the switch OFF-time. Since $B_{AC} = \Delta B/2$, we get

$$B_{AC} = \frac{V_{AVG} \Delta t}{2 \times NA}$$

Note that the very use of the term *AC* implies a waveform that is symmetrical. So DCM waveforms are being implicitly excluded here. Also, applying the preceding equation during the ON-time,

$$B_{AC} = \frac{V_{ON} D}{2 \times NAf}$$

And assuming CCM, we also have $T_{OFF} = (1-D)/f$, so

$$B_{AC} = \frac{V_{OFF}(1-D)}{2 \times NAf}$$

where f is the switching frequency in hertz.

At this point, if, and only if, a perfectly *square wave voltage* is applied to the core (defined as implying $T_{ON} = T_{OFF} = T/2$, i.e., $D = 0.5$), then from

$$B_{AC} = \frac{V_{AVG} \Delta t}{2 \times NA}$$

and

$$\Delta t = \frac{1}{2f}$$

we get

$$B_{AC} = \frac{B_{PP}}{2} = \frac{V_{AVG}}{4 \times NAf}$$

Manufacturers of magnetic components often prefer to express the *voltage-dependent* equation in terms of the *applied RMS voltage*. We know that by the volt-seconds law, for any topology in CCM, the following will hold true:

$$V_{OFF} = V_{ON} \frac{D}{1-D}$$

By the basic definition of RMS of any waveform, we can show that the RMS of this applied voltage waveform (calculated over the entire cycle) is

$$V_{RMS} = V_{ON} \sqrt{\frac{D}{1-D}}$$

or, equivalently,

$$V_{RMS} = V_{OFF} \sqrt{\frac{1-D}{D}}$$

Notice that if (and only if), $D = 0.5$ (square wave), we get $\mathbf{V_{RMS} = V_{ON} = V_{OFF}}$. And that's how we get one of the several *common forms* of the voltage-dependent equation—which power conversion engineers often blindly use without realizing it really only applies for $D = 0.5$, and for CCM of course.

$$B_{AC} = \frac{V_{RMS}}{4 \times NAf}$$

This applies under the following conditions: square voltage shape ($D = 0.5$) and CCM.

But let us also be clear about what the shape of the corresponding *current* waveform is under the *square wave voltage* case being discussed here. In reality, *the voltage-dependent equation says nothing about the absolute value of the current (or field)*. It only defines the *change* in current or field when a certain voltage is applied for a certain time. The actual value of the current and field in the inductor (center-of-ramp value) depends on the external application conditions (and the details of our circuit or topology). The DC level of the current waveform (center of ramp) could be almost anything, for the same ΔI.

A zero direct current level is one of many theoretical possibilities. To achieve it requires us to create appropriate external conditions, schematically and electrically (e.g., capacitive coupling). Now, we must also recognize that traditional methods of characterizing magnetic materials were ongoing even before switching power conversion was in sight. And most such methods are still in use today, though there are some new ones too. For example, vendors of magnetic materials historically tested cores using *sine waves*. Nowadays, they may use *square waves* (D = 0.5) to better support modern power converters. But one thing has not changed: They still tend to use a *symmetric excitation* as they did for sine waves. They do this *almost* implicitly, without even stating it clearly in datasheets. In other words, vendors deliberately *create external circuit conditions to keep the current symmetric* around zero, that is, bidirectional, with $B_{DC} = I_{DC} = 0$. But this is not necessarily what happens in most switching topologies. Most topologies are *unipolar* or *unidirectional*. Current flows through the coil in only one direction (especially at maximum load). This, in fact, forms a special case of a nonsymmetric excitation, one in which the current waveform is not just offset (i.e., $I_{DC} \neq 0$), but restricted *entirely to the upper* (or lower) *half* of the graph. The effect of this on how the equations are expressed is dramatic, and is often inadvertently misapplied by engineers. *We should be cognizant of how a given equation we may be planning to use was actually derived*!

Nonsymmetric excitation can significantly change the relationships between peak current (or field) and peak-to-peak values, etc. See Fig. 7.3 for an example of what can change, and what doesn't.

Only for symmetric excitation can we assume that

$$B_{AC} = B_{PK}$$

And that is how we get another *common form* seen in literature:

$$B_{PK} = \frac{V_{RMS}}{4 \times NAf}$$

This applies under the following conditions: square voltage shape (D = 0.5), symmetric excitation, and CCM. But some engineers use this for *everything*: even to find the number of turns of a single-ended Forward converter transformer, *despite the fact that the current is nonsymmetric, does not have D = 0.5, and is not even CCM!*

Note that in all cases (symmetric or not), the following equation certainly holds true: $B_{PP} = 2 \times B_{AC}$. So we can write another form of our voltage-dependent equation as

$$B_{PP} = \frac{V_{RMS}}{2 \times NAf}$$

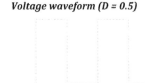

Voltage waveform (D = 0.5)

In going from symmetric to nonsymmetric excitation:
PK, DC (also RMS and AVG) values change
PP = AC+DC = Δ (also AC RMS) remain the same

For AC RMS, see Fig. 2.15

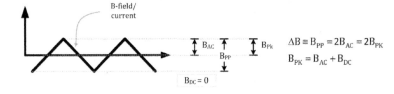

Current/field for symmetric excitation

$\Delta B \equiv B_{PP} = 2B_{AC} = 2B_{PK}$
$B_{PK} = B_{AC} + B_{DC}$

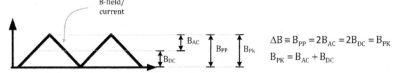

Current/field for nonsymmetric excitation (special case of CCM/DCM, i.e. BCM)

$\Delta B \equiv B_{PP} = 2B_{AC} = 2B_{DC} = B_{PK}$
$B_{PK} = B_{AC} + B_{DC}$

FIGURE 7.3 Symmetric versus nonsymmetric excitations.

This applies under the following conditions: square voltage shape (D = 0.5) and CCM. It can apply to both symmetric and nonsymmetric excitation.

Greater confusion arises if we do not recognize the fact that magnetics vendors may still only be providing data for a core under *sine wave excitation*. Here the test current will be a sine wave, and so will be the *B*-field. We can assume that the ON-time here is the first half-cycle (0 to π radians), followed by an OFF-time (from π to 2π radians). The voltage level is not *flat* in this case, and we now need to calculate the *average* voltage during the first half-cycle. The basic form of the voltage-dependent equation is

$$\Delta B = \frac{V_{AVG} \Delta t}{NA}$$

We can write the average voltage as

$$V_{AVG} = \frac{V_{AVG}}{V_{PK}} \times \frac{V_{PK}}{V_{RMS}} \times V_{RMS}$$

Realizing that V_{AVG} calculated over each half-cycle will be the same as the well-known forms applicable to a standard rectified AC sine wave (also see Fig. 2.19 in Chap. 2), we get

$$\frac{V_{AVG}}{V_{PK}} = \frac{2}{\pi}$$

and

$$\frac{V_{RMS}}{V_{PK}} = \frac{1}{\sqrt{2}}$$

So

$$V_{AVG} = \frac{2}{\pi} \times \frac{\sqrt{2}}{1} \times V_{RMS} = 0.9 \times V_{RMS}$$

The voltage-dependent equation for a *symmetric sine wave excitation* becomes

$$\Delta B = \frac{V_{AVG} \Delta t}{NA} = \frac{0.9 \times V_{RMS}}{2 \times NAf}$$

So some additional forms of the voltage-dependent equation are

$$B_{AC} = B_{PK} = \frac{V_{RMS}}{\pi\sqrt{2} \times NAf} = \frac{0.9 \times V_{RMS}}{4 \times NAf} = \frac{V_{RMS}}{4.4428 \times NAf}$$

All these apply under the following conditions: sine voltage shape and symmetric excitation.

Also, using $V_{PK} = V_{RMS} \times (2)^{1/2}$

$$B_{AC} = B_{PK} = \frac{V_{PK}}{2\pi \times NAf}$$

or equivalently, since $B_{PP} = 2 \times B_{AC}$,

$$B_{PP} = \frac{V_{RMS}}{2.222 \times NAf}$$

For a sine wave excitation, sometimes vendors use the RMS of the *B*-field in the preceding equations. Since the *B*-field is a sine wave (like the current), we can use

$$B_{PK} = \frac{B_{PK}}{B_{RMS}} \times B_{RMS} = \sqrt{2} \times B_{RMS}$$

The bottom line is that in switching converters we must be very cautious in using the myriad of common forms of Faraday's law seen in magnetics datasheets. We should go back to the basic equation if necessary, to avoid confusion.

General Design Equations	
$V = N\dfrac{d\Phi}{dt} = NA\dfrac{dB}{dt} = L\dfrac{dI}{dt}$	
Voltage-Independent Equation	
$B = \dfrac{LI}{NA}$ (connects instantaneous values, provided remanence is zero)	
Voltage-Dependent Equation	
Most general forms	$\Delta B = 2B_{AC} = \dfrac{V_{ON}T_{ON}}{NA} = \dfrac{V_{OFF}T_{OFF}}{NA}$ (always valid)
Square wave (CCM, $D = 0.5$) Symmetric or nonsymmetric excitation	$B_{AC} = \dfrac{V_{ON}D}{2 \times NAf}$ (CCM)
	$B_{AC} = \dfrac{V_{OFF}(1-D)}{2 \times NAf}$ (CCM)
	$B_{AC} = \dfrac{\Delta B}{2} = \dfrac{B_{PP}}{2} = \dfrac{V}{4 \times NAf}$ where $V \equiv V_{ON} = V_{OFF}$ (CCM, $D = 0.5$)
	$B_{AC} = \dfrac{B_{PP}}{2} = \dfrac{V_{RMS}}{4 \times NAf}$ (CCM, $D = 0.5$)
Sine wave, symmetric excitation	$B_{PK} = B_{AC} = \dfrac{B_{PP}}{2} = \dfrac{V_{RMS}}{4.4428 \times NAf}$

TABLE 7.1 Basic Design Table for Magnetics (MKS Units)

See Table 7.1 for a summary of the main equations discussed. *When in doubt, use the row marked "Most general forms."*

Units in Magnetics

Here lies another source of massive confusion. There are several systems of units in circulation. The one we have used so far (largely) is the modern, international (or *rationalized*) system based on meters, kilograms, seconds, and amperes (called the *MKS* or *MKSA* system, or the SI system, or even the Georgi system). The older, but still commonly used one is the CGS system (for centimeters, grams, seconds). There are many other engineers (mainly in the United States) who are still using the FPS system (foot, pound, second), at least in patches: inches (or mils) instead of meters or centimeters. And there are others who even happily *mix different systems into one equation*. For example, the voltage-dependent equation we have derived is actually

$$B_{AC\_Tesla} = \dfrac{V_{AVG}}{4 \times N \times A_{m^2} \times f_{Hz}}$$

where V is in volts, A is in square meters, f is in hertz, and B is in teslas. But if B is in teslas (MKS units) and we want to use cm² (CGS) instead of m², we get

$$B_{AC\_Tesla} = \dfrac{V_{AVG} \times 10^4}{4 \times N \times A_{cm^2} \times f_{Hz}}$$

That is really mixed units! But we can convert the B-field to the CGS system by using the conversions in Table 7.2.

With that, finally we get another one of the commonly seen forms of the voltage-dependent equation:

$$B_{AC\_Gauss} = \dfrac{V_{AVG} \times 10^8}{4 \times N \times A_{cm^2} \times f_{Hz}}$$

Property	CGS units	MKS units	Conversion
Magnetic flux (ϕ)	Line (li) (or Maxwell)	Weber (Wb) (or volt-seconds, V·s)	1 weber = 10^8 lines
Flux density (B)	Gauss (G)	Tesla (T) (or Wb/m²)	1 tesla = 10^4 gauss
Magnetomotive force	Gilbert	Ampere-turn	1 gilbert = $10/4\pi$ = 0.796 ampere-turn
Magnetizing force field (H)	Oersted (Oe)	Ampere-turn/meter	1 oersted = $1000/4\pi$ = 79.577 ampere/meter
Permeability	Gauss/Oersted*	Weber/ampere-turn-m (or henry/m)	$\mu_{MKS} = \mu_{CGS} \times (4\pi \times 10^{-7})$

*Both gauss and oersted are $cm^{-1/2} \cdot g^{1/2} \cdot s^{-1}$, so permeability is dimensionless in CGS units.

TABLE 7.2 Magnetic Units Conversions

For symmetric excitation we saw in Fig. 7.3 that $B_{AC} = B_{PK}$. So we could also write

$$B_{PK\_Gauss} = \frac{V_{AVG} \times 10^8}{4 \times N \times A_{cm^2} \times f_{Hz}}$$

In general, when in doubt about units (and that will be often), we should stick to one selected system of units (preferably MKS), and *only at the very end should we use conversions if we need to*, as provided in Table 7.2. We can then convert the final results, for example, from tesla into gauss. That is the final step in navigating the minefield of magnetics in power conversion.

The Magnetomotive Force (mmf) Equation

In Fig. 7.4 we present an exploded view of a toroidal core with an air gap. The flux lines are also shown. The coil that creates this flux has N turns and carries a current I. Applying Ampere's circuital law, we get by integrating over the entire inner path indicated

$$H_c l_c + H_g l_g = NI$$

where the subscript c stands for core (material) and g refers to the (air) gap, also sometimes just subscripted by 0. Note that by integrating as per Ampere's law, we get the current *enclosed* (by the path we integrate over), and that equals NI (ampere-turns).

Writing $H = B/\mu$ and $B = \Phi/A$, we get $H = \Phi/A\mu$, so

$$\frac{\Phi_c l_m}{A_c \mu_c} + \frac{\Phi_g l_g}{A_g \mu_g} = NI$$

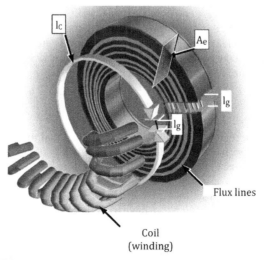

FIGURE 7.4 Exploded view of toroid with gap.

Note that usually (for small gaps) A_c is almost the same as A_g. But sometimes it may be quite different. We will take that up in the section on fringing flux later in this chapter. However, flux lines are always continuous. So, assuming we have adjusted the effective area if necessary, so that we are not excluding any flux lines (as we move from core to gap to core), then $\Phi_c = \Phi_g$ and

$$NI = \Phi \times \left(\frac{l_c}{\mu_c A_c} + \frac{l_g}{\mu_g A_g} \right)$$

In general, this is written out as the *magnetomotive force* (mmf) equation

$$\Phi = \frac{\text{mmf}}{\mathfrak{R}}$$

where *mmf*, the *magnetomotive force*, equals NI, and $\mathfrak{R}$ is the *reluctance* (of the entire magnetic circuit), given by

$$\mathfrak{R} = \frac{l_c}{\mu_c A_c} + \frac{l_g}{\mu_g A_g} = \mathfrak{R}_c + \mathfrak{R}_g$$

We observe the analogy with $V = IR$ (Ohm's law) where V, the voltage, is also called the electromotive force (emf), similar to NI, the magnetomotive force.

$$V = IR \quad \Leftrightarrow \quad \text{mmf} = \phi \mathfrak{R}$$

So in a *magnetic circuit*, flux plays the same role as current does in an electric circuit. $\mathfrak{R}$ is analogous to resistance in a circuit. We interpret this as saying that the *magnetic voltage* (the magnetomotive force NI) is responsible for producing a *magnetic current* (the flux ϕ), whose magnitude depends on the *magnetic resistance* (the reluctance $\mathfrak{R}$). Note that all the current flowing into an electrical node must leave. Flux behaves that way too.

In our case, we have two materials in series (core material and air), so the total reluctance is the series sum of two separate reluctances, one for each material. As mentioned, reluctance behaves in a completely analogous manner to electrical resistance, which is

$$\text{Resistance} = \frac{1}{\text{Conductance}} = \frac{1}{\sigma A}$$

where σ is the electrical conductivity (inverse of resistivity) and is a property of the material itself. So the opposite (inverse) of reluctance, in magnetics, is called P (permanence).

Effective Area and Effective Length in Toroids

To a very close approximation, the average area available for the flux lines is simply the *geometric* cross-sectional area of the core. It is called the effective area A_e. Now, that is not just approximately, but *completely* valid for toroids, because that is how A_e is defined, and tested by magnetic vendors, in the first place. But what we are also saying here is: it is also *almost* valid for the typical *E-type* cores we use, such as EE, ETD, and EFD.

The effective length (l_e) of a toroidal core (without any gap) is *almost* exactly equal to its average geometrical circumference. Since the core has an outer diameter (OD) and an inner diameter (ID) we take an arithmetic average of the two. So, using the equation for the circumference of a circle,

$$l_e = \pi \times \text{Diameter} = \pi \times \left(\frac{OD + ID}{2} \right)$$

Note that the reason l_e is *almost, but not exactly*, equal to the preceding amount (even for a toroid), is that we have already fixed two parameters out of three: the effective area of the core and the effective volume V_e (the latter happens to be the actual volume of the core material used). So l_e is also already fixed, and it can be derived based on the obvious relationship: $l_e = V_e / A_e$. The result of this, as it turns out, is *almost* equal to the average circumference given previously, but not exactly. Vendors have tried to create more accurate equations for calculating l_e from the core dimensions, even for a toroidal core.

Better agreement, both with real data and the more complicated equations for l_e can be obtained if the denominator of our really simple preceding equation, is made 2.1 instead of 2. With that slight change, the approximate equation becomes a little more accurate, and in fact seems very

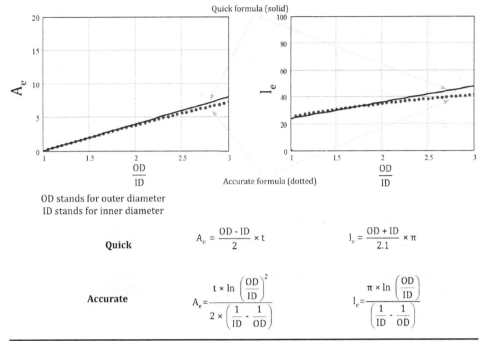

FIGURE 7.5 Effective area and effective length equations for a toroid.

acceptable, right up until the point where the outer diameter starts getting to be more than twice the inner diameter (which would be a rather unusual core). See Fig. 7.5.

Effective Area and Effective Length in E-cores

Historically, a vendor testing magnetic materials typically only characterizes them shaped in the form of toroids and then publishes their characteristics. When we try to use our familiar magnetics equations, we are also implicitly assuming toroids. But then we try to use E-cores, and we expect to use the same (toroid-based) magnetic equations and also the (toroid-based) material data from the vendor. We hope to still accurately predict, say, the peak field in the E-core, or the inductance if we wind say N turns on it, and so on. In other words, we really need to know, and *accurately* so, the A_e, l_e, and V_e of the E-core's *equivalent toroid*.

Nowadays, vendors do test E-cores and publish the E-cores' A_e, l_e, and V_e data, for our convenience. That works fine. Their numbers correspond to the *equivalent toroid* on which our equations are all based. So we can use them directly.

But it is also instructive to know how to rather accurately estimate l_e, V_e, and A_e geometrically: by just looking at the physical dimensions of the E-core that we are planning to use. In Fig. 7.6, we first see how a regular E-core can be mentally mapped into an equivalent toroid and how the l_e, A_e, and V_e can be approximated. For example, when it comes to E-cores, we first must realize that the flux in the center limb splits *equally* into the outer limbs—by sheer symmetry. And A_e is very close to the geometric cross section of the *center limb*.

It turns out that most E-cores are designed such that the sum of the cross-sectional areas of the two outer limbs equals the area of the center limb. In very rare cases (some nonstandard cores from Asia in particular, which the author actually saw years ago), we will find that for some reason, the sum of the outer limbs is less than the center limb. In such cases, and in general, we can take A_e as the area of the center limb *or* the sum of the areas of the outer limbs, *whichever is less*.

NOTE *We can get a swinging inductance by using nonuniform core shapes. As we increase current, the portions with higher flux densities (outer limbs in the preceding case) will saturate first and the inductance will drop (but not too low). The center limb has a lower flux density and will continue to provide inductance up to higher currents.*

Note that in all cases, toroids or E-cores, the effective volume of the core V_e is always the actual volume of core material used. And this is also completely true always:

$$V_e = A_e \times l_e$$

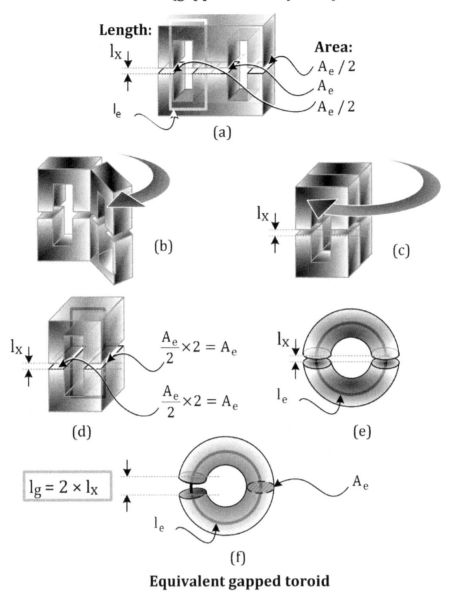

FIGURE 7.6 Mapping an E-core into a toroid.

Note *When in doubt about core geometry, we can just turn to the published values for A_e and l_e. But we should be cautious about the units. Some manufacturers give these values in mm^2 and in mm, respectively (almost as a default sometimes, without writing out the units explicitly), and some may give these in m^2 and m. Remember, if we are sticking to MKS units, we will need to convert the former to m^2 and m.*

The Effect of the Air Gap

Returning to the mmf equation

$$NI = \Phi \times \left(\frac{l_c}{\mu_c A_c} + \frac{l_g}{\mu_g A_g} \right)$$

we see that we can write $l_c = l_e$ (the effective length of the core before gapping). Also, μ_g, which is basically the permeability of air, is also μ_0. In MKS units its value is (see Fig. 7.1 too)

$$\mu_0 = 4\pi \times 10^{-7} \quad \text{W/A} \cdot \text{m} \text{ or } \text{H/m} \text{ J/A}^2 \cdot \text{m}$$

The Holy Grail: An Overview of Magnetics in Power Conversion

We also define the *relative permeability* of the material as

$$\mu = \frac{\mu_c}{\mu_g} \equiv \frac{\mu_c}{\mu_0}$$

(The permeability of the gap is the same as of air, so we set $\mu_g = \mu_0$.)

NOTE *Do not confuse the μ given here with the μ we presented in the equation $B = \mu H$. In that equation, μ was written symbolically. In reality, it referred to the material's absolute permeability (not its relative permeability as explained later, for which we are now using μ). For example, in the core we should correctly write $B_c = \mu_c H_c$ and in the air gap; $B_g = \mu_g H_g$. B_c can also be written as $B_c = \mu \mu_0 H$, where μ is the relative permeability of the core material.*

NOTE *The symbolic equation $B = \mu H$, connects the B at any point in space to the H at that point. As mentioned previously, B and H are proportional (at any given point) to each other—in linear materials—and their proportionality constant is the (absolute) permeability (at that point). Since H is proportional to current (free space being linear always), we find that B too is proportional to current (in linear materials).*

If we also assume that we have a *uniform* area $A_e = A_c$ for the entire structure (no fringing), we get from the mmf equation

$$\Phi = \frac{NIA_e\mu_c}{l_e + \mu l_g}$$

that is,

$$B = \frac{\mu_c NI}{l_e + \mu l_g}$$

We can also write

$$\Phi = \frac{NIA_e\mu_e}{l_e} \quad \text{where } \mu_e = \frac{\mu_c}{(1+\mu l_g/l_e)}$$

This means, with a gap present, we get in the core and gap

$$B = \frac{\mu_e NI}{l_e}$$

Had there been no air gap, we would have gotten (in the core and gap)

$$\Phi = \frac{NIA_e\mu_c}{l_e} \qquad B = \frac{\mu_c NI}{l_e}$$

We are implying here that B is continuous—with no breaks—and is always the same in the core material as in the air gap.

So if we increase the air gap for example, we will reduce the field in both the core material and in the air gap—but they both remain the same in the core and gap. This is natural, because we are assuming constant cross-sectional area (in core and gap) and B is always related to flux $B = \Phi/A$ (that is why it is called flux density!). Knowing that flux lines are continuous (with no breaks), so is the B-field. Then, the H-field follows from the equation $H = B/\mu$. We can see that as we move from one material to another (say core to gap or gap to core), we will get *breaks* and *jumps* in H. The *H-field is not continuous, the B-field is.* That is why we suggest calculating B first, not H. In fact, we need not even bother with what H is.

Based on the preceding, we realize that we can look at the air gap in two ways:

1. We can think of the air gap as having *caused the B-field to decrease because of an increase in the magnetic path length*—from l_e to $l_e + \mu l_g$. So

$$\Phi = \frac{NIA_e\mu_c}{l_e} \quad \Rightarrow \quad \frac{NIA_e\mu_c}{l_e + \mu l_g}$$

This means that though the gap introduced was only l_g, it contributed μ times the geometric gap length into the effective magnetic path length. Note that the factor μ is the relative permeability of the material *surrounding the gap*! So, it helps "weight" the air gap into becoming something very powerful in the overall reckoning.

2. Alternatively, we can think of the effect of the gap as having caused the B-field to decrease by changing *the permeability of the core material* from μ_c to μ_e. The latter is *effective permeability*, applicable to the entire core, treated homogenously (no longer a separate gap and core, quite like powdered iron with a distributed air gap). So

$$\Phi = \frac{NIA_e\mu_c}{l_e} \Rightarrow \frac{NIA_e\mu_e}{l_e}$$

Note that the two preceding approaches are equivalent but *alternative* ways of looking at the effects of the gap. *Therefore only one or the other should be utilized, not both simultaneously.* Further, this little mathematical trick should be used only for finding B, because B is the same in the core and gap. After that, H can be *derived*, using $H = B/(\text{absolute permeability})$ as mentioned.

In a short while we will tabulate these changes to make all this even clearer.

The Gap Factor z

We had defined an effective (lumped) permeability as

$$\mu_e = \frac{\mu_c}{(1+\mu l_g/l_e)}$$

This can be rewritten as

$$\mu_e = \frac{\mu_c}{(l_e + \mu l_g)/l_e}$$

So

$$\mu_e = \frac{\mu_c}{z}$$

Permeability of the material of the core fell by the factor $1/z$, where z is

$$z = \frac{l_e + \mu l_g}{l_e}$$

This air-gap factor z is the ratio of the *new* path length $l_g + \mu l_e$ to the old (ungapped) length l_e. It is always equal to or greater than 1. A value of $z = 1$ means an ungapped core.

Since, typically, $\mu l_g \gg l_e$, we can write

$$z = \frac{l_e + \mu l_g}{l_e} \approx \mu \frac{l_g}{l_e}$$

This means that roughly, *if we double the air gap, we double z.*

Note that as mentioned, we can use the z factor *once*—either to adjust the permeability or to adjust the length, but not both.

Applying this rule to the equation for B with no gap, we can easily get the equation with an air gap present, as follows

$$B = \frac{\mu_c NI}{l_e} \Rightarrow \frac{\mu_c NI}{z \times l_e} \equiv \frac{(\mu_c/z)NI}{l_e} = \frac{\mu_e NI}{l_e}$$

NOTE *For E-cores, if only the center limb is ground to create an air gap, and so no gap exists on the outer limbs, the gap length is equal to the physical dimensions of the gap in the center limb. Alternatively, if spacers are used on the outer limbs, the gap length is twice the gap set on each limb. See Fig. 7.5.*

The Origin and Significance of z

Flux lines are continuous. Therefore, the B-field is the same everywhere in the core and gap. But $B = \mu H$, and since μ changes from core to gap, the H-field is not the same. That is uncomfortable to some who are trained to think of H as a *driving field*, and B its response. But actually, *when a gapped core is present, it is more helpful (and easier) to think in terms of* B *instead of* H.

Let us calculate the B- and H-fields in the core and in the gap.

B in the core and gap:

$$B_c = B_g = \frac{\mu_c NI}{z l_e} \equiv \frac{\mu_e NI}{l_e}$$

H in the core:

$$H_c = \frac{B}{\mu_c} = \frac{NI}{z l_e}$$

H in the gap:

$$H_g = \frac{B}{\mu_g} = \left(\frac{\mu_c}{\mu_g}\right) \times \frac{NI}{z l_e} = \frac{\mu NI}{z l_e}$$

We can see that in this case

$$H_g = \mu \times H_c$$

Since ferrite, for example, can have a relative permeability between 1000 to 5000, *the H-field in the air gap is 1000 to 5000 larger than the H-field in the magnetic core!* We keep in mind that B is continuous in the core and gap.

Question How does the H-field compare to the field that would have been present if the magnetic core was not present (i.e., core replaced by free air)? We assume that the coil and current remain the same.

Applying Ampere's law to an air-cored coil we would get

$$H = \frac{NI}{l_e}$$

So we can conclude the following:

- The H-field in the core of the gapped magnetic core is significantly less than the air-core case by the factor $1/z$.
- The H-field in the gap of the gapped magnetic core is much greater than the air-core case by the factor μ/z, where μ is the relative permeability of the surrounding core material.

As for the B-field, for the air-core case we get

$$B = \mu_0 H = \frac{\mu_0 NI}{l_e}$$

compared with the gapped-core case

$$B_c = B_g = \frac{\mu_e NI}{l_e}$$

We can see that B *has increased substantially in the gap and core as compared to the air-core coil.*

MKS units	B			H
Ungapped core $\mu_e = \mu_c$, $z = 1$	$\dfrac{\mu_c NI}{l_e}$			$\dfrac{NI}{l_e}$
Gapped core μ_e, $z > 1$	Core	$\dfrac{\mu_c NI}{zl_e} \equiv \dfrac{\mu\mu_0 NI}{zl_e} \equiv \dfrac{\mu_e NI}{l_e}$	$\equiv \dfrac{\mu_e NI}{l_e}$	$\dfrac{NI}{zl_e}$
	Gap			$\dfrac{\mu NI}{zl_e}$
Air-cored coil $\mu_e = \mu_0$, $z = 1$	$\dfrac{\mu_0 NI}{l_e}$			$\dfrac{NI}{l_e}$
z is $(l_e + \mu l_g)/l_e$ μ is relative permeability $= \mu_c/\mu_0$, μ_e is effective permeability $= \mu_c/z$.				

TABLE 7.3 *B*- and *H*-Fields Inside the Core (and Gap)

See a comparison in Table 7.3 for a summary of the cases just discussed. In Fig. 7.7 we have tabulated how z affects all relevant magnetic quantities. In Fig. 7.8 we have listed specifically, more clearly, what the fields are, in gapped and ungapped cores.

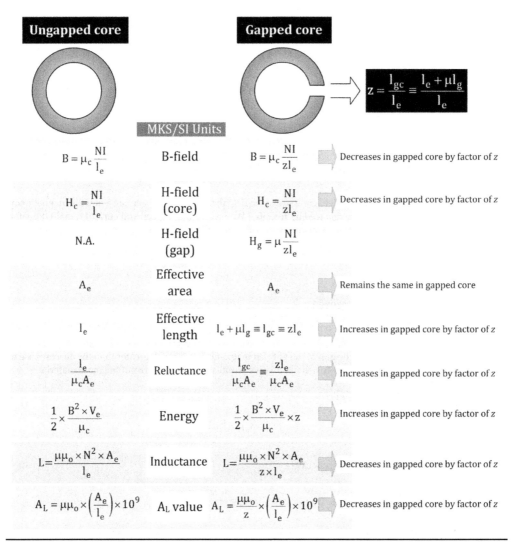

FIGURE 7.7 Comparing gapped and ungapped cores.

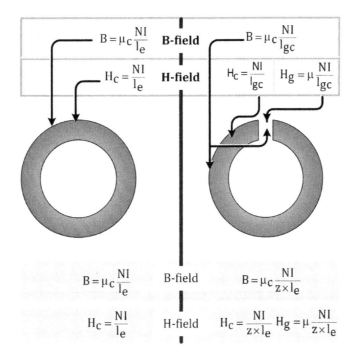

FIGURE 7.8 B- and H-fields inside the core (and gap).

Relating B to H

The equation $B = \text{permeability} \times H$ applies to a *point* in space, not to the *entire* structure as sometimes mistakenly interpreted by engineers. In fact, no equation of the symbolic form $B = \mu H$ can be written out as somehow being an average relationship valid for the entire gapped core. That is because though B is a constant as we go through the core and gap, H changes in discrete steps, and here is no "average H". Yes, we can symbolically write

$$B = \mu_e H$$

but we must remember that from Table 7.3, rather coincidentally, this can mean either:

$$B_{\text{gapped\_core}} = \mu_e \times H_{\text{air\_core}} \quad \text{OK}$$

or

$$B_{\text{gapped\_core}} = \mu_e \times H_{\text{ungapped\_core}} \quad \text{OK}$$

But not

$$B_{\text{gapped\_core}} = \mu_e \times H_{\text{gapped\_core}} \quad \text{Not OK. This makes no sense, what is } H \text{ here?}$$

Gapped E-Cores

E-cores are topologically equivalent to the toroid, as we saw in Fig. 7.6. The area of the center limb of the E-core becomes equivalent to the effective area of the toroid. The effective length measured through the center limb of the E-core passing through either side-limb becomes the effective length of the toroid. The total gap in the *center* limb of the E-core becomes the gap length of the toroid.

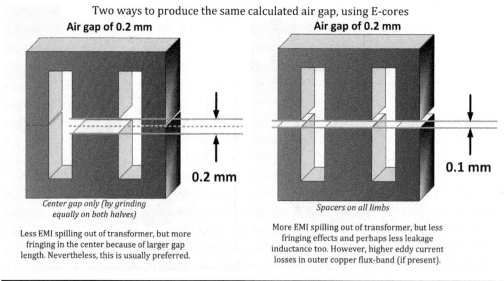

FIGURE 7.9 How to implement correct gapping in E-cores.

However, we must take note of the following:

- If both halves of an EE-core (consisting of two identical E-shaped sections) are equally ground to achieve a certain air gap length l_g, then each half must be ground to $l_g/2$.
- If only one-half of an EE-core is being ground to achieve a certain air gap length l_g, then it must be ground to l_g.
- If the center limbs of the EE-cores are not going to be ground, and the method of achieving a certain gap length l_g is by means of identical spacers on the outer limbs, then the spacer thickness must be set to $l_g/2$. See Fig. 7.9.

Energy Storage Considerations: How to Vary the Gap in Practice, Optimally

The energy stored in the inductor is the same as the energy associated with its field. That is,

$$\text{Energy} = \frac{1}{2} BH \times \text{volume}$$

Since in general, we write $B = \mu H$ (where μ here is the absolute permeability of the material), we get the following energy terms for the gapped toroid. For the core,

$$\text{Energy}_c = \frac{1}{2} \frac{B^2}{\mu_c} \times A_e l_e \quad \text{J}$$

and for the gap,

$$\text{Energy}_g = \frac{1}{2} \frac{B^2}{\mu_o} \times A_e l_g \quad \text{J}$$

So the total stored energy is

$$\text{Energy} = \frac{1}{2} B^2 A_e \left(\frac{l_e}{\mu_c} + \frac{l_g}{\mu_o} \right) \quad \text{J}$$

So,

$$\text{Energy} = \frac{1}{2} B^2 A_e l_e \frac{z}{\mu_c} \quad \text{J}$$

Or, since by definition, $V_e = A_e \times l_e$, we get

$$\text{Energy} = \frac{1}{2} B^2 V_e \frac{z}{\mu_c}$$

Rewriting it slightly to make it more intuitive

$$\text{Energy} = z \times \frac{1}{2} \frac{B^2}{\mu_c} \times V_e$$

We can guess the energy without an air gap, by simply putting $z = 1$. In other words, *the air gap helps multiply the total stored energy by the factor z*. See Fig. 7.7.

The *limit* of the core is

$$\text{Energy}_{SAT} = z \times \frac{1}{2} \frac{B_{SAT}^2}{\mu_c} \times V_e$$

Let us see how we can use this information in the lab as we attempt vary the gap optimally.

First Attempt (*N* Constant)

This is the most common and most obvious path. Though it *works*, it is not optimum at all. This is how it unfolds.

We find our core slightly saturating, so we adjust the air gap (introducing spacers), keeping all else unchanged. Note we can also do this *virtually*, using magnetics *expert software*. Yes, we do find the core is no longer saturating. But the effect of this is a bit *too extreme*. Because not only are we increasing Energy$_{SAT}$ (the energy limit) by a factor of z, but we are also lowering the amount of energy we are storing in the core by a factor $1/z$. Intuitively, we can look upon this as desiring to store a certain amount of food in a container, increasing the size of the container, but also *unnecessarily throwing away some of the food*.

The reason for the preceding statement is: As we increase z, B collapses in the core. How does that happen? The current (i.e., center-of-ramp value) depends on the application (load and line, which has not changed). So, it has a comparatively small ramp around this center value, and though that will change, it will not be by a big amount. So the ampere-turns can be considered almost constant (we saw this in the early part of this chapter too, in Example 7.1). So, in that case, B will *fall in inverse proportion to the increase in z* as we increase the gap, based on the relevant equation for B in Table 7.3.

$$B = \frac{\mu \mu_o N I}{z l_e}$$

Now looking at our energy equation,

$$\text{Energy} = z \times \frac{1}{2} \frac{B^2}{\mu_c} \times V_e$$

If B goes as $1/z$, stored energy will also go down, as $1/z$. Yes, the *size of the container*, as reflected by the limiting equation

$$\text{Energy}_{SAT} = z \times \frac{1}{2} \frac{B_{SAT}^2}{\mu_c} \times V_e$$

is increasing by the factor z, because B_{SAT} is fixed as per Fig. 7.10, but this is certainly not the optimum way to go. We are making space anyway, so why do we need to throw away food?

Second Attempt (*N* Proportional to *z*: *B* Kept Constant)

We examine our equation for B

$$B = \frac{\mu \mu_o N I}{z l_e}$$

To prevent the unnecessary collapse of B, we can make N proportional to z. Then B *will remain almost the same (unchanged)*. However, if we look at the equation for energy, we have

$$\text{Energy} = z \times \frac{1}{2} \frac{B^2}{\mu_c} \times V_e$$

So even if we keep B constant, the energy we are storing in the core is now increasing by a factor z. But this is another extreme, because though we are increasing the energy capability (size of container), we are also trying to put too much energy (food) into it. And this obvious overdesign leads to the problem that the *headroom (in energy) does not increase*. If the core was almost saturating, it still is.

We have to be conscious always of indiscriminately adding turns to a transformer close to saturation. It will then *definitely* hasten saturation, because as we saw above, *despite increasing the air gap* while adding turns, we still had no improvement. So if we do not change the air gap, while increasing the turns, there is simply no hope of ever introducing any *headroom*. In fact quite the contrary.

Third Attempt (*N* Proportional to $\sqrt{z}$: Energy Kept Constant)

Looking at

$$\text{Energy} = z \times \frac{1}{2}\frac{B^2}{\mu_c} \times V_e$$

we see that if we make B proportional to $1/\sqrt{z}$, we will keep the stored energy constant. That is what we want to achieve here (no change in the amount of food to be stored).

By increasing z, we will increase Energy_{SAT} (increase the container size). So the headroom before the onset of saturation will increase optimally.

How do we make B proportional to $1/\sqrt{z}$? From the equations, since B is proportional to N/z, if we allow N to increase by the factor $\sqrt{z}$, then B will be proportional to $1/\sqrt{z}$, as desired. That is how we achieve this. For example, if we double z (almost double the air gap), we need to increase the number of turns by 1.414 (because $\sqrt{2} = 1.414$).

Why do we consider this the optimum, and not the previous attempts? Because

1. In the first attempt, we decreased the energy being stored, but at the expense of making all currents very "peaky," thus increasing losses throughout the converter.

2. In the second attempt, though we increased the energy storage capability of the core, we also increased the energy it needed to store (by the same amount). We can look upon this as increasing the *inductance* indiscriminately, which was responsible for increasing the actual energy [$(1/2) \times L \times I^2$], thus negating the anticipated headroom. If the core was close to saturation, it still remained so.

3. In the third attempt, we varied N proportional to $\sqrt{z}$. For example, if z went from 2 to 8 (factor of four, almost four times the air gap), we doubled the number of turns.

In the last case, since inductance for a gapped core, as per Fig. 7.7, is

$$L = \frac{\mu\mu_o \times N^2 \times A_e}{z \times l_e}$$

we see that if N is set proportional to $\sqrt{z}$, then *L will remain constant*. That is the key! Because, since current ripple ratio r is inversely proportional to L, by a constant L, we will maintain r constant too. And that is the fundamental reason why the *energy we were storing in the core remained the same*. That also means there will now be no *domino effect* across the converter by increasing the air gap. Currents will not get more *peaky*; the RMS of various currents will remain unchanged. We will retain our optimal r (of about 0.4).

So finally, *with our third attempt we have reached an optimum solution for tweaking the gap*. We increased the energy capability of the core, yet did not change the energy we attempted to store in it, nor did we degrade converter efficiency in the process. Further, by keeping r unchanged, we can choose to either increase the headroom, or simply decrease the size of the core, as we will soon see.

In Chap. 10 we will learn that the energy we put into a core is almost fixed for a given topology and application. The rest is *shape dependent*. For a Buck-boost we get the required core volume to be for any application

$$V_{e\_cm3} = \left[\frac{31.4 \times P_{IN} \times \mu}{z \times f_{MHz} \times B^2_{SAT\_Gauss}}\right] \times \left[r \times \left(\frac{2}{r}+1\right)^2\right] \equiv X \times Y$$

This has a term X which for a given material, frequency, and air-gap factor only depends on power (which we assume has not changed here), multiplied by a *shape factor Y*, which is purely r-dependent. So the required volume for a given power level, all else unchanged, only depends on r. If we keep L and thereby r constant, we keep the energy constant too.

As a side note, we will see from the preceding energy equations, that *decreasing the core material permeability*, not increasing it, *will give us an increase in Energy$_{SAT}$* (for a given B_{SAT}). That may be counterintuitive to some, but that is exactly what we end up doing by air-gapping the core too: We *lower* the overall *effective permeability*.

So why not just use an air-core coil? That does have *as low a permeability as it gets*! In fact, an air-core coil would have excellent energy-storage *capabilities* (because it cannot saturate), but it is a capability that we can hardly succeed in using. Because, as per the energy equation, we have to create a huge B-field in it to achieve any significant energy storage. In the case of a gapped ferrite, it was easier: The ferrite acted much like a "conduit" to channel the flux lines into the air gap. That is why a small l_g appeared in the equations as a big $\mu_c \times l_g$. The core was a multiplier. So, without that sort of help, air by itself just won't work as the complete *core material*. Further, to create a large field we would need to place an unrealistically large number of turns in the coil, with correspondingly exorbitant copper losses. And that is also why we often end up evaluating the power capability of a given core *by examining its available window area*, which will tell us how much copper we can actually put into the coil practically.

How Much Energy Is in the Gap and How Much Is in the Core?

Let us now eliminate B from the energy equations. We can simplify the previous equations to get (*using "E" for energy here, not electric field*):

$$\frac{E_g}{E_c} = z - 1 = \frac{\mu l_g}{l_e}$$

We can see that the energy in the gap can rival the energy in the core simply because, though its dimensions are small, from the magnetics viewpoint it gets multiplied by the relative permeability of the *surrounding* material. That makes it effectively big. We can thus write

$$\frac{E_g}{E_c + E_g} \equiv \frac{E_g}{E} = \frac{\mu l_g}{l_e + \mu l_g}$$

where E is the total energy stored in the structure (i.e., $E_c + E_g$). Then using $z = (l_e + \mu l_g)/l_e$ we get

$$\frac{E_g}{E} = \frac{\mu l_g}{l_e + \mu l_g} = 1 - \frac{1}{z}$$

This is the ratio of the energy stored in the gap to the total energy. But what is the total energy? We can simplify the preceding equations to get

$$E = \frac{\mu \mu_o N^2 I^2 A_e}{2 l_e} \times \frac{1}{z}$$

Note that this is the "food" we want to store, *not the size of the container*. The size of the container corresponds to Energy$_{SAT}$, and it increases with z, unlike what the above equation seems to suggest. Yes, in this equation, if we vary N as per $\sqrt{z}$ as mentioned earlier (our third attempt), the food remains constant.

Similarly, the energy stored in the core alone (excluding the gap) is

$$E_c = \frac{\mu \mu_o N^2 I^2 A_e}{2 l_e} \times \frac{1}{z^2}$$

Using

$$E_g = \frac{\mu l_g}{l_e} E_c$$

or

$$E_g = \left(1 - \frac{1}{z}\right) \times E$$

we get

$$E_g = \frac{\mu\mu_o N^2 I^2 A_e}{2l_e} \times \frac{1}{z}\left(1 - \frac{1}{z}\right)$$

If we plot this out, we will see that *for constant ampere-turns (but not necessarily otherwise)*, the energy in the total structure starts falling *the moment we increase z (our first previous attempt)*. The energy in the core actually falls off at a much faster rate, but it is partly compensated by the rapidly increasing energy in the gap. However, the ratio of the energy in the gap to the total energy reaches 50 percent by the time z = 2, and from that point onward the energy *in the gap also starts to decrease*. Note that z = 2 corresponds to $l_g = l_e/\mu$. This applies only to the case of no change in amperes (i.e., the same application) and no change in turns—by just adding spacers to increase the air gap. In other words, for our previous *first attempt*, we will get some advantage up until the z = 2 threshold. After that it is downhill!

Optimum Design Target Values for z

Too much of an air gap requires more turns for the same field (or same energy), and this causes a progressive increase in copper losses, and also higher leakage inductance. So for transformers, using *E-shaped* cores, *the most optimum value is about z = 10 in Flyback converters*. That becomes our initial design target at least.

For chokes, such as in PFC stages, we typically target about z = 40, since there is only one main high-current winding, so it has all the window area available to it.

In Forward converters, theoretically, we need no air gap, because the core is not being used to store any of the output energy. We will learn more about this in Chaps. 11 and 12. However, the transformer core of the Forward does store some energy related to the magnetization current (this is discussed later in this chapter in the section titled "(Real) Transformers"). So, to avoid wide variations related to the magnetization current (since that can also impact the tolerance of the set switch current limit, etc.), we do want to add a small air gap, with a target z of about 1.5 to 2 at most.

Keep in mind that a very important property of the air gap is that the *effective permeability* of the core becomes increasingly dependent on the air gap itself (on air to be precise), with lesser impact from the core material characteristics. So the air gap starts dominating overall performance and behavior and makes the final characteristics of the magnetic structure far more *predictable*. The truth is that ferrites have significant mechanical and electrical tolerances, related to the very tricky process of their manufacture. Also, because of that, no exact "equivalents" really exist from vendor to vendor. So it makes commercial sense to literally "swamp out" differences in ferrite lots and sources, by introduction of at least a small air gap in Forward converters (not to exceed z = 2 though). In other cases, as we have seen, just for sheer energy storage purposes, we do need air gaps (z = 10 to 40). But now we also realize a beneficial side effect of any air gap (z = 2 to 40) is that we can switch rather easily between different vendors too.

The *BH* Curve

One intuitive reason why we add an air gap is that air never saturates. So by *mixing* some air into the structure, the structure starts taking on some of the useful properties of air. The *BH* curve shown in Fig. 7.10 *flattens out*. Though *this does nothing to change the saturation flux density* B_{SAT} *of the structure*, it does "soften" the tendency to enter saturation *abruptly*—in the sense that now a larger *change* in current (ampere-turns), at a larger current, is required to cause the same change in B. This feature helps a lot, because it gives enough time for most controllers and switches to react (by means of their current sense, for example) and to be able to turn off fast enough to prevent switch destruction. It also lowers effective permeability as is obvious from Fig. 7.10.

For gapped cores, the final A_L (inductance index expressed in nH/turns²) value also starts to become virtually independent of the permeability of the material and starts to depend almost completely on the amount of air gap. Thus the design becomes relatively

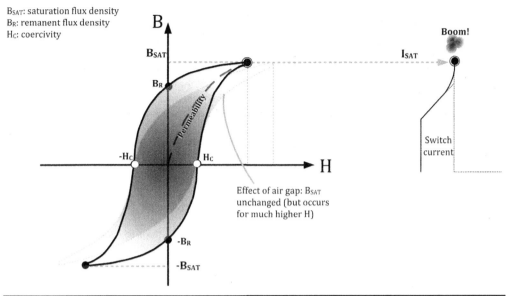

FIGURE 7.10 The *BH* curve and how the air gap affects it.

insensitive to variations in the material. So long as we can *control the air-gap tolerance well*, we can control the tolerance of our final inductance.

Careful Design Principles Help Decrease Core Size

As we saw, the air gap affects the energy storage capability. Let us recap some of the equations that went into the preceding analyses and look carefully at what ensued. It will also help in learning to manipulate the key equations of magnetics to our advantage.

We have the following five general equations (*E* is energy here, not electric field):

1. $E = \dfrac{\mu\mu_o N^2 I^2 A_e}{2 l_e} \times \dfrac{1}{z}$

2. $B = \dfrac{\mu\mu_o NI}{z l_e}$

3. $E = \dfrac{1}{2} B^2 V_e \dfrac{z}{\mu\mu_o}$

4. $L = \dfrac{1}{z}\left(\dfrac{\mu\mu_o A_e}{l_e}\right) N^2$

5. $\mu_e = \dfrac{\mu_c}{z}$

If, for example, we have set an air gap such that the (relative) permeability falls from 2000 to 200, i.e., by 10 times, then from the fifth equation, *z* must have gone from 1 (no air gap) to 10. From the second equation, we see that to keep the operating *B*-field unaltered (say, as when we are operating close to B_{SAT}), we can safely increase the ampere-turns *NI* by 10 times (i.e., the number of turns, since current is virtually predetermined). From the first equation, we see that the energy stored in the core increases by a factor of $10^2/10 = 10$ times. From the fourth equation, inductance has also been increased in the process by $10^2/10 = 10$ times. So in this case $N \propto z$, energy $\propto z$, B = constant, $L \propto z$ (our *second attempt*).

However, we know that for an optimum converter design, we want to fix a certain *r* (usually at 0.4). At a given frequency and for a fixed application condition, this means we have a specific *L* that we want to achieve. No more, no less. So in power supply design, what we really want to do is described next (corresponding to our *third attempt*).

From the fourth equation we see that to be able to keep the inductance fixed as *z* goes from 1 to 10, we only need to increase *N* by $10^{1/2} = 3.2$ times. Therefore, from the second equation we can see that if *z* went from 1 to 10, but *NI* was increased only by a factor of 3.2

(so as to keep L fixed), then the operating B-field would be *reduced* to one-third of its original value. From the first equation, the energy stored in the core has remained unaltered in the process, though from the third equation we can see that its *overload capability* (i.e., measured up to a certain B_{SAT}) has certainly increased 10 times. So any *headroom* as measured from the operating B-field value to the saturation level (B_{SAT}), or from the operating energy storage level to the peak energy handling capability, must have increased considerably—even though inductance has been kept a constant in this case. All this could translate to a *much higher field reliability where the converter will likely encounter severe abnormal or transient line/load conditions.*

Here is a recap of our three gapping attempts and how they affect the key parameters:

- If N = constant, then energy $\propto 1/z$, $B \propto 1/z$, $L \propto 1/z$.
- If $N \propto z$, then energy $\propto z$, B = constant, $L \propto z$.
- If $N \propto z^{1/2}$, then energy = constant, $B \propto 1/z^{1/2}$, L constant.

However, *if all the bells and whistles are present* in the design of the control circuitry (such as voltage feedforward, primary/secondary current limit, and duty cycle clamp), and they serve to protect the converter adequately against any abnormal conditions, this gives us a great opportunity to select a smaller core for the same power level. In doing so we would be essentially returning to the point of optimum core size where the operating peak field, B_{PK}, is set close to B_{SAT} (minimum headroom). That is the perhaps the biggest advantage accruing from the presence of an optimally set air gap—*reduction of core size, provided we have carefully set current limit, duty cycle limit, line feedforward, and undervoltage lockout (UVLO)*. A good systems designer can pull it off, temperatures permitting of course.

Understanding *L* Better

We had distinguished between maximum energy storage capability (the "container") and the energy we are trying to put into it (the "food"). Here we will describe what affects *our ability to transfer* that energy storage requirement (based on our application) into the core and how the concept of inductance enters the picture.

We saw that in the core and gap,

$$B = \frac{\mu \mu_o N I}{z l_e}$$

That means that B falls as $1/z$, just as E *(energy)* does. So, though we write E as being apparently proportional to B^2, the energy in the entire magnetic structure is actually proportional to B. We can see this by combining the following two equations:

$$E = \frac{\mu \mu_o N^2 I^2 A_e}{2 l_e} \times \frac{1}{z}$$

and

$$B = \frac{\mu \mu_o N I}{z l_e}$$

We get

$$E = \frac{1}{2} \times B A_e I N$$

So, for a given coil (fixed N, I, A_e), *energy is proportional to B. It does not even depend directly on the permeability.* The permeability is important only indirectly, because it will affect our ability to develop a sufficiently large B-field. However, note that if even we cannot develop a large enough B-field, we can still develop enough energy if we simply increase N. Our ability to store more energy in the magnetic structure is therefore based on the following:

- The permeability affects our ability to increase B to the desired level.
- If the material saturates, we cannot achieve the theoretically calculated B-field from Table 7.3.

- We can try to compensate the permeability by putting in more turns, but that may be limited by the available window. In addition, we could also end up making the structure lossy due to an increase in copper losses in the windings.

Note that most engineers already know that $E = (1/2)(LI^2)$, and since I is essentially determined by the application conditions, not by design, we expect that *all the previous three statements must be merely equivalent to our ability to achieve a high-enough inductance.*

And they are. That is the importance of defining a quantity called L. It virtually summarizes all we know about the core and gap that matters to us macroscopically. Equating the equation for E as expressed in terms of A_e, l_e, etc., to the standard $E = (1/2)(LI^2)$, we can actually derive L:

$$E = \frac{\mu\mu_o N^2 I^2 A_e}{2l_e} \times \frac{1}{z} = \frac{1}{2}LI^2$$

$$L = \frac{1}{z}\left(\frac{\mu\mu_o A_e}{l_e}\right) \times N^2 \quad \text{H}$$

We see that since A_e is being assumed to equal A_g, the reluctance

$$\Re = \frac{l_c}{\mu_c A_c} + \frac{l_g}{\mu_g A_g}$$

Becomes

$$\Re = \frac{l_e}{\mu\mu_o A_e} + \frac{l_g}{\mu_o A_e} = \left(\frac{l_e}{\mu\mu_o A_e}\right)\left[1 + \frac{\mu l_g}{l_e}\right]$$

But the term in square brackets is z. So

$$L = \left(\frac{1}{\Re}\right) \times N^2 \equiv P \times N^2$$

where P is defined as the *permeance* (the magnetic analog of the electrical parameter known as conductance), mentioned earlier.

The inductance is thus the product of several terms. The first is dependent on core geometry and material. This is multiplied by a term in z, which includes the effect of the air gap. Finally, we have the N^2 term. This comes from the self-inductance effect. As previously mentioned, the number of flux lines produced is proportional to the number of turns in the coil, N. These flux lines then create flux linkages that are also proportional to N, because the flux essentially links with the same coil (itself), and it has N turns as we know. This gives (self) inductance its overall N^2 dependency.

Since we had previously defined L in terms of the inductance index (A_L value) as

$$L = A_L \times N^2 \times 10^{-9} \quad \text{H}$$

we get

$$A_L = \frac{1}{z}\left(\frac{\mu\mu_o A_e}{l_e}\right) \times 10^9 \quad \text{nH/turns}^2$$

We also get

$$A_L = P \times 10^9$$

So, A_L is physically speaking, just the permeance. If A_L is in nanohenrys per turns squared, permeance can be thought of as being henrys per turns squared.

Scaling: Difference between an Inductor and (Flyback) Transformer

Keep in mind that the transformer of a Flyback converter is a multiwinding inductor in effect, since it too is used for energy storage purposes. It is instructive to compare them.

In Fig. 7.11, typical voltage and current waveforms are presented to indicate more clearly how they will appear for an inductor, as compared to transformers with different turns

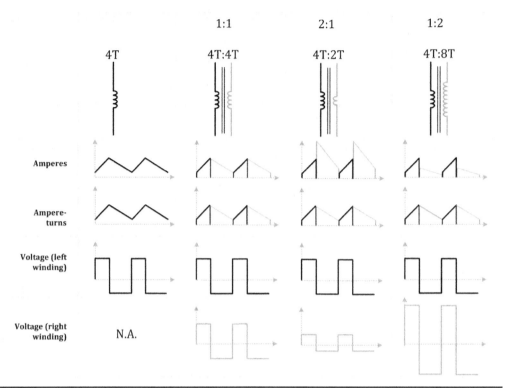

Figure 7.11 How voltages and currents will look like in multiwinding structures.

ratios. The horizontal axis is time. We must remember that as far as the core is concerned, it is totally "unaware" of whether it is has several windings or just one. If the current in one of its windings stops, the other takes over, and so continuity in ampere-turns is maintained. *And that is all the core cares about.* So what is the practical difference between an inductor and a Flyback transformer? The difference is in the shape of the *current through each winding*. If there was only one winding, the current through it would be smooth and undulating, and would be considered almost pure DC. But if there are several windings (being switched around), the current in each winding is necessarily going to be a "choppy" waveform—one that will have a very high frequency content. So, considering *skin depth*, we will now get high *AC-resistance* losses. Thus in general, core loss, which depends only on the excitation of the core (ΔB or ΔI), will not be affected in a transformer versus an inductor, but the copper loss will be much more for a transformer.

To understand "scaling" better, see also Chaps. 8 and 10.

(Real) Transformers

Flyback transformers are multiwinding inductors. When the primary winding conducts, the secondary windings do not conduct, and vice versa. Therefore, the basic purpose of a Flyback transformer is identical to that of any inductor—storage of energy during the switch ON-time for subsequent delivery to the output when the switch turns OFF. The turns ratio provides step-up or step-down as an added bonus. And we can also get primary to secondary isolation. However, when it comes to the Forward converter and similar topologies, we use a "real" transformer, that is, truly as a *transformer*, to literally transform voltages and currents by means of a step-up or step-down ratios. The transformer's main purpose in that case is not energy storage. In modern AC-DC power converters, if we didn't step down using transformer action, we would need an extremely small and impractical duty cycle to go from the highly rectified mains voltage to our typical DC output voltage levels.

We take help from both the duty cycle and the turns ratio. So in CCM for example, the input-to-output transfer function for a Forward converter is

$$V_O = V_{IN} \times \frac{D}{n}$$

where n is the turns ratio N_P/N_S, where N_P is the number of turns on the primary (input-side) winding and N_S is the number of turns on the secondary (output-side) winding. We can identify such *real-transformer* based topologies by the fact that when the primary winding conducts, so does the secondary winding. The transformer carries out the "*n*-step conversion" and applies, in effect, an effective input voltage of $V_{INR} = V_{IN}/n$ at the switching side of the output inductor (*choke*) of the Forward converter. This inductor essentially forms a Buck cell which carries out final "*D*-step conversion," that is, $V_O = V_{INR} \times D$.

Though the energy storage ability of an inductor is, and will always be, its main selection criterion, the energy stored in the *transformer*, based on its magnetization current, is purely an activation or excitation energy. It is not related to the load current at all and varies only when the input voltage changes (discussed shortly). It therefore suits us to try to keep this energy term small, since all we are going to do with it anyway is dump it back into the input bulk capacitor every cycle. For that reason too, the transformer, unlike the choke, *is invariably operated in discontinuous conduction mode*. Why put in more energy if we are going to recover and recycle it anyway? There is a also a problem with ensuring reset of the transformer, which is why it is kept to duty cycles less than 50 percent. This is further explained in Chap. 11. Also, as indicated previously, in principle, a Forward converter transformer has no air gap, since energy storage function is not required. But *a small gap may be introduced to make the design more stable and more tolerant of variations in the permeability of the material and its other tolerances*.

These factors make the selection criteria and design of a Forward converter transformer very different from that of a Flyback transformer.

Let us try to understand the transformer model more clearly here. In a real transformer, since the secondary winding conducts at exactly the same moment as the primary winding, we get the situation shown in Fig. 7.12. Here at time $t = 0$, we suddenly apply a step voltage across the Primary. We get a cause V_{PRI}, and its effect V_{SEC}, and they obey

$$V_{PRI} = -N_P \frac{d\Phi}{dt}$$

$$V_{SEC} = -N_S \frac{d\Phi}{dt}$$

What is Φ above? This is not yet clear. But it is the (net) flux in the core.

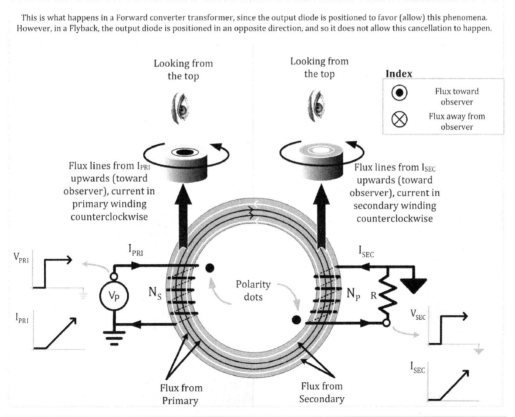

Figure 7.12 Understanding transformer action.

Whatever it is, *it connects (via its change) the voltages between the primary winding and the secondary winding*. All that is fully determined despite our lack of knowledge (so far) of the absolute value of Φ!

We can eliminate $d\Phi/dt$ from the preceding equations by equating them to get

$$\frac{N_P}{N_S} \equiv n = \frac{V_{\text{PRI}}}{V_{\text{SEC}}}$$

We thus get the familiar transformer scaling rule for voltages.

We see that voltage appears across the secondary winding irrespective of the value of R. In other words, even if the secondary winding is open (R infinite, zero current), there is a voltage present across the secondary winding. We see this despite not even knowing what the (net) flux was! But once we know that, the transformer mystery will become clearer.

NOTE *To emphasize, if R is infinite, despite zero secondary current, we do have an induced voltage across the secondary winding. This induced voltage has a polarity such that* had there been an available path *for current to flow (R not infinite), it* would *have produced an induced current, which* would *then be in such a direction that this current* would *have produced a flux, which in turn* would *have been in such a direction as to oppose the* change *causing it. So in that sense, induced voltage is "anticipative."*

NOTE *The direction of the induced-flux component is such that it opposes the cause: which is a change. The induced-flux component will try to maintain status quo: by keeping the previously existing flux level in the core unchanged. This induced-flux component does not necessarily need to be in an opposite direction to the primary-side flux. It only needs to oppose the change. So if the original flux tended to increase, the induced flux correspondingly increases in magnitude in the opposite direction or decreases in the same direction! This distinction is important in cases of preexisting DC bias in the transformer.*

Let us start thinking in terms of the flux from each winding separately. Because that constitutes the net flux. We have (from the magnetic circuit)

$$\Phi_{\text{PRI}} = \frac{N_P \times I_{\text{PRI}}}{\mathfrak{R}}$$

and

$$\Phi_{\text{SEC}} = -\frac{N_S \times I_{\text{SEC}}}{\mathfrak{R}}$$

We are saying that the secondary winding will start to conduct (pass current) at $t = 0$ as per Fig. 7.12, in an effort to oppose the changing flux caused by the primary winding. But there is something odd about this that we need to analyze further: *How can the secondary winding oppose a changing flux produced by the primary winding if the current through it is zero, as it is for the corner case of R (i.e., load resistance) infinite, for which all the preceding equations must nevertheless apply?* The answer is: It *doesn't (oppose)*. A certain component of primary-side flux is always present the moment we connect the voltage source to the primary winding. That flux changes (because that is the reason why we get reflected voltages across the boundary in the first place), but the secondary winding does not (and in the case of R being infinite, cannot) oppose this specific component of flux, because in effect, the Secondary does not "see" this part of the flux, despite the fact that the rate of change of this flux component was the root cause for the induced voltage across the Secondary in the first place. It is in reality, a rather difficult mental picture to comprehend, but surprisingly accepted without deeper thought (virtually prima facie) even by experienced power conversion engineers.

Perhaps we can mentally visualize this flux as being "unlinked" to the Secondary, quite like leakage, except that this particular flux is totally contained *in the core material*, not in the surrounding air (as in the case of leakage inductance). In that sense this is an *excitation* flux, arising from what we call the *magnetization* component of current on the primary side. This serves to "set up" the core for *further* flux variations—and the Secondary responds to the flux changing *beyond this baseline level*. To emphasize: *This baseline flux is the flux created by the Primary with the Secondary open (R infinite)*. It is the magnetization component—responsible for literally magnetizing the core, creating a reflected voltage across the boundary, and thus allowing the Secondary to conduct current and oppose any *excess* (additional) flux change beyond this baseline level.

The "baseline" is *not* flat? That is because there is yet another equation that needs to be satisfied:

$$V_{PRI} = L_{PRI} \frac{dI_{PRI}}{dt} \quad V_{SEC} = L_{SEC} \frac{dI_{SEC}}{dt}$$

So for a constant V_{PRI} applied across the Primary, the magnetization current ramps up. We see the baseline *does* change with time. It has to. Because only due to its change do we get a voltage reflected onto the secondary side. Otherwise we wouldn't.

But we have two unanswered (related) questions remaining:

1. How does the secondary current I_{SEC} relate to the primary current I_{PRI}?
2. What is the absolute value of the flux in the core in general (not just for the infinite R case)?

Here is the *problem*: In general, for a given R, the current in the secondary winding is fixed—because the voltage across it is fixed by the voltage scaling law, which applies irrespective of the value of R. And that depends only on the magnetization current. The secondary current is predetermined very simply by the equation V_{SEC}/R. We ask: How can the flux from a *fixed* (predetermined) current oppose a supposedly arbitrarily varying primary-side flux? The math does not seem to work out unconditionally. In fact, the final answer to this puzzle, one which satisfies all the equations, is that the primary-side flux measured *above the baseline level* is exactly equal (and opposite) to the secondary-side flux, as indicated in Fig. 7.12. These components cancel out completely in the core, and therefore, irrespective of the current in the Secondary (R of any value), the net flux in the core (measured above the baseline) is zero. Putting it another way, the next flux in the core is *always* just due to the magnetization current component, irrespective of R (or I_{SEC}).

To check consistency, let us first consider the case where R is not connected (i.e., R is infinite). For this corner case, we have applied "primes" to the parameters that follow.

For an open Secondary, since I'_{SEC} is always zero (no conduction), for all practical purposes the secondary winding does not exist. It produces no flux and cannot in any way affect what happens elsewhere. In essence we just have a simple inductor constituted around the primary winding, and so all the basic laws of magnetics presented earlier apply directly here too. In particular, the flux at any moment in the core is related to the instantaneous current by

$$\Phi'_{PRI} = \frac{N_P \times I'_{PRI}}{\mathfrak{R}}$$

$$\Phi'_{SEC} = \frac{N_S \times I'_{SEC}}{\mathfrak{R}} = 0$$

The net flux is

$$\Phi' = \Phi'_{PRI} + \Phi'_{SEC} = \Phi'_{PRI}$$

The voltages are

$$V_{PRI} = -N_P \frac{d\Phi}{dt} = -N_P \frac{d}{dt}\left(\frac{N_P \times I'_{PRI}}{\mathfrak{R}}\right)$$

$$V_{SEC} = -N_S \frac{d\Phi}{dt} = -N_S \frac{d}{dt}\left(\frac{N_P \times I'_{PRI}}{\mathfrak{R}}\right)$$

where $\mathfrak{R}$ is the reluctance of the magnetic core. As mentioned, let us consider this as the baseline:

$$\Phi' = \Phi'_{PRI} = \frac{N_P \times I'_{PRI}}{\mathfrak{R}}$$

Now suppose R is lowered to a finite value; current then flows in the Secondary and produces flux. Assume that the Primary also creates a flux above the baseline; then we get

$$\Phi_{PRI} = \Delta\Phi_{PRI} + \frac{N_P \times I'_{PRI}}{\mathfrak{R}}$$

$$\Phi_{SEC} = \Delta\Phi_{SEC} = \frac{N_S \times I'_{SEC}}{\mathfrak{R}}$$

The net flux is

$$\Phi = \Phi_{PRI} + \Phi_{SEC} = \Delta\Phi_{PRI} + \frac{N_P \times I'_{PRI}}{\mathfrak{R}} + \Delta\Phi_{SEC}$$

Note that this has a certain $d\Phi/dt$, which connects the voltages.

Compare this with the corner case of infinite R.

$$\Phi' = \Phi'_{PRI} = \frac{N_P \times I'_{PRI}}{\mathfrak{R}}$$

Note that this also has a certain $d\Phi/dt$, which relates to the voltages. Further, in both cases, R infinite or R finite, the voltages are the same, so the two $d\Phi/dt$'s above must be the same. That is trivially satisfied if $\Delta\Phi_{PRI} + \Delta\Phi_{SEC} = 0$. And that is exactly what happens.

In other words, as we reduce R from infinite to finite, the increase in flux *above the baseline* created by the primary winding is equal and opposite to the (increase in) flux in the secondary winding. The net flux and its rate of change both remain exactly the same as for the case of no load across the secondary winding. In effect, the secondary winding has successfully opposed the change in flux in the primary winding, *above the baseline value*.

With that information, from the magnetic circuit equations, we can easily see that the currents must scale as

$$\left| \frac{I_{PRI} - I'_{PRI}}{I_{SEC}} \right| = \frac{N_S}{N_P} = \frac{1}{n}$$

I'_{PRI} is the current in the Primary with the Secondary unloaded. We give it a new symbol I_{MAG} for magnetization current:

$$I_{PRI} = I_{MAG} + I_{SEC}/n$$

The equivalent transformer diagram (model) is thus represented as in Fig. 7.13. It separates the current into an "unlinked" component, I_{MAG}, the rest going through the "ideal transformer" where it receives a step-up or step-down, based on the turns ratio. The primary- and secondary-side leakage inductances, L_{LKP} and L_{LKS}, are also shown in the figure.

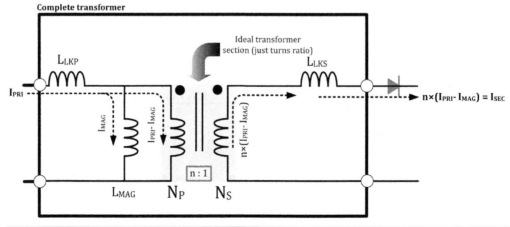

Figure 7.13 Transformer model.

Fringing Flux Correction

We have assumed so far that the area available to the flux lines in the magnetic material is the same as the area in the gap. But in the gap, the flux "balloons out." The bigger the gap, the more the ballooning. Yes it does cause an increase in eddy current losses in the windings, flux-band, Faraday shield, and so on (see Fig. 7.14)—in fact in any surface that is electrically conductive. This is one argument against the normal practice of concentrating the air gap in the center limb (see Fig. 7.9), because the air gap is bigger. However, people argue that gaps in the side limbs are worse because they run alongside the flux band.

How does this ballooning of flux change result? For one, it certainly lowers the reluctance. This is obvious from the electrical analogy: If we increase the cross-sectional area of a conductor, its resistance falls, allowing more current for a given voltage. Looking at the magnetic circuit, for a given ampere-turns (mmf) we will get more flux lines (analogous to current).

To model this, we introduce a fringing flux (FF) correction that virtually averages out the effect of this and applies it to the entire gapped structure. It turns out that this fringing flux term is equivalent to *increasing the effective area by a dimensionless factor FF (always greater than 1)*. So, the effective area gets modified to a new effective area A_{eff}:

$$A_{eff} = A_e \times FF$$

where FF is

$$FF = 1 + \frac{l_g}{\sqrt{100 \times A_e}} \times \ln\left(4 \times \frac{D + 0.5}{l_g}\right)$$

where l_g is the air gap in mm, A_e is expressed in cm², and D is the dimension (in mm) shown in Fig. 7.15.

As mentioned, by introducing fringing flux correction, reluctance decreases by the factor FF:

$$\mathcal{R} = \frac{l_c}{\mu_c A_c} + \frac{l_g}{\mu_g A_g} \Rightarrow \frac{l_c}{\mu_c A_{eff}} + \frac{l_g}{\mu_g A_{eff}}$$

(we assumed $A_c = A_g$). Note that reluctance would also decrease had we interpreted the fringing flux correction as having *decreased the lengths involved* rather than increasing the area. But air-gap length is geometrically fixed. It is the area available for the flux lines that

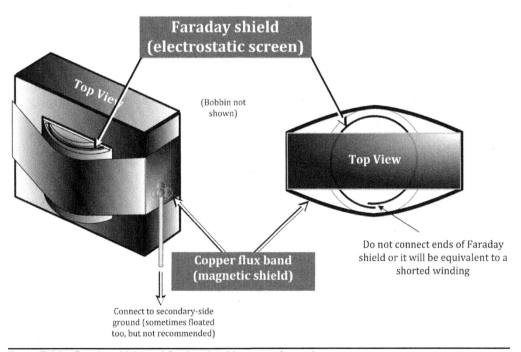

FIGURE 7.14 Faraday shield and flux band (eddy current losses).

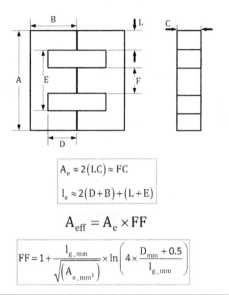

Figure 7.15 Effective area, effective length, and fringing flux estimates.

can, and do change due to fringing. With this new effective area we can go back to all the earlier equations (for inductance and so on) and simply replace A_e with A_{eff}.

We see that by introducing the fringing flux correction (A_e to A_{eff})

- B- and H-fields are not going to change (from Table 7.3, they depend on l_e, not A_e). Intuitively we have more flux lines and more area, so flux density is almost unchanged.
- z does not change.
- μ_e does not change from its value of z.
- Inductance is $1/\Re$ so it *increases* by factor FF.
- Energy ∝ area × length, so this increases by factor FF ("size of container" is bigger).

If L increases, this means that the value of L we would calculate while ignoring fringing flux would be an underestimation, because in reality when we wind the calculated number of turns on the given gapped core, we will measure a *higher inductance* than we expected. This in fact leads to an interesting error if we "check the core for saturation"—as soon explained anecdotally.

Let us consider the basic voltage-independent equation:

$$B = \frac{LI}{NA}$$

So without fringing flux considered:

$$B = \frac{L_{calculated} \times I}{N \times A_e}$$

With fringing flux considered:

$$B_f = \frac{L_{measured} \times I}{N \times A_{ef}} = \frac{(FF \times L_{calculated}) \times I}{N \times (FF \times A_e)} = B$$

So the B-field does not change, as we expected from Table 7.3 too. However, what we do in practice is we typically calculate the number of turns, and then when it is built, we measure the inductance and use it to set production limits. But if we now use the *measured* value of inductance and try to see if our core is saturating by using the *published value* of effective area of the core, we get a mismatch error:

$$B_{check1} = \frac{L_{measured} \times I}{N \times A_e} = FF \times B \quad \text{(wrong: seems too high)}$$

Since FF can even be as large as 1.4 for large gapped cores, we will end up thinking that our B-field is not say 3000 G as we had calculated, but 4200 G, which will throw everyone into a state of needless panic. We need to be consistent: Are we using FF or not? We can't do mix and match. So if we had checked B using the *calculated* value of L (with FF not considered) along with the published value of A_e, we would have got the correct value of B in the core:

$$B_{\text{check2}} = \frac{L_{\text{calculated}} \times I}{N \times A_e} = B$$

where we used calculated L (ignoring FF).

But this is also correct:

$$B_{\text{check3}} = \frac{L_{\text{measured}} \times I}{N \times A_{\text{eff}}} = B$$

In reality, this is how it played out. Around 1998, the author had created a very detailed Mathcad spreadsheet for designing a PFC choke. It had used the correction factor FF. The spreadsheet gave the number of turns required on the core for the target measured inductance. On the bench, using the recommended number of turns, and recommended air gap, the author had verified that the predicted inductance was indeed what was measured. All this indicated the validity and accuracy of the spreadsheet (including FF). The author was testing it, monitoring the PFC choke current carefully, and not seeing the slightest hint of saturation (i.e., anything other than a steady linear slope of ramp). But the author then got called up to explain an "issue" to higher-ups. A senior engineer had whispered to them that this author's PFC choke was incorrectly designed and was "exceeding 4200 G" as per his calculations. What the engineer had really done was use the formula for B_{check1}, thereby overestimating the B-field by a factor of 1.4 (check: 3000 G × 1.4 = 4200 G). The FF value was 1.4 as per the author's spreadsheet. The mistake that engineer made was to use the author's published value of L in his schematic, along with his published number of turns, with the core's datasheet value for A_e (which in effect ignored FF). It is not consistent. Had this engineer used the standard textbook equation for inductance (not involving fringing effects), instead of using its measured value, he would have at least not landed in a self-contradiction. So, he should have computed L based on the published value of A_e of the core and the number of turns on it. Then he would have arrived at the theoretical value of inductance. If he had used that computed value of inductance in the equation $B = LI/NA$, he would have calculated the correct value of B as 3000 G, not 4200 G. Because he would have used the theoretical value for A (i.e., A_e) too. Of course, after this was all over, he may then have wondered why the computed value of L was 1.4 times less than what was measured on the bench (and stated in the author's schematics). That difference was due to the fringing flux. In other words, to get the right value of B, we need to be consistent: use measured values throughout (invoking fringing effects consistently), or theoretically computed values throughout. We just can't mix and match, or we will end up over-designing the core by wrongly overestimating the B field in it, as this very senior engineer had done.

Worked Example Using Fringing Flux Correction

We have an EE 42/42/15 core set. The material is 3C85. We wind 40 turns on it. We introduce an air gap by means of two spacers of 1-mm thickness each on both sides. What is the value of L? Validate the energy equations. Assume we are passing an instantaneous current of 1.2 A through it at the moment.

Note that the core is two halves joined together, so it is sometimes referred to in terms of the size of only one half and is thus often called 42/21/15. However, note that all vendors state its relevant parameters (like V_e and l_e) with the *implicit* assumption that two halves are joined together in a set. In Chaps. 11 and 12 we have provided tables for some popular core sizes. We have for this material $\mu = 2000$, and for this core set $V_e = 17.6$ cm^3, $A_e = 1.82$ cm^2, $l_e = 97$ mm, and $2 \times D = 29.6$ mm.

Note that by geometry, if the area is 1.82 cm^2, and the depth is 15 mm as is obvious from the part number of the core (EE xx/xx/15), the center limb is 1.82/1.5 = 1.21 cm wide. EE-cores are usually made such that the center limb width splits up into the two equal side limb widths. So each limb is 1.21/2 = 0.61 cm wide. If we subtract this from the length along each side (i.e., 21 mm) we should get D. So our quick estimate of D is 21 − 6.1 = 14.9 mm. So $2 \times D = 29.8$ mm, which is almost in complete agreement with the data provided by the vendor. *So we don't really need the vendor to tell us $2 \times D$ usually.*

Therefore, our calculations proceed as follows:

$$FF = 1 + \frac{l_g}{\sqrt{100 \times A_e}} \times \ln\left(4 \times \frac{D+0.5}{l_g}\right)$$

$$= 1 + \frac{1}{\sqrt{100 \times 1.82}} \times \ln\left(4 \times \frac{14.8+0.5}{1}\right) = 1.305$$

The air gap factor is

$$z = \frac{l_e + \mu l_g}{l_e} = \frac{97 + 2000 \times 1}{97} = 21.62$$

The A_L value with FF correction is (higher inductance)

$$A_L = \frac{1}{z}\left(\frac{\mu\mu_o A_e}{l_e}\right) \times 10^9 \times FF \quad nH/turns^2$$

$$= \frac{1}{21.62}\left(\frac{2000 \times 4\pi \times 10^{-7} \times (1.82/10^4)}{97/10^3}\right) \times 10^9 \times 1.305 = 284.6 \quad nH/turns^2$$

Inductance is

$$L = A_L \times N^2 \times 10^{-9} \quad H$$

$$= 284.6 \times 40^2 \times 10^{-9} \times 10^6 = 455.4 \ \mu H$$

The energy being stored (for the given peak current I) is

$$Energy = \frac{\mu\mu_o N^2 I^2 A_e}{2l_e} \times \frac{1}{z} \times FF \quad J$$

$$= \frac{2000 \times 4\pi \times 10^{-7} \times 40^2 \times 1.2^2 \times (1.82/10^4)}{2 \times (97/10^3)} \times \frac{1}{21.62} \times 1.305 \times 10^6 = 327.9 \ \mu J$$

Comparing with the alternate form of stored energy

$$E = \frac{1}{2} \times L \times I^2 = \frac{1}{2} \times 455.4 \times 1.2^2 = 327.9 \ \mu J$$

We have complete agreement, thus validating our equation for energy too. All this is consistent with simply replacing A_e with $A_e \times FF$.

Note that in a typical power conversion application, we will actually fix L. So to account for FF we would need to decrease N from the theoretical value (with FF not considered).

CHAPTER 8
Tapped-Inductor (Autotransformer-Based) Converters

Introduction

If we are hitting a very unreasonable duty cycle, a tapped-inductor topology can be the answer. It uses a transformer, or more precisely an autotransformer, to provide one more valuable degree of freedom—the *turns ratio*. For example, if we want to go from 10 to 100 V using a DC-DC Boost, we need a duty cycle of roughly 0.9. Note that, whether it is considered advisable or not to operate at such a high duty cycle, it may in fact be impossible to do so, because most Boost controller ICs are, rightfully so, not designed to operate at such an extreme duty cycle. Most chip architects know that in a Boost and Buck-boost, we need to leave time for energy to freewheel into the output, because energy enters the output capacitors *only* during the OFF-time, unlike in a Buck. So, if we block that "moment of opportunity," we can get locked into a situation where the output never rises and the duty cycle stretches out to 100 percent to try and bring it up, but the output therefore still never rises, and so on. One such erroneously designed IC is the LM3478 Boost controller, which "proudly" features 100 percent maximum duty cycle. Yes, a 100 percent duty cycle is certainly acceptable and in fact desirable, for a Buck topology (unless of course, there is a bootstrap drive circuit requirement, which ends up restricting the maximum duty cycle). But 100 percent duty cycle is anathema for a Boost or a Buck-boost topology, because as mentioned, energy is delivered to the output only during the OFF-time in these two topologies. We must leave some available OFF-time for the energy-transfer process to complete. No wonder there are so many rather unsatisfactorily explained complaints about the LM3478 on its online support forum. In general, a much better way to handle extreme step-up or step-down ratios is to use tapped-inductor topologies instead of extreme duty cycles. For example, tapped-inductor topologies can be very useful in photovoltaic (solar) applications which often involve conversion from a few millivolts to hundreds of volts, as in modern solar microinverters.

Tapped inductors are in effect autotransformers. Autotransformers are in turn, simply transformers without isolation. We use transformers in a Forward converter not only for isolation, but for exactly the same purpose as we are now suggesting to use tapped-inductor topologies for: to avoid extremes of duty cycle in either stepping up or stepping down of voltage. For example, to go from 500 volts to 5 volts using just an ordinary Buck topology would require a theoretical duty cycle of 5/500, or one percent. That is almost impossible to achieve, and certainly impractical in any case. But if we use a transformer, or an autotransformer (tapped inductor), with a Primary to Secondary turns ratio of 25, we can first go from 500 V to 500/25 = 20 V with only the help of the turns ratio. Then using duty cycle control in the effective Buck stage that follows, we can go from the virtual input of 20 V to 5 V, using a far more practical duty cycle of 5/20 = 0.25, or 25 percent. That happens to be exactly how we compute the duty cycle of a Forward converter. Note that isolation is just an added advantage here. The turns ratio is the actual key to operation. And by the same logic, we can opt to create non-isolated Flyback converters too, using autotransformers instead of conventional transformers. We can also avail of the additional advantage of polarity correction afforded by the use of a transformer or autotransformer, because the

Buck-boost topology on which the Flyback is based, unfortunately suffers from inherent and undesirable input-output polarity inversion, as we know all too well.

We will see that there are two ways in general, to implement any tapped-inductor topology: one that uses the autotransformer to create a step-up action, and one that helps create a step-down action. Either of these can be superimposed on top of the main step-up or step-down action arising from the underlying topology. We will refer to these as the "$n < 1$" and "$n > 1$" cases, where $n = N_P/N_S$. So for a Buck we have two cases, and for a Boost we have another two: four schematics in all. In this book, for perhaps the first time in literature, both these ways for expanding each topology have been brought into the ambit of the same duty cycle equation, by a unique way of designating primary and secondary sections, similar to what we do in transformer-based topologies. The boundary case corresponding to $n = 1$ (between the two tapped-inductor schematics of each topology) is identical to our standard DC-DC topology (non-tapped inductor). So mentally, for either the Buck or the Boost, we have actually one *super-schematic* with three possibilities as we *slide* the value of n from say 10 to 0.1.

- $n < 1$: step-up turns ratio ($n = 0.1$ would give a 1:10 step-up ratio from the autotransformer).
- $n = 1$: standard DC-DC converter (single-winding inductor in effect).
- $n > 1$: step-down turns ratio (as in an AC-DC Forward converter, or Flyback: $n = 10$ would give a 1:10 step-down ratio from the autotransformer).

Note that in a Buck we would typically use the autotransformer to aid the overall step-down function of the converter, so we would prefer to use the $n > 1$ (step-down turns ratio). That is what we do in a classic Forward converter too. For a Boost, we are usually interested in aiding the overall step-up function, so we usually want to use the case of step-up turns ratio, that is, $n < 1$.

The Tapped-Inductor Buck

The best way to understand how this works is by a numerical example, as shown in Fig. 8.1. Here we have a tapped-inductor Buck. Let us say it has an input of 34 V and its output is set to 10 V.

- The current flows through all eight turns when the switch is ON. We call this the *primary* winding here.
- The current flows through five turns when the switch turns OFF. These turns are said to constitute the *secondary* winding.
- *The (algebraic) sum of the ampere-turns (summed over all windings present on the same core) must be preserved at any given moment. The total ampere-turns cannot change in a discontinuous fashion at any moment.* The reason for that becomes clearer if we write out the voltage-independent equation for a multiwinding inductor as

$$B = \frac{LI}{NA} = \frac{A_L NI}{A} \times 10^{-9} \Rightarrow \left(\frac{A_L}{A} \times 10^{-9}\right) \times \sum N_i I_i$$

The term in brackets is a characteristic, not of the windings, but of the core geometry and its material. The summation is over the index i where each winding is arbitrarily numbered $i = 1, 2, 3$, etc. So a sudden jump in the net ampere-turns would imply a sudden change in the field. The field, however, corresponds to the energy stored in the core, and energy cannot suddenly disappear and appear. Therefore, we get

$$I_{SEC} = I_{PRI} \times \frac{N_P}{N_S}$$

- According to this rule, we can conclude that *at* the moment of either crossover transition, the ampere-turns of the inductor must be preserved as the current changes its route from primary to secondary (or the other way around). So if for example, 1 A was flowing through the Primary, we have 8 ampere-turns just before

Tapped-Inductor (Autotransformer-Based) Converters

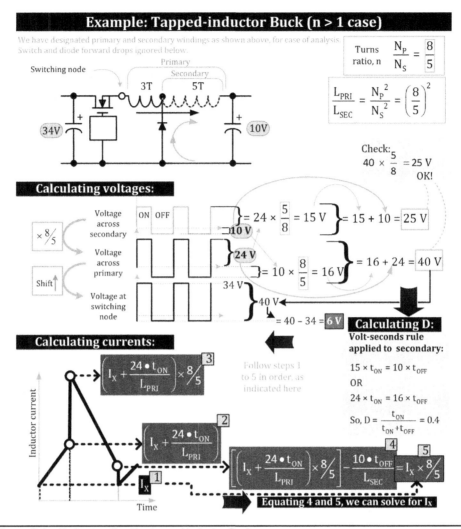

FIGURE 8.1 Understanding the tapped-inductor Buck.

the turn-off transition. Therefore, the current that develops in the Secondary just after the transition must have the same ampere-turns and thus and thus the secondary current is $8/5 = 1.6$ A.

- The voltage across the windings also scales according to the turns ratio, though in opposite manner to the current. So

$$V_S = V_P \times \frac{N_S}{N_P}$$

This follows from Faraday's law:

$$V = N \frac{d\Phi}{dt} = NA \frac{dB}{dt}$$

when written for a multiwinding choke, the equation is

$$\frac{V_i}{N_i} = \frac{d\Phi}{dt}$$

where the right-hand side is again a characteristic, not of the windings, but of the magnetic core on which the windings are wound. Therefore, the left-hand side must apply to each and every winding (each considered separately). We can now state a general rule: At any given moment *the volts per turn existing across any (and every) winding present on a core must be equal* (excluding any spikes due to uncoupled/ leakage inductance elements present, which we are ignoring here).

- *Transformer scaling* for voltages is the same as the volts/turn statement. And the (inverse) transformer scaling for currents is the same as the ampere-turns statement.

- We must, however, remember that scaling applies to the voltages present *across* windings, not to voltages present at any single *point* in the primary or secondary subcircuits. Similarly (inverse) scaling applies only to currents passing *through* windings, not elsewhere.

- In the figure, we can ask: What do we already know about the voltages present across the windings? For one, we know that the voltage across the primary winding (eight turns) is 34 − 10 = 24 V *when the switch is ON*. We also know that the voltage across the secondary winding (five turns) is 10 V when the switch is OFF. Note that these are highlighted in Fig. 8.1.

- Applying voltage scaling to each of the above, we can get the voltage across the secondary winding when the switch is ON, and across the primary winding when the switch is OFF. Therefore, we get the total peak-to-peak voltage swing across each winding. These are 40 V and 25 V, respectively, as shown in the figure.

- The voltage across the primary winding is basically the voltage at the switching node, but measured with respect to the output voltage rail. Therefore, if we want to know the voltage on the switching node (which is with reference to ground), we just need to vertically shift the primary voltage waveform by 10 V. Note that this is in conformity with the observation that when the switch is ON, the secondary winding switching node must be equal to the input voltage (34 V).

- We find that if each winding is considered separately, *the volt-seconds law applies to any chosen winding*. That is what can be seen from the highlighted block in the figure. We will calculate D = 0.4, whichever winding we consider.

- For analyzing the current waveforms, let us assume that the primary inductor current ramp starts at a reference value designated I_X on its waveform. From there, it rises according to $V = LdI/dt$, where V is the voltage across the chosen winding and L is the inductance of that winding (assuming that the other windings are open-circuited).

- The primary-side current therefore ramps up to the value

$$I_X + \frac{24 t_{ON}}{L_{PRI}}$$

- Having ramped up, the current jumps at the transition according to the scaling law. So the ramp-down of the secondary current starts at

$$\left(I_X + \frac{24 t_{ON}}{L_{PRI}}\right) \times \frac{8}{5}$$

- It ramps down according to $V = LdI/dt$, where V is the voltage during the OFF-time across the secondary winding and L is its inductance. We will see that the inductance scales according to the square of the turns ratio. Finally, the secondary current ramps down to

$$= \left[\left(I_X + \frac{24 t_{ON}}{L_{PRI}}\right) \times \frac{8}{5}\right] - \frac{10 t_{OFF}}{L_{SEC}}$$

- Then a crossover transition occurs again, and the current in the Primary must be (as per the scaling rule)

$$\left[\left[\left(I_X + \frac{24 t_{ON}}{L_{PRI}}\right) \times \frac{8}{5}\right] - \frac{10 t_{OFF}}{L_{SEC}}\right] \times \frac{5}{8}$$

But this must return the current to the starting value it had. That is the definition of a *steady state*. So by using

$$\frac{L_{\text{PRI}}}{L_{\text{SEC}}} = \frac{N_P^2}{N_S^2} = \left(\frac{8}{5}\right)^2$$

and simplifying, we get

$$t_{\text{ON}} = t_{\text{OFF}} \times \frac{2}{3}$$

This is the same statement as $D = 0.4$, which we got inside the figure by simply applying the volt-seconds law.

- A more generic derivation for the tapped-inductor Buck gives the following equation for duty cycle:

$$D = \frac{(V_O + V_D) \times N_P}{[(V_{\text{IN}} - V_{\text{SW}}) \times N_S] + [V_O \times (N_P - N_S)] + (V_D \times N_P)}$$

where V_D and V_{SW} are their forward drops across the diode and switch, respectively. Ignoring the diode and switch drops and writing

$$n = \frac{N_P}{N_S}$$

we get a form that is easier to remember:

$$D = \frac{nV_O}{V_{\text{IN}} + V_O(n-1)}$$

- The average inductor current (primary and secondary) is the load current (as for any Buck type of converter). Actually, the expression is simpler for the center of the ramp (COR value). For a standard Buck converter we know that the center is the load current. For a tapped-inductor the center of the primary current ramp is

$$I_{\text{COR\_PRI}} = \left[\frac{N_S}{D \times (N_S - N_P) + N_P}\right] \times I_O$$

The center of the secondary current ramp is by the current scaling rule

$$I_{\text{COR\_SEC}} = I_{\text{COR\_PRI}} \frac{N_P}{N_S}$$

The swing ΔI for the Primary side is of magnitude

$$\Delta I_{\text{PRI}} = \left(\frac{V_{\text{IN}} - V_{\text{SW}} - V_O}{L}\right) \times \frac{D}{f}$$

where L is the inductance of the entire coil (eight turns) and f is the frequency in hertz. The secondary-side swing is

$$\Delta I_{\text{SEC}} = \Delta I_{\text{PRI}} \frac{N_P}{N_S}$$

The same rules that apply to the tapped-inductor also apply to transformers.

NOTE *The leakage inductance has been ignored so far. In reality, to avoid damaging the switch from spikes arising from this leakage, we would usually need an RC or zener snubber/clamp for all tapped-inductor configurations. We should endure tight coupling between primary and secondary windings in any case.*

236 Chapter Eight

> *Note* *Most switcher ICs are, by design, unable to deal with tapped-inductor topologies because they do not allow the switching node to go more than about 1 V negative. For the tapped-inductor Buck topology, the negative voltage is of magnitude $V_O \times [(N_P/N_S) - 1]$. One IC that can handle this negative swing is LT1074 from Linear Tech. But usually, a controller IC, instead of a switcher IC, with some modification in its drive, is easily able to implement a tapped-inductor topology.*

Other Tapped-Inductor Stages and Duty Cycle

As mentioned earlier, there are different ways to construct tapped-inductor topologies, as shown in Fig. 8.2. The input-to-output transfer function for each is presented in the respective parts of the figure. In Fig. 8.3, we see how the duty cycle changes from the standard DC-DC converter, as we try to garner the additional advantage that comes from the turns ratio.

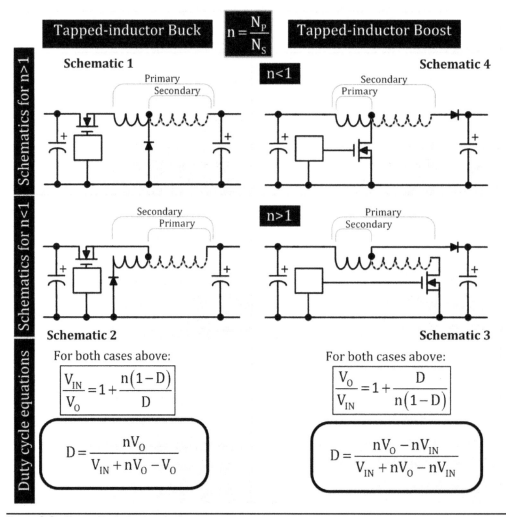

Figure 8.2 Tapped-inductor topologies.

Tapped-Inductor (Autotransformer-Based) Converters 237

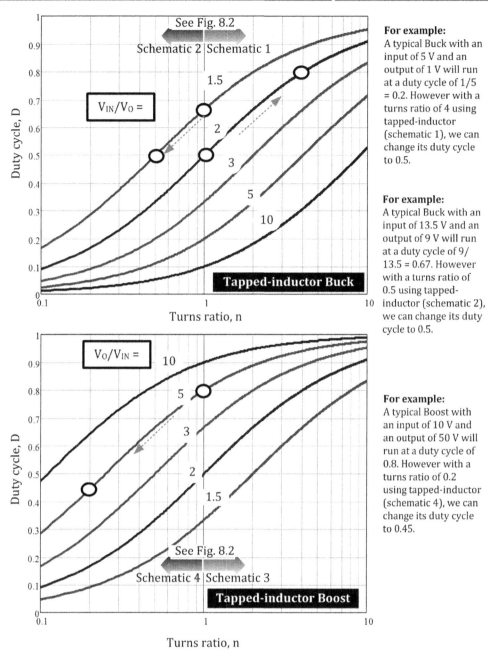

FIGURE 8.3 Comparing duty cycles for Buck and Boost tapped-inductor topologies.

CHAPTER 9
Selecting Inductors for DC-DC Converters

Introduction

Power supply engineers rarely "design" an inductor for DC-DC converter applications. The preferred approach seems to be to just use an easily available off-the shelf inductor from the nearest components cabinet! The first cut is to choose the inductance, hopefully based on a target value of current ripple ratio r. Then the engineer calculates the worst-case inductor current, hopefully cognizant of what part of the line variation leads to the maximum inductor current. Then he or she tries to find an inductor with the required inductance and current rating. That's it.

But we can ask: What is the expected peak temperature of the inductor in that application? What are its losses? The efficiency? Is the inductor really meant for our switching frequency? And so on. Unfortunately, inductor vendors don't answer all these questions in the datasheets of their parts, because there are just too many scenarios and potential applications possible, for every such part. So, if our almost casually selected inductor is found acceptable, it may either just be a plain coincidence, or simply because we have inadvertently overdesigned, by picking a much more expensive or heftier part than necessary for our application. Or maybe we just need to thank the vendor for making a part with a very wide acceptable region of operation—the equivalent of a one-size-fits-all part. In this chapter, we are going to try and minimize this chance factor as we attempt to more optimally select and characterize the inductor in our current application. In effect we are going to wrest more engineering control from what is typically considered a mundane and almost unrewarding task: selecting inductors for DC-DC converters.

The Basics

By definition, for any topology, the current ripple ratio is

$$r = \frac{\Delta I}{I_{COR}}$$

where I_{COR} is the center of ramp (average inductor current waveform) at maximum load and ΔI is the swing (peak to valley). r is defined only for continuous conduction mode (CCM) and that is our assumption too in this chapter. It can vary between 0 for an extremely large inductance to 2 for *critical* conduction mode, that is, the inductor operating at the boundary between continuous conduction and discontinuous conduction modes. A calculated value of r greater than 2 implies discontinuous conduction mode (DCM).

The worst-case inductor current depends on the topology. So for a Buck, we know that the average value (geometric center) of the inductor current waveform equals the load current, irrespective of the input voltage. But for Boost or Buck-boost the center shifts upward as we increase the duty cycle (decrease V_{IN}). So

$$I_{COR} = I_O \quad \text{(Buck)} \qquad I_{COR} = \frac{I_O}{1-D} \quad \text{(Boost/Buck-boost)}$$

These equations follow from the fact that in a Buck, energy is being transferred to the load during the entire cycle, so the average inductor current must equal the load current. For a Boost or a Buck-boost, energy is pushed into the output only when the diode conducts. So the average diode current $I_{COR} \times (1-D)$ must equal the load current.

For all topologies, a high duty cycle implies a low input voltage. So for a Boost or a Buck-boost, we get the maximum (average) inductor current at low input condition V_{INMIN}.

The reader can brush up on his or her knowledge of basic waveform analysis from Chaps. 2 and 3. We see that the peak current is

$$I_{\text{PK}} = I_{\text{COR}} \times \left(1 + \frac{r}{2}\right)$$

The reader can also prove from the method indicated for an arbitrary waveform (Fig. 2.14) that the RMS of the inductor current waveform (for all topologies) is

$$I_{L\_\text{RMS}} = I_{\text{COR}} \sqrt{1 + \frac{r^2}{12}}$$

Note that the inductor RMS current is very close to the DC (average) value, that is, I_{COR}. Even if r has its maximum value of 2, the RMS value is only 15 percent higher than the average value. For our typical design goal of $r = 0.4$, the difference is less than 1 percent. Therefore, we should not be too concerned when looking at a vendor's datasheet whether it has specified the rating in terms of I_{RMS} or I_{DC}. In all cases this refers to the continuous *heating current* that we can pass through the inductor. Clearly it will depend on the expected temperature rise when mounted on a printed-circuit board (PCB). So manufacturers will specify the DC/RMS current rating but at a specified temperature rise over ambient ΔT in degrees Celsius. This is usually in the range 30 to 50°C.

The temperature rise refers basically to the copper losses. Core loss may be only 5 to 10 percent of the total inductor losses when we are dealing with ferrites, but with powdered iron it could be as high as 20 to 30 percent. Core loss depends on the *flux swing* (see Chap. 11). So we need to be able to estimate the flux swing too, if we are going to be able to accurately predict the actual temperature rise in a real converter application. But talking about flux swings and B-fields, we also *do not want to saturate the inductor*. After all, we can simply pass direct current in the coil and measure its heating current capability, but what was the magnetic state of the core in the process? Was it offering any inductance? To know that we may need to pass something other than just direct current. The heating test ignores this, but in a real converter we do care deeply about saturation. So I_{DC} does not tell the whole story.

Beside the flux swing, we also want to know the *peak* flux density (i.e., the peak B-field). Vendors sometimes provide this figure, but in terms of the "safe" current or I_{SAT} that we can pass without saturating the core. Some of the more astute designers and vendors of standard magnetic components probably realize that there is no point using thicker copper if the core is saturating at that point, nor putting less copper which will simply be tantamount to failing to fully exploit the energy storage capability of the core. So *they usually set $I_{\text{DC}} = I_{\text{SAT}}$*. This means that at the current at which the heating test is done, the core is just at the point of saturation. In any case, if two values are provided by the vendor, and they are disparate, the power designer should ignore the higher of the two and use the *lesser* value as the practical limit of the component. That becomes its effective current rating in our first-pass selection procedure based on L and I.

NOTE *There is justification in having some headroom available before saturation occurs. This is equivalent to having an I_{SAT} higher than the continuous rating. But in the author's experience below input voltages of 40 V, we need not worry about saturation, especially with fast-acting FET-based integrated switchers with accurate internal current limiting. From 40 to 63 V, using integrated bipolar switches, some failures were seen under output short-circuit testing if the inductor had been sized only according to the maximum continuous load current, instead of the maximum possible current (i.e., the current limit). Therefore, in high-voltage applications, e.g., off-line power supplies, transformers are always sized according to the worst-case (upper tolerance limit) value of the current limit. But in DC-DC converters below 40 V, we need to size the inductor only as per the continuous load current.*

Specifying the Current Ripple Ratio *r*

The current ripple ratio r is the starting point of any converter design. By definition, r is a constant for a given converter or application (even though the actual load on the output varies). That is because, by definition, it is set at the *maximum* load. The input voltage point we set r at depends on the topology. So for a Buck we set it at the V_{INMAX} simply because though the average inductor remains constant as we vary the input voltage, the peak current does get higher at higher input voltages. For a Boost or a Buck-boost we know that the worst case for the inductor current is at the lowest input voltage, because the average current goes as $1/(1-D)$, so we set r at V_{INMIN} for these.

A high inductance reduces ΔI and results in lower r (and lower RMS current in the input and/or output capacitors) but may result in a very large and impractical inductor. So typically, for most Buck regulators, r is chosen to be in the range of 0.3 to 0.5 (at the maximum rated load, and at V_{INMAX}). Once the inductance is selected, as we decrease the load on the converter (keeping input voltage constant), ΔI remains fixed but the DC level decreases, and so r increases. Ultimately, at the point of transition to DCM, the DC level of the inductor current is $\Delta I/2$. We conclude that

- The current ripple ratio at the point of transition to discontinuous mode is 2. Therefore, for continuous conduction mode, r ranges from 0 (an extremely large inductance) to 2 (*critical inductance*).
- The load at which transition happens can be shown by simple geometry to be $r/2$ *times the maximum rated load*. So, for example, if the inductance is chosen to be such that r is 0.4 at a maximum load of 2 A, the transition to discontinuous mode of operation will occur at 0.2 times 2 A, which is 400 mA. This may be an additional consideration when choosing inductance.

Mapping the Inductor

The practical problem in evaluating an off-the-shelf component is that a vendor would have used certain test conditions to evaluate the part. The vendor would have applied a certain voltage and used a waveform at a certain frequency. The odds that these are the same as the user's conditions are extremely small. So in this chapter our main purpose is to take the vendor's test conditions and "map" them to our application. In doing so we will know more precisely how that inductor will behave as far as we are concerned. We might even find that we can change the frequency or other conditions the inductor was originally designed for and not lose much in the process. This will help us pick the most cost-effective inductor from a much wider array of candidates. The only obstacle may be that the vendor has provided insufficient data for our exercise. But the vendor can usually be prodded to produce much more information than is available on the datasheet of the part.

Volt-seconds

From the basic inductor equation $V = L\,dI/dt$ we get

$$\text{Vsec}_{ON} = L\,\Delta I_{ON} = \text{Vsec}_{OFF} = L\,\Delta I_{OFF}$$

where Vsec is the volt-seconds, and ON and OFF refer to the ON-time and OFF-time, respectively. So generically speaking

$$\text{Vsec} = L\,\Delta I$$

Sometimes engineers prefer to talk in terms of "voltµsecs" instead of volt-seconds as this gives a more *manageable number*. This is simply the voltage across the winding of the inductor times the duration *in µsecs* for which it is applied. This calculation too, as expected, will give the same result if performed for the ON-time as for the OFF-time. If it doesn't, the engineer needs to recheck the equation being used for the duty cycle.

Volt-seconds completely defines the *current swing*. Together with I_{COR}, it completely defines the application from a magnetics viewpoint. Vsec determines the AC component of the current waveform (the swing), whereas I_{DC} is the DC component (average value) of the inductor current, which depends not on volt-seconds but on the load current (and perhaps D).

As a corollary, *all applications with the same Vsec and same I_{DC} (I_{COR}) can be considered to be the "same" application from the viewpoint of inductor design and selection. An inductor designed for a given Vsec and current can be cross-utilized across all applications with the same Vsec and I_{COR} without question.* Even frequency doesn't matter, except to a second order, since it indirectly enters only into the core loss calculation. But we can be assured that the core certainly will not saturate by keeping the above I_{COR} and Vsec rule in mind.

For example, it doesn't matter if we are applying 5 V for 2 µs or 10 V for 1 µs. For a given inductance, we see from the preceding equation that ΔI will remain the same. So the core loss will be the same too (at least that part of it that depends on flux/current swing). And if I_{COR} is the same, the I^2R losses will also not change. And neither will the peak current since that is simply $I_{COR} + \Delta I/2$. The "suitability question" arises only when we attempt to use the inductor for a different Vsec, or I_{COR}, than its original design. For that we need the formal mapping procedure presented in this chapter.

Note that the volt-seconds depends only on the input and output voltages for a given topology. As long as it is in continuous conduction mode, changing the load current, L or even r, does not affect Vsec.

Choosing *r* and *L*

In the Appendix we have the relevant parameters for the three topologies. First, we start with the equation for the required peak energy-handling capability E:

$$E = \frac{1}{2} L I_{PK}^2 \quad J$$

Then using the basic equations in the Appendix, we can carry out a simple derivation as follows:

$$E = \frac{1}{2} L I_{COR}^2 \left(1 + \frac{r}{2}\right)^2$$

$$= \frac{1}{2} \frac{Vsec}{\Delta I} I_{COR}^2 \left(1 + \frac{r}{2}\right)^2$$

$$= \frac{1}{2} \frac{I_{COR} Vsec}{r} \left(1 + \frac{r}{2}\right)^2$$

$$= \frac{1}{2} I_{COR} \times Vsec \times \frac{1}{r} \times \left(1 + \frac{r^2}{4} + r\right)$$

$$= \frac{1}{8} I_{COE} \times Vsec \times \frac{r^2}{r} \times \left(\frac{4}{r^2} + 1 + \frac{4}{r}\right)$$

And finally,

$$E = \frac{I_{COR} Vsec}{8} \left[r \left(\frac{2}{r} + 1\right)^2 \right] \quad J$$

This expression is valid for all topologies, though the relationship of I_{COR} to load current may be different. But the term in the square brackets has a particular shape with respect to r—and this provides the clue for describing an optimum setting of r. To gauge the significance of the energy term vis-à-vis the other parameters in a power supply, we plot the results out for a Buck in Fig. 9.1. Here we are varying r to see what happens. This is tantamount to changing the inductance for a given frequency. We observe the following:

- The energy curve has a "knee" at around $r = 0.4$.
- Too low of an r will significantly increase the physical size of the core, because the core volume is proportional to its energy handling capability.
- Too high an r will not significantly decrease the size of the core, since we will reach a point of diminishing returns (flattening of the energy curve).
- As we increase r, the related currents in the capacitors and switch increase. This in turn may demand larger capacitors, etc.
- However, what we are showing in Fig. 9.1 are *normalized* values. So, a large variation may not be very significant if we realize that the absolute value of the concerned parameter may be very small to start with. For example, the RMS current through the output capacitor is actually very small for a Buck converter even though its variation with respect to r is the most noticeable. Further, its basic criterion of selection is usually not the RMS current, but the output voltage ripple and/or the open-loop gain (stability) characteristics (see Fig. 12.18). So, for example, with voltage-mode control, the ESR (equivalent series resistance) of the output capacitor provides the required zero in its loop response, and we know that selecting a very low ESR capacitor may cause instability.

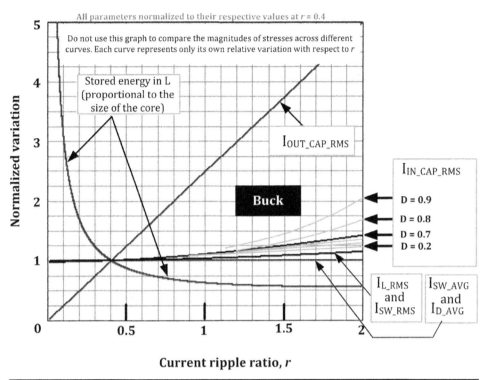

FIGURE 9.1 Variation of parameters for a Buck converter as *r* varies.

- Of greater significance in a Buck converter is the input capacitor. Its physical size may often be determined purely by the ripple current through it. So an increase in the RMS current even by a factor of two, will typically increase its size by four times.
- But generally speaking, the optimum setting is $r \approx 0.4$ for all topologies as it coincides with the knee of the energy curve.

B in Terms of Current

For full details refer to Chap. 7. Here we just use Faraday's law (in MKS units) expressed in terms of volt-seconds:

$$\Delta B = \frac{L \times \Delta I}{N \times A} = \frac{\text{Vsec}}{N \times A} \quad \text{T}$$

A is in square meters. Writing this in terms of square centimeters, gauss (G), and voltμseconds (Vμs), we get

$$\Delta B = \frac{100 \times \text{V}\mu\text{s}}{N \times A_{\text{cm}^2}} \quad \text{G}$$

This equation also applies to instantaneous values. Thus at peak current we can write (in MKS units this time)

$$B_{\text{PK}} = \frac{L \times I_{\text{PK}}}{N \times A} \quad \text{T}$$

Note also that the preceding relationships seem to indicate that we cannot change the flux density swing ΔB by any other means except by changing the area of the core, the number of turns, or the volt-seconds. But volt-seconds cannot change if the input and output voltages are held fixed (frequency being fixed too). So for a given application condition, we conclude that though we may change inductance (and thereby ΔI), since $L \Delta I$ equals Vsec, we cannot change their product. So if we lower L, simultaneously ΔI goes up by the same amount, and so the volt-seconds do not change as expected. Therefore, the only way to change the flux swing ΔB is to change the number of turns and/or the area of the core. For fixed frequency

$$\Delta B \propto \frac{1}{NA}$$

Since the applied volt-seconds is inversely proportional to frequency, we get for the general case with variable switching frequency

$$\Delta B \propto \frac{1}{NAf}$$

In Chap. 7 we had derived the following relationship

$$L = \frac{1}{z}\left(\frac{\mu\mu_o A_e}{l_e}\right) \times N^2 \quad \text{H}$$

Omitting the subscript *e* for convenience, we get

$$L \propto \frac{N^2 A}{l}$$

So, using

$$\Delta B = \frac{L \times \Delta I}{N \times A}$$

we can plug in L to simplify and get an *alternative, but equivalent proportionality statement* for B—one that should also help in selecting optimum components

$$\Delta B \propto \frac{N \Delta I}{l}$$

In this entire discussion, A is the same as the effective area of the core A_e and l is the same as its effective length l_e.

Counterintuition in Magnetics

We must be wary of applying intuition in magnetics. Here is an example. The size of the inductor is related primarily to its energy handling capability. The energy handling capability is $(1/2) \times L \times I_{PK}^2$ as this is the peak energy in the inductor. For a given application, if we reduce inductance, we may think that this would cause an increase in ΔI and in I_{PK}, and since E depends on the square of I_{PK}, it would dominate and cause E to increase. However, an actual calculation shows the converse is true. I_{PK} is equal to $I_{COR} + \Delta I/2$ and so the impact of varying ΔI on I_{PK} is much less. Thus, in the energy equation, the variation of L dominates, and the energy handling requirement actually decreases as we reduce L. This is also clear from Fig. 9.1. Reality seems counterintuitive here. It is magnetics after all!

Core-Loss Optimization by Tweaking Geometry at a Fixed Frequency

Here we will test some popular *myths* associated with magnetic components and also carry out some thought experiments that will give us a better feel for the subject of core losses, and thereby help us in selecting more optimum cores in high-efficiency, high-frequency applications in particular. In this first part we are keeping a fixed frequency, trying to reduce core losses.

Case 1: Varying the Volume of the Core (*N* Fixed)

Core loss is sometimes expressed as a power loss density, that is, in W/cm³. On this basis, some engineers feel that a larger volume will automatically increase the core loss. Let us check it out. Here we are increasing the dimensions of the core, keeping N unchanged so far. Suppose we increase *each* dimension of the core by a factor of two (double each). So effective length l_e doubles, but effective area A_e increases four times. If we keep the number of turns unchanged, since L is proportional to A_e/l_e, the inductance will double. Since r is inversely proportional to L, ΔI will get halved.

Core loss typically varies as $B^{2.5}$ for ferrites. As already discussed, ΔB is proportional to $1/NA_e$, and so, if A_e increases four times, ΔB will decrease four times (if N is unchanged). So the net effect on the total core loss is

$$P_{\text{CORE}} \propto \Delta B^{2.5} \times \text{volume} \Rightarrow \left(\frac{1}{4}\right)^{2.5} \times 8 = 0.25$$

The total core loss has actually *decreased* by a factor of four, not increased. So larger cores may not be cost effective, but neither are they going to hurt us in this manner of variation at least.

NOTE *If all dimensions are changed by a factor x, the inductance too changes by the same factor, because $A_e/l_e = x^2/x = x$.*

Case 2: Varying the Volume of the Core (*L* Fixed)

Once again we increase the volume eight times (each dimension doubled, A_e quadrupled, l_e doubled). Since L is proportional to A_e/l_e, the inductance will tend to double (case 1). But now we want to keep L fixed (to keep to our design target for r, say). Since L actually depends on turns$^2 \times A_e/l_e$, if we decrease the number of turns by the factor $2^{0.5} = 1.414$, inductance will be held constant (if we quadruple A_e and double l_e). Keeping L constant would keep ΔI unchanged. Further, ΔB is proportional to $1/NA_e$. So now we will get

$$P_{\text{CORE}} \propto \Delta B^{2.5} \times \text{volume} \Rightarrow \left(\frac{\sqrt{2}}{4}\right)^{2.5} \times 8 = 0.595$$

So *core loss has decreased*. Note that the exponent of B in the core-loss equation is 2.5 here. Had it been 2 (as for Kool Mu®), we would have gotten a result of 1, implying *no change in core loss*. If the exponent was less than two (implying a "better" material), in fact we would have got *more core loss* with larger core volumes (for the same inductance). Consult Table 11.5 (in Chap. 11) for a table of core loss coefficients (exponents). The material decides whether this type of variation is beneficial or not.

The same situation arises if an unsuspecting engineer tries the following: In an effort to increase the current rating of her inductor, say 10 μH, he or she goes to the cabinet and pulls out two inductors of 4.7 μH each, with each of higher current rating, and places them in series to get almost the same inductance with higher current rating. In effect, for almost the same net inductance, the engineer now has a much higher core volume, and for that reason alone, may in fact notice an unexpected drop in efficiency, especially if the magnetic material has a small B-related core loss exponent, as explained above.

Case 3: Varying the Volume of the Core by Varying Area (Length and *N* Fixed)

There are other ways to increase the volume of the core. So, suppose we had increased the volume four times but *only by increasing the effective area*, not the effective length, and also with no change in the number of turns. Now the inductance would increase four times and ΔI would reduce by four. Then using $B \propto 1/NA$

$$P_{\text{CORE}} \propto \Delta B^{2.5} \times \text{volume} \Rightarrow \left(\frac{1}{4}\right)^{2.5} \times 4 = 0.125$$

So core loss has decreased.

ANALYSIS *Note that here the volume increased four times but the core loss fell $1/0.125 = eight$ times (twice the increase in volume). Compare this to case 1 where volume increased eight times and core loss fell four times, a much smaller improvement. So, we conclude that it is far better in practice to increase "volume" by increasing the area, not all the dimensions. That helps achieve higher efficiencies where core loss is playing a major part, as in high-frequency switchers. In other words, all commercial inductors of the same inductance, material, and supposed current rating may not be equivalent in terms of their core loss terms. A better choice of the physical profile of the inductor may go a long way in maximizing efficiency. This aspect is almost completely overlooked in related literature. We therefore continue our core loss analysis next.*

Case 4: Varying the Volume of the Core by Varying Area (Length and *L* Fixed)

This is similar to case 2. We want to keep the inductance constant (r constant), but we also want to increase the volume by changing only the area (quadrupling). Once again, we need to decrease the number of turns, this time by a factor of two only (L depends on A_e/l_e, and we are quadrupling A_e only). So now, since B depends on $1/NA_e$, we get

$$P_{\text{CORE}} \propto \Delta B^{2.5} \times \text{volume} \Rightarrow \left(\frac{2}{4}\right)^{2.5} \times 4 = 0.707$$

The *core loss has decreased*.

ANALYSIS *Note that here, the volume increased four times but the core loss fell 1/0.707 = 1.4 times. Compare this to case 2 where volume increased eight times and core loss fell by a factor 1/0.595 = 1.68 times, a much larger improvement but at the cost of a much larger core.*

ANALYSIS *Note that the exponent of B in the core-loss equation is 2.5 here. As mentioned under case 2, had the exponent of B in the core-loss equation been 2, we would again have got a result of 1, implying no change in core loss. If the exponent were less than 2, in fact we would have got more core loss with larger core volumes (for same inductance).*

Case 5: Varying the Volume of the Core by Varying Length (Area and *L* Fixed)

Here we increase volume four times, by only increasing the length four times. *L* depends on $N^2 \times A_e/l_e$, so to keep it constant, we need to increase N^2 by a factor of 8, i.e., *N* multiplied by $\sqrt{8} = 2.83$. Since *B* depends on $1/NA_e$, *B* will fall be the same factor. So we get

$$P_{\text{CORE}} \propto \Delta B^{2.5} \times \text{volume} \Rightarrow \left(\frac{1}{2.83}\right)^{2.5} \times 4 = 0.3$$

This is over two times better in terms of core loss, than case 4. It says that if we keep *L* fixed (as we would like to do in a real application), *we are better off looking for a larger core with the same area but a larger length*.

ANALYSIS AND CONCLUSION *If we increase core volume (at a fixed frequency), keeping L and r constant, we should prefer to achieve this by varying length of core not area. That will lower core losses dramatically (provided we have a suitably high exponent for B).*

Core-Loss Optimization by Tweaking Geometry as We Vary Frequency

Now we will see the effects (on core loss) of varying frequency too.

Case 6: Varying the Frequency (Core Volume and *L* Fixed)

Now we start to vary the frequency to try and minimize core-loss increase, if not actually lower core losses. Since the core-loss equation is the product of two terms, one related to flux swing and one to *frequency*, with the exponent of the frequency term being typically around 1.5, it is sometimes believed that increasing the frequency (for a given core and inductance) will automatically increase core losses. Here we use the full form of the *B* proportionality equation, i.e., $\Delta B \propto 1/NAf$. We also note that volt-seconds is inversely proportional to frequency. So if we double the frequency, the variation of the core loss is

$$P_{\text{CORE}} \propto \Delta B^{2.5} \times f^{1.5} \times \text{volume} \Rightarrow \left(\frac{1}{1 \times 1 \times 2}\right)^{2.5} \times 2^{1.5} \times 1 = 0.5$$

So *core loss will actually decrease*. But note that we have kept inductance constant here.

Case 7: Varying the Frequency (Core Volume and *r* Fixed)

We know from a design optimization viewpoint that it is advisable to keep *r* at around 0.4, irrespective of topology, frequency, or application conditions. So let us decrease the number of turns accordingly, *while keeping to the same core size for now*. Therefore, if, for example, we increase frequency four times, we need to reduce the inductance by the same factor, and that means that the number of turns has to be halved. So now, for the variation in core loss we get (using $\Delta B \propto 1/NAf$)

$$P_{\text{CORE}} \propto \Delta B^{2.5} \times f^{1.5} \times \text{volume} \Rightarrow \left(\frac{1}{(1/2) \times 1 \times 4}\right)^{2.5} \times 4^{1.5} \times 1 = 1.414$$

The core loss has increased.

Case 8: Varying the Frequency (*r* Fixed, *N* Fixed, Each Dimension Changed)

The whole idea of using high switching frequencies is to be able to use *smaller* inductors. What if we increased frequency four times but achieved the reduction in *L* not by changing the number of turns, but by making the core *physically smaller*? The volume requirement of a core in power conversion is inversely proportional to switching frequency. So, here we do achieve reduction in volume by altering all *three* dimensions of the core. We know that to achieve a reduction of four in its inductance (for a fourfold increase in frequency), we need to change each dimension by the same factor (it depends on A_e/l_e). So A_e will now be reduced by a factor of $4 \times 4 = 16$ (l_e by a factor of four too). Applying this result to our core-loss equation we get

$$P_{\text{CORE}} \propto \Delta B^{2.5} \times f^{1.5} \times \text{volume} \Rightarrow \left(\frac{1}{1 \times (1/16) \times 4}\right)^{2.5} \times 4^{1.5} \times \frac{1}{4 \times 4 \times 4} = 4$$

So core loss has increased tremendously.

Case 9: Varying the Frequency (*r* Fixed, *N* Fixed, Only Area Changed)

Now we reduce the inductance by four, not by changing the number of turns or changing l_e, but only by changing A_e by a factor of four (volume reduced by the same factor). Then

$$P_{\text{CORE}} \propto \Delta B^{2.5} \times f^{1.5} \times \text{volume} \Rightarrow \left(\frac{1}{1 \times (1/4) \times 4}\right)^{2.5} \times 4^{1.5} \times \frac{1}{4 \times 4 \times 1} = 0.5$$

So core loss has finally decreased at higher frequencies.

ANALYSIS AND CONCLUSION *As we increase frequency, it is not just enough to decrease inductance. We need to actually reduce the physical size of the magnetic core (volume inversely with respect to frequency). And we should try to achieve this reduction in volume by changing the area, not the length of the core.*

We must emphasize that if we change inductance in inverse proportion to frequency, which is what we need to do to maintain *r*, then the energy requirement of the core also decreases roughly in the same proportion. This is fortuitous, because at no time do we want to select a core size based only on optimum core loss, because that would compromise the required energy handling capability. We must be careful to choose an A_e and l_e combination that will have enough core volume $V_e = A_e \times l_e$ to be able to do the job in all senses.

Magnetics vendors are constantly putting out new parts. Systems designers know that just as they often set disparate and sometimes meaningless current ratings (e.g., one for heating and one for saturation), they don't necessarily always pick the most optimum combination of core characteristics and winding to achieve the most optimum results. *Not all inductors are created equal.* Especially in high-frequency, high-efficiency designs, a careful evaluation of any off-the-shelf part is well deserved.

A Walk-Through Example

The input DC voltage is 24 V into a Buck converter. The output is 12 V at a maximum load of 1 A. We wish to allow a slightly more conservative current ripple ratio of 30 percent (at maximum load). We assume $V_{\text{SW}} = 1.5$ V, $V_D = 0.5$ V, and $f = 150{,}000$ Hz.

For a Buck regulator, the duty cycle *D* is

$$D = \frac{V_O + V_D}{V_{\text{IN}} - V_{\text{SW}} + V_D}$$

So the switch ON-time is

$$t_{\text{ON}} = \frac{D}{f} = \frac{(12 + 0.5) \times 10^6}{(24 - 1.5 + 0.5) \times 150{,}000} \ \mu\text{s}$$
$$= 3.62 \ \mu\text{s}$$

So the Vμs for the application is

$$V\mu s = (V_{IN} - V_{SW} - V_O) \times t_{ON} = (24 - 1.5 - 12) \times 3.62$$
$$V\mu s = 38.0$$

$$L = \frac{V\mu s}{r \times I_O} \quad \mu H$$
$$= \frac{38.0}{0.3 \times 1.0} \quad \mu H$$
$$= 127 \quad \mu H$$

The required energy handling capability is next calculated. Every cycle, the peak current is

$$I_{PEAK} = I_O + \frac{V\mu s}{2L} = 1.0 + \frac{38}{2 \times 127} \quad A$$
$$= 1.15 \ A$$

The required energy handling capability E is

$$E = \frac{1}{2} \times L \times I_{PEAK}^2 \quad \mu J$$
$$= \frac{1}{2} \times 127 \times 1.15^2 = 84 \quad \mu J$$

Choosing an Inductor

Note that for voltμseconds, the vendor we are going to look at, uses the symbol ET. To avoid confusion we too will do the same from this point on.

Our first-pass selection is based upon the inductance we calculated earlier and the DC rating (maximum load). We tentatively select a part from Pulse (at www.pulseelectronics.com) simply because its L and I_{DC} are close to our requirements, even though the rest does not seem to fit our application very well, at least not at first sight (see Table 9.1). In particular the frequency for which the inductor was designed is 250 kHz, but our application is 150 kHz. Note that we will dispel another intuitive myth: that since we are decreasing the frequency of operation of the inductor, our flux swing will increase and our core loss will therefore go up, along with the peak flux density and energy. In fact, the reverse happens in our case, and that is one example why it is important to follow the full mathematical procedure presented next.

- For core-loss equations it is conventional to use half the peak-to-peak flux swing. So, as do most vendors, the B that Pulse Electronics uses in its datasheets actually refers to $\Delta B/2$. This must be kept in mind in the calculations that follow.

- The astute designer can recognize the exponents of B and f in the core-loss equation in Table 9.1 as corresponding to a ferrite (which this is). Powdered Iron material 52 (from Micrometals Inc. at www.micrometals.com) would, for example, have given us an equation of the form $\propto B^{2.11} \times f^{1.26}$).

- Most off-the-shelf inductors are designed for a temperature rise of 30 to 50°C.

Part number	Reference values			Control values	Calculation data
	I_{DC} (A)	L_{DC} (μH)	ET (Vμs)	DCR (nom) mΩ	ET_{100} (Vμs)
P0150	0.99	137	59.4	387	10.12
- The inductor is such that 380-mW dissipation corresponds to a 50°C rise in temperature.					
- The core-loss equation for the core is $6.11 \times 10^{-18} \times B^{2.7} \times f^{2.04}$ mW where f is in hertz and B is in gauss.					
- The inductor was designed for a frequency of 250 kHz.					
- ET_{100} is the Vμs at which B is 100 G.					

TABLE 9.1 Specifications of Choke to Be Evaluated for Application

Evaluating the Inductor in Our Application

We have the inductor operating under its test conditions. We now map its performance to our specific application conditions. *We follow the summary in Table 9.2.* So, set unprimed parameters to be the *design values*, and the corresponding primed parameters as the *application values*. The overall mapping procedure is described next.

The following are the *design conditions* of the inductor

- I_{DC}
- ET
- f
- $T_{AMBIENT}$

The *application conditions* we will use the chosen inductor in are

- I'_{DC}
- ET'
- f'
- $T'_{AMBIENT}$

Design parameters	Design conditions $I_{DC}, ET, f, T_{AMBIENT}$	Application conditions $I'_{DC} \equiv I_C, ET', f', T'_{AMBIENT}$
Current swing (A)	$\Delta I = \dfrac{ET}{L}$	$\Delta I' = \Delta I \cdot \left[\dfrac{ET'}{ET}\right]$
Current ripple ratio	$r = \dfrac{ET}{L \cdot I_{DC}}$	$r' = r \cdot \left[\dfrac{ET' \cdot I_{DC}}{ET \cdot I'_{DC}}\right]$
Peak current (A)	$I_{PEAK} = I_{DC} + \dfrac{ET}{2 \cdot L}$	$I'_{PEAK} = I_{PEAK} \cdot \left[\dfrac{(2 \cdot L \cdot I'_{DC}) + ET'}{(2 \cdot L \cdot I_{DC}) + ET}\right]$
RMS current in inductor (A)	$I_{RMS} = \left[I_{DC}^2 + \dfrac{ET^2}{12 \cdot L^2}\right]^{1/2}$	$I'_{RMS} = I_{RMS} \cdot \left[\dfrac{(12 \cdot I'^2_{DC} \cdot L^2) + ET'^2}{(12 \cdot I_{DC}^2 \cdot L^2) + ET^2}\right]^{1/2}$
Flux density swing (G)	$\Delta B = \dfrac{ET}{ET_{100}} \cdot 200 = \dfrac{100 \cdot ET}{N \cdot A_e}$	$\Delta B' = \Delta B \cdot \left[\dfrac{ET'}{ET}\right]$
Peak flux density (G)	$B_{PEAK} = \dfrac{200}{ET_{100}} \cdot \left[(I_{DC} \cdot L) + \dfrac{ET}{2}\right]$	$B'_{PEAK} = B_{PEAK} \cdot \left[\dfrac{2 \cdot L \cdot I'_{DC} + ET'}{2 \cdot L \cdot I_{DC} + ET}\right]$
Copper losses (mW)	$P_{CU} = DCR \cdot \left[I_{DC}^2 + \dfrac{ET^2}{12 \cdot L^2}\right]$	$P'_{CU} = P_{CU} \cdot \dfrac{(12 \cdot I'^2_{DC} \cdot L^2) + ET'^2}{(12 \cdot I_{DC}^2 \cdot L^2) + ET^2}$
Core losses (mW)	$P_{CORE} = a \cdot \left[\dfrac{ET}{ET_{100}} \cdot 100\right]^b \cdot f^c$	$P'_{CORE} = P_{CORE} \cdot \left[\left(\dfrac{ET'}{ET}\right)^b \cdot \left(\dfrac{f'}{f}\right)^c\right]$
Energy in core (µJ)	$E = \dfrac{1}{2} \cdot L \cdot \left[I_{DC} + \dfrac{ET}{2 \cdot L}\right]^2$	$E' = E \cdot \left[\dfrac{(2 \cdot L \cdot I'_{DC}) + ET'}{(2 \cdot L \cdot I_{DC}) + ET}\right]^2$
Temperature rise (°C)	$\Delta T = R_{TH} \cdot \dfrac{P_{CU} + P_{CORE}}{1000}$	$\Delta T' = \Delta T \cdot \left[\dfrac{P'_{CU} + P'_{CORE}}{P_{CU} + P_{CORE}}\right]$

ET is in Vµs, DCR in mΩ, L in µH, f in Hz, A_e in cm², N is number of turns

TABLE 9.2 Implementing Mapping Procedure for an Inductor (Go from Left to Right: Design Conditions to Application Conditions)

In going from the *design conditions* to the *application conditions* the following are considered virtually constant:

- L
- DCR
- Rth
- The core-loss equation

Finally, to "approve" the inductor for our application we should certify that

1. r is acceptable (i.e., choice of L is OK).
2. B_{PK} is OK (important in core-saturation–limited inductors).
3. $I_{PK} < I_{CLIM}$ (or the controller will be unable to provide the required power).
4. ΔT is OK (evaluate $P_{CU} + P_{CORE}$).

We assume the vendor has provided all the following inputs:

- ET (Vμs)
- ET_{100} (Vμs per 100 G)
- L (μH)
- I_{DC} (A) (maximum rating)
- DCR (mΩ)
- f (Hz)
- The form for core loss (in mW) as $a \times B^b \times f^c$, where B is in gauss, f is hertz, and B is half the peak-to-peak flux swing
- Thermal resistance of inductor in free air (°C/W)

If any of these are unknown, the vendor should be contacted.

We now proceed with the specific example (see Table 9.2). The design conditions of the inductor are

- ET = 59.4 Vμs
- f = 250,000 Hz
- I_{DC} = 0.99 A

Our *application conditions* are

- ET′ = 38 Vμs
- f' = 150,000 Hz
- I'_{DC} = 1 A

(We assume that $T_{AMBIENT}$ is unchanged so we ignore it.)

a) Current Ripple Ratio

By design

$$r = \frac{ET}{L \times I_{DC}}$$
$$= \frac{59.4}{137 \times 0.99}$$
$$= 0.438$$

In our application

$$r' = r \times \left(\frac{ET' \times I_{DC}}{ET \times I'_{DC}}\right)$$

Selecting Inductors for DC-DC Converters

$$r' = 0.438 \times \left(\frac{38 \times 0.99}{59.4 \times 1}\right)$$
$$= 0.277$$

We expected r to be slightly lower than 0.3 since the chosen inductor has a higher inductance than we required (137 µH instead of 127 µH).

b) Peak Flux Density

By design

$$B_{PK} = \frac{200}{ET_{100}} \times \left[(I_{DC} \times L) + \frac{ET}{2}\right] \text{ G}$$
$$= \frac{200}{10.12} \times \left[(0.99 \times 137) + \frac{59.4}{2}\right] \text{ G}$$
$$= 3267 \text{ G}$$

In our application

$$B'_{PK} = B_{PK} \times \left[\frac{(2L \times I'_{DC}) + ET'}{(2L \times I_{DC}) + ET}\right] \text{ G}$$
$$= 3267 \times \left[\frac{(2 \times 137 \times 1) + 38}{(2 \times 137 \times 0.99) + 59.4}\right] \text{ G}$$
$$= 3084 \text{ G}$$

which is less than B_{PK} and is therefore acceptable.

c) Peak Current

To ensure that the regulator will deliver the rated load, we need to ensure that the peak current is less than the internal current limit of the switcher IC.

By design

$$I_{PK} = I_{DC} + \frac{ET}{2 \times L}$$
$$= 0.99 + \frac{59.4}{2 \times 137} \text{ A}$$
$$= 1.21 \text{ A}$$

This corresponds to a *B*-field of 3267 G as calculated earlier.

Note that gives us an energy handling capability of

$$E = \frac{1}{2} \times L \times I_{PK}^2$$
$$= \frac{1}{2} \times 137 \times 1.21^2 \text{ µJ}$$
$$= 100 \text{ µJ}$$

whereas we required at least 84 µJ. So the inductor seems sized about right, and we can proceed with the analysis.

In our application

$$I'_{PK} = I_{PK} \times \left[\frac{(2L \times I'_{DC}) + ET'}{(2L \times I_{DC}) + ET}\right]$$
$$= 1.21 \times \left[\frac{(2 \times 137 \times 1.0) + 38}{(2 \times 137 \times 0.99) + 59.4}\right] \text{ A}$$
$$= 1.14 \text{ A}$$

Chapter Nine

This corresponds to a B-field of 3084 G as calculated earlier. We check that the minimum value of the current limit over temperature and tolerance is 2.3 A. So since the peak value I'_{PK} is less than 2.3 A, the controller will be able to provide the desired output power (without hitting the current limit).

d) Temperature Rise

By design

$$P_{CU} = DCR \times \left(I_{DC}^2 + \frac{ET^2}{12 \times L^2} \right)$$

$$= 387 \times \left(0.99^2 + \frac{59.4^2}{12 \times 137^2} \right) \text{ mW}$$

$$= 385 \text{ mW}$$

$$P_{CORE} = a \bullet \left(\frac{ET}{ET_{100}} \bullet 100 \right)^b \bullet f^c \text{ mW}$$

where the vendor has provided that $a = 6.11 \times 10^{-18}$, $b = 2.7$, and $c = 2.04$. So

$$P_{CORE} = 6.11 \times 10^{-18} \times \left(\frac{59.4}{10.12} \times 100 \right)^{2.7} \times f^{2.04} \text{ mW}$$

$$= 18.7 \text{ mW}$$

So

$$\Delta T = R_{TH} \times \frac{P_{CU} + P_{CORE}}{1000} \text{ °C}$$

The vendor has stated that 380-mW dissipation corresponds to a 50°C rise in temperature. So the thermal resistance of the inductor is

$$R_{TH} = \frac{50}{380/1000} = 131.6 \text{ °C/W}$$

$$\Delta T = 131.6 \times \left(\frac{385 + 18.7}{1000} \right) = 53°C$$

In our application

$$P'_{CU} = P_{CU} \times \frac{(12 \times I'^2_{DC} \times L^2) + ET'^2}{(12 \times I_{DC}^2 \times L^2) + ET^2}$$

$$= 385 \bullet \frac{(12 \times 1^2 \times 137^2) + 38^2}{(12 \times 0.99^2 \times 137^2) + 59.4^2} \text{ mW}$$

$$= 389 \text{ mW}$$

$$P'_{CORE} = P_{CORE} \bullet \left[\left(\frac{ET'}{ET} \right)^b \bullet \left(\frac{f'}{f} \right)^c \right]$$

$$= 18.7 \times \left[\left(\frac{38}{59.4} \right)^{2.7} \times \left(\frac{150,000}{250,000} \right)^{2.04} \right] \text{ mW}$$

$$= 2 \text{ mW}$$

So,

$$\Delta T' = \Delta T \times \left[\frac{P'_{CU} + P'_{CORE}}{P_{CU} + P_{CORE}} \right]$$

$$= 53 \times \left[\frac{389 + 2}{385 + 18.7} \right] = 51°C$$

which was deemed to be acceptable in this application.

This completes the qualification analysis for the short-listed inductor. The summary is

- The inductor was designed for about a 50°C rise in temperature over ambient at a load of 1 A. We will get 51°C in our application.
- The copper loss (385 mW) predominates, and the core losses are relatively small. We will get 389 mW in our application.
- The peak flux density is about 3200 G, which occurs at a peak instantaneous current of 1.2 A. We will get 3080 G at 1.15 A.
- The rated energy handling capability of the core is 100 µJ. We only require 84 µJ.

CHAPTER 10
Basics of Flyback Transformer Design

Refer to Chap. 7 for magnetics concepts before starting this one. Also read Chap. 6 for a complete introduction to the Flyback. We also have a more detailed, step-by-step example in Chap. 12 to follow. We are using MKS (SI) units, unless otherwise stated.

The Voltage-Dependent Equation: A Practical Form

We discussed several forms of the *voltage-dependent* equation, also known as Faraday's law, in Chap. 7. Here is a more practical form as applied to switching converters.

The basic form of the equation is

$$B_{AC} = \frac{V_{AVG} \times \Delta t}{2 \times N \times A_e}$$

or in easy-to-remember form,

$$\text{Volt-second} = NAB$$

Just keep in mind that by convention, B_{AC} is twice the flux swing B (or ΔB). As we mentioned, the voltage-dependent equation just given only refers to the *change* in the B-field. However, we can easily relate it to the peak field too (for most common magnetic materials). Let us first assume that the gapped inductor has finally been designed fairly optimally, so that its peak field is very close to the saturation value B_{SAT}. Then realizing that *current and B are proportional to each other* (we are assuming "linear" magnetic materials, though the proportionality is not entirely true, especially for powdered iron, for example), the relationship for the current ripple ratio r must apply to the B-field too. And since $\Delta B = 2 \times B_{AC}$, we get a field-expression of our key term r:

$$\frac{2 \times B_{AC}}{B_{DC}} = r$$

But by definition, $B_{PK} = B_{DC} + B_{AC}$, i.e., $B_{DC} = B_{PK} - B_{AC}$, so

$$\frac{2 \times B_{AC}}{B_{PK} - B_{AC}} = r$$

Simplifying, and combining with the voltage-dependent equation, we get

$$N = \left(\frac{2}{r} + 1\right) \times \frac{V_{ON} D}{2 \times B_{PK} A_e f}$$

This will, for example, give us the number of turns of the primary winding if we have a Flyback transformer, but the same can be used for any inductor too.

Note that this is a very interesting relationship. *It does not depend directly on the air gap, the effective length, or effective volume of the core, and not even on the permeability!* It does not depend directly on load current either, and that is because, as implied by our use of r (which is defined only for CCM), we are talking only of continuous conduction mode. It does depend on effective area of the core though and on the field.

Also, do keep in mind that the swing in B and its peak value are related, just as current is, by virtually the same equation (see Fig. 7.2):

$$\Delta B = \frac{2r}{2+r} B_{PK}$$

Energy Stored, as Related to Core Volume in an Ungapped Core

We have shown in Chap. 7 that the energy in the core E_c is

$$E_c = \frac{1}{2} \frac{B^2}{\mu_c} \times A_e l_e \quad J$$

that is,

$$E_c = \frac{1}{2} \frac{B^2}{\mu \mu_0} \times V_e \quad J$$

Here μ_c is the absolute permeability of the core, μ_0 is the (absolute) permeability of air, and μ is the relative permeability of the core material.

Note that it is often more useful to talk in terms of energy per unit volume, that is,

$$\frac{E_c}{V_e} = \frac{1}{2} \frac{B^2}{\mu \mu_0} \quad J/m^3$$

For a typical ferrite, assuming the relative permeability is about $\mu = 2000$ and the saturation flux density $B_{SAT} = 0.3\ T$ (3000 G), we get for most *ungapped* ferrite cores a typical power density of

$$\frac{E_c}{V_e} = \frac{1}{2} \times \frac{0.3^2}{2000 \times 4\pi \times 10^{-7}} = 17.91\ J/m^3$$

$$= 17.91\ J/m^3 \quad \text{or simply } 18\ \mu J/cm^3 \quad \text{or} \quad 18\ J/m^3$$

This is for ferrites, assuming a maximum B-field of 0.3 T (3000 G) and a relative permeability of 2000.

It is the energy stored inside the core (the magnetic material), for either a gapped or an ungapped core. It is also the total overall energy stored in an ungapped core. But in a gapped core, there is typically nine times the preceding value of stored energy (inside the core), now stored separately in the air gap too—giving a total stored energy of about 10 times the preceding figure. We look into that now.

General Energy Relationships for a Gapped Core

From Chap. 7, here are the summarized relationships of the energy components in terms of z (the air gap factor):

$$\frac{E}{E_c} = z \qquad \frac{E_g}{E_c} = z - 1 \qquad \frac{E_g}{E} = 1 - \frac{1}{z}$$

Here, E is the total energy, E_c is the energy in the core, E_g is the energy in the gap, and z is defined as

$$z = \frac{l_e + \mu l_g}{l_e}$$

In a typical Flyback transformer, we set z to about 10. So we get

$$E = zE_c \approx 10 E_c$$

that is, $E \approx 180\ J/m^3$ for ferrite-based Flyback transformers

General Relationships for A_L and μ

The inductance index A_L varies as $1/z$:

$$A_L = \frac{1}{z}\left(\frac{\mu\mu_0 A_e}{l_e}\right) \times 10^9 \quad \text{nH/turns}^2$$

We can solve for the relative permeability of the material μ:

$$\mu = \frac{A_{L\_nogap} \times l_{e\_mm} \times 10}{4\pi \times A_{e\_sqmm}}$$

Duty Cycle of Universal-Input Flyback with V_{OR} = 100 V

Now we will take up a typical design case for an off-line Flyback and come up with equations to help speed up the design process.

The inductor design must proceed at the *lowest input* for this topology (as for the Buck-boost and Boost too), since the currents are highest when D is closer to 1. Assuming 100 percent efficiency, the average inductor current (in the *equivalent primary-side Buck-boost* model, see Fig. 6.6) is

$$I_{COR} = \frac{I_O/n}{1-D}$$

where n is the turns ratio N_P/N_S. The duty cycle is

$$D = \frac{nV_O}{nV_O + V_{IN}}$$

Then 90 Vac gets rectified to a peak of $90 \times 1.414 = 127$ V. Call this V_{IN} (ignore normal input ripple). So

$$D = \frac{nV_O}{nV_O + V_{IN}} = \frac{20 \times 5}{20 \times 5 + 127} = 0.44$$

In other words, we get $D = 0.44$ for the case where we have 90 Vac and $V_{OR} = 100$ V.

The Area × Turns Rule

Since a Flyback transformer design is always done at minimum V_{IN}, then, for the case of *any* worldwide input (90 to 270 Vac) power supply with a reflected output voltage V_{OR} (i.e., nV_O) of 100 V, working at a frequency of 100 kHz, with an optimum r of 0.4, we get from the preceding equation (and solution)

$$NA_e = \left(\frac{2}{r}+1\right) \times \frac{V_{IN}D}{2 \times B_{PK} \times f} = \left(\frac{2}{0.4}+1\right) \times \frac{127 \times 0.44}{2 \times 0.3 \times 10^5}$$

$$\cong 5.588 \times 10^{-3} \text{ m}^2$$

So irrespective of the core shape, core size, air gap, or even power, we need to set the following design target for any universal input Flybacks, which are "optimally designed" ($r = 0.4$), and use a ferrite transformer:

$$(\text{Area}_{mm^2} \times \text{turns}) \approx 5600 \text{ mm}^2$$

Keep in mind this is for $r = 0.4$, $V_{OR} = 100$ V, $B_{SAT} = 0.3$ T, $f = 100$ kHz and 90 Vac.
Or equivalently, for any general frequency f in hertz, in terms of square meters:

$$(\text{Area}_{m^2} \times \text{turns} \times f_{Hz}) \approx 560 \text{ m}^2$$

Worked Example (Part 1)

A 90-to 270-Vac (universal-input) Flyback with an output of 5 V @ 5 A and with a transformer turns ratio of 20 needs a core. Suggest a suitable candidate. The efficiency is 70 percent, and the switching frequency is 100 kHz.

Required L

We know from Chap. 9 that assuming perfect efficiency, the center of the current ramp is

$$I_{COR} = \frac{I_{OR}}{1-D}$$

We now increase the current to account for the less than perfect efficiency (70 percent). Note that the center of the switch current ramp is the same as the average inductor current of the *equivalent primary-side Buck-boost model*. So

$$I_L \equiv I_{COR} = \frac{5/20}{1-0.44} \times \frac{1}{70\%} = 0.64 \text{ A}$$

We set the current ripple to ±20 percent (i.e., $r = 0.4$). So the required AC ramp is

$$\Delta I = r \times I_L = 0.4 \times 0.64 = 0.26 \text{ A}$$

The required inductance is derived from $V = L\,dI/dt$ (conveniently applied for the ON-time):

$$L = \frac{V_{IN} \times D/f}{\Delta I} = \frac{127 \times (0.44/100,000)}{0.26}$$
$$L = 2.15 \text{ mH}$$

This is the inductance calculated from the required r.

Required Energy

The peak current is

$$I_{PK} = I_{COR} + \frac{\Delta I}{2} = 0.64 + 0.13 = 0.77 \quad \text{A}$$

So the required energy handling capability of the core is

$$E_{PK} = \frac{1}{2}LI_{PK}^2 = \frac{1}{2} \times 2.15 \times 10^{-3} \times 0.77^2 \quad \text{J}$$
$$= 6.37 \times 10^{-4} \quad \text{J}$$

A little later we will qualify this and try to indicate why we should be very cognizant of the set current limit while selecting the core.

Pick a Core (Energy-Storage Considerations)

Let us first just examine a popular candidate for this level of power, the E25/13/7 (EF25) core set. Later we will discuss what makes such a core a suitable choice. Its key parameters are given by its datasheet as $A_e = 52.5 \text{ mm}^2$, $l_e = 57.5 \text{ mm}$, and $A_L = 2000 \text{ nH/turns}^2$.

For this core $V_e = A_e \times l_e = 3.0 \text{ cm}^3$. So its energy handling capability (if ungapped), is

$$E_c = 18 \times \frac{3}{100^3} = 5.37 \times 10^{-5} \quad \text{J}$$

We have calculated that we need to store 6.37×10^{-4} J in our application (to avoid saturation at 90 Vac). So the balance of the energy, i.e., $6.37 - 0.537 \Rightarrow 5.833 \times 10^{-4}$ J must be able to reside in the air gap. If we cannot make this happen for any reason, for example, too much copper required to be accommodated in the available window, or an air gap that is clearly an absurdly large (impractical) value, we will certainly need to move to another core choice. For now

$$z = \frac{E}{E_c} = \frac{6.37}{0.537} = 11.86$$

But this is still a reasonable and practical value, since z can *typically vary between the values of 1 and 10*, and in some cases it is even set as high as 25.

N from Voltage-Dependent Equation

Our frequency is 100 kHz, and our $V_{OR} = V_O \times n = 5 \times 20 = 100$ V, so for an optimal design we can simply use our typical equation:

$$(\text{Area} \times \text{turns}) \approx 5600 \text{ mm}^2$$

$$N = \frac{5600}{52.5} = 107 \text{ turns} \quad \text{(Primary)}$$

Required A_L

We have calculated that the required value L is 2.15 mH. Since N is known, the calculated A_L is

$$A_L = \frac{2.15 \times 10^6}{107^2} = 188 \text{ nH/turns}^2$$

NOTE *To meet holdup time requirements the peak energy handling capability (possibly the core size) may have to be increased to prevent saturation under dropouts (or even during a normal power-down sequence). But another option is to simply set r to 0.4 at the lowest operating voltage (say 60 V as discussed in Chap. 5). The reason is the Flyback transformer only has to store all the energy going out to the load, no more, no less. But there is a shape factor consideration (based on r) superimposed on the average energy throughput requirement that affects the peak energy storage requirement. So if r remains the same at lowest voltage, there is no reason for the core to grow in size unless we increase the output power.*

Required z from Inductance Considerations

We know that A_L varies as $1/z$, so z can be calculated from this too:

$$z = \frac{A_{L\_nogap}}{A_{L\_gapped}} = \frac{2000}{188} = 10.64$$

We notice that we have different values of z from two calculations. Which one is right? The energy relationship $E_c/V_e = 17.9$ J/m³ was clearly a slight approximation and was provided only to speed up initial core selection. But also, in particular, it was based on an assumption that $\mu = 2000$. As a matter of fact, we can reverse-calculate the permeability from the more accurate value of z just calculated, and we will see that the calculated permeability is close to, but not quite equal to 2000. That explains the slight discrepancy between the z values. In the calculations, we finally used the vendor's actual (measured) value for A_L.

Size of Air Gap

$$l_g = \frac{l_e(z-1)}{\mu} = \frac{57.5 \times (10.64 - 1)}{1743} = 0.32 \text{ mm}$$

Note that the gap length is $0.32/57.5 \cong 0.6$ percent of the effective length. This is a reasonable number, since up to 2 to 3 percent is considered practical.

Permeability of the Core

$$\mu = \frac{A_{L\_nogap} \times l_{e\_mm} \times 10}{4\pi \times A_{e\_sqmm}} = \frac{2000 \times 57.5 \times 10}{4\pi \times 52.5} = 1743$$

If the designer is surprised that we have first calculated the air gap and now we are finding the permeability, he should remember that the permeability was in fact implicit in the A_L value provided for the core and used by us. So this is just a check on our calculations.

We can now see that $\mu = 1743$ is a little different from the value of 2000 we had assumed for the quick energy estimate equation. But we can see that E_c/V_e is inversely proportional to μ. So had μ been taken as 1743 instead of 2000, the energy estimate equation would have been

$$\frac{E_c}{V_e} = 17.91 \times \frac{2000}{1743} = 20.55 \text{ J/m}^3$$

This is for the actual $\mu = 1743$ and $B_{SAT} = 0.3$ T. And for our example, the corresponding z would be proportionally less from the earlier calculated value, that is,

$$z = 11.86 \times \frac{1743}{2000} = 10.34$$

The slight remaining discrepancy between the z calculated here from energy considerations, and the z calculated previously using inductance is explained by the fact that the equation used to find N was a little approximate too. Plus we had also rounded up to the nearest integral value of turns. But, either way, the calculated values are well within normal tolerances on air gap and A_L anyway. There is no real inconsistency here.

Some Finer Points of Optimization

We bullet out some useful points here to think about.

- We observe that the number of (primary) turns required varies as A^{-1}; that is, it is inversely proportional to the effective area of the core.
- We can also show that the total length of the primary winding then varies as $A^{-1/2}$ (copper losses will vary similarly).
- We have been assuming that it is possible to set $r = 0.4$ and B_{PK} close to 0.3 T (3000 G), *simultaneously*. That may, however, not be possible in practice.
- In a Flyback, with a turns ratio of say 20, we get a reflected output voltage V_{OR} of 100 V for a 5 V output. We know from Chap. 6, that for this topology, V_{OR} is a major design goal and is usually set in the range of 70 to 130 V for worldwide input off-line power supplies. But since the secondary number of turns must be an integer, the corresponding number of primary turns will vary in discrete steps of *20 turns for each additional secondary turn we decide to use*, thus denying us a smooth "continuum" of possible values. The inductance too will jump similarly in steps related to its A_L. Therefore, from the practical form of the voltage-dependent equation we can conclude the following:
  - For a given core, if we try to set B_{PK} close to B_{SAT}, we can expect that r will have to be allowed to jump in corresponding steps, and therefore we may or may not be able to set it close to the "optimum" of 0.4.
  - If we try to set r to say 0.4, B_{PK} will necessarily jump in steps, and we may or may not be able to set it to the optimum value (close to B_{SAT}).
  - Or we need to be a little flexible about the desired V_{OR}, as this will allow us to use slightly different values for the turns ratio. This is the most practical option.
- The voltage-dependent equation (its practical form) has provided us the required number of primary turns merely on the basis of the A_e of the core. The A_L, l_e, and air gap have not even entered the picture.
- As a corollary, since ultimately we will always be trying to operate close to B_{SAT}, the number of primary turns is essentially fixed for a given core in a given application, (for a target r). This asks us to put *neither more, nor less turns* than calculated.
- Note that though the number of turns required does not depend directly on the inductance, we will need to adjust the air gap to get the inductance to be such that r is close to our design target of about 0.4.
- If we allow for a different r, we can vary the number of primary turns. In fact, sometimes we may even be forced to do this if for example 107 turns cannot be physically accommodated in the available window area of the chosen core (and for some reason we don't want to go to a core with a larger window). We must, however, remember that if we do so, we have side effects from the higher r on the rest of the converter (slightly nonoptimum value).
- Once r is fixed, so is the required inductance for a given frequency (and given application). Therefore, from the basic definition of L given previously, we can then use our knowledge of N, μ, l_e, and A_e to calculate the required z (and air gap).
- We must double-check that the required number of turns can be accommodated in the available core window area.
- It is interesting to note that if r is fixed by design, and A_e is also fixed (i.e., we are sticking to a given core), then if we *also* try to fix N, we will see that the value of B_{PK} is no longer directly in our hands. In such a case we may end up not even operating close to B_{SAT}.
- Therefore, the bottomline is: We just don't have so many available *degrees of design freedom*. We have to compromise somewhere, either on the design goal for V_{OR}, on the r, or on the B_{PK}. We just cannot have everything here. That is why transformer design is so tricky (and rewarding at the same time). It is a tradeoff too.

Rule of Thumb for Quick Selection of Flyback Transformer Cores

We saw that for ferrites, the typical energy density in the core material (with certain assumptions) is

$$\frac{E_c}{V_e} \approx 18 \text{ J/m}^3$$

By adding an air gap, the energy increases as per

$$\frac{E}{E_c} = z$$

We have seen that a practical value of z is about 10. So

$$\frac{E}{V_e} \approx 180 \ \text{J/m}^3$$

This is also 1.8×10^{-4} J/cm^3. Or 180 µJ/cm^3. However we must remember that this is *not* the same parameter that most magnetics vendors provide when they draw *Hanna* curves. What they use is actually LI^2/V_e, whereas we are working with the actual physical energy density term $(1/2) \, LI^2/V_e$. However, the conversion is clear. So our rule of thumb can also be written in a form that can be better compared with a vendor's datasheet.

$$\frac{LI^2}{V_e} \approx 3.6 \ \text{J/cm}^3$$

We thus see that our rule of thumb is actually very close to the official guideline of 3.5 J/cm^3 provided for ferrite material-77 from Fair-Rite, on the basis of its Hanna curves.

Now relating the peak energy $(1/2) \times LI^2$ to the input watts $(V_{IN} \times I_{SW\_AVG} \times D)$ for a Flyback, we get a rather complicated expression:

$$P_o = \frac{8r \times (E/V_e) \times f \times V_e}{(2+r)^2} \ \text{W}$$

where E is the total energy stored in the entire gapped structure (in joules), V_e is the volume in m^3, and f is the frequency in hertz. Note however that *both D and V_{IN} have gotten cancelled out* in this relationship.

For ferrites we just saw that the maximum energy density E/V_e is about 180 J/m^3, so setting $r = 0.4$, we get a very simple relationship for predicting a ballpark figure for the power capability of a typical ferrite-based Flyback transformer:

$$P_o \approx 100 \times f \times V_e$$

Correcting this for the case of nonideal efficiency, we get

$$P_{IN} = 100 \times f \times V_e$$

This was in MKS units. Alternatively expressed, and in a slightly different manner

$$V_{e\_cm^3} = \frac{0.01 \times P_{IN}}{f_{MHz}}$$

This has been derived in Chap. 5 of *Switching Power Supplies A-Z*, 2d ed. It was also pointed out that the simplicity of this equation belies the fact that it is actually an exact equation. The only assumptions are the permeability is 2000, the peak saturation flux density (B) is 3000 G, the current ripple ratio is set to 0.4, and the air gap factor is 10. All these are are very reasonable entry points. We usually set this at the worst-case design point which is 90 Vac.

Worked Example (Part 2)

Let us use our newly derived quick-selection equation too. The EF 25 has a volume of 52.5 × 57.5 = 3019 mm^3. So at 100 kHz we get from the preceding equation set

$$P_o \approx 100 \times 10^5 \times 3019 \times 10^{-9} = 30.2 \ \text{W}$$

This means that the chosen core (EF 25) is almost certainly OK up to about 30-W input or output. Thereafter, if we choose copper wire with a thickness just enough to give us a current density of about 400 cmils/A, we will probably never run into a situation where we cannot accommodate the required copper in the available window area (for most commercial transformer cores), or where we have an overheated transformer. A discussion about current density follows next. Much more on that topic is in Chap. 12.

A common error is assuming that the core for DCM or BCM is larger than in CCM. If we increase r to 2 (boundary/critical conduction), we can actually almost double the throughput

262 Chapter Ten

power capability of a given core. This happens because the term $8r/(2+r)^2$ changes from 0.56 at $r = 0.4$ to 1 at $r = 2$—an increase of almost 100 percent. How does this correlate to our understanding that the size of the core is related to I_{PK}^2? If we decrease L, we increase r and I_{PK}. However, energy is proportional to LI_{PK}^2, so since L is proportional to N^2, energy comes down despite the fact that the peak current has gone up. We can also think in terms of ampere-turns. This is related to the energy in the core. Though we increase the (peak) amperes, the turns have to be reduced at a faster rate, so the net ampere-turns actually decrease. But remember, we may have to oversize other components on the board. That is why an optimum of $r = 0.4$ is usually suggested.

Impact of Current Limit on Core Size

In typical DC-DC converter applications, we traditionally calculate the maximum operating current and pick the inductor. For example, for a 12-to 5-V Buck, with a 5-A maximum load we pick any inductor of over 5-A rating. Simple enough. But when it comes to high-voltage applications, any momentary core saturation can cause such a steeply rising current, that even a fast-acting current limit may not be effective in arresting the rise of current and saving the FET. See the top half of Fig. 10.1. So though we may pick our core based on a maximum of 3000 G at our peak *operating* current as we did in the solved examples, is that completely right? The key question to really ask is: *What happens during an overload or even a simple start-up scenario?* Or

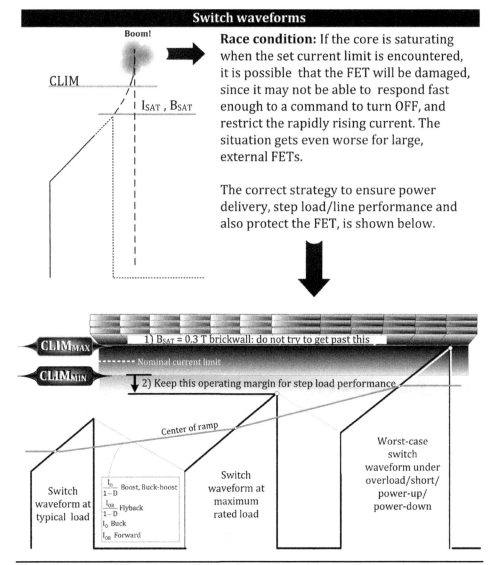

Figure 10.1 How a FET can blow up if saturation occurs, and the best way to select the core and set the current limit.

worse, during *power-down*, where we usually set our undervoltage lockout (UVLO) at around 60 Vdc to achieve required holdup time as discussed in Chap. 5—and therefore demand the converter not only keeps switching, but *regulating* too (with no foldback in output power). We must recognize that we *will* regularly hit the set current limit, perhaps momentarily, but the duration is completely immaterial here. We must ask: *Is the core designed to accept that high current (and field) for even one cycle?* As mentioned, in DC-DC converters, because of the *low voltages* involved, we literally "get away" every time the core saturates for a few cycles. But in AC-DC applications we have to be very clear what is the worst-case *B*-field in the core *under any condition*, even for one cycle, as that is potentially destructive.

The safest approach is simply to *not use the peak operating flux density* (*B*-field) but the maximum flux density related to the *maximum of the current limit spread*. The operating current just does not matter anymore, except for setting the wire gauge and current density. We should, therefore, set I_{PK} in our preceding calculations as the maximum of the current limit spread, which we call $CLIM_{MAX}$ here. We would ensure that *B* does not exceed 3000 G at $CLIM_{MAX}$. See the bottom half of Fig. 10.1.

The importance of $CLIM_{MIN}$ is that we must ensure, say at the desired UVLO point of 60 Vdc, that our calculated peak operating current is less than that of $CLIM_{MIN}$; otherwise, we will not be able to guarantee full power.

In other words,

1. We should set our current limit precisely so that the peak operating current is a little less than the set $CLIM_{MIN}$—over the entire desired input operating range (down to the set UVLO, typically 60 Vdc, at maximum load).

2. We should pick the core size based on $CLIM_{MAX}$. Use a suitable worst-case peak *B*-field. Is it 3000 G? More on this shortly.

3. We can certainly pick wire gauges based on slower-acting *thermal considerations*, that is, based on maximum load at nominal line (90 Vac).

The first two points here indicate it is very important to have *tight current limiting* (with minimum spread). Because if we think about it, we will realize that *the region between $CLIM_{MIN}$ and $CLIM_{MAX}$ actually represents wasted and unnecessary core material*. So for the same power, if the spread in the current limit is too much, we would still ensure $CLIM_{MIN}$ is where it is shown in Fig. 10.1. Unfortunately, $CLIM_{MAX}$ would move up much higher, demanding a larger core.

Vendors of some popular off-line integrated switcher ICs have often used the preceding logic to mask a major limitation of their devices, perhaps unknowingly. They do advertise their very tight current limit for exactly the reasons we have stated. But the problem is that they can only offer a family of devices with a certain limited set of discrete choices for the current limit (internally set), as opposed to controllers where we can set virtually any current limit that we like. This means that if, for example, the vendor has a 2-A (current limit) device and the next higher device is a 3-A device, then we are forced into using the latter device even for an application where the peak current is, say, only 2.2 A. How does a tight current limit help us here anyway? In effect, the current limit spread is very wide for the 2.2-A application.

How do we expect core size to scale in general? We know core size is proportional to energy and thus power. So, for a given output voltage we expect core size to be proportional to peak current. Therefore, since 2.2 A/3 A = 0.73, we expect the core size for the 2.2-A application to be 27 percent smaller than that required for a 3-A application (assuming all else is unchanged such as *r* and V_{OR}). But our inability to set the current limit precisely at 2.2 A means we cannot reduce the core size from the 3-A level either. One way to do that, however, is by changing our basic assumptions on *r* and V_{OR}. Therefore, we can force the 2.2-A application have a smaller transformer, but that is cheating in a sense. Because all we may have done to achieve this is to have increased *r* and/or V_{OR}, and thereby simply transferred the problem over into the capacitors (larger RMS currents) and so on. We should be cognizant about all these subtleties in transformer design and in component selection.

Design software from the IP-switch company does some brinksmanship here. The software permits a *peak flux density* of up to 4200 G for a few (undefined) number of cycles, say as the Flyback converter powers down. This is, however, in our opinion, not really a good design practice, especially for Flybacks with external FETs, which are a little slower in turning OFF under a rapidly rising current overload. One option is to simply design for a lower flux density at 90 Vac, so that at 60 Vdc we would still not exceed 3000 G. This would require a larger core quite naturally. Another option is to set *r* = 0.4, not at 90 Vac rectified, but at 60 Vdc (or whatever is the set UVLO level for meeting holdup). With that approach, the core would be no bigger than a core designed for *r* = 0.4 at say 90 Vac (for the same output power). For a general *r*, we can use the core selection equation provided further up, the one involving *r*.

Chapter Ten

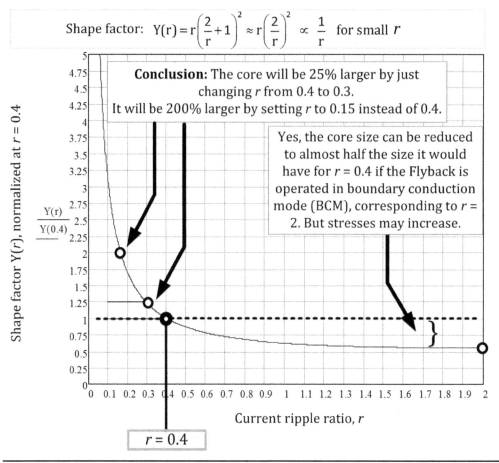

FIGURE 10.2 How current ripple ratio affects core size in a Flyback (or any topology).

The reason for this is that, as derived in *Switching Power Supplies A-Z*, 2d ed, the general equation for a Buck-boost is

$$V_{e\_cm3} = \left(\frac{31.4 \times P_{IN} \times \mu}{z \times f_{MHz} \times B^2_{SAT\_Gauss}}\right) \times \left[r \times \left(\frac{2}{r}+1\right)^2\right] \equiv (X) \times (Y)$$

This has a term X, which for a given material, frequency, and air gap factor only depends on power (which we assume has not changed here), multiplied by another *shape factor*, Y, which is purely r-dependent. So the required volume for a given power level, all else unchanged, only depends on r. What is the shape of Y above? We plot this out in Fig. 10.2. We see that $r = 0.4$ is the "knee" of the energy curve and represents an optimum of sorts. Going to BCM has the potential of halving the core size, but slight reductions in r (by increasing L) will cause a drastic change in core size. Of course, if we do not size the core correctly, as indicated by the figure, we will saturate the core. Compare with Fig. 10.2.

Circular Mils (cmil)

For fixing the wire gauge we must be familiar with circular mils. We know that 1000 mils = 1 in = 25.4 mm. So 1 mil is also 25.4 µm (µm is also called a *micron*).

A circular mil (cmil) is a term used to define conductor cross-sectional areas using an arithmetic shortcut, in which the area of a round wire is taken as d^2 in units of cmils, rather than the "correct" form $\pi d^2/4$ in units of square mils (mil^2) (d is the diameter here). Thus *1 cmil is equivalent to $\pi/4$ mil^2*. See Fig. 10.3. The rationale behind introducing this *shortcut* is that, for example, when we consider the case where we stack wires on top of each other to form a winding arrangement, the net area occupied by a wire is in effect the entire square cross-sectional area bounding that wire strand, as shown in the figure. The rest is wasted space. The area allocated to each turn is, however, d^2. So in this case, *the cross-sectional area*

Basics of Flyback Transformer Design

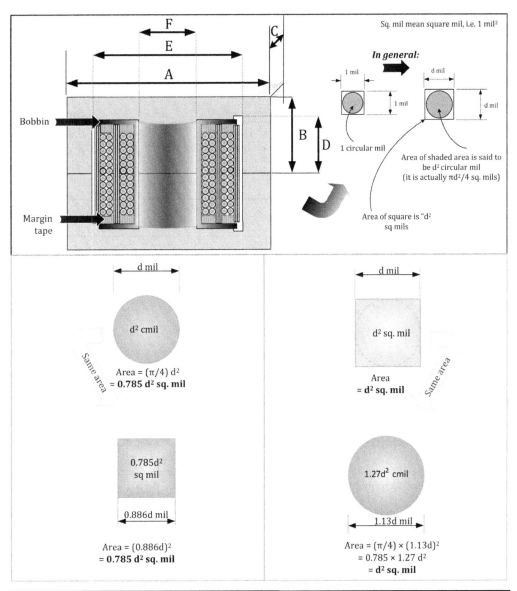

FIGURE 10.3 Circular mil explained.

expressed in cmils is numerically the same as the square mils occupied. For almost all other considerations this shortcut can become rather confusing, and we should be cautious. Conversions are also provided within Fig. 10.3 and listed out in Table 10.1.

NOTE *Some turns do settle into the spaces between turns on the layer below them, but that is unpredictable and so we are ignoring it here. Besides we usually have interlayer insulation to prevent that from happening.*

Current Carrying Capacity of Wires

The following two expressions are often used for calculating the diameter in mils for AWG (American Wire Gauge, also called the Brown & Sharpe or simply B&S wire gauge). The first form is easier to remember; the second is accurate and is the one we have used for generating the tables:

$$d_{\text{mils}} = \frac{1000}{\pi} \times 10^{-\text{AWG}/20}$$

Length	mil	in	ft	mm
mil	1	0.001	0.000083	0.0254
in	1,000	1	0.083	25.4
ft	12,000	12	1	304.8
mm	39.37	0.03937	0.0033	1
Area	**cmil**	**square mil²**	**in²**	**cm²**
cmil	1	$\pi/4 = 0.7854$	7.854×10^{-7}	5.067×10^{-6}
square mil²	$4/\pi = 1.274$	1	10^{-6}	6.452×10^{-6}
square mm²	1,973	1,550	0.00155	0.01
Current density*	**cmil/A**		**A/cm²**	**A/in²**
x cmil/A	x		$197,353/x$	$1,273,000/x$
y A/cm²	$197,353/y$		y	$6.45y$

*440 cmil/A ≅ 440 A/cm², 1000 cm/A ≅ 1270 A/in²

TABLE 10.1 Conversions for Diameter/Length, Area, and Current Density

or

$$d_{\text{mils}} = 5 \times 92^{(36-\text{AWG})/39}$$

For the range of wire diameters in common use for power supplies, there is almost no difference between these two expressions. But we must remember that either way, what we get here is the diameter of the *bare* copper wire, *excluding* any insulation or coating.

NOTE *It is possible to buy half-integral AWG sizes too for critical applications. However, many power supply companies just stock either even or odd number AWG sizes for logistical ease and lower inventory costs.*

There are several issues related to a suitable choice of AWG in switching converter transformers and inductors as we now consider:

- In inductors (single winding) the current is relatively smooth and has fairly low high-frequency content. So technically speaking, we could, for example, use a single strand of say AWG 10 (very thick wire) for a large current. There are, however, some (design for manufacturability) (DFM) issues to consider. Thick wire is harder to wind, so it will usually be replaced by several strands twisted together.

- In transformers we can have one winding suddenly stop carrying current, while one or more windings freewheel. As far as the core is concerned, it does not "know" the difference, since all it demands is that we *allow* the total ampere-turns (summed over all the windings on the given core) to remain continuous (no sudden step changes). But as far as each particular winding is concerned, the amperes (or ampere-turns) of that winding can certainly jump discontinuously. This gives us the commonly seen trapezoidal or rectangular current waveforms seen in the windings of a transformer. This also means that we now have a large high-frequency Fourier content, and therefore we have to be concerned about *skin depth*, something that doesn't bother us much in (single-winding) inductors. What this means is that as we increase the diameter of the wire beyond a certain point (twice the skin depth), the high-frequency current stays restricted in an annular surface region of the wire. Though the cross-sectional area "available" for conduction does increase thereafter, it no longer varies as d^2 but as d. So it is a situation of rapidly diminishing returns. What we need to do for higher currents is to use several strands of insulated wire, each strand diameter equal to roughly twice the skin depth. Standard Litz wire used for radio-frequency applications is not suited for commercial switching power supplies because of the wasted space due to the silk or textile covering the insulation, etc. Instead we can either make in-house, or order directly from several vendors, multistrand twisted or braided bundles of standard AWG magnet wire (also increasingly called Litz wire nowadays).

Skin Depth

Skin depth is defined as the distance from the surface where the current density has fallen by a factor $1/e$ from the value at the surface (continuing to fall exponentially as we go deeper). But the high-frequency resistance (and the loss) is the same as if the entire current was distributed uniformly up to a depth equal to the skin depth, falling abruptly to zero thereafter. We can integrate an exponential curve to see that this is true. This description of skin depth as an annular region of uniform current density, equal to the density we actually have on the surface of the inductor, leads to much easier computations.

Skin depth in mils for copper is typically presented as $2837/f^{1/2}$, and this is the equation we too have used for generating the design aids that follow. Note that in this equation f (in hertz) is rather loosely taken to be the switching frequency, and the diameter of the wire then set to *twice* this value. However, in reality we should consider *all the harmonics* that go into making up the rectangular/trapezoidal current waveform. Recommendations on what the diameter of the wire should really be vary somewhat in literature.

A more complete form of skin depth of copper expressed in milimeters is

$$\text{Skin depth} = \frac{66.1 \times [1 + 0.0042(T - 20)]}{\sqrt{f}} \text{ mm}$$

where T is the temperature of the winding in degree Celsius and f is in hertz. Note that the 0.0042 comes from the fact that the resistance of any copper trace or winding increases by 4.2 percent every 10°C rise in temperature.

For example, at 70 kHz, the skin depth is 0.27 mm. From Table 10.1 we see that 1 mm is 39.37 mil. So 0.27 mm is $39.37 \times 0.27 = 10.6$ mil. We should not choose a wire of a diameter, or a foil of thickness, more than twice the skin depth (20 mil in this case), because that is an inefficient use of copper. From Table 10.2 we see that AWG 24 has a diameter of 20 mil and is the right choice

Gauge No.	AWG (in.)	AWG (mm)	SWG (in.)	SWG (mm)	Gauge No.	AWG (in.)	AWG (mm)	SWG (in.)	SWG (mm)
0	0.3249	8.25	0.324	8.23	23	0.0226	0.574	0.024	0.61
1	0.2893	7.35	0.3	7.62	24	0.0201	0.511	0.022	0.559
2	0.2576	6.54	0.276	7.01	25	0.0179	0.455	0.02	0.508
3	0.2294	5.83	0.252	6.4	26	0.0159	0.404	0.018	0.457
4	0.2043	5.19	0.232	5.89	27	0.0142	0.361	0.0164	0.417
5	0.1819	4.62	0.212	5.38	28	0.0126	0.32	0.0148	0.376
6	0.162	4.11	0.192	4.88	29	0.0113	0.287	0.0136	0.345
7	0.1443	3.67	0.176	4.47	30	0.01	0.254	0.0124	0.315
8	0.1285	3.26	0.16	4.06	31	0.0089	0.226	0.0116	0.295
9	0.1144	2.91	0.144	3.66	32	0.008	0.203	0.0108	0.274
10	0.1019	2.59	0.128	3.25	33	0.0071	0.18	0.01	0.254
11	0.0907	2.3	0.116	2.95	34	0.0063	0.16	0.0092	0.234
12	0.0808	2.05	0.104	2.64	35	0.0056	0.142	0.0084	0.213
13	0.072	1.83	0.092	2.34	36	0.005	0.127	0.0076	0.193
14	0.0641	1.63	0.08	2.03	37	0.0045	0.114	0.0068	0.173
15	0.0571	1.45	0.072	1.83	38	0.004	0.102	0.006	0.152
16	0.0508	1.29	0.064	1.63	39	0.0035	0.089	0.0052	0.132
17	0.0453	1.15	0.056	1.42	40	0.0031	0.079	0.0048	0.122
18	0.0403	1.02	0.048	1.22	41	0.0028	0.071	0.0044	0.112
19	0.0359	0.912	0.04	1.02	42	0.0025	0.064	0.004	0.102
20	0.032	0.813	0.036	0.914	43	0.0022	0.056	0.0036	0.091
21	0.0285	0.724	0.032	0.813	44	0.002	0.051	0.0032	0.081
22	0.0253	0.643	0.028	0.711	45	0.0018	0.046	0.0028	0.071

TABLE 10.2 Bare Copper Wire Diameters for SWG and AWG

268 Chapter Ten

AWG	Current @ 1000 cmil/A or 197 A/cm²	Current @ 900 cmil/A or 219 A/cm²	Current @ 800 cmil/A or 247 A/cm²	Current @ 700 cmil/A or 282 A/cm²	Current @ 600 cmil/A or 329 A/cm²	Current @ 500 cmil/A or 395 A/cm²	Current @ 400 cmil/A or 493 A/cm²	Current @ 250 cmil/A or 789 A/cm²	Current @ 200 cmil/A or 987 A/cm²
10	10.383	11.537	12.979	14.833	17.305	20.766	25.958	41.532	51.915
11	8.2341	9.149	10.293	11.763	13.724	16.468	20.585	32.936	41.171
12	6.5299	7.2555	8.1624	9.3285	10.883	13.06	16.325	26.12	32.65
13	5.1785	5.7539	6.4731	7.3978	8.6308	10.357	12.946	20.714	25.892
14	4.1067	4.563	5.1334	5.8667	6.8445	8.2134	10.267	16.427	20.534
15	3.2568	3.6186	4.071	4.6525	5.428	6.5136	8.142	13.027	16.284
16	2.5827	2.8697	3.2284	3.6896	4.3046	5.1655	6.4569	10.331	12.914
17	2.0482	2.2758	2.5603	2.926	3.4137	4.0964	5.1205	8.1928	10.241
18	1.6243	1.8048	2.0304	2.3204	2.7072	3.2486	4.0608	6.4972	8.1215
19	1.2881	1.4313	1.6102	1.8402	2.1469	2.5763	3.2203	5.1525	6.4407
20	1.0215	1.135	1.2769	1.4593	1.7026	2.0431	2.5538	4.0861	5.1077
21	0.8101	0.9001	1.0126	1.1573	1.3502	1.6202	2.0253	3.2405	4.0506
22	0.6424	0.7138	0.8031	0.9178	1.0707	1.2849	1.6061	2.5698	3.2122
23	0.5095	0.5661	0.6369	0.7278	0.8491	1.019	1.2737	2.0379	2.5474
24	0.404	0.4489	0.5051	0.5772	0.6734	0.8081	1.0101	1.6162	2.0202
25	0.3204	0.356	0.4005	0.4577	0.534	0.6408	0.801	1.2817	1.6021
26	0.2541	0.2823	0.3176	0.363	0.4235	0.5082	0.6353	1.0164	1.2705
27	0.2015	0.2239	0.2519	0.2879	0.3359	0.403	0.5038	0.8061	1.0076
28	0.1598	0.1776	0.1998	0.2283	0.2663	0.3196	0.3995	0.6392	0.799
29	0.1267	0.1408	0.1584	0.181	0.2112	0.2535	0.3168	0.5069	0.6337
30	0.1005	0.1117	0.1256	0.1436	0.1675	0.201	0.2513	0.402	0.5025
31	0.0797	0.0886	0.0996	0.1139	0.1328	0.1594	0.1993	0.3188	0.3985
32	0.0632	0.0702	0.079	0.0903	0.1053	0.1264	0.158	0.2528	0.316
33	0.0501	0.0557	0.0627	0.0716	0.0835	0.1003	0.1253	0.2005	0.2506
34	0.0398	0.0442	0.0497	0.0568	0.0663	0.0795	0.0994	0.159	0.1988
35	0.0315	0.035	0.0394	0.045	0.0525	0.063	0.0788	0.1261	0.1576
36	0.025	0.0278	0.0313	0.0357	0.0417	0.05	0.0625	0.1	0.125
37	0.0198	0.022	0.0248	0.0283	0.033	0.0397	0.0496	0.0793	0.0991
38	0.0157	0.0175	0.0197	0.0225	0.0262	0.0314	0.0393	0.0629	0.0786
39	0.0125	0.0139	0.0156	0.0178	0.0208	0.0249	0.0312	0.0499	0.0623
40	0.0099	0.011	0.0124	0.0141	0.0165	0.0198	0.0247	0.0396	0.0494
41	0.0078	0.0087	0.0098	0.0112	0.0131	0.0157	0.0196	0.0314	0.0392
42	0.0062	0.0069	0.0078	0.0089	0.0104	0.0124	0.0155	0.0249	0.0311
43	0.0049	0.0055	0.0062	0.007	0.0082	0.0099	0.0123	0.0197	0.0247
44	0.0039	0.0043	0.0049	0.0056	0.0065	0.0078	0.0098	0.0156	0.0196
45	0.0031	0.0034	0.0039	0.0044	0.0052	0.0062	0.0078	0.0124	0.0155

TABLE 10.3 AWG versus Current Carrying Capacity for Different cmil/A

at 70 kHz. From Table 10.3 we see that for a recommended current density of 400 cmil/A, it is acceptable for up to 1 A.

Example 10.1 A 90- to 270-Vac (worldwide input) Flyback with a turns ratio of 20 needs a transformer. Suggest suitable primary and secondary wire gauges for the several cases listed here:

Case 1. Single output of 5 V @ 5 A (25 W)
Case 2. Single output of 5 V @ 8 A (40 W)
Case 3. Single output of 5 V @ 16 A (80 W)
Case 4. Dual outputs of 5 V @ 10 A and 12 V @ 2.5 (50 + 30 = 80 W)

Basics of Flyback Transformer Design

In all cases assume the efficiency is 70 percent and the switching frequency is 70 kHz.

As we calculated previously, for case 1, that is, 5 V @ 5 A, the switch current ramp (COR value) is 0.64 A at 90 Vac (using the *flat-top* approximation). For case 2, that is, 5 V @ 8 A, we can scale it easily and guess the COR value as $8 \times 0.64/5 = 1.02$ A. Similarly, for case 3, that is, 5 V @ 16 A, it is $16 \times 0.64/5 = 2.05$ A. Case 4 is identical to case 3 as far as the primary side is concerned. As mentioned, from Tables 10.2 and 10.3, we see that AWG 24 is the right choice for 70 kHz.

Note that in all the preceding cases, the duty cycle is 0.44 at 90 Vac as calculated earlier, since we are keeping the turns ratio fixed, and the V_{OR} is clearly 100.

The input RMS current in each case is

1. $0.64 \text{ A} \times \sqrt{0.44} = 0.425$ A. From Table 10.3, a single strand of AWG 27 or 28 is adequate for the Primary.
2. $1.02 \times \sqrt{0.44} = 0.68$ A. From Table 10.3, a single strand of AWG 26 is adequate for the Primary.
3. $2.05 \times \sqrt{0.44} = 1.36$ A. From Table 10.3, two strands of AWG 26 will be adequate for the Primary.
4. Same as case 3 for primary side. Secondary windings will need to use (actual) current values.

For the secondary windings, the (secondary-side) COR value must be such that $I_{COR} \times (1 - D) = I_O$. Further, the RMS value is $I_{COR} \times \sqrt{1-D}$ in this case. So the secondary-side RMS is $[I_O/(1-D)] \times \sqrt{1-D} = I_O/\sqrt{1-D}$. Therefore,

1. For a load current of 5 A, the RMS is 6.682 A. We can take seven strands of AWG 24 (preferably) twisted together.

If we take standard copper foil of *t*-mil thickness, then if the width of the foil is *w* mil, the available square mils is $t \times w$ mil². If we are aiming for close to, or slightly better than, 400 cmil/A,

$$\frac{w \times t}{I_{RMS\_SEC}} \geq 0.785 \times 400 \text{ cmil/A}$$

So, we need a foil of width greater than

$$w \geq \frac{0.785 \times \text{cmil/A} \times I_{RMS\_SEC}}{t} = \frac{0.785 \times 400 \times 6.682}{20} = 105 \text{ mil}$$

where we have used a standard 20-mil foil thickness as per the skin depth recommendations. The term in 0.785 is from Fig. 10.3 since we know that d^2 cmil equals $0.785 \times d^2$ mil²—the allowed current density is expressed in cmil, but since this is a foil, not a round wire, we need to convert that to allowed current density in terms of mil²/A.

We need a foil of width

$$w \geq 0.105 \text{ in}$$

From Table 10.1, we have 1 in is 25.4 mm, so this is $0.105 \times 25.4 = 2.7$ mm

$$w \geq 2.7 \text{ mm}$$

But to this foil width we need to typically add $4 + 4 = 8$-mm margin tape (4 mm on either side). Also, we have a bobbin of typical thickness 1.5 mm, so we should add another 3 mm to this. So we are demanding that the core provides a total window width of 2*D* (see *D* in Fig. 10.3) equal to

$$2D \geq w + 11 = 13.7 \text{ mm} \quad D > 6.85 \text{ mm}$$

The selected core (EF 25) has a *D* of 8.7 mm (see www.epcos.com/inf/80/db/fer_07/e_25_13_7.pdf). So it will work. Though it is somewhat ridiculous to look for a foil of 3-mm width! So we would rather stay with $4 \times$ AWG 24 wire here.

The problem we want to point out here, however, is that the large margin tape requirement may lead us for higher current outputs with foil windings, to *elongated profiles*: that is, core families with large *D* (like EEL/EER cores). Also see Table 12.1 in Chap. 12, for different core families.

2. We can similarly go through the remaining steps.

Graphical Solutions and a Useful Nomogram

In Fig. 10.4 we have provided a useful nomogram for the selection of round wires in particular (though foil thickness information can be gleaned from it too). A numerical example is also embedded in the figure. Note that this nomogram is supposedly for

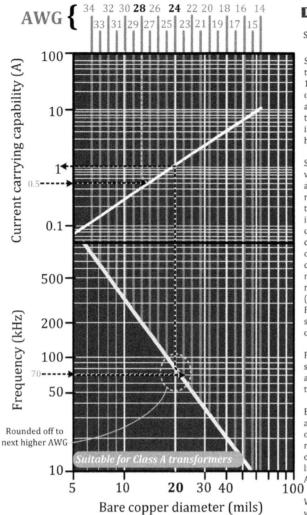

Figure 10.4 Quick estimate of area and diameter at different frequencies.

400 cmil/A, but it uses the *center of ramp* (COR) current value for simplicity in its use. But in terms of RMS, it is actually closer to 565 cmil/A-RMS (for $D = 0.5$). Some have opined that 565 cmil/A-RMS is much too conservative, and they would recommend going down by a wire gauge number or two. This entire topic is well thrashed out in Chap. 12. But keep in mind at this point that it is wiser to err on the conservative side. Thermal issues are one major cause of huge delays in release to production (EMI is another). You *don't* want that. Transformers commonly used in power supplies for information technology equipment are classified either as Class A (less than 65°C temperature rise and that would need to be confirmed by thermocouples inserted into the windings under the rated maximum load), or Class B (similarly measured to have less than an 85°C temperature rise). Note that whereas Class A transformers just require individually approved materials, for Class B transformers all the materials have to be evaluated and approved *together by safety agencies*, thus forming a designated *insulation system*. The latter adds cost, and it is no surprise that many magnetic vendors try to push this on customers by using much more aggressive current densities (saving their costs but producing much higher temperatures). If we really wish to decidedly stick to the lowest-cost Class A transformer, it is a good idea to insist on 400 cmil per amp of COR (center of ramp) current value (565 cmil/A-RMS, not

400 cmil/A-RMS), which though a bit conservative, is a much safer design target. It also leaves some margin for cheaper ferrite materials with higher core losses, rather than shoe-horning us into state-of-the-art ferrites.

A Feel for Wire Gauges

ASTM B-258 specifies that AWG is based on geometric interpolation between gauge 0000, which is 0.46 in exactly, and gauge 36 which is 0.005 in exactly. We must realize that ASTM B-258 also specifies rounding rules that seem to be ignored by makers of most tables (for convenience we will do the same too!). Actually, gauges up to 44 are to be specified with up to four significant figures, but no closer than 0.0001 in Gauges from 44 to 56 are to be rounded to the nearest 0.00001 in.

Note that $92^{1/39} = 1.123$ and is very close to $2^{1/6} = 1.122$, so *diameter is approximately halved for every six-gauges*. In a similar manner we get the following quick rules for the diameter of bare (unclad) copper wire:

- Diameter of no. 36 is 5 mils.
- Diameter changes approximately by a factor of 2 every 6 gauges.
- Diameter changes approximately by a factor of 3 every 10 gauges.
- Diameter changes approximately by a factor of 4 every 12 gauges.
- Diameter changes approximately by a factor of 5 every 14 gauges.
- Diameter changes approximately by a factor of 10 every 20 gauges.
- Diameter changes approximately by a factor of 100 every 40 gauges.

Using the definition of AWG we also see that the diameter of no. 10 is 10,380 cmil. Take this to be almost 10,000 cmil. Then we have the following rules for area (resistance per unit length varies in the same manner, though inversely):

- The area of no. 10 is 10,000 cmil.
- Area changes approximately by a factor of 2 every 3 gauges.
- Area changes approximately by a factor of 10 every 10 gauges.

We can also remember that

- Diameter increases 12 percent every decrease in wire gauge number.
- Area increases 26 percent every decrease in wire gauge number.
- Resistance/length increases 26 percent every increase in wire gauge number.

We now give some examples that give acceptable accuracy for most purposes.

Example 10.2 What is the area of no. 27 wire?

No. 10 → 10,000 cmil ⇒ no. 20 → 1000 cmil ⇒ no. 30 → 100 cmil ⇒ no. 27 → <u>200 cmil</u>

Example 10.3 What is the area of no. 28 wire?

No. 10 → 10,000 cmil ⇒ no. 13 → 5000 cmil ⇒ no. 16 → 2500 cmil ⇒ no. 19 → 1250 cmil
No. 22 → 625 cmil ⇒ no. 25 → 312 cmil ⇒ no. 28 → <u>156 cmil</u>

Example 10.4 What is the area of no. 26 wire?

No. 10 → 10,000 cmil ⇒ no. 20 → 1000 cmil ⇒ no. 23 → 500 cmil ⇒ no. 26 → <u>250cmil</u>

Diameter of Coated Wire

Typical magnet wire used in converter magnetics is coated with polyurethane plus polyamide type of insulation. It is most commonly available in *single* or *heavy* (double) insulation depending on the number of protective coatings applied on the bare copper conductor. The insulation *build* is by definition twice the thickness of the coating deposited. On the

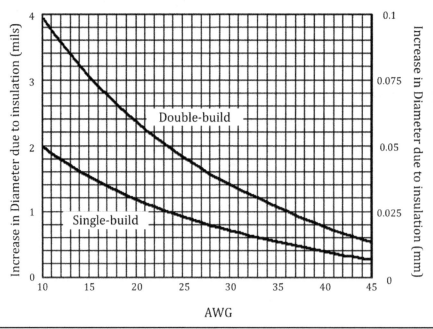

FIGURE 10.5 How insulation coating changes bare copper diameter.

high-voltage (primary) side, it is unusual not to use double insulation, whereas on the secondary side we can usually use single insulation wire. Note that from the viewpoint of safety regulations, the coating does not correspond to anything other than *functional insulation*. But to us it is more: a potential reliability issue, related to the scratches the wire may receive in production during the winding process and the resulting possibility of flashovers while in operation later.

A well-known vendor is PD Wire and Cable (a unit of Phelps Dodge Corporation) at www.pdwcg.com. It makes the popular Nyleze and Thermaleze brands. These are solderable/tinnable varieties meant for easy production flow.

We need to know the build thickness, because only then can we accurately predict how many turns we can lay on a certain bobbin width. A closed-form equation closely approximating typical overall wire thicknesses for AWGs falling between 14 and 29 is

$$d = d_{Cu} + 10^{\alpha - AWG/44.8} \text{ mil} \qquad \text{(AWG 14 through 29)}$$

where d_{Cu} is the diameter of the bare copper in mils and $\alpha = 0.518$ for single build and 0.818 for double build. For AWGs between 30 to 60 we should use

$$d = d_{ref} \times \beta \times \left(\frac{d_{Cu}}{d_{ref}}\right)^{\gamma} \text{ mils} \qquad \text{(AWG 30 though 60)}$$

where d_{ref} is an arbitrarily defined diameter used to get the dimensions right. Here it is set to be the diameter of AWG 40 wire. $\beta = 1.12$ for single-build wire and 1.24 for double-build wire. $\gamma = 0.96$ for single-build wire and 0.926 for double-build wire. On this basis we have generated Fig. 10.5.

SWG Comparison

In the *rest of the world* the standard wire gauge (SWG, also called Imperial or British wire gauge) is in more common use. In Table 10.2 we had provided (bare) copper wire diameters for SWG too. In Fig. 10.6 we have also provided a simple graphical way of quickly picking the closest SWG or AWG equivalent as required. For example, we have shown how AWG 32 gives us the closest equivalent: SWG 36 (and vice versa).

Basics of Flyback Transformer Design

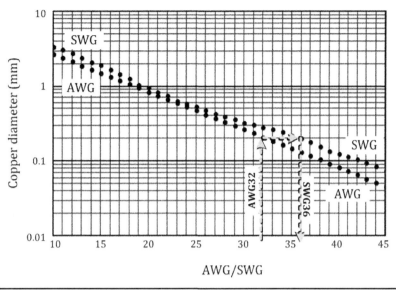

FIGURE 10.6 Graphical comparison of SWG and AWG.

For diameters of SWG the following quick rules apply within a limited range:

- Diameter of no. 19 is 40 mil (1 mm).
- From nos. 19 through 23, diameter decreases 4 mil each number.
- From nos. 23 through 26, diameter decreases 2 mil each number.
- The closest match with AWG is that AWG 24 (or 25) is almost exactly the same diameter as SWG 25 (or 26). This gives us a convenient *bridge point* to go from one to the other system.

CHAPTER 11
Basics of Forward Converter Magnetics Design

Introduction

Refer to Chap. 7 for magnetics concepts before starting this one. Also read Chap. 6 for an introduction to the Forward converter. We also have a more detailed, step-by-step example in Chap. 12 to follow. We are using MKS (SI) units, unless otherwise stated.

When it comes to the Forward converter, there are actually two magnetic components to consider. The good news is: Its transformer is *really* a transformer, unlike a Flyback where the transformer is best thought of as a multiwinding inductor. In the Forward, we also have the output choke (inductor) to consider, but since its primary purpose is energy storage, we can design it in virtually the same way as we handle any other inductor. Besides, the secondary side is not *high voltage*, so all the fuss about fast-acting current limits and so on, as mentioned in reference to the Flyback in Chap. 10, is not really an issue here. Also the Forward converter's transformer core does not "see" the high current associated with the load current, as discussed in Chap. 6. So its design complexity is mainly related to the proximity effect, which we will introduce in this chapter. A lot more on the proximity effect actually follows in Chap. 12.

The Transformer and Choke (Inductor) Compared

A Forward converter choke is usually always operated in continuous conduction mode (CCM) as any energy storage magnetic component. Its current ripple ratio r is also thus set to an optimum value of around 0.4.

A Forward converter *transformer* is, however, always operated in discontinuous conduction mode (DCM). In a Forward converter, the duty cycle D is normally set around 0.3 to 0.35 at low-line (with an input voltage doubler being typically present in universal-input applications as discussed in Chap. 5), and that is partly attributable to the need for supporting a certain holdup time. The applied voltage in the voltage-dependent equation during the ON-time is V_{IN} (the rectified and possibly voltage-doubled AC input).

For a Forward converter we have a simple relationship arising from its input-to-output transfer function (where $n = N_P/N_S$):

$$V_O = D \times V_{INR} = D \times \frac{V_{IN}}{n}$$

that is,

$$V_O (= \text{constant}) \propto D \times V_{IN}$$

We see that $D \times V_{IN}$ is a constant. Which means we can multiply the duty cycle at low-line with the input voltage at low-line, or the duty cycle at high-line with the input at high-line, and we will get the same result.

Also, since $D = T_{ON}/T = T_{ON} \times f$, this basically means that the *volt-seconds applied to the transformer does not depend on the input voltage*! We expect, therefore, that unlike for a Flyback, the peak (magnetization) current in the transformer is constant with respect to input voltage. So, if we ensure that the B-field in the transformer is say, 1500 G at high-line, it would be 1500 G at low-line too. We do not have to worry about saturating the transformer during power-up or power-down in particular, as in a Flyback.

Let us confirm the behavior of the peak field. The voltage-dependent equation is

$$B_{AC} = \frac{V_{AVG} \times \Delta t}{2 \times N \times A_e}$$

In this equation V_{AVG} is just V_{IN}. In the worst case of DCM (converter just entering DCM) we can set $B_{PK} = 2 \times B_{AC}$. So simplifying, the preceding equation becomes

$$B_{PK} = \frac{V_{IN} \times D}{N \times A_e \times f}$$

Since the numerator is a constant, and all else is fixed too, then B_{PK} is fixed too, whatever the input is.

Our unexpected stroke of luck, in terms of input independence, came about because, though the transformer is in DCM, its duty cycle is CCM based. That is, in turn, because the duty cycle is being dictated, not by the transformer, but by the output choke, which is in CCM (typically). And that, as we saw earlier; is responsible for the fact that for a Forward converter transformer, we can actually use either the V_{INMIN} or the V_{INMAX} to solve the voltage-dependent equation. So, both the following equations (for either input end) are equivalent and will give the same number of (Primary) turns when solved:

$$B_{AC} = \frac{B_{PK}}{2} = \frac{V_{INMIN} D_{MAX}}{2 \times NA_e f}$$

$$B_{AC} = \frac{B_{PK}}{2} = \frac{V_{INMAX} D_{MIN}}{2 \times NA_e f}$$

In these equations, we can set B_{PK} equal to B_{SAT} (typically 3000 G) as a worst-case scenario, but usually we keep it much less (about half of that) so as to reduce core loss (discussed later). We also note that since core loss depends on ΔI (or ΔB), *the core loss in the Forward converter transformer is also independent of line voltage. The same is true for the peak current (since I and B are almost proportional to each other in ferrites).* Of course, the underlying assumption in all the preceding equations is that the (CCM) duty-cycle equation applies to the converter.

Note that the duty-cycle equation can be written as

$$D = \frac{V_O}{V_{INR}}$$

where V_{INR} is the reflected input voltage, equal to V_{IN}/n. Applying the voltage-dependent equation to the output choke now, it becomes

$$B_{AC} = \frac{V_{ON} D}{2 \times NA_e f} = \frac{(V_{INR} - V_O) \times D}{2 \times NA_e f}$$

or equivalently (using the OFF-time instead of the ON-time)

$$B_{AC} = \frac{V_{OFF} \times (1-D)}{2 \times NA_e f} = \frac{V_O \times (1-D)}{2 \times NA_e f}$$

From this equation we can see that B_{AC} (or ΔI) increases if D decreases (line increases). Keep in mind that for all topologies, if input voltage increases, D decreases, and vice versa. *Therefore, the core losses (and peak current) are higher at high-line for the Forward converter choke. This is unlike the case of the Forward converter transformer, but identical to what we would expect in any Buck stage.*

Thus the design of the output choke should be done at high-line, just the way we design an inductor for a Buck converter application. The basic design of the Forward converter transformer (flux, number of turns etc.) can be done at either high-line or low-line.

But what exactly is the duty cycle we should use? Doesn't it depend on holdup time considerations? The answer to this follows.

In a Forward converter, *we actually do need to perform at least part of the transformer design at minimum input*, but for an altogether different reason, i.e., not related to the core saturation possibility at low inputs as in Flybacks. Here the reason is that we need to carefully calculate the *turns ratio at minimum input*. Because we have to ensure that at the lowest intended DC input, say 250 V, the converter does not hit the maximum duty-cycle limit (D_{MAX} is theoretically 50 percent, but in reality is closer to 45 percent in most practical controller ICs). The calculated turns ratio then determines the nominal duty cycle, and indirectly the volt-seconds across the transformer and its peak field too.

Turns Ratio Calculation (at Minimum Input)

We return to our previous equation and apply it at minimum input:

$$D_{MAX} = \frac{nV_O}{V_{INMIN}} \quad \text{so } n = \frac{V_{INMIN}D_{MAX}}{V_O}$$

For example, if we have a 12-V output, we need to set (ignoring output diode drop)

$$n = \frac{V_{INMIN}D_{MAX}}{V_O} = \frac{250 \times 0.45}{12} = 9.4$$

So, for example, at 220 Vac rectified (311 Vdc) input, the duty cycle will be

$$D = \frac{nV_O}{V_{IN}} = \frac{9.4 \times 12}{311} = 0.363$$

This is the nominal D of the converter. When this is plugged into the voltage-dependent equation, we can accurately estimate B_{PK}. Because otherwise, we really did not know the value of D to use, other than a rough design target.

That summarizes why minimum input voltage is of great concern in Forward converter transformer design too, just like in a Flyback, but for altogether different reasons as we have seen.

Note that in all the preceding equations, we have been using "ideal equations." To recalculate accurately, based on the estimated efficiency η, we need to simply do the following transformation (as discussed in Chap. 2, see Fig. 2.13):

$$V_{IN} \rightarrow \eta V_{IN}$$

That will give us a better estimate of the nominal D, but will also more accurately ensure that we do not hit D_{MAX} at higher input voltages than intended (which could easily lead to the holdup time requirement not being met).

Summary of Observations

At this point, see Chap. 6 too, because the following pretty much recapitulates and summarizes some of the material from that chapter. Analyzing Fig. 11.1 we see:

- On the secondary side, in effect we have a Buck converter with an effective input voltage equal to V_{INR}. The center of the inductor current is the load current as for a Buck.

- On the primary side the Forward converter (almost) "thinks" that it is a Buck with an output voltage of V_{OR} and a load current of I_{OR} (just a way to visualize it). We are following the technique for a Flyback, as introduced in Chap. 6. In a Flyback, the component that didn't help map the isolated converter to a nonisolated equivalent was the leakage inductance. In the case of the Forward, it is the magnetizing inductance.

- The difference on the primary side of the Forward is the magnetization current I_{MAG} which adds to the reflected secondary current. This sum of the reflected secondary current and the magnetization current forms the primary winding (and switch) current. We have $I_{SEC} = n \times (I_{PRI} - I_{MAG})$ as indicated in Fig. 11.1. Note that I_{MAG} is *not* dependent on power transfer taking place across the isolation boundary.

- I_{MAG} is simply freewheeled back through the tertiary winding (T) as I_T, and the associated magnetization energy of the transformer returns to the input bulk capacitor during the switch OFF-time. This is clearly a *circulating current* and some of its energy is dissipated in the tertiary diode and tertiary winding resistance.

- When the switch is ON, the voltage across the transformer is V_{IN}. When the switch is OFF, the tertiary winding will reflect an equal and opposite voltage across the primary winding (provided it has the same number of turns as the primary winding). Thus the peak voltage on the drain of the switch is $2 \times V_{IN}$ (see Chap. 6 too). This voltage stress lasts until the transformer core gets deenergized, and that is the moment I_T falls to zero.

- The magnetization current ramp is determined by the primary inductance of the transformer, the applied input voltage, etc. But as we see, it rides on a ramp or pedestal being reflected from the secondary side. That pedestal is determined completely by the load current and the inductance of the output choke.

- The inductance of the choke should be so chosen that (as for any Buck converter) r is about 0.4 at V_{INMAX}. Note that we can use all the standard Buck converter equations, including the selection criteria for inductors, if we just imagine that the effective input applied to the output step-down (Buck) stage of the Forward converter is V_{INR}. In other words, we can mentally visualize the Forward converter as a Buck converter on the secondary side, with an input DC voltage rail equal to $V_{INR} = V_{IN}/n$ (created by transformer action).

- Besides isolation, of course, the only major difference is that in a DC-DC Buck converter, we can in principle go up to 100 percent duty cycle, but in a Forward converter, we need to allow the *transformer* to reset too (besides the output inductor). The (magnitude of) the slopes of the ramp-up and ramp-down of the magnetization current are the same (= V_{IN}/L_{PRI}). So the worst case occurs at 50 percent duty cycle where the magnetization current is in "critical conduction" (the boundary between continuous and discontinuous conduction modes). If the ON-time ever exceeds this theoretical/geometrical limit, then every cycle there will be a net increase in the magnetization current. Since the feedback loop connects only to the output choke, it will have no way of realizing and correcting this buildup of transformer magnetization current. Primary-side current limiting won't help either since that is set as per the maximum load current, and the magnetizing component is a smaller (and variable) part of it. Therefore, we cannot easily sustain CCM in the single-ended Forward converter transformer. The easy option to ensure the transformer resets is to always allow *enough time*—for the magnetization current to return to zero (every cycle). Knowing what the slopes of magnetizing and tertiary winding currents are (they are equal and opposite), this simply means restricting the maximum duty cycle to less than 50 percent in any single-ended Forward converter (or even in the asymmetric two-switch Forward—see Fig. 6.3 in Chap. 6).

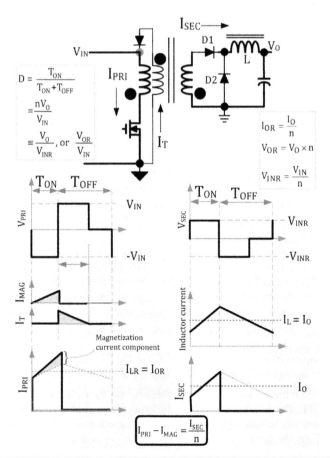

Figure 11.1 Forward converter currents.

- In an active clamp implementation, the Forward converter transformer can be (and in fact is) in CCM. This is discussed in Chap. 16.
- We reiterate: As far as the Forward converter transformer core is concerned, it "thinks" that the only current flowing around it in the windings is the magnetization current. But the *temperature* of the copper windings is a *dead giveaway*: From it we can know the actual primary and secondary currents. Yes, if we could separate sources of temperature easily, we may learn that the windings are usually hot, though the core itself is quite cool.
- One important difference is that unlike the choke, which also sees a high current, the primary and secondary windings have a *chopped* current waveform. So the high-frequency losses (AC resistance) can be very high because of the high-frequency harmonic content. It is important to try and reduce the copper losses in the windings of a Forward converter. And for that we need to understand *proximity effects*.

Introducing the Proximity Effect

Whereas the skin depth considerations that we discussed in Chap. 10 still apply, here we must go further. The skin depth represents a single wire actually. We didn't consider the fact that the field from the nearby windings (laid adjacent to it and on the winding layers around it) may be affecting the current distribution significantly, and so the annular area on the surface of the wire, which we had imagined (and hoped) was available for the high-frequency current to flow through, is actually not. In effect, we cannot really "get away" with simply saying that the diameter of the wire (or thickness of copper foil) needs to be twice the skin depth. In fact, because of this *proximity effect*, thicknesses and diameters need to be much less, because otherwise there is in effect, wasted copper (for the high-frequency Fourier components), which in turn means the *AC resistance* has gone up. An increase of AC resistance implies that, in effect, the resistivity of copper is higher at high frequencies. It is almost like a new material. We will see in this chapter and in Chap. 12 that even with the best measures, we will eventually end up with effective AC resistance ratio (all harmonics included) of about 2, so it is almost like we changed over from copper to a material of double the resistivity. Our losses also double as compared to DC estimates. And that is with the best winding approaches. It can get much worse.

More about Skin Depth

In a stand-alone conductor, the current distribution falls as e^{-1} inside the conductor. Ignoring the sinusoidal time variation here, its variation in space has the form

$$J = J_O e^{-x/\delta}$$

Looking at the upper part of Fig. 11.2, we have arbitrarily set the surface current density J_O to unity and the skin depth δ to 1. We can eyeball the curve and also confirm by simple mathematics that, by the properties of the exponential function, the area under the entire exponential curve is equal to the area of the rectangular shaded portion. This implies that we can in effect replace the entire exponential current distribution with a constant (uniform) current density equal to the actual current density on the surface, but now confined to a distance δ under the surface. This is the definition of skin depth δ.

This also leads to the concept of AC resistance. In the lower part of Fig. 11.2 we have shown a round conductor, and we can see that if direct current is passed, the current will spread out uniformly throughout the entire cross-sectional area, that is, 4^2. But if a time-varying (sinusoidal) current tries to make it through the conductor, it is confined to an annular area on the surface as shown (with the preceding exponential to rectangular simplification). The skin depth is assumed to be unity here too. So as far as the alternative current is concerned, the area available to it is less than if it was direct current. Since resistance is inversely proportional to area, the AC resistance must be related to the DC resistance by

$$F_R \equiv \frac{R_{AC}}{R_{DC}} = \frac{4^2}{4^2 - 2^2} = 1.3$$

where we have also introduced the term F_R. Clearly, F_R can be as low as unity, but as we can see, it can be very high if the diameter is increased.

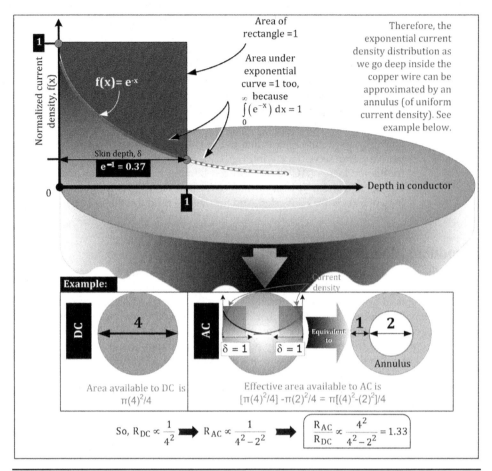

Figure 11.2 Understanding current density and skin depth.

Note *For large wire diameters, the annular region is basically the circumference of a circle, so we can still get an increase in the available area for the current. In other words, $R_{AC} \propto 1/d$, whereas we know that $R_{DC} \propto 1/d^2$. So, $F_R = R_{AC}/R_{DC}$ goes as $d^2/d = d$. In a copper foil, the width is usually fixed. So the only option is to increase the thickness of the foil (t). But that will not affect R_{AC} unfortunately, since more area does not become available to the current by doing so. However, $R_{DC} \propto 1/t$. So we get $F_R = R_{AC}/R_{DC} \propto t$. So as we increase the thickness or diameter, F_R goes up in both cases (wire and foil), but in fact that is deceptive since R_{AC} certainly falls in the case of wires. So lowering F_R is not necessarily the target; we need to lower R_{AC}. This is discussed in great detail in Chap. 12. Here we should just remember that the only way to reduce the R_{AC} in foils is to increase the* width *of the foil. This would normally mean a larger core size. However, one way out is to look for special EER cores like the EER35, for achieving high output currents when using secondary foil windings. These are long (extended) versions of comparably sized cores, so it should not affect the cost as much as increasing the core volume normally will.*

Now, if we put similar current-carrying conductors next to each other, they will affect even the availability of the annular region displayed in Fig. 11.2. This can lead to a very steep increase in F_R.

Dowell's Equations

Dowell successfully reduced a very complex three-dimensional field problem into a manageable and accurate one-dimensional calculation. But first, we must define a *portion* as per Dowell. As we look at how the magnetomotive force (mmf, i.e., ampere-turns) varies through different winding arrangements in Fig. 11.3, we realize that the key to reducing eddy current losses is to reduce the *local fields*, and we can do that by reducing the peak value of the (local) mmf. So split windings will always help. In principle, additional levels of primary-secondary interleaving will help further, with some added cost and complexity. Most practical AC-DC, medium-power designs will use a split Primary (two sections in series) and a sandwiched Secondary as shown. With this in mind, a portion is defined as the layers between a maxima and a zero of mmf. Clearly, we want to increase the number of portions if possible, thereby reducing the number of layers per portion.

Basics of Forward Converter Magnetics Design

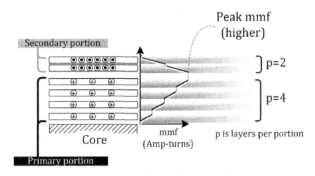

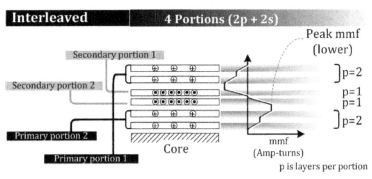

FIGURE 11.3 Magnetomotive force and *portions* for two winding arrangements.

NOTE *In Fig. 11.3, by a split winding we have reduced the peak mmf and field by a factor of two. This translates into four times less energy in this leakage field, and four times less leakage inductance too.*

According to Dowell, the F_R of a portion with an integral number of layers (p being an integer) is

$$F_R(p, X) = A(X) + \frac{p^2 - 1}{3} B(X)$$

where p is the number of layers in that portion, X is h/δ (penetration ratio) with h being the thickness of the *equivalent foil*, and

$$A(X) = X \frac{e^{2X} - e^{-2X} + 2\sin(2X)}{e^{2X} + e^{-2X} - 2\cos(2X)}$$

$$B(X) = 2X \frac{e^{X} - e^{-X} - 2\sin(X)}{e^{X} + e^{-X} + 2\cos(X)}$$

For half-integral layers per portion (p being a half-integer), we can use

$$F_R(p, X) = \left(1 - \frac{0.5}{p}\right) \times [F_R(p - 0.5, X)] + \frac{C(X)}{4p} + \frac{(p - 0.5) \times B(X)}{2}$$

where

$$C(X) = X \frac{e^{X} - e^{-X} + 2\sin(X)}{e^{X} + e^{-X} - 2\cos(X)}$$

We have plotted these out in Fig. 11.4 (in two different ways). These equations apply to foils very naturally, since we transform even layers of round wires into equivalent foils. But we have also assumed that round wires are spread out in a layer with *no space between successive (adjacent) windings*. For wires, we will describe the equivalent foil transformation in more detail later.

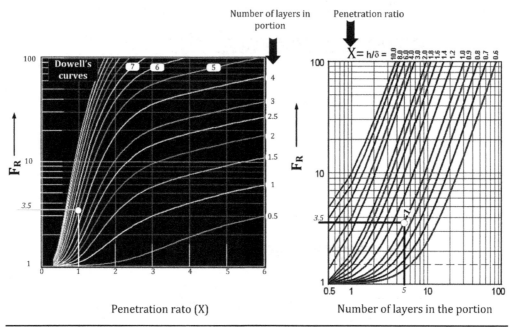

FIGURE 11.4 Dowell's curves in alternative representations.

Note that in related literature the same curves are presented with X on the horizontal axis. But we have preferred to use the number of layers in the portion p on this axis, with X as the parameter for the set of curves. That way we can present the curves for many more layers and that comes in handy when we use bundled/Litz wire as we will see.

Note that Dowell's original equations talk in terms of δ, which depends on frequency, so clearly *the curves are applicable to a sine wave only*. The actual switching current waveform is, however, usually a unidirectional (*unipolar*) rectangular or trapezoidal waveform, so *we must use Fourier analysis to split the waveform into harmonic components* (each of amplitude $|c_n|$), evaluate the AC resistance of each, and then find the effective F_R (F_{R_eff}) for the composite waveform. Taking the DC level of the current waveform as the *zeroth harmonic* c_0, we get the effective winding resistance to be

$$R_{AC\_eff} \times I^2_{RMS} = R_{DC} \times \sum_{n=0}^{40} |c_n|^2 F_{Rn}$$

where I_{RMS} is the usual RMS of the current waveform ($\cong I_{SW} \times D^{1/2}$). Simplifying

$$F_{R\_eff} \equiv \frac{R_{AC\_eff}}{R_{DC}} = \frac{\sum_{n=0}^{40} |c_n|^2 F_{Rn}}{I^2_{RMS}}$$

or

$$R_{AC\_eff} = F_{R\_eff} \times R_{DC}$$

where F_{Rn} is the F_R of the nth harmonic, and by definition, $F_{R0} = 1$. We are summing only up to the 40th harmonic.

Optimization starts by varying h (or equivalently the X) and seeing where we get the least R_{AC_eff}. But as we saw, for a foil, R_{DC} too will vary in the process, and will vary as 1/thickness (~1/h). So *we need to find the minima of the function F_R/X where X is now h/δ_1*, that is, *referred to the fundamental frequency* (the switching frequency). This function is plotted out in Fig. 11.5 from a Mathcad file, and we see that for each p, we get an optimum value for X. We have set both the rise and fall times to a typical 0.5 percent of the switching cycle time period. The duty cycle is assumed to be 50 percent.

In Figs. 11.6 and 11.7 we have collated the *optimum foil thickness* X/δ values thus generated. We have two design curves (different zoom levels and linear and log scales) to help us either with layers using a single conductor (small p) or with bundled/Litz wire (large p). In Fig. 11.8 we

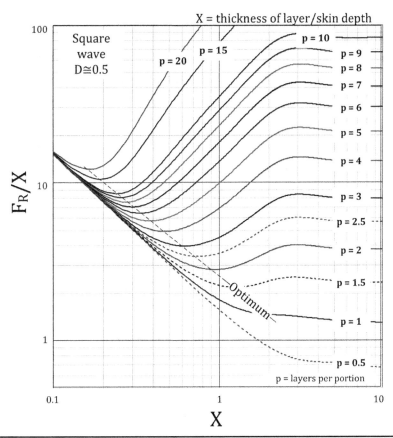

FIGURE 11.5 AC resistance as a function of thickness for a square waveform.

have the results of a mathematical iteration for the effective F_R we can expect *if we use the optimum* X/δ. We can see that *anything more than 5 to 6 layers (per portion) is not going to reduce the F_R much*. Even for bundled/Litz wire, *we are not going to be able to reduce F_R much lower than about* 2. That is *the best achievable*, proximity effects considered.

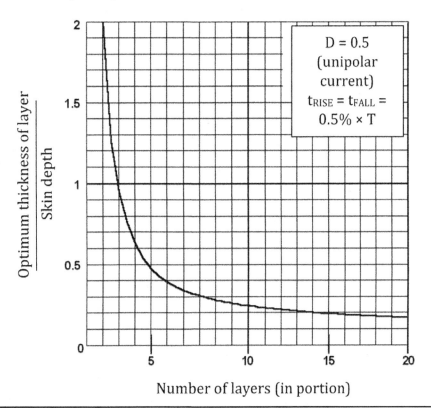

FIGURE 11.6 Optimum layer thickness for a square waveform (one view).

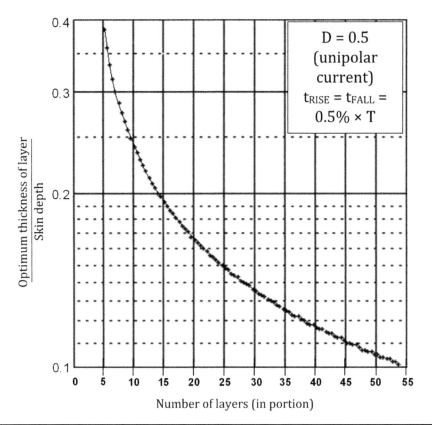

FIGURE 11.7 Optimum layer thickness for a square waveform (another scale).

In Fig. 11.6 notice how for just one layer, we can go up to twice the skin depth as suggested previously. So it is the proximity of adjacent layers that demands we decrease X to much smaller values than 2δ as the numbers of layers in the portion increases.

NOTE *According to Bruce Carsten the optimum foil thickness is roughly proportional to $D^{1/2}$. (depending on the duty cycle; we assumed $D = 0.5$) He also recommends that for bipolar current waveforms (e.g., bridge topologies), we should* halve *the unipolar current foil thickness so obtained. However, as per the author of this book, do not expect to be able to lower the F_R much from the numbers indicated in Fig. 11.8.*

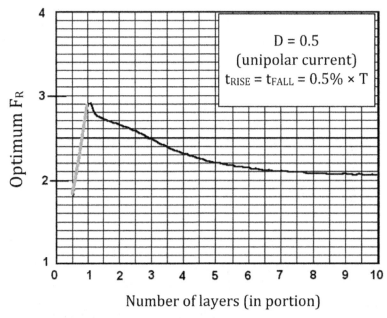

FIGURE 11.8 Lowest achievable F_R if optimum thickness is used.

From Figs. 11.6 and 11.7, we can see that $X = h/\delta$ seems roughly inversely proportional to the number of layers. A good curve-fit to our mathematically generated results is

$$h = \delta \times \left(\frac{2}{p^{1.2}} + 0.095 \right)$$

where δ is the skin depth at the switching frequency and p is the number of layers in the portion. This is roughly equivalent to saying X is inversely proportional to p. So if we increase the number of layers per portion, we need to reduce X.

If this is indeed so, it implies that the *total thickness of the entire portion, that is*, $p \times h$, *will tend to (for larger p) become virtually constant, irrespective of the number of layers*. In Fig. 11.9 we have plotted this out more accurately. We see that it does vary somewhat within the accuracy of the program used to generate these results, but starts flattening out to the right, as expected. This curve will help us a great deal especially when we start a design based on bundled/Litz wire. We can be quite clear as to how thick the primary winding portion is going to be. We can double-check the available window area too (see Chap. 12 for more detail on that).

Example 11.1 We have six layers in the Primary, and it is split into two sections. What is the optimum foil thickness if the switching frequency is 100 kHz?

Using the skin depth equation for a typical 60°C temperature rise, at 100 kHz we get a δ of 0.24 mm. Using the preceding equation for three layers per portion we get

$$h = 0.24 \times \left(\frac{2}{3^{1.2}} + 0.095 \right) = 0.15 \text{ mm}$$

Note that we can see that $h \times p = 0.15 \text{ mm} \times 3 \cong 0.45 \text{ mm}$. But this is almost equal to $2 \times 0.24 \text{ mm} = 2\delta$. This is the required *thickness of the entire portion*. We can see how the proximity effect is affecting us. If we had ignored this effect, we would have normally chosen each layer to be 2δ thick, giving us a total primary portion thickness of 6δ. Now we require it to be a third of that.

But that also means we have less bulk copper cross-sectional area to meet any required current density target! All this says is that we have less wasted copper if we use this particular fine thickness (X). But the selected X is valid for any load (we have not even stated load in this example). So, we still do not know, one way or another, whether we have enough copper cross-sectional area to meet the required load current with the target current density, even assuming all the copper area is 100 percent utilized. This is the reason the problem is best approached by the *subdivision method* discussed in Chap. 12, where we start with a certain occupied copper area and subdivide it into finer and finer strands to see when we get the lowest F_R.

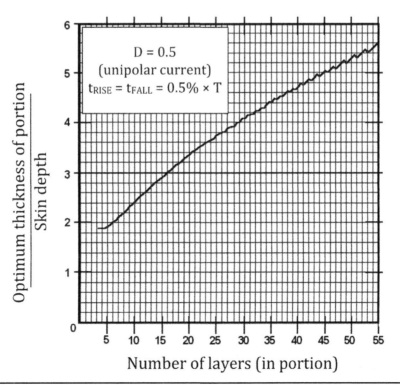

FIGURE 11.9 Optimum thickness of entire portion (not layer).

Keeping this perspective in mind, we can see from Fig. 11.8, that with this optimum thickness, the effective F_R for the entire Primary will be about 2.5. So if we know the DC resistance, we can easily compute the AC resistance and then, based on load current, we can estimate the losses in the primary winding. We can repeat the same process for the secondary winding and thus get the total copper loss in the transformer. In principle, at least for now.

In the following section we relate these results, apparently applicable only to a foil, to the case of round wire, as Dowell initially did.

The Equivalent Foil Transformation

To be able to apply Dowell's equations and the results of the preceding section, the round wire is mentally replaced by an equivalent "square wire" of the same copper area. Thus, if we have found out a certain desirable h from the previous curves or equations, the equivalent wire diameter is obtained by

$$d \Leftarrow \frac{2}{\sqrt{\pi}} \times h$$

We therefore just *need to multiply* h *obtained previously by the factor 1.13, to get* d. See Fig. 10.3 (in Chap. 10) in this regard too.

So in Example 11.1 we got an equivalent wire diameter of 0.15 mm, or AWG 33. Of course, we know that this can give us the lowest losses, but we still don't know yet if that is good enough, considering the output power and efficiency of the converter (as suggested earlier). We will come to this topic again later.

For now we can see from Fig. 11.8 that the best F_R with three layers is going to be 2.5. Why not use bundled wire? The rule for generating this transformation is also shown in Fig. 11.10. Similarly, we will see that, for example, a bundle of 16 strands can be considered to be stacked as 4×4, and so where each bundle was, we now get four effective layers. We merge them, and so, for three initial layers we now get a portion with $p = 4 \times 3 = 12$ layers (this is as per Dowell's equivalent foil transformation; these are not actual layers of winding). From Fig. 11.8 we can see that the F_R can be reduced to about 2, *provided we go to about 12 layers (per portion), by choosing an equivalent foil thickness of*

$$h = \delta \times \left(\frac{2}{p^{1.2}} + 0.095\right) = 0.24 \times \left(\frac{2}{12^{1.2}} + 0.095\right) = 0.047 \text{ mm}$$

The equivalent optimum wire diameter is $0.047 \times 1.13 = 0.053$ mm. This gives us AWG 43 for each strand.

When we come to bundled or Litz wire, if we have n strands in the bundle, we should replace these with equivalent square wire strands, and then stack them together in a square. *Ultimately we get* $n^{1/2}$ *layers from one*. See Fig. 11.10. Note that the number of turns has not really changed. In fact, the actual number of turns is not directly relevant in Dowell's analysis at all.

In Chap. 12 we will use the subdivision method to perform a full analysis and calculation, based on what we have learnt here so far.

These are equivalent foil transformations. Each effective layer shown, is not a physical layer, but a layer as per Dowell's analysis.

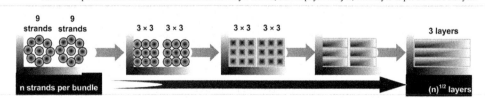

FIGURE 11.10 Replacing bundled wire with equivalent foils for using Dowell's curves.

Some Useful Equations for Quick Selection of Forward Converter Cores

For a transformer, the limitation on throughput power is essentially due to (1) the effective area A_e of the core, which determines how many turns we need (using the voltage-dependent equation), and (2) the available window area which determines how much copper we can actually squeeze in. Therefore, we do expect $P_o \propto A_e \times A_w$, where A_w is the available window area. The product $A_e A_w$ is called the area product (AP) of the core. This too is discussed in great detail in Chap. 12.

For now let us firm up the concepts. We know that the entire "available" window area will not have copper in it. We need space for the wire insulation, bobbin, tape layers, etc., and, in addition, round copper wire will have wasted air spaces between windings, however closely packed it is. Therefore, in general, in literature, a window utilization factor K_u is used, which expresses how much of A_w will be actually occupied with copper. However, out of this, only part of the area will be occupied by the primary winding. In addition, generalizing further, there is going to be a dependence on the topology too (all being Forward converter derivatives). The most commonly used equation for selecting a core is given by

$$AP = \frac{11.1 \times P_{IN}^{1.32}}{K \times \Delta B \times f} \quad cm^4$$

where $K = K_t \times K_u \times K_p$ (K_u is window utilization factor $A_w'/A_w \approx 0.4$, K_p is primary area factor A_p/A_w', and K_t is topology factor). ΔB is the flux swing in tesla (in discontinuous conduction mode for unipolar current this equals B_{PK} and is typically set to 0.15 to 0.2 for ferrites). f is in hertz. The K factors are given in Table 11.1.

The reader can refer to the conversion tables in Chap. 10. Empirically, it is seen that the maximum current density is 400 to 450 A/cm² (about 450 cmil/A) for a core with AP = 1 cm², and a 30°C temperature rise attributable to the copper losses.

So the current density for a 30°C rise is written as

$$J = 420 \times AP^{-0.24} \quad A/cm^2$$

This was the relationship that was used to calculate the preceding AP relationship. But if the core losses are almost the same as the copper losses (that is not necessarily always true), and we do not want to exceed 30°C, we will have to allocate only 15°C for the copper losses. In that case the current density should be reduced to about 300 A/cm² (about 650 cmil/A).

$$J = 297 \times AP^{-0.24} \quad A/cm^2$$

Note that

$$AP \propto \frac{1}{J}$$

so the area product should be increased accordingly.

Smaller cores have a large relative surface area for cooling. That is because V_e varies as l^3 but surface area goes as l^2. So, the thermal resistance R_{TH} of large cores is higher than for small cores. The empirical relationship for IEC qualified safety transformers (for ≈40°C rise in temperature) is stated as

$$R_{TH} = \frac{53}{V_e^{0.54}} \quad °C/W$$

The empirical relationship between V_e and AP is

$$V_e = 5.7 \times AP^{0.68}$$

where V_e is in cm³.

	K	K_t	K_u	K_p
Forward converter	0.141	0.71	0.4	0.5
Bridge/Half-Bridge	0.165	1.0	0.4	0.41
Full-wave center tap	0.141	1.41	0.40	0.25

TABLE 11.1 Selecting the K-Factors for Different Topologies for Use in the Area Product Equation

So we also get the following relationship:

$$R_{TH} = \frac{20.7}{AP^{0.367}} \quad °C/W$$

We can also rewrite the AP relationship in terms of the easier-to-find V_e parameter, to aid a quick first selection of core. This is

$$V_e = \frac{29.3 \times P_{IN}^{0.9}}{(K \times \Delta B \times f)^{0.68}} \quad cm^3$$

Example 11.2 For a 200-W Forward converter, what core should we select? The switching frequency is 100 kHz and the expected efficiency is 80 percent.

$$V_e = \frac{29.3 \times (200/0.8)^{0.9}}{(0.141 \times 0.15 \times 10^5)^{0.68}} = 23.1 \ cm^3$$

From Table 11.2, we see that the EE 42/21/20 (two halves making 42/42/20 size) will suffice for this application. One size smaller (the EE 42/42/15) may also suffice depending upon the application (how may secondary windings, etc.).

Keep in mind that the mean length per turn of ETD cores is smaller than E-cores, because of the round central limb. This lowers the resistance for a given number of turns. Also see Table 12.1 in Chap. 12 for a lot of information on popular core sizes.

Stacking Wires and Bundles

When we stack round wires next to each other, it is trivial to calculate how many turns can be accommodated per inch in a layer (transverse, say t direction) and how many can be stacked on top of each other, i.e., height-wise (say, h direction). With insulation between each layer, both are expected to be the same. We just need to know the diameter with the coating. That is provided in Chap 10. In the t direction, we will need to allow for any margin tape requirement.

When we take bunched wire (bundles twisted or *preferably braided* together), the picture is not so obvious. Some relevant empirical results are presented in Table 11.3 and have been verified

Core	V_e (cm³)	A_e (cm²)	l_e (mm)	2 × D*
EE 13/7/4	0.384	0.13	29.6	9
EE 16/8/5	0.753	0.201	37.6	11.4
EE 20/10/6	1.5	0.335	44.9	14
EE 25/13/7	3.02	0.525	57.5	17.4
EE 30/15/7	4	0.6	67	19.4
EE 32/16/9	6.18	0.83	74	22.4
EE 42/21/15	17.6	1.82	97	29.6
EE 42/21/20	23.1	2.36	98	29.6
EE 55/28/21	43.7	3.54	123	37
EE 55/28/25	52	4.2	123	37
EE 65/32/27	78.2	5.32	147	44.4
ETD 29	5.47	0.76	72	22.0
ETD 34	7.64	0.97	78.6	23.6
ETD 39	11.50	1.25	92.2	28.4
ETD 44	17.80	1.73	103	32.2
ETD 49	24.0	2.11	114	35.4
ETD 54	35.50	2.80	127	40.4
ETD 59	51.50	3.68	139	45.0

*2 × D is as per the nomenclature presented in Fig. 10.3. (2 × D is twice the length of any limb.)

TABLE 11.2 Dimensions and Effective Parameters for Various Popular Core Sizes

	Strands/Bundle						
	4	5	6	7	8	9	10
N_t	2.45	2.94	2.98	3.11	3.61	3.89	4.34
N_h	2.31	2.69	2.93	2.93	3.16	3.16	3.36

TABLE 11.3 Stacking of Bundled Wire: Transverse and Vertical (Height)

to be providing good stacking predictions. So, for example, if the diameter of each strand of a 4-strand bundle is d, then the bundle can be considered to be looking like an effective "diameter" of $2.45 \times d$ when wound side by side (i.e., along a layer). But in the *vertical direction*, since the bundles on top tend to slip slightly into the spaces of the preceding layer, the effective diameter of each bundle looks more like $2.31 \times d$. We can then very clearly estimate how the available window space is filling up with the bundles and whether the transformer suffices.

NOTE *Sometimes in the lab we just take several AWG strands and twist them into a bundle. This is not going to help reduce proximity/eddy current losses much since the same strand(s) tends to remain on the surface. What we really want to do is braid them so that they take turns on the surface and inside. It is therefore better to order correctly bunched bundles from vendors with automatic machines to do the job. The other option in the lab is to twist subbundles with lesser strands and then twist several of these bundles together.*

NOTE *For a chosen diameter, the F_R increases significantly if the number of layers increases. Further, even if a few turns are left over and wound on the last layer, from Dowell's point of view that counts as a complete layer. Therefore, if just a few turns are left over, it is better not to create an additional layer. This we can do either by omitting these extra turns entirely (i.e., reduce N_P slightly and incur slightly higher flux swing and core losses). Or we can reduce the diameter of the wire slightly (i.e., keep number of turns constant but incur slightly higher DC resistance). Both ways are usually preferable to increasing the F_R.*

NOTE *As we can see from Table 11.3, a bundle with seven strands has an N_h equal to that of the bundle with six strands. This is a correct indication of the fact that a 7-strand bundle is the most effective; that is, it has more copper within a given space. One strand tends to stay in the middle with six strands distributed evenly around its circumference. The strands can take turns, but this basic arrangement remains consistent at any cross-sectional point along its length.*

Core-Loss Calculations

Core loss depends on the flux swing ΔB and the frequency f (and also temperature, but we are ignoring this here). Note, however, that by convention, core loss is usually always quoted in terms of B_{AC} instead of ΔB. This is *half* the actual swing as indicated earlier. Unfortunately, there are also several units in which core loss is expressed in related literature, and the designer can get really confused. Therefore, they will all be touched upon here, including their relative transformations.

In general,

$$\text{Core loss} = \text{constant}_1 \times B^{\text{constant}_2} \times f^{\text{constant}_3} \times V_e$$

or simply

$$\text{Core loss} = \text{constant} \times B^{\text{exponent of } B} \times f^{\text{exponent of } f} \times V_e$$

We write the loss *per unit volume* as

$$P_{\text{CORE}} = \frac{\text{Core loss}}{V_e}$$

In Table 11.4 we have indicated the three main systems of units used. In Table 11.5 we have the values for the constants, and, therefore, core loss can be calculated for any application. We, of course, need to know the ΔB from the magnetics equations.

	Constant	Exponent of B	Exponent of f	B	f	V_e	P_{CORE}
System A	C_c $= \dfrac{C \times 10^4 \times p}{10^3}$	C_b $= p$	C_f $= d$	T	Hz	cm³	W/cm³
System B	C $= \dfrac{Cc \times 10^3}{10^4 \times Cb}$	p $= Cb$	d $= Cf$	G	Hz	cm³	mW/cm³
System C	K_p $= \dfrac{C}{10^3}$	n $= p$	m $= d$	G	Hz	cm³	W/cm³

TABLE 11.4 Converting between Different Systems of Core Loss

Material	c	d (Exponent of f)	p (Exponent of B)	μ	≈ B_{SAT} (G)	≈ max freq. (Hz)
Powdered Iron 8	4.3E-10	1.13	2.41	35	12500	1E+8
Powdered Iron 18	6.4E-10	1.18	2.27	55	10300	1E+8
Powdered Iron 26	7E-10	1.36	2.03	75	13800	2E+6
Powdered Iron 52	9.1E-10	1.26	2.11	75	14000	1E+7
Kool Mu 60	2.5E-11	1.5	2	60	10000	5E+5
Kool Mu 75	2.5E-11	1.5	2	75	10000	5E+5
Kool Mu 90	2.5E-11	1.5	2	90	10000	5E+5
Kool Mu 125	2.5E-11	1.5	2	125	10000	5E+5
MolyPermalloy 60	7E-12	1.41	2.24	60	6500	5E+6
MolyPermalloy 125	1.8E-11	1.33	2.31	125	7500	3E+6
MolyPermalloy 200	3.2E-12	1.58	2.29	200	7700	1E+6
MolyPermalloy 300	3.7E-12	1.58	2.26	300	7700	4E+5
MolyPermalloy 550	4.3E-12	1.59	2.36	550	7700	2E+5
HighFlux 14	1.1E-10	1.26	2.52	14	15000	1E+7
HighFlux 26	5.4E-11	1.25	2.55	26	15000	6E+6
HighFlux 60	2.6E-11	1.23	2.56	60	15000	3E+6
HighFlux 125	1.1E-11	1.33	2.59	125	15000	1E+6
HighFlux 160	3.7E-12	1.41	2.56	160	15000	8E+5
Ferrite Magnetics F	1.8E-14	1.62	2.57	3000	3000	1.3E+6
Ferrite Magnetics K	2.2E-18	2	3.1	1500	3000	2E+6
Ferrite Magnetics P	2.9E-17	2.06	2.7	2500	3000	1.2E+6
Ferrite Magnetics R	1.1E-16	1.98	2.63	2300	3000	1.5E+6
Ferrite Philips 3C80	6.4E-12	1.3	2.32	2000	3000	1E+6
Ferrite Philips 3C81	6.8E-14	1.6	2.5	2700	3000	1E+6
Ferrite Philips 3C85	2.2E-14	1.8	2.2	2000	3000	1E+6
Ferrite Philips 3F3	1.3E-16	2	2.5	1800	3000	1E+6
Ferrite TDK PC30	2.2E-14	1.7	2.4	2500	3000	1E+6
Ferrite TDK PC40	4.5E-14	1.55	2.5	2300	3000	1E+6
Ferrite FairRite 77	1.7E-12	1.5	2.3	2000	3000	1E+6

Note: Philips Ferrites is now Ferroxcube. Powdered iron grades are from Micrometals. High Flux, and Kool Mu are registered trademarks of Magnetics Inc.

TABLE 11.5 Typical Core-Loss Coefficients of Common Materials

Note that the table is only a rough initial guide. For example, the core-loss coefficients are not really constants, but vary with temperature. *Refer to vendors' datasheets for more accurate information including frequency and B_{SAT}.*

NOTE *Optimum results (in terms of overall losses) are said to be attained if*

$$\frac{\text{Core loss}}{\text{Copper loss}} = \frac{2}{\text{exponent of } B}$$

But treat this only as a general guideline. For example, in most off-the-shelf inductors for DC-DC converters, the copper loss is over 90 percent of the total loss. For transformers, we may be closer to 50-50. The exponent of B for ferrites tends to be around 2; for frequency it is around 1.5.

CHAPTER 12

Forward and Flyback Converters: Step-by-Step Design and Comparison

Introduction

This is a follow-up chapter to everything we have learned in previous chapters on AC-DC design in particular. It brings it all together in a top-down numerical example. Certainly a level of expert knowledge obtained from previous chapters will be required, but we are trying to keep it very accessible, so we may repeat previous learnings.

In a Flyback topology, the selection of the transformer core is fairly straightforward. The Flyback transformer has a dual function: It not only provides a step-up or step-down ratio based on the Primary-to-Secondary turns ratio, but it also serves as a medium for energy storage. The Flyback is a derivative of the Buck-boost and shares its unique property that not just part but *all* of the energy that is delivered to the output must have previously been stored (as magnetic energy) within the core. This is consistent with the fact that the secondary winding conducts only when the primary winding stops, and vice versa. We can intuitively visualize this as the windings being *out of phase*. So we have an endless sequence of energy stored-and-released followed by stored-and-released, and so on. The core-selection criterion is thus very simply as follows: The core must basically be capable of storing each packet of energy (per cycle) passing through it. That packet is equal to $P_{IN}/f = \Delta \varepsilon \approx \varepsilon_{PEAK}/1.8 = (L \times I_{PEAK}^2)/3.6$, in terms of joules. Here f is the switching frequency and ε is energy (see Fig. 5.6 of *Switching Power Supplies A-Z*, 2d ed, for a derivation of the preceding). Equivalently, we can just state that the peak current I_{PEAK} should not cause *core saturation*, though that approach gives us no intuitive understanding of the fact that if we double the switching frequency, the energy packets get reduced in half, and so in effect the same core, designed properly, can handle twice the input-output energy. But that is indeed always true whenever we use an inductor or transformer as an energy-storage medium in switching power conversion.

Coming to a Forward converter, at least two things are very different right off the bat.

1. Not all the energy reaching the output necessarily needs to get stored in a magnetic energy storage medium (core) along the way. Keep in mind that the Forward converter is based on the Buck topology. We realize from p. 208 of *Switching Power Supplies A-Z*, 2d ed, that only $1 - D$ times the total energy gets cycled through the core in a Buck topology. So, for a given P_O, and a given switching frequency, the Buck or Forward core will be roughly half the size of a Buck-boost or Flyback core, handling the same power (assuming $D \approx 1 - D \approx 0.5$).

2. Further, in a Forward converter, the energy storage function does not reside in the transformer. The storage requirement, however limited, is fulfilled entirely by the secondary-side choke, not the transformer. So we can well ask: What does the transformer do in a Forward converter anyway? It only provides *transformer action*, i.e., voltage step-up corresponding to current step-down or voltage step-down corresponding to current step-up function, based on the turns ratio—which is, in a way, half the function of a Flyback transformer. Once it provides that step-up or step-down ratio, there is an additional step-down function provided by simply running the secondary-side choke in a chopped-voltage fashion, as in any regular (nonisolated) Buck. That is why we always consider the output rail of a Forward

converter as having been derived from the input rail, with two successive step-down factors applied, as shown

$$V_O = (D \times V_{IN}) \times \frac{N_S}{N_P}$$
$$\Uparrow \qquad\qquad \Uparrow$$
$$\text{Buck} \qquad \text{Transformer action}$$

The perceptive will notice that the Forward converter's transformer action could be such that we use the transformer turns ratio to give an intermediate step-up instead of a step-down function, and then follow it up with a step-down function accruing from the inherent Buck stage based around the secondary-side choke. That could in effect give us another type of (overall) Buck-boost converter—but not based on the classic inductor-based Buck-boost anymore. And that is what we, in effect, usually do in the LLC resonant topology (see Chap. 19).

The secondary-side choke selection criterion is straightforward too: It is sized so that it does not saturate with the peak current passing through it (typically 20 percent more than the load current). We see that it is the same underlying criterion as in a Flyback, Buck, Buck-boost, and even a Boost. So this leaves us with the basic question: How do we pick the Forward converter *transformer*? What does its size depend on? What are its selection criteria?

There are two major factors affecting the Forward converter transformer selection. First, we need to understand that the primary and secondary windings conduct *at the same time*. So they are intuitively "in phase." The observed *transformer action*, that is, the simple turns-ratio–based current flow of the secondary winding, is in fact just a direct result of induced electromotive force (EMF, i.e., voltage) based on Faraday's and Lenz's laws. The induced EMF in the secondary winding in response to the changing flux caused by the changing current in the primary winding tries to oppose the change of flux, and since both windings can conduct current at the same time in a Forward converter, the two voltages (applied and induced) lead to simultaneous currents in the windings, which create equal and opposite flux contributions in the core, cancelling each other out. Yes, completely so! In effect, the *core* of the Forward converter's transformer does not "see" *any* of the flux associated with the transfer of power across its isolation barrier. Note that this flux-cancellation "magic" was physically impossible in a Flyback, simply because, though there was induced EMF in the Secondary during the ON-time, the output diode was so pointed that it blocked any current flow arising from this induced voltage—so there was no possibility of having two equal and opposite flux contributions occurring (at the same time).

This leads to the big question: If the *core* of the Forward converter's transformer does not see any of the flux related to the ongoing energy transfer through the transformer, can we transfer limitless energy through the transformer? The answer is no, because the DC resistance of copper gets in the way. This creates a *physical limitation* based on the available window area W_a of the core. We just cannot stack endless copper windings in a restricted space to support any power throughput. Certainly not if we intend to keep to certain thermal limits; because though the core may be totally "unaware" of the actual currents in the windings (because of flux cancellation), the windings themselves do see I^2R (ohmic) losses. So eventually, for thermal reasons, we have to keep to within a certain acceptable *current density*. This in effect restricts the amount of power we can transfer through a Forward converter transformer. We intuitively expect that if we double the available window area W_a, we would be able to double the currents (and the power throughput) too, for a given (acceptable) current density. In other words, we expect roughly (intuitively)

$$P_O \propto W_a$$

Truth does in fact support intuition in this case. But there is another key factor too: A transformer needs a certain excitation (magnetization) current to function to be able to provide transformer action in the first place. So there is a certain relationship to the core itself, its *ferrite-related* dimensions, not just the window area (air dimensions) that it provides. A key parameter that characterizes this aspect of the core is the area of its center limb, or A_e (often just called A in this chapter). Finally we expect the power to be related to both factors: the air-related component W_a and the ferrite-related component A_e:

$$P_O \propto W_a \times A_e$$

The product $W_a \times A_e$ is generically called AP, or the area product of the core. See Fig. 12.1.

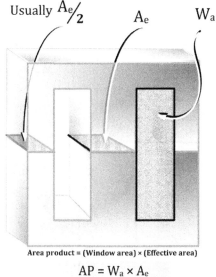

FIGURE 12.1 Basic definition of area product.

As indicated, we intuitively expect that doubling the frequency f will double the power too. So we expect

$$P_O \propto \mathrm{AP} \times f$$

Or better still, since in the worst case (losses after the transformer) the transformer is responsible for the entire *incoming* power, it makes more intuitive sense to write

$$P_{IN} \propto \mathrm{AP} \times f$$

Finer Classes of Window Area and Area Product (Some New Terminology)

As we can see from Figs. 12.2 and 12.3, we can actually break up the window area into several windows (with associated area products). We should try to distinguish between them for the subsequent analysis, since typically this becomes a source of major confusion in literature, with innumerable equations and fudge factors (generically called K_x usually) being used

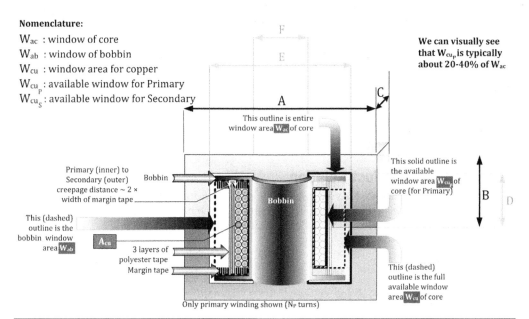

FIGURE 12.2 Finer divisions of window area and area product.

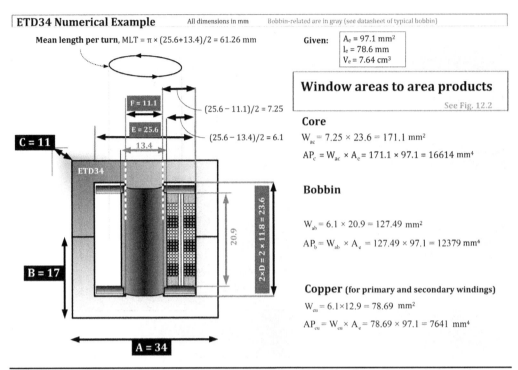

FIGURE 12.3 Numerical example showing the nomenclature of popular dimensions and also various window areas and area products.

apparently to fit equations somehow to empirical data, rather than deriving equations from first principles and then seeing how they match the data. So we create some descriptors here.

W_{ac}. This is the core window area. Multiplied by A_e, we get AP_c.

W_{ab}. This is the bobbin window area. Multiplied by A_e, we get AP_b.

W_{cu}. This is the window available to wind copper in (both primary and secondary windings). Multiplied by A_e, we get AP_{cu}.

NOTE *In a safety-approved transformer for AC-DC applications, we typically need 8-mm creepage between primary and secondary windings (see Fig. 12.2 and Fig. 17.1), so a 4-mm margin tape is often used (but sometimes only 2.5 to 3 mm wide nowadays). For telecommunication applications, where only 1500 Vac isolation is required, a 2-mm margin tape will suffice and provide 4 mm of creepage. The bobbin, insulation, etc., significantly lower the available area for copper windings—to about 0.5 × (or half) the core window area W_{ac}.*

W_{cu_p}. This is the window available for the primary winding. Multiplying it by A_e, we get AP_{cu_p}. For a safety-approved AC-DC transformer, for example, this area may be only 0.25 times W_{ac} (typically assuming W_{cu} is split equally between the primary and secondary windings).

W_{cu_s}. This is the window available for the secondary winding. Multiplying it by A_e, we get AP_{cu_s}.

Power and Area Product Relation

We remember that since the voltage across the inductor during the ON-time, V_{ON}, equals the input rail V_{IN} in *almost* all topologies (though not in the half-bridge, for example), from the original form of the voltage-dependent (Faraday) equation

$$\Delta B = \frac{V_{IN} \times t_{ON}}{N_P \times A} \quad \text{T}$$

Here A is the effective area of the core (same as A_e), expressed in square meters. (To remember try this: "volt-seconds equals NAB"). Note that

$$N_P \times A_{cu} = 0.785 \times W_{cu_p}$$

This is because a round wire of cross-sectional area A_{cu} occupies only 78.5 percent (i.e., $\pi^2/4$) of the space (square of area D^2) that it physically occupies within the layer (see Fig. 10.3). Here W_{cu_P} is the (rectangular) physical window area available to wind copper in, but is reserved only for the primary turns. We are typically assuming that the available copper space W_{cu} is split equally between primary and secondary windings. That is a valid assumption mostly.

Solving for N_P, the number of primary turns is

$$N_P = \frac{0.785 \times W_{cu_P}}{A_{cu}}$$

Using this in the voltage-dependent equation, we get

$$\Delta B = \frac{V_{IN} \times t_{ON} \times A_{cu}}{0.785 \times W_{cu_P} \times A_e} \quad T$$

Performing some manipulations

$$\Delta B = \frac{V_{IN} \times t_{ON} \times A_{cu}}{0.785 \times W_{cu_P} \times A_e} = \frac{V_{IN} \times (I_{IN}/I_{IN}) \times (D/f) \times A_{cu}}{0.785 \times W_{cu_P} \times A_e}$$

$$= \frac{P_{IN} \times D \times A_{cu}}{I_{IN} \times 0.785 \times W_{cu_P} \times A_e \times f} = \frac{P_{IN} \times D \times A_{cu}}{(I_{SW} \times D) \times 0.785 \times W_{cu_P} \times A_e \times f}$$

$$= \frac{P_{IN}}{(I_{SW}/A_{cu}) \times 0.785 \times W_{cu_P} \times A_e \times f} = \frac{P_{IN}}{J_{A/m^2} \times 0.785 \times AP_{cu_P} \times f}$$

where J_{A/m^2} is the current density expressed in A/m^2 and AP_{cu_P} is the *area product* for the copper allocated to the primary windings ($AP_{cu_P} = A_e \times W_{cu_P}$). Note that I_{SW} *here is the center of ramp (COR) of the switch current (its average value during the ON-time). The current density is, therefore inherently based on that, not on the RMS current,* as is often erroneously interpreted. Let us now convert the preceding into CGS units for convenience (writing units explicitly in the subscripts to avoid confusion). We get

$$\Delta B_G = \frac{P_{IN}}{J_{A/cm^2} \times 0.785 \times AP_{cu_{P\_cm^4}} \times f_{Hz}} \times 10^8$$

where AP_{cu_P} is expressed in square centimeters now. Finally, converting the current density into cmil/A (see Table 10.1) by using

$$J_{cmil/A} = \frac{197,353}{J_{A/cm^2}}$$

we get

$$\Delta B_G = \frac{P_{IN} \times J_{cmil/A}}{197,353 \times 0.785 \times f_{Hz} \times AP_{cu_{P\_cm^4}}} \times 10^8 \quad G$$

Solving for the Primary copper area product

$$AP_{cu_{P\_cm^4}} = \frac{645.49 \times P_{IN} \times J_{cmil/A}}{f_{Hz} \times \Delta B_G} \quad cm^4$$

Let us do some numerical substitutions here. Assuming a typical target current density of 600 cmil/A (based on COR current value as previously explained), and a typical allowed ΔB equal to 1500 G (to keep core losses down and to avoid saturation), we get the following core selection criterion:

$$AP_{cu_{P\_cm^4}} = 258.2 \times \frac{P_{IN}}{f_{Hz}} \quad cm^4 \quad \text{(for 600 cmil/A, based on center of current ramp)}$$

Keep in mind that so far this is an exact relationship. It is based on the window area available for the primary winding, because, along with the target current density in mind (600 cmil/A), this determines the ampere-turns and thus the flux.

In p. 153 of *Switching Power Supplies A-Z*, 2d ed, we derived the following relationship in a similar manner to what we have done here:

$$P_{IN} = \frac{AP_{cm^4} \times f_{Hz}}{675.6}$$

Equivalently

$$AP_{cm^4} = 675.6 \times \frac{P_{IN}}{f_{Hz}}$$

This too was based on a COR current density of 600 cmil/A. The real difference with the equation we have just derived is that the area product in the A-Z book used the entire core area. In other words we had derived this:

$$AP_{c_{cm^4}} = 675.6 \times \frac{P_{IN}}{f_{Hz}}$$

Compared to what we just derived (based on estimated area reserved for the primary winding)

$$AP_{cu_{P\_cm^4}} = 258.2 \times \frac{P_{IN}}{f_{Hz}}$$

In effect we had assumed in the A-Z book that $AP_{cu_P}/AP_c = 258.2/675.6 = 0.38$. (*Note*: The reason it seems to be set to 0.3 in the A-Z book is this: $0.3/0.785 = 0.38$! Think about it. There is no contradiction. The factor 0.785 was not factored into the current density.) But, in the A-Z book, as in most literature, the utilization factor K was just a fudge factor, applied to make equations fit data (with some physical reasoning to satisfy the critics). But in our ongoing analysis, we are trying to avoid all inexplicable fudge factors. So we should assume the equation we have just come up with is accurate.

Keep in mind that though the maximum flux swing of 1500 G is still a very fair assumption to still make in most types of practical Forward converters (to limit core losses and avoid saturation during transients), the current density of 600 cmil/A (COR value) needs further examination. And until we do that, let us stick to the more general equation connecting area product and power (making no assumptions yet).

$$AP_{cu_{P\_cm^4}} = \frac{645.49 \times P_{IN} \times J_{cmil/A}}{f_{Hz} \times \Delta B_G} \quad cm^4 \quad \text{(Maniktala, most general)}$$

In terms of A/cm², this is

$$AP_{cu_{P\_cm^4}} = \frac{645.49 \times P_{IN} \times 197,353}{f_{Hz} \times \Delta B_G \times J_{A/cm^2}}$$

or

$$AP_{cu_{P\_cm^4}} = \frac{12.74 \times P_{IN}}{f_{kHz} \times \Delta B_T \times J_{A/cm^2}} \quad \text{(Maniktala, most general)}$$

Keep in mind that J here is based on the COR value.

Current Density and Conversions Based on D

At the very start of the preceding derivation, when we set $I_{IN} = I_{SW} \times D$, in effect the current density was a *COR* current density, not an RMS value. That is how we *eliminated D* from the equation. However, heating does not depend directly on the COR value, but on its RMS. So, in effect, looking at it the other way, our area product equation actually implicitly *depends on D* through the COR current density value we picked. If we know D, we can convert the COR-based current density to an equivalent RMS current density value.

The 600-cmil/A value we used to plug in numerically into the equation should perhaps be written out more clearly as 600 cmil/A_{COR}, where A_{COR} is the center-of-ramp value of the current in amperes. We ask: What is 600 cmil/A in terms of RMS current? As indicated, that actually depends on the duty cycle. Assuming a ballpark nominal figure of $D = 0.3$ for a Forward converter, a current pulse of height 1 A leads to an RMS of $1 A \times \sqrt{D} = 1 A \times \sqrt{0.3} = 0.548$ A. In other words, 600 cmil/A_{COR} means that 600 cmil is being allocated for 0.548 A_{RMS}. In other words, this is equivalent to allocating $600/0.548 = 1095$ cmil per A_{RMS}. So we get the following conversions:

$$\frac{600 \text{ cmil}}{A_{COR}} \equiv \frac{600}{0.548} = \frac{1095 \text{ cmil}}{A_{RMS}}$$

In other words, 600 cmil/A_{COR} can be expressed as 1095 cmil/A_{RMS}, or

$$\frac{197{,}353}{600} = \frac{330 A_{COR}}{cm^2} \quad \text{(in terms of COR current, see Table 10.1)}$$

or

$$\frac{197{,}353}{1095} = \frac{180\, A_{RMS}}{cm^2} \quad \text{(in terms of RMS current, for } D = 0.3\text{)}$$

Note that we were in effect asking for a current density of 180 A/cm², which is rather lower (more conservative) than usually accepted. But let us discuss this further below.

Optimum Current Density

What really is a good current density to target in an application? Is it 600 cmil/A_{COR} (i.e., 180 A_{RMS}/cm² for $D = 0.3$) or something else? Actually, 600 mil/A_{COR} is a tad too conservative as we too will agree here. But, in general, this is a topic of great debate, much confusion, and widely dissimilar recommendations in the industry. We need to sort it out.

As a good indication of the industrywide dissonance on this subject, see the 40-W Forward converter design from an engineer at Texas Instruments at www.ti.com/lit/ml/slup120/slup120.pdf. He writes,

> The transformer design uses the Area Product Method that is described in [3]. This produced a design that was found to be core loss limited, as would be expected at 200 kHz. The actual core selected is a Siemens-Matsushita EFD 30/15/9 made of N87 material. The area-product of the selected core is about 2.5 times more area-product than the method in [3] recommended. We selected the additional margin with the intention of allowing additional losses due to proximity effects in a multi-layer foil winding that is required for carrying the large secondary currents.

In this extract, the engineer refers to reference [3], which is: Lloyd H. Dixon's "Power Transformer Design for Switching Power Supplies," Rev. 7/86, *SEM-700 Power Supply Design Seminar Manual*, Unitrode Corporation, 1990, section M5.

This means that *Unitrode [now Texas Instrument (TI)] has a recommendation on core size of Forward converters that was almost 250 percent off the mark, as reported by another TI engineer who actually tried to follow his own company's design note to design a practical converter.*

It therefore seems it is a good idea to stay conservative here, as no one in the commercial arena will appreciate or reward a thermal (or EMI) issue holding up safety approvals and production at the very last moment.

Let us start with the basics: It has been stated and seemingly widely accepted that for most E-core–type Flyback (not Forward) transformers, a current density of 400 cmil/A_{RMS} (equivalent to $197{,}353/400 \approx 500\, A_{RMS}/cm^2$) is acceptable. This seems to have at least served engineers making evaluation boards well. But is it really acceptable in trying to achieve a maximum 55°C rise (internal hot-spot temperature), so as to qualify as a commercial safety-approved Class A Forward converter transformer (maximum of 105°C)?

The problem is that a current density of 500 A_{RMS}/cm² may work for low-frequency sine waveforms, as used by most core vendors, but when it comes to Forward converters in particular, because of the skin and proximity effects, as best described by Dowell historically, the ratio F_R (AC resistance divided by DC resistance) is much higher than unity. Note that Dowell used high-frequency waves for a change, but still assumed sinusoidal waves. After that, a lot of Unitrode application notes invoked the original form of Dowell's equations, with sine waves, and arrived at "achievable" F_R values slightly greater than 1, with proper high-frequency winding techniques, and so on. However, in modern days, when we include the high-frequency harmonics of the typical *square waveforms* of switching power conversion, the best achievable AC resistance ratio F_R is not close to 1, but about 2. In other words, mentally we can think of this as windings made with a new metal that has double the resistivity of copper. Now, to arrive at the same acceptable value of heating and temperature rise as regular (low-frequency) *copper transformers*, a good target in a Forward converter would be to allocate twice the area (i.e., target half the current density expressed in A/cm²). That means we really want to target 800 cmil/A_{RMS} for a Forward converter, which would be roughly comparable in temperature to 400 cmil/A_{RMS} for a Flyback. So, assuming a Forward converter with $D = 0.3$, we need to target

$$\frac{800 \text{ cmil}}{A_{RMS}} \equiv \frac{800 \times 0.548 = 440 \text{ cmil}}{A_{COR}}$$

or

$$\frac{197{,}353}{800} \approx \frac{250\ A_{RMS}}{cm^2} \quad \text{(in terms of RMS current)}$$

or

$$\frac{197{,}353}{440} = \frac{450 A_{COR}}{cm^2} \quad \text{(in terms of COR current, for } D = 0.3\text{)}$$

If the duty cycle was $D = 0.5$ (as in a Forward converter at lowest line condition), since $\sqrt{(0.5)} = 0.707$, we could write the target current density as

$$\frac{800\ \text{cmil}}{A_{RMS}} \equiv \frac{800 \times 0.707 = 565\ \text{cmil}}{A_{COR}}$$

or

$$\frac{197{,}353}{800} \approx \frac{250\ A_{RMS}}{cm^2} \quad \text{(in terms of RMS current)}$$

or

$$\frac{197{,}353}{565} = \frac{350\ A_{COR}}{cm^2} \quad \text{(in terms of COR current, for } D = 0.5\text{)}$$

We see that for both the above duty cycles, what remained constant was the following design target: a Forward converter transformer current density of $250\ A_{RMS}/cm^2$, exactly half the *widely, and blindly accepted* current density target. The underlying reason was F_R was at best 2, not 1.

We now recall our accurate equation for a Forward converter transformer:

$$AP_{cu_{P\_cm^4}} = \frac{645.49 \times P_{IN} \times J_{cmil/A_{COR}}}{f_{Hz} \times \Delta B_G}\ cm^4$$

If we plug in our recommended current density of 800 cmil/A_{RMS}, i.e., 440 cmil/A_{COR} (for $D = 0.3$), and also assume that we have a utilization factor of 0.25 (ratio of primary winding area to core winding area, see Fig. 12.2), we get our basic recommendation to be

$$AP_{c_{cm^4}} = \frac{645.49 \times P_{IN} \times J_{cmil/A_{COR}}}{f_{Hz} \times \Delta B_G} = \frac{645.49 \times P_{IN} \times 440}{0.25 \times f_{Hz} \times \Delta B_G} = 11{,}360{,}624 \times \frac{P_{IN}}{f_{Hz} \times \Delta B_G}$$

or

$$AP_{c_{cm^4}} = 113.6 \times \frac{P_{IN}}{f_{Hz} \times \Delta B_T} \quad \text{(Maniktala, for } D = 0.3,\ J = 250\ A_{RMS}/cm^2,\ K = 0.25\text{)}$$

Plugging in a typical value of $\Delta B = 1500$ G, we get

$$AP_{c_{cm^4}} = \frac{645.49 \times P_{IN} \times 440}{f_{Hz} \times 1500 \times 0.25} = 755 \times \frac{P_{IN}}{f_{Hz}}$$

Or equivalently (using kilohertz),

$$AP_{c_{cm^2}} = 0.75 \times \frac{P_{IN}}{f_{kHz}} \quad \text{(Maniktala, for } D = 0.3,\ \Delta B = 0.15\ T,\ J = 250\ A_{RMS}/cm^2,\ K = 0.25\text{)}$$

As we can see, this equation asks for a slightly larger core than we had suggested in the numerical example from the *A-Z* book. In the *A-Z* book, though, we had used a little more generous (conservative) current density, we also set a much more optimistic *utilization (fudge) factor*. We had derived

$$AP_{c_{cm^4}} = 675.6 \times \frac{P_{IN}}{f_{Hz}}$$

or equivalently

$$AP_{c_{cm^4}} = 0.676 \times \frac{P_{IN}}{f_{kHz}} \quad \text{(Maniktala, for } D = 0.3,\ \Delta B = 0.15\ T,\ J = 180\ A_{RMS}/cm^2,\ K = 0.38\text{)}$$

We conclude that the new equation we have now derived

$$AP_{cm^2} = 0.75 \times \frac{P_{IN}}{f_{kHz}} \quad \text{(Maniktala, for } D = 0.3, \Delta B = 0.15 \text{ T}, J = 250 \text{ A}_{RMS}/\text{cm}^2, K = 0.25)$$

is a tad more realistic (and conservative in terms of available window area) than the older one in the A-Z book. This one asks for slightly higher area product (for a given power).

This slight modification of the A-Z book recommendation is a little more helpful for designing a safety-approved Class A Forward converter transformer running at a nominal $D = 0.3$.

Note that the underlying assumptions in our new equation include a *maximum flux swing of 1500 G, a current density of 250 A_{RMS}/cm^2, and a utilization factor (ratio of primary winding area W_{cup} to the full core window area W_{ac}) of 0.25.*

If we have a core with a certain core-area product, we can also flip it to find its power capability as follows:

$$P_{IN} = \frac{AP_{cm^4} \times f_{Hz}}{754} = 1.33 \times 10^{-3} \times \left(AP_{cm^4} \times f_{Hz} \right)$$

$$P_{IN} = 1.33 \times AP_{cm^4} \times f_{kHz} \quad \text{(Maniktala, for } D = 0.3, \Delta B = 0.15 \text{ T},$$

$$J = 250 \text{ A}_{RMS}/\text{cm}^2, K = 0.25)$$

For example, at $f = 200$ kHz, the ETD-34 core-set, with a core-area product of 1.66 cm^4, is suitable for

$$P_{IN} = \frac{1.66 \times 200000}{754} = 440 \text{ W} \quad \text{(recommendation example based on Maniktala)}$$

With an estimated efficiency of say 83 percent, this would work for a converter with $P_O = 365$ W.

Having understood this, we would like to compare with the equations others are espousing in related literature to see where we stand vis-à-vis their recommendations. Here is a list of other "similar" equations. It is a jungle where angels have feared to tread.

Industry-Recommended Equations for the Area Product of a Forward Converter

Fairchild Semi Recommendation

For example, see "The Forward-Converter Design Leverages Clever Magnetics by Carl Walding" at http://powerelectronics.com/mag/Fairchild.pdf:

$$AP_{mm^4} = \left(\frac{78.72 \times P_{IN}}{\Delta B \times f_{Hz}} \right)^{1.31} \times 10^4$$

This was alternatively expressed in Application Note AN-4134 from Fairchild as

$$AP_{mm^4} = \left(\frac{11.1 \times P_{IN}}{0.141 \times \Delta B \times f_{Hz}} \right)^{1.31} \times 10^4$$

But it is the same equation. It seems to be using the term area product, for the entire core. The field is in tesla. We can also rewrite this in terms of cm^4 as

$$AP_{cm^4} = \left(\frac{78.72 \times P_{IN}}{\Delta B_T \times f_{Hz}} \right)^{1.31} \quad \text{(Fairchild)}$$

Compare it to our equation:

$$AP_{cm^4} = \frac{113.6 \times P_{IN}}{f_{Hz} \times \Delta B_T} \quad \text{(Maniktala)}$$

We can simplify the Fairchild equation and set $\Delta B = 0.15$ T (the usual typical optimum flux swing to avoid saturation and keep core losses small). We get

$$AP_{cm^4} = \left(\frac{78.72 \times P_{IN}}{0.15 \times f_{kHz}} \right)^{1.31}$$

$$AP_{cm^4} = 0.43 \times \left(\frac{P_{IN}}{f_{kHz}} \right)^{1.31} \quad \text{(Fairchild, with } \Delta B = 0.15 \text{ T)}$$

We can compare this with our equation:

$$\mathrm{AP}_{c_{cm^4}} = 0.75 \times \left(\frac{P_{IN}}{f_{kHz}}\right) \quad \text{(Maniktala, with } \Delta B = 0.15 \text{ T)}$$

For example, for 440-W input power, we know at 200 kHz, we recommend the ETD-34 with $\mathrm{AP}_c = 1.66$ cm^4 (see Fig. 12.3). What does the Fairchild equation recommend? We get

$$\mathrm{AP}_{c_{cm^4}} = 0.43 \times \left(\frac{440}{200}\right)^{1.31} = 1.21 \text{ cm}^4 \quad \text{(Fairchild recommendation example)}$$

ETD-29 has an area product (core) of 1.02 cm^4. So we will still end up using ETD-34. But, in general, at least for lower powers and frequencies, the Fairchild equation can ask for up to half the area product, thus implying much smaller cores. It seems more aggressive, and unless forced into a default larger core size, it will likely require either forced air cooling, or better (more expensive) core materials to compensate higher copper losses by much lower core losses. Or the transformer will be either non-safety-approved, or Class B safety-approved.

We can also solve the Fairchild equation for the power throughput from a given (core) area product (using typical $\Delta B = 1500$ G)

$$\mathrm{AP}_{c_{cm^4}} = \left(\frac{78.72 \times P_{IN}}{0.15 \times f_{kHz}}\right)^{1.31}$$

$$P_{IN} = \mathrm{AP}_{c_{cm^4}}^{0.763} \times \frac{0.15 \times f_{Hz}}{78.72} = 1.9 \times f_{kHz} \times \mathrm{AP}_{c_{cm^4}}^{0.763}$$

$$P_{IN} = 1.9 \times f_{kHz} \times \mathrm{AP}_{c_{cm^4}}^{0.763} \quad \text{(Fairchild, for } \Delta B = 0.15 \text{ T)}$$

TI/Unitrode Recommendation

For example, see www.ti.com/lit/ml/slup126/slup126.pdf and www.ti.com/lit/ml/slup205/slup205.pdf:

$$\mathrm{AP}_{c_{cm^4}} = \left(\frac{11.1 \times P_{IN}}{K \times \Delta B_{Tesla} \times f_{Hz}}\right)^{1.143}$$

In this case K is a fudge factor related both to window utilization and topology. Unitrode asks to fix this at 0.141 for a single-ended Forward, and at 0.165 for a Bridge/half-Bridge. So with that, we get (for a single-ended Forward, assuming core area product as before):

$$\mathrm{AP}_{c_{cm^4}} = \left(\frac{11.1 \times P_{IN}}{0.141 \times \Delta B_T \times f_{Hz}}\right)^{1.143} = \left(\frac{78.72 \times P_{IN}}{\Delta B_T \times f_{Hz}}\right)^{1.143} \quad \text{(Unitrode)}$$

which is almost identical to the Fairchild equation, except that the exponent is mysteriously 1.143, leading to a much slower *rise* with power (and a *fall* with frequency), as compared to the exponent of 1.31 in the Fairchild equation. Note that this equation too (as the Fairchild equation) is said to be based on a current density of 450 A$_{RMS}$/cm^2—far more aggressive than the 250 A$_{RMS}$/cm^2 that we are espousing. But in all Unitrode application notes, the best achievable F_R was calculated to be just slightly larger than 1, *because it was based on sinusoidal waveforms*, whereas in reality, the best-case F_R is actually closer to 2, as we have assumed in our equations (see Fig. 11.8 too). That is why, our estimate seems excessively conservative, but is more accurate and realistic. However, in the TI/Unitrode recommendation, a better fit to actual data seems to have been created artificially, by introducing an arbitrary fudge factor K. Unfortunately, logically speaking, any such utilization factor should be changed or tweaked depending on the types of core being used. But that aspect is invariably overlooked.

We can also solve the Unitrode equation for the power throughput from a given (core) area product (using typical $\Delta B = 1500$ G)

$$P_{IN} = \mathrm{AP}_{c_{cm^4}}^{0.875} \times \frac{0.15 \times f_{Hz}}{78.72} = 1.9 \times f_{kHz} \times \mathrm{AP}_{c_{cm^4}}^{0.875}$$

$$P_{IN} = 1.9 \times f_{kHz} \times \mathrm{AP}_{c_{cm^4}}^{0.875} \quad \text{(Unitrode, for } \Delta B = 0.15 \text{ T)}$$

Basso/On-Semi Recommendation

For example, see www.onsemi.com/pub_link/Collateral/TND350-D.PDF:

$$AP_{c_{cm^4}} = \left(\frac{P_O}{K \times \Delta B_T \times f_{Hz}}\right)^{4/3}$$

It is suggested that $K = 0.014$ for a Forward converter. This is another unexplained fudge factor really. Simplifying, we get for a Forward converter

$$AP_{c_{cm^4}} = \left(\frac{71.43 \times P_O}{\Delta B_T \times f_{Hz}}\right)^{1.33}$$

This is indeed very close to the Fairchild equation, though this equation unconsciously and implicitly assumes 100 percent efficiency, because it uses the output power instead of the input power, whereas, the worst-case assumption is that all the losses occur after passing through the transformer, not before (see Fig. 2.13). In that case, the transformer has to throughput the full input power, not the lower output power. To try and correct for this unfortunate assumption, we now assume 90 percent efficiency. We then get

$$AP_{c_{cm^4}} = \left(\frac{71.43 \times 0.9 \times P_{IN}}{\Delta B_T \times f_{Hz}}\right)^{1.33} = \left(\frac{64.3 \times P_{IN}}{\Delta B_{Tesla} \times f_{Hz}}\right)^{1.33} \quad \text{(On-Semi corrected)}$$

Note that On-Semi says this is based on a window utilization factor of 0.4 and a current density of 420 A/cm². We assumed a 90 percent efficiency to get to the preceding equation.

The original, uncorrected On-Semi equation can also be written out for power throughput in terms of (core) area product as follows:

$$AP_{c_{cm^4}} = \left(\frac{71.43 \times P_O}{\Delta B_T \times f_{Hz}}\right)^{1.33} \Rightarrow (AP_{c_{cm^4}})^{1/1.33} = \left(\frac{71.43 \times P_O}{\Delta B_T \times f_{Hz}}\right)$$

$$P_O = \frac{\Delta B_T \times f_{Hz}}{71.43} \times AP_{c_{cm^4}}^{0.752}$$

For a flux swing of 1500 G,

$$P_O = AP_{c_{cm^4}}^{0.752} \times \frac{0.15 \times f_{Hz}}{71.43} = 2.1 \times f_{kHz} \times AP_{c_{cm^4}}^{0.752}$$

$$P_{IN} = 2.1 \times AP_{c_{cm^4}}^{0.752} \times f_{kHz} \quad \text{(On-Semi, for } \Delta B = 0.15 \text{ T, 100 percent efficiency)}$$

ST Micro Recommendation

For example, see AN-1621 at http://www.st.com/st-web-ui/static/active/cn/resource/technical/document/application_note/CD00043746.pdf:

$$AP_{c_{cm^4}} = \left(\frac{67.2 \times P_{IN}}{\Delta B_T \times f_{Hz}}\right)^{1.31} \quad \text{(ST Micro)}$$

$$P_{IN} = AP_{c_{cm^4}}^{0.763} \times \frac{0.15 \times f_{Hz}}{67.2} = 2.23 \times f_{kHz} \times AP_{c_{cm^4}}^{0.763}$$

$$P_{IN} = 2.23 \times f_{kHz} \times AP_{c_{cm^4}}^{0.763} \quad \text{(ST Micro, for } \Delta B = 0.15 \text{ T)}$$

Keith Billings and Pressman Recommendation and Explanation

For example, see *Switching Power Supply Design*, 3rd ed., by Abraham Pressman, Keith Billings, and Taylor Morey, and *Switchmode Power Supply Handbook* by Keith Billings.

Billing actually starts to derive the requisite equation in a manner identical to ours, based on basic principles, but then suddenly digresses and arrives at the exact same equation provided by TI/Unitrode previously, complete with the arbitrary fudge factor K.

This leads us to the origin of the odd exponent we are seeing in almost all the industrywide equations. Where did that come from? Almost all the equations are apparently based on an old empirical equation found in *Transformer and Inductor Design Handbook* by Colonel Wm. T. McLyman. The reason for the odd exponent stems from a *completely empirical statement* that says the optimum current density is not a constant as we assumed but is a function of area product. The paradox is that everyone (including Billings) continue to state the current density target as a fixed number anyway: 420 or 450 A/cm². But the inclusion of the odd exponent implies otherwise. Because, as indirectly explained by Billings himself in his derivation and his derivation parallels ours, except to the point that Billings plugs in McLyman's equation

$$J_{A/m^2} = 450 \times 10^4 \times AP^{-0.125}$$

So it seems that now the target current density is suddenly a function of area product, in direct contradiction to previous statements, which had declared the target to be a fixed value.

Nevertheless continuing the derivation as per Billings (ignoring fudge factors, etc., and replacing them with just an X here)

$$AP = \frac{X \times P_{IN}}{AP^{-0.125} \times \Delta B \times f}$$

$$AP^{1-0.125} = AP^{0.875} = \frac{X \times P_{IN}}{\Delta B \times f}$$

$$AP^{0.875/0.875} = AP = \left(\frac{X \times P_{IN}}{\Delta B \times f}\right)^{1/0.875} = \left(\frac{X \times P_{IN}}{\Delta B \times f}\right)^{1.143}$$

$$AP = \left(\frac{X \times P_{IN}}{\Delta B \times f}\right)^{1.143}$$

That is the underlying logic of how the strange exponent of 1.14 (or something else very close) appears in almost all equations, especially the early TI/Unitrode notes. Clearly, *the presence of this exponent implicitly assumes a variable current density, but that is not what is usually stated alongside*. Perhaps that explains the emergence of the fudge factors. It was just to get a better match to bench data. But, as mentioned, the fudge factors logically need to change with the transformer cores being used, and also their construction. For example, we may be using margin tape to comply with safety requirements, something that was ignored in the past. And so on. Besides, we realize that Dowell's equations, on which a lot of prior design equations seem to be based on, assumed sine waves. And so the AC resistance and transformer dissipation were severely under-estimated to start with. But two wrongs do not make a right.

Most prevalent equations seem to be far more aggressive at lower wattages than our recommendations. But it is possible that they will work. Keep in mind that since smaller transformers have a larger exposed surface area to volume, they cool better (smaller thermal resistance), and so inaccuracies in setting more aggressive current densities for smaller cores were perhaps not noticed, until larger cores appeared. In that case, temperatures rose much higher than expected. So now, empirically, it was decided to adjust the core size down for a given power requirement, just to get a larger surface area to allow it to cool, and of course a larger window area for allowing improved current density too. That is likely how the term −0.125 in the McLyman current density versus area product equation appeared, which in turn led to the odd exponents we see: such as 1.14, 1.31, and so on.

Disregarding where they all came from, we can certainly plot them all out for comparison to see if our guess about the historical sequence and the subsequent "equation adjustments," using fudge factors as described previously, seems plausible.

Plotting Industry Recommendations for Forward Converters

For a typical flux swing of 1500 G, we have plotted out the following recommendations:

$$P_{IN} = 1.33 \times f_{kHz} \times AP_{c_{cm^4}} \quad \text{(Maniktala, see page 301)}$$

$$P_{IN} = 1.9 \times f_{kHz} \times AP_{c_{cm^4}}^{0.763} \quad \text{(Fairchild, see page 302)}$$

$$P_{IN} = 1.9 \times f_{kHz} \times AP_{c_{cm^4}}^{0.875} \quad \text{(Unitrode/TI, see page 302)}$$

$$P_{IN} = 2.1 \times f_{kHz} \times AP_{c_{cm^4}}^{0.752} \quad \text{(On-Semi, see page 303)}$$

$$P_{IN} = 2.23 \times f_{kHz} \times AP_{c_{cm^4}}^{0.763} \quad \text{(ST Micro, see page 303)}$$

We see from these that, indeed, doubling the frequency will double the power (so we really do not need to plot out curves for 300 kHz, 400 kHz, and so on—it is obvious how to derive the results for different frequencies).

Forward and Flyback Converters: Step-by-Step Design and Comparison

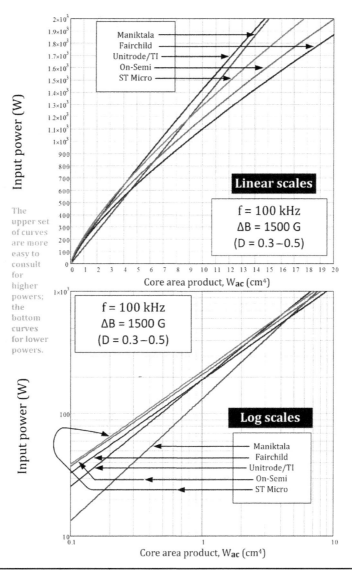

FIGURE 12.4 Comparing industry recommendations through plots of power versus core area product, assuming typical flux swing of 1500 G (at 100 kHz).

On plotting these out in Figs. 12.4 and 12.5, we see that our recommendation is more conservative for smaller output powers but is in line with others at higher power levels. We know that ours is consistently based on a constant current density target of 250 A_{RMS}/cm^2. The other recommendations do seem to be using a variable-current density target, though that is never explicitly defined in literature. They may *get away* with their more aggressive core-size recommendations *for small cores*, based on the empirical fact that smaller cores have improved thermal resistances on the bench, because of their higher ratio of surface-area to volume. And that fact may admittedly allow us also to judiciously increase the current density in small cores, say up to 350 to 400 A_{RMS}/cm^2. But it is quite clear that for larger cores, we do need to drop down to 250 A_{RMS}/cm^2 because all other recommendations do coincide with ours at high power levels, and our recommendation was based on a fixed 250 A_{RMS}/cm^2.

We can confirm from Fig. 12.5 that our recommendation is ETD34 ($AP_c = 1.66$ cm^4) for up to 440-W input power at 200 kHz, whereas the others typically allow 100 to 200 W more than that.

We can also compare with another set of curves historically available from Magnetics® at www.mag-inc.com. These are shown in Fig. 12.6 and are clearly the most aggressive. They also do not seem to spell out clearly if the topology is a single-ended Forward converter, or say, a Push-Pull (where due to symmetric excitation, some engineers claim it will give exactly twice the power reflected by the curves in Figs. 12.4 and 12.5). Keep in mind that the Magnetics Inc. curves seem to be based on low-frequency sine waves applied to test cores. Yet they were widely "referred to" in most of the prevailing Forward converter design notes around us even today.

Our conclusion is the equations proposed by us are self-consistent, derived from first principles, and less likely to run into thermally-initiated recalls.

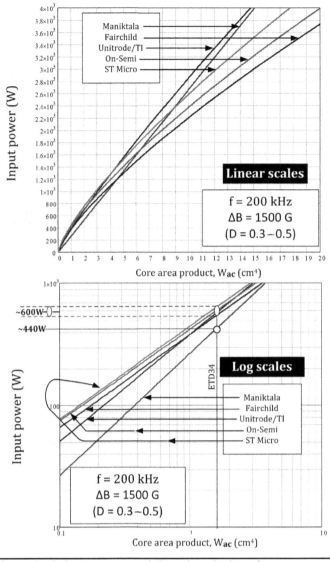

Figure 12.5 Comparing industry recommendations through plots of power versus core area product, assuming typical flux swing of 1500 G (at 200 kHz).

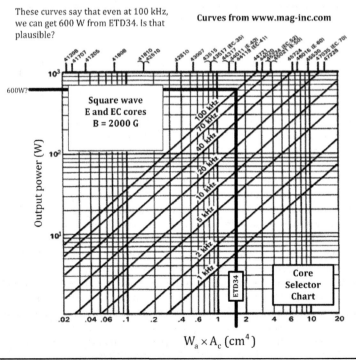

Figure 12.6 Historically available recommendations from Magnetics Inc.

Area Product for Symmetric Converters

We derived the general equation for a Forward converter

$$\Delta B_G = \frac{P_{IN}}{(J_{A/cm^2}) \times 0.785 \times AP_{cu_{p\_cm^4}} \times f_{Hz}} \times 10^8$$

The following similar equation is available at www.cedt.iisc.ernet.in/people//lums/smpc_page/pes03/pes03.pdf from Dr. Umanand at CEDT, Bangalore. This is derived in a manner similar to ours, but it does not use the COR current density to *mask* the dependence on D. Instead it uses the RMS current density and brings out $\sqrt{D}$ explicitly.

$$A_P = \frac{\sqrt{D_{max}} P_O (1 + 1/\eta)}{K_w J B_m f_s}$$

The paper also provides expressions for half-bridge, full-bridge, and Push-Pull converters. Note that it includes the efficiency η too. But we are setting that equal to 1 here (and simply changing P_O to P_{IN}). This is to retain simplicity. Also, casting the equations in terms of our ongoing nomenclature and units, we can rewrite the applicable equations as shown subsequently. Note that we have continued to include the correction factor of 0.785 for round wires so that the current density explicitly refers to the current passing through copper—though for foil windings we can actually remove it, but we will ignore that fact in our discussions for simplicity too. Note that now the current density is expressed explicitly in terms of RMS current, not the COR current value. We have

$$\Delta B_G = \frac{P_{IN} \sqrt{D_{MAX}}}{(J_{A/cm^2}) \times 0.785 \times AP_{cu_{p\_cm^4}} \times f_{Hz}} \times 10^8 \quad \text{(Forward, RMS current density)}$$

For $D = 0.5$, we get

$$\Delta B_G = \frac{P_{IN} \times 0.707}{(J_{A/cm^2}) \times 0.785 \times AP_{cu_{p\_cm^4}} \times f_{Hz}} \times 10^8 \quad \text{(Forward, RMS current density)}$$

Simplifying

$$\Delta B_G = \frac{0.90 \times P_{IN}}{(J_{A/cm^2}) \times AP_{cu_{p\_cm^4}} \times f_{Hz}} \times 10^8 \quad \text{(Forward, RMS current density)}$$

For half-bridge and full-bridge converters we have

$$\Delta B_G = \frac{P_{IN}(1+\sqrt{2})}{4 \times (J_{A/cm^2}) \times 0.785 \times AP_{cu_{p\_cm^4}} \times f_{Hz}} \times 10^8$$

$$= \frac{0.6 \times P_{IN}}{(J_{A/cm^2}) \times 0.785 \times AP_{cu_{p\_cm^4}} \times f_{Hz}} \times 10^8$$

$$= \frac{0.764 \times P_{IN}}{(J_{A/cm^2}) \times AP_{cu_{p\_cm^4}} \times f_{Hz}} \times 10^8 \quad \text{(Half- or full-bridge, RMS current density)}$$

We see that this says, in effect, for the same area product, flux swing, etc., the power throughput of the half-bridge and Push-Pull converter is greater than that for a single-ended Forward converter by the factor $0.9/0.764 = 1.18$, that is, only 18 percent more, unless we increase the flux swing, assuming that is acceptable in terms of core loss. More on that soon.

For a Push-Pull converter we have

$$\Delta B_G = \frac{P_{IN}}{\sqrt{2} \times (J_{A/cm^2}) \times 0.785 \times AP_{cu_{p\_cm^4}} \times f_{Hz}} \times 10^8$$

$$= \frac{0.71 \times P_{IN}}{(J_{A/cm^2}) \times 0.785 \times AP_{cu_{p\_cm^4}} \times f_{Hz}} \times 10^8$$

$$= \frac{0.90 \times P_{IN}}{(J_{A/cm^2}) \times AP_{cu_{p\_cm^4}} \times f_{Hz}} \times 10^8 \quad \text{(Push-Pull, RMS current density)}$$

which is the same as the Forward converter at $D = 0.5$! This requires some explanation.

Historically, the flux swing was restricted to 1500 G in a single-ended Forward converter, because it was known that the ferrite core could saturate at about 3000 G. So to avoid saturation during sudden line and load transients, a headroom of 1500 G was maintained. When engineers set to work on the Push-Pull, half-bridge, and full-bridge converters, since the core excitation was symmetric (around 0 G), the total flux swing could be increased to 3000 G (± 1500 G), and we would still have the same 1500-G safety margin. So at first sight it was felt that we could double the power throughput in most cases. It is perhaps still possible, but only at low switching frequencies (around 20 kHz).

Today, at higher switching frequencies, the flux swing is kept to 1500 G to keep core loss down to 100 mW/cm³ (for 3F3 material at 200 kHz, for example). The response of current limit circuitry, etc., is fast enough to minimize worries about hitting B_{SAT} under transients. So, in fact, *a good design (from a thermal viewpoint) may even set ΔB to 1000 or 1200 G only*. In other words, even in a *symmetric excitation converter*, we would very likely continue to keep ΔB to less than 1500 G. In that case, the preceding discussions and equations (derived from first principles), tell us there is very little to gain in terms of reduction of size of magnetic components in moving from a single-ended Forward converter to a symmetric converter. Yes, doing so may greatly help in finding components to ensure high efficiencies at high powers, and so on. But the usual *rule of thumb* that we can blindly double the power from a given core by using, say, the half-bridge instead of a Forward, is very doubtful indeed—from the thermal viewpoint in particular.

More Accurate Estimate of Power Throughput in Safety Transformers

All recommendations so far have been based on an assumption of a certain window utilization factor. All the curves we have shown in Figs. 12.4 and 12.5 have some such underlying assumption. At least in our case we have rather clearly assumed (and announced) that the primary windings will occupy exactly one-fourth of the total available *core* window area (i.e., $K = 0.25$). Most others typically provide rather vague utilization numbers, seemingly applied to somehow fit empirical data, but usually provide almost no physical explanation.

We also opined that for smaller transformers, we may be able to target higher current densities judiciously. Keep in mind that if the (exposed) area of a core was proportional to its volume, then even assuming that the coefficient of convection h was constant with respect to area (it isn't perfectly), we would expect the thermal resistance, which is assumed inversely proportional to surface area, to be inversely proportional to the volume (size of core) too. So, we would expect Rth to vary as per $1/V_e$. But that does not happen. The actual thermal resistance is much *worse than expected* for larger cores and is based on the following well-known empirical formula. See Fig. 12.7 for how a "wishful situation" was tempered with reality. So, the accepted empirical equation is

$$\text{Rth} = \frac{53}{V_e^{0.54}} \quad °C/W$$

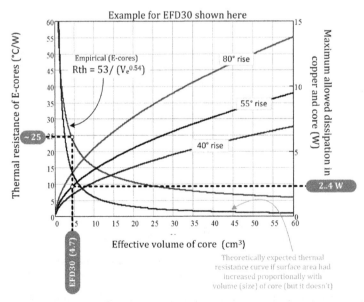

FIGURE 12.7 Thermal resistance of E-cores and maximum allowed dissipation (in windings and core).

However, we should also keep in mind that in smaller cores, less and less window utilization occurs, because the margin tape is of a fixed width (and also with a constant bobbin wall thickness), and does not decrease proportionally with core window area. So we will likely struggle even to maintain the same fixed current density. We just may not have enough winding width available, once we subtract the margin tape width on either side.

To more accurately judge what is the *real utilization factor* to plug in (instead of the default value of 0.25 we have used so far), *we need to actually compute the physical dimensions*, making some assumptions about bobbin wall thickness too. We start with some popular core sizes listed in Table 12.1, and then use that to arrive at the detailed results in Tables 12.2 to 12.5, cranked out by a spreadsheet, for the following cases: no margin tape, 2-mm margin tape (telecommunication applications), 4-mm margin tape (AC-DC with no PFC), 6.3-mm margin tape (AC-DC with Boost PFC front-end). As we can see, certain core sizes result in "NA" (nonapplicable), because after subtracting the margin tape from the available bobbin width, we get either almost no space for any winding, or worse, we have negative space. We also see that the utilization factor K_{cup}, is all over the place. Even our assumption of $K = 0.25$ was clearly a broad assumption, not really valid for small cores in particular. From these tables we can do a much more detailed and accurate calculation, as we will carry out shortly.

Basic Core Parameters (see Fig. 12.2)									
A (mm)	B (mm)	C (mm)	D (mm)	E (mm)	F (mm)	l_e (cm)	A_e (cm²)	V_e (cm³)	Core
20.00	10	5	6.3	12.8	5.2	4.28	0.312	1.34	**EE20/10/5**
25.00	10	6	6.4	18.8	6.35	4.9	0.395	1.93	**EE25/10/6**
35.00	18	10	12.5	24.5	10	8.07	1	8.07	**EE35/18/10**
42.00	21	15	14.8	29.5	12.2	9.7	1.78	17.3	**EE42/21/15**
42.00	21	20	14.8	29.5	12.2	9.7	2.33	22.7	**EE42/21/20**
55.00	28	20	18.5	37.5	17.2	12.3	4.2	52	**EE55/28/20**
28.00	14	11	9.75	21.75	9.9	6.4	0.814	5.26	**ER28/14/11**
35.00	20.7	11.3	14.7	25.6	11.3	9.08	1.07	9.72	**ER35/21/11**
42.00	22	16	15.45	30.05	15.5	9.88	1.94	19.2	**ER42/22/16**
54.00	18	18	11.1	40.65	17.9	9.18	2.5	23	**ER54/18/18**
12.00	6	3.5	4.55	9	5.4	2.85	0.114	0.325	**EFD12/6/3.5**
15.00	8	5	5.5	11	5.3	3.4	0.15	0.51	**EFD15/8/5**
20.00	10	7	7.7	15.4	8.9	4.7	0.31	1.46	**EFD20/10/7**
25.00	13	9	9.3	18.7	11.4	5.7	0.58	3.3	**EFD25/13/9**
30.00	15	9	11.2	22.4	14.6	6.8	0.69	4.7	**EFD30/15/9**
29.00	16	10	11	22	9.8	7.2	0.76	5.47	**ETD29/16/10**
34.00	17	11	11.8	25.6	11.1	7.86	0.97	7.64	**ETD34/17/11**
39.00	20	13	14.2	29.3	12.8	9.22	1.25	11.5	**ETD39/20/13**
44.00	22	15	16.1	32.5	15.2	10.3	1.73	17.8	**ETD44/22/15**
49.00	25	16	17.7	36.1	16.7	11.4	2.11	24	**ETD49/25/16**
54.00	28	19	20.2	41.2	18.9	12.7	2.8	35.5	**ETD54/28/19**
59.00	31	22	22.5	44.7	21.65	13.9	3.68	51.5	**ETD59/31/22**
74.00	29.5	NA	20.35	57.5	29.5	12.8	7.9	101	**PM74/59**
87.00	35	NA	24	67	31.7	14.6	9.1	133	**PM87/70**
114.00	46.5	NA	31.5	88	43	20	17.2	344	**PM114/93**
35.00	17.3	9.5	12.3	22.75	9.5	7.74	0.843	6.53	**EC35**
41.00	19.5	11.6	13.9	27.05	11.6	8.93	1.21	10.8	**EC41**
52.00	24.2	13.4	15.9	33	13.4	10.5	1.8	18.8	**EC52**
70.00	34.5	16.4	22.75	44.5	16.4	14.4	2.79	40.1	**EC70**

TABLE 12.1 Selection of Popular Cores with Basic Characteristics

0-mm (no) margin tape on either side.
Default values: 1.15-mm bobbin wall along direction of A, 1.35-mm bobbin wall along direction of D, additional 0.35-mm minimum clearance to the ferrite on the outside of the copper winding. See Fig. 12.2.

W_{ac} (cm²)	W_{ab} (cm²)	Width (mm)	Height (mm)	AP_b (cm⁴)	AP_c (cm⁴)	Width _tape (mm)	W_{cu} (cm²)	AP_{cu_p} (cm⁴)	K_{cu_p}	MLT (cm)	Core
0.48	0.23	9.90	2.30	0.07	0.15	9.90	0.23	0.04	0.24	4.02	**EE20/10/5**
0.80	0.48	10.10	4.73	0.19	0.31	10.10	0.48	0.09	0.30	5.42	**EE25/10/6**
1.81	1.28	22.30	5.75	1.28	1.81	22.30	1.28	0.64	0.35	7.36	**EE35/18/10**
2.56	1.92	26.90	7.15	3.42	4.56	26.90	1.92	1.71	0.38	9.36	**EE42/21/15**
2.56	1.92	26.90	7.15	4.48	5.97	26.90	1.92	2.24	0.38	10.36	**EE42/21/20**
3.76	2.97	34.30	8.65	12.46	15.77	34.30	2.97	6.23	0.40	11.96	**EE55/28/20**
1.16	0.74	16.80	4.43	0.61	0.94	16.80	0.74	0.30	0.32	5.33	**ER28/14/11**
2.10	1.51	26.70	5.65	1.61	2.25	26.70	1.51	0.81	0.36	6.16	**ER35/21/11**
2.25	1.63	28.20	5.78	3.16	4.36	28.20	1.63	1.58	0.36	7.52	**ER42/22/16**
2.53	1.93	19.50	9.88	4.81	6.31	19.50	1.93	2.41	0.38	9.56	**ER54/18/18**
0.16	0.02	6.40	0.30	0.00	0.02	6.40	0.02	0.00	0.06	2.68	**EFD12/6/3.5**
0.31	0.11	8.30	1.35	0.02	0.05	8.30	0.11	0.01	0.18	3.23	**EFD15/8/5**
0.50	0.22	12.70	1.75	0.07	0.16	12.70	0.22	0.03	0.22	4.24	**EFD20/10/7**
0.68	0.34	15.90	2.15	0.20	0.39	15.90	0.34	0.10	0.25	5.22	**EFD25/13/9**
0.87	0.47	19.70	2.40	0.33	0.60	19.70	0.47	0.16	0.27	5.89	**EFD30/15/9**
1.34	0.89	19.30	4.60	0.67	1.02	19.30	0.89	0.34	0.33	5.36	**ETD29/16/10**
1.71	1.20	20.90	5.75	1.17	1.66	20.90	1.20	0.58	0.35	6.13	**ETD34/17/11**
2.34	1.73	25.70	6.75	2.17	2.93	25.70	1.73	1.08	0.37	6.97	**ETD39/20/13**
2.79	2.11	29.50	7.15	3.65	4.82	29.50	2.11	1.82	0.38	7.85	**ETD44/22/15**
3.43	2.68	32.70	8.20	5.66	7.25	32.70	2.68	2.83	0.39	8.66	**ETD49/25/16**
4.50	3.64	37.70	9.65	10.19	12.61	37.70	3.64	5.09	0.40	9.80	**ETD54/28/19**
5.19	4.24	42.30	10.03	15.61	19.09	42.30	4.24	7.80	0.41	10.78	**ETD59/31/22**
5.70	4.75	38.00	12.50	37.53	45.01	38.00	4.75	18.76	0.42	14.03	**PM74/59**
8.47	7.32	45.30	16.15	66.58	77.10	45.30	7.32	33.29	0.43	15.87	**PM87/70**
14.18	12.66	60.30	21.00	217.80	243.81	60.30	12.66	108.90	0.45	20.94	**PM114/93**
1.63	1.12	21.90	5.13	0.95	1.37	21.90	1.12	0.47	0.34	5.43	**EC35**
2.15	1.56	25.10	6.23	1.89	2.60	25.10	1.56	0.95	0.36	6.43	**EC41**
3.12	2.42	29.10	8.30	4.35	5.61	29.10	2.42	2.17	0.39	7.65	**EC52**
6.39	5.37	42.80	12.55	14.99	17.84	42.80	5.37	7.49	0.42	9.93	**EC70**

W_{ac} is window area of core; W_{ab} is window area in side bobbin; **Width** is the width of any layer inside bobbin if no margin tape were present; **Height** is the height available for winding copper; AP_b is the area product of the bobbin; AP_c is the area product of the core; **Width_tape** is the actual width available for the copper layer with margin tape present; W_{cu} is the net window area available to wind copper (in Primary and Secondary) with margin tape and bobbin considered; AP_{cu_p} is the area product available for primary winding alone, assuming it is half the total available; K_{cu_p} is the actual utilization factor for the primary winding (ratio of AP_{cu_p} to AP_c), **MLT** is the mean (or average) length per turn with the bobbin wall thickness and required minimum clearance considered.

TABLE 12.2 Popular Cores with Area Product, Window Area, Utilization Factor with No Margin Tape

Number of Primary Turns

This is another source of confusion. Most databooks, from core vendors in particular, ask to fix the number of primary turns based on the equation for square waves shown in Fig. 12.8. Many engineers use that as a basis but do not realize that it uses the RMS value of the voltage, not the input DC rail. Besides, it assumes 50 percent duty cycle as we can see from the derivation in the figure too. We therefore do not recommend using it. The correct relationship must involve the duty cycle, just as we concluded during the core selection process too.

2-mm margin tape on either side.
Default values: 1.15-mm bobbin wall along direction of A, 1.35-mm bobbin wall along direction of D, additional 0.35-mm minimum clearance to the ferrite on the outside of the copper winding. See Fig. 12.2.

W_{ac} (cm²)	W_{ab} (cm²)	Width (mm)	Height (mm)	AP_b (cm⁴)	AP_c (cm⁴)	Width_tape (mm)	W_{cu} (cm²)	AP_{cu_p} (cm⁴)	K_{cu_p}	MLT (cm)	Core
0.48	0.23	9.90	2.30	0.07	0.15	5.90	0.14	0.02	0.14	4.02	**EE20/10/5**
0.80	0.48	10.10	4.73	0.19	0.31	6.10	0.29	0.06	0.18	5.42	**EE25/10/6**
1.81	1.28	22.30	5.75	1.28	1.81	18.30	1.05	0.53	0.29	7.36	**EE35/18/10**
2.56	1.92	26.90	7.15	3.42	4.56	22.90	1.64	1.46	0.32	9.36	**EE42/21/15**
2.56	1.92	26.90	7.15	4.48	5.97	22.90	1.64	1.91	0.32	10.36	**EE42/21/20**
3.76	2.97	34.30	8.65	12.46	15.77	30.30	2.62	5.50	0.35	11.96	**EE55/28/20**
1.16	0.74	16.80	4.43	0.61	0.94	12.80	0.57	0.23	0.25	5.33	**ER28/14/11**
2.10	1.51	26.70	5.65	1.61	2.25	22.70	1.28	0.69	0.31	6.16	**ER35/21/11**
2.25	1.63	28.20	5.78	3.16	4.36	24.20	1.40	1.36	0.31	7.52	**ER42/22/16**
2.53	1.93	19.50	9.88	4.81	6.31	15.50	1.53	1.91	0.30	9.56	**ER54/18/18**
0.16	0.02	6.40	0.30	0.00	0.02	2.40	0.01	0.00	0.02	2.68	**EFD12/6/3.5**
0.31	0.11	8.30	1.35	0.02	0.05	4.30	0.06	0.00	0.09	3.23	**EFD15/8/5**
0.50	0.22	12.70	1.75	0.07	0.16	8.70	0.15	0.02	0.15	4.24	**EFD20/10/7**
0.68	0.34	15.90	2.15	0.20	0.39	11.90	0.26	0.07	0.19	5.22	**EFD25/13/9**
0.87	0.47	19.70	2.40	0.33	0.60	15.70	0.38	0.13	0.22	5.89	**EFD30/15/9**
1.34	0.89	19.30	4.60	0.67	1.02	15.30	0.70	0.27	0.26	5.36	**ETD29/16/10**
1.71	1.20	20.90	5.75	1.17	1.66	16.90	0.97	0.47	0.28	6.13	**ETD34/17/11**
2.34	1.73	25.70	6.75	2.17	2.93	21.70	1.46	0.92	0.31	6.97	**ETD39/20/13**
2.79	2.11	29.50	7.15	3.65	4.82	25.50	1.82	1.58	0.33	7.85	**ETD44/22/15**
3.43	2.68	32.70	8.20	5.66	7.25	28.70	2.35	2.48	0.34	8.66	**ETD49/25/16**
4.50	3.64	37.70	9.65	10.19	12.61	33.70	3.25	4.55	0.36	9.80	**ETD54/28/19**
5.19	4.24	42.30	10.03	15.61	19.09	38.30	3.84	7.06	0.37	10.78	**ETD59/31/22**
5.70	4.75	38.00	12.50	37.53	45.01	34.00	4.25	16.79	0.37	14.03	**PM74/59**
8.47	7.32	45.30	16.15	66.58	77.10	41.30	6.67	30.35	0.39	15.87	**PM87/70**
14.18	12.66	60.30	21.00	217.80	243.81	56.30	11.82	101.68	0.42	20.94	**PM114/93**
1.63	1.12	21.90	5.13	0.95	1.37	17.90	0.92	0.39	0.28	5.43	**EC35**
2.15	1.56	25.10	6.23	1.89	2.60	21.10	1.31	0.79	0.31	6.43	**EC41**
3.12	2.42	29.10	8.30	4.35	5.61	25.10	2.08	1.87	0.33	7.65	**EC52**
6.39	5.37	42.80	12.55	14.99	17.84	38.80	4.87	6.79	0.38	9.93	**EC70**

W_{ac} is window area of core; W_{ab} is window area in side bobbin; **Width** is the width of any layer inside bobbin if no margin tape were present; **Height** is the height available for winding copper; AP_b is the area product of the bobbin; AP_c is the area product of the core; **Width_tape** is the actual width available for the copper layer with margin tape present; W_{cu} is the net window area available to wind copper (in Primary and Secondary) with margin tape and bobbin considered; AP_{cu_p} is the area product available for primary winding alone, assuming it is half the total available; K_{cu_p} is the actual utilization factor for the primary winding (ratio of AP_{cu_p} to AP_c), **MLT** is the mean (or average) length per turn with the bobbin wall thickness and required minimum clearance considered.

TABLE 12.3 Popular Cores with Area Product, Window Area, Utilization Factor with 2-mm Margin Tape

Other engineers (such as AN-4134 from Fairchild) ask to use this equation:

$$N_{P\_MIN} = \frac{V_{INMIN} \times D_{MAX}}{A_e \times f \times \Delta B} \times 10^6$$

We need to correct some wrong impressions here. There is actually no need to do the calculation at minimum input. The reason is that the duty cycle of a Forward converter is based on the Buck cell that follows the transformer stage, which has an effective DC input of V_{INR} (the reflected input voltage) and an output of V_O. So we have (as for a Buck)

$$D = \frac{V_O}{V_{INR}} = \frac{n \times V_O}{V_{IN}}$$

4-mm margin tape on either side.
Default values: 1.15-mm bobbin wall along direction of A, 1.35-mm bobbin wall along direction of D, additional 0.35-mm minimum clearance to the ferrite on the outside of the copper winding. See Fig. 12.2.

W_{ac} (cm²)	W_{ab} (cm²)	Width (mm)	Height (mm)	AP_b (cm⁴)	AP_c (cm⁴)	Width_tape (mm)	W_{cu} (cm²)	AP_{cu_p} (cm⁴)	K_{cu_p}	MLT (cm)	Core
0.48	0.23	9.90	2.30	0.07	0.15	1.90	NA	NA	0.05	4.02	**EE20/10/5**
0.80	0.48	10.10	4.73	0.19	0.31	2.10	NA	NA	NA	5.42	**EE25/10/6**
1.81	1.28	22.30	5.75	1.28	1.81	14.30	0.82	0.41	0.23	7.36	**EE35/18/10**
2.56	1.92	26.90	7.15	3.42	4.56	18.90	1.35	1.20	0.26	9.36	**EE42/21/15**
2.56	1.92	26.90	7.15	4.48	5.97	18.90	1.35	1.57	0.26	10.36	**EE42/21/20**
3.76	2.97	34.30	8.65	12.46	15.77	26.30	2.27	4.78	0.30	11.96	**EE55/28/20**
1.16	0.74	16.80	4.43	0.61	0.94	8.80	0.39	0.16	0.17	5.33	**ER28/14/11**
2.10	1.51	26.70	5.65	1.61	2.25	18.70	1.06	0.57	0.25	6.16	**ER35/21/11**
2.25	1.63	28.20	5.78	3.16	4.36	20.20	1.17	1.13	0.26	7.52	**ER42/22/16**
2.53	1.93	19.50	9.88	4.81	6.31	11.50	1.14	1.42	0.22	9.56	**ER54/18/18**
0.16	0.02	6.40	0.30	0.00	0.02	NA	NA	NA	NA	2.68	**EFD12/6/3.5**
0.31	0.11	8.30	1.35	0.02	0.05	0.30	NA	NA	NA	3.23	**EFD15/8/5**
0.50	0.22	12.70	1.75	0.07	NA	NA	NA	NA	NA	4.24	**EFD20/10/7**
0.68	0.34	15.90	2.15	0.20	0.39	7.90	0.17	0.05	0.13	5.22	**EFD25/13/9**
0.87	0.47	19.70	2.40	0.33	0.60	11.70	0.28	0.10	0.16	5.89	**EFD30/15/9**
1.34	0.89	19.30	4.60	0.67	1.02	11.30	0.52	0.20	0.19	5.36	**ETD29/16/10**
1.71	1.20	20.90	5.75	1.17	1.66	12.90	0.74	0.36	0.22	6.13	**ETD34/17/11**
2.34	1.73	25.70	6.75	2.17	2.93	17.70	1.19	0.75	0.25	6.97	**ETD39/20/13**
2.79	2.11	29.50	7.15	3.65	4.82	21.50	1.54	1.33	0.28	7.85	**ETD44/22/15**
3.43	2.68	32.70	8.20	5.66	7.25	24.70	2.03	2.14	0.29	8.66	**ETD49/25/16**
4.50	3.64	37.70	9.65	10.19	12.61	29.70	2.87	4.01	0.32	9.80	**ETD54/28/19**
5.19	4.24	42.30	10.03	15.61	19.09	34.30	3.44	6.33	0.33	10.78	**ETD59/31/22**
5.70	4.75	38.00	12.50	37.53	45.01	30.00	3.75	14.81	0.33	14.03	**PM74/59**
8.47	7.32	45.30	16.15	66.58	77.10	37.30	6.02	27.41	0.36	15.87	**PM87/70**
14.18	12.66	60.30	21.00	217.80	243.81	52.30	10.98	94.45	0.39	20.94	**PM114/93**
1.63	1.12	21.90	5.13	0.95	1.37	13.90	0.71	0.30	0.22	5.43	**EC35**
2.15	1.56	25.10	6.23	1.89	2.60	17.10	1.06	0.64	0.25	6.43	**EC41**
3.12	2.42	29.10	8.30	4.35	5.61	21.10	1.75	1.58	0.28	7.65	**EC52**
6.39	5.37	42.80	12.55	14.99	17.84	34.80	4.37	6.09	0.34	9.93	**EC70**

W_{ac} is window area of core; W_{ab} is window area in side bobbin; **Width** is the width of any layer inside bobbin if no margin tape were present; **Height** is the height available for winding copper; AP_b is the area product of the bobbin; AP_c is the area product of the core; **Width_tape** is the actual width available for the copper layer with margin tape present; W_{cu} is the net window area available to wind copper (in Primary and Secondary) with margin tape and bobbin considered; AP_{cu_p} is the area product available for primary winding alone, assuming it is half the total available; K_{cu_p} is the actual utilization factor for the primary winding (ratio of AP_{cu_p} to AP_c), **MLT** is the mean (or average) length per turn with the bobbin wall thickness and required minimum clearance considered.

TABLE 12.4 Popular Cores with Area Product, Window Area, Utilization Factor with 4-mm Margin Tape

where $n = N_P/N_S$. The volt-seconds across the primary winding is

$$\text{Volt-seconds} = V_{IN} \times \frac{D}{f}$$

So at minimum input we get

$$\text{Volt-seconds}_{MIN} = V_{INMIN} \times \frac{1}{f} \times \frac{N_P \times V_O}{V_{INMIN}}$$

6.3-mm margin tape on either side.
Default values: 1.15-mm bobbin wall along direction of A, 1.35-mm bobbin wall along direction of D, additional 0.35-mm minimum clearance to the ferrite on the outside of the copper winding. See Fig. 12.2.

W_{ac} (cm²)	W_{ab} (cm²)	Width (mm)	Height (mm)	AP_b (cm⁴)	AP_c (cm⁴)	Width_tape (mm)	W_{cu} (cm²)	AP_{cu_P} (cm⁴)	K_{cu_P}	MLT (cm)	Core
0.48	0.23	9.90	2.30	0.07	0.15	NA	NA	NA	NA	4.02	**EE20/10/5**
0.80	0.48	10.10	4.73	0.19	0.31	NA	NA	NA	NA	5.42	**EE25/10/6**
1.81	1.28	22.30	5.75	1.28	1.81	9.70	0.56	0.28	0.15	7.36	**EE35/18/10**
2.56	1.92	26.90	7.15	3.42	4.56	14.30	1.02	0.91	0.20	9.36	**EE42/21/15**
2.56	1.92	26.90	7.15	4.48	5.97	14.30	1.02	1.19	0.20	10.36	**EE42/21/20**
3.76	2.97	34.30	8.65	12.46	15.77	21.70	1.88	3.94	0.25	11.96	**EE55/28/20**
1.16	0.74	16.80	4.43	0.61	0.94	4.20	0.19	0.08	0.08	5.33	**ER28/14/11**
2.10	1.51	26.70	5.65	1.61	2.25	14.10	0.80	0.43	0.19	6.16	**ER35/21/11**
2.25	1.63	28.20	5.78	3.16	4.36	15.60	0.90	0.87	0.20	7.52	**ER42/22/16**
2.53	1.93	19.50	9.88	4.81	6.31	6.90	0.68	0.85	0.13	9.56	**ER54/18/18**
0.16	0.02	6.40	0.30	0.00	0.02	NA	NA	NA	NA	2.68	**EFD12/6/3.5**
0.31	0.11	8.30	1.35	0.02	0.05	NA	NA	NA	NA	3.23	**EFD15/8/5**
0.50	0.22	12.70	1.75	0.07	0.16	0.10	NA	NA	NA	4.24	**EFD20/10/7**
0.68	0.34	15.90	2.15	0.20	0.39	3.30	0.07	0.02	0.05	5.22	**EFD25/13/9**
0.87	0.47	19.70	2.40	0.33	0.60	7.10	0.17	0.06	0.10	5.89	**EFD30/15/9**
1.34	0.89	19.30	4.60	0.67	1.02	6.70	0.31	0.12	0.11	5.36	**ETD29/16/10**
1.71	1.20	20.90	5.75	1.17	1.66	8.30	0.48	0.23	0.14	6.13	**ETD34/17/11**
2.34	1.73	25.70	6.75	2.17	2.93	13.10	0.88	0.55	0.19	6.97	**ETD39/20/13**
2.79	2.11	29.50	7.15	3.65	4.82	16.90	1.21	1.05	0.22	7.85	**ETD44/22/15**
3.43	2.68	32.70	8.20	5.66	7.25	20.10	1.65	1.74	0.24	8.66	**ETD49/25/16**
4.50	3.64	37.70	9.65	10.19	12.61	25.10	2.42	3.39	0.27	9.80	**ETD54/28/19**
5.19	4.24	42.30	10.03	15.61	19.09	29.70	2.98	5.48	0.29	10.78	**ETD59/31/22**
5.70	4.75	38.00	12.50	37.53	45.01	25.40	3.18	12.54	0.28	14.03	**PM74/59**
8.47	7.32	45.30	16.15	66.58	77.10	32.70	5.28	24.03	0.31	15.87	**PM87/70**
14.18	12.66	60.30	21.00	217.80	243.81	47.70	10.02	86.15	0.35	20.94	**PM114/93**
1.63	1.12	21.90	5.13	0.95	1.37	9.30	0.48	0.20	0.15	5.43	**EC35**
2.15	1.56	25.10	6.23	1.89	2.60	12.50	0.78	0.47	0.18	6.43	**EC41**
3.12	2.42	29.10	8.30	4.35	5.61	16.50	1.37	1.23	0.22	7.65	**EC52**
6.39	5.37	42.80	12.55	14.99	17.84	30.20	3.79	5.29	0.30	9.93	**EC70**

W_{ac} is window area of core; W_{ab} is window area in side bobbin; **Width** is the width of any layer inside bobbin if no margin tape were present; **Height** is the height available for winding copper; AP_b is the area product of the bobbin; AP_c is the area product of the core; **Width_tape** is the actual width available for the copper layer with margin tape present; W_{cu} is the net window area available to wind copper (in Primary and Secondary) with margin tape and bobbin considered; AP_{cu_P} is the area product available for primary winding alone, assuming it is half the total available; K_{cu_P} is the actual utilization factor for the primary winding (ratio of AP_{cu_P} to AP_c); **MLT** is the mean (or average) length per turn with the bobbin wall thickness and required minimum clearance considered.

TABLE 12.5 Popular Cores with Area Product, Window Area, Utilization Factor with 6.3-mm Margin Tape

We can see that *the input voltage actually cancels out*. So, in fact, the volt-seconds across the transformer at high-line remain the same at low-line! So does the current swing, and therefore the flux swing. In fact, we can pick *any input voltage* (minimum, maximum, or nominal) and we will get the (same) primary turns if we use

$$N_P = \frac{V_{IN} \times D}{A_e \times f \times \Delta B} \times 10^6$$

314   Chapter Twelve

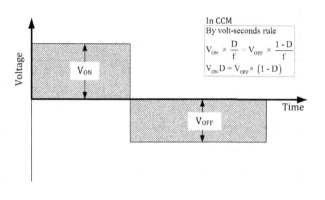

Figure 12.8 Different equations for finding primary number of turns.

We may still need to use V_{INMIN} for an entirely different reason. We need to confirm that the turns ratio is such that at minimum input, *the duty cycle does not exceed the maximum duty cycle limit of the controller IC*. See the worked example that follows.

The correct equation to use is the more basic form of Faraday's law (Volt-seconds = NAB)

$$V_{IN} \times \frac{D}{f_{Hz}} = N_P \times A_{e_{m^2}} \times \Delta B_T$$

$$N_P = \frac{V_{IN} \times D}{f_{Hz} \times A_{e_{m^2}} \times \Delta B_T} = \frac{V_{IN} \times D \times 10^4}{f_{Hz} \times A_{e_{cm^2}} \times \Delta B_T}$$

We will use this in the numerical example.

Worked Example: Flyback and Forward Alternative Design Paths

In a telecommunications application, such as power over ethernet (PoE), we have an input voltage of 36 to 57 V. We want to design a 200-kHz, 12 V @ 11 A (132 W) Forward converter (the controller is limited to a maximum duty cycle of 44 percent as in a typical single-ended type). Select the transformer core, and calculate the primary and secondary number of turns on it. Also select a secondary choke. If the same application and the same control IC are used for a Flyback, what would be the required core size and the number of turns?

Step-by-Step Forward Converter Design

Core Selection

Assume the efficiency will be close to 85 percent. So for an output of 132 W, the input will be 132/0.85 = 155.3 W. We target a flux swing of 0.15 T maximum and a COR current density of 500 A/cm² (slightly more aggressive than the 450 A/cm² we usually recommend, see pages 298 and 300). So

$$AP_{cu_{P\_cm^4}} = \frac{12.74 \times P_{IN}}{f_{kHz} \times \Delta B_T \times J_{A/cm^2}} = \frac{12.74 \times 155.3}{200 \times 0.15 \times 500} = 0.132 \text{ cm}^4$$

This is the required area product in terms of the primary winding. We expect to use 2-mm margin tape. Therefore, we look at Table 12.3. We see that AP_{cu_p} of 0.13 cm⁴ is available from EFD 30/15/9, almost exactly what we need here (0.132 cm⁴). That is the selected core.

Primary Turns

We assume the turns ratio will be fixed such that at minimum input the duty cycle is 0.44. So

$$N_P = \frac{V_{IN} \times D \times 10^4}{f_{Hz} \times A_{e_{cm^2}} \times \Delta B_T} = \frac{36 \times 0.44 \times 10^4}{200,000 \times 0.69 \times 0.15} = 7.65 \text{ turns}$$

Magnetization Inductance and Peak Magnetization Current

What is the magnetization inductance? The EFD30 with no air gap, made of 3F3 from Ferroxcube, has a datasheet A_L value of 1900 nH/turns². So if we use 8 primary turns, we get an inductance of 1900 nH × 8² = 121 μH.

Note that an alternative calculation in literature uses

$$L = \frac{\mu \mu_o N^2 \times A_e}{z \times l_e} \quad \text{(MKS units)}$$

where l_e is the effective length and A_e is the effective area of the core, as defined in Chap. 7. Plugging in our values, we get for primary inductance

$$L = \frac{2000 \times (4\pi \times 10^{-7}) \times 8^2 \times 0.69 \times 10^{-4}}{1 \times 6.8 \times 10^{-2}} = 1.63 \times 10^{-4} \quad \text{(MKS units)}$$

This is 163 μH.

The difference between the two results is based on the fact that the A_L value provided by the vendor is more practical: It includes the small default air gap since it is not possible to eliminate all air gaps when clamping two separate halves together. So in theory, if there was zero air gap (i.e., an air-gap factor z of 1, see the A-Z book), we would get 163 μH. In reality, the magnetization peak current will be higher than expected, because of the minute residual air gap, which has reduced the measured inductance to 121 μH.

So the actual peak magnetization current component in the switch will be a little higher than anticipated (though this will be the same at any input voltage as explained earlier):

$$I_{MAG} = \frac{V_{IN} \times D/f}{L} = \frac{36 \times 0.44/200,000}{121 \times 10^{-6}} = 0.655 \text{ A}$$

Turns Ratio

The turns ratio is derived from

$$D = \frac{V_O}{V_{INR}} = \frac{n \times V_O}{V_{IN}}$$

$$n = \frac{D \times V_{IN}}{V_O} = \frac{0.44 \times 36}{12} = 1.32$$

Voltage Ratings

So the maximum reflected input voltage is (see Appendix for voltage stress tables)

$$V_{\text{INRMAX}} = \frac{V_{\text{INMAX}}}{n} = \frac{57}{1.32} = 43.2 \text{ V}$$

The minimum voltage rating of the output diode is

$$V_{D1} = V_{\text{INRMAX}} + V_O = 43.2 + 12 = 55.2 \text{ V}$$

The minimum voltage rating of the catch diode is

$$V_{D2} = V_{\text{INRMAX}} = 43.2 \text{ V}$$

If the two diodes are in the same package, we may just get away with a 60-V Schottky with a slight adjustment of the turns ratio to increase headroom.

If this is a single-switch Forward converter, the maximum drain-to-source voltage is twice the input, i.e., 2×57 V. So we should look for a 150 V FET. If this is a two-switch Forward (asymmetric half-bridge) the maximum drain-to-source voltage is only 57 V.

Secondary Turns

The number of secondary turns is

$$N_S = \frac{N_P}{n} = \frac{8}{1.32} = 6.06 \approx 6 \text{ turns}$$

Sense Resistor

The peak output current is about 1.2×11 A $= 13.2$ A. This occurs at high-line actually, and the inductor is designed to give a 20 percent peak above average ($r = 0.4$). On the switch, we also need to add the peak magnetization current of 0.655 A. So the sense resistor will have to be set to permit the normal operating current of 13.2 A/n + 0.655 A = (13.2 A/1.333) + 0.655 = 10.6 A. We can set current limiting at about 12 A. So, for example, if the controller IC sense threshold is 200 mV, we need a sense resistor of $V/I = 0.2/12 = 0.017$ Ω. The RMS current through the sense resistor is about $(11 \text{ A}/n) \times \sqrt{D} = (11 \text{ A}/1.333) \times \sqrt{0.5} = 5.8$ A. So heating in the sense resistor is about $5.8^2 \times 0.017 = 0.57$ W. For adequate derating we can pick two 33-mΩ/0.5-W resistors in parallel.

Minimum Duty Cycle

We will need this shortly:

$$D_{\text{MIN}} = \frac{V_O}{V_{\text{INRMAX}}} = \frac{n \times V_O}{V_{\text{INMAX}}} = \frac{(8/6) \times 12}{57} = 0.28$$

Choke Inductance and Rating

We have to design this at maximum input because, as in any regular Buck, the maximum peak current occurs at maximum input. At that point we want a total swing ΔI equal to about 40 percent the average value (11 A). This is 20 percent above and 20 percent below the center at I_O.

We need the duty cycle at maximum input from the preceding above step. So, setting a current ripple ratio of 0.4, using the standard Buck equations:

$$L_{\mu H} = \frac{V_O}{I_O \times r \times f_{\text{Hz}}} \times (1-D) \times 10^6 = \frac{12}{11 \times 0.4 \times 200,000} \times (1-0.28) \times 10^6 = 9.82$$

So we pick an inductance of 10 µH. It must have a minimum saturation rating of 12 A.

Overall Loss Estimation in Transformer

From Table 12.3, we see that EFD30 with 2-mm margin tape has a total available area of 0.38 cm² for winding copper (both Primary and Secondary). Assuming no Litz wire (no silk covering, etc.) and ignoring wires slipping into adjacent spaces between wires (as in a bundle), the simplest assumption to make here is that 78.5 percent (i.e., $\pi/4$) of the physical space is occupied by copper. So the actual area occupied by copper in our case is $0.38 \times 0.785 = 0.2983$ cm².

Let us for now just look at the primary winding, and assume that it occupies half the available window area W_{cu}. So we have $0.2983/2 = 0.15$ cm² window reserved for primary-side copper. This has $N_p = 8$ turns. Assuming all eight turns are laid out side by side, no wasted space, each turn will have a cross-sectional area of $0.15/8 = 0.01875$ cm². The mean length per turn is also given in Table 12.3 for the EFD30 core as 5.89 cm. The length of the entire primary winding is therefore $N_p \times \text{MLT} = 8 \times 5.89 = 47.12$ cm. Using the resistivity of copper (17 n$\Omega \cdot$ m) we get

$$R = \rho \times \frac{l}{A} = 17.2 \times 10^{-9} \ \Omega\text{m} \times \frac{47.12 \times 10^{-2}\,\text{m}}{0.01875 \times 10^{-4}\,\text{m}^2} = 4.32 \ \text{m}\Omega$$

The RMS switch/primary current is about 5.8 A as shown under the sense resistor calculations given earlier. See page 188 too. So the primary side dissipation is

$$P_{cu_p} = 2 \times 5.8\,\text{A}^2 \times 4.32 \ \text{m}\Omega = 0.29 \ \text{W}$$

where we have also silently doubled the dissipation *because we have assumed that the AC resistance is, at best, twice the DC resistance.*

With the same current density all through the transformer, we can assume that the dissipation is split equally between the primary and secondary windings. So the total transformer dissipation is finally estimated to be 0.29×2, that is, about 0.6 W. But this assumes we can achieve $F_R = 2$. If F_R was closer to 3 (more likely), we would get about 1 W. Further, if copper loss equals core loss, we would then get up to 2-W total transformer dissipation. From Fig. 12.7 we see that an EFD30 transformer (with an effective volume of 4.7 cm³) has a thermal resistance of about 25°C/W. So we could expect a temperature rise of up to 2 W $\times$ 25 = 50°C. We would then be approaching the limit of a Class A transformer (55°C allowable rise).

Core Loss and Total Estimated Loss

Assuming we are using 3F3 material from Ferroxcube, the core loss equation is (see Tables 12.6 and 12.7):

$$\text{Core loss} = C \times B^p \times f^d \times V_e \quad \text{(using system B)}$$

where for 3F3, we have (valid up to 300 kHz):

$$C = 1.3 \times 10^{-16}, \quad p = 2.5, \quad d = 2$$

Here B is half the flux swing (in gauss) since vendors use symmetric sine wave excitation for testing, and B refers to the amplitude of the swing (around zero). We get for our current example (EFD30)

$$\text{Core loss} = 1.3 \times 10^{-16} \times 750^{2.5} \times 200,000^2 \times 4.7 = 376.5 \ \text{mW} \quad \text{(using system B)}$$

where we have used the core-loss coefficients provided in Table 12.6. We can also derive these as shown in Fig. 12.9. Note that our design point corresponds to (half total flux swing of) 75 mT (750 G), and this corresponds to 100 kW/m³, which is numerically the same as 100 mW/cm³. So we have targeted a conservative core loss of 100 mW/cm³, as most in the industry do (though some claim up to 200 mW/cm³ is OK). Basically, our maximum switching frequency is largely determined by this aspect (core loss).

	Constant $\times$ B(exponent of B) $\times$ f(exponent of f) (Core loss per unit volume)						
	Constant	Exponent of B	Exponent of f	B	f	V_e	Units
System A	C_c	C_b	C_f	T	Hz	cm³	W/cm³
	$= \dfrac{C \times 10^{4 \times p}}{10^3}$	$= p$	$= d$				
System B	C	p	d	G	Hz	cm³	mW/cm³
	$= \dfrac{C_c \times 10^3}{10^{4 \times C_b}}$	$= C_b$	$= C_f$				
System C	K_p	n	m	G	Hz	cm³	W/cm³
	$= \dfrac{C}{10^3}$	$= p$	$= d$				

TABLE 12.6 The Different Systems in Use for Describing Core Loss (and Their Conversions)

Material (vendor)	Grade	C	p (B^p)	d (f^d)	μ	≈ B_{SAT} (G)	≈ f_{MAX} (MHz)
Powdered iron (Micrometals)	8	4.3E – 10	2.41	1.13	35	12,500	100
	18	6.4E – 10	2.27	1.18	55	10,300	10
	26	7E – 10	2.03	1.36	75	13,800	0.5
	52	9.1E – 10	2.11	1.26	75	14,000	1
Ferrite (Magnetics, Inc.)	F	1.8E – 14	2.57	1.62	3,000	3,000	1.3
	K	2.2E – 18	3.1	2	1,500	3,000	2
	P	2.9E – 17	2.7	2.06	2,500	3,000	1.2
	R	1.1E – 16	2.63	1.98	2,300	3,000	1.5
Ferrite (Ferroxcube)	3C81	6.8E – 14	2.5	1.6	2,700	3,600	0.2
	3F3	1.3E – 16	2.5	2	2,000	3,700	0.5
	3F4	1.4E – 14	2.7	1.5	900	3,500	2
Ferrite (TDK)	PC40	4.5E – 14	2.5	1.55	2,300	3,900	1
	PC50	1.2E – 17	3.1	1.9	1,400	3,800	2
Ferrite (Fair-Rite)	77	1.7E – 12	2.3	1.5	2,000	3,700	1

Note: (a)E-(b) is the same as (a) × $10^{-(b)}$.

TABLE 12.7 Typical Core-Loss Coefficients of Common Materials (System B)

We see that we may get up to 0.6 to 1-W copper loss and 0.4 W of core loss. They are not equal, and in fact in most modern converters, the assumption of equal copper and core losses is not necessarily true. As mentioned in Chap. 11, it has been reported that a more optimum operating point (minimum core and copper losses combined) is actually defined by

$$\frac{\text{Core loss}}{\text{Copper loss}} = \frac{2}{\text{exponent of }B} \equiv \frac{2}{p} \Rightarrow \frac{2}{2.5} = 0.8 \quad \text{(for 3F3)}$$

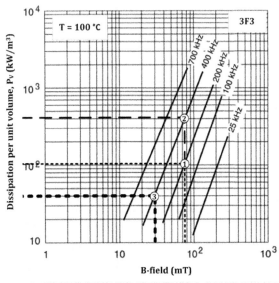

Three points selected in succession as shown: the first two at same B, the last two at same f.

Point 1: 75 mT, 200 kHz, 100 kW/m³
Point 2: 75 mT, 400 kHz, 400 kW/m³
Point 3: 400 kHz, 30 mT, 40 kW/m³

Coordinates

$P_{v1} = 100$, $P_{v2} = 400$, $P_{v3} = 40$

$f_1 = 200$, $f_2 = 400$, $f_3 = 400$

$B_1 = 75$, $B_2 = 75$, $B_3 = 30$

Equations/Calculations

$$\exp\_f = \frac{\ln(P_{v1}/P_{v2})}{\ln(f_1/f_2)} = \frac{\ln(100/400)}{\ln(200/400)} = 2$$

$$\exp\_B = \frac{\ln(P_{v2}/P_{v3})}{\ln(B_2/B_3)} = \frac{\ln(400/40)}{\ln(75/30)} = 2.5$$

Using System B (see accompanying text) (cm³, mW/cm³, Hz, G), and keeping in mind that kW/m³ is numerically equal to W/cm³, we get the constant term C as follows

See Pages 106, 107 of *Switching Power Supplies A-Z, 2e*

(kW/m³) = (constant × $B^{\exp\_B}$ × $f^{\exp\_f}$)

$$C = \text{constant} = \frac{\text{kW/m}^3}{B^{\exp\_B} \times f^{\exp\_f}} = \frac{100}{750^{2.5} \times (200000)^2} = 1.6 \times 10^{-16}$$

This is based on eyeballing the above curve.
More exact values are (in System B):

$C = 1.3 \times 10^{-16}$, exponent of B = 2.5, exponent of f = 2

FIGURE 12.9 Evaluating the core-loss coefficients from the vendor's core-loss curves, such as for 3F3.

Copper Sizing and Transformer Windings

The width available for the primary winding is 15.7 mm from Table 12.3 (EFD30 with 2-mm margin tape). The skin depth at 200 kHz is (assuming a temperature of 80°C for adjusting the resistivity of copper better)

$$\delta_{mm} = \frac{66.1[1 + 0.0042(T - 20)]}{\sqrt{f_{Hz}}} = \frac{66.1[1 + 0.0042(80 - 20)]}{\sqrt{200,000}} = 0.185 \text{ mm}$$

When dealing with round wires, we find that just decreasing F_R does not necessarily correlate with lowest F_R because F_R is a ratio. The logic is explained in Fig. 12.10 and leads us to the subdivision strategy shown in Fig. 12.11.

The subdivision strategy is further explained in the *A-Z* book (chapter 3), but as we saw there too, it is not a good idea to start off with an X ($X = h/\delta$) of greater than around 4; otherwise, we have to subdivide too much, never really getting to a low-enough F_R value. And when we do, we will end up with impracticable or nonexistent wire gauges.

To start with, let us assume just a simple winding arrangement of a primary winding followed by a secondary winding. Let us call this a *P-S arrangement*, one primary portion followed by a secondary portion.

P-S Winding Arrangement

Let us start with just a single strand for the Primary and adjust its diameter so it completely fills one layer with eight turns. The layers per portion p for this equals 1. The diameter is width/N_p = 15.7 mm/8 T = 1.963 mm. Note that using the equivalent foil transformation of Dowell, that diameter would give an equivalent foil of thickness $h = 0.886 \times 1.963 = 1.739$ mm (since $\sqrt{\pi/4}$ equals 0.886, see Figs. 10.3 and 11.10). In terms of penetration ratio X, that would be $X = h/\delta = 1.739$ mm/0.185 mm = 9.4. But that is too high a starting value for entering the subdivision process.

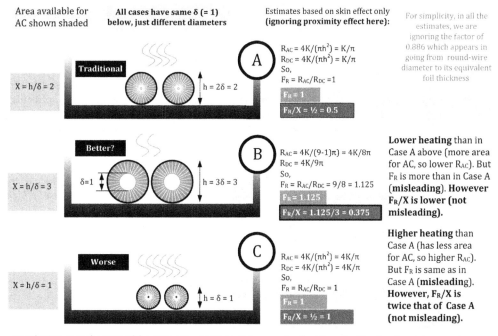

Conclusion: F_R is the ratio of the AC to DC resistance, and may not reflect the actual in-circuit resistance and thereby the dissipation. F_R does not necessarily seem to correlate well with the copper heating in the case of round windings (high-current copper foil windings are discussed in Fig. 2.15). F_R/X *seems* to reflect the heating better. So, at first sight, we may want to find the lowest F_R/X, not lowest F_R. However, if in Case B above (labeled "Better?"), increasing the diameter eventually leads to an increase in the *number of layers*, then considering proximity effects, the overall dissipation will increase, not decrease, as per Dowell's equation — because there are now *more layers to sum over*, and so, even if each layer has less dissipation, the overall dissipation would likely have increased. So, for Forward converter transformer design, the most important thing is to try and *minimize the actual physical number of copper layers first*. That is the basis of the subdivision method being used in this chapter: the layers are held constant, then we optimize the wire diameter based on the lowest F_R, not necessarily the lowest F_R/X.

FIGURE 12.10 F_R/X, not just F_R, correlate better with lowest losses, but the best way is to keep layers unchanged and subdivide strands to achieve lower AC resistance. See Fig. 12.11.

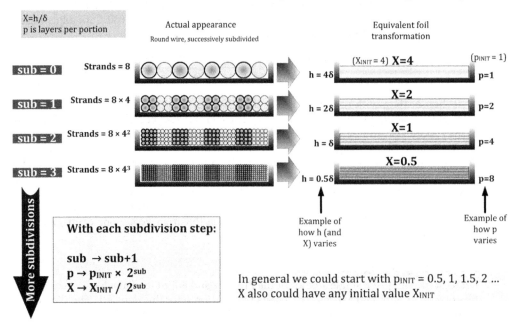

FIGURE 12.11 Subdivision strategy for round wires explained.

So instead, let us start with two paralleled strands of round wire, of diameter half of that, that is, Width/$(2 \times N_p)$ = 1.963/2 = 0.98 mm, but still laid out in one layer. So the layers per portion p still equals 1. We say p_{INIT} is 1. Similarly, the starting X value (X_{INIT}) is

$$X_{INIT} = \frac{0.886 \times \text{Dia}}{\delta} = \frac{0.886 \times 0.98}{0.185} = 4.7$$

Let us go with this for now. From Fig. 12.12 (lower half, that is, p_{INIT} = 1) we see that for X_{INIT} = 4.7 we need six subdivisions to get F_R to fall below 2. Each subdivision splits each strand into four wires, each of half the diameter.

NOTE *In Fig. 12.13, for convenience we have also provided the plots for p_{INIT} equal to 1.5 and 2. These curves are just Dowell's curves plotted out in a way that is useful for the subdivision strategy, and, of course, to which DC bias has been added (which Dowell had not included), and also summed up to 40 Fourier harmonics as shown in Fig. 12.14 (Dowell had just used a high-frequency sine wave in his analysis, as explained).*

After six subdivision steps, we will be left with a wire diameter of $d/(2)^{sub}$ = 0.98/2^6 = 0.015 mm. In mils this is

$$\text{mils} = \frac{\text{mm}}{0.0254} \Rightarrow \frac{0.015}{0.0254} = 0.591 \text{ mils}$$

The nearest AWG is

$$\text{AWG} = 20 \times \log\left(\frac{1000}{\text{mils} \times \pi}\right) = 20 \times \log\left(\frac{1000}{0.591 \times \pi}\right) = 54$$

But that is an impractically thin AWG (if it exists!). The minimum wire gauge usually available is AWG 52, and the thinnest practical value to use for multistrand bundles of magnet wire is AWG 42 to AWG 44).

So we should now start with three paralleled strands instead (in one layer, so p_{INIT} = 1). The starting diameter is 1.963/3 = 0.645 mm. The starting X is

$$X_{INIT} = \frac{0.886 \times \text{Dia}_{INIT}}{\delta} = \frac{0.886 \times 0.645}{0.185} = 3.1$$

Forward and Flyback Converters: Step-by-Step Design and Comparison

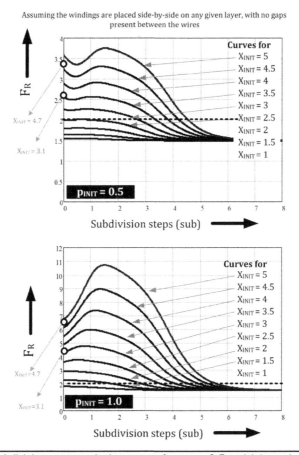

FIGURE 12.12 Subdivision strategy design curves for $p_{INIT} = 0.5$ and 1 (see also Fig. 12.16).

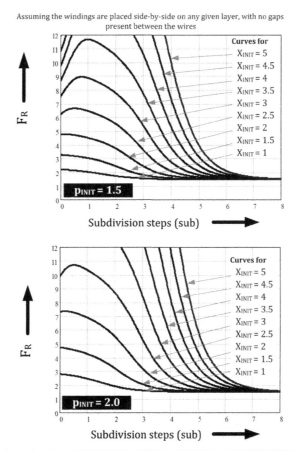

FIGURE 12.13 Subdivision strategy design curves for $p_{INIT} = 1.5$ and 2 (see also Fig. 12.16).

322 Chapter Twelve

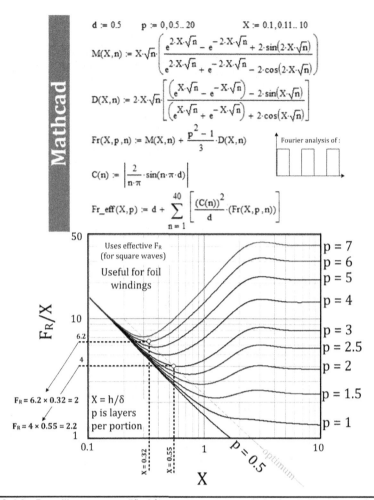

Figure 12.14 Dowell's curves modified for square waves and plotting F_R/X versus X for foil design.

p_{INIT} is still 1. From lower half of Fig. 12.12 we see that for X_{INIT} of 3.1, we need three subdivisions to get F_R to fall close to 3. So we will be left with a diameter of $0.645/2^3 = 0.081$ mm.

$$\text{mils} = \frac{\text{mm}}{0.0254} \Rightarrow \frac{0.081}{0.0254} = 3.19 \text{ mils}$$

The nearest AWG is

$$\text{AWG} = 20 \times \log\left(\frac{1000}{\text{mils} \times \pi}\right) = 20 \times \log\left(\frac{1000}{3.19 \times \pi}\right) = 40$$

This is acceptable, though the F_R is around 3, not 2.

In any subdivision step, a single strand becomes four strands. So, the number of strands after splitting each strand *sub* number of times, is $4^{\text{sub}} = 4^3 = 64$. So one possible implementation is to use a twisted bundle of magnet wire, consisting of 64 strands of AWG 40. Further, consistent with our starting assumption, three such bundles need to be laid out in parallel (all on one physical layer) to complete eight turns of the primary winding.

For the secondary winding, we look at Fig. 12.15. We see that indeed, optimizing F_R/X is a good idea for foils, especially because, unlike round wires, the layers per portion remain fixed if we increase the foil thickness. To optimize F_R/X, we can consult the lower part of Fig. 12.14. We see that with six layers (turns) per portion, as in our case, we have an optimum $F_R/X = 6.2$ for $X = 0.32$ (corresponding to $F_R = 6.2 \times 0.32 = 2$). So we need a copper foil of thickness $h = X \times \delta = 0.32 \times 0.185$ mm $= 0.059$ mm.

In mils, this is

$$\text{mils} = \frac{\text{mm}}{0.0254} \Rightarrow \frac{0.059}{0.0254} = 2.3 \text{ mils}$$

This is the optimum thickness of foil suggested. In general, 1, 1.4, 3, 5, 8, 10, 16, and 22 mils are the more commonly available foil thicknesses, but others can be ordered too.

Forward and Flyback Converters: Step-by-Step Design and Comparison

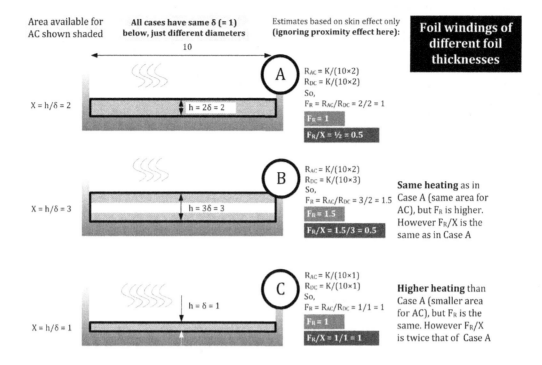

Conclusion: F_R does not correlate well with the heating in the case of foil windings either. However F_R/X reflects the heating well, and we should try to optimize F_R/X not F_R

FIGURE 12.15 Subdivision strategy for foil windings explained.

The width of this foil can be up to 15.7 mm (see "Width_tape" for EFD30 in Table 12.3). So the total copper cross-sectional area is 0.059 mm × 15.7 mm = 0.9263 mm². Our secondary-side current is 11 A (COR value), which for D = 0.5 is an RMS of $11 \times \sqrt{0.5}$ = 7.8 A. If this passes through 0.9263 mm², the current density will be 7.8/0.0093 = 838 A/cm². At a minimum duty cycle of 0.28 we will get the highest RMS value of $11 A \times \sqrt{1-D_{MIN}} = 11 \times \sqrt{0.72}$ = 9.33 A. So the worst-case RMS current density will be 9.33/0.0093 = 1000 A/cm². This is well over our target of 250 A/cm² RMS current denity, so losses will increase significantly.

So far we have ignored the possibility of interleaving. We have eight turns in the Primary, and we would like to keep that as one physical layer. Instead let us split the Secondary into two *series sections*. We can think of splitting the Secondary into two parallel sections too, but in that case we can get severe EMI due to slight imbalances in the paralleled halves, a sort of "ground loop" phenomena deep inside the transformer, unless we decide to "OR" the paralleled windings though separate output diodes. Also, in paralleled windings, the layers per portion will not decrease; in fact, the number of secondary portions will just double.

So we are now trying a series-split Secondary, sandwiching a single layer of Primary.

S-P-S Winding Arrangement

Let us start with just a single strand for the Primary and adjust its diameter so it completely fills one layer with eight turns. The layers per portion p for this, now equals 1/2, not 1, since each half Primary gets assigned to half the split Secondary on either side. The diameter is width/N_p = 15.7 mm/8T = 1.963 mm. Note that using the equivalent foil transformation of Dowell that diameter would give an equivalent foil of thickness $h = 0.886 \times 1.963 = 1.739$ mm (since $\sqrt{\pi/4}$ equals 0.886). In terms of penetration ratio X, that would be $X = h/\delta = 1.739$ mm/0.185 mm = 9.4. But that is too high a starting value for entering the subdivision process.

So instead, let us start with two paralleled strands with round wire, with diameter half of that, i.e., width/$(2 \times N_p)$ = 1.963/2 = 0.98 mm, but still laid out in one layer. So the layers per portion p still equals 1/2. We say p_{INIT} is 1/2. Similarly, the starting X value (X_{INIT}) is

$$X_{INIT} = \frac{0.886 \times \text{Dia}}{\delta} = \frac{0.886 \times 0.98}{0.185} = 4.7$$

Let us go with this for now. From Fig. 12.12 (upper half, that is, $p_{INIT} = 1/2$) we see that we need three subdivisions to get F_R to fall below 3. That could be OK, but we also look at the following case.

Suppose we start with three paralleled strands instead (in one layer, so $p_{INIT} = 1/2$). The starting diameter is $1.963/3 = 0.645$ mm. The starting X is

$$X_{INIT} = \frac{0.886 \times \text{Dia}_{INIT}}{\delta} = \frac{0.886 \times 0.645}{0.185} = 3.1$$

p_{INIT} is still 1/2. From Fig. 12.12 (upper half), we see that for X_{INIT} of 3.1, we *do not need any subdivisions* to get F_R to fall close to 3. In fact we are starting at $F_R = 2.5$ already. So no further subdivision is required.

In mils the strand diameter is

$$\text{mils} = \frac{\text{mm}}{0.0254} \Rightarrow \frac{0.645}{0.0254} = 25.4 \text{ mils}$$

The nearest AWG (for strand of diameter) is

$$\text{AWG} = 20 \times \log\left(\frac{1000}{\text{mils} \times \pi}\right) = 20 \times \log\left(\frac{1000}{25.4 \times \pi}\right) = 22$$

So *the Primary consists of three paralleled strands of AWG 22.*

Let us check whether we can accommodate this. In all, we will have $8 \times 3 = 24$ strands side by side in one layer. So it will occupy $0.645 \times 24 = 15.5$ mm. We know from Table 12.3 that we have 15.7 mm available. So this is acceptable.

NOTE *Coming to the Secondary, we could consider trying to avoid a foil winding if possible (for reasons of cost). We can then proceed as we did for the Primary. But keep in mind that p_{INIT} is no longer equal to 3 for the Secondary if we use round wires instead of a foil.*

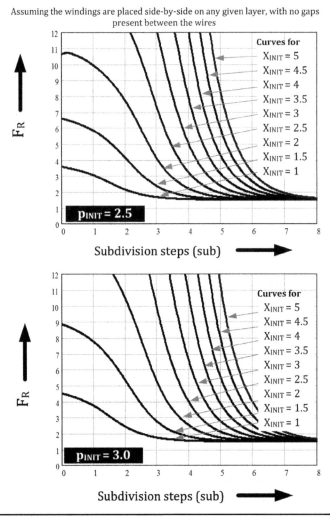

FIGURE 12.16 Subdivision strategy design curves for $p_{INIT} = 2.5$ and 3.

Let us return to three turns of foil on each side of the sandwiched Primary. We look at Fig. 12.14, along the $p = 3$ curve this time. We see that we have an optimum $F_R/X = 4$ for $X = 0.55$ (corresponding to $F_R = 4 \times 0.55 = 2.2$). So we need a copper foil of thickness $h = X \times \delta = 0.55 \times 0.185$ mm $= 0.102$ mm.

In mils, this is

$$\text{mils} = \frac{\text{mm}}{0.0254} \Rightarrow \frac{0.102}{0.0254} = 4.0 \text{ mils}$$

We look for a 4-or 5-mil-thick copper foil.

The width of this foil can be up to 15.7 mm (see "Width_tape" for EFD30 in Table 12.3). So the total copper cross-sectional area is 0.102 mm $\times$ 15.7 mm = 1.6 mm^2. Our secondary-side current is 11 A (COR value), which for $D = 0.5$ is an RMS of $11 \times \sqrt{0.5} = 7.8$ A. If this passes through 1.6 mm^2, the RMS current density will be 7.8/0.016 = 487 A/cm^2. This is within the usual industry target of 500 A/cm^2 at least (see page 299). But the losses are going to be higher than we expected, based on $F_R = 2$. Our initial target was 500 A$_{COR}$/cm^2 on page 315.

Keep in mind that in a foil winding, as explained in Fig. 12.15, even if we increase the foil thickness, since the skin and proximity effects restrict the effective cross-sectional area actually passing high-frequency current, that area remains fixed if we increase foil thickness beyond a certain point. So AC resistance will not improve. In reality because of *excess copper* exposed to higher proximity effects (higher X), we could end up worsening the situation. So how do we improve the current density for foils? The only possibility is by choosing cores with *long* profiles. For this we should also consider EER/ER and EERL/ERL cores in particular.

Input Capacitor Selection

In AC-DC applications, this is a wide topic involving holdup time, power factor correction, etc. Here we are dealing only with a DC-DC converter for telecommunications applications. So, the primary (dominant) selection criterion is the RMS current. We first select a bulk capacitor, preferably aluminum electrolytic, for cost reasons. The input RMS of a Buck (at $D \approx 0.5$) is $I_O/2$ (see page 712 of the A-Z book). In a Forward, this is reflected to the Primary through the turns ratio, as $I_O/2n$, i.e., $I_{OR}/2$. So, ignoring the small magnetizing current, we need a capacitor with an RMS rating of $I_O/n = 11$ A/(2 $\times$ 1.33) = 4.1 A. We tentatively pick UVY1J102MHD from Nichicon. This is a 105°C, 1000-h, 1000-μF, 63-V capacitor with a stated RMS current capability of 0.93 A. At high frequencies we can apply the typical frequency multiplier of $\sqrt{2} = 1.414$ (Nichicon actually allows 1.5). So its high-frequency RMS rating is, conservatively, 0.93 A $\times \sqrt{2} = 1.32$ A. If we parallel three of these, we get $1.32 \times 3 = 4$ A, which is what we need. At a maximum ambient of 45°C, we can add a worst-case additional 10°C rise from local heating (hot components, enclosure, etc.), so we estimate the surface of the capacitor can be at 55°C. We know that every 10°C below the upper category temperature (105°C here), we get a doubling of life (provided we do not exceed its datasheet value of ripple current rating at room temperature). So we expect life expectancy to be 1000 h $\times 2^{(105-55)/10} = 1000$ h $\times 2^5 = 320$ kh, which is about 3.7 years when operated 24 h a day. If we need more life, we need to pick a 2000 h at 105°C capacitor.

The ESR of each capacitor can be determined from its stated tangent of loss angle, $\tan \delta = 0.1$ (here δ is not the skin depth but the loss angle, expressed by this vendor, and most other vendors too, at 120 Hz). The relationship is

$$\tan \delta = \frac{\text{ESR}}{X_C} = \text{ESR} \times 2 \times \pi \times f \times C$$

Solving

$$\tan \delta = \frac{\text{ESR}}{X_C} = \text{ESR} \times 2 \times \pi \times f \times C$$

$$\text{ESR} = \frac{\tan \delta}{2 \times \pi \times f \times C} = \frac{0.1}{2 \times \pi \times 120 \times 1000} \times 10^6 = 0.133 \, \Omega$$

Note that this is the ESR at 120 Hz. We can assume the ESR of an aluminum electrolytic capacitor gets better by a factor of 2 at high frequencies (this is the origin of the frequency multiplier $\sqrt{2}$). In addition, we have three capacitors in parallel. So the net high-frequency ESR is 133 mΩ/6 = 22 mΩ.

We therefore have a net capacitance of 3000 µF with an ESR of 22 mΩ. This is acceptable, but where space is of concern, we would like to reduce the bulk capacitance, by paralleling several ceramic capacitors. The technique to do this will now be explained.

Paralleling Ceramic and Electrolytic Capacitors at the Input

Looking at Fig. 12.17 we see that there will be two contributions to the ESR: one from the capacitance and one from the ESR. In an electrolytic capacitor, generally the first contribution is negligible compared to the second. We have a reflected load current of $I_{OR} = 8.2$ A (I_{OR} for a Forward, I_O for a Buck), split in three capacitors, each therefore supporting 8.2 A/3 = 2.73 A load current. The high-frequency ESR of each is 22 mΩ × 3 = 66 mΩ. With a current through each capacitor of peak-to-peak value $I_{OR}(1 + r/2)/3$ (as per Fig. 12.18, but with three capacitors in parallel sharing I_{OR}), we will get a peak-to-peak ripple of [8.2 A × (1 + (0.4/2))/3] × ESR = 216.5 mV. In fact each identical capacitor produces this very ripple voltage, and these voltages are all in parallel so they do not pass current between one another (ideally). When we use several ceramic capacitors in parallel to replace one or two of the aluminum electrolytic capacitors ("Elkos"), the first thing we have to do to avoid upsetting this "apple cart" is to ensure they produce a ripple smaller than 216.5 mV. If the ripple they produce is more than this, the incoming current will prefer to shift more current through the remaining Elko(s). That will pass excess RMS current through them. But if less ripple is created in parallel, the ceramic capacitors will start taking up more and more of the current, and we may even be able to reduce the size of the last remaining Elko(s). So, let us target 150 mV for the ceramic capacitor combination.

Suppose we have three ceramic capacitors in parallel, their ESR will be typically 20 mΩ/3 = 7 mΩ, since each is about 20 mΩ. This is a rather small contribution to the voltage ripple. The main contribution to the ripple in this case will come from the fact that their net capacitance is not as large as for aluminum capacitors.

We have to now use the capacitance-based equation in Fig. 12.17 for the ceramic capacitors. *We are trying to get rid of two of the three Elkos.* The replacement ceramic combination must support at least 2 × 8.2 A/3 = 4.47 A (shared) load current. Setting a peak-to-peak ripple target of 150 mV, we get

$$150 \text{ mV} = \frac{I_{OR} \times D \times (1-D)}{f \times C_{IN}} = \frac{4.47 \text{ A} \times 0.5 \times (1-0.5)}{200,000 \times C_{IN}}$$

Solving for capacitance,

$$C_{IN} = \frac{4.47 \times 0.5 \times (1-0.5)}{200,000 \times 150 \times 10^{-3}} = 3.7 \times 10^{-5} \text{ F}$$

This is 37 µF (net value). To keep the ESR contribution low (well below 33 mΩ, i.e., whatever they are replacing: two 66-mΩ capacitors in parallel here), we should try two or three ceramic capacitors in parallel. Also, knowing that the actual value of a ceramic capacitor in any application may only be 60 percent the printed value in its datasheet, we should aim for almost twice the calculated value we obtained for choosing the printed value.

So finally, a possible solution is: One 1000-µF, 63-V aluminum Elko, in parallel with two 33-µF, 100-V ceramic capacitors. (The two paralleled 33-µF ceramic capacitors will give an in-circuit net value of about 37 µF.)

Output Capacitor Selection

The output is a simple Buck stage. We are trying to choose ceramic capacitors here. There are three main criteria we need to satisfy. This is explained in Fig. 12.18.

1. Maximum peak-to-peak output ripple to be within 1 percent (i.e., ±0.5 percent) of output rail, i.e., $V_{O\_RIPPLE\_MAX} = 12 \text{ V}/100 = 0.12$ V.
2. Maximum acceptable droop during a sudden increase in load: $\Delta V_{DROOP} = 0.5$ V.
3. Maximum acceptable overshoot during a sudden decrease in load: $\Delta V_{OVERSHOOT} = 0.5$ V.

We have minimum output capacitances based on criteria 1 to 3. For criteria 1,

$$C_{O\_MIN\_1} = \frac{r \times I_O}{8 \times f \times V_{O\_RIPPLE\_MAX}} = \frac{0.4 \times 11}{8 \times 200 \times 10^3 \times 0.12} = 2.3 \times 10^{-5} \text{ F}$$

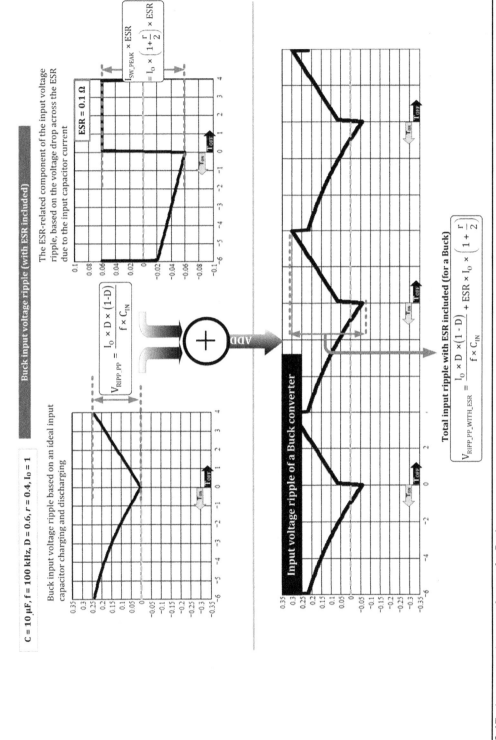

FIGURE 12.17 Input capacitor waveforms of a Buck.

Chapter Twelve

Maximum ESR: based on maximum output ripple

Ignoring ESR and ESL, purely based on capacitance, the maximum allowed output ripple determines a minimum output capacitance.

$$C_O = \frac{r \times I_O}{8 \times f \times V_{RIPPLE}} \quad \text{(see Fig. 13.5 in } \textit{Switching Power Supplies A-Z, 2e}\text{)}$$

$$C_O \geq \frac{r \times I_O}{8 \times f \times V_{RIPPLE\_MAX}}$$

Including ESR, but assuming C is large and ESL is negligible. The maximum allowed voltage ripple determines a maximum ESR

$$V_{RIPPLE} = ESR \times I_O \times r$$

$$ESR \leq \frac{V_{RIPPLE\_MAX}}{I_O \times r}$$

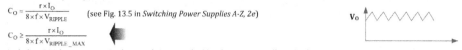

Minimum capacitance: based on maximum droop

Typically, with a well-designed loop, it takes about three switching cycles for the loop to react and start correcting the output to meet a sudden load demand. During that time we do not want the output capacitor to fall more than a certain value V_{droop}. Thus, using $I = C\, dV/dt$, we get

$$I = C\frac{\Delta V}{\Delta t} \Rightarrow C \geq \frac{I \times \Delta t}{\Delta V} = \frac{I \times 3T}{\Delta V_{droop}} = \frac{I \times 3}{\Delta V_{droop} \times f}$$

Here the droop is actually related to the <u>extra</u> load demand, since the normal load requirement is being met every cycle without any droop. So the current here is actually the load increase.

$$C_O \geq \frac{3 \times \Delta I_O}{\Delta V_{droop} \times f}$$

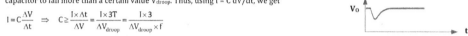

Minimum capacitance: based on maximum overshoot

There is another criterion. In case of a sudden release of load demand, say from maximum load I_o to zero, the inductor energy will all get dumped into the output capacitor. If we do not want too much of an overshoot (say, to restrict the output to some value V_x):

$$\tfrac{1}{2} \times C\left(V_x^2 - V_O^2\right) = \tfrac{1}{2} \times L\left(I_O^2\right) \Rightarrow C \geq \frac{L\left(I_O^2\right)}{(V_x + V_O) \times (V_x - V_O)} \approx \frac{L\left(I_O^2\right)}{(2V_O) \times (\Delta V_{overshoot})}$$

$$C_O \geq \frac{L\left(I_O^2\right)}{(2V_O) \times (\Delta V_{overshoot})}$$

(where we have used the approximation $V_x + V_O \approx 2 \times V_O$. Also, $V_x - V_O = \Delta V_{overshoot}$)

FIGURE 12.18 Output capacitor selection criteria for Buck (and Forward) converters.

That is 23 µF. Now based on the second criterion,

$$C_{O\_MIN\_2} = \frac{3 \times (I_O/2)}{\Delta V_{DROOP} \times f} = \frac{3 \times (11/2)}{0.5 \times 10^6} = 3.3 \times 10^{-5}$$

That is 33 µF. Now based on the third criterion,

$$C_{O\_MIN\_3} = \frac{L \times I_O^2}{2 \times V_O \times \Delta V_{OVERSHOOT}} = \frac{10 \times 10^{-6} \times 11^2}{2 \times 12 \times 0.5} = 1.00 \times 10^{-4}$$

This is 100 µF. We need to account for tolerances, etc. Let us therefore pick an output ceramic capacitance of 120 µF/16 V. This will satisfy all the criteria. This may have a large ESR, so we may want to pick paralleled capacitors.

We should double-check that the ESR of the selected capacitor is small enough. The ESR should be less than

$$ESR_{Co\_MAX} = \frac{V_{O\_RIPPLE\_MAX}}{I_O \times r} = \frac{0.12}{11 \times 0.4} = 0.027\ \Omega \quad \text{(i.e., 27 m}\Omega\text{)}$$

Most small-capacitance ceramic capacitors will have no trouble complying with this. But in our case we may prefer to pick two 56-µF or 68-µF/16-V ceramic caps in parallel, to keep ESR down.

Note that as explained in Fig. 12.19, the contributions to output voltage ripple are not *in phase*. So, for example, if the ESR-based ripple is 120 mV, and the capacitance-based ripple is 120 mV, the total ripple is still around 120 mV, not 240 mV. This is unlike the input.

Step-by-Step Flyback Converter Design

Here the requirements are the same as for the preceding Forward converter. This exercise will give us insight into how a Flyback compares with a Forward, in terms of design methodology and component selection, especially at these high power levels.

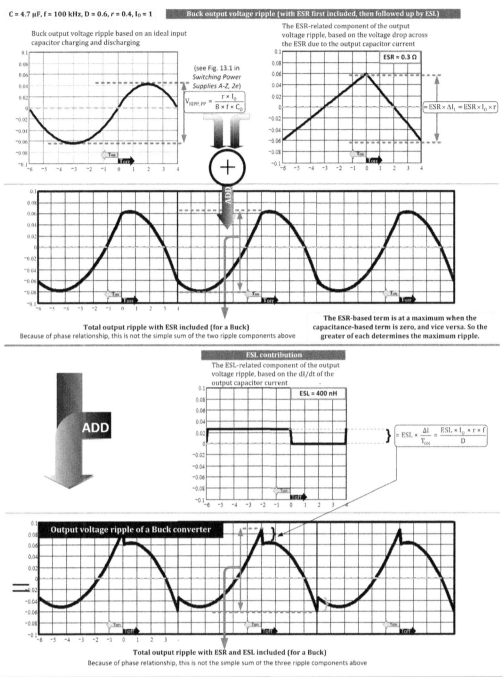

FIGURE 12.19 Output capacitor waveforms of a Buck.

Choosing V_{OR}

Once again, assume the efficiency will be close to 85 percent. So for an output of 132 W, the input will be 132/0.85 = 155.3 W. This is to compare apples to apples, though a Flyback will have much lower efficiency at these power levels, largely due to the huge pulsating current into the output capacitors and leakage inductance dissipation.

We need to set the reflected output voltage (the effective output rail as seen by the primary side). This is also based on the maximum duty cycle limit condition at low-line. We have (see the DC transfer function equation of a Buck-boost in the Appendix)

$$V_{OR} = V_{INMIN} \times \frac{\eta_{VINMIN} \times D_{MAX}}{1 - D_{MAX}} = 36 \times \frac{0.85 \times 0.44}{1 - 0.44} = 24.04 \text{ V}$$

Turns Ratio

Therefore, the turns ratio must be

$$n = \frac{V_{OR}}{V_O} = \frac{24.04}{12} = 2$$

Core Selection

$$V_{e_{cm^3}} = \frac{31.4 \times P_{IN} \times \mu}{z \times f_{MHz} \times B_{SAT_G}^2}\left[r \times \left(\frac{2}{r}+1\right)^2\right] = \frac{31.4 \times 155.3 \times 2000}{10 \times 0.2 \times 3000^2}\left[0.4 \times \left(\frac{2}{0.4}+1\right)^2\right] = 7.8$$

Here we have used the equation derived in *Switching Power Supplies A-Z*, 2d ed (page 225). We have set relative permeability to 2000, maximum saturation flux density to 3000 G (0.3 T), air-gap factor z to 10, and current ripple ratio to 0.4. We need a core volume of 7.8 cm^3. Looking at Table 12.1 we see that the EFD30 we selected for the Forward converter has a volume of 4.7 cm^3. We need almost twice that here. From Table 12.1 we see that a close fit is ETD34 with a volume of 7.64 cm^3 and an effective area of 0.97 cm^2.

Primary Turns

As derived in the *A-Z* book (page 236),

$$N_P = \left(1 + \frac{2}{r}\right) \times \frac{V_{INMIN} \times D_{MAX}}{2 \times B_{SAT_T} \times A_{e_{m^2}} \times f_{Hz}}$$

$$= \left(1 + \frac{2}{0.4}\right) \times \frac{36 \times 0.44}{2 \times 0.3 \times 0.97 \times 10^{-4} \times 200,000} = 8.2 \approx 8 \text{ turns}$$

Secondary Turns

$$N_S = \frac{N_P}{n} = \frac{8}{2} = 4 \text{ turns}$$

Note that the turns ratio is 8/4 = 2, as compared to 1.33 for the Forward converter. This helps pick lower voltage components on the secondary side since the reflected input voltage is lower.

Primary Inductance

From the Appendix, and as derived in *A-Z* book (see page 139),

$$L_{P\_\mu H} = \frac{V_{OR}}{I_{OR} \times r \times f_{Hz}} \times (1 - D_{MAX})^2$$

$$= \frac{24.04}{(11/2) \times 0.4 \times 200,000} \times (1 - 0.44)^2 = 1.714 \times 10^{-5}$$

So we need a primary inductance of 17.14 µH.

Zener Clamp

For good efficiency, the zener clamp voltage must be greater than 1.4 times the reflected output voltage. So the minimum recommended clamp voltage is $1.4 \times V_{OR} = 1.4 \times 24.04 = 33.7$ V. But see the subsequent final choice below (i.e., 58 V).

Voltage Ratings

This clamp would require a minimum FET rating of 57 V + 33.7 V = 90.7 V. But if we pick a 100-V FET, there is very little headroom (margin). So we actually pick a 150-V FET instead, and also can then use a 58-V zener clamp. That will give us a maximum drain-to-source voltage of 57 V + 58 V = 115 V. That gives us a good derating margin of 115 V/150 V = 0.77 (i.e., headroom of 23 percent).

We assume the typical leakage is 1 percent of the primary-side inductance, $L_{lk} = 171$ nH. The peak primary current at low-line is

$$I_{PK\_PRI} = \frac{I_{OR}}{1 - D_{MAX}} \times \left(1 + \frac{r}{2}\right)$$

$$= \frac{11/2}{1 - 0.44} \times \left(1 + \frac{0.4}{2}\right) = 11.8$$

The zener clamp dissipation will be

$$P_Z = \frac{1}{2} \times L_{lk} \times I_{PK}^2 \times f \times \frac{V_Z}{V_Z - V_{OR}}$$

$$= \frac{1}{2} \times 171 \times 10^{-9} \times 11.8^2 \times 200,000 \times \frac{58}{58-24} = 4.1 \text{ W}$$

In reality, if we do bench measurements, we will likely find that the peak current that actually freewheels into the clamp is less. It can be anywhere between 0.7 and 0.9 of the expected peak current, on account of the parasitic capacitances in the transformer, especially in the case of multilayered primary windings. If that happens, the zener clamp dissipation can even be $0.7^2 = 0.5$ (half) what we are expecting from the preceding equation.

The reflected input voltage is

$$V_{INRMAX} = \frac{V_{INMAX}}{n} = \frac{57}{2} = 28.5 \text{ V}$$

The minimum output diode voltage is therefore

$$V_{D1} = V_{INRMAX} + V_O = 28.5 + 12 = 40.5 \text{ V}$$

The currents are very high, so certainly a single diode will not serve the purpose. We will need paralleled diodes, and very likely low-R_{DS} synchronous FET rectification.

Input Capacitor Selection

The equation for the RMS current of an input capacitor of a Buck is

$$I_{IN\_RMS} = I_O \sqrt{D\left(1 - D + \frac{r^2}{12}\right)}$$

so for $D = 0.5$ and r small, we get

$$I_{IN\_RMS} \approx \frac{I_O}{2}$$

For a Forward converter, we just replaced I_O with $I_{OR} = I_O/n$. Similarly, for a Buck-boost we have

$$I_{IN\_RMS} = \frac{I_O}{1-D} \sqrt{D\left(1 - D + \frac{r^2}{12}\right)},$$

so for $D = 0.5$ and r small, we get

$$I_{IN\_RMS} \approx I_O$$

For a Flyback converter, we just replace I_O with $I_{OR} = I_O/n$.

We see that for duty cycles close to 50 percent, the Flyback has twice the input capacitor RMS, compared to a Forward. Another way of looking at this is the Flyback of x W, is equivalent to a Forward of $2x$ W. So, we can repeat the calculations we did for a Forward converter, but mentally visualizing it as a 12 V @ 22 A converter now. All our calculated input capacitances will double. We can similarly trade-off ceramics for some of the Elkos. We conclude that a possible solution is two 1000-μF/63-V aluminum Elkos, in parallel with four 33-μF/100-V ceramic capacitors.

Output Capacitor Selection

In a Flyback and Buck-boost we have a pulsating current into the output capacitors too. It is not *smoothened* by an inductor along the way, as in a Buck or Forward. So the dominant criterion is simply based on the need to be able to absorb this high RMS, without overheating. Any further reduction in output voltage ripple, if necessary, usually comes from a small post-LC filter placed after the initial output capacitors just after the output diode, which are the ones that really take the entire brunt of the output diode current.

The calculation is actually very similar that for to the input capacitor. The equation for the output capacitor RMS of a Buck-boost is

$$I_{O\_RMS} = I_O \sqrt{\frac{D + r^2/12}{1 - D}}$$

so for $D = 0.5$ and r small, we get

$$I_{O\_RMS} \approx I_O$$

We need a net capacitor RMS rating of 11 A! We are getting no help from the turns ratio here, as in the case of input capacitors with pulsating currents. Looking at the catalog, we see that rather than use traditional aluminum electrolytic capacitors with huge capacitance values, since we only need capacitors for less than 25 V, a very good candidate is APXC160AR-A820MH70G from Chemicon. It is a 82-μF/16-V *conductive polymer aluminum solid capacitor* with a RMS rating of 2.83 A_{RMS} (at high frequencies), due to its extremely low ESR of 25 mΩ. We need four of these in parallel for a $2.83 \times 3 = 11.32$ A_{RMS} rating.

Copper Windings

The skin depth at 200 kHz is (assuming a temperature of 80°C for setting the resistivity of copper more accurately)

$$\delta_{mm} = \frac{66.1[1 + 0.0042(T - 20)]}{\sqrt{f_{Hz}}} = \frac{66.1[1 + 0.0042(80 - 20)]}{\sqrt{200,000}} = 0.185 \text{ mm}$$

We choose a round wire of diameter 2δ. So we look for a wire of cross-sectional area $0.185 \times 2 = 0.37$ mm. In mils the strand diameter is

$$\text{mils} = \frac{\text{mm}}{0.0254} \Rightarrow \frac{0.37}{0.0254} = 14.6 \text{ mils}$$

The nearest AWG (for strand of diameter) is

$$\text{AWG} = 20 \times \log\left(\frac{1000}{\text{mils} \times \pi}\right) = 20 \times \log\left(\frac{1000}{14.6 \times \pi}\right) = 27$$

We choose AWG 27. Its cross-sectional area is

$$\text{Area}_{AWG} = \frac{\pi \times D^2}{4} = \frac{\pi \times 0.37^2}{4} = 0.11 \text{ mm}^2$$

At a target current density of 250 A_{RMS}/cm^2, we can pass $250 \times 0.11/100 = 0.275$ A_{RMS}. However, at low line, our COR current is $I_{OR}/(1 - D_{MAX}) = 11$ A$/[2 \times (1 - 0.44)] = 9.82$ A. Its RMS value is $9.82 \text{ A} \times \sqrt{D} = 9.82 \text{ A} \times \sqrt{0.44} = 6.5$ A_{RMS}. So the number of strands we need for the Primary are $9.82/0.275 = 36$ strands. If we double the current density target to 500 A_{RMS}/cm^2, we can go in for 18 strands. Also, as explained in Fig. 12.10, in the case of round wires, if we go to higher diameters than 2δ, we do get an improvement in AC resistance, even though F_R worsens. So we can in fact judiciously go in for thicker wire gauges (lesser number of strands), to fill in each layer fully, if required, and thus get a better build.

On the secondary side, we use the same wire gauge. At $D = 0.44$, the RMS of the secondary current is $I_O/\sqrt{(1-D)} = 11 \text{ A}/\sqrt{(1 - 0.44)} = 14.7$ A_{RMS}. So if we are targeting 250 A/cm^2, we know that AWG 27 is only capable of 0.275 A_{RMS}. In that case the number of strands required is $14.7/0.275 = 53$ strands. If we decided we can double the current density, we can go in for 26 strands for the Secondary.

Note that we have not touched upon the topic of proximity effects in the Flyback, since most agree it is an extremely difficult problem to tackle through closed-form equations. Instead we are just sticking to current density targets. This is discussed next.

Keep in mind that split and sandwiched windings help here too, but mainly to reduce leakage inductance and reduce zener clamp dissipation. Otherwise our underlying assumption of leakage inductance being just 1 percent of the primary inductance won't be true.

Industrywide Current Density Targets in Flyback Converters

In the *A-Z* book, we suggested 400 cmil/A as a recommended current density for the Flyback. See its nomogram and contained explanation on page 145 of the *A-Z* book. That was based on the COR value. To make that clearer here, as per our current terminology, we prefer to write it as 400 cmil/A_{COR}.

Assuming $D \approx 0.5$, we have $\sqrt{D} = 0.707$, so the conversions are

$$\frac{400 \text{ cmil}}{A_{COR}} \equiv \frac{600}{0.707} = \frac{565 \text{ cmil}}{A_{RMS}}$$

or

$$\frac{197{,}353}{400} = \frac{493 \, A_{COR}}{cm^2} \quad \text{(in terms of COR current)}$$

or

$$\frac{197{,}353}{565} = \frac{350 \, A_{RMS}}{cm^2} \quad \text{(in terms of RMS current)}$$

In other words, we were recommending somewhere between 250 (conservative) to 500 A_{RMS}/cm^2 (overly aggressive). But a lot depends on core losses too, because we should remember, the flux swing in a typical Flyback is always fixed at around 3000 G, not 1500 G as in a Forward converter. So core losses can be four times that of a Forward converter transformer (since for ferrites, we can have B^2 dependency in the core-loss equation). However, we are also using a (Flyback) core size that is twice that in a Forward converter. So it is better exposed to cooling. But at the same time, everything else is scaling too. For example, we first calculate core loss per unit volume and then multiply that with volume to get the total core loss. So if volume is doubled, for the same flux density swing, we will get double the core losses! And so on. The picture is really murky. We do need to depend a lot on industry (and our own) experience here. In the case of this author, it was 400 cmil/A_{COR}, just for achieving Class A transformer certification (and barely so). So it is probably best to target 350 A_{RMS}/cm^2 at worst. A lower density is even better (say 250 A_{RMS}/cm^2). But what do others say?

- AN-4140 from Fairchild allows 500 A_{RMS}/cm^2, suggesting up to 600 A_{RMS}/cm^2.
- Texas Instruments, www.ti.com/lit/an/slua604/slua604.pdf allows 600 A_{RMS}/cm^2.
- International Rectifier, www.irf.com/technical-info/appnotes/an-1024.pdf, suggests 200 to 500 cmil/A_{RMS}. This translates to 400 to 1000 A_{RMS}/cm^2.
- AN017 from Monolithic Power allows 500 A_{RMS}/cm^2.
- AN-9737 from Fairchild, www.fairchildsemi.com/an/AN/AN-9737.pdf, asks for 265 A_{RMS}/cm^2, very close to our conservative suggestion of 250 A_{RMS}/cm^2.
- On-Semi, www.onsemi.com/pub_link/Collateral/AN1320-D.PDF, allows 500 A_{RMS}/cm^2.
- Power Integrations recommends 200 to 500 cmil/A, *but in calculations often uses the COR value without necessarily pointing it out*, and typical values used are 500 A_{COR}/cm^2. That is 19,737/500 = 400 cmil/A_{COR}, same as what was suggested in the *A-Z* book. From page 333, that is 350 A_{RMS}/cm^2. See Fig. 10.4 too.

Keep in mind there is a big difference in making a small and *attractive* transformer for an evaluation board, and between a commercial product that meets safety approvals.

Comparison of Energy Storage Requirements in a Forward and Flyback

Irrespective of efficiency considerations, the most basic question is: By going from a Flyback to a Forward, do we end up requiring more magnetic volume or less?

We saw earlier, that when we went to the Flyback, its *transformer volume was twice that of the Forward converter*. But the Forward converter has an additional magnetic component, its energy storage element, that is, its secondary-side choke. Generally we pick an off-the-shelf inductor for that. But we can ask: If we use a gapped ferrite for the choke, how will its volume compare with the transformer of the Flyback? Keep in mind that in a Flyback, its transformer is also the energy storage element.

The answer to this is on page 225 of the *A-Z* book, where we show that for a Buck, the volume is $(1 - D) \times$ the volume of a Buck-boost, for the same energy, current ripple ratio, etc. so for a duty cycle of about 0.5, the volume of a Buck choke will be half that of a Buck-boost.

We learned that the transformer of a Forward is half the size of a Flyback, but then we need a secondary-side choke for it, equal to half the size of the Flyback transformer. The total gain or *loss is virtually zero*. Both the Forward and the Flyback need almost the *same total volume* of magnetic components. Yes in a Forward, the heat gets split into two components and their total exposed area is more than that of a single component of the same net volume. This is one of the reasons a Forward is preferred at higher powers. But the Flyback also suffers from zener clamp dissipation and high-RMS output capacitor current.

CHAPTER 13
PCBs and Thermal Management

PART 1 PCBs AND LAYOUT

Introduction

When it comes to switching regulators, it is not enough to concern oneself with just the basic routing and connectivity and related mechanical and production issues. Both the power supply designer and the CAD person need to be aware that the design of a switching power converter is only as good as its layout.

The overall area of printed-circuit board (PCB) design is an extremely wide one, embracing several testing, mechanical, and production issues and applicable compliance and regulatory issues. Most of the issues discussed in this chapter revolve around simply ensuring basic functionality. Though luckily, as the beleaguered switcher designer will be happy to know, in general, all the electrical aspects involved are related and point in the same general direction. So, for example, an *ideal* layout, that is, one that helps the IC function properly, also leads to reduced electromagnetic emissions, and vice versa. There are, however, some exceptions to this helpful trend, particularly in the practice of indiscriminate copper-filling, and this will be touched upon too.

Trace Analysis

We must first learn to identify the troublesome or critical trace sections of any topology. The following rule is simple and applies to all topologies.

During a crossover transition the current flow in some trace sections has to suddenly come to a stop, and in certain others it has to start equally suddenly. These are the critical traces for any switcher PCB layout. These should be identified up front.

In Fig. 13.1 we have trace analysis for the Buck. We have omitted most of the control traces like feedback, bootstrap, and enable, as these hardly carry any current and are *not* critical from the perspective of current and power flow. But do note that the routing of the feedback trace (which will be taken up later) is certainly important too, but mainly from the viewpoint of noise pickup, after the concerns already expressed in Fig. 1.21.

The bold traces in Fig. 13.1 show the power flow. We have shown the traces, which are passing current during the switch ON-time, followed by the traces conducting during the OFF-time, and finally in the lowermost schematic we work out the *difference* between the two previous schematics. These *difference traces* are the ones in which the current must either abruptly turn ON or turn OFF during crossover (switch transition).

But all traces have nonzero inductance. So if current is passing through them, they have a certain amount of stored energy $(1/2) \times LI^2$. When the current in such a trace is commanded by the switch to stop flowing, its stored energy cannot disappear immediately. So it "complains loudly" in the form of a voltage spike (shown with gray triangles). Similarly, the same traces generate a voltage spike in the opposite direction, when we try to force current suddenly through them. The amplitude of the spikes can be significant as per the standard equation $V = L\,dI/dt$. This means that even if L is small, if the dI/dt is very high, so will be the spike. The dI/dt depends on how fast the FET/BJT switches from the ON-state to OFF-state (and the other way around).

The problem with the spikes is that not only can they cause EMI and output ripple, but *they easily enter the control sections of the IC* through their connecting pins, causing the IC to "misbehave." Note that the noise rejection capability (or equivalently susceptibility to noise)

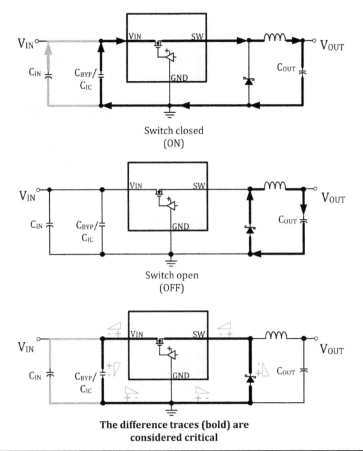

FIGURE 13.1 Trace analysis of a Buck converter to identify critical traces.

of switcher ICs is never specified in datasheets. However a *good device* is one that is fairly tolerant to these noise spikes. But, in general, we can affect the behavior of a switcher IC (more than a controller IC), if we do not reduce the inductance of the identified critical traces. This is what may lead to a constantly griping switcher IC customer rather than a satisfied repeat-order one. It's usually just the layout.

We do note that ICs with BJT switches always tend to switch more slowly than ICs that use FET switches. So it is no surprise that there are less *product complaints* from customers who are using the slower devices. In reality, those were likely not "product complaints" but admissions of bad PCB layout.

We need to highlight some more details from Fig. 13.1. C_{IC} is a bypass capacitor specifically meant to smooth out the supply rails (which are ultimately going to the control sections of the IC). If C_{IC} is very close to the IC, it provides most of the high-frequency content demanded (mainly by the edges) of the switch current waveform. C_{IN} is the input bulk capacitor, and it provides most of the remaining current waveform. It is being constantly DC-refreshed from a distant bulk capacitor, in this case belonging to the DC bench power supply.

We can see that the way to reduce the length, and thereby the inductance of the critical traces, is to bring their *associated components close to the IC*. That will automatically reduce the corresponding trace lengths. It will also automatically reduce the high-frequency *current loop* and thereby reduce EMI too. So, for example, the position of C_{IC} is critical in Fig. 13.1 (for this topology). This should thus be as close as possible to the IC. C_{IN} can be usually an inch or two away in that case. It is, therefore, not very critical and is shown in bold gray lines in the schematic. If the dedicated IC decoupling capacitor C_{IC} is omitted (and it can usually be left out if we have a BJT-based Buck switcher IC, provided we have C_{IN} of course), then C_{IN} must be brought very close to the IC. More than even 1 inch away from the IC is known to have caused problems. Similarly, for a FET-based switcher, even two series intervening vias (one on each side of the decoupling capacitor), are known to have caused erratic performance. This will happen if the decoupling capacitor is not on the same layer as the IC, but on the other side of the board. Which is why, in Fig. 13.2, we show the best way of doing IC decoupling.

In Fig. 13.3 we do the same analysis for all three topologies and thus arrive at Table 13.1, where we have summarized the critical components for the three topologies.

PCBs and Thermal Management 337

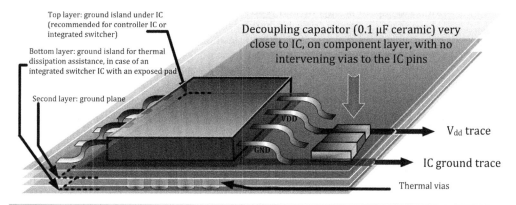

FIGURE 13.2 Recommendation for proper decoupling.

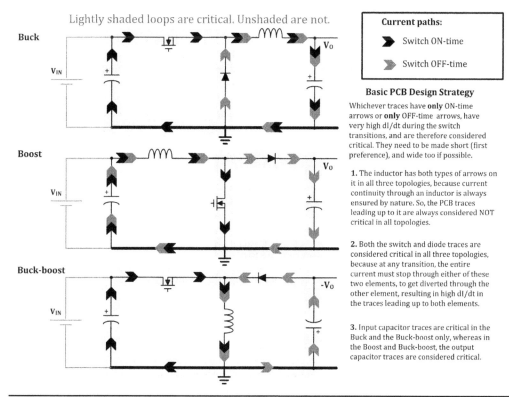

FIGURE 13.3 Trace analysis for all three topologies.

	C_{IN} (input bulk capacitor)	C_{IC} (bypass capacitor)	C_{OUT} (output capacitor)	Inductor	Diode
Buck	Critical	Critical	Not critical	Not critical	Critical
Buck-boost	Critical	Critical	Critical	Not critical	Critical
Boost	Not Critical	Critical	Critical	Not critical	Critical

TABLE 13.1 Critical Components in PCB Layout for Integrated Switchers and Controllers

Miscellaneous Points to Note

1. The oft-repeated rule of thumb is that every inch of trace length has an inductance of about 20 nH.

2. The transition time is about 10 to 30 ns for high-speed FET switchers and is about 75 to 100 ns for the slower bipolar switchers. This also means that the voltage spikes in the high-speed family can be more than twice that in the slower family,

for a comparable layout and load (because of higher dI/dt). Therefore, layout becomes all the more critical in high-speed switchers. One inch of trace, switching, say, 1 A of instantaneous current, in a transition time of 30 ns, gives 0.7 V theoretically. For 3 A, and 2 inches of trace, the induced voltage tries to be up to 4 V! These spikes may not actually be seen, however, as they easily get absorbed by parasitics, but some of this switching noise may certainly enter the IC itself (which is the root cause of its reported misbehavior).

3. None of these problems can be easily corrected, or even band-aided, once the layout is bad. So the important thing is to get the layout *right* to start with.

4. For the Buck topology, the output capacitor current is smooth (because the inductor is in series with it). In a Boost topology the situation is reversed: i.e., the input capacitor current is smooth and the current into the output capacitor is pulsed. In a Buck-boost or Flyback, both the input and the output capacitor currents are pulsed.

5. In some *not-so-horrible* Buck converter layouts, the converter was successfully band-aided by a small-series resistor-capacitor (RC) snubber across the catch diode. This consists typically of a resistor (low-inductance type preferred) of value 10 to 100 Ω and a capacitor, preferably SMD ceramic, of value 470 pF to 2.2 nF. A larger capacitance than this would lead to unacceptably higher dissipation ($= C \times V^2 \times f$), mainly in the resistor of the snubber. However, note that this RC snubber actually needs to be placed very close to and across the switching (SW) pin and ground (GND) pin of the IC, with short leads and traces.

    Sometimes the RC snubber is perceived to be across the diode, and meant for the diode, because on the schematic there is no way to tell the difference. However, the purpose of this RC snubber is to actually absorb the voltage spikes of the *trace inductances*, and therefore, its position must be such that it bypasses the critical or AC trace sections of the output side. It must *straddle* the culprit trace sections (of the fairly bad layout). Especially where the catch diode is a Schottky diode (not a regular ultrafast diode), reverse recovery or forward recovery issues are hardly present, so the RC snubber is not there because of any diode characteristics it needs to quell but to account for bad trace sections. The corollary is that RC snubbers are not usually necessary if the layout is carried out in accordance with the lessons learned via Figs. 13.1 and 13.3—unless, of course, the controller/switcher IC itself is incredibly susceptible to noise. In that case we should change the chip vendor, not add (dissipative) snubbers.

6. The traces to the critical components should be short, reasonably wide, and should not go through any vias. The inductance of a via is given by

    $$L = \frac{h}{5}\left(1 + \ln\frac{4h}{d}\right) \text{ nH}$$

    where h is the height of the via (equal to the thickness of the board) and d is the diameter, both in millimeters.

    Expressed in centimeters for easy comparison to the inductances provided for a trace or wire length, this is

    $$L = 2h\left(1 + \ln\frac{4h}{d}\right) \text{ nH}$$

    where h is now the height of the via and d is the diameter, both in centimeters.

    This inductance is *not insignificant* especially when we are confronted with the high dI/dt's of modern switchers. So, if vias have to be used in a critical trace for any reason, *several vias in parallel* will yield better results than a single via. And larger via diameters would help further. We should certainly avoid vias in the path of chip decoupling.

7. A first approximation for the inductance of a conductor having length l and diameter d is

    $$L = 2l \times \left(\ln\frac{4l}{d} - 0.75 + \frac{d}{2l}\right) \text{ nH}$$

    where l and d are in centimeters. Note that the equation for a PCB trace is not much different from that of a wire as shown here:

    $$L = 2l \times \left(\ln\frac{2l}{w} + 0.5 + 0.2235\frac{w}{l}\right) \text{ nH}$$

where w is the width of the trace and l the length, both in centimeters. For PCB traces, L of a trace hardly depends on the thickness (depth) of the copper layer (i.e., it is the same for a 1-oz or 2-oz board, these terms being explained later). Note that the preceding, equation is actually for a flat ribbon of copper (freestanding). In an actual case, the return path (PCB trace) may pass directly under the forward trace (automatically, if we have a ground plane with no cuts below the forward trace). In that case there can be a significant reduction in inductance because of mutual cancellation of the fields produced by the forward and return traces. So the preceding equation represents the worst case in a sense.

There is another lesson we can learn from the preceding equations. Note that in all cases l appears outside the log term, so it is fair to say that doubling the length will roughly double the inductance. But since w or d is within the log, we realize that to halve the inductance, we have to increase the width by a large factor (up to 10 times). This is related to the mutual inductance effects of several thin parallel strips constituting the trace. See Fig. 13.4 for a plot based on these inductance equations. We can see that standard PCB thickness of 1.6 mm will also lead to a via inductance of around 1 or 2 nH for a via of diameter 1 mm, increasing if we reduce the diameter of the via. If we parallel several vias, the inductance will fall significantly.

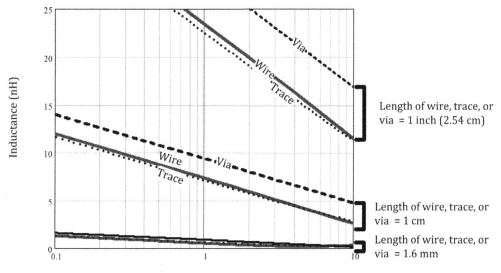

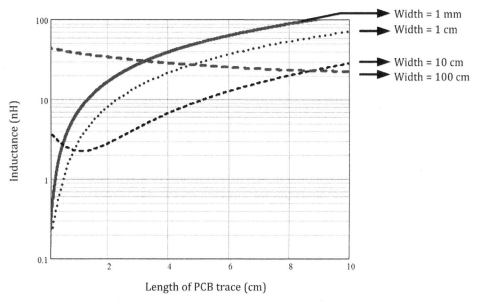

FIGURE 13.4 Trace, via, and wire inductances.

As a corollary, making traces as wide as possible may not be a very good idea. It doesn't reduce the inductance much, and it can also become a nice electric field antenna if there is a pulsating voltage across it (e.g., if it is the switching node). So at some point we must learn to resist the tendency of copper-filling, especially if the copper fill is anything other than the ground plane. Note that in a typical Buck, the cathode of the diode is the switching node of the topology. Knowing that the cathode of the diode connects thermally to the die, tab, and pad, we are tempted to increase the copper area of the switching node for thermal reasons. But we realize we need to be particularly judicious about this particular aspect alone.

Sizing Copper Traces

Commercial PCBs are often referred to as *1-oz* or *2-oz*, for example. This refers to the weight of copper in ounces per square foot deposited on the copper-clad laminate. The 1-oz PCB is equivalent to 1.4-mil copper thickness (or 35 μm). Similarly the 2-oz PCB is 2.8 mils or 75 μm.

There are complicated curves available for copper versus temperature rise of PCB traces in the now-obsolete military standard MIL-STD-275E (see Fig. 13.5). These curves have also found their way into more recent standards like IPC-2221 and IPC-2222. Engineers often try to create elaborate curve-fit equations to match these curves. But the truth is the earlier curves can be easily approximated by simple linear rules as follows.

The required cross-sectional area of an external trace is approximately

37 mil^2 per ampere of current for 10°C rise in temperature (most recommended).

25 mil^2 per ampere of current for 20°C rise in temperature.

18 mil^2 per ampere of current for 30°C rise in temperature.

The MIL curves are for traces exposed to air. It has been observed later that for the traces in inner layers, we can just multiply the calculated width of an external trace by the factor 2.6, to get the required width.

To calculate the width of a trace from the cross-sectional area, keep in mind that 1-oz copper is 1.4 mil thick and 2-oz copper is 2.8 mil thick.

So the preceding can be written as follows:
For 1-oz copper:

26 mil per ampere of current for 10°C rise in temperature.

18 mil per ampere of current for 20°C rise in temperature.

13 mil per ampere of current for 30°C rise in temperature.

For 2-oz copper:

13 mil per ampere of current for 10°C rise in temperature.

9 mil per ampere of current for 20°C rise in temperature.

6.5 mil per ampere of current for 30°C rise in temperature.

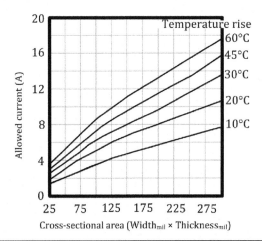

FIGURE 13.5 MIL-STD-215E curves for temperature rise of PCB traces.

It is very important to size the traces correctly, as this can also contribute significantly to the overall dissipation, efficiency, and temperature rise.

We see that the current-handling capability of traces is related to the temperature rise we permit. Though military standards (MIL standards or MIL-STD) recommend a maximum 10°C rise in PCB trace temperature, we can often go up in commercial designs to twice or thrice that.

Routing the Feedback Trace

The only critical signal trace is the feedback trace. This is more liable to pick up noise if the pin it connects to, is a high-impedance node. ICs with *fixed output voltage options* like 5 and 3.3 V require no external voltage divider, because the divider is internal to the IC. Therefore, for such parts, the feedback trace is not going to a high-impedance input pin, and it is relatively immune to noise pickup. For the *adjustable voltage parts* greater care must be taken how we route this trace. We should keep the feedback trace short *if possible* so as to minimize pickup, but it should certainly be kept away from noise sources (e.g., switch, diode and even inductor). We must not compromise the positions of the input decoupling capacitor and catch the diode in the process. If necessary we can use a via to drop into the *ground plane* and route the feedback trace through it. This keeps its surroundings *quiet*.

Routing the Current-Sense Trace

This applies particularly to AC-DC converters, especially those using current-mode control, such as the popular UC384 × controllers. It is a common pitfall to run the current-sense trace parallel to the Gate trace since both leave the controller IC and head to the FET. However, if a quiet trace runs parallel to a noisy trace even for a few millimeters, it can pick up tremendous jitter. One way out, especially in cheaper one-sided PCBs, is to *guard-band*, by running the ground trace in between as shown in Fig. 13.6. Of course, the feedback trace is not used in such cases, since the opto-coupler is connected directly to the low-impedance COMP pin (input of the PWM comparator). So the feedback trace is not an issue here.

The Ground Plane

With double-sided (two-layer) boards, it is a common practice to almost completely fill one side with ground (copper-filling). In four-layer boards, one plane is dedicated to ground. There are people who usually rightly so consider this a panacea for most problems. As we

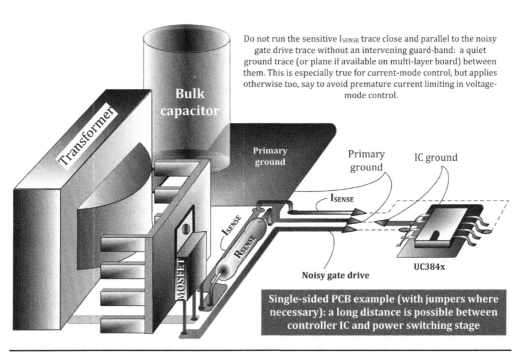

FIGURE 13.6 Routing the sense trace in AC-DC converters.

have seen, every signal has a return, and as its harmonics get higher, the return "wants" to be directly under the signal path, thus leading to field cancellation and reduction of inductance. See Fig. 13.7. It also helps thermal management as it couples some of the heat to the other side. The ground plane can also capacitively link to noisy traces above it, causing reduction in overall noise and EMI.

The ground plane can also, however, end up radiating if there is *too much* capacitive coupling from noisy traces into this plane. By its sheer area, it can be a very good antenna if we just give it the necessary degree of freedom. The author, therefore, prefers to avoid dedicating an internal copper plane (in four-layer and higher-layer boards) to the positive incoming supply rail (V_{IN} or V_{CC}), as is considered a fairly common practice. The reason is that the incoming rail is also usually derived from a switching power supply and can inject noise into the ground plane, besides becoming a good electric field radiator itself. The author recommends having only thick traces for the incoming supply rail, keeping just a ground plane. We can consider multiple ground planes too, provided they are stitched together by multiple vias spaced at short distances, avoiding inadvertent ground loops.

As shown in Fig. 13.7, at low frequencies (low-frequency harmonics), the return current flowing in the ground plane tries to return by the path of least resistance, i.e., a straight line running across the ground plane between two points. But for higher-frequency harmonics it tries to image itself directly under the forward trace as it tries to cancel the mutual field, minimize the loop area, and lower the total inductance, all at the same time. This now becomes

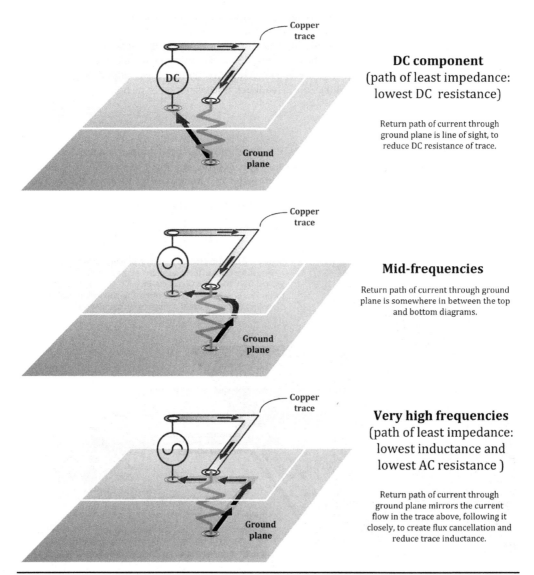

FIGURE 13.7 How a ground plane of copper helps in lowering PCB parasitics (without cuts).

the path of least reactive impedance. So, if the ground plane is partitioned in odd ways, either to create thermal islands or to route other traces, the current flow patterns in it can become "unnatural". We must learn to allow nature to help us. It always does. With various thoughtless cuts in the ground plane, we can get effects quite similar to a slot antenna.

We also note that generally, a ground plane should supplement other recommendations. It does not necessarily substitute for the correct placement of the critical components in particular.

Sometimes we forget that board parasitics are also known to inadvertently *help* us in some cases, such as in preventing unwanted interactions between different circuit blocks—the reason why we put decoupling in the first place. So the designer should weigh everything carefully before deciding on a board configuration and layout.

In multilayer boards we should try to keep the ground plane as the internal plane just below the component side. By the fact that the distance to the component side is now much less (provided the overall board thickness is still the standard 1.6 mm), the length of the interconnecting vias is much smaller and the coupling in general is better. Therefore, noise gets suppressed much better. Current-sense sections seem to work better. EMI is better. And the layout becomes rather forgiving to bad practices. So if the engineer is too busy or nonchalant to sit with the CAD guy through all this, then this is the way to go. Never mind the cost. If the cost is already inherent in the design, by the fact that the converter is just being placed on board an existing multilayer card, the ground plane becomes a no-brainer.

Some Manufacturing Issues

A 1-oz double-sided board passes through an electroless copper plating process stage (before solder mask is applied) to create vias [also called plated through-holes (PTH)], and so it may end up effectively as being considered closer to a 1.4-oz copper board. Therefore, it is a good idea to check this out with the PCB manufacturer before even starting the layout. This will help determine how wide the copper traces actually need to be for a particular current-carrying capacity and temperature rise. Also note that even a single-sided board passes through a hot-air solder level finishing stage (after solder mask), where a thin tin (or tin-lead in the past) layer is deposited on the *unmasked* (no solder mask) copper areas. This does increase the effective thickness of these traces, but doesn't help as much as copper plating, since tin-lead has 10 times higher resistivity than copper.

We are also getting so used to seeing standard *green boards* (FR-4) that we may forget that there are cheaper alternatives available. In power supplies where cents count, and the PCB could cost several dollars, we cannot afford to forget that cheaper materials could cut the overall PCB cost by almost a factor of four. The designer may, however, have to struggle since not all board laminate materials are equal, and they have limitations that need to be overcome by careful layout.

A strategy once used by the author was to move as many control components as possible onto *one* side of a *Primary-to-Secondary* double-sided SMD (or PTH) daughter board *made of FR-4*, while placing *all* the bulkier components which were *not capable* of being handled by pick and place machines onto a cheaper single-sided CEM-1 board. This was considered to be a major cost saver for the company, not only in terms of component cost but also production and test costs. It did require SMD opto-couplers though.

But first we need a small discussion about what are FR-4 and CEM-1. FR stands for *fire resistant* or *fire retardant*. Incidentally, that is not exactly the same as the issue of *flammability*, which is the key concern of safety agencies. FR-4 and also the cheaper CEM laminates (*composite epoxy material*) do typically all come with the highest flammability rating of safety agencies (94V-0, or equivalently just V-0).

> *CEM-1.* This consists of a cellulose paper core and a woven glass cloth surface, combined with an epoxy resin binder. It can be recognized by its whitish to cream color (opaque). It has good mechanical strength, is easily punchable, and drills well too. It is very cheap and not suitable for PTHs (*vias*). It is useful for single-sided boards.
>
> *CEM-2.* Considered virtually identical to CEM-1.
>
> *CEM-3.* Similar to CEM-1 except that the core is made of nonwoven glass. CEM-3 is considered virtually interchangeable with FR-4. It is recognizable by its natural light brown color (opaque). PTH is possible with this material. In Japan, for example, CEM-3 has a very large market share.
>
> *CEM-4.* Considered virtually identical to CEM-3, particularly from an electrical standpoint.

FR-2. This is a paper-based laminate with phenolic resin binder. It has a high moisture resistance and is flame retardant.

FR-3. This is a paper-based laminate with epoxy resin binder. It has high flexural strength and is flame retardant.

FR-4. This is a woven glass cloth with epoxy resin binder. It is the same as CEM-3 except that it is fire retardant.

Some general comments and observations:

- FR-4 is used for its inherent stability. Its characteristics remain constant over long periods of time, even under adverse environmental conditions. However, there are different grades and specifications even within FR-4, with special versions for chemical resistance and for use in multilayer designs. Different grades for higher temperatures are also available. The rating we cannot exceed is the T_g (glass transition temperature) of the PCB laminate.

- An important reason to use FR-4 along with SMD components is that almost all manufacturers of such components spend a great deal of effort matching their *CTE* (coefficient of thermal expansion) or *TCE* to that of standard FR-4 boards. If the thermal expansions are not well matched, components may start to crack under extended periods of thermal cycling and, also, solder joints will develop fatigue and start to give.

- For boards devoted to through-hole components, we can consider using either CEM-1 or CEM-3 to reduce costs. But note that whenever we use through-hole components, that board will pass through a wave-soldering process. We should, however, avoid immersing SMD components in hot solder as will happen during wave soldering if they are on the underside of the board. Generally, if SMD components must be on the same board as through-hole components, for various reasons we should try to restrict them to the component side only (where the power components are placed).

- If a double-sided SMD control board is being used, we may like to avoid placing components on both sides of the board. Because if the components lie only on one side, they are just placed on top of the board by the pick and place machines and pass through the reflow soldering. They thus stay in place with the help of gravity (and solder paste). However, if we have components on both sides, we will have to ensure that the components on one side do not fall off. Adhesives will become necessary. All this adds to the cost.

- A possible concern with the use of single-sided boards for power components is that under a vibration test, a single-sided board's copper traces can come undone rather easily. Double-sided boards offer much greater *holding capability*, especially when the heavy component's leads pass through plated through-holes. Here the solder seeps into the PTH creating a very firm bond. So with a single-sided board we should at least try to leave large copper areas around the legs of heavier components, like the transformer. Gluing the component to the board with room temperature vulcanizing silicone (RTV) or hot-melt will also probably be required.

- Single-sided boards will require jumpers. Most of the time these may have to be hand inserted, which can add to the cost depending on the geographic area in which the production is being done. Jumpers have a higher inductance than traces. We must, therefore, study the layout to see where currents *change direction* during a given switch transition. Those traces are considered critical, and no jumpers should ever be placed particularly on such traces.

- The *transition temperature* T_g of the PCB material is also important. We don't want to use a low-temperature board material and have to provide more expensive better cooling systems. The standard FR-4 has a T_g of approximately 130°C. The CEM boards may be rated 20 to 30°C less than FR-4 depending upon their manufacturer.

- One of the most important ratings of a PCB laminate is its *comparative tracking index (CTI)*. We should confirm what the CTI is from the specific vendor as it has a bearing on the required *creepage* in off-line power supplies. See Chap. 17 for more details on CTI.

PCB Vendors and Gerber Files

We must understand that our PCB should match the ability of the vendor if we want to reduce costs. For example, every vendor has *preferred drill sizes*. We should get that information beforehand.

Too often we also don't understand what information we need to provide to the vendor. Often the vendor just doesn't ask and simply uses default values. So another vendor's board may behave very differently, simply because that vendor's default values were different. It will take us a while to figure out why the new build is not working properly anymore. We need to avoid this situation by specifying clearly all that is important to us.

The Gerber File format is the industry-standard format for files used to generate artwork necessary for circuit board imaging. The preferred Gerber format today is *extended Gerber* or RS274X, which embeds the apertures within the specific files. The apertures assign specific values to design data (specific pad size, trace width, etc.), and these values make up a D-code list.

The current Gerber file format specification is revision I3 from June 2013. It can be freely downloaded from the Ucamco download page at www.ucamco.com/downloads.

When files are not saved as RS274X, a text file with values must be included because the values must be hand-entered by CAM operators. This slows down the process and increases the margin for human error, as well as lead time and cost.

Here is a sample of a request to a PCB vendor:

Please build a quantity of 200 pieces of this 6-layer board. We need 1-ounce base copper weight, 0.062" thick FR-4 material, with green solder mask, and white legend.

File Name: FLYBACK_70_REVA.ZIP includes:

FLYBACK70_REVA.CMP Layer 1 (component side)
FLYBACK70_REVA.gnd Layer 2 (ground layer)
FLYBACK70_REVA.IN1 Layer 3 (internal signal)
FLYBACK70_REVA.IN2 Layer 4 (internal signal)
FLYBACK70_REVA.VCC Layer 5 (power layer)
FLYBACK70_REVA.SLD Layer 6 (solder side)
FLYBACK70_REVA.CSM (component-side solder mask)
FLYBACK70_REVA.SSM (solder-side solder mask)
FLYBACK70_REVA.TSK (component-side silkscreen/nomenclature)
FLYBACK70_REVA.BSK (solder-side silkscreen)
FLYBACK70_REVA.DRL (Excellon drill file)
FLYBACK70_REVA.TOL (drill tool description)
FLYBACK70_REVA.APT (aperture information)
FLYBACK70_REVA.DWG (drawing or print)

PART 2   THERMAL MANAGEMENT

Introduction

Natural convection starts with the simple statement: "Hot air rises". So, for example, a power semiconductor mounted on a metal plate gets exposed to the natural flow of air, and this transfers heat from the power device to the surroundings. We also know intuitively that using larger metal plates helps this process, and this lowers the temperature. We probably also know that very large plates don't help much beyond a certain point. We are also perhaps aware that higher dissipations actually help the cooling process, though that doesn't mean we try to increase our dissipation to avail this so-called *advantage*. It just means that the temperature rise *per watt* is less under high-dissipation conditions. So, clearly, nature plays a helping hand, but to understand how it goes about the task, we need to come up with equations of convection.

Thermal management is not a black art. The rules are actually quite simple. We noticed that one of the major problems is *not* that there was a lack of available rules, empirical or otherwise, but that they were being cast in so many diverse forms that engineers just don't know how they compared and which one to choose. Our purpose here is not to declare which rule is the most accurate, but to show the different forms each rule can be written in. Further, if the equations are all converted into the same format, they become amenable to a simple point comparison. And then we will see how amazingly close the apparently different equations really are in their respective predictions of temperature. Then we can pick a *conservative estimate*, or a more practical estimate.

Thermal Measurements and Efficiency Estimates

One problem in thermal management is measurement. Measuring a temperature rise and relating it to the dissipation is one of the most difficult bench studies an engineer can undertake. Some issues to be aware of are as follows:

- *Where temperature should be measured.* Heatsink manufacturers frequently provide procedures they used, but often these are unfortunately impractical for a real-life converter measurement, for example, drilling a small hole through the heatsink until the thermocouple can be inserted to make contact with the tab. Some say that the best point for judging the junction temperature of a three-terminal FET, for example, is the central lead, just at the point where it exits the plastic. But this gives measurements that are slightly lower than with other methods. One thing is clear: The location that gives us the highest measurement is likely to be the most accurate. Some say the tab of a mounted transistor is the best point, and the thermocouple should be centered just below the mounting hole, at the point closest to the plastic body. The author, however, personally prefers to loosen the mounting screw slightly, insert the thermocouple under the tab, put in a lot of thermal grease to fill the air space, and then tighten the screw again. Note that in the previous methods, the thermocouple needs to be glued onto the device. Thermally conductive adhesives are available, but some take several hours to cure. A quick method, and one that is internally allowed by some major power supply vendors, is to use *super glue* (cyanoacrylate). This has been found to be acceptable, provided the thermocouple is pressed firmly against the tab for the minute or so it takes for the glue to harden. Moisture (a little dab of water) hastens the curing process for this glue. Or we can use a formal accelerator spray.

NOTE *Thermocouples are either (1) T type: these are made of copper-constantan and work over the range –270 to 600°C with a swing of 25 mV, providing 40.6 µV/°C at 25°C, or (2) J type: these are of iron-constantan and work over the range –270 to 1000°C with a swing of 60 mV, providing 51.70 µV/°C at 25°C. The ends are usually spot-welded together, but soldering often works too for quick results. The solder should be 60 percent lead and 40 percent tin as it has a higher eutectic temperature than the more commonly used 40 percent lead and 60 percent tin.*

- *Variable ambient conditions.* In a typical room, unbeknownst to us, small movements of air (draughts) keep taking place, like someone walking past or an overhead air-conditioning vent suddenly coming into play. The measurements, therefore, rarely turn out to be reproducible to within the required couple of degrees of accuracy. An enclosed box may be built, but then local pockets of hot air form inside. To keep a proper average ambient temperature inside, we need some movement of air. If a fan is used, it may be too much. We want to create natural convection, not forced air convection.
- *What is the dissipation?* Even if we do get the temperature, how do we relate it to the dissipation? A FET, for example, has an R_{DS} that is a function of temperature, as is its dissipation. So we end up with a cyclical argument. We want to know the temperature at a certain dissipation, but the dissipation itself depends on the temperature. Nature ultimately iterates to achieve a stable solution, but for us the math may be hard to figure out. One thing we can do is replace the FET with a diode in the same package. We can pass direct current through the diode and monitor the voltage across it by a kelvin measurement using a multimeter. We should increase our current slowly (and wait!) so that we get the same temperature we got with the switching FET in place. We thus now know the dissipation correctly, and we also have the corresponding temperature reading. Having characterized the heatsink, we can put in the FET again and try to validate our design estimates against the measured dissipation.
- *Measuring conduction loss.* We go back to the bench and try to measure the forward voltage drop across the FET (or diode), for example. This is in our actual switching application, and we are thus trying to validate our estimate of conduction loss, and possibly create a model for purposes of optimization. A multimeter will clearly not work here. So we use an oscilloscope and try to "see" the forward voltage across the FET. We are surprised to find that it seems too high, or too low, or even goes negative. We make sure our probes are well compensated by connecting the probe to the test signal on the front panel of the scope, and adjusting the trimpot on the probe by the

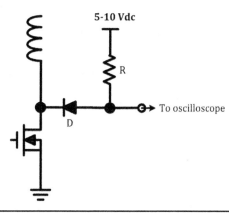

FIGURE 13.8 Measuring the forward voltage drop.

small, insulated screwdriver provided in the accompanying probe kit. Now our test signal looks *nice and* square, but when we look at the FET again, we still get absurd results. The problem here is that we have adjusted the vertical scale of the scope to a few millivolts per division as we try to zoom in on the small forward drop. So when the FET turns OFF, the voltage is high and *out of scale*. The invisible part of the signal we are trying to measure, therefore, actually ends up overdriving the internal amplifiers of the scope, and when the FET turns ON again in the next cycle, the scope just can't faithfully reproduce the signal correctly (for a certain undefined recovery time). The solution to this is to clamp the voltage to a lower level of about 5 to 10 V as shown in Fig. 13.8. We would need to characterize the *VI* characteristics of the diode beforehand since we are now going to be reading the forward drop across the FET plus the diode drop. The diode should be carefully selected to be a fast small-signal PN-type diode. *R* will also need to be adjusted, typically to around 1 to 2 kΩ, to bias the diode correctly. The DC voltage may also need to be further adjusted to keep the signal limited to the screen.

- *Measuring switching losses.* Here we look at the overlap between the voltage waveform and the current waveform. We have to remember that active current probes are not so fast and typically have a delay. Therefore, switching loss measurements can be erroneous. Passive probes (which just have a coil inside) may work much better for this measurement. Further, even the scope probes used for voltage measurement have a few nanoseconds of delay, and this can vary from probe to probe, especially if their lengths are different. We should check for these *propagation delays* and correct our readings accordingly. On some scopes we can apply this correction on-screen too. Scopes with general purpose interface bus (GPIB) output can be used to drop the data into an Excel spreadsheet, in which we can apply the necessary offsets, Δt and Δv, and then do the usual *VI* crossover calculation automatically.

The Equations of Natural Convection

If we know the temperature of the *heatsink* (which could just be the copper in the immediate vicinity of a surface-mounted device), and also the dissipation, we can estimate the junction temperature by using

$$R_{th\_JA} = R_{th\_JH} + R_{th\_HA}$$

where R_{th} is the thermal resistance in °C/W, J refers to the junction, H is the heatsink, and A stands for ambient. If P is the dissipation in watts,

$$T_J = P \times (R_{th\_JH} + R_{th\_HA})$$

Note that R_{th_JH} can be further split:

$$R_{th\_JH} = R_{th\_JC} + R_{th\_CH}$$

where C is for case. But the only parameter we probably have full control over is R_{th_HA}. So this chapter focuses on understanding this critical link. *Henceforth, R_{th} just refers to R_{th_HA}.*

There are abundant equations or empirical formulas available for the purpose. Though they actually just fall under two or three umbrellas, and these too are very close. But that is not obvious at first sight. One source of confusion is that the *area* used in these equations often refers to the area of the plate (*one* side of it) even though both sides are exposed to natural convection. But some equations recognize this up front, and use twice the value (total exposed area). This actually changes the appearance of the equation almost completely, and therefore they become hard to compare. Then still others use area in square inches, though some use square meters (or some other units). The problem is that even a small change of units makes the equation unrecognizable from its root form, because of the exponents involved.

In this chapter, A refers to the area of one side of a plate, both of whose sides are exposed to cooling A. The total exposed area is being called $\underline{A}$ (so $\underline{A} = 2A$).

Historical Definitions

We take the simplest case of a square plate made from a very good thermally conducting material, dissipating P watts. After some time, we will find that the plate stabilizes at a certain temperature rise of ΔT over the ambient.

We expect that the temperature rise will be proportional to the dissipation. The proportionality constant is called the thermal resistance R_{th} in °C/W. So

$$R_{th} = \frac{\Delta T}{P}$$

Similarly, we expect that the thermal resistance will vary inversely with the area:

$$R_{th} \propto \frac{1}{A}$$

The inverse of the proportionality constant above, is h in W/°C per unit area and is called by various names, like *convection coefficient* or *heat transfer coefficient*.

$$R_{th} = \frac{1}{h\underline{A}} = \frac{1}{2hA}$$

Finally we have the basic equation set

$$P = h \times \underline{A} \times \Delta T = 2 \times h \times A \times \Delta T = \frac{\Delta T}{R_{th}}$$

Explicitly,

$$R_{th} = \frac{\text{temperature rise}}{\text{watts}}$$

$$h = \frac{\text{watts}}{\text{total exposed area} \times \text{temperature rise}}$$

and

$$h\underline{A} = \frac{1}{R_{th}}$$

The reason we kept using the word *expect* is that historically speaking these relations were presumed to be true, and it was thought that R_{th} and h were proportionality constants. But later it was realized that this was not so. However, the preceding classical equations were still maintained for the sake of consistency (enforced), *but what changed was that no longer were* h *or* R_{th} considered constants. They were now *allowed* to depend on area and on dissipation etc., the intention being to indirectly factor in the observed deviations from the *expected* results.

Available Equations

As a first approximation, h is often stated (at sea level) to be

$$h = 0.006 \text{ W/in}^2 \cdot °C$$

If area is expressed in meters, this becomes

$$h = 0.006 \times (39.37)^2 = 9.3 \text{ W/m}^2 \cdot °C$$

(since there are 39.37 inches in a meter).

Nowadays we know that in reality, h can vary about 1:4 times from the commonly assumed precedng value. So in literature we can find the following generalized empirical equation for h and this becomes our *standard equation no. 1*:

$$h = 0.00221 \times \left(\frac{\Delta T}{L}\right)^{0.25} \quad \text{W/in}^2 \cdot {}^\circ\text{C} \quad \text{(standard equation no. 1)}$$

where L is the length along the direction of natural convection (vertical). In the case of the simple square plate, $L = A^{0.5}$, so we can write this as:

$$h = 0.00221 \times \Delta T^{0.25} \times A^{-0.125} \quad \text{W/in}^2 \cdot {}^\circ\text{C} \quad \text{(standard equation no. 1)}$$

Also observe that the preceding equation uses A, which is actually half the area exposed to cooling. So, we can equivalently rewrite it in terms of the actual area involved in the cooling process:

$$h = 0.00221 \times \Delta T^{0.25} \times \left(\frac{\underline{A}}{2}\right)^{-0.125} \quad \text{W/in}^2 \cdot {}^\circ\text{C}$$

$$h = 0.00241 \times \Delta T^{0.25} \times \underline{A}^{-0.125} \quad \text{W/in}^2 \cdot {}^\circ\text{C}$$

These are all available and published forms of the *same* equation for h.

NOTE *The preceding equation predicts that h has a specified dependency on the exposed area of the plate and also on its temperature differential with respect to ambient. This dependency (i.e., $A^{-0.125}$) implies that the cooling efficiency per unit area (i.e., h) of large plates is worse than that of small plates. However, if this sounds surprising, we note that the overall and total cooling efficiency of a plate is $h \times A$, which depends on $A^{+0.875}$. So, thermal resistance goes as $1/A^{+0.875}$ and is clearly lower for a large plate than for a small plate as we would expect. Compare this to the ideal $1/A$ variation, which was as per classical equations, what was initially expected for thermal resistance.*

In literature we often find the following *standard* formula (area in square inches), hereafter referred to as our *standard equation no. 2*:

$$R_{th} = 80 \times P^{-0.15} \times A^{-0.70} \quad \text{(standard equation no. 2)}$$

Where A is in square inches. We notice that the first equation is written in terms of h and the second in terms of R_{th}. How do we compare them?

We will now do some more manipulations on these equations to bring them to a comparable format.

Manipulating the Equations

We have provided tables for the purpose, but let us do one such manipulation to get comfortable with the process:

1. We can rewrite our standard equation no. 1 in terms of dissipation instead of temperature rise:

$$h = 0.00221 \times \left[\frac{P}{h \times A \times 2}\right]^{0.25} \times A^{-0.125}$$

So

$$h = 0.00654 \times P^{0.2} \times A^{-0.3} \quad \text{W/in}^2 \cdot {}^\circ\text{C}$$

2. Or we can write it in terms of the total exposed area:

$$h = 0.008 \times P^{0.2} \times \underline{A}^{-0.3} \quad \text{W/in}^2 \cdot {}^\circ\text{C}$$

3. We can also now try to see what this will look like in MKS (SI) units. The conversion is not obvious and so we proceed as follows: Take an imaginary plate of size 39.37 in × 39.37 in, or 1 m × 1 m. Clearly the thermal resistance of the plate is in °C/W and is therefore independent of the units used to measure area and must remain unchanged by any change in the system of units used. This means that $1/(h \times \underline{A})$ is independent of units, and so is $h \times \underline{A}$. Therefore, if in MKS units we first assume a similar form for h:

$$h = C \times \Delta T^{0.25} \times A^{-0.125} \quad W/m^2 \cdot °C$$

Equating,

$$h \times A = C \times \Delta T^{0.25} \times A_{m^2}^{-0.125} \times A_{m^2} = 0.00221 \times \Delta T^{0.25} \times A_{in^2}^{-0.125} \times A_{in^2}$$

$$C \times A_{m^2}^{0.875} = 0.00221 \times A_{in^2}^{0.875}$$

$$C = (39.37^2)^{0.875} \times 0.00221 = 1.37$$

So finally, in MKS units,

$$h = 1.37 \times \Delta T^{0.25} \times A^{-0.125} \quad W/m^2 \cdot °C$$

4. In terms of the total exposed area:

$$h = 1.49 \times \Delta T^{0.25} \times \underline{A}^{-0.125} \quad W/m^2 \cdot °C$$

5. We can also express h in terms of P instead of temperature as before:

$$h = 1.12 \times P^{0.2} \times A^{-0.3} \quad W/m^2 \cdot °C$$

6. In terms of the total exposed area:

$$h = 1.38 \times P^{0.2} \times \underline{A}^{-0.3} \quad W/m^2 \cdot °C$$

7. We can also recast standard equation no. 1 in terms of thermal resistance instead of h. We get several different forms:

$$R_{th} = \frac{1}{2hA} = 76.5 \times P^{-0.20} \times A^{-0.70} \quad \text{(area in in}^2\text{)}$$

8. Or in terms of the total exposed area:

$$R_{th} = \frac{1}{h\underline{A}} = 124.3 \times P^{-0.20} \times \underline{A}^{-0.70} \quad \text{(area in in}^2\text{)}$$

9. In MKS units

$$R_{th} = \frac{1}{2hA} = 0.45 \times P^{-0.20} \times A^{-0.70} \quad \text{(area in m}^2\text{)}$$

10. Or in terms of the total exposed area:

$$R_{th} = \frac{1}{h\underline{A}} = 0.72 \times P^{-0.20} \times \underline{A}^{-0.70} \quad \text{(area in m}^2\text{)}$$

Comparing the Two Standard Equations

Our standard equation no. 2 is

$$h = 80 \times P^{-0.15} \times A^{-0.70} \quad \text{(area in in}^2\text{)}$$

The result of our manipulations on standard equation no. 1 gives us

$$R_{th} = \frac{1}{2hA} = 76.5 \times P^{-0.20} \times A^{-0.70} \quad \text{(area in in}^2\text{)}$$

And we thus see that the two equations, one initially expressed in terms of h and the other in terms of R_{th}, are not very different at all, if brought to a similar form as we have done. This is now an apples-to-apples comparison.

h from Thermodynamic Theory

Without needing to go too deep into thermodynamic theory, here is a quick check on the equations we can derive from theory. We have the dimensionless Nusselt number, Nu, which is the ratio of the convection heat transfer to the conduction heat transfer. We also have the dimensionless Grashof number Gr, which is the ratio of buoyant flow to viscous flow. Under natural convection (laminar flow), we have the following defining equations in MKS units:

$$\text{Nu} = 3.5 + 0.5 \times \text{Gr}^{1/4}$$

where

$$\text{Gr} = \frac{g \times [1/(T_{AMB} + 273)] \times \Delta T \times L^3}{v^2}$$

where $g = 9.8$ (acceleration due to gravity in m/s²) and $v = 15.9 \times 10^{-6}$ (kinematic viscosity in m²/s). At an ambient temperature $T_{AMB} = 40°C$ it can be shown that this simplifies to

$$\text{Nu} = 3.5 + 52.7 \times \Delta T^{0.25} \times L^{0.75}$$

The coefficient of cooling is by definition

$$h = \frac{\text{Nu} \times K_{AIR}}{L}$$

where K_{AIR} is the thermal conductivity of air (0.026 W/m·°C). So we get our third standard equation:

$$h = 0.091 + 1.371 \times \left(\frac{\Delta T}{L}\right)^{0.25} \quad \text{W/m}^2 \cdot °C \quad \text{(standard equation no. 3)}$$

or

$$h = 0.091 + 1.371 \times \Delta T^{0.25} \times A^{-0.125} \quad \text{W/m}^2 \cdot °C \quad \text{(standard equation no. 3)}$$

Comparing this to the previously given empirical equations, we find that this equation too is surprisingly close, especially to the comparable form 3) of our standard equation no. 1.

Unfortunately, though this form may be more accurate because of the constant term in its equation, for that very reason it is more difficult to manipulate into all the forms the previous equations could be manipulated into. So we won't even try here. We will manipulate the previous equations into a similar form and then compare them toward the end. In effect, we are ignoring standard equation no. 3 from this point on, with the understanding that it is almost identical to standard equation no. 1.

Working with the Tables of the Standard Equations

In Table 13.2 we have the complete procedure for manipulating an equation given in terms of h into all the other forms. We have four cases each time:

Case 1. Area used in equation is half of exposed area (in square inches).

Case 2. Area used in equation is full exposed area (in square inches).

Case 3. Area used in equation is half of exposed area (in square meters).

Case 4. Area used in equation is full exposed area (in square meters).

In the same table we have shown the numbers (in gray) thus generated from our standard equation no. 1. At the bottom of the same table we show how to generate all the forms from an equation given in terms of R_{th} (such as from our standard equation no. 2). In one step we can go from an equation in terms of R_{th} to the same expressed in terms of h, that is, to the very beginning of the table, and then we can work our way down as before, generating all the other forms.

$$h = \alpha \times \frac{\Delta T^\beta}{\text{Area}^\gamma}$$

Case 1 inches, half area	$\alpha_1 \equiv \alpha = ?$ 0.00221	$\beta_1 \equiv \beta = ?$ 0.25	$\gamma_1 \equiv \gamma = ?$ 0.125
Case 2 inches, full area	$\alpha_2 = \alpha \times 2^\gamma$ 0.00241	$\beta_2 = \beta$ 0.25	$\gamma_2 = \gamma$ 0.125
Case 3 meters, half area	$\alpha_3 = \alpha \times 39.37^{2-2\gamma}$ 1.37	$\beta_3 = \beta$ 0.25	$\gamma_3 = \gamma$ 0.125
Case 4 meters, full area	$\alpha_4 = \alpha \times 39.37^{2-2\gamma} \times 2^\gamma$ 1.49	$\beta_4 = \beta$ 0.25	$\gamma_4 = \gamma$ 0.125

$$h = x \times \frac{P^y}{\text{Area}^z}$$

Case 1 inches, half area	$x_1 = \left(\dfrac{\alpha_1}{2^\beta}\right)^{1/(\beta+1)}$ 0.00653	$y_1 = \dfrac{\beta}{\beta+1}$ 0.20	$z_1 = \dfrac{\beta+\gamma}{\beta+1}$ 0.30
Case 2 inches, full area	$x_2 = (\alpha_2)^{1/(\beta+1)}$ 0.00805	$y_2 = \dfrac{\beta}{\beta+1}$ 0.20	$z_2 = \dfrac{\beta+\gamma}{\beta+1}$ 0.30
Case 3 meters, half area	$x_3 = \left(\dfrac{\alpha_3}{2^\beta}\right)^{1/(\beta+1)}$ 1.12	$y_3 = \dfrac{\beta}{\beta+1}$ 0.20	$z_3 = \dfrac{\beta+\gamma}{\beta+1}$ 0.30
Case 4 meters, full area	$x_4 = (\alpha_4)^{1/(\beta+1)}$ 1.38	$y_4 = \dfrac{\beta}{\beta+1}$ 0.20	$z_4 = \dfrac{\beta+\gamma}{\beta+1}$ 0.30

$$R_{TH} = \frac{C\alpha}{\Delta T^{C\beta} \times \text{Area}^{C\gamma}}$$

Case 1 inches, half area	$C\alpha_1 = \dfrac{1}{2 \times \alpha_1}$ 226.2	$C\beta_1 = \beta$ 0.25	$C\gamma_1 = 1 - \gamma$ 0.875
Case 2 inches, full area	$C\alpha_2 = \dfrac{1}{\alpha_2}$ 415	$C\beta_2 = \beta$ 0.25	$C\gamma_2 = 1 - \gamma$ 0.875
Case 3 meters, half area	$C\alpha_3 = \dfrac{1}{2 \times \alpha_3}$ 0.365	$C\beta_3 = \beta$ 0.25	$C\gamma_3 = 1 - \gamma$ 0.875
Case 4 meters, full area	$C\alpha_4 = \dfrac{1}{\alpha_4}$ 0.67	$C\beta_4 = \beta$ 0.25	$C\gamma_4 = 1 - \gamma$ 0.875

$$R_{TH} = \frac{Cx}{P^{Cy} \times \text{Area}^{Cz}}$$

Case 1 inches, half area	$Cx_1 = \dfrac{1}{2 \times x_1}$ 76.5	$Cy_1 = y_1$ 0.20	$Cz_1 = 1 - z_1$ 0.70
Case 2 inches, full area	$Cx_2 = \dfrac{1}{x_2}$ 124.2	$Cy_2 = y_2$ 0.20	$Cz_2 = 1 - z_2$ 0.70

TABLE 13.2 Conversion Table for Comparing Equations of Natural Convection

Case 3 meters, half area	$Cx_3 = \dfrac{1}{2 \times x_3}$ 0.45	$Cy_3 = y_3$ 0.20	$Cz_3 = 1 - z_3$ 0.70
Case 4 meters, full area	$Cx_4 = \dfrac{1}{x_4}$ 0.72	$Cy_4 = y_4$ 0.20	$Cz_4 = 1 - z_4$ 0.70
Direct ⇓ ⇓ ⇓ (Cases 1 and 3 only)	$Cx = \dfrac{1}{2} \times \left(\dfrac{2^\beta}{\alpha}\right)^{1/(\beta+1)}$	$Cy = \dfrac{\beta}{\beta+1}$	$Cz = 1 - \dfrac{\beta+\gamma}{\beta+1}$
Direct ⇑ ⇑ ⇑ (Cases 1 and 3 only)	$\alpha = \dfrac{1}{2} \times \dfrac{1}{Cx^{1/(1-Cy)}}$	$\beta = \dfrac{Cy}{1-Cy}$	$\gamma = 1 - \dfrac{Cz}{1-Cy}$

TABLE 13.2 Conversion Table for Comparing Equations of Natural Convection (*Continued*)

In Table 13.3 we have compared the numerical results in each form for our two standard equations. Sure enough, we have seen many of these (or very close) in related literature, though we probably didn't realize they were all the same equation.

We can provide a simple equation for estimating the copper area on a PCB. This is not a plate, but a copper island on a PCB, and *only one side is exposed to cooling*. This is not the same as using the area of one side of a plate, both sides of which are exposed to cooling. Here we use the equation both sides of which are exposed to cooling. For that reason we just call it "Area" here, instead of A or $\underline{A}$ used previously. We can however use standard equation no. 1 for this purpose too, to get the applicable equation as (see last row of Table 13.2 on page 352)

$$R_{th} = \dfrac{124.2}{P^{0.20} \times \text{Area}^{0.70}} \quad °C/W \quad \text{(area in square inches)}$$

Solving for Area we get

$$\text{Area} = \left(\dfrac{124.2}{P^{0.20} \times R_{th}}\right)^{1/0.70}$$

$$= 981 \times R_{th}^{-1.43} \times P^{-0.29} \quad \text{(area in square inches)}$$

Example 13.1 We have a dissipation of 0.45 W from an SMT device, and we want to restrict the temperature of the PCB to a maximum of 100°C to avoid getting too close to the glass transition of the board (which is around 120°C for standard FR-4). If the worst-case ambient is 55°C, let us find the amount of copper that should be made available to the device.

The required R_{th} is

$$R_{th} = \dfrac{°C}{W} = \dfrac{100-55}{0.45} = 100°C/W$$

So from our equation (based on standard equation no. 1) we get

$$\text{Area} = 981 \times 100^{-1.43} \times 0.45^{-0.29} = 1.707 \text{ in}^2$$

So we need a square copper area of side $1.707^{0.5} = 1.3$ in.

We also plot out the standard design equations, no. 1 and 2, in Fig. 13.9. We see that standard equation no. 2 is always more conservative than standard equation no. 1; that is, it calls for slightly larger areas and therefore may be a *safer bet* in most cases. Though not compared, standard equation no. 3 is almost coincident with standard equation no. 1, though it predicts slightly lower temperatures (less conservative).

PCBs for Heatsinking

As mentioned, commercial PCBs are often referred to as *1-oz* or *2-oz* for example. This refers to the weight of copper in ounces per square foot deposited on the copper-clad laminate. The 1-oz is actually equivalent to 1.4 mil copper thickness (or 35 µm). Similarly 2-oz is twice that (70 µm).

$$h = \alpha \times \frac{\Delta T^{\beta}}{\text{Area}^{\gamma}} \equiv \alpha \times \frac{\Delta T^{\beta}}{L^{2\gamma}}$$

	Standard equation no.	α	β	γ
Case 1	1	0.00221	0.25	0.125
inches, half area	2	0.00288	0.18	0.18
Case 2	1	0.0024	0.25	0.125
inches, full area	2	0.0033	0.18	0.18
Case 3	1	1.37	0.25	0.125
meters, half area	2	1.22	0.18	0.18
Case 4	1	1.49	0.25	0.125
meters, full area	2	1.38	0.18	0.18

$$h = x \bullet \frac{P^{y}}{\text{Area}^{z}} \equiv x \bullet \frac{P^{y}}{L^{2z}}$$

	Standard equation no.	x	y	z
Case 1	1	0.0065	0.20	0.30
inches, half area	2	0.0063	0.15	0.30
Case 2	1	0.0081	0.20	0.30
inches, full area	2	0.0077	0.15	0.30
Case 3	1	1.12	0.20	0.30
meters, half area	2	1.07	0.15	0.30
Case 4	1	1.38	0.20	0.30
meters, full area	2	1.32	0.15	0.30

$$R_{TH} = \frac{C\alpha}{\Delta T^{C\beta} \times \text{Area}^{C\gamma}}$$

	Standard equation no.	Cα	Cβ	Cγ
Case 1	1	226.2	0.25	0.875
inches, half area	2	173.4	0.18	0.82
Case 2	1	414.9	0.25	0.875
inches, full area	2	306.8	0.18	0.82
Case 3	1	0.37	0.25	0.875
meters, half area	2	0.41	0.18	0.82
Case 4	1	0.67	0.25	0.875
meters, full area	2	0.72	0.18	0.82

$$R_{TH} = \frac{Cx}{P^{Cy} \times \text{Area}^{Cz}}$$

	Standard equation no.	Cx	Cy	Cz
Case 1	1	76.5	0.20	0.70
inches, half area	2	80	0.15	0.70
Case 2	1	124.3	0.20	0.70
inches, full area	2	130.0	0.15	0.70
Case 3	1	0.45	0.20	0.70
meters, half area	2	0.47	0.15	0.70
Case 4	1	0.73	0.20	0.70
meters, full area	2	0.76	0.15	0.70

TABLE 13.3 Numerical Comparison of the Two Standard Equations of Natural Convection

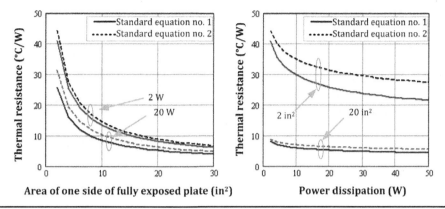

Figure 13.9 Comparing the two standard design equations of natural convection.

Larger and larger areas of copper do not help thermally, especially with thinner copper. A point of diminishing returns is reached for a square copper area of length (each side) 1 inch. Some improvement continues until about 3 inch, especially for 2-oz boards and better, but beyond that, external heatsinks are required. A practical value attainable for the thermal resistance (from the case of the power device to the ambient) is about 30°C/W.

That is not to say that heat is lost only from the copper side. The usual laminate (board material) used for SMT applications is epoxy-glass FR-4, which is a fairly good conductor of heat. So some of the heat from the device side does get to the other side where it contacts the air. Therefore, putting a copper plane on the other side (this need not even be electrically the same node, it could be the ground plane) also helps, but only by about 10 to 20 percent as compared to a copper plane on only one side. A much greater reduction of thermal resistance by about 50 to 70 percent can be produced if *thermal vias* are used to conduct heat to the other side. This *shunts* or bypasses the board material to get the heat to the other side, where there is more air waiting to act. See Fig. 13.2.

The ground or tab can be a wide copper plane since it is *quiet* and will not radiate. It can then serve to carry heat away from the device to aid convection. If a double-sided board is used, several small vias sunk right next to the IC ground can be used to connect to a *ground plane* on the other side of the PCB. These vias not only help in the correct electrical implementation of grounding, but also serve as thermal shunts. They are, therefore, called *thermal vias*. It is recommended that they be small (0.3- to 0.33-mm barrel diameter) so that the hole is essentially filled up during the plating process. Too large a hole can cause *solder wicking* during the reflow soldering process. The pitch (distance between the centers) of several such thermal vias in an area is typically 1 to 1.2 mm, and a grid of thermal vias can be created right under the tab.

However, overestimating the amount of the copper plane for device cooling is a common mistake and can lead to excessive EMI too. The switching node is the biggest culprit. We should think twice about the copper really needed here. In some topologies the cathode of the catch diode (which is usually its tab or substrate) needs to be connected to the switching node (e.g., a Buck). We should carefully estimate the dissipation before providing the required copper area for it. So, for example, we know that a typical Schottky diode has a forward voltage drop of 0.5 V. If the load current is 5 A and the duty cycle is 0.4, the dissipation is only $5 \times 0.5 \times (1 - 0.4) = 1.5$ W for the Buck. For Boost and Buck-boost, the average diode current is fixed: It is the load current.

Natural Convection at an Altitude

At sea level, over 70 percent of heat is transferred by natural convection and the rest by radiation. At very high altitudes (over 70,000 ft) the ratio inverts and the heat lost by radiation could be 70 to 90 percent of the total, even though the radiated transfer is unchanged. So by about 10,000 ft the overall efficiency of cooling typically falls to 80 percent, at 20,000 ft it is only 60 percent, and at 30,000 ft it is 50 percent.

Knowing that the coefficient of natural convection goes as $P^{1/2}$, where P is the pressure of air, a good curve fit gives us the following useful relationship:

$$\frac{R_{th\_feet}}{R_{th\_sea\text{-}level}} = [(-30 \times 10^{-6} \times \text{feet}) + 1]^{-0.5}$$

So, for example, we find that at 10,000 ft, all the R_{th}'s in Tables 13.2 and 13.3 need to be increased by 19.5 percent.

Forced Air Cooling

Fans are rated for a certain cubic feet per minute (cfm). The actual cooling, however, depends on the linear feet per minute (lfm) to which the heatsink is subjected. Two parameters are needed to find the velocity in lfm: (1) the volume of air discharged from the fan in cfm, and (2) the cross-sectional area through which the cooling air passes in square meters. So lfm = cfm/area. But finally we should derate the calculated lfm by 60 to 80 percent to account for back pressure.

At sea level the following formula gives a rough estimate of the required airflow:

$$\text{cfm} = \frac{1825}{\Delta T} \times P_{kW}$$

The ΔT is the differential between the inlet and outlet temperatures. It is typically set to about 10 to 15°C.

Note that if the inlet temperature, which is the room ambient, is 55°C, for example, then we need to add this differential ΔT as the actual local ambient inside the power supply when doing our initial calculations. However, ultimately we will be carrying out an actual temperature test by attaching thermocouples to all the components. We will thus certainly see an advantage in moving hotter components closer to the inlet during the design phase.

The linear speed is often expressed in terms of meter per second: 1 m/s is equal to an lfm of 196.85. Roughly, 1 m/s is 200 lfm.

Some empirical results are as follows: at 30-W dissipation, an unblackened plate of 10×10 cm has the following R_{th}: 3.9°C/W under natural cooling, 3.2°C/W with 1 m/s, 2.4°C/W with 2 m/s, 1.2°C/W with 5 m/s. Provided the air flows parallel to the fins, with speed > 0.5 m/s, the thermal resistance hardly depends on the power dissipation. That is because, on its own, even in static air, hot plates produce enough air movement around them to help in the heat transfer. Also note that blackening of plates has some effect under natural convection, but curves for forced convection depend very little on this aspect. Radiation is improved by blackening, but at sea level it is only a small part of the overall heat transfer. In general, black anodized heatsinks in typical forced-air designs are a waste, and they should be replaced with uncoated aluminum.

Under steady state, roughly 2-mm-thick copper is almost exactly equivalent to 3-mm-thick aluminum. The only advantage of copper is its better thermal conductivity, so it may be used to avoid thermal constriction effects when using very large areas.

The curve of thermal resistance to air flow falls off roughly exponentially, and so the improvement in thermal resistance in going from still air to 200 lfm is the same as from 200 to 1000 lfm. Velocities in excess of 1000 lfm (about 5 m/s) do not cause significant improvement.

Under forced convection the Nusselt number at sea level is

$$\text{Nu}_F = 0.664 \times \text{Re}^{1/2} \times \text{Pr}^{1/3} \quad \text{(laminar flow)}$$
$$\text{Nu}_F = 0.037 \times \text{Re}^{4/5} \times \text{Pr}^{1/3} \quad \text{(turbulent flow)}$$

Note that generally for natural convection, we should assume laminar flow. But under high dissipation, the hot air tends to rise so fast that it breaks up into turbulence. This is actually very useful in reducing the thermal resistance (increasing the h). For forced air, it is common to cut fingers on the sides of plate metal sinks and bend them alternately in and out. The purpose here is to actually create turbulent flow in the vicinity of the heatsink, thus lowering its thermal resistance. However, we do note from the formal analysis and equations that follow, turbulent flow provides better cooling (high h) under conditions of high lfm and/or large plates only. Laminar flow will provide better cooling otherwise.

We have defined the Prandtl number, Pr, which is the ratio of momentum diffusion to thermal diffusion. We can take its value at sea level to be 0.7. Re is the dimensionless Reynolds number, which is the ratio of momentum flow to viscous flow. If the plate has two dimensions L_1 and L_2 (so that $L_1 \times L_2 = A$), and L_1 is the dimension along the flow of air, then Re is

$$\text{Re} = \frac{\text{lfm}_{\text{sea-level}} \times L_{1\_\text{meters}}}{196.85 \times \nu}$$

where we already know $v = 15.9 \times 10^{-6}$ (the kinematic viscosity in m²/s). Thus we get the h under forced convection:

$$h_F = \frac{\text{Nu}_F \times K_{\text{AIR}}}{L_{1-\text{meters}}} \quad \text{W/m}^2 \cdot °\text{C}$$

where K_{AIR} is the thermal conductivity of air (0.026 W/m.°C). Putting all the numbers together, we simplify to get

$$h_{\text{FORCED}} = 0.086 \times \text{lfm}^{0.8} \times L^{-0.2} \quad \text{(turbulent flow, } L \text{ in meters, sea level)}$$
$$h_{\text{FORCED}} = 0.273 \times \text{lfm}^{0.5} \times L^{-0.5} \quad \text{(laminar flow, } L \text{ in meters, sea level)}$$

This assumes the two plate dimension L_1 and L_2 are equal, called L. At higher altitudes we need to increase the cfm calculated at sea level by the following factor, so as to maintain the same effective cooling. This is because a fan is a constant volume mover, not a constant mass mover, and at high altitudes, the air density is much lower. Therefore, the cfm has to be increased in inverse proportion to the pressure.

$$\frac{\text{cfm(feet)}}{\text{cfm(sea level)}} = \frac{1}{(-30 \times 10^{-6} \times \text{feet}) + 1}$$

For example, at 10,000 ft the calculated cfm at sea level has to be increased by 43 percent to maintain the same h_{FORCED}.

Radiative Heat Transfer

Radiation does not depend on air and can take place even in a vacuum since it is electromagnetic in nature. At high altitudes, radiative heat transfer can become a significant part of the overall heat transfer. The equation for h is

$$h_{\text{RAD}} = \frac{\varepsilon \times (5.67 \times 10^{-8}) \times [(T_{\text{HS}} + 273)^4 - (T_{\text{AMB}} + 273)^4]}{T_{\text{HS}} - T_{\text{AMB}}} \quad \text{W/m}^2 \cdot °\text{C}$$

Note that at high altitudes, under forced air cooling, the cfm falls, and so the inlet to outlet ΔT increases somewhat. Therefore, T_{AMB} goes up, and this affects h_{RAD}. So it ends up looking like radiation is getting affected too at higher altitudes. Luckily this actually improves the situation somewhat, by about 2 percent every 10,000 ft in typical applications.

The emissivity of the surface is ε. It is 1 for a perfect blackbody, but for polished metal surfaces we should take this as 0.1. If the surface is anodized, we can take it as about 0.9.

Miscellaneous Issues

- A typical power supply specification will ask for meeting an altitude requirement of 10,000 ft (3000 m). They will usually not *relax* the ambient temperature up to about 6000 ft, after which they will allow us to reduce the upper ambient limit by about 1°C every 1000 ft higher.

- A typical industry rule of thumb for testing power supplies at sea level for a certain altitude requirement is to *add 1°C every 1000 ft to the upper limit of the maximum specified operating ambient*. So if the power supply is designed for 55°C at sea level, we should test it at 65°C. However, this is not adequate. Nor do any temperature derating margins at sea level necessarily help. A key limiting factor is not the junction temperature but the temperature on the PCB where the device is mounted. We usually cannot exceed more than about 100 to 110°C on the PCB or it will degrade.

- We can sum over all the h's calculated in this chapter as follows:

$$h_{\text{total}} = h_{\text{RAD}} + \left(h_{\text{FORCE}}^3 + h_{\text{NATURAL}}^3\right)^{1/3}$$

- Extruded heatsinks are certainly very useful under forced air cooling, because then the efficiency of cooling depends on their surface area. But correlation of experimental data indicates that their cooling capabilities under natural convection conditions are

Core Sizes	°C/W
EC35/17/10	17.4
EC41/19/12	15.5
EE42/42/15	10.4
EE42/42/20	10.0
EE30/30/7	23.4
EE25/25/7	30.0
EE20/20/5	35.4
EE42/54/20	8.3
EE55/55/21	6.7
EE55/55/25	6.2
UU15/22/6	33.3
UU20/32/7	24.2
UU 25/40/13	15.7
UU 30/50/16	10.2

TABLE 13.4 Typical Thermal Resistance of Cores

a function of the volume of the space they occupy, that is, their *envelope* (ignoring the finer detail of their fin structure). That is because heat lost from one fin is largely reacquired by the adjacent fins, and so there are very small deviations with regard to the *exoticness* of their actual shape. Typical values drawn from published curves are as follows: 0.1 in³ will give about 30 to 50°C/W, 0.5 in³ will give about 15 to 20°C/W, 1 in³ will give about 10°C/W, 5 in³ will give about 5°C/W, 100 in³ will give about 0.5 to 1°C/W. The preceding data is for one device mounted on the heatsink. Roughly, there will be a further 20 percent improvement in the thermal resistance if two devices share the dissipation and are mounted slightly apart.

- For common magnetic cores (like the E-cores, and ETD cores, EFD cores), thermal resistance under natural convection can be approximated by

$$R_{th} \cong 53 \times V_e^{-0.54}$$

where V_e is in cm³. We have some typical thermal resistances based on actual empirical testing in Table 13.4. Also see Fig. 12.7.

- When using the equation for extrusions, the volume to be used is the overall space they occupy (i.e., ignoring the details of their fin structure).

- With extrusion heatsinks, if the fins are too close to each other, heat radiated from one is simply absorbed by the adjacent pins. So the overall emissivity is not as high as we may have thought.

- If the fins are too close, they also impede the flow of air. Therefore, the recommended optimum fin spacing is about 0.25 in for natural convection, at 200 lfm it is about 0.15 inch, and at 500 lfm it is about 0.1 inch. This applies for heatsinks up to 3 inches in length. We can increase the fin spacing by about 0.05 inch for heatsinks as long as 6 inches.

- Here is a quick rundown on fans. Ball-bearing fans are more expensive. They have a longer life when the temperature (as seen by the bearing system) is higher. But they can get noisier over time. If the useful life of a fan was defined as ending when the fan became noisy, the ball-bearing fan would have a shorter life than the sleeve-bearing fan. Sleeve-bearing fans are less expensive, and quieter, and easily handle any mounting attitude (angle). Their life as good as that of a ball-bearing fan provided temperatures are not very high. They can sustain multiple shocks (without impacting noise or life).

CHAPTER 14

Closing the Loop: Feedback and Stability

Basic Terminology

In power converters, we commonly refer to the steady-state ratio, output voltage divided by input voltage V_O/V_{IN}, as the *DC transfer function* of the (entire) converter. But, in general, the transfer function of any designated circuit block or component is also the ratio of its output to its input. The two may not even be like quantities. For example, the transfer function of a sense resistor, output divided by input, is equal to sense-voltage divided by current. The transfer function of the PWM stage is output divided by input, equal to duty cycle divided by error-amp-output-voltage. Duty cycle is dimensionless, of course, so the transfer function can have units or be dimensionless. In general, since a transfer function can involve imaginary numbers (to express phase angle, for example), the word *gain* refers to the magnitude of the transfer function, unless the transfer function is a real number to start with, in which case the two terms are synonymous.

In Fig. 14.1, we define terms commonly used in the area of loop stability, not just in relation to power converters. In particular, note that open-loop gain is $|GH|$, the (magnitude of the) product of G (i.e., the *plant* in general, or the *power stage* in power converters) and H, the feedback section (or *compensator*). Knowing G, we must design H so that GH reflects a stable system.

In general, any network transfer function can be described as a ratio of two polynomials:

$$G(s) = \frac{V(s)}{U(s)} = k\frac{a_0 + a_1 s + a_2 s^2 + a_3 s^3 + \cdots}{b_0 + b_1 s + b_2 s^2 + b_3 s^3 + \cdots}$$

Here $s = j \times \omega$ where $\omega = 2\pi \times f$ and $j = \sqrt{(-1)}$. This is math in the *s*-plane, that is, the imaginary (or complex) frequency domain. G (or H in this case) can be factored out as

$$G(s) = K\frac{(s-z_0)(s-z_1)(s-z_2)\cdots}{(s-p_0)(s-p_1)(s-p_2)\cdots}$$

Zeros (i.e., leading to the numerator being zero) occur at the complex frequencies $s = z_1, z_2, z_3, \ldots$ and so on. *Poles* (denominator zero) occur at $s = p_1, p_2, p_3, \ldots$ and so on.

In power supplies, we usually deal with transfer functions of the form

$$G(s) = K\frac{(s+z_0)(s+z_1)(s+z_2)\cdots}{(s+p_0)(s+p_1)(s+p_2)\cdots}$$

So the "well-behaved" poles and zeros are actually in the *left half* (of the complex frequency) plane (they are *LHP* poles and zeros). Their locations are at $s = -z_1, -z_2, -z_3, -p_1, -p_2, -p_3, \ldots$ and so on. We can also, in theory, have right-half-plane (RHP) poles and zeros that have very different behavior to normal poles and zeros and can therefore cause almost intractable instability. These occur in Boost and Buck-boost topologies (and also in their derivatives such as the Flyback).

NOTE *Sometimes the magnitude of the transfer function, the ratio output/input, is called* gain, *but sometimes $20 \times \log$ (output/input) is also called gain (in decibels). It is usually obvious which is intended, but we should be careful about this.*

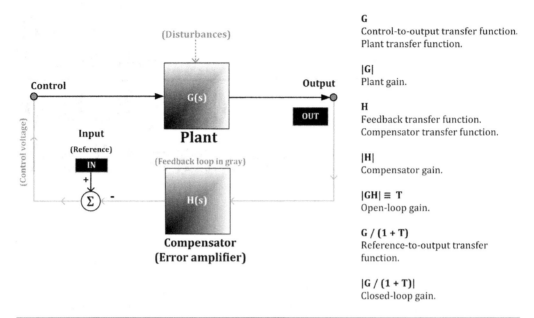

Figure 14.1 Control-loop terminology.

In a simple RC-based low-pass filter, the gain at high frequencies starts decreasing *by a factor of 10 for every 10-fold increase in frequency*. This is a single (or simple) pole by definition. Note that by the definition of a decibel, a 10:1 voltage ratio is 20 dB [check 20 log(10) = 20]. Therefore, we can say that the gain of a single pole falls at the rate of *− 20 dB per decade* at higher frequencies. A circuit with a slope of this magnitude is called a *first-order filter* (in this case a low-pass one, because it attenuates higher frequencies). Since this slope is *constant* (in the log versus log plane), the signal must also decrease by a factor of two for every doubling of frequency. Or by a factor of four for every quadrupling of frequency, and so on. But a 2:1 ratio is 6 dB, and an *octave* is a doubling (or halving) of frequency. Therefore, we can also say that the gain of a low-pass first-order filter falls at the rate of *− 6 dB per octave* (at high frequencies). Similarly, when we have filters with *two reactive components* (i.e., an inductor *and* a capacitor), we will find the slope is − 40 dB/decade (i.e., − 12 dB/octave). This is usually called a " − 2" slope or a double pole.

At an LHP pole, *both* the gain and the phase *fall* as frequency *increases.* At an LHP zero, both the gain and the phase rise as frequency increases. The phase angle shift associated with both the single pole and single zero is 90°; the accompanying sign depends on whether we are talking about a pole or a zero.

In Tables 14.1 and 14.2, we present convenient lookup tables, relating factors (i.e., ratios) to their respective decibels, and vice versa, hopefully in an easy-to-remember form. Note how

Factor	20 × log (factor)
× 1	0 dB
× 1.5	3.5 dB
× 2	6 dB
× 3 (2 × 1.5)	9.5 dB (6 + 3.5)
× 4 (2 × 2)	12 dB (6 + 6)
× 5 (10/2)	14 dB (20 − 6)
× 6 (3 × 2)	15.5 dB (9.5 + 6)
× 7	17 dB
× 8 (4 × 2)	18 dB (12 + 6)
× 9 (3 × 3)	19 dB (9.5 + 9.5)
× 10 (2 × 5)	20 dB (6 + 14)

Table 14.1 Ratios to Decibels

dB	Ratio	Easier to remember (Ratio)
1	1.122	
2 (= 12 − 10)	1.265 (= 4/$\sqrt{10}$)	
3	1.414 (= $\sqrt{2}$)	$\sqrt{2}$
4 [= (20 − 12)/2]	1.581 [= (10/4)$^{1/2}$]	
5 (= 10/2)	1.778 [(= $\sqrt{10}$)$^{1/2}$]	
6	2	2
7	2.24 (= $\sqrt{5}$)	
8 (= 20 − 12)	2.5 (= 10/4)	2.5
9 (= 6 + 3)	2.828 (= $\sqrt{8}$) (= 2 × $\sqrt{2}$)	$\sqrt{8}$
10 (= 20/2)	3.17 (= $\sqrt{10}$)	$\sqrt{10}$
11 (= 8 + 3)	3.536 (= 2.5 × $\sqrt{2}$)	
12 (= 6 × 2)	4 (= 2^2)	4
13 (= 10 + 3)	4.472 (= $\sqrt{10}$ × $\sqrt{2}$ = $\sqrt{20}$)	
14 (= 7 × 2)	5 (= $\sqrt{5^2}$)	5
15 (= 12 + 3)	5.657 (= 4 × $\sqrt{2}$)	
16 (= 8 × 2)	6.25 (= 2.5^2)	
...	...	...
20 (= 10 + 10)	10	10

TABLE 14.2 Decibels to Ratios

two numbers that get multiplied together are so much more easily handled in the log plane, as the simple sum of their respective decibels.

Loop instability occurs if any disturbance (harmonic frequency) travels around the G and H stages (plant and compensator) in succession (with a combined gain of GH) and arrives at the same point with exactly the same amplitude and phase (0° or 360° phase). This is an outcome of the formula for closed-loop gain $G/(1 + GH)$, because if GH ever equals − 1, it means the denominator "explodes" and we have instability. So knowledge of the open-loop gain is the key to defining the behavior of the closed-loop gain. Note that the term closed-loop gain in control-loop theory actually refers to the ratio of the output to the reference voltage (see Fig. 14.1). The reference is considered to be the input of the control loop, not the input rail of our power converter. The transfer function connecting the output DC rail to the DC input rail of a converter is called the *line transfer function* (or line-to-output transfer function, or audio susceptibility). It is the product of two cascaded blocks, one consisting of the basic DC transfer function of the topology (V_O/V_{IN}), and the transfer function of the *power stage* expressed generically as an *LC* filter

$$\frac{1/LC}{s^2+s(1/RC)+1/LC}$$

Note that for a Buck, this term is obvious, because the *LC* stage is really a separate block and we can multiply the transfer functions of two cascaded blocks to find out the net gain. But for a Boost and a Buck-boost, we really do not have a distinct and separate post-LC block. However, it can be shown via the *canonical model* that we can consider the *LC* of a Boost and a Buck-boost as a separate cascaded block too, provided we replace L in the above *LC* transfer function by an *equivalent inductance* $L/(1 − D)^2$. In other words, the post-*LC* filter for the latter two topologies has the following transfer function:

$$\frac{1/\underline{L}C}{s^2+s(1/RC)+1/\underline{L}C} \quad \text{where } \underline{L}=\frac{L}{(1-D)^2}$$

Note that of the 360° phase allowance mentioned earlier, we must subtract 180° right off the bat, since the error amplifier always adds 180° phase shift due to the signal anyway—because the feedback signal is almost invariably applied to the inverting terminal of the

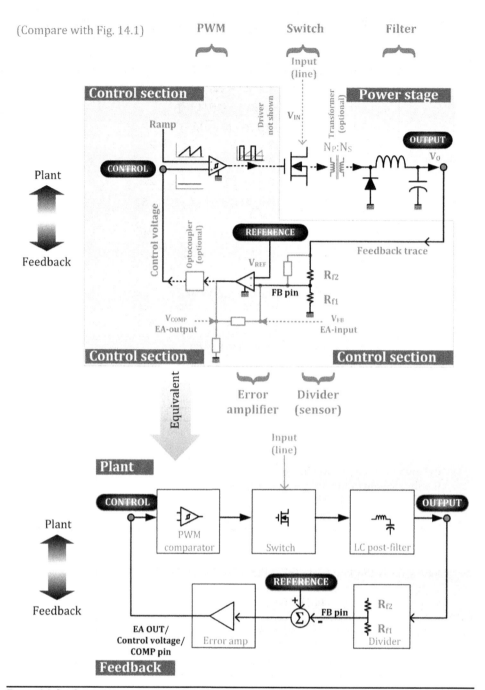

Figure 14.2 Control-loop terminology vis-à-vis a DC-DC converter.

error amplifier, to allow the gain to fall below unity for achieving a full range of correction. So, in effect, the loop stability criterion requires that the phase shift of the open-loop transfer function GH should not be equal to 180° (or −180°) *at the crossover frequency* (no other frequency is relevant actually). Because if that happens, then, combined with the 180° shift due to the error amplifier itself, this amounts to 360°, which would lead to complete reinforcement of disturbance, that is, instability.

Crossover is defined as the frequency at which we have unity open-loop gain, i.e., 0 dB. Typically, in voltage-mode control (VMC), (i.e., with a fixed clock-derived voltage ramp applied at the PWM input along with the output of the error amplifier), we aim for a crossover in the open-loop gain at $f_{SW}/10$, whereas in current-mode control (CMC), (i.e., with an inductor-current-derived voltage ramp applied at the PWM input along with the output of the error amplifier), we target a slightly higher crossover of $f_{SW}/6$. However, in the process of avoiding the potentially high quality factor Q of the *subharmonic (half-frequency) instability peak* occurring at $f_{SW}/2$, which CMC can throw up, we may have to be less ambitious in the crossover target for CMC and/or apply high levels of *slope compensation* and/or increase the inductance, as discussed later.

In general, we need to ensure a certain *margin of safety too*. This can be expressed in terms of the degrees of phase angle *short of 180°*, at the crossover frequency. This headroom is called the *phase margin*. But we could also talk almost equivalently about the safety margin in terms of the amount of gain *below the 0-dB level* at the point where we get 180° phase shift. This gives us the term *gain margin*. It is usually considered desirable to have around 45° of phase margin, since that margin accounts for typical tolerances, drifts over time, and temperature, and also gives a good compromise between output overshoots or undershoots during line and load variations and the *settling time* (i.e., the time to return to steady-state conditions at the output following any disturbance).

PART 1   STABILIZING CURRENT-MODE CONVERTERS

Background

For voltage-mode control, setting the current ripple ratio $r = 0.4$, as recommended throughout this book, will usually stand up to scrutiny. But when it comes to peak current-mode control that may not be enough. Because, if we have current-mode control, and our duty cycle is over 50 percent, and we are also in continuous conduction mode (CCM), then, as we reduce the inductance we could arrive at a strange type of oscillation. This will manifest itself as an alternating pattern of one large pulse followed by one small pulse. It is a *steady-state* of sorts too, except that the pattern repeats itself every two cycles instead of one. A system in this state may appear normal in *steady state*, not even exhibiting significantly higher output voltage ripple. But a Bode (gain-phase) plot of such a system (see Fig. 14.3) will reveal an incomprehensible plot, and further, the system, if subjected to step loads, will show very poor response time and large overshoots and undershoots. This is, therefore, alternatively called *subharmonic instability*, *alternate cycle instability*, *half-frequency instability*, *period-doubling instability*, etc. They all mean the same thing. The familiar cure is *slope compensation*. Unfortunately, as we may have inferred, the amount of slope compensation and the value of the inductance are interlinked. So, for example, a poorly designed slope compensation circuit may demand that we use a *higher* inductance (than that which comes out of the condition $r = 0.4$). This is clearly something we need to fix in the *slope compensation* circuit itself. On the other hand our desire may be to allow r to increase so as to reduce the size of our magnetics, irrespective of the possible impact on the related power components. That is also understandable, because that is the very reason why we aim for higher and higher switching frequencies in the first place. But then we may run into subharmonic instability.

The designer is cautioned that there are other reasons why we may see the same behavior. In Fig. 14.4, we present one such *other possible cause* that will mimic this symptom. Coincidentally, that is also usually associated with CMC, but its solution is altogether different (increase blanking time, *lower* the inductance). Slope compensation will not help here.

The problem with subharmonic instability is that "playing" with the usual loop compensation components is not likely to help at all, simply because the root cause is not the usual *incorrect* placement of poles and zeros. This particular instability is not even included in the usual small-signal models around us. And it applies to all topologies using peak current-mode control. Understanding the equations behind this phenomenon helps us pick the smallest possible inductor in a switching application (though not necessarily the most optimum one in terms of its overall impact on system size and cost).

The unambiguous solution is to introduce slope compensation, and we are assuming that the phrase at least is familiar to most engineers. Here we will see that we then need to keep the inductance higher than a certain amount (hitherto unspecified), and also that this *value of inductance is the highest at the lowest input voltage. So, we should design our slope compensation circuit at* V_{INMIN} *for any topology*.

Designers familiar with the popular peak current-mode controller the UC3842/3844 series know that, surprisingly, this peak current-mode control IC has no built-in slope compensation. It is like a bicycle without tires. For the UC3844 we can understand the absence, because subharmonic instability occurs only if D is greater than 50 percent, and the UC3844 being intended for single-ended Forward converters is by design limited to a maximum of D slightly less than 50 percent (though the designer is cautioned that it is now known that subharmonic instability can even show up around 45 percent duty cycle and above, with some saying it is even 30 percent and above in some cases). But what about the UC3842? We all remember putting in that mysterious 47 pF (sometimes only 10 or 22 pF, or sometimes 100 pF, all by trial and error) from the clock pin to the current sense pin. What this really was

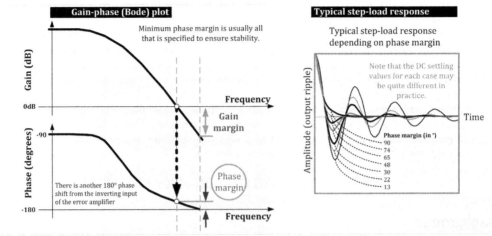
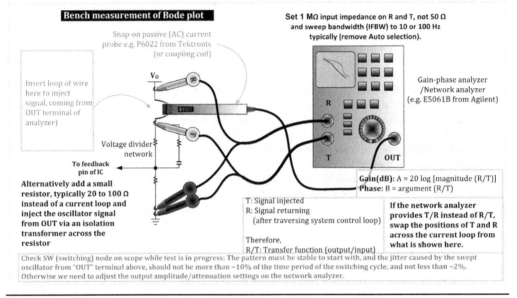

Figure 14.3 Bode plot setup.

doing was mixing some of the clock ramp with the current sense signal. In doing so, *we were actually introducing a little voltage-mode control into current-mode control*. The ramp going to the PWM comparator is now partly current-sense signal and partly fixed ramp. In fact, this is what we essentially do even in formal slope compensation. However, instead of looking at the situation as a ramp *added* to the current sense, we can equivalently look at it as the current ramp having remained the same and, instead, an *inverted ramp* being summed with the output of the error amplifier (see Fig. 14.5). *The PWM comparator is concerned only with the difference in voltage at its inputs*. The relative voltage is all that matters. So, both these methods are equivalent as far as the comparator is concerned.

NOTE *The small (~22 pF) capacitor also unexpectedly helps with the UC3844 and Forward converters in general. But that is only because the current-sense signal is a few millivolts at light loads, and the signal-to-noise ratio may be poor. So we can get a lot of jitter. By providing a small fixed ramp, we mimic full voltage-mode control at very light loads, and thus get clean pulse patterns. But this is not a case of subharmonic instability; it's only noise and jitter.*

We also note that the way we implement slope compensation in the UC3842 (and also in several older controllers and switchers) affects the current limit too, so we will have to either decrease our sense resistor somewhat or derate the power available. But in more recent controllers, the current limit circuit is kept independent of the PWM comparator section, and so it remains *flat* and unaffected, despite the applied slope compensation. Because otherwise

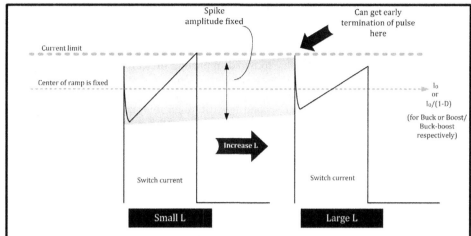

A major concern is the leading edge spike. This can cause jitter, and in severe cases, premature termination of the ON-pulse and a consequent inability to deliver full power. If we increase the inductance, we can induce premature termination, especially in current-mode controllers, because by a large L, we inadvertently raise the pedestal on which the spike is riding on. If premature termination occurs, since less than the required energy will be delivered for that (prematurely terminated) cycle, in the next cycle the converter will try to compensate by a larger duty cycle. In this process it gets some unexpected help, because after the early termination of the previous pulse, the inductor current has had a longer time to slew down, and thus the pedestal on which the leading edge spike is riding, comes down, probably enough to help it evade early pulse limiting in the next cycle. Finally, what we may see on the oscilloscope are *alternate wide and narrow pulses*, which mimic what we get under genuine subharmonic instability. We may be surprised, because we thought that a high inductance helps avoid subharmonic instability, but here it seems to be aggravating it. The leading edge spike also causes erratic responses from the current limit protection circuit for both current-mode and voltage-mode control. We can't ever set an effective current limit based on a spike, especially since we will discover that this spike varies from unit to unit, being based on uncontrolled and/or uncharacterized parasitics. We could of course set a large blanking time for current-mode control, and/or we can add some delay to the current limit detect circuit. But we also then run the danger of not being able to react fast enough to an actual abnormal load condition, especially if the inductor starts to saturate. **Ripple-phobia should thus be avoided.** Besides, we should not oversize our inductor only because of an instinctive, but unjustifiable desire to decrease the current ripple ratio *r*. An *r* of 0.4 is usually optimal.

FIGURE 14.4 Mimicking subharmonic instability by excessively large inductances.

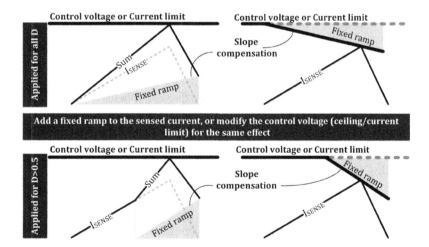

FIGURE 14.5 Applying a fixed ramp to the sensed current is equivalent to modifying the control voltage, applied over the entire period or only for duty cycles exceeding 50 percent.

the overall converter design can become tricky: Whenever we adjust the slope compensation, we will need to double-check whether we are even going to be able to deliver maximum power (without hitting the new effective current limit).

Recognizing that subharmonic instability only occurs if the duty cycle is greater than 50 percent, some controllers try to avoid impacting the current limit over the entire input voltage range by introducing slope compensation ramp only for $D > 50$ percent. See the lower half of Fig. 14.5. The applied slope compensation can eventually be expressed in terms of an equivalent amperes per second (even though internally it may be a voltage ramp, expressed as volts per second). See Fig. 14.6.

This is also a good place to point out that as switching frequencies increase, the typical 50- to 150-ns blanking time requirements and oversensitivity to layout may take a lot of the shine out of current-mode control. The subharmonic oscillation problem with peak current-mode control does not help either. So voltage-mode controllers did make a comeback. In particular, using techniques like input voltage feedforward.

Input (Line) Feedforward in Voltage-Mode Control (Wannabe CMC)

In feedforward, the up-slope of the internal ramp used in VMC is made proportional to the input voltage. We thereby replicate some of the advantages of current-mode control, in terms of its ability to react extremely fast (on a pulse-by-pulse basis) to changes in input voltage. See Fig. 14.7. Incidentally, in this figure, it is also shown that the transfer function (gain) of the PWM stage is $1/V_{RAMP}$.

This technique requires the input voltage to be sensed, and the slope of the comparator sawtooth ramp increases if the input goes up. In the simplest implementation, a doubling of the input causes the slope of the ramp to double. Then, from Fig. 14.7, we see that if the slope doubles, the duty cycle is immediately halved. In a Buck, the governing DC transfer function equation is $D = V_O/V_{IN}$. So if a doubling of input occurs, we know that the duty cycle will eventually halve. So, rather than wait for the control voltage to slowly decrease by half to lower the duty cycle (keeping the ramp unchanged), we have changed the ramp itself—in this case to double the slope of the previous ramp and thereby achieved the very same result (i.e., halving of the duty cycle) *almost instantaneously*.

It turns out that the duty cycle correction afforded by this *automatic* input-proportional ramp correction is exactly what is required for a Buck, since its duty cycle D equals V_O/V_{IN}. More importantly, this correction is virtually instantaneous—we didn't have to wait for the error amplifier to detect the error on the output (through the inherent delays of its RC-based compensation network scheme) and respond by altering the control voltage. So in effect, by input/line feedforward, we have bypassed all major delays and therefore line correction is as fast as possible. This amounts to almost *perfect* rejection of the line disturbance.

In current-mode control, the PWM ramp is basically an appropriately amplified version of the switch/inductor current. The original inspiration behind the idea of input feedforward came from *current-mode control*—in which the PWM ramp, generated from the inductor current, automatically increases if the line voltage increases. That partly explains why current-mode control seems to respond so much "faster" to line disturbances than traditional voltage-mode control, and one of its oft-touted advantages. However, the "built-in" automatic line feedforward in current-mode control is not as perfect as its supposed imitation. Because in a Buck topology, the *slope* of the inductor current up-ramp is equal to $(V_{IN} - V_O)/L$. So if we double the input voltage, we do *not* end up doubling the slope of the inductor current. Therefore, neither do we end up automatically halving the duty cycle, as we do in line feedforward applied to voltage-mode control. In other words, voltage-mode control with proportional line feedforward control, though inspired by current-mode control, provides *better* line rejection than current-mode control (for a Buck).

How Much Slope Compensation?

In Fig. 14.8 we have a peak current-mode controller and we apply a small disturbance. We can see that this disturbs the regular pulse pattern. But at least (we hope) it subsides. If it subsides, we won't get sustained oscillations, but *if the disturbance increases every cycle*, we will. On the other hand, we would also prefer that it subside *in a fairly short time*.

We can be assured that the disturbance will not increase, if the amount of slope compensation is limited as per

$$S \geq \frac{S_2 - S_1}{2} \equiv \frac{1}{2}, \text{(which is the difference between down-ramp and up-ramp)}$$

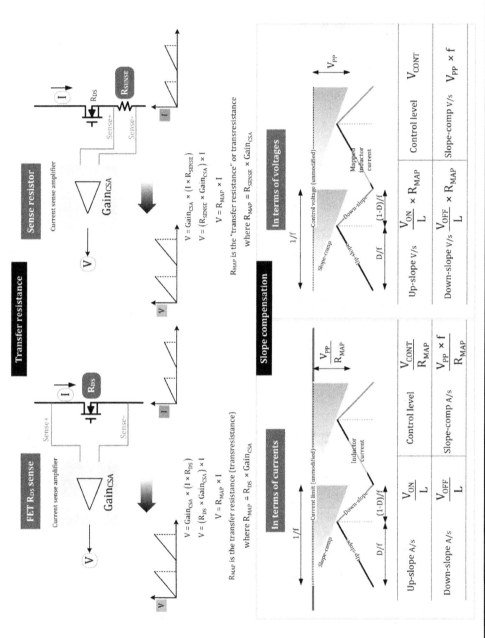

Figure 14.6 Transfer resistance, alternate ways of expressing slope compensation, and current sense methods.

Conventional pulse width modulation (PWM) in voltage-mode control explained

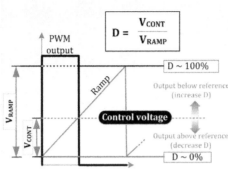

The control voltage (V_{CONT}) is the output of the error amplifier. It falls (or rises) if the output is above (or below) the set reference level. That causes the duty cycle D to increase (decrease).

(In current-mode control (CMC), the ramp is derived from the switch/inductor current waveform).

In voltage-mode control (VMC), the ramp is internally generated (from the clock).
(If the ramp is made proportional to the input voltage, we get line/input voltage feedforward; see further below).

Pulse width modulator gain (units of V^{-1})

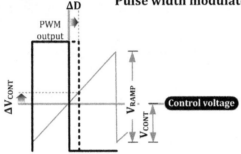

$$D = \frac{V_{CONT}}{V_{RAMP}} \quad \text{(as shown above)}$$

$$\Delta D = \frac{\Delta V_{CONT}}{V_{RAMP}}$$

$$G_{PWM} = \frac{\Delta D}{\Delta V_{CONT}} = \frac{1}{V_{RAMP}}$$

This is the transfer function/gain of the PWM comparator stage (it is the change in duty cycle due to a change in the control voltage)

Line feedforward in VMC explained

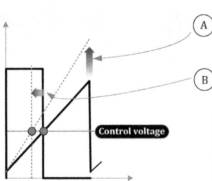

(A) If the input voltage increases, the PWM comparator ramp (its slope) is made to increase proportionally and instantly, to mimic CMC somewhat (in fact better)

(B) The duty cycle decreases immediately as a result of the intersection with the control voltage (which has not had time to re-adjust yet)

Note: The control voltage has not even had time to respond. Yet, as the line voltage increased, the duty cycle decreased instantly, as required. The slower response from the error amplifier and the LC-based power stage did not hold up the line correction.

FIGURE 14.7 Duty cycle, gain of the PWM stage, and voltage-mode control with input feedforward explained.

In related literature, *the terms used more commonly are S_n for S_1, S_f for S_2, and S_e for S.*
We have several possibilities:

- If

$$S = S_2$$

the disturbance will be "killed" in the very same cycle it started. Of course, to achieve this, we may have to either significantly increase the slope compensation or increase inductance (lower *r*). This is *overkill*. It also seriously affects the much vaunted *instantaneous line rejection* of current-mode control.

- If

$$S = \frac{S_2 - S_1}{2} \quad \text{or} \quad S_2 = 2S + S_1$$

then the disturbance tends to last forever. In practice, this tells us the *smallest* inductance threshold and/or least slope compensation. Because beyond this boundary

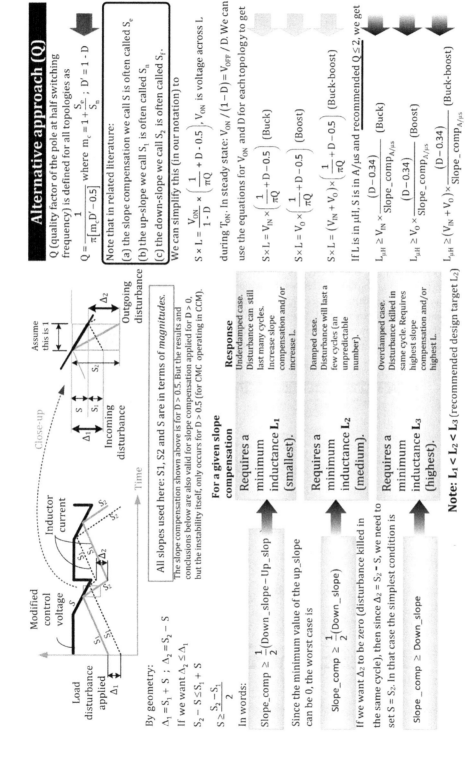

FIGURE 14.8 Amount of slope compensation required: traditional and modern approach.

(smaller inductances and/or lesser slope compensation), the disturbance will increase, not decrease.

- It is commonly stated in older textbooks that the condition to avoid subharmonic instability is

$$S = \frac{S_2}{2} \quad \text{or} \quad S_2 = 2S$$

which, for a fixed slope compensation means that we are demanding a smaller down-slope S_2 (larger inductance) than in the previous case (in which the disturbance lasted forever). This will ensure that the disturbance is fairly quickly attenuated: not instantataneously, but not too long either. It is a good compromise usually.

Note that the preceding cases of *subharmonic instability* occur in CCM, in peak current-mode control (all topologies), and theoretically only for duty cycles exceeding 50 percent. Nowadays it is increasingly said that subharmonic instability can occur even for a set (CCM) duty cycle of 40 to 45 percent. But perhaps the instability was simply instigated by a momentary excursion (perturbation) into the region $D > 50$ percent. The key question is: Does it really occur if the control IC is itself firmly clamped to less than 50 percent (like the UC3844), which would truly avoid even momentary excursions into the region $D > 50$ percent?

Yes, even in the UC3844 (maximum 50 percent duty cycle control IC), it was customary to place a very small capacitor between the clock and sense current pin (the usual way of injecting slope compomenation in the UC384 × family, such as in the UC3842). However, the key advantage of that in the UC3844 was that at very light loads, the current sense signal would be very small and noise would ride on it, causing a lot of jitter in the switching waveform. So a little clock ramp injected would really help overwhelm the noise and allow a steady switching pattern. This had nothing to do with subharmonic instability, as mentioned previously too.

Leaving the enduring *mysteries of subharmonic instability* behind, there is another problem with peak current-mode control. The output voltage is determined largely or in part (depending on the topology) by the *average* current in the inductor. If the average decreases in response to a small decrease in input voltage, it may pose a problem. See Fig. 14.9. One obvious solution

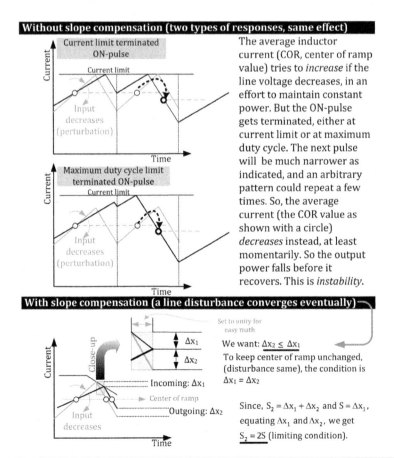

Figure 14.9 Average current problem with line disturbances, and slope compensation to resolve.

is to shun peak current-mode control altogether and opt for *average current-mode control*. Another is to apply slope compensation even for duty cycles much less than 50 percent.

Recall that in Fig. 14.8 we saw the same slopes for the disturbance, which implied that the input and output voltages had remained the same. But it can be shown that for *line disturbances*, as drawn in Fig. 14.9, in which the slopes could also change, a similar logic will apply. To avoid this particular type of disturbance from increasing over time, we need to apply slope compensation too: the mathematical condition for which happens to be the same as one of the previous conditions:

$$S = \frac{S_2}{2} \quad \text{or} \quad S_2 = 2S$$

We notice that the up-slope doesn't enter the picture for line disturbances. We emphasize that this type of temporary instability can occur starting from any duty cycle (*not just more than 50 percent*). Usually, this issue is ignored by many engineers. Line transient testing is rarely performed in the lab. It is load transient response and related issues that seem to occupy most engineers. But it is good to keep this issue in mind too. In other words, *slope compensation can help for any duty cycle*. It may be unavoidable, however, only if the controller IC allows duty cycles exceeding 50 percent.

Generalized Rule for Avoiding Subharmonic Instability

As we said, early small-signal models did not reveal subharmonic instability. However, if we do a bench measurement for the Bode plot, and we zoom in on the area around half the switching frequency with appropriate bandwidth, we will see that the gain plot that should have been continuing to fall smoothly past the crossover frequency has a *spike* at exactly half the switching frequency. This peaking is the physical manifestation of subharmonic instability. It has recently been modeled by theoretical analysis, and a quality factor Q has been assigned to it. In practical bench measurements we will see that if Q is set low, say around 0.5, then the half frequency peak is not visible at all. If Q is very large, this peak can be high enough to intersect the 0-dB (unity gain) line (x-axis of the Bode plot), and thereafter almost immediately the Bode plot will change into a rather abnormal-looking plot. That is because, though it is quasi-stable, we have entered the undesirable alternate-cycle pattern, and the control loop, as we know it, is gone. The converter will never recover without our intervention (power cycling, for example). We note that a conservative Q of around 0.5 leads to excessively high inductances, but a Q of around 2 is just right as indicated by several bench measurements the author performed. $Q = 2$ requires a fairly small minimum inductance and is assuredly stable.

What do we mean by saying, "set the Q to 2"? The required relationship is as follows and applies to all the topologies:

$$Q = \frac{1}{\pi[(1 - S/S_1) \times (1 - D) - 0.5]}$$

Note that all the slopes are only their respective *magnitudes* here. Now we solve for L, and we get for each topology the following useful equations:

$$L_{\mu H} = \frac{1/\pi Q + D - 0.5}{\text{SlopeComp}(A/\mu s)} \times V_{IN} \quad \text{(Buck, set } Q \approx 2\text{)}$$

$$L_{\mu H} = \frac{1/\pi Q + D - 0.5}{\text{SlopeComp}(A/\mu s)} \times V_O \quad \text{(Buck, set } Q \approx 2\text{)}$$

$$L_{\mu H} = \frac{1/\pi Q + D - 0.5}{\text{SlopeComp}(A/\mu s)} \times (V_{IN} + V_O) \quad \text{(Buck-boost, set } Q \approx 2\text{)}$$

In Fig. 14.10 we have plotted the minimum inductance required to avoid instability. We have set $Q = 2$, and the slope compensation value used is 0.25 A/μs. Since inductance is inversely proportional to slope compensation, we can calculate the inductance for a general condition. Note that for the curves to apply, we should be in a region of $D > 50$ percent and in CCM. If the slope compensation is affecting current limit, we must ensure that peak power is still able to be delivered. This is particularly true for the Buck converter, since it is the only topology in which the peak current increases with input voltage (decreasing D). The peak current equation for a Buck is

$$I_{PK} = I_O \times \left[1 + \frac{r}{2}\right]$$

372 Chapter Fourteen

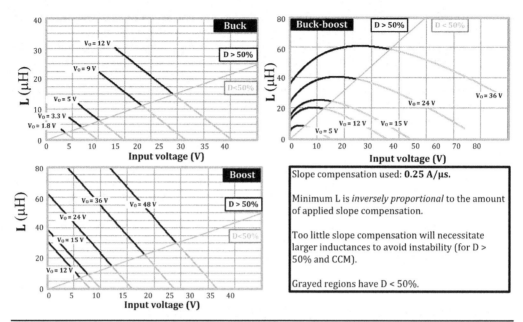

FIGURE 14.10 Minimum inductance required to avoid subharmonic instability.

where (see Appendix)

$$r = \frac{V_O + V_D}{I_O \times L_{\mu H} \times f}(1-D) \times 10^6$$

Therefore, the peak current will change at a rate dependent upon voltage, load current, inductance, etc. We have to make sure that the calculated peak current is not in excess of the effective current limit (with slope compensation included), or we will not get the output power we were hoping for. We must test this out at V_{INMAX}. But we remember that the inductor design for avoiding slope compensation is at V_{INMIN}. Therefore, our general philosophy should be:

1. Find the minimum inductance required to keep the converter in CCM.
2. Find the minimum inductance required to meet peak power requirement (at V_{INMAX}).
3. Find the minimum inductance to avoid subharmonic instability.

Take the maximum of all three *minimum inductances*. Note that the last condition comes into the picture only if allowed D > 50 percent. We may have to iterate a few times to arrive at the final and most optimum result.

PART 2 RETROSPECTIVE: VOLTAGE MODE, CURRENT MODE TO HYSTERETIC

Introduction

Switching regulators have been with us for many years. They were considered tricky to design—and still are. In 1976, Silicon General introduced the first monolithic (IC-based) switching controller, the SG1524 *Pulse Width Integrated Circuit*. A little later, this chip was improved and became the *SG3524* industry workhorse. And very soon thereafter, it was available from multiple chip vendors. Keep in mind that switching stages based on discrete designs were already gaining ground, particularly in military applications. In fact, some resourceful engineers had even made "switch-mode power supplies" by adding related circuitry around one of the highest-selling chips in history: the *555 timer* (sometimes called the *IC Time Machine*), introduced in 1971 by Signetics (later Philips, then NXP). The SG1524 was, however, only the *first* IC in which *all* the required control functionality was present on a single chip or die. With the rapidly escalating concurrent interest in switching power supplies at the time, it is no surprise that as early as 1977 the very first book on the subject, written by the late Abraham Pressman, appeared on the scene. Together, these events spurred interest in an area well beyond most people's expectations and ushered in the world of switching power conversion as we know it today.

Another well-known technique today, which has also been around since the 1980s, senses the peak current in the power switch or inductor, and turns the switch OFF at a programmed level of current. This technique is current-mode control (CMC). Keep in mind it was not *brand-new* at the time. In fact it had been discovered years ago. See US Patent Number 3,350,628 at http://www.archpatent.com/patents/3350628, from L.E. Gallaher et al, dated 1967, and also US Patent Number 4,456,872 at http://www.archpatent.com/patents/4456872 from Thomas A. Froeschle, dated 1984. However, few had realized its significance until Unitrode Corp. (now Texas Instruments) came along. It received a huge boost in popularity in the form of the world's first CMC chips: the UC1846, followed by UC1843, from Unitrode.

Brian Holland from Unitrode introduced the UC1846 in 1983 at Powercon 10, and also wrote U-93 at www.ti.com/lit/an/slua075/slua075.pdf. Quoting from this:

> The inherent advantages of current-mode control over conventional PWM approaches to switching power converters read like a wish list from a frustrated power supply design engineer. Features such as automatic feed forward, automatic symmetry correction, inherent current limiting, simple loop compensation, enhanced load response, and the capability for parallel operation all are characteristics of current-mode conversion. This paper introduces the first control integrated circuit specifically designed for this topology, defines its operation and describes practical examples illustrating its use and benefits.

But a few years into this success story, expectations got somewhat blunted. The disadvantages of CMC had surfaced slowly. That growing realization was succinctly summed up in a well-known Design Note, DN-62, from Unitrode in 1994, by Bob Mammano (see, http://www.ti.com/lit/an/slua119/slua119.pdf which said:

> There is no single topology which is optimum for all applications. Moreover, voltage-mode control—If updated with modern circuit and process developments—has much to offer designers of today's high-performance supplies and is a viable contender for the power supply designer's attention.

It also says:

> It is reasonable to expect some confusion to be generated with the introduction of the UCC3570—a new voltage-mode controller introduced almost 10 years after we told the world that current-mode was such a superior approach.

> **NOTE** *Bob Mammano is called "the father of the PWM controller IC industry," because he developed the first voltage-mode control IC, the SG1524. Later he was staff technologist in the Power IC division of Unitrode, a division that he had jointly created with two others from Silicon General.*

Plant Transfer Functions

This is often called the *control-to-output* transfer function, since it is in effect the output voltage divided by the control voltage, the control voltage being the output of the error amplifier. It is a product of three *successive* (independent) gain blocks: the PWM comparator, the switching power section, and the output LC filter. As mentioned, only in the Buck, do we have an actual LC postfilter, but by using the *canonical model*, it can be shown that an *effective* LC postfilter (a separate gain block) can be visualized *for non-Buck topologies* too, provided we use the effective inductance $\underline{L} = L/(1-D)^2$.

The important point to note is that in the non-Buck topologies, the effective inductance $\underline{L}$ varies with input voltage (for a given output), since D increases as input falls, causing a higher effective inductance, thereby tending to make the loop response more sluggish. In general, large inductors and capacitors need several more cycles to reach their new steady-state energy levels, after a line or a load transient. This sluggishness, based on larger effective L, is of greater impact in VMC, because as we will see very shortly, when using CMC, the inductance is not even part of the plant transfer function to start with. With VMC it is.

We can now go through several figures, to sum up the plant transfer functions for VMC and CMC.

Figure 14.11

This is a Buck with classic VMC. The DC gain (gain at low frequencies) changes as a function of V_{IN} in classical VMC, because V_{RAMP}, the amplitude of the sawtooth applied to the PWM comparator, is traditionally fixed. This causes a change in loop response characteristics with

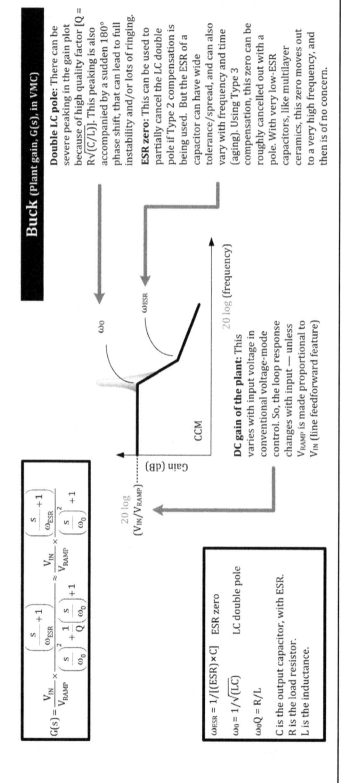

Figure 14.11 Plant transfer function for a Buck, using VMC.

respect to input. Furthermore, the *line rejection* is not good under suddenly changing conditions. The reason is that a sudden change in line is not "felt" by the PWM comparator *directly*, and so it continues with the same duty cycle for a while. But any change in input voltage requires a change in the steady-state duty cycle (as per the steady DC transfer function equation of the converter: $D = V_O/V_{IN}$). So not changing the duty cycle quickly enough leads to an output overshoot or undershoot. The system has to *wait* for the output error to be sensed by the error amplifier and for that information to be communicated to the PWM comparator as a change in the applied *control voltage*. That eventually corrects the duty cycle and the output too, but not before some swinging back and forth (ringing) around the settling value. However, if we could just change the ramp voltage directly and *instantaneously* with respect to the input voltage, we would not need to wait for the information to return via the control voltage terminal. Then the line rejection would be almost instantaneous, and furthermore, the DC gain would not change with input voltage.

To implement the preceding, we would need to do the following:

$$V_{RAMP} \propto V_{IN} \quad \text{so} \quad \frac{V_{IN}}{V_{RAMP}} = \text{constant}$$

This is called *line feedforward* as shown in Fig. 14.7. We see that CMC has similar properties very *naturally*, which was one of the reasons for its perceived superiority for a long time. Pure VMC, on its own, is certainly impaired, especially in this respect. But *VMC with line feedforward (when implemented in a Buck) actually offers superior line rejection compared to the one coming "naturally" from CMC.*

Figure 14.12

This is a Boost with VMC. Note that line feedforward is not practical here considering the complexity of the terms. We would need to somehow set V_{RAMP} proportional to $V_{IN} \times (1 - D)^2$, based on the DC gain of $G(s)$. Also note the appearance of the right half-plane (RHP) zero. It also appears in the Buck-boost. It is present for any duty cycle, and for either VMC or CMC. In a "well-behaved" (left half-plane) zero, the gain *rises* (or changes) by the amount +1 at the location of the zero, and its phase *increases correspondingly*. With an RHP zero, the phase *falls* even though the gain *rises*, making this particular zero very difficult to compensate or deal with.

The existence of the RHP zero in the Boost and Buck-boost can be traced back to the fact that these are the only topologies where an actual *LC postfilter doesn't* exist on the output. So even though we created an *effective post-LC filter* by using the canonical modeling technique, in reality there is a switch/diode connected between the actual L and C of the topology, and that is what is ultimately also responsible for creating the RHP zero. The RHP zero is often explained intuitively as follows: If we suddenly increase the load, the output dips. This causes the converter to increase its duty cycle in an effort to restore the output. Unfortunately, for both the Boost and Buck-boost, energy is delivered to the load only during the switch OFF-time. So, an increase in the duty cycle decreases the OFF-time, and unfortunately there is now a smaller interval available for the stored inductor energy to get transferred to the output. Therefore, the output voltage dips even further for a few cycles, instead of increasing as we were hoping. This is the RHP zero in action. Eventually, the current in the inductor does manage to ramp up over several successive switching cycles to the new level, consistent with the increased energy demand, and so this strange counterproductive situation gets corrected. Of course, that is provided full instability has not already occurred! As mentioned, the RHP zero can occur at any duty cycle. Note that its location moves to a lower frequency as D approaches 1 (i.e., at lower input voltages). It also moves to a lower frequency if L is increased. That is one reason why bigger inductances are *not* preferred in Boost and Buck-boost topologies. It is also perhaps the reason why some engineers feel that the RHP zero does not exist in DCM (discontinuous conduction mode). Others say that is only because of the much lower inductances involved in DCM.

The usual method to deal with the RHP zero is literally "pushing it out" to higher frequencies where it can't significantly affect the overall loop. Equivalently, we need to reduce the bandwidth of the open-loop gain plot to a frequency low enough that it just doesn't "see" this zero. In other words, the crossover frequency must be set much lower than the location of the RHP zero. In effect, *the bandwidth and loop response suffers on account of the RHP zero.*

Figure 14.13

This is a Buck-boost with VMC. As for the Boost, line feedforward is not a practical goal here. The RHP zero is also present here, though its location is a little different as compared to a Boost.

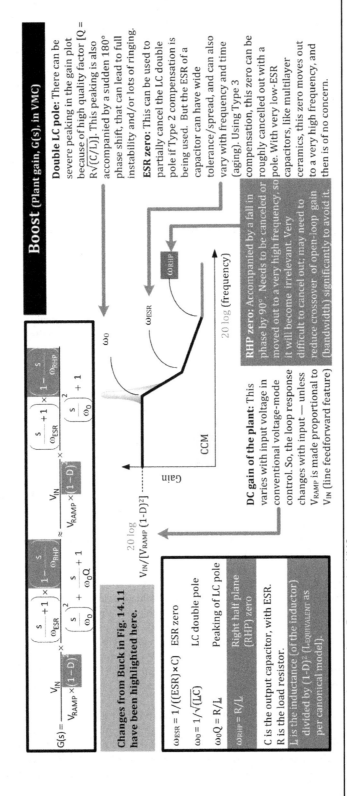

Figure 14.12 Plant transfer function for a Boost, using VMC.

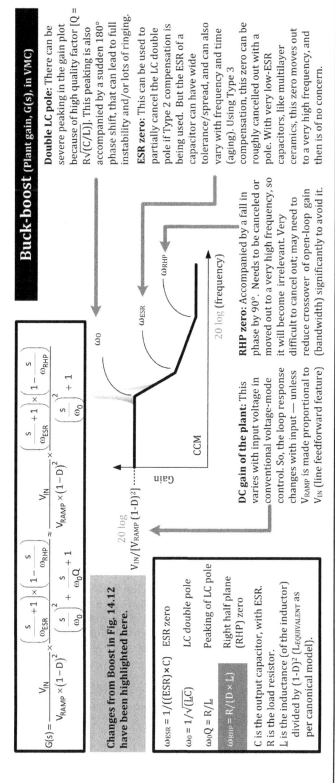

FIGURE 14.13 Plant transfer function for a Buck-boost, using VMC.

Figure 14.14

This is a Buck with CMC. We see the differences compared to VMC. The DC gain is *not* a function of input voltage (at least to a first approximation). The reason is that the PWM ramp is derived from the current ramp, and we know that the current ramp swing ΔI is a function of the voltage across the inductor during the ON-time; that is, it depends on $V_{IN} - V_O \approx V_{IN}$. So in effect, there is *pseudo* line feedforward—not as perfect as we can implement by choice in VMC. Nevertheless, CMC does offer good line rejection, which was historically one of the key reasons for its wide popularity. However, we can see that the DC gain is a function of R, the load resistance. This causes a change in the loop characteristics with changes in load. But there is an interesting property shown in the diagram because the *location* of the pole also varies with load. Therefore, the *bandwidth* remains unchanged with load (and line). That is actually the same as in VMC with line feedforward implemented.

We do see that CMC has a *single pole* at the resonant frequency of the load resistor and the output capacitor. In contrast, we saw that VMC has a double (two single) poles at the resonant frequency of the inductor and output capacitor. That by itself is not really an issue, because what it eventually means is that we need two zeros from the compensator to cancel out the double pole in VMC, but only one zero to cancel out the single pole of CMC. So the compensator can be *simpler* for CMC than for VMC. Typically that means we can use a type 2 compensator (often based around a simple transconductance error amplifier, see Fig. 14.15) for CMC, whereas we usually need a more complicated type 3 compensator for VMC (see Figs. 14.16 and 14.17). Other than that, what is the difference? The difference is that despite *cancellation* of the double pole arising from VMC, there can be a huge residual *phase shift* (rather, a huge back-and-forth phase swing) in the region around the cancellation frequency. This can lead to conditional stability issues, especially under nonlinear (large) line and load disturbances. So CMC has somewhat more predictable and acceptable responses in general.

This almost sums up our overview of the key differences between CMC and VMC. There is one last issue as discussed next. Note that all along we have restricted ourselves to continuous conduction mode (CCM), mainly for simplicity sake. In any case, discontinuous conduction mode (DCM) is encountered only for much lighter loads, and further, in many modern synchronous topologies, we may continue to stay in CCM by choice, down to zero load [as in forced continuous conduction mode (FCCM)].

Figure 14.18

This is a Boost with CMC. It also has the troublesome RHP zero.

Figure 14.19

This is a Buck-boost with CMC. It also has the troublesome RHP zero.

Conclusions on CMC versus VMC

We thus realize that when we take a closer look, CMC and VMC are just alternative ways of achieving loop stability. The type of compensation scheme we need may be simpler for CMC (type 2) than for VMC (type 3). But then, CMC also needs slope compensation, and so on. As usual, there are pros and cons.

CMC also suffers from high PCB sensitivity, since we usually depend on a small sensed current to generate the voltage ramp applied to the PWM comparator. Since a lot of noise is generated when the switch turns ON, we usually need to introduce a *blanking time* to avoid triggering the comparator for at least 50 to 200 ns after the switch turns ON. This unfortunately leads to indirectly establishing a minimum ON-time pulse width also, of 50 to 200 ns, and the corresponding minimum duty cycle can play havoc with high-voltage to very low-voltage down-conversion ratios, especially with high switching frequencies. In comparison, VMC is inherently more robust and noise resistant.

A modern preference seems to be in the direction of VMC with line feedforward, just as Bob Mammano implicitly predicted in DN-62. But another recently emerging major thrust is actually heading toward hysteretic control, as we now see.

Hysteretic Control: Energy on Demand

Looking at Fig. 14.7 we see that the basic way a PWM comparator works is by creating ON-OFF pulses from the intersection of two voltage profiles at its input terminals: one *steady voltage* level (the control voltage) and another sawtooth voltage profile (ramping up and down). We can well ask, Why not apply the reference voltage directly as the *smooth* voltage level (instead of the error voltage) on the other terminal, and use a sawtooth based

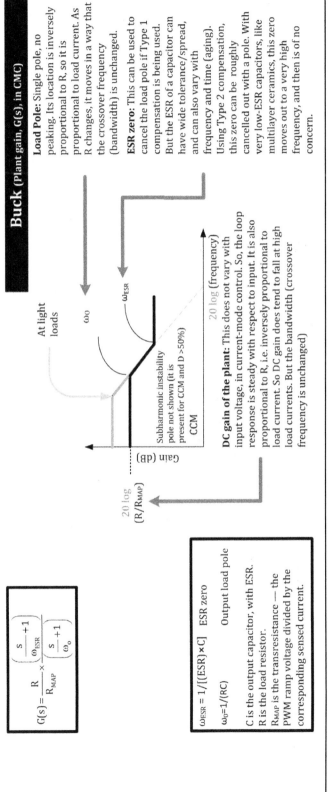

Figure 14.14 Plant transfer function for a Buck, using CMC.

Transfer functions of possible error amplifiers
(This is AC analysis. Therefore V_{REF} is being ignored below — it is just a fixed DC bias level)

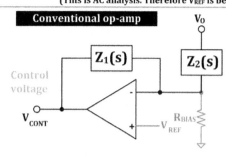

Conventional op-amp

Inverting op-amp:

$$\frac{V_{CONT}}{V_O} = -\frac{Z_1(s)}{Z_2(s)}$$

Therefore, transfer function (ignoring sign) is

$$H(s) = \frac{Z_1(s)}{Z_2(s)}$$

Magnitude of this is the gain of the error amp.

R_{BIAS} is only a DC-biasing resistor --- it does not appear in the AC analysis, and is therefore **not** included above.

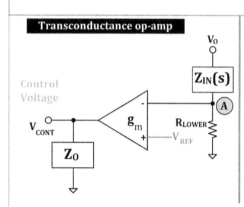

Transconductance op-amp

Voltage at point A is: $V_A = \dfrac{R_{LOWER}}{R_{LOWER} + Z_{IN}(s)} \times V_O$

By definition of transconductance, g_m

$$g_m = \frac{\text{Output current}}{\text{Input voltage}} = \frac{\frac{V_{CONT}}{Z_0(s)}}{V_A} \quad \text{(ignoring sign)}$$

Simplifying, the transfer function (ignoring sign) is

$$\frac{V_{CONT}}{V_O} = \frac{R_{LOWER}}{R_{LOWER} + Z_{IN}(s)} \times g_m \times Z_0(s)$$

Usually, $Z_{IN}(s)$ is just a resistor

The lower resistor R_{LOWER} not just a DC-biasing resistor — it *does* affect the AC signal at the input, and through g_m it affects the output current, and thereby the output voltage V_{CONT}. So R_{LOWER} is included above.

Some conclusions:
(A) If we are using a transconductance op-amp, only the ratio of the feedback resistors is important. We could for example have a combination of 1 kΩ/4 kΩ or 10 kΩ/40 kΩ, and so on. They would all create the same gain (attenuation), and the gain-phase plot would *not* change.
(B) If we are using a conventional op-amp, the upper resistor affects the gain-phase plot. If we change that, we will get entirely different results. So just keeping the ratio unchanged does *not* keep the gain-phase plot unchanged.
(C) In adjustable regulators, if we want to change the output voltage, it is therefore best to change the *lower* feedback resistor only, keeping the upper resistor the same. That way, only the DC-biasing will change, not the gain-phase (AC) characteristics of the feedback section.

FIGURE 14.15 The two main types of error amplifiers.

on the inductor current (as in CMC)? This is shown in Figs. 14.20 and 14.21. Note that as shown in the latter figure, this is not really a "PWM comparator" anymore, at least not in the sense we were used to so far. It is now a "hysteretic comparator" that terminates the ON pulse if the ramp voltage exceeds the reference voltage by a certain amount Δv ($\Delta HYS/2$) and turns the switch back ON when the ramp falls below a certain threshold slightly lower than the reference voltage ($-\Delta v$). It is therefore often called a *bang-bang* regulator. If there is a sudden line or load transient, it can react by either turning OFF completely for several pulses in succession or turning ON fully. Therefore its transient response is excellent—energy on demand in effect.

Early forms of bang-bang regulators have existed for decades, based on silicon-controlled rectifiers (SCRs) or bipolar transistors, but *without inductors*. This ancient technique has been literally reinvented in modern switching power conversion and offers tremendous advantages as can be confirmed on the bench. The bandwidth of the loop response is close to the switching frequency itself. There is no feedback or compensator to design, nor poles and zeros to manipulate. But the mathematical models of hysteretic controllers are still very complicated and just evolving. This, however, has not stopped designers from trying to eke out the full commercial advantage of hysteretic control. One of the biggest advantages is that because there is no clock and no error amplifier, nor any compensation circuitry, the quiescent

Closing the Loop: Feedback and Stability

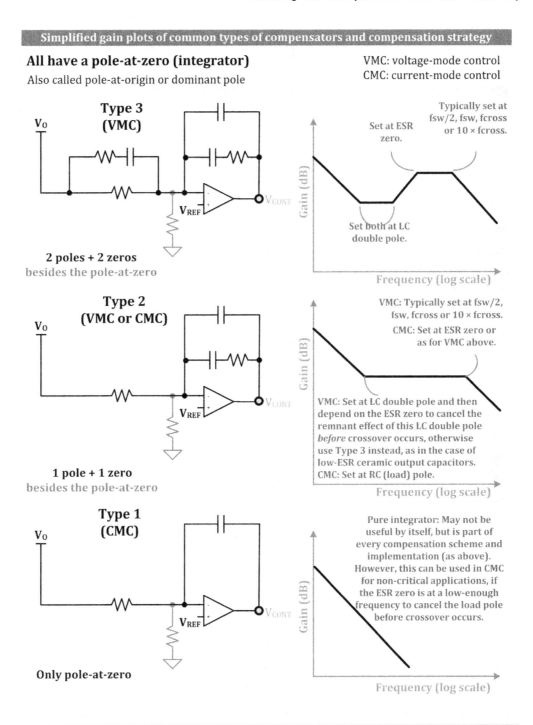

FIGURE 14.16 Types of compensation schemes.

current (I_Q, zero load, but still switching), is very low (typically less than 100 μA). This makes the hysteretic converter very suitable for modern battery-powered applications in particular. In the world of cell phones and tablets, hysteretic control has started leading the way.

Hysteretic control does have its limitations. Because there is no formal clock, it is hard to ensure constant frequency. There can also be a lot of erratic pulsing, usually accompanied by unacceptable audio noise (squealing), and also unpredictable EMI. The way to avoid erratic pulsing is to ensure the ramp waveform applied to the hysteretic comparator is an exact replica of the actual inductor current. This way we get a chicken-and-egg situation where *the duty cycle created by the comparator is exactly what the system naturally demands under the given line and load conditions*. Then there are no missed pulses accompanied by audible low-frequency harmonics. The way to adjust the frequency to an acceptable constant level is by *symmetric* variation of the comparator thresholds, as indicated in Fig. 4.22. If we do not do

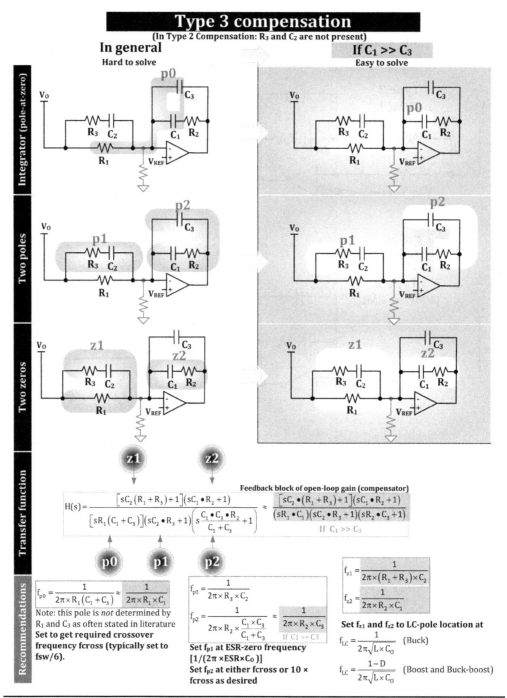

Figure 14.17 Summary of type 3 compensation.

this *symmetrically*, there will be a resultant DC offset, causing a drift or off-centering in the output voltage.

Another way to try obtaining hysteretic control with almost constant frequency is to use a constant ON-time controller (COT). We know that (for a Buck):

$$D = \frac{T_{ON}}{T} = T_{ON} \times f = \frac{V_O}{V_{IN}}$$

Therefore,

$$f = \frac{V_O}{V_{IN} \times T_{ON}}$$

In other words, if we fix the ON-time of the hysteretic converter, but also make that ON-time inversely proportional to the input, we will get a constant f. That is the underlying principle

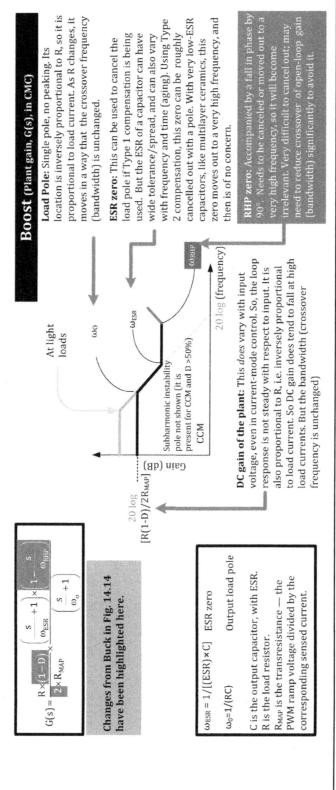

Figure 14.18 Plant transfer function for a Boost, using CMC.

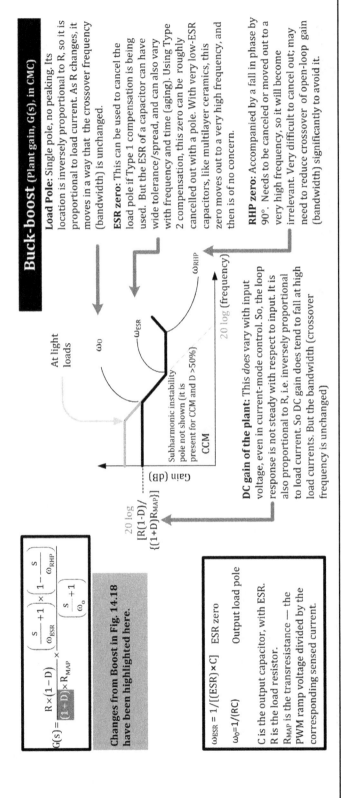

FIGURE 14.19 Plant transfer function for a Buck-boost, using CMC.

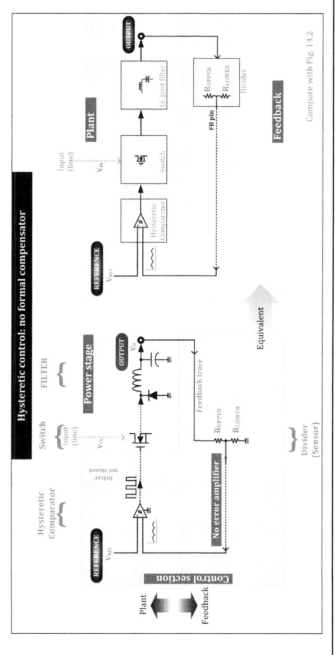

FIGURE 14.20 Functional blocks of the hysteretic converter.

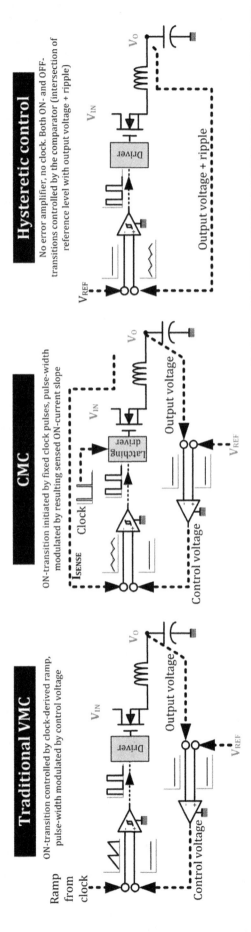

Figure 14.21 Basic changes to achieve hysteretic control.

Several techniques exist to superimpose an artificially generated ramp on to the DC level of the output voltage, without introducing any DC offset. This is usually done by sensing the voltage on both sides of the inductor, or directly across an output capacitor with significant ESR. Thereafter, this ramp-plus-DC signal is injected on to the PWM comparator. But it is hoped that there will be no external components necessary (such as DC-blocking capacitors), to eliminate any DC offset caused by mixing the ramp with the DC output level. Also, by varying the hysteresis levels for example, a pseudo-constant switching frequency can be created.

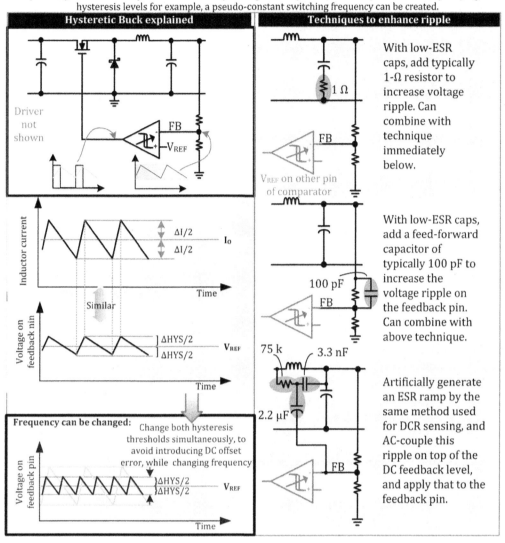

FIGURE 14.22 Hysteretic control explained in a simplified manner.

of the COT (Buck) converter. Note that the COT converter has a preordained ON-time, so there is no upper comparator threshold required anymore. But for the same reason, we can get some DC offset here (load regulation affected).

In the case of a very-high-voltage to low-voltage conversion (like 48 V to 3.3 V), we know that in traditional converters we are rather limited by the minimum (practical) pulse width of the converter, especially in the case of CMC (due to the blanking time issue as discussed earlier). But in the COT Buck, by fixing the minimum ON-time, we in effect not only lower the frequency, but also achieve smooth down-conversion without any unexpected overshoots during line and load transients as in traditional control methods.

In a similar manner it can be shown that a constant OFF-time will give a constant frequency when applied to a *Boost*. We'll remind ourselves once more of the intuitive reason for the RHP zero: Under a sudden load demand the output dips momentarily and therefore the duty cycle increases. But in the process, the OFF-time decreases. Since in a Boost (and Buck-boost) energy is delivered to the output only during the OFF-time, a smaller OFF-time leaves less time for the new energy requirement to be met, which temporarily causes the output to dip even further before things get back to normal. So we intuitively realize that fixing a certain minimum OFF-time will help in this case. And that is in fact true: *The RHP*

	CMC	VMC	Hysteretic
Rejection of line disturbances (dynamic line response)	Good (inherent)	Very good (with line feedforward)	Excellent (inherent)
Rejection of load disturbances (dynamic load response)	Good (constant bandwidth)	Good	Excellent (inherent)
Constant frequency	Excellent	Excellent	Poor, need to vary hysteresis band *or* use constant ON-time (Buck), *or* use constant OFF-time (Boost)
Predictable EMI	Excellent	Excellent	OK (with above COT techniques)
Audible noise suppression	Excellent	Excellent	OK (with above COT techniques)
Extreme down conversion (Buck)	Poor	Good	Excellent, with COT techniques
Insensitivity to PCB layout	Poor	Excellent	Good, with artificial ramp generation, otherwise poor
Excellent stability of loop responses (tolerances and long-term drifts)	Excellent	Good	Fair
Simplicity of compensation	Good	Poor	Excellent
I_Q (quiescent current)	Good	Poor	Excellent
Loop stability with use of output ceramic capacitors	Excellent (with type 3 compensation)	Very good	Good, with artificial ramp generation, otherwise poor

TABLE 14.3 Summary (Voltage Mode versus Current Mode versus Hysteretic Control)

zero is not present when operating the Boost in constant OFF-time mode. And we can get constant-frequency operation too, by setting $T_{OFF} \propto V_{IN}$.

For a Buck-boost, the relationship for achieving constant frequency is too complicated to implement easily, without sacrificing the key advantages of hysteretic controllers: simplicity and low I_Q. So we will respect this roadblock.

This concludes a brief summary of the pros and cons of hysteretic control. See Table 14.3 for a summary of pros and cons of all three control methods.

PART 3 DESIGN EXAMPLES FOR VMC AND CMC NONISOLATED, AND FOR TL431 + OPTO ISOLATED FLYBACK CASE

Design Examples for VMC and CMC, Using Type 2 and 3 Error Amps or an OTA

Our first example in Fig. 14.23 starts with a type 3 amplifier for VMC. Here, we have a 300-kHz Buck, converting 48 V to 12 V at 5 A (60 W). Its PWM ramp is 1.5 V. It has a 100-µF aluminum electrolytic capacitor of ESR 100 mΩ. We carry out its compensation in Fig. 14.23.

In particular, keep in mind that the DC gain of the plant is V_{IN}/V_{RAMP}. This was also indicated in Fig. 14.11. That was for a Buck. For other topologies, we can just use the DC gain from the figures immediately past Fig. 14.11.

Closing the Loop: Feedback and Stability

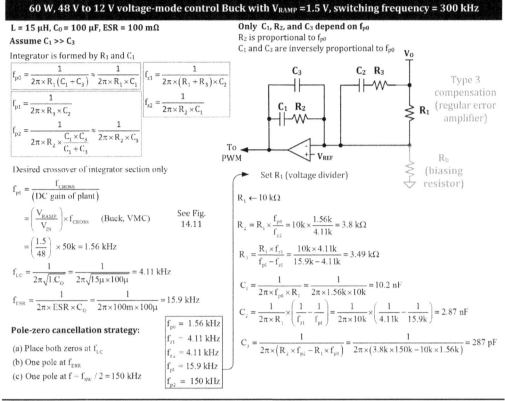

FIGURE 14.23 60-W Buck, VMC, using a type 3 error amplifier.

So, for example, in this case we have the DC plant gain (in decibels) as

$$a = 20 \log\left(\frac{V_{IN}}{V_{RAMP}}\right) = 20 \log\left(\frac{48}{1.5}\right) = 30.1 \text{ dB} \quad \text{(DC gain of plant, Buck in VMC)}$$

We want to set the crossover frequency at one-sixth the switching frequency, so

$$f_{CROSS} = \frac{f}{6} = \frac{300k}{6} = 50 \text{ kHz}$$

The inductance is chosen based on the usual guiding point of $r = 0.4$. So

$$L = \frac{V_O}{I_O \times r \times f} \times (1-D) = \frac{12}{5 \times 0.4 \times 300,000} \times \left(1 - \frac{12}{48}\right) = 1.5 \times 10^{-5}$$

We have chosen 15 μH. The double LC pole is at (in Hz)

$$f_{LC} = \frac{1}{2\pi\sqrt{LC}} = \frac{1}{2\pi\sqrt{15\mu \times 100\mu}} = 4.11 \times 10^3$$

The ESR zero is at (in Hz)

$$f_{ESR} = \frac{1}{2\pi \times ESR \times C} = \frac{1}{2\pi \times 100 \text{ m} \times 100\mu} = 1.59 \times 10^4$$

To calculate the required crossover of the integrator (which we call f_{p0}), we use the following key relationship:

$$f_{p0} = \frac{f_{CROSS}}{\text{DC gain of plant}}$$

The preceding equation is actually valid for all topologies, and for CMC and VMC—based only on our simple pole-zero cancellation strategy: creating a straight-line open-loop gain curve of slope -1 (in the log plane). We have therefore used the preceding relationship in all the design examples. In Fig. 14.24 we show how we can find f_{p0} graphically instead of using

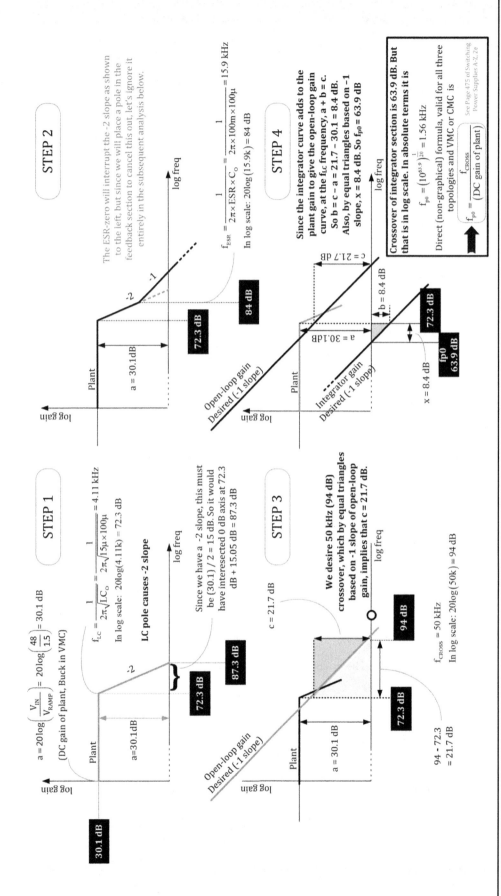

Figure 14.24 Graphical way of figuring out integrator crossover in Fig 14.23.

the preceding simple equation. It seems a lot of work, just to get the same results by "playing" in the log plane. But it is very instructive and helps us visualize the situation much better than just grinding away on equations. The reader is encouraged very strongly to go through the steps in that figure. Figure 14.24 develops intuitive skills in carrying out compensation strategies in general, and on occasion, also helps us visualize which compensation approach or strategy is not feasible, or just incorrect.

NOTE *A number X expressed in decibels can be expressed as a factor by using $10^{X/20}$.*

In Fig. 14.25 we carry out almost the same exercise for CMC. Since we do not have a double *LC* pole, but a single *LC* pole, we can use type 2 compensation. Here we are using a regular error amplifier. We should also look at Figs. 14.16 and 14.17 as it helps understand the poles and zeros. Notice that we have generated type 2 compensation from type 3 by literally blanking out the *RC* across the upper divider resistor. Note that the equations are all predictable and conform to Fig. 14.23, except for the *different plant gain*, as in CMC. In Fig. 14.26 we have compiled the CMC equations for plant and DC gain of all the three basic topologies, both in their exact form and in an approximate (but easy-to-use) form.

In Fig. 14.27 we carry out compensation of the Buck, this time using a transconductance op-amp, as described in Fig. 14.15.

Design Examples for TL431 and Opto

In Fig. 14.28 we turn our attention to the classic TL431 + opto problem. It seems there is no easy, direct, published design procedure in literature. We have in particular Basso's new book mentioned here: www.psma.com/HTML/newsletter/page11.html. It says

> For example the TL431 is a complex device and its implementation in compensation stage is often overlooked, especially if an optocoupler is added in the chain. In his new book Christopher Basso dedicates an entire 70-page chapter to the TL431, detailing internals of this popular component. The book reviews the three compensation types (1, 2 and 3) for isolated and non-isolated converters built around the 3-leg self-contained reference voltage op amp.

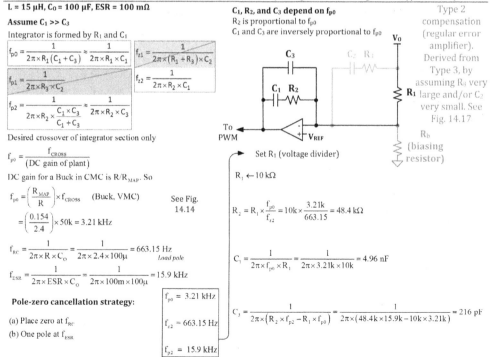

FIGURE 14.25 60-W Buck, CMC, using a type 2 error amplifier.

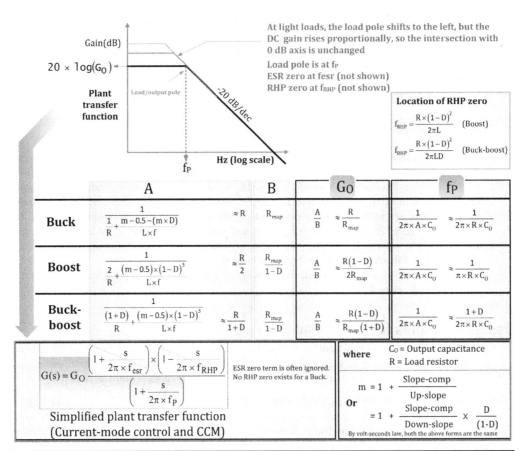

FIGURE 14.26 Load pole positions in CMC for all topologies.

60 W, 48 V to 12 V current-mode control Buck with V_{RAMP} = 1.0 V, switching frequency = 300 kHz (OTA)

Load resistor is V_O/I_O = 12 V/5 A = 2.4 Ω. Switch peak current corresponding to maximum current on sense (for Buck) is about $1.3 \times I_O$ = 1.3 × 5 A = 6.5 A. This corresponds to 0.2 V on the sense resistor in this example. Sense resistor value is V/I = 0.2 V/6.5 A = 0.031 Ω. After that, a current sense amplifier is present, of gain ×5, so the swing at the comparator input is 1 V. Therefore R_{MAP} = V/I = (0.2 V × 5)/ 6.5 A = 0.154 Ω. In general, R_{MAP} = gain of current sense amplifier × R_{SENSE} (see Page 497 in Switching Power Supplies A-Z, 2e)

L = 15 µH, C_O = 100 µF, ESR = 100 mΩ
Assume $C_1 \gg C_2$
Integrator is formed by R_1 and C_1

$$f_{p0} = \frac{1}{2\pi \times \left(\frac{C_1}{y \times g_m}\right)} \qquad f_z = \frac{1}{2\pi \times R_1 \times C_1}$$

$$f_p = \frac{1}{2\pi \times R1 \times C2}$$

Desired crossover of integrator section only

$$f_{p0} = \frac{f_{CROSS}}{(\text{DC gain of plant})}$$

DC gain for a Buck in CMC is R/R_{MAP}. So

$$f_{p0} = \left(\frac{R_{MAP}}{R}\right) \times f_{CROSS} \quad \text{(Buck, VMC)}$$

$$= \left(\frac{0.154}{2.4}\right) \times 50k = 3.21 \text{ kHz}$$

$$f_{RC} = \frac{1}{2\pi \times R \times C_O} = \frac{1}{2\pi \times 2.4 \times 100\mu} = 663.15 \text{ Hz (Load pole)}$$

$$f_{ESR} = \frac{1}{2\pi \times ESR \times C_O} = \frac{1}{2\pi \times 100m \times 100\mu} = 15.9 \text{ kHz}$$

If ESR zero is at a higher frequency than the desired crossover frequency (50 kHz here), we can omit C_2.

Pole-zero cancellation strategy:
(a) Place zero at f_{RC}
(b) One pole at f_{ESR}

f_{p0} = 3.21 kHz
f_p = 15.9 kHz
f_z = 663.15 Hz

C_1, R_1, and C_2 depend on f_{p0}
R_1 is proportional to f_{p0}
C_1 and C_2 are inversely proportional to f_{p0}

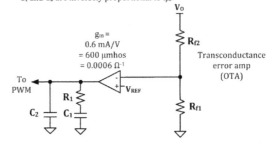

Select R_{f2} and R_{f1} (voltage divider ratio). This sets the attenuation ratio, defined as

$$y \equiv V_{REF}/V_O = 2.5 \text{ V}/12 \text{ V} = 0.208$$

(in the case of TL431 for example, where V_{REF} = 2.5 V)

$$C_1 = \frac{y \times g_m}{2\pi \times f_{p0}} = \frac{0.208 \times 0.0006}{2\pi \times 3.21k} = 6.2 \text{ nF}$$

$$R_1 = \frac{1}{2\pi \times f_z \times C_1} = \frac{1}{2\pi \times 663.15 \times 6.2n} = 38.7 \text{ k}\Omega$$

$$C_2 = \frac{1}{2\pi \times f_p \times R_1} = \frac{1}{2\pi \times 15.9k \times 38.7k} = 259 \text{ pF}$$

FIGURE 14.27 60-W Buck, CMC, using an OTA error amplifier.

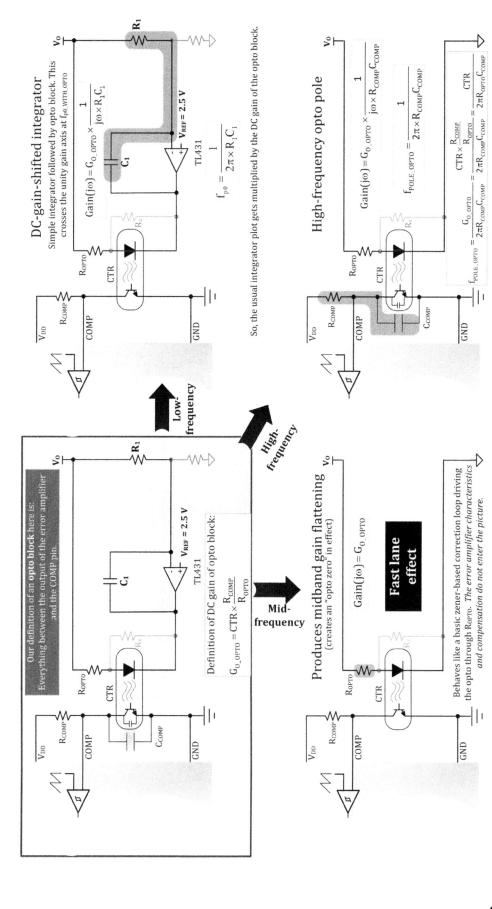

FIGURE 14.28 Low-, mid-, and high-frequency gain of TL431 with opto.

394 Chapter Fourteen

In this section we are trying to reduce an obviously complex problem to a simple one-page (albeit cramped) design procedure for engineers on the run—and in English, not in French.

The way the TL431-based network looks at low, mid, and high frequencies is described in Dr. Ridley's article found here: http://switchingpowermagazine.com/downloads/15 Designing with the TL431.pdf

Note that usually when we have an integrator consisting of R_1 and C_1, it has a pole at the origin (the *pole at zero* as we call it here), and its gain falls continuously with a -1 slope, crossing the unity gain axis (0 dB) at $1/2\pi R_1 C_1$. In our specific case, R_1 is the upper resistor of the divider and C_1 is the capacitance across the op-amp as shown. However, using a TL431 and opto, a new zero emerges, and its position and reasons for existence are not at all obvious. We are calling it the *opto-zero* here. It is created not by any *RC* resonance really, as the equation for its location seems to indicate, but by the fact that as we increase the

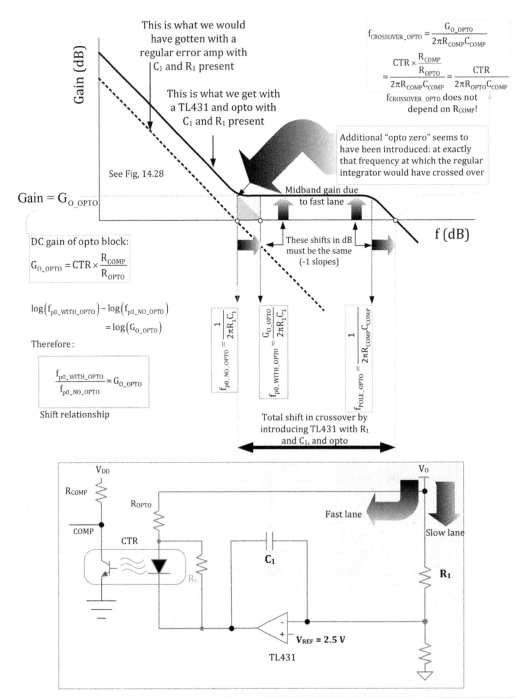

Figure 14.29 Low-, mid-, and high-frequency gain of TL431 with opto (graphically explained).

frequency of disturbance, the midband gain cannot fall below a certain value: $G_{O\_OPTO} = CTR \times R_{COMP}/R_{OPTO}$. This corresponds to the TL431 simply acting as a dead short, doing all it can, pulling in as much current as possible, through the opto, but not quite managing to reduce the output signal (disturbance) to zero. At best it gets to G_{O_OPTO}, and then plateaus. It is something similar to the gain of a noninverting op-amp never managing to go below 1 (0 dB), which is why we always prefer to inject the feedback signal to the inverting input of the error amplifier. The only way to reduce the disturbance completely from going around the loop is to (1) break the loop up entirely, say by setting CTR = 0, (2) increase R_{OPTO} to infinity (no opto-LED current to start with), or (3) set R_{COMP} to zero so that COMP just does not wiggle at all, despite any amount of current through the opto-transistor. This plateau in the gain plot, in effect, introduces a *zero*, but only because it changes the slope of the gain curve *from whatever it was* − 1 in our case) to zero (the slope of any plateau, by definition). It just turns out that the change in slope is − 1, because that was the slope just before the plateau occurred. So it looks like a *simple pole*. But in reality it is *not due to the resonance of* R_1 *and* C_1: That "resonance" only created a *pole at zero*, which tends to cross the 0-dB axis at $f_{p0\_NO\_OPTO} = 1/2\pi R_1 C_1$. Now, because the gain cannot fall below the value G_{O_OPTO}, a *zero* appears, right where it was about to cross over: see Fig. 14.29.

The problem with this zero is that it is a new *zero*, one which causes a gain plateau and makes it very hard for the open-loop-gain curve to crossover gracefully (i.e., with − 1 slope), and at a reasonably low frequency of, say, 50 kHz as we intend. The cause of this *zero* is the *fast-lane* response shown in Fig. 14.28. This plateau has nothing really to do with the error-amplifier circuitry, except that the TL431 is fully conducting through R_{OPTO}. We can attempt to close down this fast lane and thereby avoid introducing this *opto-zero* by some simple strategies, as shown in Fig. 14.30. Once we do that, we can easily adapt and use the regular

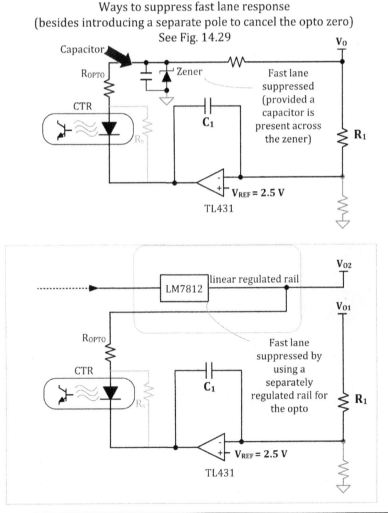

FIGURE 14.30 Ways to suppress fast lane (and midband gain plateau).

design procedure for type 2 and type 3 compensation, even for the case of the TL431 plus opto, because the TL431 is in fact just an open-collector (but otherwise regular) op-amp. The best indication of that is the lower resistor of the divider does not apparently affect the Bode plot. Yet, there are application notes, which state that the TL431 is in reality a transconductance op-amp. Yes, it does convert voltage into current, but all error amplifiers do that eventually. If it really were an OTA in the sense presented in Fig. 14.15, then the lower resistor of the divider would have entered the calculations too and affected the Bode plot (see the attenuation ratio y used in Fig. 14.27).

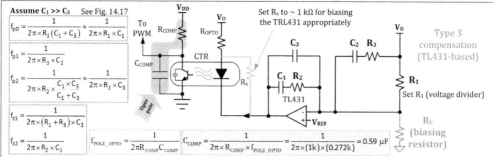

FIGURE 14.31 40-W flyback design example, VMC, using type 3 compensation.

The other option is to *cancel* the opto-zero using an entirely new pole. Let us call that the *opto-pole*. The reason is it needs to be a really *low frequency* pole. We can't get that positioning easily from the usual type 2 and type 3 compensation schemes, because each component in that network affects both a pole and a zero, so getting such a low-frequency pole from such a closely intertwined network, while still maintaining flexibility in placing the other poles and zeros at will (at a much higher frequency), is almost impossible. So we look for a new independent pole. And that is created by the R and C connected to the collector of the opto-transistor. We call these R_{COMP} and C_{COMP}. The procedure to fix this pole is shown in Fig. 14.31 for example. This particular design example is for a type 3 scheme, and assumes VMC. The steps parallel the Buck VMC calculation in Fig. 14.23, except for the change in DC plant gain and the cancellation of the opto-zero by the opto-pole. We could, of course, do a calculation even more similar to Fig. 14.23 if we use fast-lane cancellation techniques as shown in Fig. 14.29.

In Fig. 14.32, we show some aspects of Fig. 14.31 in graphical form, to feel comfortable with the compensation strategy. In Figs. 14.33 and 14.34 we carry out the final design exercise using type 2 compensation based on the TL431 and assuming CMC.

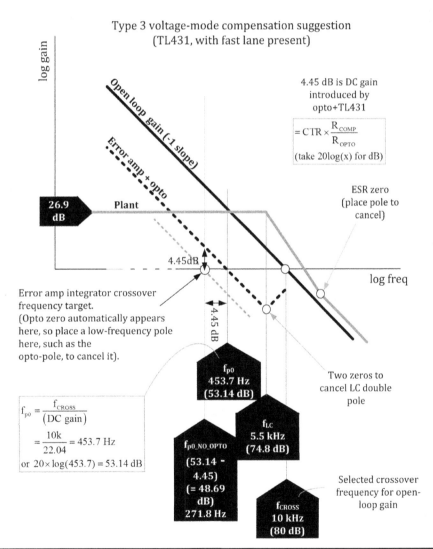

Figure 14.32 Graphical analysis of type 3 compensation (VMC) using TL431 and opto.

40 W, 48 V to 5 V current-mode control Flyback with $V_{RAMP} = 1$ V, switching frequency = 200 kHz (Part 1)

$L_{PRI} = 70$ μH, $N_P = 15$, $N_S = 3$, $C_O = 300$ μF, ESR = 25 mΩ

Assume we choose PC817B. We pick 2 to 5 mA operating point (for example $R_{COMP} = 1$ kΩ to $V_{DO} = 5$ V). We target a crossover frequency of open loop gain of 10 kHz. The CTR is assumed to be as low as 100% (to account for temperature).

On the opto-diode side from 5 V, we should plan on up to 5 mA too, since CTR is 1. So subtracting 2 V from V_O, to account for the lowest possible drop across the opto-diode and the TL431, we will place $R_{OPTO} = (5 V - 2 V)/5$ mA = 0.6 Ω. So, finally, our values are

CTR = 1, $R_{COMP} = 1$kΩ, $R_{OPTO} = 0.6$ kΩ.

DC gain shift introduced by the opto and the TL431 circuit is

$$G_{O\_OPTO} = CTR \times \frac{R_{COMP}}{R_{OPTO}} = 1 \times \frac{1k}{0.6k} = 1.67$$

In dB: $20\log(1.67) = 4.45$ dB ←

This will shift the integrator curve by this very amount.

Other calculations:

$$n = \frac{N_P}{N_S} = \frac{15}{3} = 5$$

$$D = \frac{V_{OR}}{V_{OR} + V_{IN}} = \frac{nV_O}{nV_O + V_{IN}} = \frac{5 \times 5}{(5 \times 5) + 48} = 0.34$$

RHP zero location for Buck-boost = $\frac{R \times (1-D)^2}{2\pi LD}$

So, RHP zero location for a Flyback = $\frac{R_{LOAD} \times (1-D)^2}{2\pi L_{SEC} D}$

$$= \frac{R_{LOAD} \times (1-D)^2 \times n^2}{2\pi L_{PRI} D}$$

$$= \frac{\frac{5V}{8A} \times (1-0.34)^2 \times 5^2}{2\pi (70\mu) \times 0.34} = 45.5 \text{ kHz}$$

Ensure this is at much higher frequency than the desired crossover. From the factors, the actual frequency is $(10^{6.155})^{\frac{1}{20}} = 1.2$ kHz. So, we have picked 10 kHz as target crossover (may prefer 5 kHz).

The center-of-ramp peak primary-side current is $I_{OR}/(1-D)$. We add 20% to this to account for ripple, plus 10% headroom, and this produces a 0.2 V swing across sense resistor. So

$$V_{SENSE} = R_{SENSE} \times 1.3 \times \frac{I_O}{n(1-D)}$$

So

$$R_{SENSE} = \frac{V_{SENSE} \times n(1-D)}{1.3 \times I_O} = \frac{0.2 \times 5 \times (1-0.34)}{1.3 \times 8} = 0.063$$

So current-mode transresistance R_{MAP} is found from the equation

$R_{MAP} = R_{SENSE} \times (\text{gain of CSA}) = 0.063 \times 5 = 0.317$.

However, this only relates the voltage on the PWM to the switch current. So far, we have it in the form V/I where I is the primary-side current. In reality we want to relate it to the secondary-side (load) current, which is n times greater. So the Flyback transresistance is actually

$R_{MAP} = R_{SENSE} \times (\text{gain of CSA})/n = 0.317/5 = 0.063$

Buck–boost plant DC gain = $\frac{R}{R_{MAP}} \times \frac{1-D}{1+D}$ (current mode) See Fig. 14.19

Flyback plant DC gain = $\frac{R_{LOAD}}{R_{MAP}} \times \frac{1-D}{1+D}$

$$= \frac{5V/8A}{0.063} \times \frac{1-0.34}{1+0.34}$$

$$= 4.9$$

In terms of log this is $20 \times \log(4.9) = 13.8$ dB

Desired crossover of integrator section only

$f_{p0} = \frac{f_{CROSS}}{(\text{DC Gain})} = \frac{10k}{4.9} = 2$ kHz

In terms of frequency in dB, this is $20 \times \log(2k) = 66$ dB

But this includes the actual integrator curve plus the DC gain from the opto stage (calculated to be **4.45 dB** above). So the actual integrator crossover frequency must be 66 dB - 4.45 dB = 61.55 dB.

Set $f_{p0\_NO\_OPTO} = 1.2$ kHz

$$f_{RC} = \frac{1}{2\pi \times R \times C_O} = \frac{1}{2\pi \times \frac{5V}{8A} \times 300\mu} = 848.8 \text{ Hz}$$

$$f_{ESR} = \frac{1}{2\pi \times ESR \times C_O} = \frac{1}{2\pi \times 25m \times 300\mu} = 21.2 \text{ kHz}$$

Pole-zero cancellation strategy:

(a) Place zero at f_{RC}
(b) One pole at f_{ESR}

SET
$f_{p0} = 1.2$ kHz
$f_{z2} = 848.8$ Hz
$f_{p2} = 21.2$ kHz

Opto zero appears at f_{p0} if fast lane is present

$R_1 \leftarrow 10$ kΩ

$R_2 = R_1 \times \frac{f_{p0}}{f_{z2}} = 10k \times \frac{1.2k}{848.8} = 14.1$ kΩ

$$C_1 = \frac{1}{2\pi \times f_{p0} \times R_1} = \frac{1}{2\pi \times 1.2k \times 10k} = 13.26 \text{ nF}$$

$$C_3 = \frac{1}{2\pi \times (R_2 \times f_{p2} - R_1 \times f_{p0})}$$

$$= \frac{1}{2\pi \times (14.1k \times 21.2k - 10k \times 1.2k)} = 555 \text{ pF}$$

Other components already calculated are:

CTR=1, $R_{COMP} = 1$ kΩ, $R_{OPTO} = 0.6$ kΩ

See C_{COMP} calculation and R_X suggestion in Part 2 in Fig. 14.34.

Figure 14.33 40-W flyback design example, CMC, using type 2 compensation (part 1).

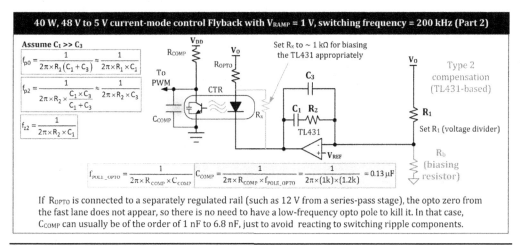

FIGURE 14.34 40-W Flyback design example, CMC, using type 2 compensation (part 2).

CHAPTER 15
Practical EMI Filter Design

The CISPR 22 Standard

The applicable standard we usually follow for IT (information technology) equipment is CISPR 22, now called EN 550022 (EN stands for European norm). In Table 15.1 we have the *conducted emission limits*. OEM specifications usually ask us to meet the more stringent Class B limits. One rather unusual addition by the author in the table provided here is that dBμV has also been expressed as mV. Hopefully that is more comfortable to some designers who have trouble thinking logarithmically. Note that by definition, dB = 20 × log (voltage ratio). So, a dBμV implies a logarithmic comparison to a reference of 1 μV. We get the corresponding mV by using

$$(mV) = (10^{(dB\mu V)/20}) \times 10^{-3}$$

The corresponding EMI requirement in the United States (FCC Part 15) does not have average limits, only quasi-peak limits, though it does accept certification to CISPR 22. For CISPR 22 (Class B) we can use the following equations for the 150 to 500 kHz range.

- *The average limit* is (almost)

$$(dB\mu V_{AVG}) = -20 \times \log(f_{MHz}) + 40$$

- *The quasi-peak limit CISPR curve (conducted emissions)* is (almost)

$$(dB\mu V_{QP}) = -20 \times \log(f_{MHz}) + 50$$

We can see that for Class B, the quasi-peak (*QP*) limit is always 10 dB higher.

The LISN

First note that "CM" or "cm" both stand for *common-mode* noise in this chapter, and "DM" or "dm" refer to *differential-mode* noise. We measure the conducted EMI emissions across a *line impedance stabilizing network* (LISN). The LISN provides the following load impedances to the noise generators (in the absence of any input filter).

- The CM load impedance is 25 Ω.
- The DM load impedance is 100 Ω.

As we flick the switch on the front panel of the LISN, we measure the following noise voltages (subscript *L* stands for *line* and N for *neutral*)

$$V_L = 25 \times I_{cm} + 50 \times I_{dm}$$
$$V_N = 25 \times I_{cm} - 50 \times I_{dm}$$

Since we have an AC line input, both lines are essentially symmetric at the input of the power supply, and so we don't usually see anything more than minor differences in the two signals above. Special LISNs are available (e.g., from Laplace Instruments) which can provide separated DM and CM components to help in troubleshooting.

	CLASS A (industrial)							
	FCC Part 15				CISPR 22			
Freq (MHz)	Quasi-peak		Average		Quasi-peak		Average	
	dBμV	mV	dBμV	mV	dBμV	mV	dBμV	mV
0.15 – 0.45	NA	NA	NA	NA	79	9.0	66	2.0
0.45 – 0.5	60	1.0	NA	NA	79	9.0	66	2.0
0.5 – 1.705	60	1.0	NA	NA	73	4.5	60	1.0
1.705 – 30	69.5	3.0	NA	NA	73	4.5	60	1.0
	CLASS B (residential)							
	FCC Part 15				CISPR 22			
Freq (MHz)	Quasi-peak		Average		Quasi-peak		Average	
	dBμV	mV	dBμV	mV	dBμV	mV	dBμV	mV
0.15 – 0.45	NA	NA	NA	NA	66-56.9	2.0-0.7	56-46.9*	0.63-0.22*
0.45 – 0.5	48	0.25	NA	NA	56.9-56	0.7-0.63	46.9-46*	0.22-0.2
0.5 – 5	48	0.25	NA	NA	56	0.63	46	0.2
5 – 30	48	0.25	NA	NA	60	1.0	50	0.32

*This is a straight line on the standard dBμV versus log f plot.
† NA stands for not applicable.

TABLE 15.1 Conducted Emissions Limits

Fourier Series

For a function $f(x)$ with time period expressed as an angle (2π) we can write

$$f(x) = \frac{1}{2}a_o + \sum_{n=1}^{\infty}(a_n \cos nx + b_n \sin nx)$$

$$a_n = \frac{1}{\pi}\int_0^{2\pi}[f(x)\cos nx]dx$$

$$b_n = \frac{1}{\pi}\int_0^{2\pi}[f(x)\sin nx]dx$$

Alternatively,

$$f(x) = \frac{1}{2}a_o + \sum_{n=1}^{\infty} c_n \cos(nx - \phi_n)$$

$$c_n^2 = a_n^2 + b_n^2$$

$$\tan\phi_n = \frac{b_n}{a_n}$$

Here the period is expressed as 2π, but in power supplies we know that the period we are interested in is in units of time, not angle, that is, $T = 1/f$. The way to convert angle θ to t is to use the equivalence

$$\frac{\theta}{2\pi} \rightarrow \frac{t}{T}$$

or

$$\theta \rightarrow 2\pi \times \frac{t}{T}$$

The first term ($\frac{1}{2} \times a_o$) of the Fourier expansion really doesn't matter, and it can even be expressed differently. But it is simply the arithmetic average of the waveform (a pure direct current). The sign of the c_n is similarly not important, either from the viewpoint of the measured EMI spectrum or our EMI suppression methods.

The Trapezoid

If we take a rectangular wave with nonzero rise and fall times, we can show that (for the case of equal rise and fall times)

$$c_n = A \times \frac{2 \times (t_{ON})}{T} \times \left[\frac{\sin\left\{\frac{n \times \pi \times t_R}{T}\right\}}{\frac{n \times \pi \times t_R}{T}} \right] \times \left[\frac{\sin\left\{\frac{n \times \pi \times (t_{ON})}{T}\right\}}{\frac{n \times \pi \times (t_{ON})}{T}} \right]$$

where $t_{RISE} = t_{FALL} = t_R$, and A is the amplitude (*actually peak-to-peak*). We are ignoring any signs as they are essentially irrelevant.

We have two "break points" of slope (in the log versus log plot). The first occurs at

$$\frac{n \times \pi \times t_{ON}}{T} = 1$$

that is,

$$n_1 = \frac{T}{\pi \times t_{ON}}$$

Since n = frequency of harmonic/fundamental frequency, i.e., $n = f \times T$, we get the corresponding break frequency to be

$$f_{BREAK\_1} = \frac{1}{\pi \times t_{ON}} = \frac{0.32}{t_{ON}}$$

The second break point is at

$$n_2 = \frac{T}{\pi \times t_R}$$

that is, the break frequency is

$$f_{BREAK\_2} = \frac{0.32}{t_R}$$

We know what to expect too, that after the second break point, the net roll-off will be at $20 + 20 = 40$ dB per decade. See Fig. 15.1. Note that n must be an integer to have any physical meaning. The first break point, therefore, may not be visible. What we will apparently perceive is that the envelope ramps down almost from the lowest frequency at the rate of 20 dB per decade.

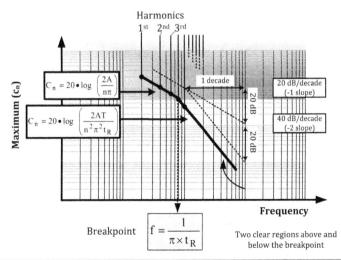

FIGURE 15.1 Envelope of amplitudes of harmonic of rectangular waveform of amplitude A and $t_R = t_F$

Below the first break frequency, the envelope of the harmonics actually becomes flat. The first break point should therefore be calculated and the envelope should be *guillotined* (truncated) below this frequency. We can use the following equations to describe the c_n. Note that in these equations, the c_n are no longer the actual coefficients of the Fourier expansion, but rather they represent the *envelope* (since that is all that matters from the standpoint of the EMI filter design).

$$c_n = 20\log\left(\frac{2A}{n\pi}\right)$$

$$c_n = 20\log\left(\frac{2A}{n^2\pi^2 t_R f_{SW}}\right)$$

The first of these two equations is valid between the first and second break points, and the second equation is valid for all frequencies higher than the second break point. Note that the switching frequency is $f_{SW} = 1/T$.

Practical DM Filter Design

Differential-mode noise is caused by the voltage drop produced across the equivalent series resistance (ESR) of the input bulk capacitor by the high-frequency switching current.

The voltage across the ESR of the capacitor is

$$v = \text{ESR} \times I_{SW} \quad \text{V}$$

If there was no filter present, the switching noise current received by the LISN is

$$I_{LISN} = \frac{v}{100} = \frac{\text{ESR} \times I_{SW}}{100} \quad \text{A}$$

since the LISN has an impedance of 100 Ω for DM noise. However, the analyzer measures the noise across one of the two effective series 50-Ω resistors in the LISN. So the measured level of noise is

$$V_{LISN\_DM\_NOFILTER} = I_{LISN} \times 50 = \frac{\text{ESR} \times I_{SW}}{2} \quad \text{V}$$

We have assumed that C_{BULK} is very large and that it has no ESL, and also that its ESR is much less than 100 Ω.

Example 15.1 What is the DM noise spectrum measured at the LISN for a 5 V @ 15 A Flyback at an input of 265 Vac, with a transformer ratio of 20? We are using an aluminum electrolytic bulk capacitor whose datasheet states that it has a capacitance of 270 µF, a dissipation factor (tangent of loss angle) of tan δ = 0.15 measured at 120 Hz, and a frequency multiplier factor of 1.5 at 100 kHz.

First the ESR is to be computed at the 120-Hz test frequency. By definition

$$\text{ESR}_{120} = \frac{\tan \delta}{2\pi f \times C} = \frac{0.15 \times 10^6}{2 \times 3.142 \times 120 \times 270} = 0.74 \, \Omega$$

At a high frequency, the ripple current is allowed to increase by the frequency multiplicative factor of 1.5. Therefore, since the heating ($I^2 \times \text{ESR}$) must still be the same, it means that the ESR at a high frequency must be $1/(1.5)^2$ times the ESR at low frequency. Therefore, for our purpose

$$\text{ESR} = \frac{1}{1.5^2} \times 0.74 = 0.33 \, \Omega$$

$$V_{LISN\_DM\_NOFILTER} = \frac{\text{ESR} \times I_{SW}}{2} = 0.17 \times \frac{n_s}{n_p} \times I_O = 0.13 \text{ V}$$

This is the amplitude of the measured signal. It is the A in the corresponding Fourier series. In terms of dBµV its value is $20 \times \log(0.13/10^{-6}) = 102$ dBµV. We can thus predict that the spectrum is as shown in Fig. 15.2.

We have also shown how the spectrum relates to the CISPR 22 Class B quasi-peak emission limits (bold line). We are showing a sample case where the switching frequency is just below

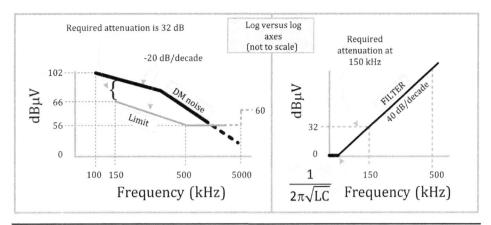

FIGURE 15.2 Worked example for DM filter design.

where the CISPR 22 limits start. Let us assume that it is 100 kHz here. For simplicity, we are also assuming that the common-mode noise is not a major contributor at these low frequencies. In practice, we should allocate some dB of margin for the CM noise (which will be estimated in the next section). We are also not explicitly accounting for any other required margins. But since we are comparing peak values (which are always higher than quasi-peak values) with quasi-peak limits, this automatically provides us with some natural headroom or margin.

The equation of the envelope at a frequency f kHz is

$$\mathrm{dB\mu V}(f) = -\mathrm{slope}\{[\log(f) - \log(100)]\} + 102$$

Since the (magnitude of the) slope is 20 dB per decade, at 150 kHz we get

$$\mathrm{dB\mu V}(f) = -20[\log(150) - \log(100)] + 102 = 98 \text{ dB}$$

This is 98 – 66 = 32 dB higher than it should be. So, we need to attenuate the noise by 32 dB. We, therefore, need to pick a low-pass LC filter which provides this attenuation at 150 kHz. We can thus calculate its break frequency. For example, if we are using an LC low-pass filter, it has an *attenuation characteristic of about 40 dB per decade* above its break frequency [(i.e., $1/2\pi(LC)^{0.5}$]. So, from Fig. 15.2 we can see that the equation it needs to satisfy is

$$32 = \mathrm{slope} \times [\log(f) - \log(f_{\mathrm{BREAK}})] = 40 \times [\log(150) - \log(f_{\mathrm{BREAK}})]$$

Solving

$$\log(f) = \log(150) - \frac{32}{40} = 1.38$$

$$f = 10^{1.38} = 24 \text{ kHz}$$

We therefore need a filter that has an LC of

$$LC = \left(\frac{1}{2\pi \times 24,000}\right)^2 = 4.4 \times 10^{-11} \text{ s}^2$$

Therefore, if the net X-capacitance (line-to-line capacitance) C is say 0.22 μF, we get the corresponding L to be 200 μH.

Since the break point associated with the *rise and fall times* (crossover) didn't enter the picture, does this mean that it doesn't matter how fast we turn ON and turn OFF the FET? Yes, from the DM-noise viewpoint it really doesn't matter much. However, there are parasitics that we ignored, chiefly the ESL and trace inductances. Since, unlike the ESR, these will produce frequency-dependent voltage spikes, it is in our interest not to keep the FET crossover times too small.

Practical CM Filter Design

We are assuming that the FET heatsink is tied to the chassis and that at one end of the parasitic capacitor C_p between the (earthed) case and chassis we are applying a trapezoid (describing the drain waveform). This causes a CM noise current I_{cm} to flow through the earth wire. We assume that the parasitic capacitances and/or X-caps cause this injected current to split *equally* between the L and N wires (that is CM noise by definition). So, we have $I_{cm}/2$ flowing in each of these two wires.

We already know the Fourier components of the trapezoid, and so we can treat them individually and determine the injected current due to each harmonic. As an example, we take the Forward converter. Here the peak-to-peak amplitude of the drain to source waveform (V_{max} or A) is twice the supply rail (V_{IN}).

$$c_n = A \times \frac{2 \times (t_{ON})}{T} \times \left[\frac{\sin\left\{\frac{n \times \pi \times t_R}{T}\right\}}{\frac{n \times \pi \times t_R}{T}} \right] \times \left[\frac{\sin\left\{\frac{n \times \pi \times (t_{ON})}{T}\right\}}{\frac{n \times \pi \times (t_{ON})}{T}} \right]$$

Since $\sin x / x \sim 1$ if x is very small, we get

$$c_n \approx 2A \times \left[\frac{\sin\left\{\frac{n \times \pi \times (t_{ON})}{T}\right\}}{n \times \pi} \right]$$

So, assuming duty cycle is about 50 percent, the amplitude of the fundamental (first harmonic) is

$$c_1 \cong \frac{2A}{\pi} = \frac{4 \times V_{IN}}{\pi} \quad V$$

We again realize that as for DM noise, from the viewpoint of the noise envelope and the required attenuation, it is the fundamental harmonic (only) which really counts. The current caused by this is

$$I_{cm} = \frac{V_{DS}}{25 - j\frac{T}{2\pi \times C_p}}$$

since the LISN presents an impedance of 25 Ω to I_{cm}. The voltage measured across the 50-Ω resistor of the LISN is due to $I_{cm}/2$ flowing through it. So,

$$V_{cm} = \frac{I_{cm}}{2} \times 50 = I_{cm} \times 25 \quad V$$

Simplifying

$$V_{cm} = \frac{4 \times V_{IN}}{\pi - \frac{j}{50 \times C_p \times f_{SW}}} \quad V$$

$$|V_{cm}| = \frac{200 \times V_{IN} \times C_p \times f_{SW}}{\sqrt{(50\pi \times C_p \times f_{SW})^2 + 1}} \quad V$$

In terms of dBµV this is

$$V_{LISN\_DM\_NOFILTER} = 20 \log\left(\frac{|V_{cm}|}{10^{-6}}\right) = 120 + 20 \log(|V_{cm}|) \text{ dBµV}$$

So, if, for example, $V_{IN} = 100$ V ($A = 200$ V), $C_p = 200$ pF, $f_{SW} = 100$ kHz, we get $V_{cm} = 0.4$ V or 112 dBµV (for the first harmonic). We can follow a similar procedure as for the DM filter to calculate the LC of the common-mode filter. Thereby we can calculate L_{cm}, and the net Y-capacitance (line to earth).

We can make the following key observations with regard to common-mode noise as borne out of an even more detailed computation:

- The envelope is flat and fixed at 100 $V_{max} C_p f_{SW}$. At the break point described by $f = 1/(\pi t_{RISE})$, the envelope rolls off at 20 dB per decade.
- The pedestal (flat part) does *not* depend on rise or fall times *contrary to popular perception*. So, the envelope does change, but not at the low-frequency end. For EMI

purposes it is that end which is our starting point for the filter design. So, any subsequent roll-off is not going to affect the filter design.

Since the pedestal of the common-mode noise envelope is independent of the rise and fall times, does this mean that it doesn't matter how fast we turn ON and OFF the FET? Yes, it doesn't. But see the last paragraph on DM filter design.

In Chap. 17 we have provided reasons why the amount of Y-capacitance we can use is restricted due to IEC safety regulations. Since C is necessarily small, L has to be much larger for CM filter stages. For DM filters, we usually put several X-caps in parallel (each of which is usually a maximum of 0.22 µF, that being based only on various manufacturing constraints). So the L for a DM stage can in turn be made much smaller. The usual practice for low to medium power off-line converters is *not* to use any actual choke for the DM filtering. Instead, typically, we just use one (sometimes two) standard *common-mode chokes* in which the *leakage inductance* between the two windings provides the small amount of necessary DM filter inductance. Suppose L_{cm} is the inductance of *each* winding of the CM choke, and L_k the leakage inductance present in *each* limb of the CM choke, then effectively, the LC filter inductance of the CM stage is L_{cm}, and the LC filter inductance of the DM stage is $2 \times L_k$. If the line-to-line capacitance is C_x, then the effective LC filter capacitance of the DM stage is C_x. If we have two Y-caps (each connected between one of the two line inputs and earth), and each has a capacitance C_y, then the effective LC filter capacitance of the CM stage is $2 \times C_y$. We can thus relate our theoretical calculations above to a real-life filter.

We should be aware of some key factors impacting our filter design in the low-frequency region (150 to 500 kHz).

- We know that the limit lines allow for progressively *higher emissions* below 500 kHz.
- But the sensitivity of the LISN *decreases* as the frequency falls off. This effectively allows us *more noise*. Roughly, we can say that the LISN impedance falls from 50 Ω at about 500 kHz to about 5 Ω at very low frequencies, at an approximate rate of 10 dB per decade below 500 kHz.
- However, we note that the EMI filter becomes less effective at low frequencies, since it is naturally a low-pass type. Typically, its attenuation rolls off at the rate of 40 dB per decade.

Let us see what all this nets us. Suppose we have, by suitable design, achieved compliance at the lowest frequency. If the switching frequency is less than 150 kHz, it would mean that we have about 2 mV (66 dBµV) of noise emissions at 150 kHz (Class B). Now let us go in the *reverse direction*, that is, from low to high frequency. This is what we find:

1. The LISN sensitivity *increases* (at the rate of ~ 10 dB per decade). So we would start getting higher and higher noise readings.
2. But the EMI filter starts becoming more and more effective, attenuating the signal at a typical rate of 40 dB per decade.
3. This swamps out the increasing LISN sensitivity, and so our measured noise actually *falls* at the rate of 40 − 10 = 30 dB per decade.
4. But the limit lines are asking us to decrease the noise level at the rate of only 20 dB per decade.

Therefore the measured noise level continues to fall below the limit lines with an *increasing headroom* of 30 − 20 = 10 dB per decade. This is why we try to go about first trying to achieve compliance at the lowest frequency, since then we usually have automatic compliance at higher frequencies (unless EMI spikes due to parasitic resonances, inadequate grounding and/or radiation problems are present).

CHAPTER 16

Reset Techniques in Flyback and Forward Converters

The first thing we need to understand is that *reset* techniques may look similar from a circuital point of view and appearance, but the motivation behind them is quite different in the case of a Flyback converter versus a Forward converter, as explained here.

PART 1   FLYBACK CONVERTER TRANSFORMER RESET (OF LEAKAGE INDUCTANCE)

In this section we look at the zener and RCD clamps, commonly used in Flyback converters.

Zener Clamp

See Fig. 16.1. In a Flyback, the magnetization current (its associated energy) is delivered to the output during the OFF-time by virtue of the polarity of the secondary winding. The only component of primary-side energy that is not linked to the Secondary is the leakage inductance energy. There is *no place for it to go*. But the leakage inductance has significant energy in it, prior to turning the FET OFF. We know that we cannot ever afford to interrupt the current in an inductor without providing an alternative path for it to flow in; otherwise we will get a huge inductive voltage spike. So the simplest technique used in a *cheap-and-dirty* Flyback, as it is often unfairly called, is the zener clamp.

What is the energy that gets delivered into the clamp? It is *not* just the energy associated with the leakage, as sometimes stated in literature. In other words (f being the switching frequency), this equation is *not* true:

$$P_{CLAMP} = \frac{1}{2} \times L_{LK} \times I_{PRI\_PK}^2 \times f \quad \text{Not true!}$$

The reason this is not true is that until the leakage inductance is *reset* (i.e., it reaches zero current), it forces current into the zener clamp (the alternative path to flow in as provided by us). Unfortunately, the leakage inductance L_{LK} is in series with the magnetization inductance L_{MAG} (i.e., the primary inductance L_{PRI}), so some energy is expended by the input source in moving current across the voltage gradient present across L_{MAG}, and this energy is an additional term that finds its way into the zener clamp. When we do the math, we actually get two terms for dissipation in the clamp:

$$P_{CLAMP} = \frac{1}{2} \times L_{LK} \times I_{PRI\_PK}^2 \times f + \frac{1}{2} \times L_{LK} \times I_{PRI\_PK}^2 \times f \times \frac{V_{OR}}{V_Z - V_{OR}}$$

where V_Z is the zener clamp voltage and V_{OR} is the reflected output voltage ($V_{OR} = V_O/n$, where n is the turns ratio N_P/N_S). Simplifying, we get

$$P_{CLAMP} = \frac{1}{2} \times L_{LK} \times I_{PRI\_PK}^2 \times f \times \frac{V_Z}{V_Z - V_{OR}}$$

This is the same as writing more generically (for clamps)

$$P_{CLAMP} = \frac{1}{2} \times L_{LK} \times I_{PRI\_PK}^2 \times f \times \frac{V_{CLAMP}}{V_{CLAMP} - V_{OR}}$$

In any case, we are simply burning up the energy here, and this is not really good for efficiency.

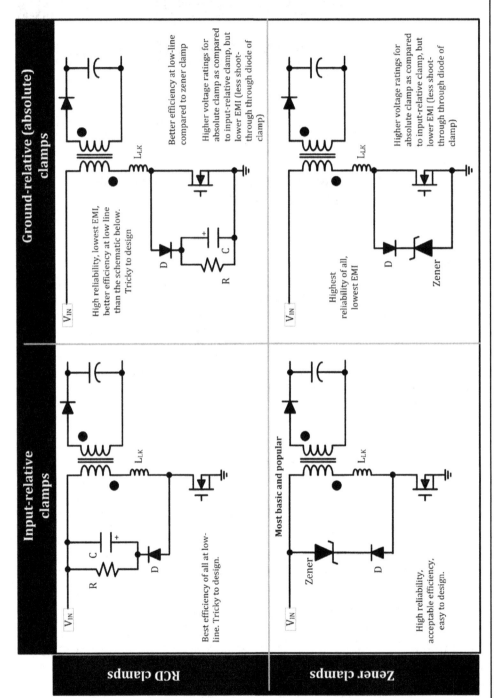

Figure 16.1 Types of RCD and zener clamps.

RCD Clamp

See Fig. 16.2. Engineers often use the RCD clamp instead of the zener clamp, thinking it somehow (magically) improves efficiency. But the fact is that if the input voltage is *almost fixed*, there is *no difference in the dissipation* compared to a zener clamp. We can't fool physics!

When we design the RCD clamp, by a suitable choice of the R of the RCD, we actually (must) ensure that the capacitor charges up to exactly the same value as a zener clamp in that position. Because ultimately, the intention of any clamp is to protect the semiconductor, the FET in this case. So with a certain stated absolute maximum of V_{DS}, and a certain design derating target, we will set the clamping level, either by a zener clamp or an RCD clamp, to *exactly the same level*, just to protect the switch. Hence, the dissipation will be the same.

We should also not set the clamp level lower than just about necessary, because we see from the preceding equations for clamps, applicable to the RCD clamp too (by replacing V_Z with V_{CLAMP}), that there is a term $V_Z - V_{OR}$ (or $V_{CLAMP} - V_{OR}$) in the denominator, and if we decrease V_{CLAMP} and bring it closer to V_{OR}, we will get a huge increase in dissipation. So we need to set V_Z (or V_{CLAMP}) as high as possible *usually*. Why usually? Because there are known adverse implications on EMI and also on *cross-regulation in multi-output Flyback converters* by setting the clamping level too high.

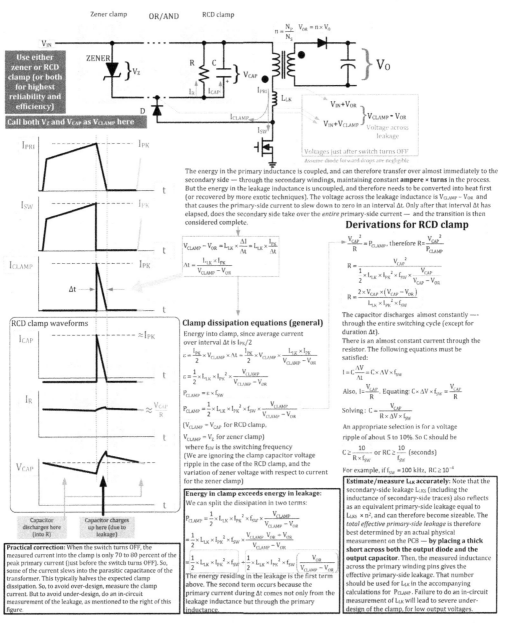

FIGURE 16.2 RCD and zener clamps analyzed (in Flyback converters).

The advantage of the RCD clamp emerges in wide-input converters. Because in a typical Flyback, the peak switch current is the lowest at V_{INMAX}, which is the point at which we will actually choose V_Z of a zener clamp, or set V_{CLAMP} in an RCD clamp to the safe voltage rating of the FET. However, when we lower the input, a zener clamp still clamps at the same level, so $V_Z - V_{OR}$ does not change (remember that V_{OR} is fixed by regulation too). However, because of the higher peak currents, the C of an alternative RCD clamp charges up to a higher level at low-line. In other words, we actually have a low-line clamping level V_{CLAMP_LO}, which is higher than the clamping level at high-line V_{CLAMP_HI} (the latter we fixed to be the same as V_Z to protect the semiconductor at high-line). So because the clamping level moves higher at low-line, $V_{CLAMP} - V_{OR}$ is not fixed in an RCD clamp, unlike in a zener clamp, and *increases at low-line*. From the preceding dissipation equation, we see that this will help lower the clamp dissipation—but lower, not as compared to the dissipation in the clamp at high-line, but as compared to a zener clamp in the same position and application. This means that an RCD clamp, for example, in a universal input AC-DC application, will give roughly 20 percent less dissipation in the clamp and an improved overall low-line (worst case) efficiency by typically 2 to 3 percent. For a detailed line variation calculation, see page 306 of *Switching Power Supplies A-Z*, 2d ed.

Lossless Snubbers

See Fig. 16.3. Here we are doing two things: (1) trying to use the energy in the leakage, say to power our control IC (will need a zener though, not shown, to prevent cumulative buildup due to the energy difference term, and (2) trying to store the energy in an inductor and cycle it back to the input rail during the OFF-time. One way is to try to combine techniques as also shown. This was first reported perhaps in the article, "Lossless Snubber Circuit in Flyback Converter and Its Utilization for a Low Operating Voltage" by In-Hwan Oh, *Fairchild Semiconductor*, July 28, 2005.

All such implementations will require a lot of bench optimization, and no straightforward equations seem to be available in related literature. In fact, in the case of the simple inductor-based snubber, this is what Motorola/On-Semi's *Switchmode Handbook* has to say:

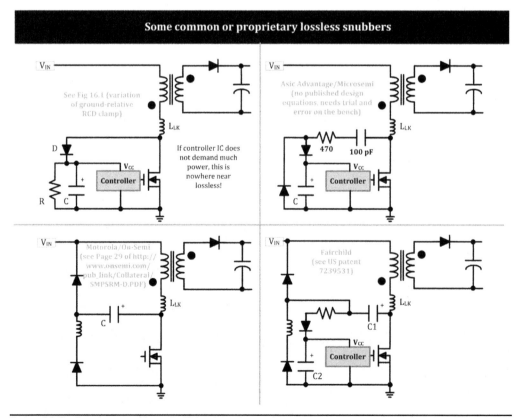

Figure 16.3 Lossless snubbers (in Flyback converters).

A lossless snubber is a snubber whose trapped energy is recovered by the power circuit. The lossless snubber is designed to absorb a fixed amount of energy from the transition of a switched AC voltage node. This energy is stored in a capacitor whose size dictates how much energy the snubber can absorb ... The design for a lossless snubber varies from topology to topology and for each desired transition. Some adaptation may be necessary for each circuit. The important factors in the design of a lossless snubber are: 1. The snubber must have initial conditions that allow it to operate during the desired transition and at the desired voltages. Lossless snubbers should be emptied of their energy prior to the desired transition. The voltage to which it is reset dictates where the snubber will begin to operate. So if the snubber is reset to the input voltage, then it will act as a lossless clamp which will remove any spikes above the input voltage. 2. When the lossless snubber is "reset," the energy should be returned to the input capacitor or back into the output power path ... Returning the energy to the input capacitor allows the supply to use the energy again on the next cycle. Returning the energy to ground in a boost-mode supply does not return the energy for reuse, but acts as a shunt current path around the power switch. Sometimes additional transformer windings are used. 3. The reset current waveform should be band limited with a series inductor to prevent additional EMI from being generated. Use of a 2 to 3 turn spiral PCB inductor is sufficient to greatly lower the di/dt of the energy exiting the lossless snubber.

Keep in mind, no equations were provided here either! So we will bypass this approach too.

PART 2 FORWARD CONVERTER TRANSFORMER RESET (OF MAGNETIZATION INDUCTANCE)

Introduction

There is a steady buildup of current (and stored energy) every cycle within the magnetization inductance of any transformer, just as in any inductor. It is this energy that gets delivered to the secondary side in a Flyback topology. But in a Flyback, that transfer of energy event could occur only because the secondary winding's polarity was such that it conducted when the primary winding was turned OFF. We remember that in a (single-winding) inductor, we cannot force sudden discontinuities in the amperes passing through its winding, or we will destroy the semiconductors due to huge resulting inductive voltage spikes. Very similarly, in a multiwinding structure (i.e., a transformer), we have to maintain the *net* ampere × turns (proportional to net flux in the shared core)—summed over all the windings placed on the core. No sudden discontinuities are allowed in the *net* ampere × turns, or once again, we will similarly destroy the semiconductors. But we can certainly turn OFF one winding and let the other take up the ampere-turns relinquished (immediately)—provided the following conditions hold: (1) the polarity (with the connected diode considered) is such that it allows current to actually flow in the other winding, and (2) the two windings are fully coupled, i.e., they share "space" in the same core, in terms of their flux. In a Flyback, the net ampere-turns lost (relinquished) by the primary winding when the switch turns OFF is immediately picked up by the secondary winding, because the polarity allows it—so condition 1 is satisfied. Further, since $N_P \times I_P$ must equal $N_S \times I_S$ (no change in net ampere-turns allowed), we get $I_S = I_P \times n$, where n is the turns ratio N_P/N_S. This is *transformer action*, Flyback-style. Note that in the process we have managed to transfer magnetization energy across the isolation boundary. The only inductance in a Flyback that does not *share space* on the same core is the leakage inductance. Its flux is primarily through air, not the core. Condition 2 is therefore not being satisfied *for the leakage*. This inductance is consequently considered *unlinked* to the Secondary, and we therefore need to allow the current through to freewheel and then either burn the associated energy up in a clamp or try to recover it using a lossless snubber, as explained earlier. Because if we do not *reset* it (i.e., bring the core to the condition it was in at the start of the cycle), there would be an eventual buildup of energy in the core and clamp, and sooner or later, some component will fail.

In a Forward converter, even condition 1 remains unsatisfied—for the magnetization inductance (due to the polarity of the windings) and also for the leakage (primarily because of condition 2). The good news is that whatever we do to account for the magnetization energy, and find some solution for, works for the leakage inductance too, since they are effectively both in series, and, in this case, both are equally unlinked to the Secondary. We can consider them part of the same problem, in fact the same problem.

The magnetization energy is much larger than the leakage energy. Because, typically, leakage inductance is only 1 percent of the magnetizing inductance in a transformer.

We can try RCD clamps or zener clamps for a Forward converter too, but clearly the impact on efficiency will be unacceptable. The conventional way to handle this is to use a third *energy recovery* winding. This thin *tertiary*, or *third* or *reset*, winding is tightly coupled to the primary winding (usually wound together, i.e., "bifilar") to minimize leakage between the primary winding and the energy recovery winding, so that the energy recovery winding can take up all the ampere × turns relinquished by the primary winding whenever the switch turns OFF. The diode direction and polarity of the winding is such that the winding conducts just like a (very tightly coupled) Flyback winding. Since this new winding typically connects through a diode into the input capacitor (of the input rail), therefore, during the OFF-time, it places exactly V_{IN} across the primary winding, but as expected, this voltage is in an opposite direction to the V_{IN} impressed across the same winding during the ON-time. So the magnetization current component of the primary-side current ramps up with a slope of V_{IN}/L_{MAG} during the ON-time and ramps down with exactly $-V_{IN}/L_{MAG}$ slope during the OFF-time. To ensure it can complete the ramp down to zero (and thus get *reset*), we have to allocate enough time. Suppose the ON-time is 60 percent of the time period (T), and the OFF-time is only 40 percent, there will be a bigger ΔI (increase) during the ON-time and a smaller ΔI (decrease) during the OFF-time. So we will end the cycle with a net *increment* in current. The core would not be *reset* and there will be staircasing every cycle until destruction occurs. Therefore, we must allocate at least 50 percent of the time period as T_{OFF} just to ensure reset. This time symmetry is purely because of the equal slopes mentioned earlier. It is for this reason that all controller ICs intended for single-ended Forward converters are limited to 50 percent duty cycle (to leave margin, they may actually be set to a nominal of 0.47 ± 0.03 duty cycle, that is, 0.44 to 0.5 duty cycle. So the minimum of the duty cycle limit should be taken as 44 percent for design purposes. In particular, we must ensure this maximum duty cycle limit (brickwall) is not "hit" at the lowest input voltage at which we want to have a regulated output. Otherwise the output will start to droop. In both the Forward and Flyback converters, since we do not usually have any control of the maximum duty cycle of the converter, we need to adjust our turns ratio very carefully for this very reason. *The turns ratio is invariably determined by the duty cycle limit of the controller and the lowest input voltage under consideration.*

This 50 percent duty cycle limit is conducive to also eliminating subharmonic instability in current-mode control (with continuous conduction mode), and therefore turns out to be useful even in Flyback controllers (using current-mode control).

General Tertiary Winding

Yes, in the case of the single-ended Forward converter with tertiary winding, we can set the duty cycle limit of the controller to say $0.75 \times T$ (where T is the time period, i.e., $1/f$) and *still ensure transformer reset occurs*. How do we do that? By changing the turns ratio between the primary and the tertiary winding. Note that we can't change either the ON-time or the OFF-time voltages across the primary winding, so we can't change the slope of the up and down magnetization current ramps either. But we can change the *number of turns*, as shown in Fig. 16.4. We have shown a more general case here. But to understand it easily, think of a specific case if we halve the windings in the energy recovery winding. We can show this allows the current to return to zero in a very short time indeed, in fact, in much less than the remaining $0.25 \times T$. This is how that happens: The initial value of the current in the energy recovery winding (i.e., just as the FET turns OFF) is now double what it was, because we have halved the number of turns of the energy-recovery winding, and since ampere × turns needed to be maintained across the switch transition. So the current is actually starting its down-ramp at a much higher value. We first think it will take even longer to return to zero. But the opposite happens. Because, when we halved the turns of the tertiary winding, the inductance associated with this winding dropped to *one-fourth* of its value—since L is proportional to N^2. So now, the down-slope is four times greater. Therefore, even though the peak current doubled, because of this much higher down-slope, the current will return to zero in less than $0.25 \times T$. Transformer reset will still occur. The general relationship helps us to correctly set the turns ratio n_{RESET}, based on the maximum duty cycle of the controller (not the other way around!).

$$\frac{T_{ON}}{T_R} = n_R, \quad \text{where } n_R = \frac{N_P}{N_T}$$

Reset Techniques in Flyback and Forward Converters

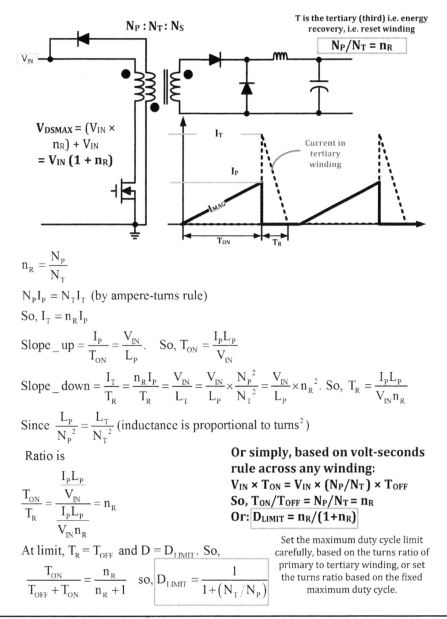

FIGURE 16.4 General variation of the tertiary (third) winding turns ratio in a Forward converter.

where T_R is the reset time (which should be less than T_{OFF} or staircasing will occur). The maximum duty cycle of the controller corresponds to $T_R = T_{OFF}$ and must be, therefore,

$$D_{LIMIT} = \frac{n_R}{n_R + 1}$$

The voltage stress on the FET is

$$V_{DSMAX} = V_{INMAX} \times (n_R + 1)$$

So for the case of $D_{LIMIT} = 0.5$, $n_R = 1$. If we want to go to a higher duty cycle, we can solve the preceding equations for n_R:

$$n_R = \frac{D_{LIMIT}}{1 - D_{LIMIT}}$$

Plugging in, say, $D = 0.67$, we get

$$n_R = \frac{D_{LIMIT}}{1 - D_{LIMIT}} = \frac{0.67}{1 - 0.67} = 2$$

In other words, the tertiary winding is half the number of turns of the primary winding.

During the OFF-time the voltage across the tertiary winding V_{TER} equals V_{IN}. This voltage reflects to the primary winding as per the volts per turn rule, so we get $V_{PRI}/N_P = V_{TER}/N_T$ where $V_{TER} = V_{IN}$. Solving, $V_{PRI} = V_{IN} \times n_R$. This voltage further reflects onto the secondary winding as per volts per turn rule too. So we get $V_{PRI}/N_P = V_{SEC}/N_S$. Solving, $V_{SEC} = V_{PRI}/n = V_{IN} \times n_R/n$ where $n = N_P/N_S$. In the OFF-time the voltage sees this reflected secondary voltage plus the voltage on its anode (equal to V_O).

We can therefore conclude that

1. The maximum voltage across the FET is $V_{PRI} + V_{IN} = V_{IN} \times n_R + V_{IN} = V_{IN} \times (n_R + 1)$. Check this at maximum input.

2. The maximum voltage on the output diode is $V_{IN} \times (n_R/n) + V_O$. Check this at maximum input.

So, if we decrease the number of turns on the tertiary winding from its starting value (equal to N_P turns), n_R will increase, so the maximum voltage on the FET will increase. From the preceding equations, the voltage stress on the output diode increases too. If all the stresses are more, why should we even bother with this different tertiary winding turns ratio? The reason is that *since the turns ratio is determined by the maximum duty cycle limit in general*, we are now in a position to change the turns ratio too. The turns ratio is always fixed at the minimum input voltage, since that corresponds to the maximum duty cycle D_{MAX} demanded by the control loop. We thus have:

$$\text{Set } D_{MAX} = D_{LIMIT} \Rightarrow \frac{V_{OR}}{V_{INMIN}} = \frac{V_O \times n}{V_{INMIN}} = \frac{n_R}{n_R + 1}$$

where n is N_P/N_S as usual. Solving

$$n = \frac{n_R V_{INMIN}}{V_O(n_R + 1)}$$

For example, consider the case of a 5-V output from a 20-V to 30-V input (nominal 25 V). If n_R was set to 1 (as is the usual case in conventional single-ended Forward converters), we know that the maximum duty cycle (theoretically) is 50 percent. So if we want the Buck stage that follows to run at a maximum duty cycle of 50 percent, and the desired output to be 5 V, we must have V_{INMINR} (the reflected input voltage at minimum input) to be 10 V. Then we get the duty cycle as 5 V/10 V = 0.5. Therefore, if the lowest input was 20 V, we would use a turns ratio of $V_{INMIN}/V_{INMINR} = 20/10 = 2$. This is what the preceding equation also tells us:

$$n = \frac{n_R V_{INMIN}}{V_O(n_R + 1)} = \frac{1 \times 20}{5 \times (1+1)} = 2$$

But if n_R goes from 1 to 2 (halving the number of tertiary winding turns), we get

$$n = \frac{n_R V_{INMIN}}{V_O(n_R + 1)} = \frac{2 \times 20}{5(2+1)} = 2.67$$

So the Primary-to-Secondary turns ratio n has increased too!

The voltage across the FET is greater because so is n_R,

$$V_{DSMAX} = V_{INMAX} \times (n_R + 1) = 30 \times (2+1) = 90 \text{ V}$$

If the turns ratio of the reset winding was 1, we would have gotten only

$$V_{DSMAX} = V_{INMAX} \times (n_{RESET} + 1) = 30 \times (1+1) = 60 \text{ V}$$

What about the diode voltage now? This equals $V_{IN} \times (n_R/n) + V_O$. In our case, n_R is increasing, but so is n. Which one dominates? We now have $n_R/n = 2/2.67 = 0.74$. If n_R was 1, we would have gotten the diode stress as $(V_{IN} \times n) + V_O$ with $n = 2$, i.e., $(0.5 \times V_{IN}) + V_O$. Now we have $(0.74 \times V_{IN}) + V_O$. So the diode voltage has gone up too. Table 16.1 summarizes the equations for a general tertiary winding case.

Note that the asymmetric half-bridge (two-switch) Forward converter is identical to the single-ended Forward with a 1:1 ($n_R = 1$) tertiary winding, from the viewpoint of the Secondary. But on the primary side, the voltage stays clamped to the input rail (V_{IN}).

Design table for third (energy recovery) winding with arbitrary turns ratio	
Turns ratio N_P/N_S	n
Turns ratio N_P/N_T	n_R
Reset time	T_R
Reset time T_R	$\dfrac{T_{ON}}{n_R}$
Duty cycle	$D = \dfrac{nV_O}{V_{IN}}$
Maximum controller duty cycle limit to ensure no staircasing D_{LIMIT}	$\dfrac{n_R}{n_R + 1}$
Recommended turns ratio n	$n = \dfrac{n_R V_{INMIN}}{V_O(n_R + 1)}$
Voltage stress on FET	$V_{DSMAX} = V_{INMAX} \times (n_R + 1)$
Voltage stress on output diode	$V_{DMAX} = V_{INMAX} \times \left(\dfrac{n_R}{n}\right) + V_O$

TABLE 16.1 Design Table for Forward Converter with Tertiary Winding Using Arbitrary Turns Ratio

Active Clamp Reset

Before we discuss this, we may need to understand synchronous topologies a little better.

1. They are always in forced continuous conduction mode (FCCM) since the current is allowed to reverse direction in the inductor. The average of the inductor current waveform (its center of ramp value) equals the load current. So if the load current is zero, the converter continues passing current back and forth, charging up the output capacitor and then discharging it back into the input capacitor, and so on. This is constantly circulating energy.

2. The Boost and the Buck-boost are not much different actually, as shown in Fig. 16.5. The difference is that one uses a capacitor on the low-side and one on the high-side. This leads to the oft-quoted difference in their DC transfer functions:

$$V_{O\_BOOST} = V_{IN} \times \frac{1}{1 - D_{BOOST}}$$

$$V_{O\_BUCK\text{-}BOOST} = V_{IN} \times \frac{D_{BUCK\text{-}BOOST}}{1 - D_{BUCK\text{-}BOOST}}$$

Now looking at Fig. 16.6, we see that the *high-side* active clamp is just a *parasitic synchronous Buck-boost*. From Fig. 16.7, we see that the *low-side* active clamp is just a *parasitic synchronous Boost*. In both cases, these are not *regulated* converters but are instead *slaved* at the duty cycle determined by V_{IN} and V_O (by the main switching stage). So the voltage on the clamp capacitors can vary significantly, and it is obvious that we will get the following steady voltages on the clamping capacitors:

$$V_{CLAMP\_LO} = V_{IN} \times \frac{1}{1 - D}$$

$$V_{CLAMP\_HI} = V_{IN} \times \frac{D}{1 - D}$$

The detailed calculations and waveforms for the high-side active clamp are shown in Fig. 16.8 as an example. The design table (derivations are tedious but obvious) is presented in Table 16.2. Note our terminology here:

LAC stands for low-side active clamp.

HAC stands for high-side active clamp.

ERW stands for energy recovery winding case.

Positive-to-positive Boost OR Negative-to-positive Buck-boost

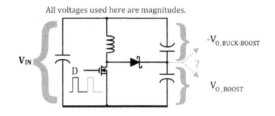

All voltages used here are magnitudes.

Start with DC transfer function of a Boost topology:

$$\frac{V_{O\_BOOST}}{V_{IN}} = \frac{1}{1-D}$$

Using $V_{O\_BOOST} - V_{IN} = V_{O\_BUCK-BOOST}$ (from above schematic)

$$V_{O\_BUCK-BOOST} + V_{IN} = \frac{V_{IN}}{1-D}$$

Simplifying,

$$\frac{V_{O\_BUCK-BOOST}}{V_{IN}} = \frac{D}{1-D}$$

We get the DC transfer function of a Buck-boost topology, starting from a Boost topology! That is by just drawing power from across either of the two caps shown above.

FIGURE 16.5 Boost and Buck-boost are very similar.

Analysis and Conclusions (Figs. 16.9 to 16.11)

Note that for a given turns ratio n, the curve for the active clamp (high-side or low-side, HAC or LAC) intersects the energy recovery winding (ERW, 1:1) case at exactly $D = 0.5$ (at this point the duty cycle is for active clamp or energy recovery winding cases). The ERW curve does not go to lower inputs because the duty cycle limit of the controller is 50 percent.

It is commonly stated in literature that the advantage of the active clamp is it "allows" us to go to duty cycles greater than 50 percent. It certainly does, but as we see in Fig. 16.9, there is no point in doing that. The FET voltage stress goes up significantly above $D = 0.5$, as does the diode stress as per Fig. 16.10. The reason is obvious: If we apply V_{IN} across the primary winding for larger T_{ON}, the voltage across the Primary during the OFF-time (its so-called reset voltage) must go up, as also indicated in Fig. 16.11. This is because *for any winding on a transformer the volt-seconds law must be satisfied (in steady state)*. Again there is no escaping physics here.

So though the active clamp enables us to go to $D > 0.5$, *we should not*. In fact it is advisable to firmly clamp the duty cycle to 50 percent (if not lesser), just as in the case of the conventional Forward converter with a 1:1 energy recovery winding.

So how do we benefit? Assuming the turns ratio is still the same, we see from Fig. 16.9 that at duty cycles less than 50 percent (higher input voltages than their intersection, and corresponding to the shaded area), the ERW curve and the HAC/LAC curve diverge, so the active clamp helps lower the voltage stress. This is because the clamp voltage is determined no longer by us, or the Forward converter per se, but by whatever output voltage the (active clamp) Boost or Buck-boost develops, when slaved to the master duty cycle (based on the Buck stage of the Forward converter). It just so happens that the "output rail" (clamping capacitor voltage) is less than V_{IN} (in the case of the parasitic Buck-boost, i.e., high-side active clamp) or $2 \times V_{IN}$ (in the case of the parasitic Boost, i.e., low-side active clamp).

The active clamp therefore reduces the voltage stresses (FET and diode) as compared to the energy recovery winding, for $D < 0.5$. The down-side is that since the active clamp is sloshing current back and forth continuously, as in the case of any synchronous topology at almost zero load, the circulating current is relatively higher in an active clamp case, compared to the energy recovery winding case.

Figures 16.9 and 16.10 also tell us that if for a given input and output voltage, and for duty cycles less than 50 percent, a further reduction in voltage stress occurs if we lower the

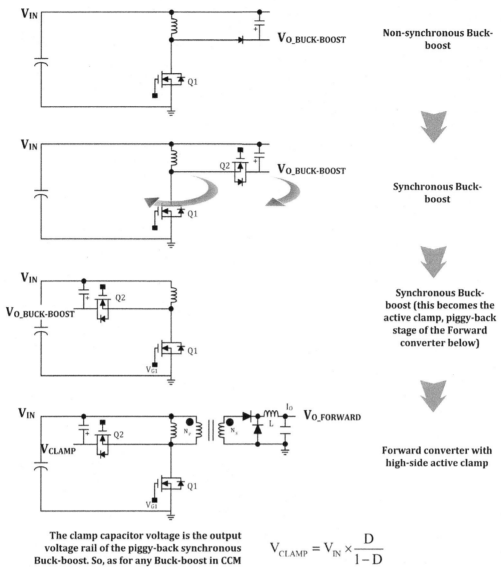

FIGURE 16.6 The *high-side* active clamp is just a parasitic (slave) synchronous Buck-boost.

turns ratio (not increase it as sometimes stated in literature). The physical reason for this is easy to understand. Because by lowering the turns ratio n, we are making the reflected input voltage V_{INR} bigger. So the duty cycle $D = V_O/V_{INR}$ will decrease. Then, since T_{ON} of the main FET is now less, the input voltage V_{IN} applied across the Primary is present for a smaller time. Then by the volt-seconds law, the voltage on the Primary during the OFF-time (the reset voltage, V_{RESET}) will be less. This is what gets reflected to the output diodes as $V_{RESET}/n + V_O$. So it tends to decrease. Though, if n had decreased faster than V_{RESET} had decreased, the diode stress would go up! But we see from Fig. 16.10 that this does not happen. So we can conclude that *lowering the turns ratio would benefit us from the viewpoint of diode and FET stresses*. The down-side is that because of the smaller duty cycle, we will tend to run into discontinuous conduction mode (DCM) in the Buck stage for higher than light loads, and we may therefore need to increase the inductance of the output choke, for example. Also, as we lower the input, we may more readily hit the often unspecified but real D_{MIN} limit of the controller IC, especially at lighter loads.

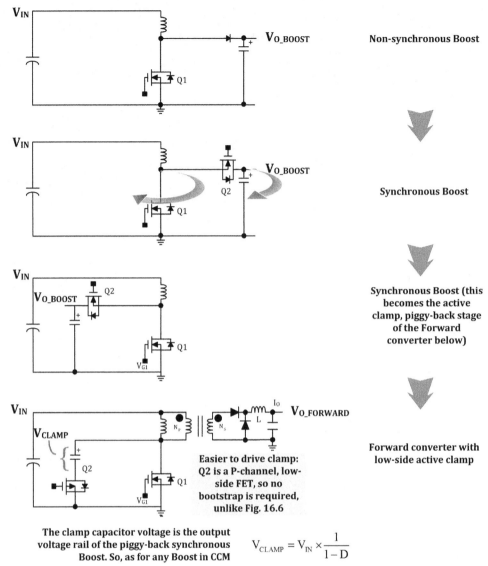

$$V_{CLAMP} = V_{IN} \times \frac{1}{1-D}$$

Conclusion: The low-side active clamp is in effect, simply a synchronous Boost, operating at almost zero average load current (since no load is physically connected across the clamp capacitor, which happens to be the output capacitor of the embedded/hidden Boost). So, energy constantly cycles back and forth between the clamp capacitor and the input rail. Note that the output of this embedded Boost stage (active clamp) is not being regulated to any set value, rather, it is being driven with the duty cycle of the Forward converter D = V_O/V_{IN} (to keep the output V_O constant). So the clamp voltage will vary with D. But like any Boost, the highest output (clamp capacitor) voltages will result at the maximum duty cycle limit of the controller.

FIGURE 16.7 The *low-side* active clamp is just a parasitic (slave) synchronous Boost.

Which Is Better: High-Side or Low-Side Active Clamp?

Note that from the upper half of Fig. 16.10, we see that the low-side active clamp has much higher voltage stresses. Diode stresses and reset voltages are compared for all the cases, in Figs. 16.10 and 16.11. The high stresses of the low-side active clamp compared to the high-side version is actually analogous to the situation that arose when we compared the low-side (absolute) RCD and zener clamps, with their high-side versions, as shown in Fig. 16.1. So, the high-side clamp has lower stresses. Other than that, the voltage and diode stresses on the Forward converter components and the transformer windings are all identical. High-side or low-side does not matter there.

One problem here, especially in high-voltage applications, is that the low-side clamp position is most suited to using a P-channel FET. But this is also the position in which we are getting the highest voltage stresses. It may be difficult to even find P-channel FETs with ratings greater than 400 V.

The high-side clamp is suited for an N-channel FET, but we will need a Gate driver transformer to translate the voltage to its source (its Gate reference level). On the other hand,

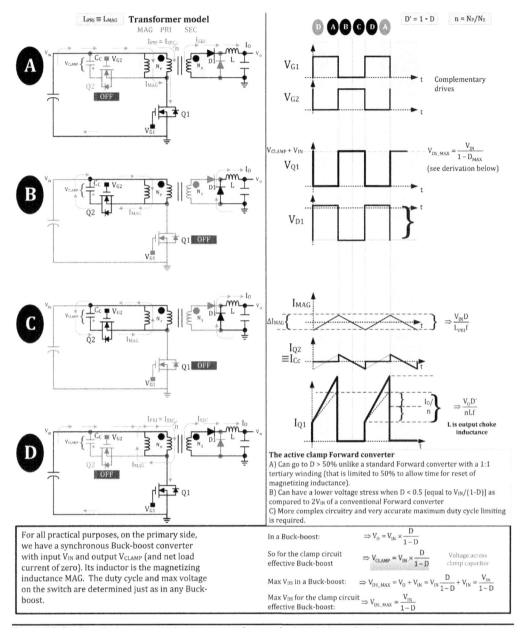

FIGURE 16.8 Detailed current and voltage waveforms of a high-side active clamp.

the low-side clamp uses a P-channel FET and can be driven with a simple capacitor coupling circuit directly from the controller IC. This translates the signal to provide negative pulses to turn ON the P-channel FET.

The polarity of the voltages must be kept in mind. To turn ON a P-channel FET, we need the controller IC to produce a LOW level when the Gate of the main FET goes LOW. Whereas to drive the N-channel FET, it needs to produce a HIGH pulse when the Gate of the main FET goes low. Simple buffer or inverting circuits, derived from the main Gate pin signal, do not work well because we need a certain deadtime before we turn on any active clamp FET after the main FET turns OFF, and similarly before we turn it OFF, a little before the main FET turns ON. So, for example, a simple complementary signal from the main FET drive pin is not enough to drive the N-channel FET of the high-side clamp: We definitely need deadtime. So a dedicated pin is usually provided on a control IC for that. It may also feature an inverted (but similarly deadtime gated) signal for driving the P-channel FET of a low-side active clamp. For the high-side active clamp, *do not use a floating driver IC!* This usually has hundreds of nanoseconds of propagation delay, and that will annihilate any deadtime planning and will thereby cause shoot-through in the main FET and active clamp FET, destroying either or both (if one FET goes, the other will follow very shortly!).

Low-side clamp	High-side clamp	Tertiary (Energy-Recovery) Winding 1:1
\multicolumn{3}{c}{$D = \dfrac{V_O}{V_{IN}} \times n$ (where $n = N_P/N_S$) $\quad V_{IN} = V_O n D$}		
$V_{CLAMP\_LO} = V_{IN} \times \dfrac{1}{1-D}$ $= \dfrac{V_O n D}{1-D}$ $= \dfrac{V_{IN}^2}{V_{IN} - n V_O}$ (clamp capacitor average, add typical 10 % for ripple component)	$V_{CLAMP\_HI} = V_{IN} \times \dfrac{D}{1-D}$ $= \dfrac{V_O n D^2}{1-D}$ $= \dfrac{n V_O V_{IN}}{V_{IN} - n V_O}$ (clamp capacitor average, add typical 10 % for ripple component)	Not applicable
$V_{DS} = V_{CLAMP\_LO}$ $V_{DS} = \dfrac{V_O n D}{1-D} \equiv \dfrac{V_{IN}}{1-D}$ $V_{DS} = \dfrac{V_{IN}}{1 - n V_O/V_{IN}} = \dfrac{V_{IN}^2}{V_{IN} - n V_O}$ (Drain-source FET voltage, same as minimum voltage rating of active clamp FET)	$V_{DS} = V_{CLAMP\_HI} + V_{IN}$ $= \dfrac{V_O n D}{D(1-D)} + \dfrac{V_O n}{D}$ $V_{DS} = \dfrac{V_O n D}{1-D} \equiv \dfrac{V_{IN}}{1-D}$ $V_{DS} = \dfrac{V_{IN}}{1 - n V_O/V_{IN}} = \dfrac{V_{IN}^2}{V_{IN} - n V_O}$ (Drain-source FET voltage, same as minimum voltage rating of active clamp FET)	$V_{DS} = 2 V_{IN}$ $V_{DS} = 2 V_O n D$ (Drain-source FET voltage)
\multicolumn{2}{c}{$V_{DS} = \dfrac{V_{IN}}{1-D}$ $\quad V_{DS} = \dfrac{V_O n D}{1-D}$ $\quad V_{DS} = \dfrac{V_{IN}^2}{V_{IN} - n V_O}$ (Drain-source FET voltage)}		$V_{DS} = 2 V_{IN}$ $V_{DS} = 2 V_O n D$
$V_{RESET} = V_{CLAMP\_LO} - V_{IN}$ (voltage across Primary during OFF-time)	$V_{RESET} = V_{CLAMP\_HI}$ (voltage across Primary during OFF-time)	$V_{RESET} = V_{IN}$ (voltage across Primary during OFF-time)
\multicolumn{2}{c}{$V_{RESET} = \dfrac{V_O V_{IN} n}{V_{IN} - n V_O}$ (voltage across Primary during OFF-time)}		$V_{RESET} = V_{IN}$ (voltage across Primary during OFF-time)
$V_D = \dfrac{V_{RESET}}{n} + V_O$ $= \dfrac{V_O}{1-D}$ (diode rating: check at maximum V_{IN})	$V_D = \dfrac{V_{RESET}}{n} + V_O$ $= \dfrac{V_O}{1-D} \times (D^2 - D + 1)$ (diode rating: check at maximum V_{IN})	$V_D = \dfrac{V_{RESET}}{n} + V_O$ $= V_O(1 + D)$ $= \dfrac{V_{IN} + n V_O}{n}$ (diode rating: check at maximum V_{IN})
\multicolumn{2}{c}{$V_D = \dfrac{V_O(2 V_{IN} - n V_O)}{V_{IN} - n V_O}$ (diode rating: check at maximum V_{IN})}		

TABLE 16.2 Design Table for Active Clamp (and Comparison to Conventional Third-Winding Case)

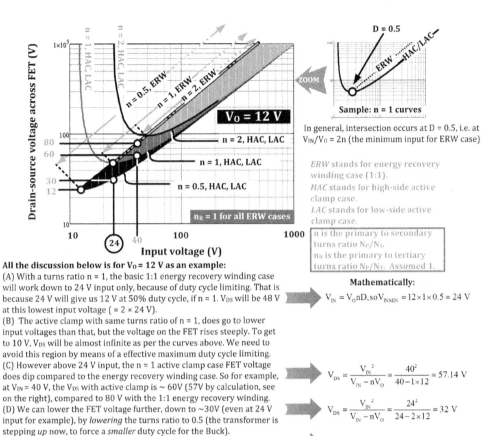

All the discussion below is for $V_O = 12\ V$ as an example:

(A) With a turns ratio n = 1, the basic 1:1 energy recovery winding case will work down to 24 V input only, because of duty cycle limiting. That is because 24 V will give us 12 V at 50% duty cycle, if n = 1. V_{DS} will be 48 V at this lowest input voltage (= 2 × 24 V).

(B) The active clamp with same turns ratio of n = 1, does go to lower input voltages than that, but the voltage on the FET rises steeply. To get to 10 V, V_{DS} will be almost infinite as per the curves above. We need to avoid this region by means of a effective maximum duty cycle limiting.

(C) However above 24 V input, the n = 1 active clamp case FET voltage does dip compared to the energy recovery winding case. So for example, at V_{IN} = 40 V, the V_{DS} with active clamp is ~ 60V (57V by calculation, see on the right), compared to 80 V with the 1:1 energy recovery winding.

(D) We can lower the FET voltage further, down to ~30V (even at 24 V input for example), by *lowering* the turns ratio to 0.5 (the transformer is stepping *up* now, to force a *smaller* duty cycle for the Buck).

(E) By lowering the turns ratio to n = 0.5, we can also go down to 12 V input (with a duty cycle still less than 50%). Above D = 0.5, V_{DS} increases.

Mathematically:

$$V_{IN} = V_O nD, so\ V_{IN\,MIN} = 12 \times 1 \times 0.5 = 24\ V$$

$$V_{DS} = \frac{V_{IN}^2}{V_{IN} - nV_O} = \frac{40^2}{40 - 1 \times 12} = 57.14\ V$$

$$V_{DS} = \frac{V_{IN}^2}{V_{IN} - nV_O} = \frac{24^2}{24 - 2 \times 12} = 32\ V$$

$$V_{IN} = V_O nD, so\ V_{IN\,MIN} = 12 \times 2 \times 0.5 = 12\ V$$

Conclusions about the active clamp:

(1) Though the active clamp does allow you to go to duty cycles greater than 0.5, the FET voltage stress increases dramatically for D > 0.5. So we do *not* want to operate in that region anyway. The purpose of the active clamp is *not* to somehow or the other, just go to higher than 50% duty cycles. That only increases V_{DS}.

(2) The biggest advantage in voltage stresses occurs for duty cycles less than 0.5, when using the active clamp.

(3) Lowering the duty cycle below 0.5, i.e. by decreasing (not increasing) the turns ratio, will help significantly lower FET voltage stress. But if the maximum duty cycle limit of the control IC is still 50% or higher, we still have to pick a FET with a much higher V_{DS} rating, for transient conditions. Even at low D, the RMS switch current waveforms are more peaky. So losses will increase.

It is a systems-level tradeoff. There isn't a sweet spot or optimum turns ratio. In general, despite popular notions, lowering (not increasing) the turns ratio, reducing (not increasing) the duty cycle, will help reduce at least the FET voltage stresses.

FIGURE 16.9 Drain-source voltage on main FET, comparing active clamps and energy recovery winding cases. Also a graphical and mathematical analysis based on these graphs and equations in Table 16.2.

424 Chapter Sixteen

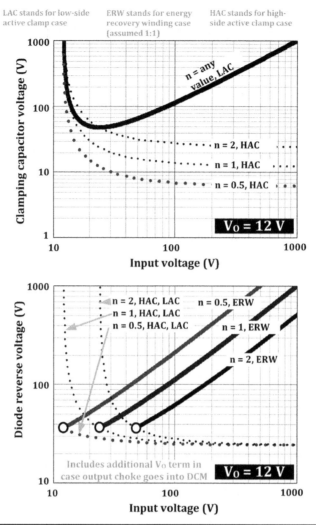

Figure 16.10 Clamping and diode voltage stresses compared.

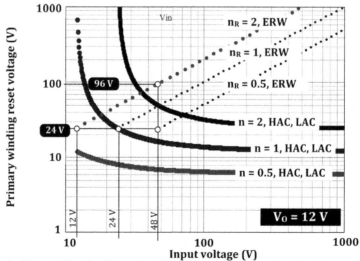

Example: For ERW, $n_R = N_P/N_T = 0.5$, at 48 V input, V_{ON} = 48 V across N_P for a certain amount of time (depending upon D and n). During the OFF-time, N_T is clamped to the input rail of 48 V. But $N_T = 2 \times N_P$. So, N_T reflects 48/2 = 24 V onto N_P during a part of T_{OFF}. By volt-seconds law applied to N_P, the set duty cycle limit must be 0.33. Similarly, for $n_R = 2$, with 48 V input, N_T reflects 48/0.5 = 96 V onto the Primary during a part of T_{OFF}. It's D_{LIMIT} must therefore be 0.667. See also, Fig. 16.4.

Figure 16.11 Reset voltages compared.

CHAPTER 17

Reliability, Testing, and Safety Issues

Introduction

Power supply engineers have several performance goals that they typically weigh against cost. One of these performance goals is reliability. As we will see, a few cents saved by a bad design choice is not worth it if we are going to see excessive numbers of field returns. That is why, among others, a *demonstrated reliability test* (DRT) is considered a must. Here we will briefly cover the terms in frequent use in the power supply industry as regards testing and qualification. We will also touch upon safety issues and show how they can be impacted by seemingly innocuous design choices.

Reliability Definitions

Reliability is the probability that a device will perform its specified function in a given environment for a specified period of time. In other words, reliability is quality over time and environmental conditions.

The exponential distribution is one of the most common distributions used to describe reliability. We assume that the reliability exponentially decays with time as

$$R(t) = R(0) \times e^{-\lambda t}$$

where $R(0)$ is the reliability at time $t = 0$ and is assumed maximum [that is, $R(0) = 1$]. Then the time constant is the point where the number of units still functioning is $1/e$ of the original number, that is, 36.8 percent are left. This time constant is called the *mean time between failure* (MTBF). 1/MTBF is the failure rate. Failure rate can be expressed as the percentage failures per 1000 device-hours of operation (usually called λ), or as the total number of failures in a million device-hours (expressed as parts per million, or ppm), or as the number of failures per billion device-hours (expressed as failures in time, or FITs). See Table 17.1 for the respective definitions and their relationships.

NOTE *We can have 1000 devices operating for 1 h, or one device operating for 1000 h. They are statistically equivalent and both correspond to 1000 device-hours. For power supplies we prefer to talk in terms of power-on-hours (POH). This is the cumulative hours of operation of several power supplies working together. Keep in mind that hereafter, h stands for hour.*

NOTE *The MTBF is almost the same as mean time to failure (MTTF), and the two are used almost equivalently. Technically, however, MTBF should be used only in reference to repairable items while MTTF should be used for nonrepairable items. However, MTBF is commonly used for both cases.*

Now, since we have 8760 h in 1 year, a typical power supply MTBF of 250,000 h is equal to about 30 years. Yes, we could theoretically have one power supply working for 30 years, but we know it really doesn't (or can't). The reason is that we have wear-out failures starting to occur after about 3 to 5 years. We know that the chief culprits are the aluminum electrolytic capacitors (or the fan). So what an MTBF of 250 kilohour *does not mean* is that the power supply will work 30 years before we see a failure. We could however have 100 units in the field

MTBF (hours)	Failure rate (per POH) (h⁻¹)	λ (% per 1000 h)	ppm	FITs (failures per billion POH)
10^9	10^{-9}	10^{-4}	10^{-3}	1
10^8	10^{-8}	10^{-3}	10^{-2}	10
10^7	10^{-7}	10^{-2}	10^{-1}	10^2
10^6	10^{-6}	10^{-1}	1	10^3
10^5	10^{-5}	1	10	10^4
10^4 ⇐1/x	10^{-4} ×10^5 ⇒	10 ×10 ⇒	10^2 ×10^3 ⇒	10^5
10^3	10^{-3}	10^2	10^3	10^6

TABLE 17.1 Failure Definitions and Their Conversions (POH stands for power-on-hours)

and they would accumulate 30 power-on-years in 0.3 calendar years only. That is when we will see the first failure (*on an average*). And that is what MTBF generally indicates. See also Fig. 11.2 for an important clue.

The definition of MTBF applies only in the region where life or wearout issues are not present. In fact the decay in the number of functioning units is exponential on the basis of the fact that the *failure rate is a constant*. So, if we start off with say 1000 units and 10 percent fail in the first year, we are left with 900 units. In the next year, another 10 percent fail so we are left with 810 units. The next year yet another 10 percent, that is, 81 fail, so we are left with 729 units and so forth. This goes on until wearout starts occurring and the failure rate then climbs steeply. However, if we plot the values before wearout—1000, 900, 810, 729, and so on—we get an exponential curve. The time constant of this curve is the MTBF.

In the above example we took the failure rate as 10 percent in 8760 h; that is, the percentage in 1000 h is 10/8.76 = 1.142. From Table 17.1 we see that this is λ by definition. To get the failure rate per POH we need to divide this by 10^5. Thus we get the latter to be 1.142×10^{-5}. From the same table we see that to get MTBF we have to take the reciprocal of this. Therefore, MTBF is $10^5/0.001142 \approx 9 \times 10^7$ h.

Similarly, a typical power supply MTBF of about 30 years (250 kilohour) means a failure rate of $1/250\,k = 4 \times 10^{-6}$. To get λ we have to multiply this by 10^5 (see Table 17.1) and so we get 0.4 percent per 1000 h.

Chi-Square Distribution

As mentioned, to verify MTBF we could power up one unit and wait for about 30 years for a failure to confirm that the MTBF is 250 kilohour. But the problem with that is that (1) within the test time we are not allowed to have any wear-out failures, (2) we need to verify the MTBF *before* we release the product, not 30 years later, and finally (3) how do we know that that particular power supply was indeed a representative sample, or "typical"? We want to increase the number of units being tested so as to include several more production batches to see the effect of variations/spreads/tolerances, and also to increase the POH instead of the calendar years. Since the analysis clearly becomes statistical at this point, a typical power supply specification may ask for "an MTBF of 250 kilohour at 60 percent confidence level, at an ambient temperature of 55°C." This would mean that we should be able to assert with 60 percent confidence that indeed the MTBF is better than 250 kilohour. The means to do this is the χ^2 (Chi-square) distribution and its table. This is described in the United States Military Handbook Mil-Hdbk 781A. The engineer doesn't need to know statistical theory in detail here but only to understand how to use it to compute the MTBF. Here are the steps

1. By definition, failure rate is

$$FR = \frac{\chi^2(\alpha, 2f+2)}{2 \times POH}$$

or

$$MTBF = \frac{2 \times POH}{\chi^2(\alpha, 2f+2)} \quad h$$

No. of failures	χ^2 at 60% CL	χ^2 at 90% CL
0	1.833	4.605
1	4.045	7.779
2	6.211	10.645
3	8.351	13.362
4	10.473	15.987
5	12.584	18.549

TABLE 17.2 Chi-square Lookup Table

where α is the *significance level* (or the acceptable risk of error) and is related to the confidence level by

$$CL = 100 \times (1 - \alpha)\%$$

f is the number of failures.

While the engineer can look up a book on statistics to get the χ^2 table, we have provided the results most frequently used for estimating reliability of power supplies directly in Table 17.2. The confidence levels usually used are the 60 percent and 90 percent levels. Further, the testing usually never goes for more than zero or one failure, as it would involve an even larger number of power supplies. Let us do some sample calculations.

Example 17.1 How many POH are required to demonstrate an MTBF of 250 kilohour at 90 percent confidence level (temperature specified: usually 55°C)?
We should have zero failures when operated for

$$POH_0 = \frac{\chi^2 \times MTBF}{2} = \frac{4.605 \times 250,000}{2} = 575,625 \text{ h}$$

And we should have had only one failure when operated for

$$POH_1 = \frac{\chi^2 \times MTBF}{2} = \frac{7.779 \times 250,000}{2} = 972,375 \text{ h}$$

On completion of these many *power-on-hours* we can expect the second failure to occur.

Example 17.2 How many units are required for a 4-week test to demonstrate an MTBF of 250 kilohour at 60 percent confidence level, with one failure?
For one failure we need

$$POH_1 = \frac{\chi^2 \times MTBF}{2} = \frac{4.045 \times 250,000}{2} = 505,625 \text{ h}$$

In 4 weeks we have 672 h. Thus we need

$$\frac{505,625}{672} = 752 \text{ units}$$

Note that these units will all need to be operated simultaneously at maximum load or 80 percent of maximum load (as specified) and at maximum ambient of 55°C (or as specified). Typically, some of these will be run at the customer's location and some at the power supply manufacturer's location.

An alternative form does not require the Chi-square table. This says that the demonstrated MTBF is

$$MTBF = \frac{1}{1 - (1 - CL)^{\frac{1}{POH}}}$$

where CL is the confidence level and POH the number of power on hours *before* the first failure occurs (i.e., for zero failures).

Chargeable Failures

When doing the demonstrated reliability test, some failures may be discounted. For example, if the failure is clearly a result of user mishandling, it may not be considered a potential cause of future field failures, and therefore is not relevant to a reliability estimate.

Failures such as these are considered *nonchargeable*. On the other hand, *chargeable* failures are those which if not corrected *will occur in the field*. "Failure" may just mean that the power supply has gone out of its stated specifications. Mechanical dents and the like are not chargeable. Note that a chargeable failure may be the result of a quality-of-work problem during production, and if so, it has to be corrected by process improvement and/or training so that we can be sure that the failure will not occur in the field. When the supplier has provided adequate evidence that a particular failure mode has been understood and a fix implemented, only then will it become a *nonchargeable* failure (and will be eliminated from being counted in the ongoing demonstrated reliability test). Note that this may also involve a design change. Finally, it must be agreed to by all concerned as constituting a corrective action.

Here is a summary of these failure classifications

1. *Chargeable failures*
   - Any failure that degrades the performance and effectiveness beyond acceptable limits
   - Any failure that could result in significant system damage, such as to preclude mission accomplishments
   - Any failure that will need repair activities
2. *Nonchargeable failures*
   - Any failure that does not degrade the overall performance and effectiveness of the system beyond acceptable limits
   - Any failure that has been corrected and proven to be fixed
   - Any failure that occurs during run-in test or incoming inspection

Warranty Costs

Most engineers are surprised to find out how much it can cost to get back one unit, repair it, and then have it reinstalled. This was in fact posed as a prize-winning question for employees in the German factory the author worked in, a few years ago. The correct answer was around DM180 (approximately $120 at that time). But it was interesting to note that almost no guess from the hundreds of employees even made it into the ballpark.

Engineers should also know the 10× rule which goes as follows: if a failure is detected at the board level and costs $1 to fix, then if discovered at a system level it will cost $10, and if it goes to the field and then failure occurs, it will cost $100 to repair and so forth.

We should know how the figures relating to warranty add up as a function of MTBF. Here is a sample calculation.

Example 17.3 If there are 1000 units with an MTBF of 250 kilohour, how many are expected to fail over 5 years?

Let us assume that the failed unit is immediately replaced with a unit of similar reliability. Then the number of failures over 5 years (43.8 kilohour) is

$$\text{No. of failed units} = \frac{1000 \text{ units} \times 43{,}800 \text{ h/unit}}{250{,}000 \text{ h/failure}} = 175.2 \text{ failure}$$

If it takes $100 estimated to repair one unit in the field, this will cost $17,520. So cost per unit is $17.52.

This calculation is often performed in terms of *annualized* (or *annual*) *failure rate* (AFR) and gives almost the same result. Starting with N power supplies, with a certain MTBF, we know that the sample size shrinks as

$$N(t) = N \times e^{-t/\text{MTBF}}$$

So at the end of 1 year (i.e., $t = 8760$ h) we are left with

$$N \times e^{-8760/\text{MTBF}} \quad \text{units}$$

The number of units failing every year is therefore described by

$$\text{Annual failure rate} = 1 - e^{-8760/\text{MTBF}}$$

Over 5 years the cost is estimated to be

$$\text{Cost/unit} = (\text{annual failure rate}) \bullet (\text{years of product life}) \bullet (\text{per unit repair cost})$$

So, if MTBF is 250 kilohour, and the cost to bring back, repair, and return one failed unit is $100, over 5 years we get the following cost/unit

$$\text{Annual failure rate} = 1 - e^{-8760/250,000} = 0.034$$

So,

$$\text{Cost/unit} = (0.034) \bullet (5) \bullet (\$100) = \$17$$

This is going to increase the selling price dramatically. Unless of course the warranty is drastically reduced (90 days?)!

Calculating Reliability

Until fairly recently the United States Military Handbook Mil-Hdbk 217F was commonly used to calculate reliability based on either a simple *part count* or a *part stress* analysis.

This particular standard may not be in common use today, but in the author's opinion the underlying philosophy should still be taken note of. This philosophy is in fact shared by many other reliability prediction methodologies, many of which are still in frequent use. In all of these the failure rate of the equipment is taken to be the sum of the failure rates of all the individual components. Each component has a specified *base failure rate*. This is then adjusted by multiplying it with a number of factors (called the *pi factors*) which are related to the environment π_E, application π_A, quality level π_Q, secondary stresses (e.g., voltage stress π_V and the like. The π *factors* in the Mil-Hdbk were based on statistical data gathered over several years, and so they weren't always reflecting the recent and rapid improvements in component technology. Therefore it was common to get a calculated MTBF for a power supply of the order of only 100 kilohour by part stress analysis, and around 150 to 200 kilohour by part count analysis (under *ground benign conditions*). But neither of these numbers came close to what demonstrated reliability (and field) tests showed (typically 400 kilohour for an off-line power supply for example). So, over the last decade or so most companies had established an unofficial and internal thumb-rule multiplicative factor (of around 4) with which to multiply the MTBF prediction from the part stress analysis of Mil-Hdbk 217F.

Though Mil-Hdbk 217F is now rarely used, *stress analysis must always be performed*. This involves testing and profiling *each* component inside the power supply. Voltage levels, current, and temperature should be recorded, and this exercise is also helpful in identifying the weak links in the design. Note that the part count analysis of Mil-Hdbk 217F always seemed somewhat of a "play with numbers." Some prominent industry personalities had even laid out their "pet peeve" as follows: if you remove several components corresponding to a certain protection block, the reliability would fall, though part count analysis would say it has improved because there are now fewer components which can fail. However, it was always understood that part count analysis was to be carried out in the initial bidding phases only since there was no prototype on hand yet to carry out a formal part stress analysis. But it has to be acknowledged that even the latter analysis would likely not account for most *abnormal* operating conditions, i.e., those in which the protection block would act to protect the power supply. Therefore engineering judgment should always be considered of prime importance when designing power supplies, especially when carrying out reliability enhancements.

As mentioned, there are several other reliability calculation methods, and some companies (e.g., Siemens) even have their own internal prediction methodologies. All are, however, very similar to Mil-Hdbk 217F in philosophy, though they do end up with very different final numbers, mainly because their π factors are numerically different.

Demonstrated reliability is still clearly the best way to go, despite the fact that a very large number of units need to be tested. But we must also realize that even this test is usually done only under steady operating conditions. Thus, a good design engineer will also carry out several bench tests to confirm reliability under abnormal conditions.

Testing and Qualifying Power Supplies

A power supply may go through several tests before it is considered mature. Besides the obvious functional checks, thermal scans, and thermocouple measurements (and safety and EMI compliance tests) some of the other tests typically required fall into the following general categories.

HAST/HALT, HASS, and ESS

HAST/HALT, HASS, and ESS stand for *highly accelerated stress/life test*, *highly accelerated stress screen*, and *environmental stress screening*, respectively. HALT is performed during the development phase of the product. It is intended to stress a few samples to find the product's limits before destruction occurs. Thermal soak, thermal cycling, vibration, and other stresses may be applied that will sometimes exceed the specified operating range of the product. Thus HALT will give us a quantitative idea of the design margins.

Typically, HALT will be performed several times on successive design iterations to measure improvement. On the other hand, HASS is conducted during the subsequent production cycles to measure the effect of normal process variations on product reliability. But we note that the stress limits set for HASS are sufficiently reduced below the maximum limits indicated by the HAST test so as to avoid causing premature ageing during the production cycle itself. In HASS we will apply all selected stresses simultaneously, with the product functioning and being constantly monitored during that time.

In ESS, various stresses are applied to each outgoing unit. The idea is to weed out early life failures (i.e., those with "infant mortality") in production rather than hear about it from the field. Various forms of stress can be applied. The most common are thermal soak (*burn-in*), thermal cycling, and vibration. Sometimes shock and input voltage margining are used, though humidity may also be increased (e.g., to 85 percent).

Some specific examples of power supply/module tests are

1. *Burn-in*. A typical commercial power supply burn-in setup is actually very simple. Several power supplies are powered up for several hours in a small room with resistive loads on their outputs. The heat from this combined resistive load heats the room up. A thermostat operates an exhaust fan when the temperature rises above 55°C typically, and this constitutes a crude ambient temperature control. Each output has a light emitting diode (LED) across it so that the operator can periodically check to see if any unit has failed. Some limited form of thermal cycling, and simple tests like a timer operated power-on/power-off cycling may also be carried out.

2. *Overstress tests (e.g., on 4 units)*. The purpose of all this is to determine if it is cost effective to take steps to improve reliability. For example, if the stress margins can be improved significantly by just raising the wattage of a certain resistor, it would be usually well worth it. This test is thus done during the design phase and is essentially exploratory in nature.
   - *Thermal*. One unit will run at full load in a thermal chamber at maximum rated temperature and nominal line. Thermal protection, if fitted, will be deactivated. The temperature is increased in 10°C steps with a 30-min operation at each temperature until failure occurs. Failure analysis and repair are performed. The test is repeated (total 2 runs) and during the second run, the temperature on the suspected weak link is monitored.
   - *Line voltage*. One unit will run at full load in a thermal chamber at maximum-rated temperature and maximum line voltage. The line is increased in steps of 10 Vac, until failure occurs. Failure analysis and repair are performed. The test is repeated (total 2 runs) and during the second run the voltage and current of the suspected weak link are monitored.
   - *Load stress*. One unit will run at full load, nominal line, in a thermal chamber at maximum-rated temperature and maximum load. The load is then increased in steps of 10 percent of maximum load (each output simultaneously), at 30-min intervals until failure occurs. Failure analysis and repair are performed. The test is repeated (total 2 runs) and during the second run the voltage and current of the suspected weak link are monitored. We again repeat the test, but now with load on only one output at a time ("corner conditions"). Note that any current limit protection on the outputs is to be deactivated for the purpose of this test.

4. *Thermal ESS (e.g., on 4 units)*. Here normal operation is expected through 10 cycles of −30°C to 85°C at a rate of 20°C/min. The thermal profile is provided. Typically, there will be three dwell levels per cycle of 15 min each. One will be at the highest temperature, after which the unit is taken straight down to the lowest temperature where it remains for 15 min, and thereafter the unit is brought back to room temperature where it stays for 15 min before it starts the next thermal cycle. The chamber temperature rate may be varied from 20°C/min to 60°C/min. After this test, the random vibration test is to be done so as to provoke failures due to any damage during this test.

Reliability, Testing, and Safety Issues

4. *Vibration test (e.g., on 4 units)*. The test is conducted with the system energized, fixed onto the vibration table in its normal configuration. The frequency band is set from 5 to 500 Hz. Each system is run for 10 min at 3, 4, 5, and 6 g (RMS), 5 min at 7 g (RMS) in the x axis, and then the system integrity is checked. The test is repeated for the y and z axes. Plots of the control accelerometer and the response accelerometers are recorded for the starting and maximum operating test levels.

Safety Issues

Here we will focus mainly on the issue of "clearances" and "creepages" as they relate to our design and topology. Some basic definitions are as follows:

Clearance distance. This is the shortest distance between two conductive parts (or between a conductive part and the bounding surface of the equipment) measured through air. Clearance distance helps prevent dielectric breakdown between electrodes caused by the ionization of air. The dielectric breakdown level is further influenced by relative humidity, temperature, and *degree of pollution* (or *pollution degree*, defined below) in the environment.

Creepage distance. This is the shortest path between two conductive parts (or between a conductive part and the bounding surface of the equipment) measured along the surface of the insulation. An adequate creepage distance is meant to protect against *tracking*. This is a process that produces a partially conducting path of localized deterioration on the surface of an insulating material. The degree of tracking required depends on two major factors. One is the *comparative tracking index* (CTI) of the electrical insulating material (expressed in volts) and the other is the *pollution degree* of the environment.

To see how these need to be computed see Fig. 17.1. Note that IEC 60950-1 Ed. 2.2 is the commonly followed internationally recommended norm (originally IEC 950). This provides tables giving clearances and creepages depending on pollution degree and CTI (as applicable) and voltage. Here are some points we should remember and double check:

The peak voltage between two points affects the required clearance and therefore it must be measured. Creepage depends only on *working voltage* which is basically the RMS of the voltage between the points.

For worldwide input power supplies for information technology (IT) equipment, a popular rule of thumb is to allow 8-mm creepage between primary and secondary circuits and 4 mm between primary side and chassis ground. Clearance usually gets automatically taken care of in the process. Thus 4-mm *margin tape* was historically used in safety transformers as this produced 8-mm creepage between primary and secondary windings. If the above spacings are used, there is a high probability (over 90 percent) that the power supply will pass the relevant safety test. However, with power factor

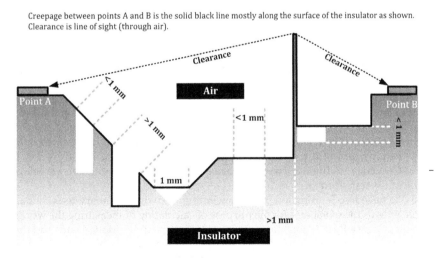

Figure 17.1 How clearances and creepages should be measured.

correction [where the high voltage DC (HVDC) rail is around 400 V], we have to be very careful as this rule may not hold. We should compute the required spacings (and margin tape width) in the manner indicated later in this chapter.

Triple insulated wire is an emerging choice, and it may be used in a transformer if cost permits. Its advantage is that no margin tape is required. But the wire must comply with Annex U in IEC 60950-1.

The optocouplers (especially surface mount packages) should be checked carefully to see if they comply with the mandatory separation requirements. As per IEC, for optocouplers, a working voltage measurement can be taken to determine creepage, but if the measured voltage is less than the mains voltage, the latter is to be used. If the working voltage of the optocoupler is more than 50 V_{RMS}, a minimum distance of 0.4 mm through the insulation is needed. But also, see http://www.vishay.com/docs/83743/83743.pdf.

There is no requirement for the distance through insulation of sheet materials. We always use three layers of Mylar (i.e., polyester) tape at the isolation boundary of a transformer, and to reduce leakage this can even be 0.5 mil thick if cost permits. Though usually, three layers of 1-mil- or 2-mil-thick polyester are more common.

Manufacturers often introduce slots in the PCB to enhance the creepage between primary and secondary circuits. This depends on the CTI of the PCB material. But note that the slots must be wider than 1 mm to count (see Fig. 17.1).

The most commonly used PCB materials and polyester insulators have a CTI of 200 to 400 V. They thus fall into *Material Group III*. We should calculate creepages based on this assumption, unless we know better.

As for pollution degree, the most common assumption in commercial power supplies is *Pollution degree 2*. This corresponds to a typical office environment where only temporary conductivity may occur by condensation. *Pollution degree 1* is for dry, nonconductive pollution, *Pollution degree 3* corresponds to a typical heavy industrial environment, and *Pollution degree 4* corresponds to persistent conductivity as by rain or snow.

In testing, every component can be subjected to a 10-N force, and the required spacings should still be maintained. This is one reason why we should prefer to lay components flat on the PCB, especially those near the edges of the PCB (near the chassis) and those near the isolation boundary.

Mandatory separations also need to be maintained if, for example, a single solder joint (anywhere) comes undone. This can cause a two-lead component to rotate, thereby possibly bridging the primary and secondary sides. Therefore, room temperature vulcanizing silicone (RTV) or hot melt glue is usually liberally applied on the PCB (this being also useful for clearing a typical vibration test especially when using single-sided PCBs.).

Calculating Working Voltage

The impact of topology on correctly calculating working voltage is now taken up. This is required to calculate the width of the margin tape, for example. We will focus on the case of a single-ended Forward converter with a PFC front-end and an HVDC of around 400 V. This case is special because the 4 mm/8 mm rule of thumb does not work and we actually need more than that. Rather than take it to the test lab and discover that the transformer needs to be completely redesigned, it is important that we know exactly what the RMS voltage we are talking about here is.

What exactly are we measuring? The transformer is isolated; so what do we mean by RMS voltage *across* the transformer? To answer this question we have to trace the path backward as in Fig. 17.2. We can see that since *protective earth* (PE) and *neutral* (N) are connected at the service entrance and the secondary ground is the system ground, the HVDC (i.e., bus voltage) has the shape shown. Note that the black part of the AC waveform occurs when *L* goes below *N*. During this time, the HVDC also falls with respect to system ground. On the gray part of the waveform, *L* is higher than *N*, so the HVDC is flat. Besides this low-frequency undulation, we also have a high-frequency switching waveform on either side of the transformer.

We have two termination pins per winding. So, which primary-to-secondary combination gives us the worst case for the purpose of calculating or measuring the working voltage? The worst-case RMS is actually recorded between the drain (dotted end) and the nondotted end of the Secondary. That is because the nondotted end has opposite phase. So when the drain goes high, this nondotted end of the Secondary goes low and the *difference* between the two is the maximum. In Fig. 17.3 we have shown what the voltage waveforms look like.

Reliability, Testing, and Safety Issues 433

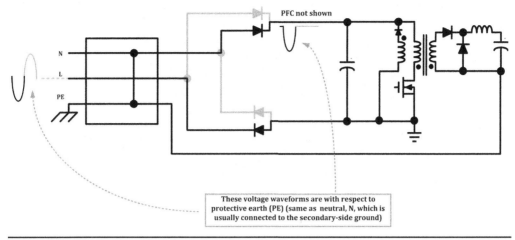

FIGURE 17.2 How the bus voltage looks with respect to the secondary ground.

What if we have multiple outputs? For calculating the maximum RMS voltage across the transformer we need to consider the winding which corresponds to the *highest output voltage* (magnitude).

We have provided a Mathcad file to do the calculations in Fig. 17.4. For simplicity, we have avoided performing a summation over vectors (which is actually the best way to prevent Mathcad convergence problems) and preferred a simple integration instead. But because integration is then required over the two very different frequencies involved, for certain output voltages the Mathcad file will not give desired results. In that case, the reader may then have to reduce the set PWM frequency down to 1 kHz just to make the program work. But once it does, we always get the right result, since the voltage does not really depend on the switching frequency. This program has actually been compared by the author to a far more detailed file (one that does involve a summation over vectors, and which also uses the exact switching frequency of the PWM), and the results from the simpler file were shown to be equally accurate. Note that we can enter four schematic options as displayed in Fig. 17.5. The voltage values shown boxed next to each schematic are the result of the attached Mathcad file. We see that if we want a positive output voltage rail, option 4 provides the lowest voltage across the transformer and thus possibly the narrowest margin tape. Mandatory creepage distances as a function of working voltage are displayed in Table 17.3. We have also provided an example of an interpolated creepage calculation within the same figure. So for an estimated working voltage of 450 $_{RMS}$ we need margin tape of width 4.6 mm. *Note that the worst case occurs at 90 Vac not 270 Vac (when PFC is present)*!

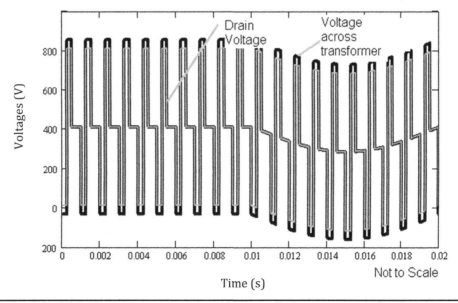

FIGURE 17.3 Voltage on drain and across transformer for Forward converter (Option 2).

> Select := 1
> **Enter 1** here if this is conventional Forward Converter (with Output Diodes in Common Cathode configuration), and output being $+Vo$ (return connect to PE)
> **Enter 2** here if this is conventional Forward Converter (with Output Diodes in Common Cathode configuration), and output being $-Vo$ (return connect to PE)
> **Enter 3** here if this is reversed Forward Converter (with Output Diodes in Common Anode configuration), and output being $-Vo$ (return connect to PE)
> **Enter 4** if here this is reversed Forward Converter (with Output Diodes in Common Anode configuration), and output being $+Vo$ (return connect to PE)
> $k := 10^3 \quad m := 10^{-3} \quad u := 10^{-6} \quad t := 0, 1 \cdot u .. 20 \cdot m$
> Enter inputs $Vac := 90 \quad Vdc := 385 \quad Vo := 12$
> Turns Ratio $n := \frac{44}{5} \quad D := \frac{n \cdot Vo}{Vdc} \quad D = 0.274$
> Switching Frequency
> (enter $2k$ to avoid non-convergence) $f := 2 \cdot k \quad \omega := 2 \cdot \pi \cdot f \quad T := \frac{1}{f}$
> Line Frequency $fl := 50 \quad \omega l := 2 \cdot \pi \cdot fl \quad Tl := \frac{1}{fl}$
> Construct Waveforms:
> $Vrect(t) := Vac \cdot \sqrt{2} \cdot |\sin(\omega l \cdot t)|$
> $Vpwm1(t) := \text{if } [\text{mod}(t, T) < D \cdot T, Vdc, 0]$
> $Vpwm2(t) := \text{if } [\text{mod}(t, T) < (T - D \cdot T), Vdc, 0]$
> $Vpwm(t) := Vpwm1(t) + Vpwm2(t)$
> $Vsec1(t) := \text{if}\left[\text{mod}(t, T) < D \cdot T, \frac{Vdc}{n}, 0\right]$
> $Vsec2(t) := \text{if}\left[(1 - D) \cdot T < \text{mod}(t, T) < T, \frac{-Vdc}{n}, 0\right]$
> $Vsec(t) := Vsec1(t) + Vsec2(t)$
> According to reference of measurement introduce required voltage difference
> $vt := \begin{cases} Vsec(t) & \text{if Select} = 1 \\ Vsec(t) + Vo & \text{if Select} = 2 \\ 0 & \text{if Select} = 3 \\ (-Vo) & \text{if Select} = 4 \\ \text{"Error"} & \text{otherwise} \end{cases}$
> $Vacross(t) := \begin{cases} [Vpwm(t) - Vrect(t) + v(t)] & \text{if } t > \frac{Tl}{2} < t < Tl \\ Vpwm(t) + v(t) & \text{otherwise} \end{cases}$
> Worst-case maximum rms voltage
> $Vworking := \left(\frac{\int_0^{Tl} Vacross(t)^2 dt}{Tl}\right)^{\frac{1}{2}}$
> $Vworking = 471.016$ Volts

FIGURE 17.4 Working voltage of single-ended Forward converter transformer.

Estimating Capacitor Life

With 500-V electrolytic capacitors becoming available, no other capacitor has the CV (capacitance times its voltage rating) capability of this popular capacitor. For cost reasons too, it is virtually indispensable in commercial off-line power supplies. People may not always want it, but usually can't do without it either. Even in modern DC-DC converters, where the "all ceramic" solution is much the buzzword nowadays, after finding unexpectedly severe input voltage overshoots (under hard application of input power) and the consequent instability and ringing, a common emerging recommendation is to put a *high-ESR* capacitor like an aluminum electrolytic in parallel to the ceramic input capacitor, in an effort to damp out the input resonances (lower the Q-factor). See Chap. 17 of author's A-Z book, 2d ed.

In most commercial power supplies the aluminum electrolytic capacitors determine the eventual life of the product under operation. It is therefore important to get a much better

Reliability, Testing, and Safety Issues 435

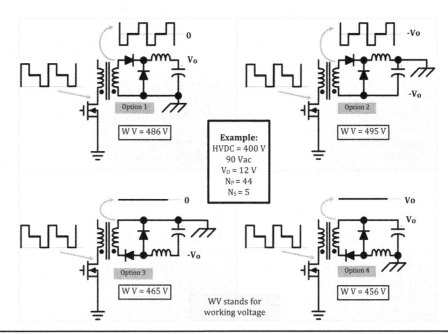

Figure 17.5 Different options referenced in the Mathcad file, for a single-ended forward converter.

understanding of this critical (and almost unavoidable) component. The most important aspect of designing with this capacitor is estimating its life.

The first parameter in a typical aluminum electrolytic capacitor's datasheet that we must understand is the *dissipation factor* (DF) or tan δ. It is related to the *equivalent series resistance* (ESR) by

$$\text{ESR} = \frac{\tan \delta}{2\pi f \times C}$$

DF is thus the ratio of the resistance to the reactance $1/2\pi C$. Note that the ESR used here must be the ESR at the frequency f (usually stated at 120 Hz). Clearly a low DF means a low ESR.

Working voltage V (RMS or DC)	Pollution Degree 2		
	Material group		
	I CTI > 600 V	II 600 > CTI ≥ 400	III 400 V > CTI ≥ 100 V
≤50	0.6	0.9	1.2
100	0.7	1.0	1.4
125	0.8	1.1	1.5
150	0.8	1.1	1.6
200	1.0	1.4	2.0
250	1.3	1.8	2.5
300	1.6	2.2	3.2
400	2.0	2.8	4.0
600	3.2	4.5	6.3
800	4.0	5.6	8.0
1000	5.0	7.1	10.0
Linear interpolation permitted (round off to next higher 0.1 mm increment) e.g., 450 V requires $\frac{450-400}{600-400} \times (6.3-4) + 4 = 4.6$ mm			
This is for basic insulation (i.e., primary to safety ground creepage, also width of margin tape). Double this distance for reinforced insulation (i.e., primary to secondary creepage).			

Table 17.3 Creepage Distances (in mm) as per the xx950 Safety Standards

What is defined as "end of life"? Typically this is said to occur when the change in capacitance measured from the *initial* value is greater than ±20 percent and/or the tan δ has become more than 200 percent of its *initial* value. Considering the fact that the initial specified value of the capacitor lies within a certain standard tolerance band (usually ±20 percent), we need to include another 20 percent to account for that. For example, if our theoretically calculated holdup time requirement is 100 μF, we should start with a capacitor value greater than 156 μF *nominal* (check: 156/(0.8 × 0.8) = 100). This is really a more than worst-case estimate. Nevertheless, *several manufacturers of reliable commercial power supplies do add a flat 40 percent to the theoretically calculated value to account for ageing and initial tolerance.*

How bad can the ESR really be at the end of the useful life? Since DF is allowed to double, we see that the change in ESR can be

$$\text{ESR} \propto \frac{\tan \delta}{C} \Rightarrow \frac{200\%}{80\%} = 250\%$$

A higher ESR generally plays a helpful rule in the feedback loop, but we should still do a Bode plot to confirm the phase margin. We can hand-select (using actual measurements) lower capacitance, and higher ESR capacitors from the used parts bin in an effort to mimic end-of-life characteristics on our bench prototype.

We should also be aware that as the ESR increases, the heating in the capacitor will also increase toward end of life, raising its temperature further. However, manufacturers of quality capacitors like the Chemicon group (www.chemi-con.com) have clearly accounted for this in their life predictions, so we usually do not need to worry about this particular issue.

The next important datasheet parameter is the ripple current rating. It is typically stated in amperes RMS at 120 Hz and 105°C. This essentially means that if the ambient temperature is at the maximum rated of 105°C, we can pass a low-frequency current waveform with the stated RMS, and in doing so we will get the stated life. The life figure is typically 2000 to 10,000 h under these conditions. Yes, there are lower-grade 85°C capacitors available, but they are rarely used as they can hardly meet typical life requirements at high ambient temperatures. Remember that h stands for hour here.

The datasheet also provides certain *temperature multipliers*. For example, for the LXF series from Chemicon, the numbers are

1. At 65°C the temperature multiplier is 2.23.
2. At 85°C the temperature multiplier is 1.73.
3. At 105°C the temperature multiplier is 1.

This is easy to understand if we realize that such (long-life) capacitors are typically designed for a 5°C differential from ambient to *core* (deep inside the capacitor), and about 5°C differential from ambient to case. The temperature inside the capacitor is thus slightly higher than the case. But it is the temperature of the electrolyte that determines how quickly it evaporates and how long the capacitor will last. Misestimation of a few °C of this temperature could mean thousands of hours subtracted from the life of the product.

Now, realizing that the amount of heating and the temperature rise are proportional to I^2_{RMS}, we can see that at 85°C

$$\frac{T_{\text{CORE}} - 105}{T_{\text{CORE}} - 85} = \frac{I^2_{105}}{I^2_{85}} = \frac{1}{1.73^2} = \frac{1}{3}$$

Solving for T_{CORE}, we get

$$T_{\text{CORE}} = 115°C$$

Using this value we can confirm the 65°C multiplier

$$\frac{115 - 105}{115 - 65} = \frac{10}{50} = \frac{I^2_{105}}{I^2_{65}}$$

So, the multiplier must be $5^{0.5} = 2.236$, which agrees with the specified value. Therefore from the vendor's ripple current multipliers (for temperature) we can easily deduce the *design core temperature for this series of capacitors*. Then, with a little help from an empirical relationship toward the end of this section, we can also figure out the most important capacitor design parameter—the *difference from case to core* (or *can to core* as more frequently called).

Out of the 10°C rise in temperature, roughly 5°C is from ambient to case and another 5°C from case to core. Let us call these differentials symbolically as $\Delta T_{\text{case\_amb}}$ and $\Delta T_{\text{case\_core}}$.

Note that these are the *design values* used by the manufacturer for the specific family of capacitors. They are not our actual measurements or estimates in an application. *If, however, we pass the rated ripple current (as specified at 105°C) through the capacitor, we do get exactly these design differentials in practice, and that is true whatever our actual ambient is.*

Example 17.4 A typical 85°C-rated capacitor has the following temperature multipliers: 1 at 85°C and 1.3 at 70°C. What is the designed core temperature?

$$\frac{T_{CORE} - 85}{T_{CORE} - 70} = \frac{I_{85}^2}{I_{70}^2} = \frac{1}{1.3^2}$$

Solving

$$T_{CORE} = 85 + 15 \times \left(\frac{1}{1.3^2 - 1}\right) = 107°C$$

So in this case, typically, $\Delta T_{case\_amb} \approx \Delta T_{case\_core} \approx 11°C$

We see that 85°C capacitors do have a much higher allowed temperature differential. This is an advantage, but not quite as much as it seems. The problem is that temperature multipliers provided in the datasheet are never used in practice since what they amount to is increasing the current so as to bring the core temperature back up to its maximum value. But we know that if the core temperature is at its maximum, the life we can expect is only the specified 2000 to 10,000 h. But we want much more.

A typical power supply design requirement is 5 years or 44,000 h at maximum load (or 80 percent of maximum load) operating at a room ambient of 40°C. Note that *the power supply is typically tested at a maximum of 55°C, but for life expectancy, a reduced temperature is usually specified*. But how can we achieve even that from a 2000-h capacitor? We use the doubling rule as derived from Arrhenius' theory

$$L = L_O \times 2^{\frac{\Delta T}{10}}$$

This effectively states that the life of a capacitor doubles every 10°C fall in temperature (of the core). L_O is the guaranteed life (2000 to 10,000 h) when passing the maximum specified ripple current at 105°C that is, when its core is at 115°C, as calculated above.

HINT *For semiconductors a similar rule of thumb prevails—the failure rate doubles every 10°C rise in temperature. Life and failure rate are actually separate issues. Life is a wearout effect at the end of the familiar "bathtub" curve, whereas failure rate is measured between the regions of infant mortality and wearout.*

NOTE *The standard 44,000 h life requirement is equivalent to 5-year operation at 24 × 7 × 365. In reality this may never be the case. It is better to discuss this with the customer as it adds greatly to the cost of the power supply. A well-known PC-market competitor usually specified a life of only 15,000 h for most products. This amounts to roughly 8 h a day for 5 years, and is probably a far more realistic goal, one which can also often be met with cheaper 85°C capacitors at certain key locations in the power supply. But we do have to use extremely good quality capacitor manufacturers. No cheap substitutes please! See http://www.theguardian.com/technology/blog/2010/jun/29/dell-problems-capacitors.*

Example 17.5 If we pass the rated ripple current through a 2000-h capacitor (no temperature multipliers applied) at an ambient of 55°C, what is the expected life (first pass estimate)?

At the rated current we can expect that the core is at 55°C + ΔT_{core_amb}. So the temperature "advantage" we have gained (measured from the maximum-rated temperature) is (105°C + ΔT_{core_amb} minus (55°C + ΔT_{core_amb} i.e., 50°C. Since this capacitor provides 2000 h at the maximum temperature, at the reduced ambient we may get a life of

$$2000 \times 2 \times 2 \times 2 \times 2 \times 2 = 64,000 \text{ h}$$

Note that in the above analysis ΔT_{core_amb} eventually got canceled out. So, this amounts to writing the following simple equation for life

$$L = L_O \times 2^{\frac{T_{core\_rated} - T_{core\_application}}{10}} = L_O \times 2^{\frac{T_{rated} - T_{amb}}{10}}$$

In our example, ΔT_{core_rated} is 115°C, and T_{rated} is the maximum-rated ambient of 105°C. T_{amb} is the actual ambient in our application. $\Delta T_{core\_application}$ is the temperature of the core in our application, which in our example is 65°C. But note, however, that we would have got the same life prediction had the capacitor manufacturer used any other designed ΔT_{core_amb}. As we saw, it got canceled out, but that is only because in the example, we *followed the manufacturer's recommendations* and passed only the maximum-rated current through the capacitor.

In practice, we don't have a good way of knowing the local ambient (in the immediate vicinity) of the capacitor. Nearby components may also be heating the capacitor. Therefore *a common and conservative industry practice is to cut the outer sleeving of the capacitor and to insert a thermocouple under the sleeve in contact with the metal case*. This way, small air draughts don't affect the results. We then take this case temperature as the effective ambient, *unless we know better*. Suppose the case temperature is measured to be 70°C. Then the estimate of the capacitor life is now

$$L = L_O \times 2^{\frac{T_{rated} - T_{application}}{10}} = 2000 \times 2^{\frac{105-70}{10}} = 22{,}600 \text{ h}$$

However, we have to be clear what the *source* of this heating is. If it is *not* heat from nearby components, the ΔT_{case_core} may be actually much higher than we think. The life cannot be the same as compared to the case where the heat is purely from external sources, since that won't produce the harmful internal temperature differential (from case to core). Therefore a case temperature measurement is simply not enough. *We have to measure the ripple current* through the capacitor too, to at least confirm that we have not exceeded the maximum ripple current rating of the capacitor (which is equivalent to not exceeding the designed case to core delta). The relevant points are summarized below.

Capacitor manufacturers recommend that in general we don't pass any more current than the maximum-rated ripple current. This ripple current rating is the one specified at the worst-case ambient (e.g., 105°C). But even at lower temperatures we should not exceed this current rating. No temperature multipliers are to be used to buttress this rating. Only then is the case to core temperature differential within the design specifications of the part. And only then are we allowed to apply the simple doubling rule for life, since the core temperature rise is then considered factored into the life prediction figures provided by the manufacturer.

If the measured ripple current is confirmed to be within the rating, only then can we take the case temperature measurement as the basis for applying the doubling rule, even if the heat is coming from adjacent sources. Again, this is because the case to core temperature differential is within design expectations.

However, in one-on-one communications, Chemicon has in the past allowed a higher ripple current than rated, but the life calculation method given is then slightly different. This amounts to a special doubling rule which we will describe below using a practical example.

Example 17.6 We are using a 2200 µF/10 V capacitor from Chemicon. Its catalog specifications are 8000 h at maximum-rated 1.69 A, stated at 105°C and 100 kHz. The measured case temperature in our application is 84°C and the measured ripple current is 2.2 A. What is the expected life?

The life calculation provided by Chemicon was

$$L = L_O \times 2^{\frac{105-84}{10}} \times 2^{\frac{5-\Delta T}{5}} \text{ h}$$

where

$$\Delta T = 5 \times \left(\frac{2.2}{1.69}\right)^2 = 8.473°C$$

So,

$$L = L_O \times 2^{\frac{105-84}{10}} \times 2^{-0.695} = 21{,}000 \text{ h}$$

Let us understand the terms involved here. The ΔT calculation above essentially says what we already know

$$\frac{\Delta T}{\Delta T_{case\_core}} = \left(\frac{I_{application}}{I_{rated}}\right)^2$$

We know from the vendor's data that this family of capacitors was designed for a 5°C differential between case and core, and that differential is caused by passing the rated 1.69 A through it. So this ΔT calculation gives us the temperature differential when we pass 2.2 A through it. We then get a rise of 8.473°C rather than the designed 5°C.

The term $(5 - \Delta T)$ in the exponent of the life calculation gives us the temperature *in excess of the designed 5°C*. Let us call this ΔT_{excess}. So the life equation is

$$L = L_O \times 2^{\frac{T_{rated} - T_{case}}{10}} \times 2^{\frac{-\Delta T_{excess}}{5}} \text{ h}$$

The first term with the positive exponent causes the life to increase above L_O and the second term exerts the opposite effect. We can also see that a temperature differential from case to core *in excess of the designed value* is considered more harmful than a normal temperature differential (i.e., one that is caused by staying within the current rating). Chemicon models this excessive temperature rise conservatively as *causing a halving of life every 5°C increase, rather than the usual 10°C*.

Note *This equation should not be used to predict life if the ripple current is less than rated. ΔT_{excess} is not allowed to be negative here.*

Note *Capacitor manufacturers typically don't guarantee life under forced air cooling. The designer should either measure the capacitor without forced air cooling if possible, or add a judicious safety margin.*

Rather than take the case temperature as the ambient temperature of the capacitor, which is more of a worst-case calculation, we could try to actually measure its local ambient. Assume that the general ambient is T_{amb_ext}. The local ambient near the capacitor is T_{amb}. The procedure to deal with radiation from nearby components is as follows:

1. Take the capacitor from the circuit board putting it on the underside, but still connected to the circuit. In this position we can measure the temperature on its case T_{case_1}. This is

$$T_{case\_1} = T_{amb\_ext} + T_{self\text{-heating}}$$

2. At the same time we place an exactly similar capacitor at the position where the original capacitor was, but this has one lead "missing," so it is in effect not connected to the circuit. We measure its case temperature T_{case_2}. This is

$$T_{case\_2} = T_{amb\_ext} + T_{ext\_heating} \equiv T_{amb}$$

3. Therefore having measured the ambient in the surrounding air which is T_{amb_ext}, we know all the required components of the temperature buildup.

4. Also note that the following equation is recommended for a more careful analysis of the ratio that exists between ΔT_{core_case} and ΔT_{case_amb} (which was earlier stated to be ≈ 1)

$$\frac{\Delta T_{core\_case}}{\Delta T_{case\_amb}} = 0.0231 \times \text{Case Dia}_{mm} + 0.845$$

This curve-fit equation was derived by the author from data provided by Chemicon. It is accurate to within 6 percent for capacitor outer diameters in the range of 10 to 76 mm. Above $D = 40$ mm, the error from the use of this formula is less than 1 percent.

Caution *A temperature measurement is not enough, nor is a ripple current measurement. Measuring ripple current with a view to at least confirm that it is below the rated ripple current of the component is certainly advisable, though the designer is cautioned against using the ripple current alone to estimate heating on the basis of some assumed coefficient of convection h (see Chap. 13 for more on h), because that would ignore heating from nearby components, and we will thus overestimate the life expectancy.*

HINT *When measuring ripple current through a capacitor, the normal procedure is to lift the lower terminal (the one going to ground) and to insert a loop of wire for inserting a current probe. But this reading is extremely hard to do without affecting the current in the capacitor even as we are attempting to measure it. This may become an invasive measurement. An alternative is to insert a small noninductive and calibrated sense resistor (e.g., made of manganin or constantin wire) and measure the voltage across it. However, we should not place reliance on a direct RMS reading of the sensed voltage since the noise will likely skew the results. We should record the waveform on an oscilloscope and then do a calculation based on the vertices as shown is Fig. 2.14. Also, when measuring the RMS current through paralleled capacitors (e.g., those on an output rail) it is not a good idea to lift the lead of only one of them to do a current measurement since the current will just happily redistribute into the remaining capacitors. We should cut a common return trace and then insert either the current probe or a sense resistor.*

Lastly, the vendor may have directly provided a ripple current rating at 100 kHz in addition to the 120 Hz number. If not, the vendor would certainly have provided *frequency multipliers*. A typical frequency multiplier is 1.43 at 100 kHz. That means that if we are allowed 1 A ripple current at 120 Hz, then at 100 kHz we are allowed 1.43 A. This by design will produce the same heating as 1 A causes at 120 Hz. Therefore this is also equivalent to saying that the ESR at 100 kHz is related to the ESR at 120 Hz by the following equation

$$\left(\frac{I_{100\,kHz}}{I_{120\,Hz}}\right)^2 = \frac{\text{ESR}_{120\,Hz}}{\text{ESR}_{100\,kHz}} = (1.43)^2 = 2.045$$

Thus the high-frequency ESR is about half the low-frequency ESR. Clearly, since the case to core temperature differential is unaffected in the process, *frequency multipliers can and should be used*. In Chap. 5, we can see how to apply these frequency multipliers to our advantage when the current waveform has both low-frequency and high-frequency components.

Safety Restrictions on the Total Y-Capacitance

In off-line power supplies Y-caps are usually connected from line to safety ground (that is, PE, or protective earth). The purpose is to bypass the high-frequency common-mode noise. But these don't just bypass noise, they also conduct some of the low-frequency line current. That is what the line-to-line capacitors (X-caps) also do; the difference is that the Y-caps carry this current into the protective earth (chassis). If the earthing is not good for any reason, the user could get electrocuted on touching the chassis (or housing). Therefore, international safety agencies limit the total RMS current introduced into the earth by the equipment to a maximum of typically 0.25, 0.5, 0.75, or 3.5 mA (depending on the type of equipment and its *installation category*—its enclosure, its earthing, and its internal isolation scheme). Somehow, 0.5 mA seems to have become the industry default design value, even in cases where 0.75 mA or 3.5 mA may have been allowed. *It is important to know how high one can actually go in terms of ground leakage current, as this dramatically impacts the size and cost of the line filter, in particular the choke.*

Keeping the discussion here at a theoretical level, we can easily calculate that we get *79 μA per nF at 250 Vac/50 Hz*. This gives us a maximum allowed capacitance of 6.4 nF for 0.5 mA, or 44.6 nF for 3.5 mA. Typical configurations in off-line power supplies are four Y-caps, each being 1 or 1.2 or 1.5 nF, or only two Y-capacitors, each of value 2.2 nF. Note that there may be other parasitic capacitances or/and filter capacitances present, which should be accounted for in computing the total ground leakage current, and thereby correctly selecting the Y-caps of the line filter. However, we must keep in mind that if for improved EMI performance (CM noise rejection), a Y-cap is connected from the *rectified* input DC rails to earth (or from the output rails to earth), there is no ground leakage current through these capacitors. Therefore, in principle, there is no limit on their capacitance in this position either. However, we may need to put *two* Y-caps in series at these positions to comply with worldwide safety standards and regional deviations.

Safety and the 5-cent Zener

In Fig. 17.6 we have presented several issues which should be kept in mind when driving FETs with controllers (in off-line applications).

1. R_1 is a Gate pull-down resistor that is not just indispensable, it needs to be of low enough resistance too. The reason is that under a sudden (hard) application of input power at high-line conditions, as the voltage on the drain of the FET suddenly

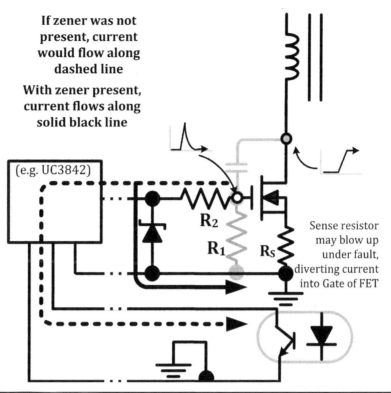

FIGURE 17.6 Things to consider when driving FETs.

ramps up, it injects a pulse through the drain-to-Gate (Miller) capacitance. This follows the simple equation $I = C \times dV/dt$. This injected current charges up the floating Gate and thus has the potential to spuriously turn the FET ON. This is further aggravated by the fact that the controller is not likely to have had time to be powered up fully at this moment. Most controllers (like the 3842/3844 series) have in effect a tristate output until the reference voltage on the pin of the IC becomes available. So, the IC cannot effectively provide a pull-down during this time. In actual tests, R_1 had to be decreased to between 4.7 and 10 kΩ to ensure a safe input power application under all conditions (in offline power supplies).

a) When abnormal tests are conducted by safety agencies, they can short or open any single component in the power supply. This is expected to usually lead to failure, but that is ok, provided the power supply "fails safe." This means that at no time should a hazardous voltage appear across the accessible output terminals. The problem here is that when the FET fails, it almost invariably causes a large momentary surge of failure current flowing internally first from drain to source. But the metal oxide sense resistor R_s invariably fails open shortly thereafter (before even the fuse can blow). However, the energy in the inductor is not yet spent, and it still demands a freewheeling path to flow through. So, now it "knocks" at the Gate of the FET. A huge surge current finally goes through the Gate, and follows a path through the IC (destroying it) and then onward into the optocoupler. In actual tests the opto package cracked open sometimes, thus potentially breaching the "sacred" Primary-to-Secondary boundary. That is certainly unacceptable to safety agencies. But, if an 18-V Gate zener is placed as shown, the zener almost invariably fails in a shorted condition. It thus diverts the FET failure current away from the IC (until the fuse blows). This helps pass safety testing, but is also invaluable during prototyping because though FETs can be quickly replaced, desoldering tiny control ICs a few times destroys the fine copper traces on the PCB, rendering the board unusable in a short while. With the zener in place, the IC usually never gets damaged (nor any of the components connected to its other pins).

2. Some high-end designs used to put a transient voltage suppressor diode (TVS) in parallel to R_s for the same reason. The TVS is basically just a higher peak energy zener that is guaranteed to fail in a shorted condition.

R_2 is usually not required, nor used. But a few years ago there was some suspicion (still unsubstantiated) that the small body capacitance of the zener was creating a resonant *C- L- C* type of tank circuit with the inductance of the PCB trace going to the Gate (including the internal bond wires) and the Gate capacitance, thus leading to oscillations and "inexplicable" failures. To play safe, cagey designers add a 10-Ω resistor in between the zener and the Gate with an intent to lower the *Q* of the tank circuit, and to thus damp out any oscillations. We should certainly mount the zener as close to the FET as possible to avoid introducing more parasitic inductance.

CHAPTER 18

Watts in It for Us: Unraveling Buck Efficiency

PART 1 BREAKUP OF LOSSES AND ANALYSIS

Introduction

The first step we have taken toward systematically understanding the topic of Buck efficiency is to write a detailed Mathcad model of virtually each significant loss. We are deliberately ignoring smaller losses; those that we know barely affect overall efficiency. For example, we know that the RMS current in the output capacitors is very small in a Buck, and we have disregarded that specific loss component. We have also disregarded core losses, since up to 1 MHz, they are usually relatively very small in a Buck. We have also made some "convenient" assumptions, since we do not want to be guilty of not seeing the forest for the trees. So, for example, the transition (crossover) time is set *equal* for turn-ON and for turn-OFF transitions. We have disregarded smaller losses from the charging and dumping of parasitic capacitances, such as those present across the FETs and the inductor. We have also assumed a Schottky of diode drop 0.6 V across each FET during the deadtime. In reality, the body diode may be conducting instead (if the Schottky is not present, or it is not placed properly on the PCB directly across the FET with very low inductance traces). The R_{DS} used and stated in our calculations and graphs is not the nominal datasheet value, or some arbitrarily scaled temperature-compensated value, but the actual value present. We have included controller losses however, as this can rather dramatically affect efficiency at very light loads. We are ignoring pulse-skip modes as they are very implementation-dependent, and difficult to model, and not universally applicable either. But we have included the possibility of running the converter either in forced continuous conduction (i.e., full synchronous) mode at light loads (which we call *FCCM*), or in diode emulation mode (we call it *DCM*, for discontinuous conduction mode).

Note that all our familiar converter design equations typically involve the parameter r, the current ripple ratio, defined in general as $\Delta I/I_{COR}$, where ΔI is the entire current swing (not *half* of the swing as sometimes used in literature) and I_{COR} is the *center-of-ramp* current value, which for a Buck is simply the load current I_O. We know that when $r = 2$, we are in critical conduction mode. We may not realize it, but in fact, all our usual CCM (continuous conduction mode) equations apply even when r exceeds 2, provided we are in FCCM. So all the usual CCM equations were extrapolated down to very light loads ($r > 2$) in the FCCM/CCM case. However, in diode emulation mode, we enter DCM after the $r = 2$ boundary is reached. For DCM, we do have to use the correct DCM equations as provided in Chap. 4. In this manner we can finally describe the performance of the converter under changes in load or line for either operating mode.

But we can also "assemble" each loss one by one, starting from the "ideal converter," at 100 percent efficiency. In this manner we can examine how each loss affects the "shape" of the efficiency curve. This in turn leads us to understanding exactly how to raise the efficiency curve at different load or line conditions.

In the last part of this chapter we present a rather remarkable validation of the efficiency file we are using. Most evaluation boards use low-enough ESR and DCR, so those hardly affect efficiency. The key losses are just the switching (crossover) losses and the switch-related conduction losses. If so, we show how knowing just three points on a typical datasheet efficiency plot, two taken at the same load (but different line conditions), and two at the same line condition (but different loads), allows us to *predict the efficiency at any other line or load condition*. We show this can be used even for older devices like the LM2592 (which uses bipolar transistor switches). We show almost a perfect match in estimating the efficiency at another line and load condition. Clearly, this simple discovery can make life very easy for the rushed (but hopefully not lazy) applications engineer. So, three data points, one Mathcad file, and everything important is known! In fact from such three points *we can even figure out three unknowns: the R_{DS_TOP}, R_{DS_BOT}, and t_{CROSS}* (the R_{DS} of the top and bottom FETs, and the crossover time). We have provided the closed-form equations for that purpose in this book, for the very first time.

Only One Loss Term at a Time: Understanding Each

Our base example is a 5- to 1.8-V converter, with maximum load I_{OMAX} equal to 10 A. We start with no losses. In Fig. 18.1 we introduce only crossover loss. The first thing we vary is the crossover loss itself, by varying the crossover time t_{CROSS}. We also vary frequency, input voltage, load, and r_{SET}. This is the set r at maximum load and at maximum line. Of course, if we change application conditions, by slowly reducing the input voltage for example, or reducing the load, r will vary from the set point. But we are interested in knowing what will be the difference in the efficiency curves if we set r to 0.5 versus, say, 0.2 at the design entry point of the maximum line and load. And that set design value is r_{SET} here.

Note that, in general, *efficiency curves are plotted in two ways: (1) efficiency versus load current (for various constant input voltages), and (2) efficiency versus line voltage (for various constant loads)*. In most of the curves we are following the former method of display. Later we will see the second type appear too. We also use the log scale for load current (*x*-axis) for greater *visibility* of the situation at light loads, so the curves may *look different*. But they actually have the same basic shape as standard datasheet efficiency curves as we will soon see.

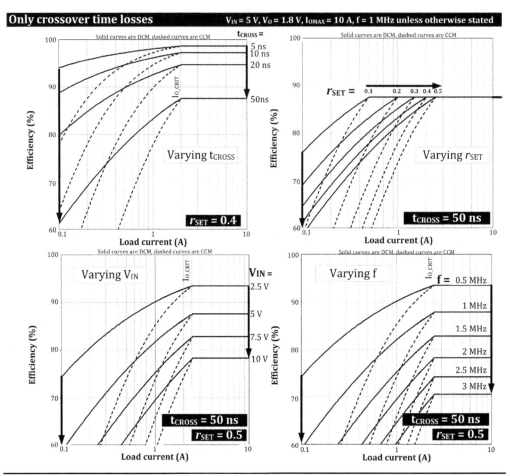

Figure 18.1 Effect of crossover time only (starting with ideal converter).

Crossover Losses

From Fig. 18.1 we see that these losses all stay flat with respect to load current, right up to the point where r exceeds 2 and the curves suddenly fall off. So switching losses seriously affect efficiency below the critical conduction point, and more so, for FCCM/CCM rather than DCM. This realization allows us to significantly reduce the sudden dip in efficiency at light loads. But it is important we do not increase DCR in the process of going to larger inductances to reduce r. Otherwise, the improvement in switching losses at light loads will be swamped out by the increased conduction loss due to higher DCR. In Table 18.1 we summarize our observations and provide detailed suggestions to reduce the crossover losses.

Deadtime Losses

These are actually a mix of what we may call switching losses and conduction losses. They are proportional to frequency, but also depend on the deadtime itself, and the (assumed 0.6 V) drop across each FET during the deadtime. In Fig. 18.2 we plot the efficiency versus load current for only deadtime losses. The findings, and suggestions for improvement, are tabulated in Table 18.2. Note they are very similar to the crossover losses, except for one notable exception: Deadtime losses do not depend on input voltage. That would be one way of trying to gauge on the bench, whether the poor efficiency at light loads is due to crossover losses or deadtime losses (or controller IC and driver losses).

Input Capacitor ESR Losses

In Fig. 18.3 we plot the variations of these and summarize the conclusions in Table 18.3. Note that, as for all conduction losses, changing the underlying resistance (the ESR_IN in this case)

Crossover time profile (one change at a time)		
Parameter	**Effect**	**Suggestion and impact**
Increasing t_{CROSS}	Efficiency will fall as expected, but the fall is equal in the region from I_{OMAX} to critical load I_{O_CRIT}. Below I_{O_CRIT} it will have an increasingly significant effect on efficiency, but more for FCCM than for DCM.	Reduce t_{CROSS} if possible. Will improve efficiency for all loads.
Increasing V_{IN}	Efficiency will fall as expected, but the fall is equal in the region from I_{OMAX} to critical load I_{O_CRIT}. Below I_{O_CRIT} it will have an increasingly significant effect on efficiency, but more for FCCM than for DCM.	Reduce V_{IN} if possible. Will improve efficiency for all loads.
Increasing f	Efficiency will fall as expected, but the fall is equal in the region from I_{OMAX} to critical load I_{O_CRIT}. Below I_{O_CRIT} it will have an increasingly significant effect on efficiency, but more for FCCM than for DCM.	Reduce frequency if possible. Will improve efficiency for all loads.
Increasing r_{SET} (r_{SET} is the set r at maximum load, maximum input)	Changing r_{SET} (different inductance, but maintaining a low-enough DCR as we change r_{SET}) will not affect the efficiency between I_{OMAX} to I_{O_CRIT}. But since efficiency drops below I_{O_CRIT} simply on account of crossover loss in general, a higher r_{SET} will have an increasingly bad effect on efficiency at light loads (both for DCM and FCCM). So, decreasing r_{SET} will reduce the impact of crossover loss significantly at light loads, even for the same crossover time. To reduce r_{SET}, we need a higher inductance. So long as this is not accompanied by an increase in DCR, then from the DCR efficiency curves we see that reducing r_{SET} will not affect efficiency at maximum loads, but will cause great improvement in light-load efficiency on account of the DCR versus r_{SET} plots too. So lowering r_{SET} *without increasing DCR* will cause a great improvement in light load efficiency on account of the profiles of crossover loss and DCR loss.	Reduce r_{SET} if possible. Will improve efficiency significantly at light loads.

TABLE 18.1 Effect of Crossover Time Only and Suggestions to Improve Efficiency

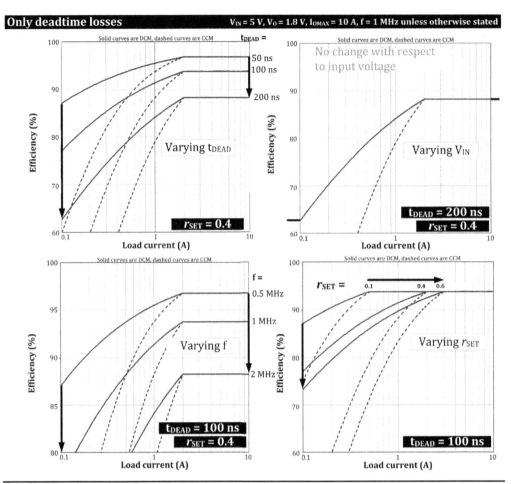

FIGURE 18.2 Effect of deadtime only (starting with ideal converter).

causes the greatest impact on efficiency at high loads. Changing V_{IN} in this case has a U-turn effect: It "maxes" out at $V_{IN} = 2 \times V_O$, corresponding to the highest RMS currents at $D = 0.5$. We know this is true for a Buck input capacitor. (For a Boost, the input capacitor RMS is very small, so we can ignore it usually. For a Buck-boost, the input capacitor RMS increases dramatically and steadily for all duty cycles from 0 to 1.)

Note that changing r_{SET} will not noticeably affect the efficiency at maximum loads, and also not at light loads for DCM, but in the midcurrent range, there is some effect. Since this is a pure conduction loss term, it is unaffected by frequency as expected.

Conduction Losses (R_{DS} and DCR)

In Fig. 18.4 we show that DCR and R_{DS} variations behave similar to ESR_IN conduction losses. One key point to note is that if we set R_{DS_BOT} as zero, then as we increase the input, efficiency improves. That is understandable because as input increases, the duty cycle pinches OFF, and so less time is spent by the inductor current in the dissipative element (the top FET). Similarly, if we only have R_{DS} present in the bottom FET (R_{DS_TOP} is zero), then as we increase the input, efficiency falls as more time is spent in the dissipative element (the bottom FET). In practice, when both R_{DS} terms are present, what happens to overall efficiency with respect to V_{IN} depends on which R_{DS} is bigger: the top FET or the bottom FET.

But there is another important lesson here. If we have a system with $D < 0.5$ (say 5 V to 1.8 V), and we want to *distribute* a net R_{DS} (fixed die cost) appropriately between the top and bottom FET positions, we are better off allocating the lower R_{DS} to the *bottom* FET, since the current spends more time in the lower FET. However, if we have a case where $D > 0.5$ (such as 5 to 3.3 V), we design our system more optimally by allocating the lower R_{DS} to the *top* FET. In fact, in general, we can proportion the two R_{DS}'s to be inversely proportional to the conduction time of each FET, so the losses will be well distributed (and minimum overall). In Tables 18.4 to and 18.6 we have summarized the trends for the R_{DS} terms and DCR for completeness sake. In Fig. 18.5, we compare the relative effect of these conduction loss terms too.

Deadtime profile (one change at a time)		
Parameter	**Effect**	**Suggestion and impact**
Increasing t_{DEAD}	Efficiency will fall as expected, but the fall is equal in the region from I_{OMAX} to critical load I_{O_CRIT}. Below I_{O_CRIT} it will have an increasingly significant effect on efficiency, but more for FCCM than for DCM.	Reduce t_{DEAD} if possible. Will improve efficiency for all loads.
Increasing V_{IN}	Efficiency does not depend on V_{IN} since the drop across the FET during the deadtime (V_{DEAD}) is fixed (we have assumed a default of 0.6 V for the curves). Only changing that voltage drop will affect efficiency results.	Reduce V_{DEAD} if possible. That will improve efficiency for all loads.
Increasing f	Efficiency will fall as expected, but the fall is equal in the region from I_{OMAX} to critical load I_{O_CRIT}. Below I_{O_CRIT} it will have an increasingly significant effect on efficiency, but more for FCCM than for DCM.	Reduce frequency if possible. Will improve efficiency for all loads.
Increasing r_{SET} (r_{SET} is the set r at maximum load, maximum input)	Changing r_{SET} (different inductance, but maintaining a low-enough DCR as we change r_{SET}) will not affect the efficiency between I_{OMAX} to I_{O_CRIT}. But since efficiency drops below I_{O_CRIT} simply on account of deadtime loss in general, a higher r_{SET} will have an increasingly bad effect on efficiency at light loads (both for DCM and FCCM). So, decreasing r_{SET} will reduce the impact of deadtime loss significantly at light loads, even for the same deadtime. To reduce r_{SET}, we need a higher inductance. So long as this is not accompanied by an increase in DCR, then from the DCR efficiency curves we see that reducing r_{SET} will not affect efficiency at maximum loads, but will cause great improvement in light-load efficiency on account of the DCR versus r_{SET} plots too. So by lowering r_{SET} *without increasing DCR*, will cause a great improvement in light-load efficiency on account of the profiles of deadtime loss and DCR loss.	Reduce r_{SET} if possible. Will improve efficiency significantly at light loads.

TABLE 18.2 Effect of Deadtime Only, and Suggestions to Improve Efficiency

Controller IC Losses

We are assuming the controller IC draws a fixed current I_{CONT}, irrespective of input voltage. We see that this has a gradually increasing significant effect at light loads as expected. We have seen that *not* all the conduction loss terms cause the efficiency to fall below the $r = 2$ boundary unless we are in CCM/FCCM. If the chip is DCM-enabled, the drop in efficiency below the $r = 2$ boundary occurs only due to switching losses. And it occurs right from the $r = 2$ boundary. However, if we minimize all switching losses, there will still be a *hump* in efficiency for very light loads—and this is due to I_{CONT} losses. The hump due to this is not related at all to r_{SET} (or where exactly the $r = 2$ boundary is). This becomes clearer when we cumulate the losses next.

Cumulating Losses: Adding Them up One by One

In Fig. 18.6 we now cumulate the loss terms one by one, showing at each step what the impact on efficiency is. So we are in effect constructing the efficiency onion (reverse peeling). We also plot the same without log scales to show the familiar shape of published efficiency curves. We learn that the fall in efficiency at maximum load regions is primarily due to conduction losses, whereas the fall at lighter loads is more due to switching losses, which occurs most significantly below the $r = 2$ boundary. However, by decreasing r_{SET}, from the usual "optimum of $r = 0.4$," to say 0.2 or even 0.1, *but without increasing DCR losses*, will cause a

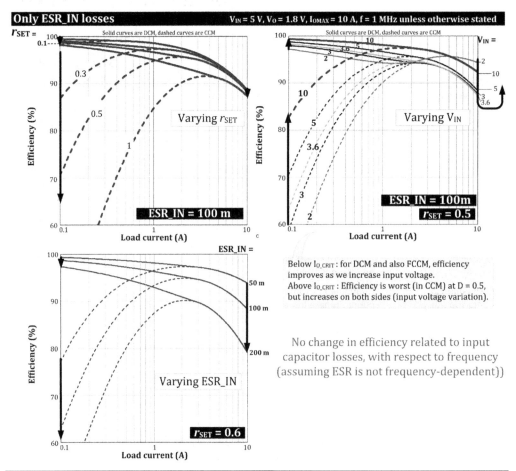

FIGURE 18.3 Effect of ESR_IN only (starting with ideal converter).

Input capacitor ESR profile (one change at a time)		
Parameter	**Effect**	**Suggestion and impact**
Increasing ESR_IN	Efficiency will fall as expected, but the fall is most at maximum loads. For DCM there is almost no effect on efficiency at light loads. For FCCM, there is an increasingly bad effect on efficiency at light loads, but the least effect of ESR_IN is in the region of I_{O_CRIT}.	Reduce ESR_IN to improve high-load efficiency in any mode and light-load efficiency in FCCM.
Increasing V_{IN}	Efficiency does not depend on V_{IN} at light loads in DCM. In FCCM at light loads, increasing input voltage improves efficiency (lower input current). In CCM, at maximum loads, the effect actually depends on duty cycle. When input voltage is twice the output voltage ($D = 0.5$), there is maximum impact on efficiency on account of ESR_IN, the effect decreasing on either side of the input range.	Increase V_{IN} to improve the efficiency at light loads in FCCM and to improve efficiency at maximum loads in CCM too for input voltages greater than $2 \times V_O$.
Increasing f	Efficiency does not change	No effect
Increasing r_{SET} (r_{SET} is the set r at maximum load, maximum input)	Increasing r_{SET} (different inductance, but maintaining a low-enough DCR as we change r_{SET}) will barely affect maximum load efficiency for CCM. It will barely affect light-load efficiency in DCM either, but in CCM/FCCM, it will cause significant worsening of light-load efficiency, and to a lesser extent, some efficiency loss in the region of I_{O_CRIT}.	Reduce r_{SET}, if possible, to improve efficiency at mid and light loads in CCM/FCCM.

TABLE 18.3 Effect of ESR_IN Only, and Suggestions to Improve Efficiency

Watts in It for Us: Unraveling Buck Efficiency

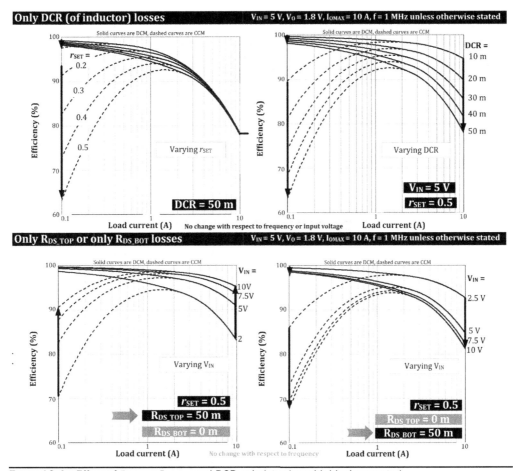

Figure 18.4 Effect of R_{DS_TOP}, R_{DS_BOT} and DCR only (starting with ideal converter).

	Inductor DCR profile (one change at a time)	
Increasing DCR	Efficiency will fall as expected, but the fall is most at maximum loads. For DCM there is almost no effect on efficiency at light loads. For FCCM, there is an increasingly bad effect on efficiency at light loads, but the least effect of DCR is in the region of I_{O_CRIT}.	Reduce DCR to improve high-load efficiency in any mode and light-load efficiency in FCCM.
Increasing V_{IN}	Efficiency does not depend on V_{IN}.	No effect
Increasing f	Efficiency does not change.	No effect
Increasing r_{SET} (r_{SET} is the set r at maximum load, maximum input)	Increasing r_{SET} (different inductance, but maintaining a low-enough DCR as we change r_{SET}) will barely affect maximum load efficiency for CCM. It will barely affect light-load efficiency in DCM either, but in CCM/FCCM, it will cause significant worsening of light-load efficiency, and to a lesser extent, some efficiency loss in the region of I_{O_CRIT}.	Reduce r_{SET}, if possible, to improve efficiency at mid and light loads in CCM/FCCM.

Table 18.4 Effect of DCR Only, and Suggestions to Improve Efficiency

dramatic increase in the maximum efficiency, simply because the switching-loss related hump moves to lower and lower load currents, and that just allows the conduction loss rising curve (for currents to the right of the $r = 2$ boundary) to naturally keep rising more and more before $r = 2$ is encountered and the efficiency falls off.

This effort continues to Fig. 18.7 where we learn to look at the efficiency curves and immediately figure out if the losses are primarily conduction loss related (curves drooping

Upper MOSFET R_{DS} (R_{DS_TOP}) profile (one change at a time)		
Increasing R_{DS_TOP}	Efficiency will fall as expected, but the fall is most at maximum loads. For DCM there is almost no effect on efficiency at light loads. For FCCM, there is an increasingly bad effect on efficiency at light loads, but the least effect of DCR is in the region of I_{O_CRIT}.	Reduce R_{DS_TOP} to improve high-load efficiency in any mode and light-load efficiency in FCCM.
Increasing V_{IN}	Efficiency improves at maximum load (CCM) and at light loads in FCCM/CCM.	Increase V_{IN} to improve the efficiency at light loads in FCCM and to improve efficiency at maximum loads in CCM too.
Increasing f	Efficiency does not change.	No effect
Increasing r_{SET} (r_{SET} is the set r at maximum load, maximum input)	Increasing r_{SET} (different inductance, but maintaining a low-enough DCR as we change r_{SET}) will barely affect maximum load efficiency for CCM. It will barely affect light-load efficiency in DCM either, but in CCM/FCCM, it will cause significant worsening of light-load efficiency, and to a lesser extent, some efficiency loss in the region of I_{O_CRIT}.	Reduce r_{SET}, if possible, to improve efficiency at mid and light loads in CCM/FCCM.

TABLE 18.5 Effect of R_{DS_TOP} Only, and Suggestions to Improve Efficiency

Lower MOSFET R_{DS} (R_{DS_BOT}) profile (one change at a time)		
Increasing R_{DS_BOT}	Efficiency will fall as expected, but the fall is most at maximum loads. For DCM there is almost no effect on efficiency at light loads. For FCCM, there is an increasingly bad effect on efficiency at light loads, but the least effect of DCR is in the region of I_{O_CRIT}.	Reduce R_{DS_BOT} to improve high-load efficiency in any mode and light-load efficiency in FCCM.
Increasing V_{IN}	Efficiency worsens at maximum load (CCM), and at light loads in FCCM/CCM.	Reduce V_{IN} to improve the efficiency at light loads in FCCM and to improve efficiency at maximum loads in CCM too.
Increasing f	Efficiency does not change.	No effect
Increasing r_{SET} (r_{SET} is the set r at maximum load, maximum input)	Increasing r_{SET} (different inductance, but maintaining a low-enough DCR as we change r_{SET}) will barely affect maximum load efficiency for CCM. It will barely affect light-load efficiency in DCM either, but in CCM/FCCM, it will cause significant worsening of light-load efficiency, and to a lesser extent, some efficiency loss in the region of I_{O_CRIT}.	Reduce r_{SET}, if possible, to improve efficiency at mid and light loads in CCM/FCCM.

TABLE 18.6 Effect of R_{DS_BOT} Only, and Suggestions to Improve Efficiency

at maximum load) or switching loss related (curves drooping at mid to light loads). One particular case, where we see a constantly rising efficiency curve right up until I_{OMAX}, actually indicates an excessively large r_{SET} value. The solution is to increase the inductance (decrease r_{SET}) *but without significantly increasing DCR*. That will yield big benefits to efficiency.

In Fig. 18.8, we take the "onion" and show the direct impact on this curve by various maneuvers and stratagems to increase overall efficiency. We show what happens as we change r_{SET}, or frequency, and so on.

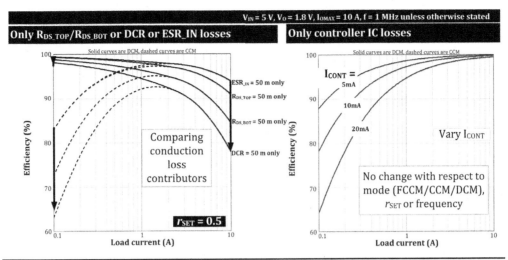

FIGURE 18.5 Comparing effect on efficiency of conduction loss contributors and the IC (controller) current.

The Underlying Buck Spreadsheet

In Fig. 18.9 we share all the equations used in the spreadsheet for all the efficiency curves presented above. These incorporate the relevant DCM equations extracted from Chap. 4.

PART 2 PREDICTING EFFICIENCY AND REVERSE-ENGINEERING TRICKS

In Fig. 18.10 we present the derivations of equations for predicting efficiency. In Fig. 18.11 we apply these equations to the published curves of LM2592 at www.ti.com. We show how by three points, (1, 2, and 3) we can not only predict the fourth point labeled "4," but also accurately figure out the transition time and R_{DS} (or drop across BJT).

In fact, we can go much further. Having figured out the conduction losses and switching losses, and disregarding other losses, we can use our Buck efficiency spreadsheet of Fig. 18.9, to actually regenerate all the curves of the LM2592, as shown in Fig. 18.12.

As closing thoughts, we see that the LM2592 curves were drawn with line voltage on the *x*-axis, not load current. So, we have to alert the reader to be watchful of the differences and not draw the wrong conclusions. Now compare Fig. 18.13 with Fig. 18.14 and notice how the switching loss and conduction loss dominated areas get swapped if we draw the efficiency curves in the two different ways.

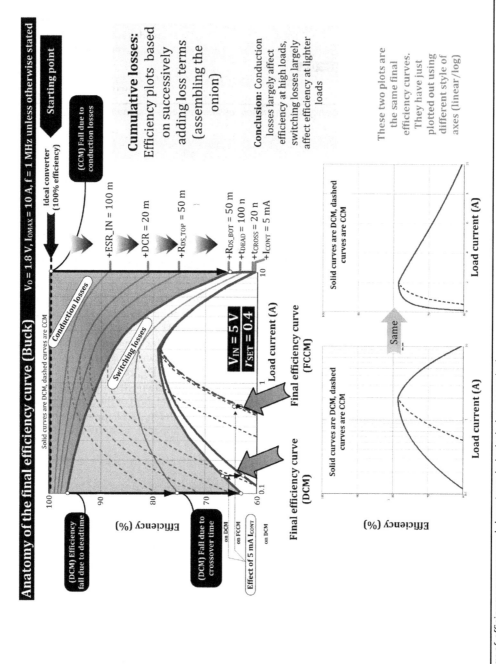

Figure 18.6 Anatomy of efficiency curves: each loss term is included successively.

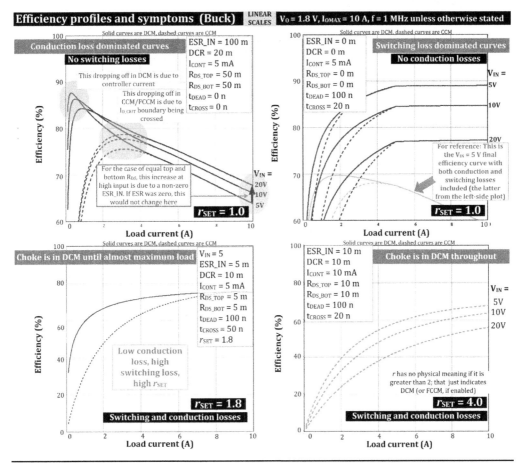

Figure 18.7 Recognizing profiles of measured efficiency and knowing what to fix.

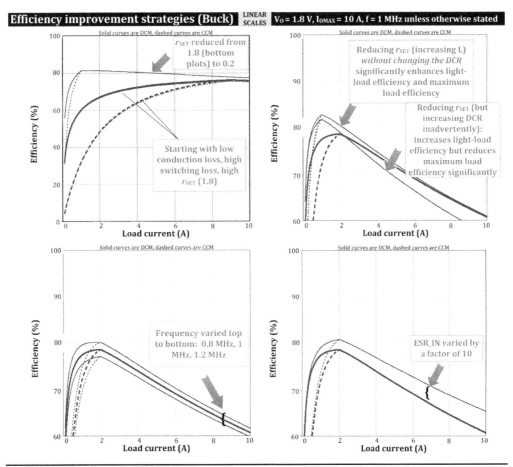

Figure 18.8 Suggestions for improvement in efficiency without major redesign.

Buck efficiency calculations (full Buck model)

CCM/FCCM Efficiency

$$D = \frac{V_O}{V_{IN}}; \quad D' = 1-D$$

$$L = \frac{V_O}{I_{OMAX} \times r_{SET} \times f}(1-D) \quad \text{(find L by setting r to } r_{SET} \text{ at maximum load and maximum input)}$$

$$r = \frac{V_O}{I_O \times L \times f}(1-D) \quad \text{(actual variation of r with load and D)}$$

Conduction losses (in CCM/FCCM)

$$I_{RMS\_TOP} = I_O \sqrt{D\left(1 + \frac{r^2}{12}\right)} \quad \text{(RMS current in top FET)}$$

$$I_{RMS\_BOT} = I_O \sqrt{D'\left(1 + \frac{r^2}{12}\right)} \quad \text{(RMS current in bottom FET)}$$

$$I_{RMS\_IND} = I_O \sqrt{1 + \frac{r^2}{12}} \quad \text{(RMS current in inductor)}$$

$$I_{RMS\_CIN} = I_O \sqrt{D\left(1 - D + \frac{r^2}{12}\right)} \quad \text{(RMS in input cap)}$$

$$P_{TOP} = I_{RMS\_TOP}^2 \times R_{DS\_TOP}; \quad P_{BOT} = I_{RMS\_BOT}^2 \times R_{DS\_BOT}$$

$$P_{IND} = I_{RMS\_IND}^2 \times DCR; \quad P_{CIN} = I_{RMS\_CIN}^2 \times ESR_{IN}$$

Switching (crossover) loss:

$$P_{SW} = \frac{V_{IN} I_O}{2} \times \left(1 + \frac{r}{2}\right) \times (f \times t_{CROSS}) + \frac{V_{IN} I_O}{2} \times \left|1 - \frac{r}{2}\right| \times (f \times t_{CROSS})$$

(for FCCM, allow r to exceed 2, so we need to use magnitude sign above).

Deadtime loss:

Assume Schottky across both FETs, so $V_{DEAD} = 0.6$ V typ

$$P_{DEAD} = \frac{V_{DEAD} I_O}{2} \times \left(1 + \frac{r}{2}\right) \times (f \times t_{DEAD}) + \frac{V_{DEAD} I_O}{2} \times \left|1 - \frac{r}{2}\right| \times (f \times t_{DEAD})$$

t_{DEAD} is deadtime (assumed same for ON and OFF transitions).

Controller IC loss: assume constant current drawn I_{CONT}

$$P_{CONT} = V_{IN} \times I_{CONT}$$

Total loss in CCM/FCCM:

$$P_{CCM} = P_{TOP} + P_{BOT} + P_{IND} + P_{CIN} + P_{SW} + P_{DEAD} + P_{CONT}$$

Efficiency in CCM:

$$\eta_{CCM} = \frac{V_O I_O}{V_O I_O + P_{CCM}}$$

Efficiency when system moves into DCM

$$D_{DCM} = \left(\frac{2 \times I_O \times L \times f \times V_O}{(V_{IN} - V_O) \times V_{IN}}\right)$$

$$I_{PK\_DCM} = \frac{(V_{IN} - V_O) \times D_{DCM}}{L \times f} \quad \text{(peak current)}$$

$$I_{RMS\_TOP\_DCM} = I_{PK\_DCM} \times \sqrt{\frac{D_{DCM}}{3}} \quad \text{(RMS current in top FET)}$$

$$D'_{DCM} = \frac{2 \times I_O}{I_{PK\_DCM}} - D_{DCM} \quad \text{("diode" duty cycle)}$$

$$I_{RMS\_BOT\_DCM} = I_{PK\_DCM} \times \sqrt{\frac{D'_{DCM}}{3}}$$

$$I_{RMS\_IND\_DCM} = I_{PK\_DCM} \sqrt{\left(\frac{D_{DCM}}{3} + \frac{D'_{DCM}}{3}\right)} \quad \text{(RMS in inductor)}$$

$$I_{AVG\_TOP\_DCM} = \frac{I_{PK\_DCM}}{2} \times D_{DCM} \quad \text{(average current in top FET)}$$

$$I_{RMS\_CIN\_DCM} = \sqrt{I_{RMS\_TOP\_DCM}^2 - I_{AVG\_TOP\_DCM}^2} \quad \text{(RMS input cap)}$$

$$P_{TOP\_DCM} = I_{RMS\_TOP\_DCM}^2 \times R_{DS\_TOP}$$

$$P_{BOT\_DCM} = I_{RMS\_BOT\_DCM}^2 \times R_{DS\_BOT}$$

$$P_{IND\_DCM} = I_{RMS\_IND\_DCM}^2 \times DCR$$

$$P_{CIN\_DCM} = I_{RMS\_CIN\_DCM}^2 \times ESR_{IN}$$

Switching (crossover) loss:

$$P_{SW\_DCM} = \frac{V_{IN} I_{PK\_DCM}}{2} \times (f \times t_{CROSS})$$

Deadtime loss:

Assume Schottky across both FETs, so $V_{DEAD} = 0.6$ V typ

$$P_{DEAD\_DCM} = V_{DEAD} \times I_{PK\_DCM} \times (f \times t_{DEAD})$$

Controller IC loss: assume constant current drawn I_{CONT}

$$P_{CONT} = V_{IN} \times I_{CONT}$$

Total loss in DCM:

$$P_{DCM} = P_{TOP\_DCM} + P_{BOT\_DCM} + P_{IND\_DCM} + P_{CIN\_DCM} + P_{SW\_DCM} + P_{DEAD\_DCM} + P_{CONT}$$

Efficiency in DCM:

$$\eta_{DCM} = \frac{V_O I_O}{V_O I_O + P_{DCM}}$$

FIGURE 18.9 Equations for CCM/FCCM and DCM for use in "full Buck model" spreadsheet.

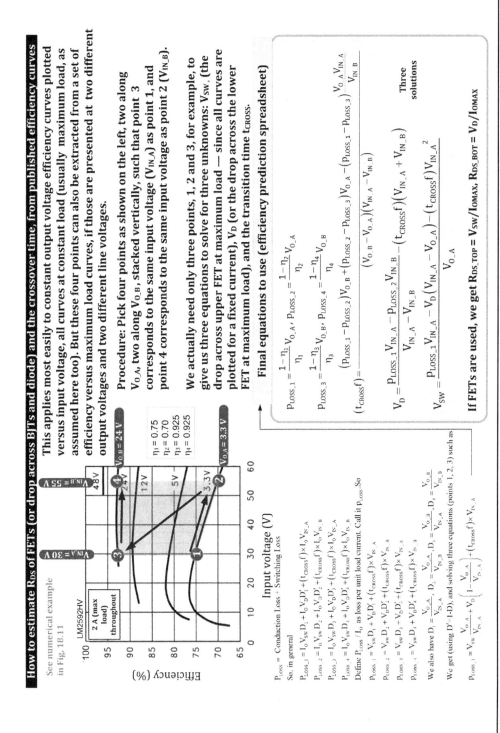

FIGURE 18.10 Solving equations for the efficiency prediction spreadsheet.

LM2592 Calculation
Mathcad-based efficiency prediction worksheet

Io := 2   All selected points are at this load current (usually max load)

VinA := 30   VinB := 55   Select two input voltages

VoA := 3.3   VoB := 24   Select two output voltages

Input data (three points, 1, 2 and 3)

Measured efficiencies (points 1, 2, 3 and 4):

$\eta 1 := 0.75$ $\eta 2 := 0.7$ $\eta 3 := 0.925$ $\eta 4 := 0.925$

$$\text{ploss1} := \frac{(1-\eta 1)}{\eta 1} \cdot \text{VoA} \quad \text{ploss2} := \frac{(1-\eta 2)}{\eta 2} \cdot \text{VoA} \quad \text{ploss3} := \frac{(1-\eta 3)}{\eta 3} \cdot \text{VoB} \quad \text{ploss4} := \frac{(1-\eta 4)}{\eta 4} \cdot \text{VoB}$$

(Equations for loss per unit load current as a function of efficiency)

$$k := \frac{1}{(\text{VoB} - \text{VoA}) \cdot (\text{VinA} - \text{VinB})} \cdot \left[(\text{ploss1} - \text{ploss2}) \cdot \text{VoB} + (\text{ploss2} - \text{ploss3}) \cdot \text{VoA} - (\text{ploss1} - \text{ploss3}) \cdot \frac{\text{VinA}}{\text{VinB}} \cdot \text{VoA} \right]$$

$$\text{Vd} := \frac{\text{ploss1} \cdot \text{VinA} - \text{ploss2} \cdot \text{VinB}}{\text{VinA} - \text{VinB}} - k \cdot (\text{VinA} + \text{VinB})$$

$$\text{Vsw} := \frac{\text{ploss1} \cdot \text{VinA} - \text{Vd} \cdot (\text{VinA} - \text{VoA}) - k \cdot \text{VinA}^2}{\text{VoA}}$$

Equations from Fig. 18.10 (in simple Mathcad notation)

Vsw = 1.74   Volts   Switch drop

Vd = 0.514   Volts   Diode drop

Solutions

Transition time $f := 150 \cdot 10^3$ $\text{tcross} := \frac{k}{f}$ $\text{tcross} = 1.002 \times 10^{-7}$

Predicting efficiencies at point 4 for example

$$D4 := \frac{\text{VoB}}{\text{VinA}} \quad \text{Ploss4\_calc} := \text{Io} \cdot [\text{Vsw} \cdot D4 + \text{Vd} \cdot (1 - D4) + k \cdot \text{VinB}]$$

$$\eta 4\_\text{calc} := \frac{\text{VoB} \cdot \text{Io}}{\text{VoB} \cdot \text{Io} + \text{Ploss4\_calc}}$$

$\eta 4\_\text{calc} = 0.928$ $\eta 4 = 0.925$

Calculated versus measured

Efficiency at any point can now be predicted very accurately — from three initial points!

Final Results
LM2592 example

Diode drop   Vd = 0.514
Switch drop   Vsw = 1.74
Crossover time $\text{tcross} = 1.002 \times 10^{-7}$

All forward drops calculated. Transition time also now known.

In case of Synchronous version

$$\text{Rds\_top} := \frac{\text{Vsw}}{\text{Io}} \quad \text{Rds\_top} = 0.87 \text{ ohms}$$

$$\text{Rds\_bot} := \frac{\text{Vd}}{\text{Io}} \quad \text{Rds\_bot} = 0.257 \text{ ohms}$$

Instead of BJT's, if FETs are used, they can be handled in this manner

FIGURE 18.11 The efficiency prediction spreadsheet in action on the LM2592.

Watts in It for Us: Unraveling Buck Efficiency

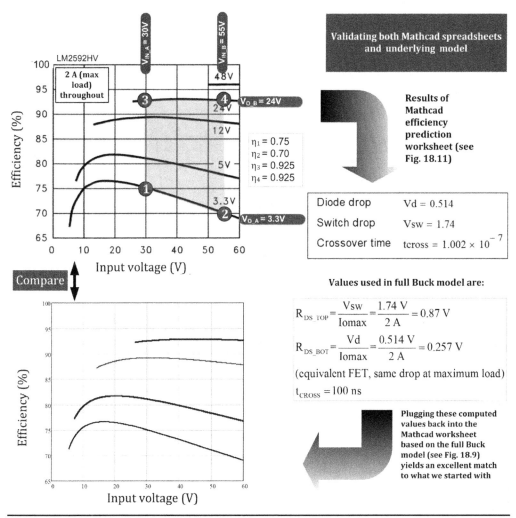

FIGURE 18.12 Validating both the Buck spreadsheets by generating the same measured curve we started off with.

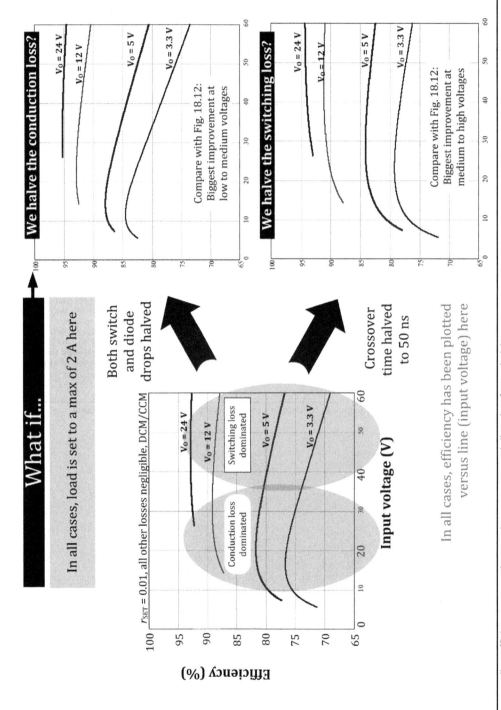

FIGURE 18.13   Understanding efficiency versus input voltage curves and direction for optimizing.

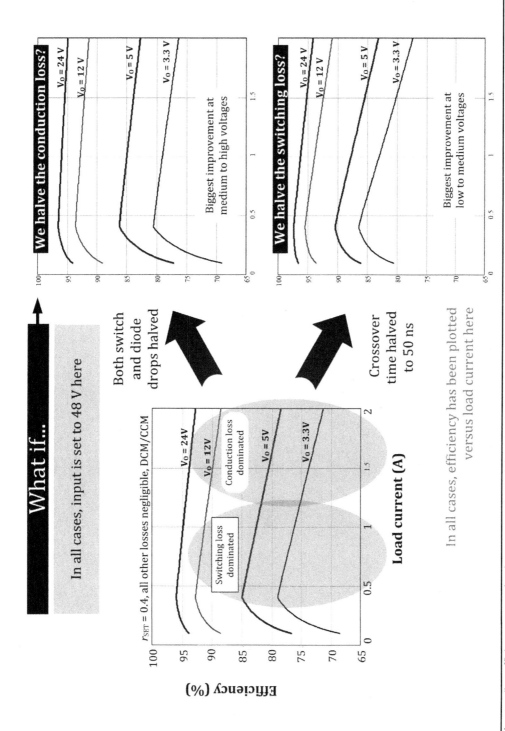

FIGURE 18.14 Understanding efficiency versus load current curves and direction for optimizing.

CHAPTER 19

Soft-Switching and Designing *LLC* Converters

PART 1 OVERVIEW OF THE TRANSITION FROM CONVENTIONAL PWM POWER CONVERSION TO RESONANT TOPOLOGIES

The reader is cautioned that this chapter involves a unique design methodology of creating an *LLC* converter discovered by this author. It offers very wide input voltage variation compared to existing *LLC* converters. This unique method also allows fixing the minimum and maximum of frequency operation precisely. It may be covered by various intellectual property and/or trade secret protections.

Introduction

At an early moment during the development of conventional *square-wave* switching power conversion, it was said that nature abhors any attempt to suddenly turn OFF the current passing through an inductor. That fear prompted attempts at creating *resonant topologies*. But for various reasons, in particular a very wide, almost uncontrolled switching frequency and resulting EMI spread, resonant topologies did not become mainstream, and *still aren't*. But recently, there is a sudden resurgence of interest, primarily through the "*LLC* topology." This unique combination of two inductors and one capacitor ("L-L-C"), offers a relatively narrow range of switching frequencies, which are much easier to design a standard EMI filter for, combined with the capability of producing zero-voltage switching (soft switching) through careful design, which can significantly improve EMI and efficiency over a wide load range. But the overall topic and analysis is still considered complex, almost mystical, and is relatively poorly understood. A majority of successful *LLC* designs are truthfully still based on the age-old method called *trial and error*, that is, the art of tweaking. The difference is that nowadays it may be happening in a real lab environment, but also in a virtual lab, using SPICE-based simulators such as PSpice, LTSpice, and Simplis. This trial-and-error phase has significantly hindered their adoption in a modern design-and-development, flowchart-based, commercial product environment.

To tackle the underlying *fear of resonance*, in these pages, emphasis is laid on first understanding the guiding principles of resonance and soft switching. Only after acquiring the underlying intuition and analytical depth, do we start to build the *LLC* converter, and we do that in steps: assembling the building blocks *one by one*, at each stage analyzing the performance, providing the necessary simplified math, and validating each step via Mathcad and Simplis. In the process, a very unique method is being proposed on these pages perhaps for the very first time. It starts with what we are calling an *LLC* seed. This is quite literally a well-analyzed resonant *seed*, from which we can grow, literally generate, almost any desired *LLC* converter—through *power and frequency scaling* techniques, which we sometimes forget, or don't realize, but in fact use almost intuitively every day, in conventional "square-wave" switching power conversion. (Scaling is a well-known technique, as discussed in the author's *Switching Power Supplies A-Z*, 2d ed, pp. 512–523). Finally, some of the fear of using *LLC* topologies for a *wide-input* application is also tackled by generating a universal Mathcad-based lookup chart, which tells us very clearly how we can trade off power capability to attain a broader input range—so, in effect we overdesign, but not the *entire* LLC converter (power stage), *only its constituent LLC resonant cell* for achieving that purpose.

Finally a complete top-down design for a 25-W PoE powered device (PD) is presented. So far, the LLC topology has not really been seen in PoE applications, perhaps because in PoE, we typically need a rather wide-input, low-voltage input operating range of 32 to 52 V (mandated is actually 36 to 57 V) for the switching converter stage. Most LLC designs around us today are in the high-voltage region, being run off the well-regulated HVDC (high-voltage DC) output rail of a conventional PFC front-end in an AC-DC application, so they "see" very little input voltage swings. But this input swing restriction is not really necessary, once we learn the techniques presented here.

Soft and Hard Switching

To close loose ends, we need to revisit some of the originally expressed anxiety against interrupting inductor current in *square-wave* switching power conversion. That fear too is admittedly somewhat true, because energy is stored in an inductor as $(1/2) \times L \times I^2$, which depends on the current passing through it at any given moment. Since by the law of conservation of energy, energy cannot just be *willed away* in an instant, we should never try to interrupt the inductor current by simply placing a switch (or FET) in series with it and turning that OFF. That attempt *will* destroy the switch—on account of a huge voltage spike (induced voltage) which will suddenly appear. But there is a way out as people soon had discovered: *We can consciously provide an alternate "freewheeling" path for the inductor current—to keep it flowing smoothly (without any discontinuity) once the switch turns OFF.* That logic led to the *catch diode* seen in conventional switching power conversion. The provision of this diversionary diode path, a detour in effect for inductor current, is a much wiser option than trying to contain the current in a cul-de-sac of sorts, leaving it no way out (and then it *will* fight back!). The catch diode is the reason we can, by now, manage to get away with turning the switch (FET) of any conventional power converter ON and OFF repetitively, hundreds of thousands of times every second, and reliably so. Nature is clearly not complaining. But there *was a price to pay* for this inductor-driven brinksmanship of sorts. Diverting the current at the very last moment, and we do mean *very last moment* here, does create *penalties*. Because whenever we try to turn OFF inductor current, by say dragging the Gate of the series FET to ground (or to its source terminal as applicable), we do *not* manage to *instantaneously* reduce the current in the FET as we had perhaps hoped to do. On closer analysis we see that there needs to be a full rail-to-rail *voltage swing completed* (across the FET), just to be able to forward-bias the catch diode to get it to even *start* taking up some of the current away from the FET. But before that happens, we cannot afford to somehow force the FET to relinquish its current, for fear of the induced voltage effect. In other words, there is a transitory moment where there is a full voltage swing taking place across the FET. Though the average value of this voltage swing is half its peak-to-peak value, it is present with the *full* preexisting inductor current continuing to pass through the FET. There is *no* reduction in FET current *yet*, because the diode is still in the process of being "set up" to *start* taking up slack. And that does not even happen until the *entire* voltage swing is *completed* and the diode gets *forward-biased*. Geometrically speaking, we see an "overlap" of voltage and current across the FET as shown in Fig. 19.1, and that constitutes "crossover" (switching) losses in the FET. Mathematically speaking, there is a non-zero $V \times I$ product across the FET during the switch transition. Yes, there are other subtle contributors to this crossover loss component too, such as the forcible discharging (burning up of energy) of the parasitic capacitance across the FET, and also the drain-to-Gate *Miller effect* causing an increase in switching (transition) time and causing even more V-I crossover energy to be wasted, etc. Therefore, as we go to higher and higher switching frequencies, this useless *work* done by the input source (in the form of dissipation) leads to several percentage points of degradation in overall efficiency. It is for these reasons that the small, but extremely significant moment of switch transition is garnering a lot of attention today, piquing renewed interest in resonant topologies that offer hope in this regard.

The reduction of switching losses during the few nanoseconds of switch state transition (i.e., crossover of the FET) is critical to improving power-conversion efficiency. But we are also hoping to reduce EMI by the softer "resonant transitions," since we know that conventional "hard transitions" are the main source of most of the EMI in conventional converters. We are almost intuitively expecting that by using resonant topologies, the voltage will be softly reduced by self-resonant action, so that the V-I "overlap" will be likely reduced, or become *almost zero*, as also displayed in Fig. 19.1. If so, that should help reduce EMI too.

Soft-Switching and Designing LLC Converters 463

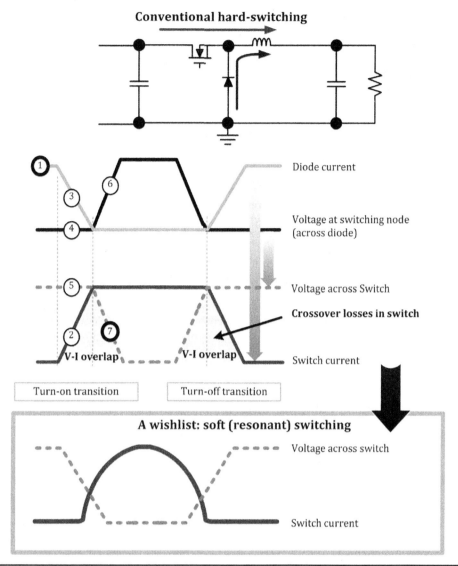

FIGURE 19.1 Hard switching compared to resonant (soft) switching.

NOTE *Another way out, suitable for low-power applications, is to use discontinuous conduction mode (DCM), because we can thereby ensure the current is zero whenever we turn the FET ON. We can implement this using the well-known ringing choke converter (RCC) principle, which basically senses the ringing voltage of the inductor/transformer to gauge when there is no residual current left (i.e., core is deenergized) and turns the FET ON at that moment. This is a variable-frequency PWM, in the form of boundary-conduction mode (BCM). It was actually used on a very wide scale in the 1980s in the form of the ubiquitous and historic television power supply Flyback controller IC, the TDA4601, originally from Philips, and still available from Infineon. The modern iPhone charger seems to be using the L6565 from ST microelectronics, which though based on the same old ringing choke principle of the TDA4601, prefers to call itself a QR (quasi-resonant), ZVS (zero voltage switching) topology SMPS controller chip, in keeping with the times. But yes, there are still switching losses when we turn the FET OFF.*

Keep in mind, however, that the resonant soft-switching sketched intuitively in Fig. 19.1 is only a "wish list" so far. It is all much easier said than done. On deeper examination, resonant topologies are not similar or identical in *all* respects. Not all "naturally" offer soft-switching, for example. In the well-known words of Bob Mammano (quoted from his presentation titled "Resonant Mode Converter Topologies," Topic 6, in the 1988/89 series of Unitrode Power Supply Design Seminars): "While basically simple, this principle can be applied in a wide variety of ways, creating a bewildering array of possible circuits and operating modes."

In fact not all resonant topologies and modes are necessarily even conducive to the task of reducing switching losses, reducing stresses, improving efficiency, or even reducing EMI. In fact, some do not even lend themselves to a proper *control strategy* as we will see. This last aspect is extremely important but often overlooked in a virtual lab. The entire topic of resonant power conversion turns out to be an extremely complex one to (1) understand and (2) implement effectively.

But do we at least fully understand *conventional* power conversion well enough? We may realize we need to do better in that regard, because some *critical hints for the successful analysis and implementation of resonant topologies are contained in conventional power conversion*. We may have missed the signs. For example, a potentially puzzling question in a conventional converter is: Why do we *not* have any overlap of voltage and current *across the catch diode*? As indicated in Fig. 19.1, there is in fact no significant *V-I* overlap across the diode, which is the reason we typically assume almost zero switching losses for the diode in an efficiency calculation. This does remain a valid assumption to make, even when we move to synchronous topologies—in which we replace (or supplement) the catch diode with *another FET*. So, consider this puzzle: We now *have a totem pole of nearly identical FETs (in a synchronous Buck, for example), yet we somehow still disregard the switching losses in the lower FET, but not in the upper FET*. How come? In what way is the *location* of the lower FET so different from that of the upper FET? Why does nature seem to favor the top location from the bottom one?

Even more surprisingly, if we delve deeper we will learn that in one particular operating mode, for just *part of the cycle*, even the synchronous Buck exhibits almost no crossover loss in the top (control) FET, and actually shifts those losses to the lower (synchronous) FET! How did this role reversal happen? *Once we understand all this, we will understand the intuitive direction we need to take in designing good resonant topologies too*.

Two Key Concerns (Guiding Criteria)

These form our guiding criteria toward identifying *suitable* resonant topologies. The most basic questions we need to ask of any proposed circuit are as follows.

Question 1. Does it *really* reduce switching losses? If so, in which *region of its operation* (or operating mode)? We need to know the line and load boundaries of any such soft-switching region, and thus try to ensure that at least at maximum load, we can significantly reduce switching losses (and preferably continue to do so at light loads too). Finally, we hope we can do this, without asking for, or somehow creating, too *wide a variation in the switching frequency*. Because, though we may have soft switching, if we have a very wide variation in switching frequency, that too cannot be good from the viewpoint of either the economics, or the size, of the EMI filtering stage. We do not gain by simply shifting the cost and dissipation from one circuit block to another (the EMI stage in this case).

Question 2. The simplest, seemingly most obvious, question can in fact become the hardest to answer: How do we "regulate" the output voltage (of a resonant converter)? Do we do pulse-width modulation (PWM) or something else? And is there a simple way to implement an unambiguous, almost "knee-jerk" (rapid) response to disturbances, *analogous to what we do in conventional power conversion?* We remember that in conventional converters, we pretty much blindly *increase* the duty cycle in response to a *fall* in the output below a set or reference level, for *any reason whatsoever*. Similarly, we decrease the duty cycle if the output rises. That is the heart of basic PWM implementation. We also know that, luckily, for all conventional topologies, a *falling input rail* or an *increasing load both* tend to cause the output to *fall*, both therefore demanding an *increase* in duty cycle (pulse width) to correct. Though that rapid response does in effect lead to the complicated area of loop stability (overcorrection versus undercorrection imbroglio), the good news is that we do at least have a *simple, immediate, and unambiguous response to line and load disturbances*, one which is guaranteed to always be in the *right "direction" for achieving correction*. After all, it makes no sense that we start to increase the duty cycle, only to find the output falls even more (somewhat like the right half-plane zero issue in Boost and Buck-boost converters). We cannot, for example, demand a complicated control scenario that asks of us something like: Increase the duty cycle if the input rail falls to 80 percent of its set value, but thereafter *decrease* the duty cycle (to correctly regulate the output). This algorithm would be impossible to implement practically. But unfortunately, *in resonant topologies, we can get exactly this odd situation*. So, if we do not define a very clear region of operation (the allowable and expected line and load variation, related to component selection), we will end up in an area of operation where we are *expecting* the output to get corrected by our response, but in fact the output veers away in the *opposite direction* (hopefully collapsing at least, not overshooting). This becomes another additional, and very tricky, part of the resonant topology puzzle that we seek to uncover in these pages.

We have listed the two key concerns or guiding criteria that will be addressed as we go along. But before we do that, as mentioned, first we need to understand a little more about *conventional converters* and understand, in particular, why the switching losses are lopsided, that is, losses in one FET, not in the other, in a synchronous Buck converter, for example. That will take us to the next step of development of resonant topologies.

Note that we are eventually going to split our total task ahead into *four basic steps*:

1. Understand the behavior of modern conventional (square-wave) synchronous topologies and the physical conditions that can help create soft-switching effects in them. Carry these lessons over to resonant circuits.

2. Understand resonant *components*, and build resonant *circuits* out of them, at first driven by pure AC sources, thereby identifying the particular LC network ("tank circuit") that "works" in all desired aspects, especially as related to the guiding criteria presented earlier.

3. Replace the AC source (the excitation to the tank circuit) with "square-switched" transistors, as in conventional power conversion, to reduce conduction-loss related dissipation in the switches, and see how the square switching approximates a sine AC source from the viewpoint of the tank circuit and its response.

4. Then add output diodes and output filtering capacitors to create a rectified DC rail, and see the effect of these on the previous circuit.

In this manner we hope to create a usable resonant converter at the end of the road, based on the 'guiding criteria' listed previously.

Switching Losses in a Synchronous Buck and Lessons Learned

In Fig. 19.1, we first consider the *turn-ON* transition (on the left side). Prior to this, the diode is observed carrying the full inductor current (circled "1"). It is obviously forward-biased. But then the switch starts to turn ON, trying to take away the inductor current (circled "2"). Correspondingly, the diode current starts to fall (circled "3"). However, the important point is that while the switch current is still changing and has not yet taken over the full inductor current, the diode needs to continue to conduct the remaining current, i.e., the difference between the inductor current and the switch current, at any given moment. However, to conduct even *some* current, the diode must remain *fully* forward-biased. So, there is no change in the voltage across the diode (circled "4"), or therefore, across the FET (circled "5"), as the current through the FET swings from zero to maximum, and in the diode from maximum to zero (crossover). Finally, only when the *entire* inductor current has shifted over to the switch, does the diode "let go" of the voltage. The switching node is released, and it flies up very close (a little higher) to the input voltage rail (circled "6"). So now, the voltage across the switch is allowed to fall (circled "7") by Kirchhoff's voltage law.

We therefore see that *at turn-ON, the voltage across the switch does not change until the current waveform has completed its transition*. We thus get a significant *V-I* overlap in the switch (FET). That is "hard switching" by definition.

If we do a similar analysis for the turn-OFF transition (right side of Fig. 19.1), we will see that for the switch current to start decreasing by even a small amount, the diode must *first* be "positioned" (in terms of voltage) to take up *any* current coming its way (relinquished by the switch). So the voltage at the switching node must *first* fall close to zero (a little below ground), so as to forward-bias the diode. That also means the voltage across the switch must first transition *fully*, *before* the switch current is even allowed to decrease slightly.

We therefore see that at turn-OFF, the current through the switch does not change until the voltage waveform has *completed* its transition. We thus get a significant *V-I* overlap in the switch (FET).

But in neither transition, turn-ON or turn-OFF, is there any significant V-I *overlap across the diode.* We therefore typically assume that there are negligible switching losses in the diode in a conventional topology—it is the switch (FET) that gets hard-switched during *both* transitions.

466 Chapter Nineteen

> **NOTE** *In a Boost topology, despite no measurable V-I overlap term present across the diode, the diode can surprisingly cause significant additional switching losses to appear in the FET, rather than in itself. For example, we can get losses in the FET due to shoot-through (poor reverse recovery characteristics of the catch diode, or the relatively "poor" body diode of a synchronous boost converter's synchronous FET). In other words, the diode may technically never "see" any significant V-I crossover loss across itself, but it can certainly cause, or instigate, the switch (FET) to experience significant, perhaps even externally invisible, losses. This particular situation is often tackled by trying to achieve zero-current switching (ZCS), rather than the more common ZVS (zero-voltage switching) which we focus on in these pages. One simple way out, especially for low-power applications, is to run the converter in DCM, because if the Boost diode is no longer carrying current, it has no reverse recovery issues either (it has already recovered when the switch tries to turn ON).*

However, now let us look at the synchronous Buck in Fig. 19.2 more closely, at what really happens when the inductor current is *momentarily negative* (below zero, i.e., flowing from top to bottom, through the lower FET). Suppose during this negative current phase, the lower FET turns OFF, while the top FET is forced ON. It turns out we have to pay close attention to what happens *during the deadtime*, that is, during the small interval between one FET turning OFF and the other turning ON. Most engineers are aware that the primary purpose of deadtime is to avoid any brief possibility of both top and bottom FETs being ON at the very same moment, which would cause efficiency loss due to cross-conduction/shoot-through, and possible destruction of the FETs too. But there is another, subtle advantage that deadtime provides, as explained in Fig. 19.2.

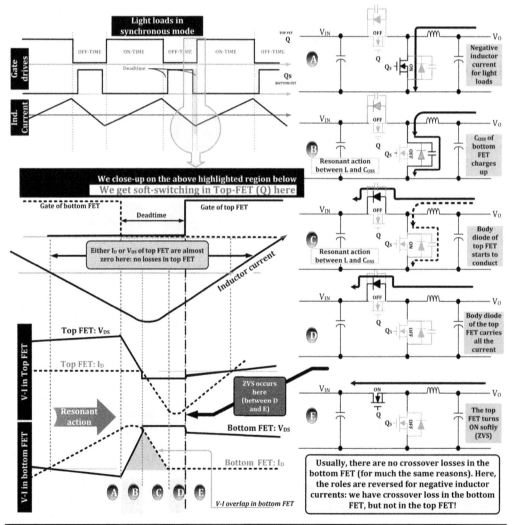

FIGURE 19.2 We do get ZVS naturally, in some cases or modes of the synchronous Buck too.

From Fig. 19.2 we can conclude that if the current is momentarily *negative*, and we turn OFF the lower FET at that moment, the inductor will in effect resonate with the drain-source parasitic capacitances of the FETs, causing the voltage at the switching node to rise, eventually causing a flow of current back into the input rail via the body diode of the top FET. Keep in mind that to cause the body diode to conduct, it must be forward-biased. So the voltage on the switching node must swing. Now when the top FET turns ON, it does so with near-zero voltage across it (a forward-biased body diode drop across it). That is *ZVS* by definition. We see the top FET being "soft-switched" for a change. Though this effect is occurring only during part of the switching cycle, it is exactly the way it can be made to happen in both (or more) FETs in resonant power conversion too.

NOTE *Hypothetically, if the average of the inductor current of the Buck falls completely below zero (possibly a small part of the inductor current waveform is positive), we would be now constantly drawing current from the "output" (now really an input) and delivering it to the "input" (now really an output). If no part of the waveform is above zero, we would have in effect a full-fledged (conventional) Boost topology, not a Buck topology anymore. And we also know that in a Boost, it is the lower FET that is the "control FET" and the upper FET is the synchronous FET. We therefore intuitively expect to see only the lower (control) FET to have switching losses in a Boost. And, quite expectedly, all the crossover losses of this reversed Buck (which actually makes it a Boost) will now shift to the lower FET over the entire switching cycle. We will have no crossover loss anymore in the top FET. So the "role reversal" now makes sense. During the time when part of the inductor current is negative (in a synchronous Buck), a switch transition during that interval will produce switching losses in the synchronous (bottom) FET, not in the upper (control) FET. Only when a switching transition occurs during the positive-current part of the cycle do we have switching loss in the control (top) FET, as is usually assumed to be true always.*

We have learned that we can achieve ZVS for whichever FET has current flowing *through its body diode* (or an antiparallel diode placed externally across it) during the *preceding* deadtime (just before transition). So, the *direction* of inductor current at the moment of transition is the key that identifies (or distinguishes) which FET position receives soft switching and which one gets hard-switched. Keep in mind that what constitutes a "positive" or a "negative" current depends completely on what we are calling the "input" and what is the "output" and which direction we consider "normal" energy flow.

Note that in either of the preceding scenarios, we *certainly need current passing through an inductance to try and "force matters" to achieve ZVS*. And of course, we also need to leave a small (but not too small) intervening deadtime for the induced voltage to be able to act. We also need enough inductance to force a *full* voltage swing across the parasitic drain-source capacitances in particular. Very similarly, in resonant topologies too, whatever the type of simple or complex L-C network is being switched, we learn that *the tank circuit needs to appear "inductive" to the input source, for being able to achieve ZVS*. This is a basic condition for ZVS.

Summarizing, the two most basic prerequisites for achieving ZVS are:

1. We need the current to "slosh" *back and forth*—something that occurs naturally in *resonant* topologies but is also enforced in conventional half-bridge, full-bridge wave, and push-pull topologies. All these are suitable candidates for ZVS, subject to the following stated condition.

2. The tank circuit (network) impedance must appear *inductive* to the input voltage source, because that is what is responsible, eventually, for trying to brute-force the current through (using induced voltage), and in the process creating zero-voltage switching across the FETs—though as mentioned, we also must have *enough inductance* (vis-à-vis available deadtime and parasitic FET output capacitances) so that the voltage on the swinging node can be forced to swing in a *timely* manner (before the deadtime runs out, so that in turn it can reduce the voltage across the FET *before* the FET is turned ON).

NOTE *Even in "ZVS-eligible" conventional topologies such as the full-bridge, we can enforce the second requirement and produce "quasi-resonant (QR) ZVS." These topologies are all based on the Forward converter topology using a transformer. We need to add a small primary-side inductance (often just the transformer leakage) for this purpose.*

Building Basic Resonant Circuits from Components

In resonant topologies, just as in conventional switching power conversion, we try to pick *reactive* components (inductors and capacitors). Because we know that ideally, they store energy, but cannot dissipate any. We always require *resistance*, either in the form of intervening parasitics, or in the form of the load itself, to dissipate any energy, either usefully or wastefully. But in addition, the L and C can pass energy back and forth *between each other* on account of their *complementary phase angles*. We can try to use that useful property to transport energy on their shoulders, from the input (source) to the output (load), somehow creating *DC regulation too along the way if possible,* which is a fundamental requirement of any type of practical power converter.

That is the general direction for implementing any "lossless" (high-efficiency) power conversion methodology. The key difference between resonant topologies and conventional switching topologies is primarily related to the actual values of the L and C components used. In conventional power conversion, we use relatively "large" capacitors at the input and output, so the natural LC frequency happens to be very large relative to the switching frequency. We just do not "see" resonant effects anymore, because *we do not wait that long* before we switch. But if we reduce the L and C values significantly, we will start to see resonant effects even in conventional power conversion. Keep in mind though that just because they are *resonant* doesn't make them *acceptable* or useful. Their eventual usefulness is judged mainly on the basis of the two "guiding criteria" described previously.

NOTE *Modifying conventional topologies by reducing their L and C values is usually not the best way to create resonance, except, for example in the "ZVS phase-modulated full bridge," which is best described as a "crossover converter" (like a crossover vehicle), literally bridging conventional power conversion with resonant effects (though resonance occurs only during the transition deadtimes). This maintains the desirable characteristics of constant clock frequency of conventional PWM topologies but uses resonance during deadtime for inducing soft switching (in all the FETs). But it also has four FETs, and typically an extra inductor too.*

To create proper resonant circuits, let us start with a simple inductance and a simple capacitance. But they are *not connected to each other yet*. To make this "real," let us pick some numerical values. We choose $L = 100$ mH and $C = 10$ µF. These may seem rather big, but as a result, our switching frequency is also going to be low, and that is quite helpful, initially at least, for ease of discussion and for cleaner and faster circuit simulations, etc. So, everything is initially being scaled to a lower switching frequency for convenience.

Let us connect each separated L and C component to *identical* AC sources, of 30 V (amplitude, or half peak-to-peak value) with a frequency of 300 Hz, and just see the currents through each. We run this through a Simplis simulator. The results are presented in Fig. 19.3.

As expected, the input and output voltages on either component are the same, at 30 V. The currents are different. We expect the corresponding current amplitudes to be:

$$Z_C = \frac{1}{2 \times \pi \times f \times C} = \frac{1}{2 \times \pi \times 300 \times 10^{-5}} = 53.05 \; \Omega \quad \text{So } I_C = \frac{V}{Z_C} = \frac{30}{53.05} = 0.565 \text{ A}$$

$$Z_L = 2 \times \pi \times f \times L = 2 \times \pi \times 300 \times 0.1 = 188.5 \; \Omega \quad \text{So } I_L = \frac{V}{Z_L} = \frac{30}{188.5} = 0.159 \text{ A}$$

This is exactly what we got through the simulations shown in Fig. 19.3. But via that figure, we see another possibility, one that we can hopefully exploit: The peak of the inductor current comes a little *later* (exactly one-fourth of a cycle) after the peak of the inductor voltage (which is the same as the input voltage in this case). That is why we usually say that the *current lags the voltage in an inductor by 90°*. We can confirm from conventional switching power conversion too, that when we apply a voltage across an inductor, the current ramps up slowly. So current does lag the voltage in that sense, though unfortunately, we can't really visualize or define what the "lag" is for nonsinusoidal waveforms, as in conventional power conversion. Now, looking at Fig. 19.3 once again, we see that the cap voltage peak *lags* the cap current peak by exactly the same amount, that is, 90°. In other words, the *capacitor and inductor currents are relatively exactly 180° out of phase*. They are *complementary*, because 180° is *just a change of sign* or direction (with respect to each other). Note that we have implicitly chosen and assumed the convention that current *into* the component (L or C) is "positive," whereas current coming out of it is "negative". So a 180° relative phase shift simply means that when 159 mA

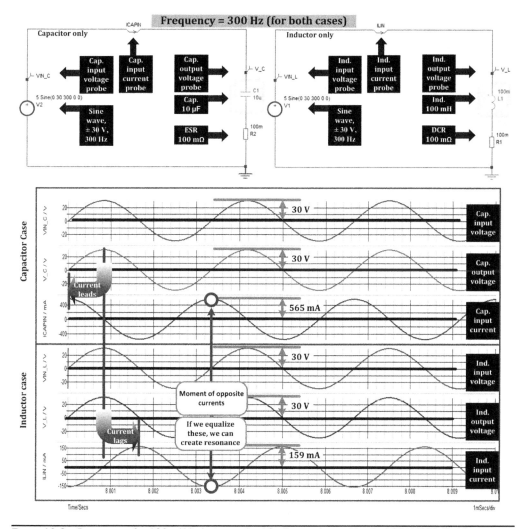

FIGURE 19.3 Response of a 100-mH inductor and a 10-μF capacitor to identical 30 V/300 Hz AC sources.

is coming *out* of the inductor, *exactly at that moment*, 565 mA is going *into* the capacitor. We can confirm this from Fig. 19.3.

The preceding logic also leads us to the following thought process: Could we have X mA coming out of the inductor and exactly X mA going into the capacitor at the same moment? Yes, by varying the frequency we can always do that—because in one case (inductor) the impedance increases with frequency, whereas in the other case (capacitor) it decreases with frequency. So certainly, we can get the two values to converge at some "intersection" point at which their impedances would be equal *in magnitude* (opposite in sign, though). See Fig. 19.4. That intersection then forms a "natural (or resonant) frequency" for the chosen L and C. The only question is: At what frequency? That is just the point at which the inductor's impedance $2\pi f \times L$, equals the capacitor's impedance $1/(2\pi f \times C)$. Equating the two, we solve to get the well-known equation for *resonant frequency*: $f_{RES} = 1/(2\pi \times \sqrt{LC})$. We will describe resonance more clearly now, based on these observations.

We could connect the two components (L and C) in parallel across each other and inject some energy into the "tank," say by *tuning* the applied AC source to the natural frequency of the two. Thereafter, *theoretically speaking*, once injected into this tank, energy could just slosh back and forth forever between the L and C. It would be self-sustaining. See Fig. 19.4. See also Fig. 19.5 for a more detailed explanation of the phenomenon.

Note that once the *LC* "tank" has been excited and *set in motion*, we could even remove the AC source completely and the oscillations would continue indefinitely (ideally). But note that if we had continued to keep the AC source connected to this tank circuit, it would make no difference at all really (unless of course the AC source tried to drive the tank circuit at some frequency *other* than its natural frequency). At the resonant (natural) frequency, provided the *LC* circuit has reached the voltage level of the AC source, the AC source does *not* need to

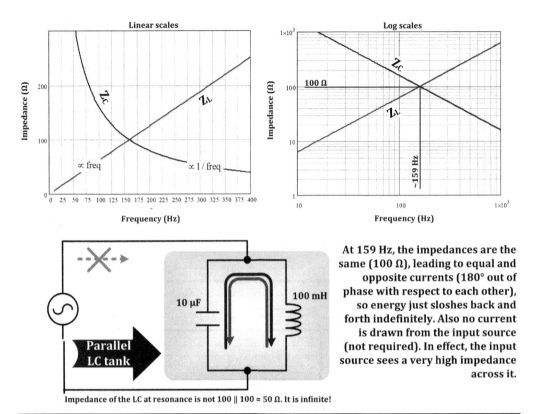

FIGURE 19.4 Explaining the energy storage capabilities of the parallel resonance tank circuit when driven at its resonant (natural) frequency.

replenish the energy in the tank (assuming no parasitic series resistances are present). In other words, the AC source will thereafter provide *no* current at all, even if connected—simply because it does *not need to*. But we also know from our usual electrical definitions that if the current drawn from the input source is zero, then the *impedance (of the LC tank connected across the source), that is, as seen by the source, is infinite (just like an open circuit, though only true at that specific frequency)*.

We have just created our first resonant circuit: a pure *LC* parallel circuit (with no parasitic resistances). It is our first building block. We have intuitively also just learned that this pure parallel *LC* tank circuit has an infinite impedance at its resonant frequency f_{RES}. As mentioned, we have the result: $f_{RES} = 1/(2\pi \times \sqrt{LC})$.

It is interesting to point out that at resonance, the overall impedance of the tank circuit is not half the impedance of each equal limb as we expect in the case of paralleled resistors. It is because of the *complementary phase angles* between voltage and current in the paralleled components that the net impedance becomes infinite in magnitude, at the resonant frequency.

In reality, any small resistances in series with the *C* and the *L* (such as ESR and DCR) will cause the oscillations to decay exponentially. And in that case, to "replenish" the tank, the AC source will need to provide a small amount of current, *even at resonance*. Which implies that though the impedance is still very high, it is not "infinite" anymore, not even at the resonant frequency.

We have simulated an (almost unloaded) parallel *LC* circuit—with a 30 Vac source set exactly to the natural (computed) frequency of 159.155 Hz for the selected components (10 μF and 100 mH). The results are presented in Fig. 19.6. Compare that with Fig. 19.3 where we still had "unmarried" and separate components. If we had not loaded the circuit a bit, the Simplis simulator would have "complained" due to infinite numbers.

The impedance of the parallel *LC* falls off on either side of the resonant frequency, irrespective of parasitics being present or not. At very low frequencies, we can assume the inductor is just a short (piece of wire), and so the AC source too will see a short (zero impedance), whereas at very high frequencies, the capacitor becomes a short (high-frequency bypass), so again, the AC source will see almost zero impedance.

We sum this up by saying that to the *left of the resonant frequency of a parallel* LC *circuit, the* LC *network appears "inductive"* (current lagging the voltage, but not necessarily by a full 90°). Whereas to the right of the resonant peak, the network appears "capacitive" (voltage lagging the current, but not necessarily by a full 90°). Since we have learned in the previous sections that a

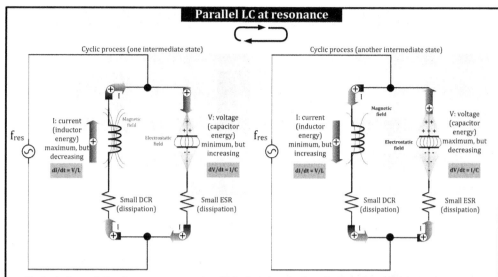

FIGURE 19.5 Explaining the energy storage of the parallel LC in more detail.

resonant network or circuit should appear "inductive" to the source, to be able to create ZVS, we realize that *the parallel LC circuit is useful to us only provided we operate to the left of its resonant peak.*

Any small resistances in series with the L and C (their typical parasitics like ESR and DCR) can be transformed, or modeled, into an equivalent large resistance placed in parallel to the paralleled (pure) L and C. In any proposed resonant converter based on this type of tank circuit, the load too will be connected in parallel to the paralleled L and C. So the effect of all resistances is to cause an eventual decay of the stored energy if the AC source is suddenly disconnected. If the AC source continues to be connected, it now has to constantly provide current and energy into the system to keep it "topped up"—in the form of (1) useful energy delivered to the load, and (2) wasted energy dissipated in the series parasitics (or the equivalent parallel resistance). That is, any energy leaving the LC tank.

NOTE *The "series-to-parallel equivalent transformation" indicated earlier, in effect, makes the parallel resistance a function of the AC frequency. For any given frequency we have to recalculate it.*

Finally, keep in mind that though the AC source may be providing only a tiny current at resonance, the current sloshing back and forth in the parallel LC can be *very high*. It is limited only by the impedances of the individual L and C, as we can see from Fig. 19.6. The AC

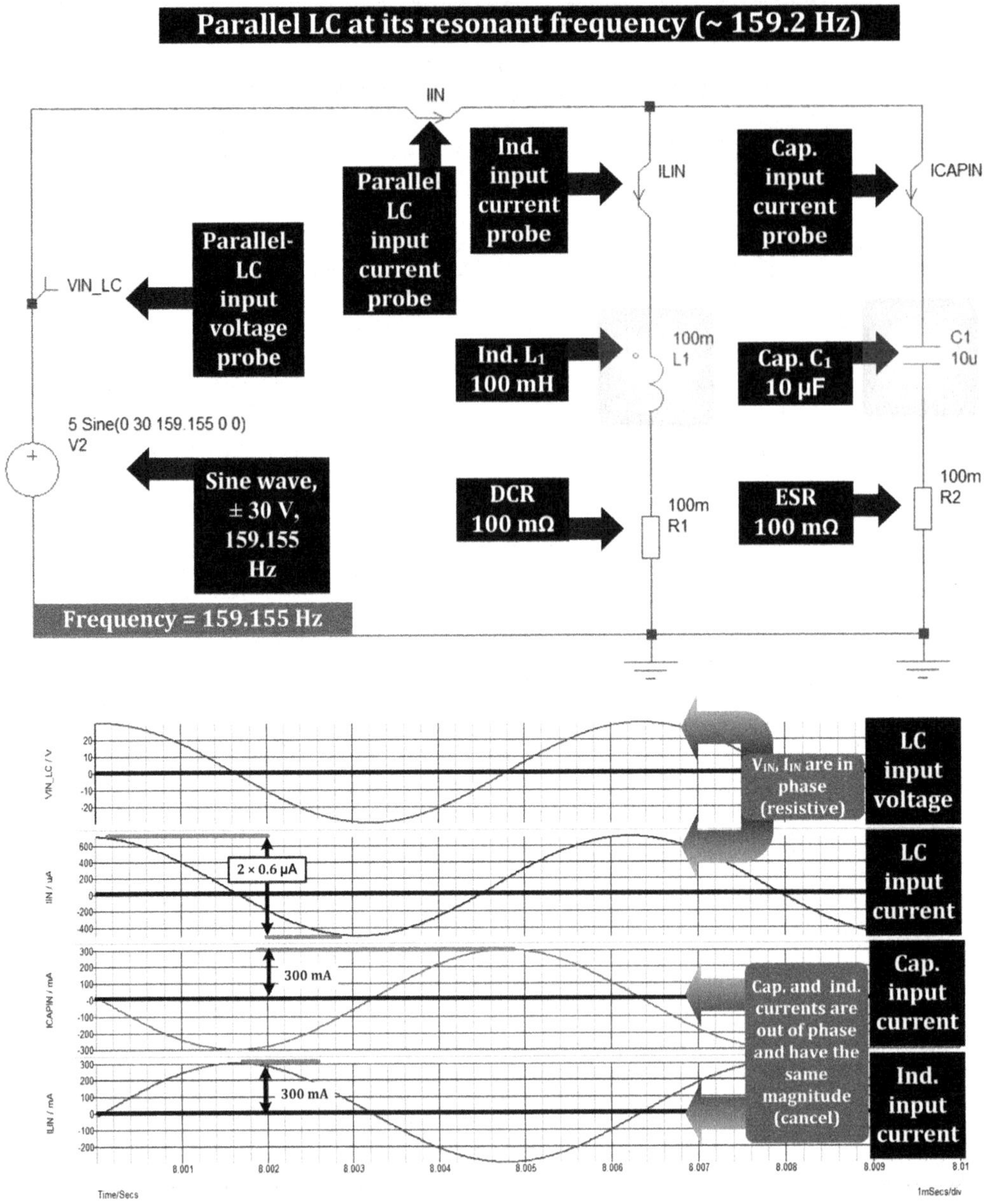

FIGURE 19.6 The (almost unloaded) parallel LC at resonance, showing the high circulating currents in L and C.

source may never "know" about these high currents since it is only delivering a very small current, but in a practical converter, where we would have transistors in the path of the circulating current, we are in danger of damaging our parallel resonant converter (PRC, based on this building block tank circuit). That is one limitation. Another limitation of this very basic PRC is that there is no obvious way to regulate the output! The load is connected in parallel to the input source, so it has exactly the *same* voltage as the incoming rail. What use is that if we can't offer load and line regulation? Yes, we can add other L and C's to try to get this to work, but at this point we prefer to move on to another, more promising resonant converter variation, which was in fact quite popular in older resonant converters: the SRC (series resonant converter). It is based on the series resonant LC cell, discussed next.

NOTE *We can prove that the real world actually depends on their being "parasitics" present almost everywhere. Very strange things would happen in our "real world" if there were no resistive parasitics present in particular. It turns out, it is also a very good idea to include small resistive parasitics during simulation, just as we have done in Fig.19.6. To avoid mysterious simulation stalls, we should also try putting in initial conditions for the L and C, for the voltage across the capacitor or the initial current through the inductor.*

PART 2 BUILDING RESONANT TANK CIRCUITS FOR FUTURE CONVERTERS

After having completed the initial transition from "PWM-thinking" to a better understanding of resonant behavior, we now start putting the LC components together to create some "trial balloons" first. Eventually we will then understand the best direction to take for practical converters.

Series Resonant Tank Circuit

This formed the basis of several commercial resonant converters. Let us get familiar with its pros and cons.

In Fig. 19.7, we have simulated the series resonant tank, driven at its resonant frequency, almost fully unloaded (i.e., small parasitic resistances too). We see very *high currents* at resonance, clearly *limited only by the parasitics*. Then using $V = I \times Z$, because of the very high currents, we also get *very high voltages* across the L and C. However, these high voltages are out of phase and almost fully cancel out from the viewpoint of the input source. So we are just left with 30 V (AC amplitude) to satisfy Kirchhoff's voltage law. The AC source does not "see" the high voltages, but does see the very high currents, because they pass through it.

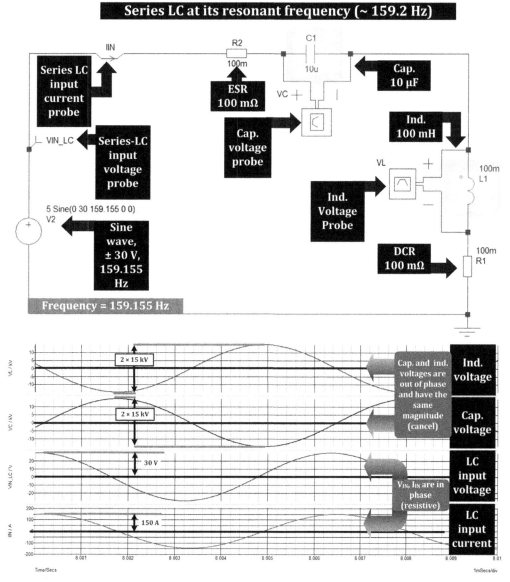

FIGURE 19.7 The (almost unloaded) series *LC* at resonance, showing the high voltages across *L* and *C*.

In a practical converter based on this series *LC* principle, we would typically connect the load across the *L*. Now we can use a "parallel-to-series equivalent (frequency dependent) transformation" and visualize this parallel load resistor as appearing in series with the *L* and *C*, just like the resistive parasitics (ESR and DCR) do. Or we could, in fact, really place the load in series with the *L* and *C* as shown in Fig. 19.8. Whichever way, the load helps significantly reduce the high voltages and currents of the series *LC* [in any proposed series resonant circuit (SRC), that is, series resonant converter]. This is called *damping*. It does make the series *LC* tank a possible practical choice, *provided we do not try to run it with no load across L*—because damping would be lost in that case, and we could easily damage it due to the high peaking at resonance of the series *LC*. That is one limitation of the SRC (based on this tank circuit).

Can we at least use the SRC to provide a regulated output rail, something we couldn't do easily with a proposed PRC? As shown in Fig. 19.8, we in fact use the basic voltage divider principle we traditionally use in setting the output voltage of a typical conventional PWM regulator, to produce a voltage rail (always) lower than the input. Now we have to *vary the impedance of the* LC appropriately to produce whatever output we want. In other words, *in a series resonant circuit (in general, many types of resonant circuits), we need to vary the "switching" (driving) frequency to create output regulation, just as in a conventional converter we need to vary the pulse width to create regulation.*

Can we ensure ZVS in a series *LC*-based converter? For that we need the tank circuit to appear *inductive* to the AC source as explained. So, without bothering to plot it out so far here, we can intuitively visualize that at low frequencies, the inductance of the series *LC* would appear as a piece of wire in series with a capacitor, so the AC source would only see a capacitance at low frequencies. At very high frequencies, the capacitor would appear shorted to an AC signal, so the tank would now appear inductive. We conclude that in a practical SRC, *we would need to operate to the right of the resonant frequency* for achieving ZVS.

One famous "last-but-not-least" type of question remains: If the output falls, do we need to *increase* or *decrease* the frequency? To figure that out, in Fig. 19.9, we plot out the gain of the series *LC* (we refer to the gain as "conversion ratio" at various places, but it happens to be just the output divided by the input). We see that in the *ZVS-valid* region to the right of the resonant peak, we need to reduce the frequency as load increases (to get the gain to increase). Conversely, at light loads, we have to significantly increase the frequency to regulate. For the case of $R = 100\ \Omega$, we see we need to go from 300 to 170 Hz to regulate the output (keeping the same conversion ratio, since input, i.e., line, has not changed here). For $R = 1000\ \Omega$, we are already in big trouble: Theoretically, we have an *almost infinite switching frequency spread* in an SRC, just to regulate to a set level (at approximately zero load). We also know that at very light loads, the voltages and currents can be extremely high in a series *LC* tank, which not only can damage the switches of an SRC but can result in very poor efficiency at light loads. These are all the major limitations of the series *LC* and its practical form, the SRC.

Introducing the *LLC* Tank for Creating Future *LLC* Converters

Here all we do is take the series LC tank and add a *relatively large inductor in parallel to the load*, as shown in Fig. 19.10. So, now we have two inductors and one capacitor. We call this *LLC* for *L-L-C* (two inductors, one capacitor).

Historically, the *LLC* was not really an unknown topology. But its full significance was not clearly understood until just a few years ago when engineers started looking rather keenly once again at resonant topologies in an effort to reduce switching losses.

NOTE *It is paradoxical though, that the design of LLC converters is still so poorly understood that even though LLC converters are based on a principle that should help achieve very high efficiencies at very high frequencies (where the last stumbling block was always the* switching loss *term), yet most commercial LLC converters are still operating only in the range of 80 to 200 kHz. Technology will certainly improve with a better understanding. However, there have been 1-MHz LLC prototypes reported (see "Optimal Design Methodology for LLC Resonant Converter," by Bing Lu, Wenduo Liu, Yan Liang, Fred C. Lee, Jacobus D. van Wyk, at Center for Power Electronics Systems Virginia Polytechnic Institute and State University, Blacksburg, VA 24061, USA).*

We need to understand very clearly how to properly and optimally design an *LLC* converter—*based on physical principles and deeper understanding*, rather than just relying on

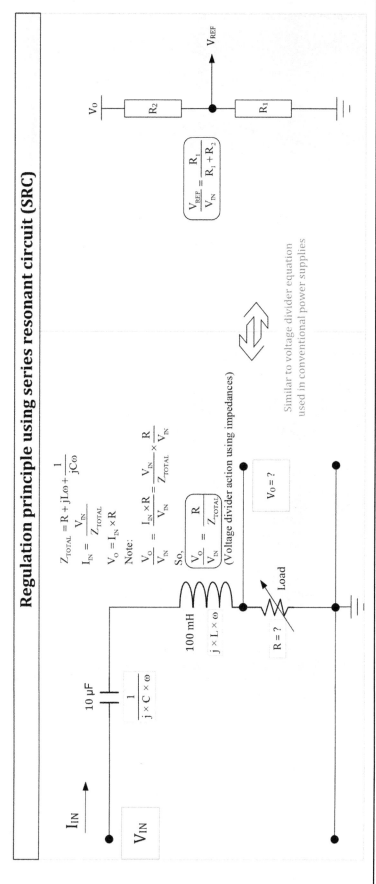

FIGURE 19.8 Using the series resonant circuit to create a regulated voltage in principle.

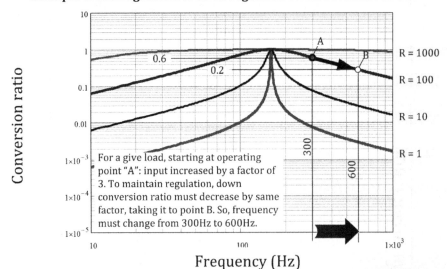

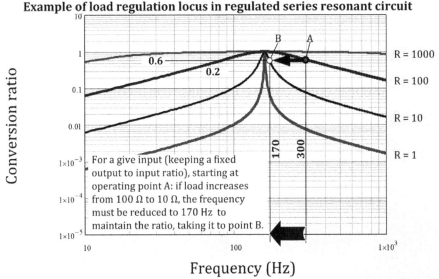

FIGURE 19.9 The series resonant *LC* tank, showing gain (conversion ratios), and how line and load regulation can possibly be carried out in a practical series resonant converter (SRC).

trial and error in a real lab, or on simulations in a virtual lab. It is all quite tricky. As even Bob Mammano admitted: It can become "bewildering." We hope to overcome some, if not all, the mystique behind the *LLC* converter in this penultimate chapter.

We expect two resonances, because we have one *C*, which can "resonate" with not one, but two inductors. Note that two *L*'s cannot "resonate" with each other, because we need *complementary* phase angles to resonate! We have to see how this additional *L* modifies the series *LC* circuit, and if it helps overcome some of the previously discussed limitations of the SRC.

In Fig. 19.10, we show how the voltage divider principle can be made to work here too, this time similar to the case of a *three*-resistor voltage divider that, in principle, can be used for setting the output in any conventional PWM converter. Then, in Fig. 19.11, we have used Mathcad

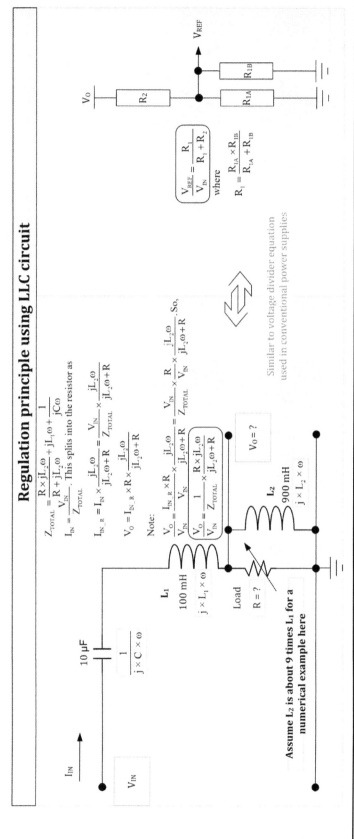

FIGURE 19.10 Using the *LLC* resonant circuit to create a regulated voltage in principle.

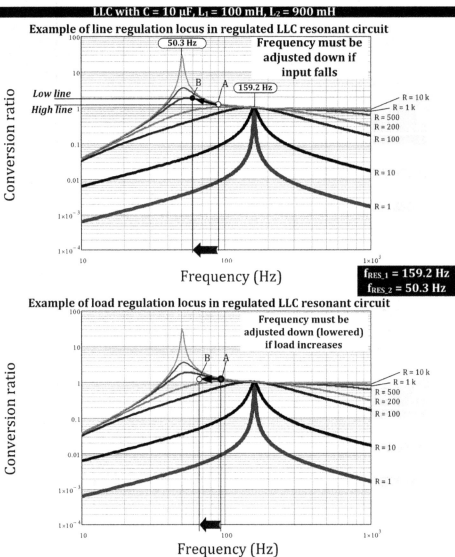

Increasing the load or reducing the input, both tend to cause the output to fall, and thus the regulation loop will need to try and correct this by increasing the conversion ratio (output/input). (A) If the system is operated *between the two fixed resonant frequencies* at all loads, we can set the logic such that the control loop will *always decrease the switching frequency in response to a falling output.* (B) We can also restrict ourselves completely to the right side of the right-hand resonant peak at all loads. But that has all the disadvantages of conventional series resonant circuits (e.g., huge, uncontrolled swing in frequency at light loads). (C) We can also decide to restrict ourselves *completely to the left side of the left-hand resonant peak*, and set the logic such that the control loop will *always increase the switching frequency in response to a falling output.* That will work in terms of regulation, but it will not be ZVS. (D) But we can never set an easy control algorithm if we allow operation *across* either resonant peak, since the conversion ratio slope changes on either side of a peak.

FIGURE 19.11 The gain (conversion ratio) of the *LLC* resonant circuit for different loads.

to plot the equation introduced in Fig. 19.10. But before we get to fully discussing Fig. 19.11 (the gain), let us first analyze the *phase* relationships accruing from the basic equation in Fig. 19.10, so we can be sure to identify the region of operation *in which the circuit appears "inductive"* and is therefore conducive to establishing ZVS. We have to admit we cannot analyze all this very intuitively anymore and need to plot it out mathematically as shown in Fig. 19.12. Note that as for the series *LC* circuit, we have chosen L_1 to be 100 mH, and C is still 10 μF. *The newly introduced inductance is* L_2, and we have used a seemingly arbitrary value of *900 mH* (a factor of nine higher, as compared to L_1). Actually, that factor is almost optimally set here. If the ratio is larger, the frequency variations are more, and if it is made too small (similar to L_1), the currents can be much higher, and the efficiency lower.

In the next section, we start by analyzing Fig. 19.12 and then Fig. 19.11 again, going back and forth to draw pointers toward designing a practical *LLC* switching converter. Note that

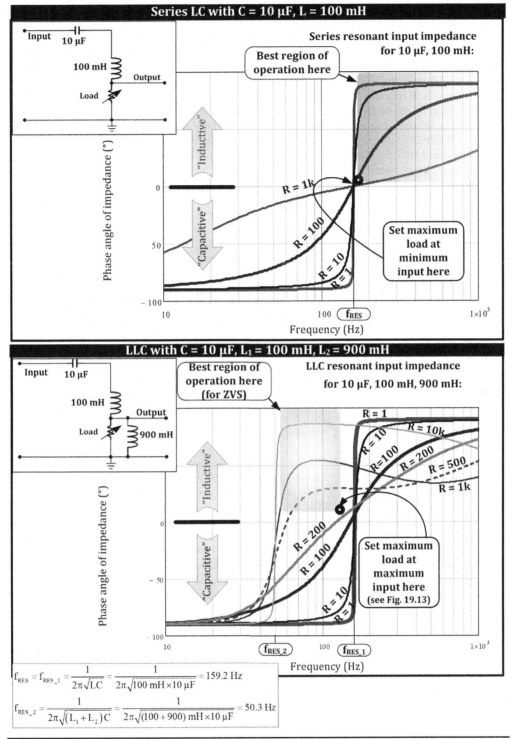

FIGURE 19.12 The phase of the *LLC* resonant circuit for different loads, compared to the series *LC*.

for now, we are only discussing how the *LLC tank circuit behaves* and only under a *sine wave* stimulus (AC source). Later we will build a practical switching converter out of it in several steps.

Analyzing the Gain-Phase Relationships of the *LLC* Tank Circuit

Looking at Fig. 19.12, we see that in a series resonant *LC*, for any load, we always had to be to the right of the resonant peak (~159 Hz) for the phase to be greater than 0° ("inductive," i.e., for current to lag the voltage as desired). That would mean ZVS was possible only for >159 Hz (potentially up to very high frequencies). In the *LLC* tank, we get a "low-frequency phase boost" arising from the lower resonant peak, which in turn is related to L_2, and this makes a wide range of load resistances with ZVS possible *even at lower frequencies*. In fact when we do the full math, we see that we get two resonant peaks, one formed by C and L_1 (as for series resonant tank), and another one formed by C and $(L_1 + L_2)$. Hence we have the resonant frequencies designated f_{RES_1} and f_{RES_2} shown at the bottom of Fig. 19.12. Numerically, for the values chosen, the second peak appears very close to 50 Hz. So, we can typically operate to the *right of 50 Hz and up to 159 Hz* for all loads ranging from open circuit to $R = 200\,\Omega$ (it seems so far), and achieve ZVS. For loads greater than that (say $R < 200\,\Omega$), we have to operate above 159 Hz; otherwise we cannot get ZVS because the phase angle is negative ("capacitive").

Regarding whether we can operate down to lower frequencies for a range of loads, or not, you might say: "So what"? That by itself doesn't seem to matter! In fact we usually want to operate at *high* frequencies anyway, and want to try and ensure ZVS at *high* loads primarily. So what is the advantage of the *LLC*? The real advantage of the second peak in the *LLC* tank shows up only in Fig. 19.11 (in the *gain* plots). In this figure, we have plotted out the gain, which we often call the *conversion ratio* (i.e., the output voltage of the *LLC* stage, divided by the input voltage, where output is just the voltage across the inductor in our simple AC case so far) versus frequency (for different loads).

For very high loads (e.g., $R = 1$), we get a conversion ratio *always less than 1*. That is just like a series resonant *LC* in which we only get step-down ratios. If we decide to operate in that region with our *LLC* tank, we can do that. But there is a major stumbling block: For very light loads (such as 1 kΩ or 5 kΩ, as shown in Fig. 19.11), we cannot even get much less than gain = 1, unless we move to infinite frequencies. We could end up in a situation where we can't even regulate anymore over the full load range. Or worse, maybe we designed our entire converter to step down, but at light loads the control loop moves in the "wrong direction"— and ends up "stepping up" (perhaps because the shape of the curve, in particular its slope, had its sign flipped as we moved from one side of the resonant peak to the other side). Yes, maybe we can "preload" the converter (at a huge expense to light-load efficiency), so our tank circuit never sees very light loads. In fact, we always need to do that in the SRC just to avoid the huge frequency variation we mentioned previously, but we could end up in the same situation with the *LLC* too, if we are not careful in designing our region of operation and our control strategy.

Therefore looking at things very practically, we decide we should *not try and operate an* LLC *based converter to the right of the higher-frequency resonant peak* (159 Hz in our case).

Keep in mind that designing our system to be "step-up" or "step-down" is actually *our initial design decision*, whatever the relationship of the input and output voltage levels. Because eventually, *we will use a transformer* with an appropriate turns ratio to correct for that. For example, we could design our *LLC* for a 1:1.1 ratio (an output 10 percent higher than the input), and then use a transformer with a Primary-to-Secondary turns ratio of, say, 2:1, to bring the output down to about half the input voltage. That is very similar to what we do, or can do, with the conventional Flyback topology too. For example, in a typical universal-input off-line (AC to DC) Flyback power supply, we create an "invisible" *intermediate primary-side regulated rail* of about 100 V. This is, for all practical purposes, the output rail as "seen" by the primary side, and it regulates (and sets the duty cycle) to keep this level fixed at 100 V. But after that, a 20:1 transformer ratio is inserted via the transformer to produce 5 V from the 100-V intermediate rail. Likewise, in *transformer-based resonant topologies* too, we have an additional degree of design freedom coming to us from the turns ratio, and we can voluntarily pick whether we want to operate the primary side to act as a step-up or step-down stage. As a corollary, since we will use a transformer anyway (also for creating the magnetic elements of the *LLC*), isolation is a natural advantage that accrues from an *LLC* converter (whether we like it and/or need it, or not).

We have decided not to operate to the right of 159 Hz in Fig. 19.12. What about to the left of the second (lower-frequency) peak, i.e., below 50 Hz? Unfortunately, looking at the phase diagrams in Fig. 19.12, we realize that the phase is below 0° in that region, and therefore the network will be capacitive to the input source. We know that is not going to lead us to ZVS operation. In other words, finally *we are left only with the region between 50 and 159 Hz to operate over the entire line and load variation*. We have no other practical and unambiguous choice really, given the guiding criteria discussed previously and the shape of the gain curves. But one question remains: Does this (available) region (fully) meet *all* our needs? Do we need to look more closely?

Let us look again at Fig. 19.11 and see how we intend to implement *line and load regulation*. We can see that, depending on the load, we can get either step-up or step-down (i.e., a conversion ratio greater or less than unity). Most of the curves are in fact with a step-up ratio. We can actually look at a close-up of this vital region, as we do in Fig. 19.13 and realize that

Soft-Switching and Designing *LLC* Converters   481

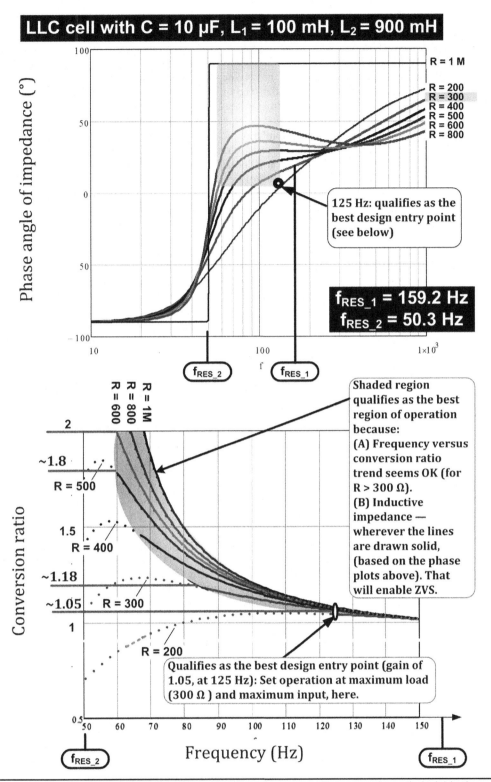

FIGURE 19.13 Close-up of phase for different loads, and corresponding conversion ratio (gain) in the usable area of the *LLC* tank circuit between the two resonant frequencies.

the curve of $R = 200\ \Omega$, which on the basis of its positive (inductive) phase had been previously deemed "ZVS-capable" and therefore acceptable, won't *work for an entirely different reason*. This is related to one of the two guiding criteria we talked about earlier: Do we have a "simple way to implement an unambiguous, almost "knee-jerk" (rapid) response to disturbances, *analogous to what we do in conventional power conversion*"? For the 200-Ω curve we are failing this requirement for line regulation. We can see from Fig. 19.13 that the 200 Ω curve slopes *downward* as frequency decreases from f_{RES_1} to f_{RES_2} (159 to 50 Hz). This is *opposite* to all other neighboring curves here. In all the other cases, based on Fig. 19.11, we have obviously designed the converter in such a way that if input falls, we lower the frequency (in knee-jerk fashion), to correct for that. But for the 200-Ω case (as per the closeup in Fig. 19.13), decreasing the frequency will cause the conversion ratio to fall even more, so the output *will collapse*. We conclude: This is not in the "right direction" for correction, commensurate with neighboring trends and/or our control strategy.

So, *resistive loads less than 200 Ω are definitely considered "overloads" (based on the LLC tank circuit values chosen so far)*. Output foldback will occur in those cases. These overload values cannot be supported, even though the phase may have seemed right.

This "dropping off" of gain is coincidentally almost commensurate with the phase just going negative (below 0°, i.e., capacitive), as indicated in Fig. 19.13. While plotting these particular curves, the Mathcad spreadsheet was set to "test the gain versus the phase condition," and "appear dotted" if the phase was reported to be negative. In general we can safely conclude that whenever the gain starts "heading down," the phase simultaneously is close to dropping below the 0° line (it becomes negative, and thus capacitive). Therefore, for *two* reasons, not one, the dotted part of the curves in Fig. 19.13 is considered no-man's land from now on. If the system enters that region, it won't be destroyed, but its efficiency will suddenly worsen (no ZVS), and also, any attempt at output correction will most likely cause output foldback. On this basis, we rule out 200 Ω—we can see from Fig. 19.13 that it just doesn't go far enough into the lower frequency region (starting from f_{RES_1}) before it turns "dotted."

We also conclude that based on the tank circuit values we have chosen in this example, the maximum usable load for the corresponding *LLC* converter is 300 Ω. But to maintain some design margin, we set the maximum usable load to be a little less, at 300 Ω, not 250 Ω. In other words, any resistance value smaller than 300 Ω is now considered an overload by definition.

- At very light loads, the frequency will shift close to 159 Hz, not more (certainly unlike an SRC that tries to go to infinite frequencies to regulate at light loads).
- At very heavy loads, the frequency will shift closer to 50 Hz.

We can easily sense, based on the equations for f_{RES_1} and f_{RES_2}, that increasing L_2 significantly will move f_{RES_2} toward much lower frequencies. Though that may help in some way, it is clear that there is a penalty for making L_2 much larger than L_1. For one, the frequency variation spread, going from minimum to maximum load, and from low-line to high-line, will become much larger.

We have decided we need to *design for conversion ratios higher than unity*, so we can cover the full load range from 250 Ω (peak load) to higher resistance values (up to unloaded). In particular, to ensure ZVS, we will set the *LLC* converter at peak load, at high-line, to be at the encircled point at the lower right tip of the shaded area in the LLC phase diagram of Fig. 19.12. At that particular design entry point (shown more clearly and precisely with an ellipse or circle in Fig. 19.13), we will have a frequency closer to the higher resonant peak (~125 Hz in our case here) and a corresponding gain (conversion ratio) of about 1.05.

This *gain target of 1.05 at maximum load and high-line is actually our universal recommended entry point for all LLC converter designs*, because otherwise the gain curve "droops" and the correction loop will take it in the "wrong direction" (foldback) as discussed previously. If we set it closer to 1, say at 1.02, our design entry point will shift closer to the higher-frequency resonant peak. But that is a bit too marginal.

The full optimum region of operation is also shown (shaded) in the gain curves of Fig.19.13. This area is the part of the gain curve *for which the phase angle is positive (inductive)*. As indicated, we have "AND-ed" the gain and phase curves using Mathcad, so the curves become dotted if unacceptable. As also indicated, this does not mean the system will necessarily stay in this optimum region, or we are somehow constraining it to. That locus of movement is determined solely by the regulation loop, as it relates to the specific line or load variations. So certainly, our selected *design entry point must lie within this shaded region (at the circle or ellipse)*, so that we get ZVS at maximum load and at high-line, but we *may just lose the advantage of ZVS if we leave this shaded area*. For example, if we set the maximum load as 300 Ω, we will be able to reach a conversion ratio of about 1.18, as per Fig. 19.13, before we leave the ZVS region. That seems to equate to an allowed 18 percent reduction in input (but we will soon show it is only 12 percent actually). Either way, we can produce *derating curves* to show how we can *trade off maximum load versus input operating range, weighing it against the price of overall expected efficiency, provided of course the range can be achieved or is acceptable in terms of our guiding criteria*.

The fear of predicting and/or really losing efficiency at low-line is perhaps one reason *LLC* converters are still largely being used only where the input is relatively stable (and comfortably so), such as where the *LLC* converter is driven off the output HVDC rail of a front-end PFC stage. We can see that if we really want wide-input variation, we have to *overdesign* the (*LLC* tank of the) converter (300-Ω maximum load as in our example will only get us a 12 percent input voltage variation factor as explained more carefully later). Holdup time can also be a major issue in AC-DC designs in which the *LLC* converter stage typically follows after a conventional PFC stage.

Two Resonances in the *LLC* Tank

In related literature, it is often said that the *LLC* converter has two resonant peaks: one a series resonance between C and L_1, and another "parallel" resonance between C and $(L_1 + L_2)$. So, people call it by many names: f_r, f_s, f_p, f_m, f_o, f_∞, and so on. It can become confusing, which is why we have just preferred to call them f_{RES_1} and f_{RES_2}, based on the simple order in which they actually appeared in our discussions. Yes, f_{RES_1} does come about from the resonance between C and L_1, whereas the lower frequency peak f_{RES_2} comes from the resonance between C and $(L_1 + L_2)$. But are these *series* or *parallel LC* resonances?

We should not get confused by the fact that the conversion ratio (gain) at the first (higher) frequency is less than 1, as is typical of a series *LC* circuit, whereas the gain at the second (lower) frequency is greater than 1 (which is perceived as true for parallel resonant circuits, in tuned circuits for radio applications, etc.). To really find out whether the peaks are series or parallel resonances, we need to plot out the input impedance (as seen by the source). This is presented in Fig. 19.14.

It is interesting that in going from a shorted load to no load, the impedance presented to the AC input shifts between what are clearly *two resonant peaks* of series resonant characteristics, because in parallel *LC* resonance, the impedance becomes very high at resonance, not low as indicated by Fig. 19.14.

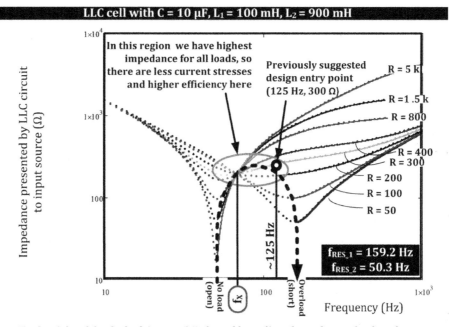

To the right of the dashed, inverted U-shaped locus line above, for any load, we have an inductive phase angle (helpful for achieving ZVS). Those areas are indicated by solid impedance lines. The dotted lines indicate capacitive phase angles (keep-off areas). We should thus choose the switching frequency such that we consciously stay only to the *right* of this inverted U, at least at maximum load.

Caution: These curves only tell us the input impedance for a given load and frequency, so we can check whether the network appears inductive (for ZVS), or not. They do *not* provide frequency response as we change the load. They don't help us pick a design entry point.

Figure 19.14 Plotting input impedance of the *LLC* tank circuit.

One of the contributors to efficiency is switching losses, which we have tried to minimize by invoking ZVS. But that advantage could be easily lost by excessively high conduction losses, as caused by *high circulating currents*. One of the contributors to that circulating current is indeed L_2, which is why typically L_2 is kept at least 5 times larger than L_1 (usually less than 10 times though, as explained further later). Especially when we insert a transformer with output diodes, we will see L_2 significantly affects the current distributions in the *LLC* tank. Therefore, it is recommended to *keep* L_2 *between 4 to 11 times* L_1 *in any practical LLC converter. In our case we have first fixed on a factor of nine.* However, the author has also built high-efficiency wireless mat chargers with leakage inductance (L_1) twice the magnetizing inductance (L_2). So clearly, the above-mentioned ratio is not necessarily restricted to a number greater than 4, but can even be set to 0.5 or less. That also gives a very tight frequency range of operation, rivaling that of conventional PWM converters.

Note that in a parallel *LC*, the circulating current sloshes back and forth between the *L* and *C*, so the input source may never "see" the high current. But in a series resonant case, the current does pass through the input source and is inversely proportional to the *impedance* the network presents at its input terminals (to the source). If the impedance is high, we will get small circulating currents (and higher efficiency). In Fig. 19.14 we see that there is a *shaded encircled area* of high impedance for almost any load. It is recommended in related literature that we should try to remain here as much as possible. But note that our previously honed choices, $R = 250\ \Omega$ (or $R = 300\ \Omega$), fall in a very flat part of the encircled region. In other words, our previous design choice is actually good even from the viewpoint of low conduction losses. We have to do nothing more to optimize.

Just for information, we mention that in related literature, the cusp labeled "f_X" in Fig. 19.14 is mentioned as some sort of *LLC* converter "design target" because it offers high impedance. Its equation is $f_X = 1/2\pi\sqrt{C \times [L_1 + (L_2/2)]}$. We have however preferred to generalize our approach by describing a certain load resistor corresponding to maximum load at high-line, one which (1) gives an unambiguous direction of correction, and (2) makes the network appear just slightly inductive to the source (for ZVS). *The desire for high input impedance is automatically taken care of*, as we see from Fig. 19.14.

PART 3   GENERALIZING VIA SCALING STRATEGIES

(This is where we "grow the *LLC* seed".) For convenience of discussion and also our simulations and certain presentation aspects so far, we had started by randomly picking some arbitrary values of L_1, L_2 and C and then seeing what we could do with our chosen *LLC* tank circuit. In practice, that approach is lopsided: In an actual design scenario, we start with a design target *first*, and *then* we pick the *L*'s and *C*'s of the *LLC* circuit. So our flowchart needs to be reversed. We will do that shortly, but in a rather unique manner.

We also chose impracticably large values for the *L*'s and *C*, and therefore ended up with seemingly low and unusable frequencies. But that is really not a problem either—if we understand the *concept of scaling as applied to reactive components*. On deeper thought, we realize we do that instinctively all the time in conventional switching power conversion. *We halve the inductance and halve the capacitance if we double our frequency. We also halve the inductance and double the capacitance if we double the power.* The very same rules apply to resonant mode converters too, because these scaling properties are not specific to any switching topology or methodology: They are keenly related to the energy storage capabilities of reactive components. The author's *A-Z* book also has an intuitive analogy on page 1 (the metro terminus analogy) which should help with visualizing this aspect better.

So, with a knowledge of scaling, we can go from our arbitrary low-frequency *LLC* tank circuit (which we have already understood very well) and adapt it to any desired switching frequency and power. We will demonstrate that deceptively trivial procedure shortly. But note that in this process we have *just anointed our simple* LLC *tank circuit consisting of 10 µF, 100 mH, and 900 mH, as an "LLC seed"*—from which we will spawn general LLC design procedures and eventually create practical LLC converters.

Let us illustrate scaling by a numerical example. We picked $L_1 = 100$ mH, $L_2 = 900$ mH, and C = 10 µF, and so we got two resonant peaks, at 159 Hz and 50 Hz, respectively. On that basis we identified that we could use our converter up to a maximum load of 250 Ω (preferably 300 Ω to leave a little design margin). We also set *an initial design target of gain = 1.05 at maximum load and high-line*. Now, we reduce all inductances and capacitances by a factor of 1000; i.e., we use $L_1 = 100$ µH, C = 10 nF, $L_2 = 900$ µH. We thus get two resonant peaks, at 159 kHz and 50 kHz, respectively. Our operation will be restricted to this new region. We have quickly shifted to a practical *high-frequency LLC* application circuit in one easy step!

What about *power capability*? We identified maximum load as 250 to 300 Ω. Now, it may seem strange to state that this specific value of resistance, as applied to our *LLC* circuit, does not depend on a specific input voltage (and power rating thereby)—it applies to *any* input voltage (and power rating as a result of that choice). The reason is we are only interested in sticking to a certain optimally (and slowly) rising gain-curve *shape*, for reasons explained earlier. And that certain *shape factor* comes about from the definition of the quality factor (Q) of the associated resonance. The Q (shape factor) is related to the *ratio* of the L and C vis-à-vis the load resistor R. It really has nothing to do with any specific input voltage or power.

But what exactly is Q here? The basic problem is we have two resonant peaks. In related literature, we find that Q is more commonly expressed in terms of the higher-frequency ("upper") resonant frequency, that is, as

$$Q_{RES\_1} = \frac{\sqrt{L_1/C}}{R}$$

However, since in reality, when we change the conversion ratio (gain), we travel along the lines of the *lower-frequency* resonant peak (see upper half of Fig. 19.11), we felt it was physically more intuitive to fix Q based on the lower-frequency ("lower") resonant peak instead. Of course, once the ratio of L_1 and L_2 is known and is fixed, as in our case, the Q of one resonant peak can be "mapped" to the Q of the other peak. It really doesn't matter eventually whatever we choose to do. They are completely linked and related.

We have thus determined that there is an *optimum Q* for our maximum load curve. For 250 Ω (this is considered the peak loading here):

$$Q_{RES\_2} = \frac{\sqrt{(L_1+L_2)/C}}{R} = \frac{\sqrt{(100\text{ mH} + 900\text{ mH})/10\,\mu\text{F}}}{250\,\Omega} = 1.265$$

For 300 Ω (preferred maximum load setting),

$$Q_{RES\_2} = \frac{\sqrt{(L_1+L_2)/C}}{R} = \frac{\sqrt{(100\text{ mH} + 900\text{ mH})/10\,\mu\text{F}}}{300\,\Omega} = 1.054$$

In other words, our final (preferred) design recommendation (for any power level, and any frequency) is to *target a Q typically 1.054 (worst case 1.265) at maximum load and high-line. Here, we are also assuming that the ratio* L_2/L_1 *is set to around 9 (but it can be anywhere between 7 and 11).* So in any general case, we fix the minimum (primary-side) load resistance (maximum load) as per this equation:

$$R_{MIN\_WORSTCASE} = \frac{\sqrt{(L_1+L_2)/C}}{1.265}$$

but preferably

$$R_{MIN\_TYP} = \frac{\sqrt{(L_1+L_2)/C}}{1.054}$$

Realize that when we scale our components for high-frequency operation, we do that by *dividing both L's and C by the same factor*. But that means the *ratio L/C* remains unchanged, and therefore *the quality factor too is unchanged and leads to the same optimal loading that we want to retain*. In other words, the 300-Ω (selected) maximum load resistor applies to any input voltage and any *switching frequency* too. The final power level of our converter depends on the input voltage with a 300-Ω load placed across the output of the *LLC* stage.

Summing up: The 300-Ω value, as a suitable loading for the *LLC* tank, "rose to the top," based on the *ratio of the* L *and* C. The L and C were (starting) values that we just happened to pick for our "*LLC* seed." Yes, we could have arrived at an equally workable seed with different L and C combinations, and that would tell us the best value of maximum load resistor for *that* combination. Indeed, we have many other possibilities for choosing L, C, and the corresponding R (for optimal maximum loading of the tank). But the guiding design rules have already been distilled, and these are now summarized as follows:

1. Set L_2 to about 9 times L_1. That gives about $1 : \sqrt{1+9} = 1 : 3.16$ frequency spread (in practice we get only 1:2.5 since we are operating somewhat within this region, not at extremes). We recall that some areas of the frequency region can cause foldback or no-ZVS and are thus ruled out by our initial design targets.

2. Our design target for the *gain (initial conversion ratio)* is set to *1.05* (slight step-up), and this is set at *maximum load and at maximum input*. We know that will ensure ZVS for most of the region of operation (though the exact boundaries will be discussed in more detail later).

3. To get the actual desired output voltage, we then *use a transformer with the appropriate turns ratio*, which will take the primary-side (virtual) output of the *LLC* converter from $1.05 \times V_{INMAX}$ down (or up) to V_{OUT}. As indicated, this is exactly the same approach we use in a conventional Flyback, where we set a *primary-side* (virtual) output voltage rail, which we call V_{OR}, for reflected output voltage, and then use a transformer to create a step-up or step-down ratio as desired—to go from V_{OR} to $V_O = V_{OR}/n$, where n is the turns ratio N_P/N_S. This is discussed in detail on pages 129 and 130 of *Switching Power Supplies A-Z*, 2d ed. See also Fig. 6.6 in this book.

4. The maximum load must be such that the *Q* (based on the low-frequency resonant peak) is *1.054* or less (though for a peak of 250 Ω instead of the maximum of 300 Ω, it may momentarily be 1.265). The way we have defined Q, we get more "peaking" at f_{RES_2} (lower peak), for smaller and *smaller values* of Q, not larger. A small Q corresponds to a larger R, that is, less output wattage. So as we decrease the load (increase R), we get higher and higher possible conversion ratios (gain), and we can use the natural peaking of the gain curve to extend our input operating region. But if we overload the tank circuit, and/or are not on the "correctly shaped" curve, we will not be able to get much input variation from our *LLC* converter, unless we significantly lower the load on it (thereby increasing R and lowering Q, and getting more "peaky"). This is discussed in more detail shortly.

A note on terminology: *Let us hereafter call the intermediate (virtual) primary-side output rail of the LLC converter V_{OR}, completely analogous to what we do in Flyback design and for much the same reasons*. Note that we are still dealing with pure AC only so far, so this V_{OR} is not a DC voltage, but represents the peak (*not* peak-to-peak) of the AC voltage appearing across L_2 (the larger inductance, see Fig. 19.10 again). The gain of the *LLC* tank is basically the V_{OR} divided by the peak AC input voltage applied across the entire tank circuit, consisting of two inductances and one capacitance in series (with R present across L_2).

Step 1 of Design Validation Process: An "AC-AC Converter"

Here we do not mean "AC" as in 50-Hz line frequency. We are just connecting a high-frequency sine-wave AC source to the input of the *LLC* tank (driving it in the region of its two natural resonances) and examining its response. As a *load*, we initially just have a resistor placed across the (larger) inductance. We will build a proper converter in several steps starting with its most basic (primitive) form here.

Here is our *design target*: Suppose we want to convert a (AC) sine wave of amplitude 52 V (*its peak-to-peak swing is 104 V*) to a (AC) sine wave output of amplitude 12 V, with a *peak* AC power output of 60 W (i.e., at the peak of the AC input), assuming 95 percent efficiency. We do not want to go below 100 kHz to avoid larger reactive components. What are the values of inductances and capacitances we need? What is the maximum switching frequency we will see? What is the turns ratio we need for the transformer? What is the final frequency spread?

We start with the "*LLC* seed" we have studied so carefully earlier and scale it for the required power level and switching frequency.

The minimum frequency we got for our seed was more accurately 50.3 Hz (using $1/2\pi \sqrt{(L_1+L_2)C}$. Now we want to take this up to 100 kHz.

The frequency scaling factor f_{SCALE} is almost 100 kHz/50 Hz = 2000. We get there by the following transformation to achieve the required frequency of operation:

$$L_1: 100 \text{ mH} \to L_1/f_{SCALE} = 100 \text{ mH}/2000 = 50 \,\mu\text{H}$$
$$L_2: 900 \text{ mH} \to L_1/f_{SCALE} = 900 \text{ mH}/2000 = 450 \,\mu\text{H}$$
$$C: 10 \,\mu\text{F} \to C/f_{SCALE} = 10 \,\mu\text{F}/2000 = 5 \text{ nF}$$

Now let us scale our *LLC* seed to the required *power* level. We learned that our LLC seed functioned well in terms of the "shape" of the curves by placing 300 Ω of load across its output. As indicated, it may seem strange at this point that this value of resistor is valid for any input voltage and any frequency. But as mentioned, the reason that is true is based on establishing a certain recommended Q for the gain curve at maximum load. That would make the system "work well": We will get a clean direction for correction (i.e., decrease frequency *always* if load increases or line voltage falls), and we will get ZVS for most of the line-load operating region. We also know that this will make the *LLC* resonant tank stage produce a conversion ratio (gain) of about 1.05, which, *for the same applied input voltage as required in our design case*, is an effective (primary side) sine output voltage (amplitude) of $V_{OR} = 52\ V \times 1.05 = 54.6\ V$ (peak-to-peak is twice that). With 300 Ω connected across this virtual (pre-transformer) output rail (V_{OR}), the *peak* AC output power is $V_{OR}^2/R = 54.6^2/300 = 9.94\ W$. That is the peak AC power level acceptable in terms of loading, to our specific (now frequency-scaled) LLC seed. But that much power is not enough!

What we need here is 60 W/0.95 = 63.2 W (assuming 95 percent efficiency). Our *power scaling factor* is 63.2/9.94 = 6.36. So *we need to be able to reduce the load resistance from 300 Ω to 300 Ω × 9.94 W/63.2 W = 300/6.36 = 47.2 Ω*. This will produce the power scaling we need (because we will still have the same gain and input and output voltage, that is, 1.05 V, 52 V, and 54.6 V, respectively).

But if we just blindly connect a lower load resistance across the inductor, we will make the Q go up much higher than 1.054, so our gain peaking curve will collapse, and we will neither get ZVS, nor an unambiguous direction of correction, that is, if we are lucky to be able to get the converter to even work! The way out is to restore the Q despite the lower load resistance value we need here, and that we do *by changing the ratio L/C, by the same ratio*, as shown subsequently.

We have learned we need *more power*, expressed by the *power scaling factor* $P_{SCALE} = 63.2\ W/9.94\ W = 6.35$. We also know from conventional power conversion that for higher power we need to *proportionally* decrease inductance and proportionally increase the capacitance. That will change Q, but will not affect the resonant frequency, since that depends on the *product* of L and C, not the ratio L/C on which Q depends. So the final values of the *LLC* tank are

$$L_1:\ 50\ \mu H \rightarrow L_1/P_{SCALE} = 50\ \mu H/6.35 = 7.874\ \mu H$$
$$L_2:\ 450\ \mu H \rightarrow L_2/P_{SCALE} = 450\ \mu H/6.35 = 70.87\ \mu H$$
$$C:\ \ \ 5\ nF\ \ \ \rightarrow C \times P_{SCALE} = 5\ nF \times 6.35 = 31.75\ nF$$

To validate our scaling principles, we first use these component values and plot the results—using the same Mathcad spreadsheet we used previously for plotting the basic *LLC* seed characteristics. In this manner we get Fig. 19.15. This looks exactly like what we were expecting. The 47.2-Ω gain curve has the same *basic shape* (and obviously Q) as the 300-Ω curve that we had previously gotten for our *LLC* seed. We see the same slow upward trend as we move away from the high-frequency peak (the peak is a little above 300 kHz in this case) toward the low-frequency peak, which is at 100 kHz. We can still unambiguously correct the output voltage, but within an apparent input variation factor of ~1.18 as before. Note that if we want higher input variation, in effect, *we have to overdesign the tank for higher than*

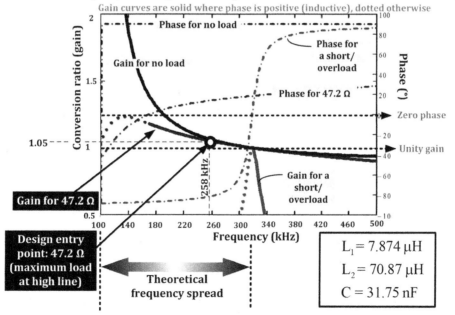

FIGURE 19.15 Plotting the final Mathcad results of our 60-W (peak AC power) AC-AC design example.

maximum load and then derate the power supply. We will discuss that shortly. In our current design example, we also confirm that the 47.2-Ω plotted gain curve has a positive phase angle (inductive) in the designated operating region, and so we expect to achieve ZVS for it.

Our original *LLC* seed (100 mH, 900 mH, 10 µF), if used in our current design case (with 104 V peak-to-peak applied input AC swing), would have provided us only 9.94 W of peak AC power and would have needed to be switched in the range of 50 to 159 Hz. By applying appropriate frequency scaling and power scaling factors, we have made it suitable for 60-W peak AC power, needing to be switched between 100 to 300 kHz.

The design entry point is, as before, shown in Fig. 19.15. But how do we expect to actually achieve that point, that is, get there and stay there? By the *turns ratio*! For example, if we want an output which is a 12-V peak-to-peak sine wave, we will use a step-down transformer with *a step-down turns ratio of* $V_{OR}/V_O = n = 54.6\ V/12\ V = 4.55$. Then we use a high-gain error amplifier, with a reference voltage set with a divider to regulate exactly at 12 V (the AC amplitude in this case, not a DC level yet). *The underlying difference from conventional PWM regulation is that the error amplifier is set to command a decrease in frequency (not change in pulse width) if the output starts to fall (such as if load increases or input decreases).* Knowing the shape of the gain curve in the chosen region of operation, we already know that that will automatically produce line and load regulation.

Note that at the design entry point itself, we will get a frequency of around 258 kHz—based on the Mathcad plot, because that corresponds to our initial assumption of gain equal to 1.05 with the calculated maximum load resistor of 47.2 Ω (a Q of 1.054, we did not change that) connected across V_{OR}. In other words, by just carefully fixing the transformer ratio, we will achieve our correct (recommended) design entry point. But if, for example, we do not set the turns ratio correctly, we will indeed have very unexpected behavior. We will then certainly find the *LLC* extremely "bewildering." *Set the turns ratio accurately*!

So 47.2 Ω is the primary-side loading of the *LLC* tank circuit. What is the load we can place across the 12 Vac output rail (i.e., through the transformer)? Very simply, that resistor will be scaled as *square of turns ratio*: we get $47.2\ /(4.55)^2 = 2.3\ \Omega$. Let's check we get the required peak AC power: $12\ V^2/2.3\ \Omega = 63\ W$. This was our target to start with.

The reflection of components from one side of the transformer to the other is tabulated on page 130 of *Switching Power Supplies A-Z*, 2d ed. See also, Fig. 6.7 in this book.

Let us also simulate this circuit to see the actual response. This is shown in Fig. 19.16. Its V_{OR} (peak amplitude AC here) turns out to be 54.6 V, indicating that we do have a gain of $54.6/52 = 1.05$, by using an *LLC* tank of 7.87 µH, 70.87 µH, 31.75 nF, and driving it at *258 kHz (2.58 times the lower peak of 100 kHz)* as suggested by the Mathcad file (Fig. 19.15). *This validates our scaling principles, and also our Mathcad file and its equations.*

Derating Maximum Power for Increased Input Range

Looking at Fig. 19.15, we see that if we lower the input, and a higher conversion ratio is required, we can only get to about 1.18 gain, before *two* adverse things happen almost simultaneously: (1) the curve turns dotted, because the Mathcad file "AND-ed" gain with phase, so this dotted line means the phase turns negative for these gain values (capacitive), and (2) the curve dips down, which means increasing the frequency as input falls is only going to make things worse. In other words, we cannot achieve any desired gain above 1.18 at maximum load (and maximum line).

NOTE *This is why typical LLC converter applications are based on using the relatively steady output of a conventional PFC front-end converter as the input rail to the LLC stage. Further, the output of the LLC stage itself is usually used only to power high-voltage LED strings, where there is in effect, no holdup time requirement, unlike digital circuits, which can get reset if their input rail droops momentarily. In other words, the widely-perceived disadvantage of the LLC topology, that it cannot handle input variations well, has typecast it in a rather limited role for LED backlighting as described above. But in reality, the LLC topology can handle input variations almost as well as the Flyback, and certainly much better than a conventional single-ended Forward converter (with energy-recovery winding). Keep in mind that the latter converter is rather severely limited in terms of its input variation, because its duty cycle is limited to 50 percent. As a result, to handle universal-input applications, a voltage doubler is required as shown in Fig. 5.6. But for an LLC converter, the doubler is not required. In fact, on June 4, 2013 the author internally reported building a 20-W nonsynchronous 12-V output, universal-input LLC converter with 88 to 90 percent efficiency over the entire input range 120 Vdc to 400 Vdc. It used a small EFD-25 transformer only. The magnetics are so small in the LLC topology.*

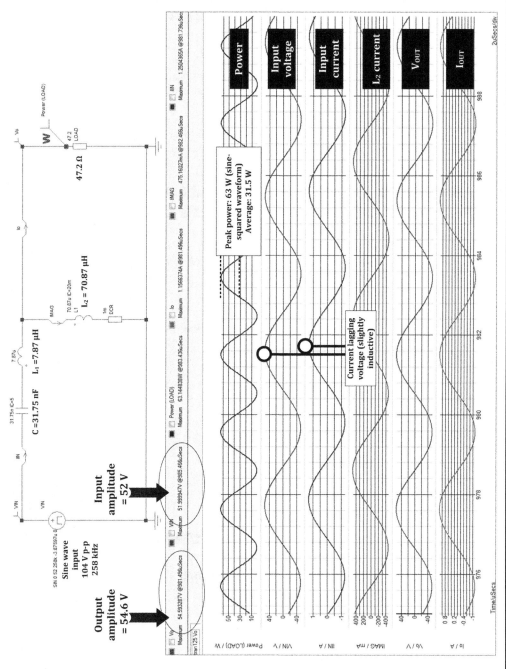

Figure 19.16 Simulating the final 52-to 12-V AC-AC design example (60-W peak AC power, 30-W average power).

On closer examination of the Mathcad file, this number is actually more precisely *1.1785*. Does that mean we can vary the input by that exact factor? Can we go down to say, 52 V/1.1785 = 44.12 V? *No we can't!* Keep in mind that our initial (starting) gain level was *1.05* (and we need to maintain that minimum gain, based on our selected and fixed turns ratio). So in reality, with a computed gain of 1.1785, we can only go down by a factor of 1.1785/1.05 = *1.122*. That is only *12 percent of allowed input variation calculated at maximum load!* Numerically, in our case, we can go down to 52 V/1.122 = 46.4 V at maximum load, and no more.

What is the conversion ratio we require to go down to, say, 36 V, on the input? That is (52 V × 1.05)/36 V = 1.517. To go down to 32 V, we need a conversion ratio of (52 × 1.05)/32 = 1.71, and so on. *In each case we need to end up with a V_{OR} of 54.6 V, to be able to get the required output voltage,* based on the preset turns ratio.

So, what does our Mathcad file predict if we want to, achieve operation down to 36 V and also down to 32 V? We can decrease the load manually (in the Mathcad file), until we get exactly to those required gain values. The numerical results are presented in Fig. 19.17.

Keep in mind that so far, we are still talking in terms of the numbers exclusively from our ongoing *AC-AC example*. But since our design rules (shape of curves, etc.) will remain the same for any proposed design (AC-AC or DC-DC), we can generalize our approach about this aspect too right now, as in Fig. 19.18. We can use these curves for *all* LLC applications (assuming it is based on our *LLC* seed, and thus retains a ratio of 9 for L_2/L_1).

Step 2 of Design Validation Process: An AC-DC Converter with Diodes and Transformer (Still No Output Capacitor)

In Fig. 19.19, we take the circuit in Fig. 19.16 and add a 1:1 transformer with two diodes. We want to rectify (but not yet filter) the output and show that *the same 47.2-Ω load resistor we used in Fig.19.16 loads the LLC tank of Fig. 19.18, by exactly the same amount*, evident by the

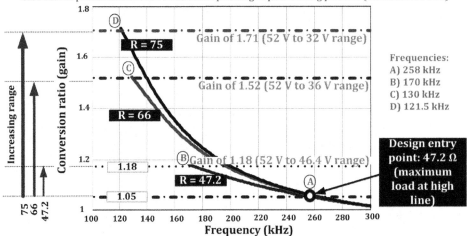

Therefore, 60 W is supported only from 52 V to 52 /1.122 = 46.4 V.
To go down to 36V, we need to derate maximum power to 43 W.
To go down to 32V, we need to derate maximum power to 38 W.
Note: These are all AC waves, and voltages here refer to sine amplitudes (half peak-to-peak), and power is peak power (the average power is half the stated peak value).

Figure 19.17 How reducing load helps achieve higher input operating range.

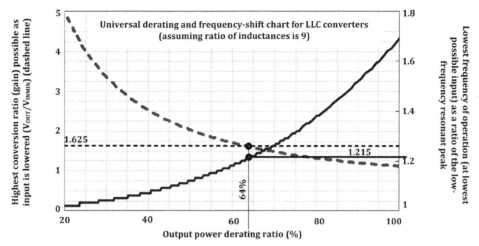

Example: From 52 Vdc maximum input, if we want to go down to 32 Vdc minimum input, the (additional) conversion ratio required is 52/32 = 1.625. From above, the frequency at minimum input is 1.215 times the lower resonant frequency (if the latter is 100 kHz, we will get 121.5 kHz at 32 Vdc). Further, we see from above that the power needs to be derated by 64%. So, if our desired maximum power is 38 W, we will need to plan the LLC tank to be able to deliver a virtual power of 38/0.64 = 60 W at maximum input, if we want to be able to deliver up to 38 W down to 32 V input. If our calculated load resistor was 47.2 Ω at maximum input, we need to make this 47.2 Ω / 0.64 = 74 Ω at minimum input. This agrees very closely with the numbers in Fig. 19.17.

FIGURE 19.18 Derating chart for LLC converters for achieving wide-input operation.

same response (gain/power). If that is true, in effect, the *LLC* circuit does not "see" the diodes or transformer (which is what we want).

Note that we have used two diodes, one ideal and one "real," in the simulation, just to show the commutation spikes (encircled in Fig. 19.19). This causes some power loss, because V_{OR} (voltage across inductor) is seen to be slightly less than what we were expecting, i.e., 54.6 V—which is the input amplitude of 52 V multiplied by the target gain of 1.05. But in the same circuit, if we put in two ideal diodes, we do get almost exactly 54.6 V, proving the non-ideal diode was responsible for the slight power loss. Note that the peak power is still a little above 60 W in Fig. 19.19, and the average power of this converter is still a little above 30 W, just as it was in Fig. 19.16.

NOTE *The built-in Simplis model for "ideal diode" is not so "ideal" either: It includes a forward voltage drop that increases rapidly with current and is set to about 600-mV drop at 10 mA of current. Even at a couple of μA, its diode drop is above 350 mV.*

This proves that the transformer and diodes are virtually transparent to the *LLC* tank circuit, yet we have managed to generate a rectified (but unfiltered) DC output with a peak value very close to 54 V (which is simply V_{OR}, the voltage across L_2, now rectified by the diodes).

Step 3 of Design Validation Process: A DC-DC Converter with Diodes and Transformer (and Also an Output Capacitor)

In Fig. 19.20, we have a *potentially* practical DC-DC converter stage. Why potentially? The output is now rectified and filtered. So that is a pure DC voltage. The input is in this case a *chopped square wave*. In an actual converter, by switching transistors connected to a steady DC rail, we will in fact be injecting a high-frequency chopped square wave at the input of the *LLC* tank circuit. That is why this design is amenable, and, in fact, just one step short of becoming a full-fledged practical DC-DC converter (based on *LLC* topology). That we will do in the next step. Keep in mind that to reduce losses, we always prefer to switch the FETs ON and OFF. Therefore, in practice, we will never really apply a sine-wave voltage excitation to the half-bridge and its *LLC* tank, but a square-wave voltage instead. However, as we will see shortly, *the first harmonic of that applied square wave voltage, which is naturally a sine wave, eventually ends up driving a sine-wave current into the load*. So, the energy-transfer process is eventually based on a resonant sine-wave action, not a square-wave action as in conventional PWM-based power conversion.

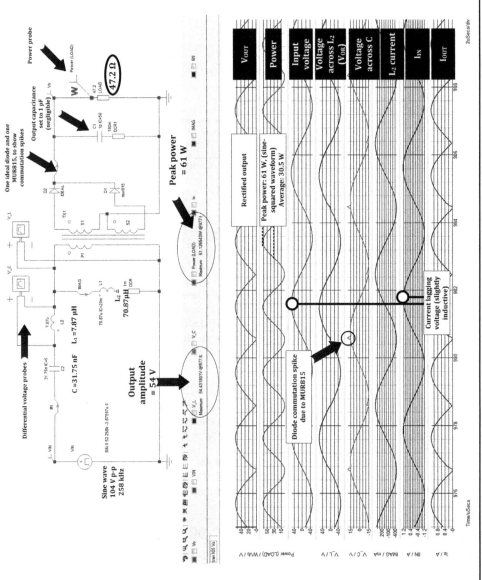

FIGURE 19.19 AC to (unfiltered) DC converter.

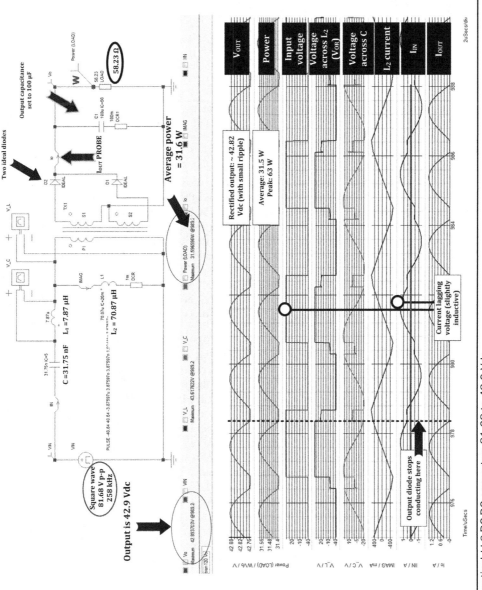

FIGURE 19.20 Potential 30-W practical *LLC* DC-DC converter: 81.68 to 42.9 Vdc.

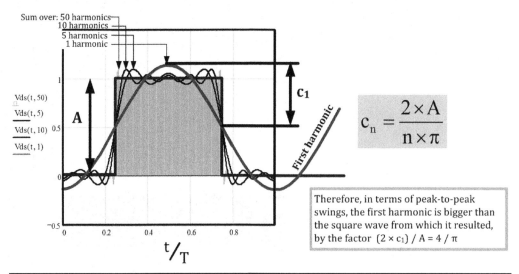

Figure 19.21 The first harmonic is bigger (peak-to-peak) than the square wave from which it came about, by the factor $4/\pi$.

Let us recap: We still are switching at 258 kHz, based on the fact that we wanted a gain of 1.05 at the design entry point. We have retained the previously scaled (frequency and power) *LLC* seed that we set out to design for 60-W peak power, but initially with a sine-wave input source (with an average output power of ~ 30 W). In Fig. 19.20, we see significant changes in input/output voltage levels and also the load resistor. *Yet we are claiming this schematic is for all practical purposes very similar to what we had in Figs. 19.16 and 19.19, from the viewpoint of the LLC tank circuit, its loading and resulting responses, etc.* We now explain that claim carefully in steps below. Because once we understand that, we will be able to design an *LLC* converter starting with a design requirement, rather than going in the opposite direction of exploring the performance and capabilities of a given *LLC* cell.

1. The input voltage was changed in Fig. 19.20 from a sine wave of peak-to-peak swing of 104 V to a *square wave of peak-to-peak swing of 81.68 V*. The reason is explained in more detail in chapter 18 of *Switching Power Supplies A-Z*, 2d ed. But we have also inserted Fig. 19.21 to show that the first harmonic (fundamental) in the Fourier decomposition is bigger than the square wave it came from, *by the factor $4/\pi$*. So, we are trying to tackle the case of a square wave excitation of our tank circuit by "thinking" (only) in terms of the first (sine) harmonic of that wave. This is called *FHA* (for first harmonic approximation) in related literature. It is a fairly effective way of tackling square waves, because if the resonant tank is not too lossy or loaded, it will tend to *reject* other frequencies and harmonics. Knowing this, if we had applied a square wave of peak-to-peak value 104 V, its first harmonic would have far exceeded our design target of 104-V peak-to-peak sine-wave amplitude, because the "equivalent" sine wave of an applied 104-V peak-to-peak square wave, based on the first harmonic approximation, would have been $104\ \text{V} \times 4/\pi = 132.4$ V. So we backed off on the peak-to-peak value of the square wave by the inverse factor: $104\ \text{V} \times \pi/4 = 81.68$ V. And that is why the square wave input (corresponding to the DC rail of a proposed equivalent DC-DC converter) is set to 81.68 V in the simulation shown in Fig. 19.20.

2. The load resistor was changed from 47.2 to 58.23 Ω in Fig. 19.20. The reason is that it has been shown in related literature that when we put a resistor after a filtering capacitor and diodes, its effective AC loading on the tank circuit is derived from

$$R_{\text{EFFECTIVE}} = \frac{8}{\pi^2} \times n^2 \times R_{\text{OUT}}$$

where "n" is the turns ratio (N_P/N_S) set to unity so far in our discussions, and R_{OUT} is just what we have been simply calling R so far (the resistor connected as a load across output terminals). The factor $8/\pi^2 = 0.811$. We want $R_{\text{EFFECTIVE}}$ to be 47.2 Ω (load as "seen" by the tank), so the load resistor we need to place across the rectified and filtered

DC output rails, based on a 1:1 transformer turns ratio so far, is $R_{\text{EFFECTIVE}}/0.811 = 47.2\,\Omega/0.811 = 58.23\,\Omega$. That is the value used in Fig. 19.20.

3. Finally, the DC output we measured is 42.9 V. If the loading of the *LLC* tank is "correct," we should be getting a steady 31.5-W out of it, because we started off by trying to get 63 W from an AC-AC converter (and so average power was exactly half the peak AC power of 63 W, i.e., 31.5 W). Therefore,

$$V = \sqrt{P_{\text{OUT}} \times R_{\text{OUT}}} = \sqrt{31.5 \times 58.23} = 42.8\text{ V}$$

This agrees with the simulation result of 42.9 V, and also validates our assumptions and procedure.

We conclude that we have understood and learned fully about how to design an *LLC* converter. So *now we reverse the process. We start with a design requirement*, and then design the entire converter from scratch.

PART 4 DESIGNING A PRACTICAL *LLC*-BASED PD

Let us design a 12-V output (25.5 W) converter as a DC-DC stage in a powered device (PD). In PoE applications, we typically need a fairly wide input DC-DC converter. Note that now we are talking of DC inputs and outputs, not AC-AC converters.

1. *Derating and required power.* We have learned that without output power derating, on a maximum load setting, we will only manage to reduce input by 12 percent from its maximum input value. Assume that the maximum value is 52 V here. In that case we will go to 52 V/1.12 = 46 V only. However, in PoE applications, we need to support voltages down to 36 V at least. In fact we target 32 V here, which is a down-conversion factor of 52/32 = 1.625. We know from Fig. 19.18 that this correction capability accrues only if we derate maximum power by 64 percent. In other words, to achieve a 25.5-W power output, we need to design the converter upfront for a maximum load at a maximum line of 25.5 W/0.64 = 40 W. That becomes our design target here in terms of selection of tank circuit components for providing desired performance (25.5 W) at low line (32 V). Note that this 40-W design target is only for determining the most appropriate Q of the *LLC* tank, for achieving wide-input operation. With suitable current and power limiting, we do not ever intend the *LLC* converter to actually work at 40-W output level. So, none of its power components need to be sized for that level. For that reason, we hereby give the 40-W design target (meant solely for achieving a wide-input range), a new name altogether: *peak virtual power*, to distinguish it from the usual maximum or peak AC power of the converter.

2. *Frequency selection.* We know that using inductors (L_1 and L_2) with a ratio of 1:9, we will get a frequency spread between the two resonant peaks separated as per $1{:}\sqrt{10}$, i.e., 1:3.16. This means the *high-frequency resonant peak will always be 3.16 times the frequency of the lower-frequency resonant peak*. Suppose we desire to keep the operation of the *LLC* below a maximum of 140 kHz for EMI reasons. In that case, the lowest frequency needs to be set to 140 kHz/3.16 = 44 kHz. Correspondingly, the highest frequency will be set to 140 kHz.

3. *Frequency scaling (of LLC seed).* Starting with our basic building block, the one we characterized so carefully initially, we chose $L_1 = 100$ mH, $L_2 = 900$ mH, $C = 10\,\mu\text{F}$. This gave a low-frequency resonant peak of 50 Hz. So to get to our current application, we need to apply a frequency scaling factor of 44 kHz/50 Hz = 880. All L and C are scaled accordingly. We get

$$L_1: 100\text{ mH} \rightarrow L_1/f_{\text{SCALE}} = 100\text{ mH}/880 = 113.6\,\mu\text{H}$$
$$L_2: 900\text{ mH} \rightarrow L_1/f_{\text{SCALE}} = 900\text{ mH}/880 = 1022.7\,\mu\text{H}$$
$$C: 10\,\mu\text{F} \rightarrow C/f_{\text{SCALE}} = 10\,\mu\text{F}/880 = 11.36\text{ nF}$$

4. *Input voltage.* At maximum input, which is a *DC level* of 52 V, in the final version of the *LLC* switching converter, we will be injecting a square wave of peak-to-peak value 52 V at the input of our *LLC* tank. Note that we must continue to work in terms of amplitudes as we did earlier. So the corresponding square-wave amplitude (*half* of the peak-to-peak swing) is actually 26 Vdc in our application. Further, as per the first harmonic approximation (FHA), this is effectively a sine wave of amplitude

26 V × 4/π = 33.1 V (the peak-to-peak swing is 66.2 V). Note that in going from the square wave of amplitude 26 V to its equivalent sine wave (as per FHA), we have multiplied the voltage by a factor of 33.1/26 = 1.273, which is just 4/π. We can remember this, it is always true: *We have to increase the square wave (DC input rail in this case) by 27.3 percent to get the equivalent peak-to-peak sine* (first harmonic). We can divide that by 2 to get the amplitude of the sine wave. Alternatively, we can go straight from a DC rail of V_{IN} to the amplitude of the equivalent sine wave by multiplying the DC rail by the factor 1.273/2 = *0.636*.

5. *Power scaling.* With the recommended 300-Ω maximum load resistor (as per our previous analysis of the chosen *LLC* seed) at our current input level (amplitude of the sine wave input being 33.1 V), the basic *LLC* seed is only good for a *peak power* of $(33.1 \text{ V} \times 1.05)^2/300 \, \Omega = 4.026 \text{ W}$. The term in *brackets*, i.e., 33.1 V × 1.05, is actually V_{OR} (i.e., V_{IN} × gain). As mentioned, so long as Q remains the same (i.e., for the same load resistor and same L and C too), the input is whatever it is (and that determines the power capability of the *LLC* tank). Here Q has not changed under frequency scaling either, because in frequency scaling we change *both L and C* by the same factor, keeping the ratio L/C unchanged—so, for the same desired Q of 1.054, R remains the same, irrespective of applied input voltage.

In our present case, we want a virtual peak power capability of 40-W (steady DC value). But in the equivalent AC-AC converter, since the output power of that is a sine-squared ($\sin^2$) curve (see Fig. 19.19), if we want the average of that to be 40 W, the virtual peak AC power we must plan on must be exactly twice that, i.e., 80 W in our case. That is our power target for the *LLC* tank. In other words, our *LLC* seed, even after frequency scaling, was only capable of 4.026 W of peak AC power, but we actually need 80 W from our equivalent AC-AC stage. Therefore, the power scaling factor required is 80 W/4.026 W = 19.87. Applying this factor to the frequency-scaled values, we get

$$L_1: 113.6 \, \mu\text{H} \to L_1/P_{SCALE} = 113.6 \, \mu\text{H}/19.87 = 5.72 \, \mu\text{H}$$
$$L_2: 450 \, \mu\text{H} \to L_1 \times 9 = 5.2 \, \mu\text{H} \times 9 = 51.47 \, \mu\text{H}$$
$$C: \ \ 11.36 \, \text{nF} \ \to \ C \times P_{SCALE} = 11.36 \, \text{nF} \times 19.87 = 225.8 \, \text{nF}$$

Note that we are consistently keeping L_2 as nine times L_1. Also we have used a Mathcad spreadsheet for the final L and C numbers, so those are more accurate than it seems by simple multiplication (that multiplies the rounding errors of the previous numbers).

6. V_{OR}. This is by our definition the amplitude (peak value) of the sine voltage that appears across the inductor in the *equivalent AC-AC converter* (with the recommended loading for a Q of 1.054). It is, in effect, the *output voltage* of the equivalent AC-AC converter. Now, if we have performed the scaling and loading correctly, we expect to get a gain of 1.05 at maximum load (when the LC network is driven at the frequency of the recommended design entry point, which we describe next). So V_{OR} is 33.1 V × 1.05 = 34.755 V. If we design the regulator carefully, it will in effect be able to regulate and keep V_{OR} constant, even if the input starts to fall (within a specified range) or the load decreases down to zero.

7. *Frequency at maximum load at high-line.* In our most recent design example, we approached the design target of gain = 1.05 by switching at 258 kHz, when our lowest frequency was 100 kHz. This value of 258 kHz was determined by careful examination on the Mathcad plot. The ratio of this frequency with respect to the low-frequency resonant peak was therefore 258 kHz/100 kHz = 2.58. In fact we can keep the same ratio in all cases assuming we have set the right Q, etc. Since our low-frequency peak is 44 kHz in this PoE design example, the desired switching frequency to achieve the optimum design entry point (of maximum load at high-line) is 44 kHz × 2.58 = 113 kHz. This is the frequency we will plug into the Simplis simulator to test our circuit's response at the maximum load, high-line setting (design entry point).

8. R_{EFF}. (Same as the $R_{EFFECTIVE}$ we used earlier). As mentioned, we always need to target a gain of 1.05. So the V_{OR} became 34.755 here. For the required virtual peak AC power of 80 W, the effective resistor to place across the larger L of the *LLC* stage (in the equivalent AC-AC schematic) is

$$R_{EFF} \equiv \frac{V_{OR}^2}{P_{PK}} = \frac{34.755^2}{80} = 15.1 \, \Omega$$

This will give us the desired steady virtual peak power of 40 W, and thus the desired maximum output power of 25.5 W over the entire input range of 32 to 52 Vdc. And this R_{EFF} is therefore the required "loading," corresponding to the maximum load, as seen by our newly designed LLC tank circuit. This value of R is expected to yield the optimum "shape" of the gain curve. We could also have found out R_{EFF} from the Q formula, using the target value of $Q = 1.054$ as before. That calculation would give exactly the same result for R_{EFF}. Note that this 15.1-Ω resistor takes the place (in this application) of the 300-Ω resistor that we had identified for our original LLC seed (that too gives us a Q of 1.054, as in the current design too).

9. R_{LOAD}. We have learned that when we put in a large filtering capacitor after rectifier diodes, there is a factor of 0.811 that we have to account for in going from the actual resistor placed across the output capacitor to its effective load across the first-harmonic sine wave resonant LLC tank. Assuming a 1:1 turns ratio for the transformer so far, if we want the LLC tank to "experience" a loading resistor of 15.1 Ω, we need to increase the load resistor placed at the (rectified and filtered) output terminals to a *larger* value: $R_{EFF}/0.811 = 15.1\,\Omega/0.811 = 18.62\,\Omega$. Keep in mind that 0.811 comes from $8/\pi^2 = 0.811$.

   To get 40 W out of this resistor (virtually), the output voltage (DC rail) must be

   $$V_{O\_INTERMEDIATE} = \sqrt{P_{OUT} \times R_{OUT}} = \sqrt{40\,W \times 18.62\,\Omega} = 27.3\,V$$

10. *Turns ratio*. Note that so far we are still only with a 1:1 transformer. In reality we need to go down to 12 Vdc. So the required turns ratio is $27.3\,V/12.7\,V = 2.15$. In terms of the reciprocal, that is, N_S/N_P which is "turns ratio" in the Simplis simulator, this is actually 0.465. Here we have included a 0.7-V diode drop since even the so-called ideal diode of Simplis has a typical forward drop of around that value at even 10 to 50 mA of current.

    Note that the turns ratio used here refers to the ratio of the primary winding to *one* secondary winding of two identical windings. In a center-tapped transformer, with one primary winding and one secondary winding, the equivalent turns ratio is actually half.

11. *Actual maximum load resistor*. If we have 12 V on the output, to get 40 W, the required load resistor is

    $$R_{LOAD} = \frac{V_{OUT}^2}{P_{OUT}} = \frac{12^2}{40} = 3.6\,\Omega$$

    This is for 40 W of (DC steady virtual) power.

12. *Derated maximum load resistor*. To get to lowline (32 Vdc), we had to derate power by 64 percent as per Fig. 19.18. It is the same for this resistor. So in the 32-V simulations we need to change it back to $3.6/0.64 = 5.625\,\Omega$.

13. *Frequency at low-line*. To get to low-line (with derated power), we also need to increase gain by reducing frequency. From the derating chart in Fig. 19.18, we can see the frequency will be higher than the low-frequency resonant peak by the factor 1.215. So in the 32-V simulations, we have fixed the frequency to $44\,kHz \times 1.215 = 53.46\,kHz$.

We now have the full proposed solution. This should give us a ~ 25-W LLC converter working over the PoE input rail of 32 to 52 V. Let us see how this fares in simulations right off the bat (eventually there will be an active feedback loop to correct our predictions anyway). Keep in mind that in Power over Ethernet, 25.5 W is the maximum power that can be drawn at the input of the powered device (PD). It is not its output power. So we do not really have to account for less than the 100 percent converter efficiency implicitly assumed above. But in general applications we would need to do that, by simply increasing the desired output power by dividing it by the efficiency.

PART 5 VALIDATING OUR THEORETICAL PD DESIGN VIA SIMULATIONS

We perform simulations at high-line and low-line to check how the converter behaves. We collect the key design inputs from the preceding design procedure:

1. In all cases our LLC tank is

   $$L_1 = 5.72\,\mu H \quad L_2 = 51.47\,\mu H \quad C = 225.8\,nF$$

2. Turns ratio (each secondary half-winding with respect to primary winding): 0.465

3. Design entry point: 40-W average output (at 113 kHz)
4. Low-line frequency: 53.46 kHz by calculation (needed to increase to 59.1 kHz)
5. Output load resistor for 40-W operation: 3.6 Ω
6. Output load resistor at low-line (32 V): 5.625 Ω.
7. Input DC voltage at high-line: 52 V
8. Input DC voltage at low-line: 32 V
9. Equivalent AC sine-wave input voltage at low-line: 40.744-V peak-to-peak

Simulation Run 1 (52 V, Square-Wave Driven, See Fig. 19.22)

This was run at the predicted frequency of 113 kHz, and we can see it very naturally produced what we designed for: 12-V output with 40W.

Simulation Run 2 (32 V, Sine-Wave Driven, See Fig. 19.23)

This was run at the predicted frequency of 53.46 kHz. The equivalent AC input applied was 40.744 V peak-to-peak instead of 32 Vdc (as per first harmonic approximation). We can see it very naturally produced *almost* what we designed for: 12-V output with 25-W output. Any slight output correction would of course be quickly carried out by the regulation loop (not part of our simulations), by tweaking the frequency.

Simulation Run 3 (32 V, Square-Wave Driven, See Fig. 19.24)

This was run at the predicted frequency of 53.46 kHz, but the output was noticeably higher and so was the output power. But we "pretended" to be the feedback loop and manually dialed in a higher and higher frequency to reduce the gain. Finally at 59.1 kHz we once again got very close to our expectations of 12 V with 25 W. Note that the slight breakdown of prediction compared to simulation run 2, was obviously attributable to the first harmonic approximation (FHA) breaking down somewhat. Because when we put back the AC (sine wave) source of appropriate amplitude, we once again get very good agreement as in simulation run 2.

Conclusion

This excellent matching up of our underlying intuitive reasoning, scaling strategies, design entry approach guidelines, Mathcad plots, and simulation results indicate that we have a validated and effective design procedure.

PART 6 RATIO OF INDUCTANCES (PROS AND CONS)

Why are high ratios of L_2/L_1 preferred? The answer revolves around a deeper understanding of the current waveforms observed. See Fig. 19.25 closely from this viewpoint. The first thing we observe is that of all the currents, I_O, which is the rectified but unfiltered current coming out of the diodes, has the most "regular" shape. In other words, looking at I_O, we can sense that useful energy (to the load) is being transferred in a nice, resonant-mode fashion, across the transformer. This explains why, despite the seemingly distorted waveforms in Fig. 19.22 for example, the final results matched our assumptions almost perfectly, even though our initial estimates were based on sine-wave excitations only, and also with no diodes or output filtering capacitor present (just a pure "AC-AC converter"). Note that we had set the frequency of operation based on our calculations and then discovered that despite there being no "regulation loop" to help us in our simulation schematics, *exactly the 12-V output we had predicted, and with almost exactly 40 W of load*, did happen. On the face of it, the waveforms on the schematic barely looked "resonant." So how did we get such good results? The answer is buried in the shape of I_O. It is very close to a rectified sine wave. That part of the circuit "sees" only the *LLC* tank, being excited by a sine-wave source. Unfortunately, there is another part of the circuit that is not so straightforward. The primary-side waveforms are a composite of two, as we will see.

One thing is sure: The LLC tank energy is delivered in a *Forward converter style* into the load, i.e., without being stored first in the core and then being pumped out, as in a Flyback. Observe the polarity of the windings in all the simulation schematics, which is not the way Flybacks work. We also know that in Forward converters, since (most of the) energy is not stored in the core, we can use comparatively small cores as compared to a Flyback, and in fact the biggest restriction is just the available *window area* for windings. If we can pack the required turns and copper thickness (to keep temperatures down) in the available window area of the core, the core would work, almost irrespective of the power level. For that reason,

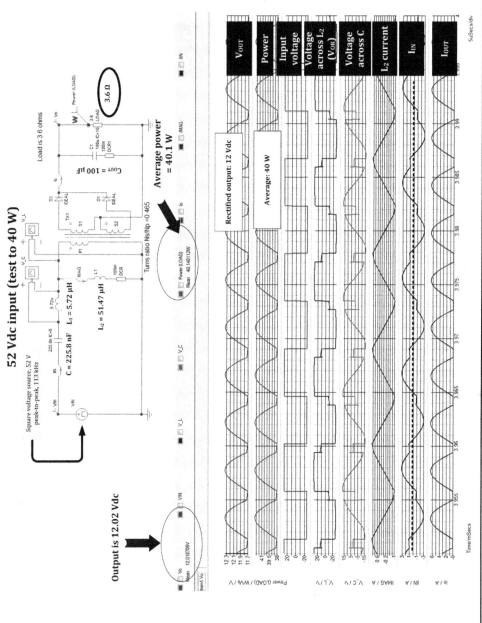

Figure 19.22 Validation of PD design through simulation, at peak load (40 W) and high-line.

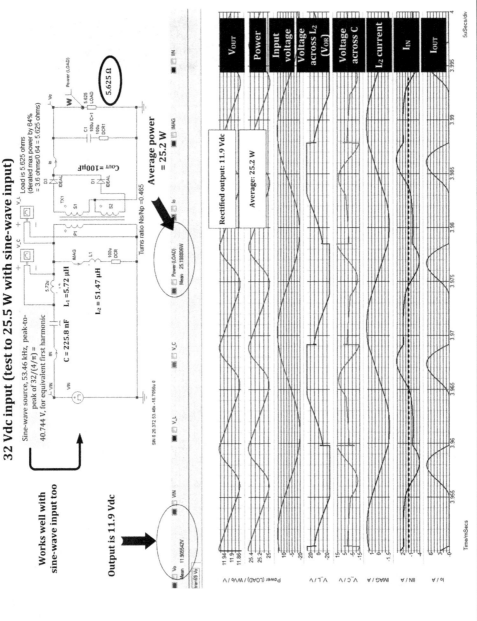

FIGURE 19.23 Validation of PD design through simulation, at maximum load (25 W) and low-line, with sine wave.

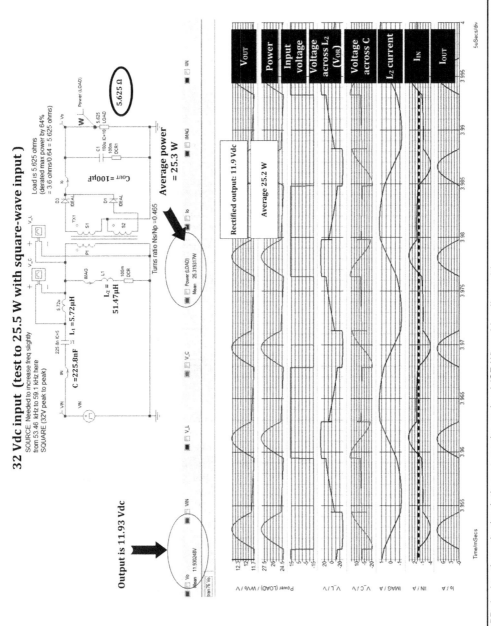

FIGURE 19.24 Validation of PD design through simulation, at maximum load (25 W) and low-line, with square wave.

502 Chapter Nineteen

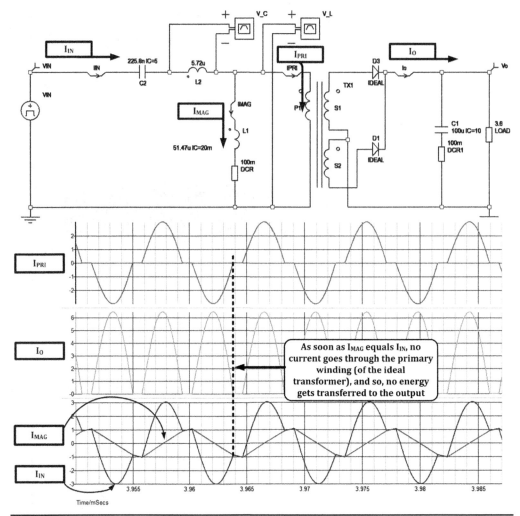

FIGURE 19.25 Understanding the current waveforms of the LLC stage.

even for picking *LLC* transformers, we can use the same straightforward core selection procedure we use for Forward converters. The reader can find such a procedure in chapter 4 of *Switching Power Supplies A-Z*, 2d ed. Also see Chaps. 11 and 12 of this book.

There is another commonality that *LLC* converters share with Forward converters, and this is not so helpful: the problem of the magnetization (core excitation) current. Yes, for a given core, we can fix the number of primary-side turns based on the allowed swing of the B-field (and that depends on the material of the core), the applied voltμseconds (*Et*), and the effective area of the core (*A*). The commonly used equation is

$$\Delta B = \frac{100 \times Et}{N \times A}$$

Note that this says nothing about the actual *inductance*. So if, for a given core, we keep the number of turns *N* constant, but change the inductance, say by varying its air gap, we will definitely get higher peak currents, but surprisingly, the B-field in the core *does not change*! And so it won't suddenly enter core-saturation either, which was the purpose of setting the number of turns we did. In effect, we have several possible magnetizing inductances that we can pick. But *we don't go in for very low inductances*. The reason is, by definition (and polarity of windings), *the magnetizing inductance in a Forward converter is not linked to the output*. Its energy really has nowhere to go, so sometimes we might just burn (dissipate) the associated energy. An acceptable way is to use an energy recovery ("tertiary") winding, for example, to pump the retrieved magnetization energy back into the input supply. In a resonant converter, even though we are delivering energy to the output in much the same style as a Forward converter, we do not need an energy recovery winding, because as we have learned: in a resonant converter, energy sloshes back and forth naturally—so we can easily hope to recover it. But still, the process is not very efficient: Some energy will likely need to be dissipated since the magnetization current passes through a diode (or body diode) along the way. So, some of the "circulating" energy is never recovered. Which is why in a Forward converter, too, we try to keep the magnetizing inductance reasonably high, so that it proportionally decreases the associated *magnetization current*. Keep in mind that energy depends on $L \times I^2$, so an x percent reduction in I will have a much greater effect in lowering energy than an x percent increase in L. Therefore, by keeping magnetization inductance high, the total associated energy that is being circulated (and partially wasted) is significantly reduced. The same concept holds true for resonant converters. As in Forward converters, the switch current is a sum of a reflected output current plus the magnetization current: so, there is one part related to useful energy (going to the load) and another part that is related to circulating energy, which we wish to minimize.

Returning to Fig. 19.25, we see that "I_{IN}," the current coming in from the input source, has a part equal to the magnetization current. So $I_{IN} = I_{MAG} + I_{PRI}$. Why is I_{MAG} ramping up in a *straight line*? That is so similar to a Forward converter, yet we have resonant voltage waveforms here in general. In fact, even if we replace the square-wave input source with an AC (sine-wave) source, we will still get a *straight ramp* for the magnetization current (for the most part). The reason is there is an almost square voltage waveform present across L_2. We see that in Fig. 19.22 for example. Where does that come from? It is simply because the voltage across L_2 (the magnetization inductance) is being clamped through one of the diodes at any time by the big *output filter capacitor*. This resulting "square" voltage across the magnetization inductance causes a straight and rising current ramp. At some point this rising current (I_{MAG}) will become equal to the input current ($I_{MAG} = I_{IN}$). So I_{PRI} becomes zero. Note that I_{PRI} is the resonant part of the input current and is related only to output power requirements. It is responsible for delivering I_O to the secondary side. Once I_{PRI} goes to zero, the output diodes no longer conduct. The input current then "follows" the magnetization current for a short while, till the applied input voltage flips again, driven by the source. At that point the magnetization current starts ramping down in a straight line with the same slope (in magnitude), because it is now clamped through the *other* secondary-side diode to the very same already-charged output capacitor.

How do all these current waveforms vary with load? In Fig. 19.26, we have superimposed the results of several simulation runs, using a *range of load resistors*. Since the output voltage is almost steady for a very wide range of loads (one of the advantages of operating at the design entry point, as is obvious from Figs. 19.11 and 19.13), the shape of I_{MAG} hardly changes, as we can see from Fig. 19.26. We can also discern that the input current has two highlighted parts: one, the near-constant magnetization current component (ramp up and down), the other the one truly involved in the resonant power delivery process—and this component expectedly has a wide variation with load.

We have learned that in a typical resonant-mode transformer, the magnetization inductance plays the part of L_2 in its *LLC* tank circuit. So by keeping the ratio of L_2/L_1 high, in effect we keep the circulating magnetization current low, and achieve higher efficiency too. If we make the L_2/L_1 ratio too high, the frequency spread will be much more than the roughly 1:3 variation we achieved by keeping L_2 as $9 \times L_1$. And also, the gain curves, slope between the two resonant peaks will likely *sag*, potentially violating our own initial "guiding criteria"—that of requiring an unambiguous direction of correction.

PART 7 HALF-BRIDGE IMPLEMENTATION OF LLC CONVERTER AND SOLVED EXAMPLES

It is clear that the sine-wave excitation we used in places earlier was necessary to explain the underlying principles of resonance, but in reality, we want to turn the switches ON and OFF between fully-conducting and fully-nonconducting states to minimize conduction loss. So only the examples we discussed earlier with "square-wave" inputs really apply to practical implementations. The easiest way to do this, and therefore one of the most popular, is to use a half-bridge (or "H-bridge"), because when the top FET turns ON (bottom FET OFF), it excites the tank circuit in one direction, and when the bottom FET turns ON (top FET OFF), it excites it in the opposite direction. It is similar to two parents positioned on opposite sides of a child's swing, each pushing slightly just when the swing approaches and then *starts to move away*. That would be the analog of soft switching, because if the pushing was "out of phase" to the swinging, and the swing was pushed when it was still approaching, the child in the swing would have every reason to complain (do not try this at home).

So the symmetric excitation possibility, along with the soft-switching capability, makes the H-bridge popular in general even in conventional power conversion. By reducing the

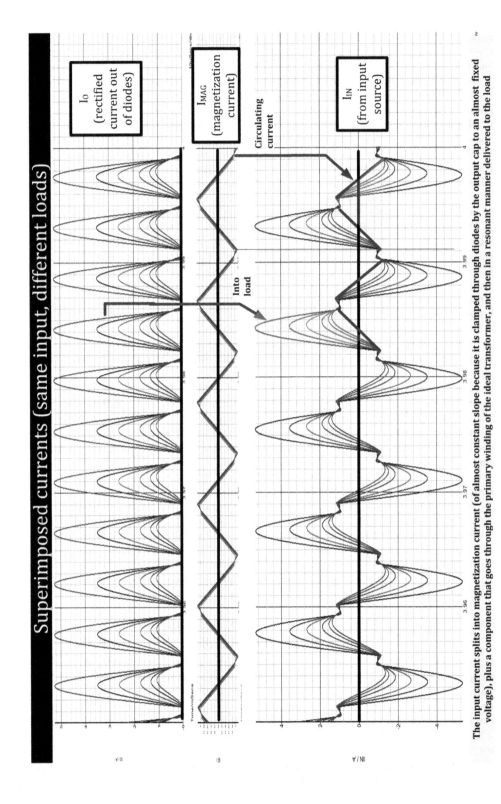

FIGURE 19.26 The current waveforms of the LLC stage, as load is varied.

usual coupling (DC-blocking) capacitor, we can partake in the resonant-mode behavior. In an *LLC* converter, since the transformer is clearly part of the parcel, we get free isolation (whether we need it or not), but also the possibility of integrating both the inductances L_1 and L_2 into the transformer. Because in all our simulations so far (those using "transformers") we ultimately used an *ideal* transformer, one which is not inductive, but simply performs a step-up or step-down operation as per the turns ratio. In reality, a real transformer has a magnetization inductance that can play the part of L_2. In addition, if we wind the primary and secondary windings, not on top of each other as in a conventional transformer, but alongside (using a split bobbin), as shown in Fig. 19.27, we will generate enough leakage inductance in both the primary and secondary sides. This can play the part of L_1 in our *LLC* tank. Of course any primary-side leakage is "seen" by the secondary side, and vice versa, because leakage reflects to the other side as per the turns ratio squared. This aspect was also carefully explained in chapter 4 of *Switching Power Supplies, A-Z*, 2d ed, in the context of how secondary-side PCB traces can cause a huge primary-side zener-clamp dissipation term in the Flyback. See also, Chap. 10 of this book.

Finally, we connect a half-bridge to the square-wave *LLC* resonant circuit we had in our earlier PoE example (see Fig. 19. 22). The result is Fig. 19.28. Note that we still get *40 W and 12-V output*.

We have thus completed our full development of an LLC *converter for PD applications, based on the initial design requirements.*

In Fig. 19.29, we show how the input current splits up in the two FETs. In Fig. 19.30, we show how zero voltage switching (ZVS) is occurring, and this is very similar to how the upper FET gets switched softly in a synchronous Buck, for negative inductor current excursions, as discussed previously.

Efficiency Estimates

A quick check of efficiency follows: The following equation has been quoted in related literature. For example in "Optimal Design Methodology for LLC Resonant Converter," by Bing Lu, Wenduo Liu, Yan Liang, Fred C. Lee, Jacobus D. van Wyk, Center for Power Electronics Systems, Virginia Polytechnic Institute and State University:

$$I_{RMS\_IN} = \frac{1}{8} \times \frac{V_O}{n \times R_{LOAD}} \sqrt{\frac{2 \times n^4 \times R_{LOAD}^2}{L_2^2 \times f^2} + 8\pi^2}$$

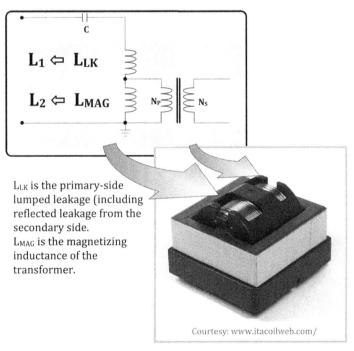

L_{LK} is the primary-side lumped leakage (including reflected leakage from the secondary side.
L_{MAG} is the magnetizing inductance of the transformer.

Courtesy: www.itacoilweb.com/

FIGURE 19.27 Construction of a typical *LLC* transformer.

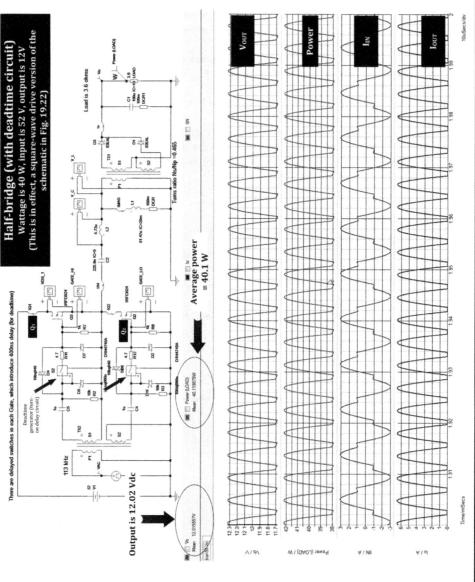

Figure 19.28 Final completed H-bridge LLC converter for PD applications (40 W at 52 Vdc, 25 W at 32 Vdc).

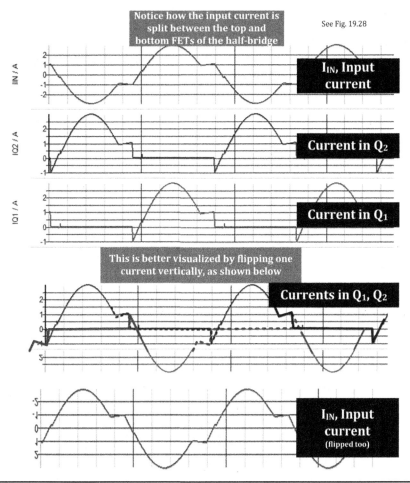

Figure 19.29 How currents in two FETS "add up" to produce the input current waveshape.

Plugging in the values for our PD design, when operated at full power (40 W) at 52 V:

$$I_{RMS\_IN} = \frac{1}{8} \times \frac{12}{2.15 \times 3.6} \sqrt{\frac{2 \times 2.15^4 \times 3.6^2}{51.47\mu^2 \times 113k^2} + 8\pi^2} = 1.89 \text{ A}$$

The Simplis simulation yielded 1.96 A, which is within 5 percent of the closed-form equation. If both the chosen FETs have an RDS of 100 mΩ, the combined FET dissipation is

$$P_{FET} = (I_{RMS\_IN})^2 \times R_{DS} = (1.89)^2 \times 0.1 = 0.36 \text{ W}$$

The dissipation in diodes, assuming a diode drop of 0.6 V and an average current of 40 W/12 V = 3.33 A is 3.33 A × 0.6 = 2 W. So total dissipation (assuming negligible losses in the transformer and ignoring small switching losses) is 2.36 W. The first efficiency estimate is therefore 40/42.36 = 94.4 percent.

Solved Example for LLC Selection with Maximum 1:3.16 Frequency Spread

We need a universal input AC-DC *LLC* converter that can output 12 V, 20 W. (Using an *LLC* seed of 100 mH, 900 mH, 10 μF, 300 Ω).

1. A 400 Vdc input, when switched, will produce a square-wave excitation on the tank circuit, of amplitude (half peak-to-peak value) 200 Vdc. The corresponding AC amplitude is more: $4/\pi \times 200$ V = 254.6 Vac (its amplitude, not peak-to-peak).

   Note the shortcut we had indicated earlier: Multiply the DC rail by 0.636 to get the sine wave amplitude. *Check:* 400 V × 0.636 = 254.4 V. Close enough.

2. With the right Q (loading), the gain will be 1.05 times. So V_{OR} will then be 254.6 V × 1.05 = 267.33 V (peak value of equivalent sine wave). With a "300-Ω" optimum resistance

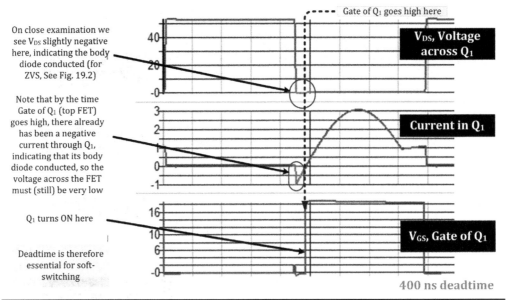

FIGURE 19.30 Zero-voltage switching occurring in an *LLC* converter.

across it (decided earlier from our seed generation study), we will get a peak AC power of $267.33^2/300 = 238.22$ W. *Note*: This is the power measured at the maxima of the sine-squared power waveform curve associated with the equivalent AC converter. This is not DC power!

3. We need 20 W of average DC output power over the entire input range. Assuming worst-case 85 percent efficiency, we should plan on $20\,\text{W}/0.85 = 23.53$ W maximum DC power. That is equivalent to twice, that is, 47 W of peak AC power (at maxima of sine-squared waveform) from the equivalent AC-AC converter.

4. But from Fig. 19.18 we see that we need derating to handle low input and must therefore plan for five times more *LLC* tank circuit power capability. That is $47\,\text{W} \times 5 = 235$ W peak AC virtual power (at maximum line, measured at maxima once again).

5. Therefore, our power scaling factor P_{SCALE} is $235\,\text{W}/238.22\,\text{W} = 0.99$. Coincidentally very close to unity.

6. Suppose we want to set our maximum operating frequency to 140 kHz, say for EMI reasons.

7. *Frequency selection.* We know that using inductors L_1 and L_2 with a ratio of 1:9, we will get a frequency spread between the two resonant peaks separated as per $1{:}\sqrt{10}$, that is, 1: 3.16. This means the high-frequency resonant peak will always be 3.16 times the frequency of the lower-frequency resonant peak. Suppose we desire to keep the operation of the LLC below a maximum of 140 kHz for EMI reasons. In that case, the lowest frequency needs to be set to $140\,\text{kHz}/3.16 = 44$ kHz. If we do so, then correspondingly, the highest frequency will not exceed 140 kHz.

8. *Frequency scaling (of LLC seed).* Starting with our basic building block, the one we characterized initially, we chose $L_1 = 100$ mH, $L_2 = 900$ mH, $C = 10$ μF. This gave us a low-frequency resonant peak of 50 Hz. So to get to our current application, we need to apply a frequency scaling factor of $44\,\text{kHz}/50\,\text{Hz} = 880$. The L's and C are scaled accordingly. We get

$$L_1: 100\,\text{mH} \rightarrow L_1/f_{SCALE} = 100\,\text{mH}/880 = 113.6\,\mu\text{H}$$
$$L_2: 900\,\text{mH} \rightarrow L_2/f_{SCALE} = 900\,\text{mH}/880 = 1022.7\,\mu\text{H}$$
$$C:\ \ 10\,\mu\text{F} \ \rightarrow\ C/f_{SCALE} = 10\,\mu\text{F}/880 = 11.36\,\text{nF}$$

9. Applying the power scaling factor to the frequency-scaled values, we get

$$L_1: 113.6\,\mu\text{H} \rightarrow L_1/P_{SCALE} = 113.6\,\mu\text{H}/0.99 = 115\,\mu\text{H}$$
$$L_2: 1.023\,\text{mH} \rightarrow L_2 \times 9 = 115\,\mu\text{H} \times 9 = 1.035\,\text{mH}$$
$$C:\ \ 11.36\,\text{nF} \ \rightarrow\ C \times P_{SCALE} = 11.36\,\text{nF} \times 0.99 = 11.5\,\text{nF}$$

10. The preceding frequency-scaled, power-scaled *LLC* seed, is supposedly good for 235 W of peak AC power at high-line. The value of the loading resistor to make this happen is $V_{OR}^2/P = 267.33^2/235 = 304.1\ \Omega$.

11. If we were using a 1:1 transformer with a rectified DC output, we know that the load resistor across the output capacitor, which will produce 304.1-Ω effective load across the *LC* tank circuit, needs to be larger by the amount $\pi^2/8 = 375.2\ \Omega$.

    Note the shortcut we had indicated earlier: going from the DC output resistor to the corresponding *LC* loading, we need to multiply the load resistor by 0.811. *Check:* $375.2\ \Omega \times 0.811 = 304.3\ \Omega$. Close enough.

12. The peak AC power as per our definition is calculated at the top of a sine-squared waveform. So the average (DC) power is actually half of that: 235 W/2 = 117.5 W. Assuming a rectified DC output with a 1:1 transformer, this "intermediate rectified DC rail" must be

$$V_{O\_INTERMEDIATE} = \sqrt{P_{OUT} \times R_{OUT}} = \sqrt{117.5\ W \times 375.2\ \Omega} = 210\ V$$

Note that there is a simple shortcut to get to this intermediate DC rail. It is very simply

$$V_{O\_INTERMEDIATE} = \frac{V_{INMAX}}{2} \times \text{gain} = \frac{400\ V}{2} \times 1.05 = 210\ V$$

13. Turns ratio. We want to go from this intermediate DC rail to the actual output DC rail by the step-down turns ratio. So very simply

$$\frac{N_S}{N_P} \equiv \frac{1}{n} = \frac{V_O}{V_{O\_INTERMEDIATE}} = \frac{12.7}{210} = 0.06 \qquad n = 16.535$$

(including 0.7 V for the output diode). So if, for example, the Primary has about 33 turns, we need 2 + 2 turns for the Secondary (center-tapped). Note that when we use Simplis or PSpice, the turns ratio used by the software is N_S/N_P instead. So we need to set it as 0.06.

Final Selection

We can use a leakage inductance of 115 µH. The primary inductance must be 1.035 mH, and the capacitor must be 11.5 nF. Turns ratio is 16.54 (in a simulator we need the inverse of this, i.e., 0.06). We can use standard Forward converter design criteria for selecting the core and number of turns.

The lowest expected frequency is 44 kHz. The maximum frequency spread is 1:3.16. The frequency at the design entry point is 2.58 × lowest frequency, that is, 44 kHz × 2.58 = 113 kHz (test frequency for simulator at design entry point).

Alternate *LLC* Seed

Sometimes, we may want to narrow the frequency variation by lowering the ratio L_2/L_1. This will produce larger circulating currents, but the spread will be narrower, and also, the transformer is usually easier to wind (with split bobbins). We arrive at a worst-case 1:2 frequency variation by the following seed (ratio of inductances is 1:3):

1. Leakage inductance of 300 mH
2. Primary inductance of 900 mH
3. Capacitance of 10 µF
4. Load resistance $R = 270\ \Omega$

As before, we want this *LLC* seed to fix the maximum load plus maximum-line operating point on a curve that has a gain of 1.05 (on the basis of the turns ratio and a regulation loop to enforce it, of course). This *LLC* seed produces soft switching up to higher gains than with the previous seed, even at maximum load and maximum-line condition as we lower the input (the gain goes from 1.05 to 1.25, compared to only up to 1.18 with the previous *LLC* seed).

The operating frequency point in this case is, however, only *1.72* times the lower resonant frequency, that is, at 79 Hz, since the lower resonant peak is at *46 Hz*. Note that for the previous 1:9 inductance ratio *LLC* seed, this design entry point was about *2.58* times the lower resonant frequency of 50.3 Hz, that is, at 129 Hz. Also, with the previous seed, the upper resonant frequency was $\sqrt{1+9} = 3.16$ times the lower resonant frequency, that is, 3.16×50.3 Hz = 159 Hz. With our new *LLC* seed, the upper resonant frequency is $\sqrt{1+3} = 2$ times the lower resonant frequency, that is, 2×46 Hz = *92 Hz*. With this new seed, we can once again design our worldwide input 20 W AC-DC stage as follows.

NOTE *The key to this new seed was establishing 270 Ω as the preferred (optimum) maximum load. Why this value? On the basis of a calculation based on Q of 1.054, we actually get a recommended R of 328 Ω (for the chosen L's and C). Alternatively, if we try to pick a curve that offers the same gain correction of 1.18 at maximum load, we actually end up with an R of 240 Ω. Neither of these values served our purpose. We wish to be able to offer the same gain derating curve as in Fig. 19.18, for ease of use. On that basis, using our Mathcad file, 270 Ω emerged as the optimum choice for this seed, which would completely mimic the 300-Ω choice for the previous seed. Now, the gain correction curve of Fig. 19.18 is still applicable to this new seed, though the right-hand axis of the figure is obviously not applicable, since the range of frequency variation has been "squished" by the choice of L's and C. But the required power derating can still be gleaned.*

Solved Example for LLC Selection with Maximum 1:2 Frequency Spread

We need a universal input AC-DC *LLC* converter that can output 12-V, 20 W (using an *LLC* seed of 300 mH, 900 mH, 10 μF, 270 Ω). Use the shortcuts from the preceding example.

1. Multiply the DC rail by 0.636 to get the sine-wave amplitude: 400 V × 0.636 ≈ 254.6 V.

2. With the right Q (loading), the gain will be 1.05 times. So V_{OR} will then be 254.6 V × 1.05 = 267.33 V (peak value of equivalent sine wave). With a 270-Ω optimum resistance across it (decided from our *LLC* seed generation study), we will get a peak AC power of $267.33^2/270 = 264.7$ W.

3. We need 20-W average DC output power over the entire input range. Assuming worst-case 85 percent efficiency, we should plan on 20 W/0.85 = 23.53 W. That is equivalent to twice, that is, 47 W of peak AC power, from the equivalent AC-AC converter. But from Fig. 19.18 we see that we need derating to handle low input and must therefore plan for *five times* more *LLC* tank circuit power capability, that is, *235-W* peak AC virtual power (at maximum line).

4. Therefore, our power scaling factor P_{SCALE} is 235 W/264.7 = 0.89.

5. Suppose we want to set our maximum operating frequency to 170 kHz.

6. *Frequency selection.* We know that using inductors L_1 and L_2 with a ratio of 1:3, we will get a frequency spread between the two resonant peaks separated as per $1:\sqrt{4}$, i.e., 1:2. This means the high-frequency resonant peak will always be two times the frequency of the lower-frequency resonant peak. Suppose we desire to keep the operation of the *LLC* below a maximum of 170 kHz. In that case, the lowest frequency needs to be set to 170 kHz/2 = 85 kHz. If we do so, then correspondingly, the highest frequency will not exceed 170 kHz.

7. *Frequency scaling (of LLC seed).* Starting with our basic building block, the one we characterized initially, we chose $L_1 = 300$ mH, $L_2 = 900$ mH, $C = 10$ μF. This gave us a low-frequency resonant peak of 46 Hz. So to get to our current application, we need to apply a frequency scaling factor of 85 kHz/46 Hz = 1848. The L's and C are scaled accordingly. We get

$$L_1: 300 \text{ mH} \rightarrow L_1/f_{SCALE} = 300 \text{ mH}/1848 = 162 \text{ μH}$$
$$L_2: 900 \text{ mH} \rightarrow L_2/f_{SCALE} = 900 \text{ mH}/1848 = 487 \text{ μH}$$
$$C: 10 \text{ μF} \rightarrow C/f_{SCALE} = 10 \text{ μF}/1848 = 5.4 \text{ nF}$$

8. Applying the power scaling factor to the frequency-scaled values, we get

$$L_1: 162 \text{ μH} \rightarrow L_1/P_{SCALE} = 162 \text{ μH}/0.89 = 182 \text{ μH}$$
$$L_2: 487 \text{ μH} \rightarrow L_2 \times 9 = 182 \text{ μH} \times 3 = 546 \text{ μH}$$
$$C: 5.4 \text{ nF} \rightarrow C \times P_{SCALE} = 5.4 \text{ nF} \times 0.89 = 6.1 \text{ nF}$$

9. The frequency-scaled, power-scaled *LLC* seed, is supposedly good for 235-W peak AC power at high-line. The value of the loading resistor to make this happen is $V_{OR}^2/P = 267.33^2/235 = 304.1\ \Omega$.

10. If we were using a 1:1 transformer with a rectified DC output, we know that the load resistor across the output capacitor, which will produce 304.1-Ω effective load across the *LC* tank circuit, needs to be larger by the amount $\pi^2/8 = 375.2\ \Omega$.

    Note the shortcut we had indicated earlier: going from DC output resistor to *LC* loading, we need to multiply by 0.811. *Check:* $375.2 \times 0.811 = 304.3\ \Omega$. Close enough.

11. The peak AC power is calculated at very top of a sine-squared waveform. So the average DC power is $235\ W/2 = 117.5\ W$. Assuming a rectified DC output with a 1:1 transformer, this intermediate rectified DC rail must be

$$V_{O\_\text{INTERMEDIATE}} = \sqrt{P_{\text{OUT}} \times R_{\text{OUT}}} = \sqrt{117.5\ W \times 375.2\ \Omega} = 210\ V$$

Note that there is a simple shortcut to get to this intermediate DC rail. It is very simply

$$V_{O\_\text{INTERMEDIATE}} = \frac{V_{\text{INMAX}}}{2} \times \text{gain} = \frac{400\ V}{2} \times 1.05 = 210\ V$$

12. Turns ratio. We want to go from this intermediate rail to the actual output rail by the step-down turns ratio. So very simply

$$\frac{N_S}{N_P} \equiv \frac{1}{n} = \frac{V_O}{V_{O\_\text{INTERMEDIATE}}} = \frac{12.7}{210} = 0.06 \qquad n = 16.535$$

(including 0.7 V for the output diode). So if, for example, the Primary has about 33 turns, we need 2 + 2 turns for the Secondary (center-tapped). Note that when we use Simplis or PSpice, the turns ratio used by the software is N_S/N_P instead. So we need to set it as 0.06.

Final Selection

We can use the leakage inductance of 182 µH. The primary inductance must be 546 µH and the capacitor must be 6.1 nF. The turns ratio is 16.54 (in the simulator we need the inverse of this, i.e., 0.06).

We can use standard Forward converter design criteria for selecting the core and number of turns.

The lowest expected frequency is 85 kHz. The maximum frequency spread is 1:2. The frequency at the design entry point is 1.72 × lowest frequency, that is, 85 kHz × 1.72 = 146 kHz (test frequency for simulator at design entry point).

Breakdown of First Harmonic Approximation and Other Subtleties

When the preceding results were simulated, it was clear they were right because when using AC sources, the results were as expected, with AC output or rectified DC output (with appropriate loading resistors as suggested). Wide input variation was also OK. However, when we switched to the "equivalent" square wave, the maximum 20-W power at low-line was somewhat in question. The inverse turns ratio had to be changed in simulation from 0.06 to about 0.07. That restored operation.

The other thing we have to keep in mind is that the "resonant frequencies" we calculated were based on ideal *LC* circuits (no load). In reality, when we load the *LC* tank, the resonant frequency shifts lower. So, for example, in the preceding example, we may need to test its response at low-line to slightly below 85 kHz to get the required gain correction. On the controller IC too, we must allow it to go to a slightly lower frequency than the "ideal" low resonant peak.

CHAPTER 20
Things to Try

Introduction

Too often we grope within our minds to remember a circuit we may have seen somewhere, or even built ourselves, which at that time had seemed "neat." Too bad we threw it away while relocating. We could have used it now, but it is history. The author actually did manage to hold on to some of the most interesting circuits he either worked with, saw, or built, some of which are presented here with suitable comments. Most of them are well-tried circuits, not design ideas, and may even have been built in extremely large commercial volumes. But readers must satisfy themselves that there are no patent protections in force before using any of these, though most likely they are only clever, not patentable tricks.

Synchronizing Two 3844 ICs

It is often believed that it is possible to synchronize two 3842 ICs but not two 3844 ICs because the latter includes a frequency doubler. The 3844 is intended for a Forward converter in which we are not allowed to exceed 50 percent duty cycle. So the 3844 simply omits every alternate cycle generated by its clock to achieve an effective D_{MAX} of (a little less than) 50 percent. Therefore if we synchronize the clocks of the two 3844 ICs, we cannot guarantee that the two ICs will synchronize *in-phase* or *out-of-phase*. The clock simply doesn't know which pulse we are omitting inside the IC, and there is no pin coming out from this low-cost eight-pin IC that gives us that information either. The idea in Fig. 20.1 was created by the author, and it solves the problem (assuming continuous conduction mode) in a very simple way—by exploiting the rule that synchronization is only possible if the "master" has a slightly *higher* frequency than the "slave." So we have a phase detector on the outputs of the 3844 ICs which basically interrupts the synchronization if it detects that the outputs are out of phase. It does this by injecting an additional charging current through the 21.5-kΩ resistor into the timing capacitor of the slave. Finally, after a few unsynchronized cycles, the outputs will get into phase again (by sheer chance), at which point synchronization would be immediately allowed again by turning OFF the upper transistor. The outputs would then lock in-phase again. In-phase synchronization is required for implementing the PFC-PWM synchronization scheme discussed in Chap. 5 if there are *two* PWM power-trains running from the output rail of the PFC preregulator. But if there is no PFC present, we should use *out-of-phase* synchronization to reduce the input capacitor ripple current. With a very simple modification of this circuit we can get the two 3844 ICs to synchronize out-of-phase.

A Self-Oscillating Low-Cost Standby/Auxiliary Power Supply

Most practical power supplies have a separate on-board low-power auxiliary power supply for various reasons. For example, we almost invariably need a current limit on each output, particularly for meeting safety regulations—specifically those limiting the maximum energy we can derive from outputs designated "SELV" (safety extra low voltage). We also need output overvoltage/undervoltage protection (OVP/UVP), input undervoltage lockout (UVLO), and the like. Most often, by customer specification, the overcurrent protection (OCP) needs to be *self-recovering* on removal of the overload. In contrast, if there is an overvoltage, we will usually require the power supply to latch-off, necessitating the mains input be "cycled" before it tries to restore the output rail. In OCP we would normally sense each output, and then pass the OR-ed information to the primary side through the *fault optocoupler* (i.e., the one placed in addition to the main *regulation optocoupler*). So, if there is an overcurrent on *any* output, we usually get a situation where all outputs will collapse. Now, if *hiccup mode* is allowed under OCP, then there may be no need for any auxiliary power supply, since the

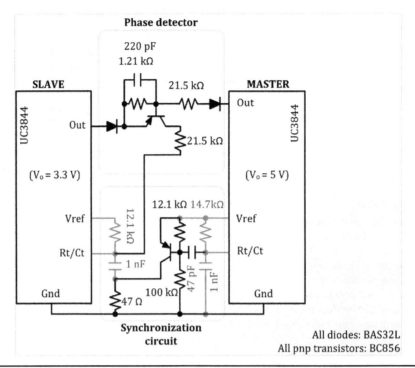

Figure 20.1 Synchronizing two xx3844 controllers.

power supply can just keep resetting and trying. But if hiccup mode is not allowed, *we need to be able to continue various activities like monitoring the current during an overload, maintaining switching action, and maintaining all the fault logic even as the main outputs power down.* Therefore, now the supply rails going to the secondary- and primary-side logic sections cannot be allowed to collapse when the main outputs do. We, thus, need an auxiliary/standby power supply to provide the internal supply rails for functioning under abnormal conditions. But why make it self-oscillating? That is considered quite advantageous since such power supplies are not only cheaper, they are inherently self-protecting by their ability to change their frequency (naturally) under overload conditions. But self-oscillating power supplies always seem to have a poorer efficiency due to their sluggish turn-ON and turn-OFF. The circuit in Fig. 20.2 therefore has several waveshaping zener diodes to make the Gate drive "sharper." It provides power to both the primary-side and secondary-side logic, and also to the PWM controller IC. It can also provide a standby output rail of 5 V which is accessible outside the power supply box. Here we should remember that any rail coming out of the box could be

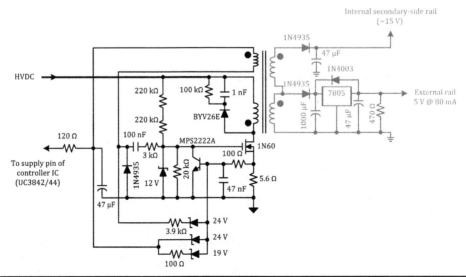

Figure 20.2 A self-oscillating auxiliary power supply for 90 to 270 Vac.

subject to safety regulations if it becomes accessible to the user. However, if it comes from an LDO stage (like the 7805), it is considered inherently protected and is not required to have any separate current-limiting circuit.

An Adapter with Battery Charging Function

In Fig. 20.3, we see on the top left side the required characteristics. We have a CV (constant voltage) region, followed by a CP (constant power) region, and then a CC (constant current) region. Actually a constant power region will give $V \propto 1/I$ which happens to be a rectangular hyperbola (not a straight line), but the approximation to a straight line is still very good over a limited region. This circuit was created by the author and it works under the following intuitive principle: if as the current increases we "fool" the feedback circuit into thinking that the voltage has gone up, it will reduce the output voltage. So we may end up getting the product VI to be fairly constant. The interesting part about the circuit is that as we increase current, first the usual regulation reference (LM431) is active (CV section, grayed out), then at higher currents the CP section automatically takes over, and finally at higher currents, the CC section takes control. Once the reader has understood the principle he or she can actually get rid of the divider on the CC section by adjusting the gain of the op-amp.

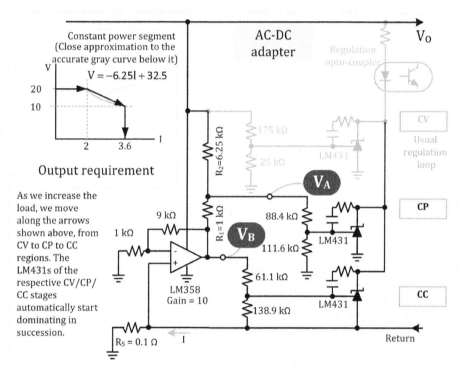

In CC region, we regulate to V_B.

$$V_B = I \cdot R_s \cdot \text{Gain (voltage proportional to current)}$$

With 3.6 A, 0.1 Ω and Gain of 10, we get   CC: $V_B = 3.6$ V

In CP region, we regulate to V_A. This is a deliberate mix between CC and CV stages, by means of the divider with R_1 and R_2. So, by choosing the right resistors, we get the right slope, one that approximates CP behavior closely (see inset above). The equations are:

$$V_o = -\text{Gain} \cdot R_S \cdot \left(\frac{R_2}{R_1}\right) \cdot I + \left(1 + \frac{R_2}{R_1}\right) \times V_A$$

Comparing with generic equation $V_o = (\text{Slope} \times I) + \text{Constant}$

$$\frac{R_2}{R_1} = \frac{\text{Slope}}{\text{Gain} \times R_S} \qquad V_A = \frac{\text{Constant}}{1 + \frac{R_2}{R_1}}$$

$$\text{So, } V_A = \frac{V_o \cdot R_1}{R_1 + R_2} + \frac{V_B \cdot R_2}{R_1 + R_2} \qquad \text{CP: } V_A = 4.48\text{ V}$$

Once R_2, R_1, V_A & V_B are known, the CC and CP dividers can be set for 2.5 V on the LM431 Reference pin as shown above.

Figure 20.3 A CV/CC/CP adapter control.

Paralleling Bridge Rectifiers

Ever needed to increase the dissipation and current in an input bridge rectifier without looking for exotic packages? In Fig. 20.4 we show how two bridge rectifier packs can be successfully paralleled using lower-cost packages. Two diodes inside the same package can usually be assumed to be well matched and will therefore share current well. This is a standard technique used on high-power supplies.

Self-Contained Inrush Protection Circuit

The basic inrush protection circuit shown in Fig. 20.5 is fairly standard, though it assumes that PFC is present. There are two high-power wire-wound resistors which charge the bulk capacitor and are then bypassed by the silicon controlled rectifier (SCR). The problem with driving an SCR at this position is that if the voltage to do that is provided by the auxiliary power supply, then we will need to do level shifting to reference the signal to the cathode of the SCR. But since the drive current for an SCR is not insignificant, we will actually lose a great deal of efficiency in the process through the dissipation in the level-shifting circuit. If we try to put a few turns around the PFC choke for the purpose, we may have a problem because the PFC varies from low duty cycle to high. Therefore, a unidirectional winding may not work over the entire range. So here we have a charge pump and voltage doubler to do the job. It works well over 90 to 270 Vac. It has been built in very large volumes.

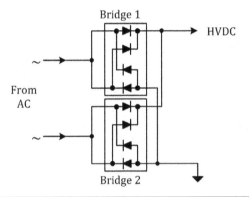

Figure 20.4 Paralleling input bridge rectifiers.

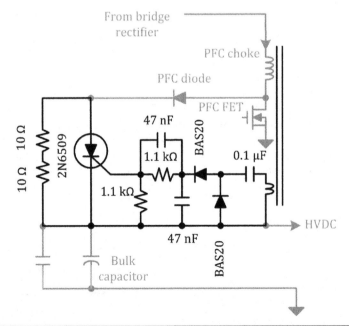

Figure 20.5 Inrush protection circuit.

Cheap Power Good Signal

A simple three terminal device at the output can provide power good indication. This is a fairly unique device from Mitsumi (at www.mitsumi.co.jp). Several variations of the PT series (e.g., PST591 to 595) are available for different applications. See Fig. 20.6.

An Overcurrent Protection Circuit

In Fig. 20.7 we have thrown a thinner gauge wire but with the same number of turns on the forward converter choke. Thus the two windings are magnetically equivalent and there should have been no voltage difference between them. But there is. The *DC resistance* (DCR) of the main winding creates a slight differential voltage which depends on the load current. This is detected by the op-amp, and by adjusting R_X we can set a current limit. Remember that this limit will be fairly crude because the resistance of copper increases by 4 percent every 10°C. However, this circuit has been used in large volumes.

Another Overcurrent Protection Circuit

In Fig. 20.8, R is the DCR of a post LC filter of a Flyback, or could be an actual sense resistor, or even just a length of manganin (or constantin) wire. It has been built in extremely large volumes. It uses the drop across a diode like the 1N4148 (≈ 0.6 V) as a reference to compare the drop across the sense resistor with. The resistor in series with the diode is calculated to

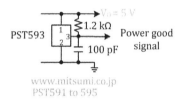

Figure 20.6 Easy power good indication.

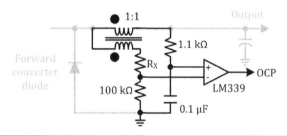

Figure 20.7 Overcurrent protection for a forward converter.

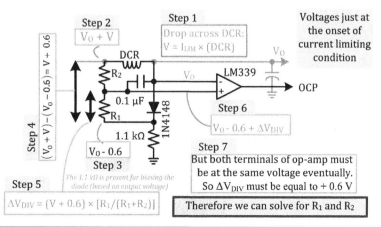

Figure 20.8 A cheap overcurrent protection circuit for Flybacks.

operate the diode close to the "knee" in its *VI* characteristics. The voltage divider R_1/R_2 is adjusted to set the current limit. Note that to avoid common-mode noise or interference problems, most experienced designers of Flybacks try not to put any sense resistor in the return rail (ground). They prefer to put it on the positive output rail itself.

Adding Overtemperature Protection to the 384x Series

The popular 3842/3844 series of controllers do not have a built-in *over temperature protection* (OTP). In Fig. 20.9 we use the fact that in typical off-line applications, the error amplifier of the controller is not being used because the error amplifier function is being carried out by the LM431 on the secondary side. So the feedback is usually directed to Pin 1 rather than Pin 2. Pin 2 is therefore vacant. Luckily, the output of the 3842/3844 error amplifier is an open-collector type, and so it is possible to perform OTP too, besides regulation. The *negative temperature coefficient* (NTC) thermistor should be actually attached to the plastic body of the switching FET by a thermal glue for the most effective protection under overloads and short circuits on the outputs. This circuit has also been used in very large volumes.

Turn-On Snubber for PFC

In Fig. 20.10 we have a popular PFC turn-on snubber, using four additional diodes, two toroidal turn-on chokes, and a large electrolytic capacitor. This works by resisting the reverse recovery shoot-through that occurs in the PFC diode when the switch turns ON. Then when the switch turns OFF, the energy gets recycled into the bulk capacitor after being temporarily stored in the large electrolytic capacitor. So it is not wasted. The author

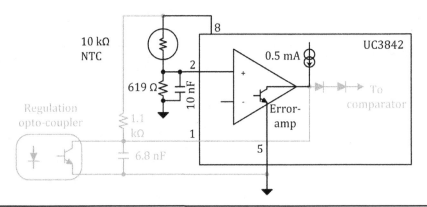

Figure 20.9 A cheap overtemperature protection add-on to the 3842/3844.

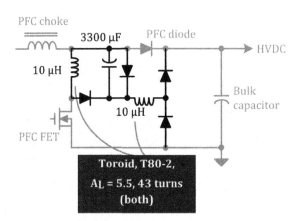

Figure 20.10 Turn-ON snubbers for PFC stages.

further modified this original (proprietary) circuit by reasoning that since the voltages across the two chokes are almost exactly equal (that can be confirmed by analysis or measurement), why not wind them on the same core? So, in the schematic we can omit the two diodes shown with gray fill, but we should ensure that the dots on the windings are as shown. We can also halve the capacitance to about 1500 µF. The result is a nonproprietary circuit that works well, is also much cheaper, but perhaps needs to be more thoroughly tested.

A Unique Active Inrush Protection Circuit

This was developed by the author for a low-volume universal input 300-W telecom power supply. Since space was at a premium (it had to be in a 3U rack-mount profile), the two large 10-W resistors of more conventional inrush protection circuits were not possible. Here the inrush control is by means of a FET (see Fig. 20.11). But the FET is *not* operated linearly and need not be mounted on a heatsink either. This circuit waits until there is a zero-crossing in the input AC voltage waveform and only then allows the FET to turn ON. So, the inrush current starts rising *with* the rising voltage, rather than turning ON and finding the instantaneous voltage at any arbitrary level. The inrush current therefore reaches a well-controlled peak (independent of parasitics like ESR) of about 42 A. The required specification on the inrush current was 45 A maximum (cold or hot power-up).

Floating Drive from a 384x Controller

In Fig. 20.12, we have a method of floating the output of the 384x controller. This was implemented for an external DC-DC converter module providing 3.3 V from a power supply which had only 12 V and 5 V outputs. Note that slope compensation is also required for such a topology or it just won't seem to work. In a 384x this normally takes the form of a simple capacitor (around 33 to 100 pF) between the Clock pin (i.e., R_T/C_T) and the Isense pin.

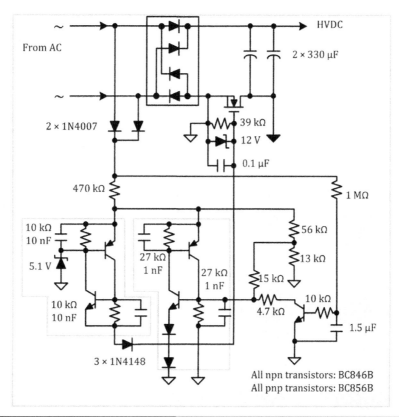

Figure 20.11 An active inrush protection circuit that works on the zero-crossing principle.

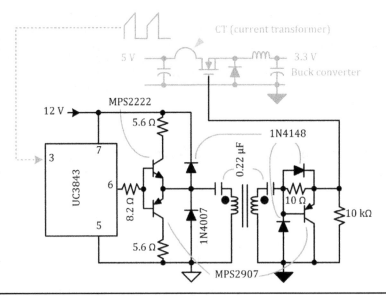

FIGURE 20.12 Floating drive from a 384x controller output.

Floating Buck Topology

This is shown for the simplest case of turns ratio 1:1 in Fig. 20.13. Here the IC floats on an auxiliary rail it creates, which also happens to settle down to exactly half the voltage of the main output rail. This decreases the voltage stress on the IC since even the SW pin of the IC sees a voltage less than V_{IN} under steady operating conditions. The input pin of the IC also sees a voltage of only $V_{IN} - V_{AUX}$. The operating principle is as follows: during the switch conduction time the voltage across the main winding is $V_{IN} - V_O$. When the switch stops conduction, the voltage across the main winding is $V_O - V_{AUX}$. But the latter must be equal to the voltage across the auxiliary winding, which is clamped to V_{AUX}. Equating, we get $V_{AUX} = V_O/2$. The auxiliary rail can be used to deliver about 1/10th the load current without disturbing the energy balance. This was developed by the author and bears a U.S. patent number.

Symmetrical Boost Topology

In Fig. 20.14 we have a true AC-DC topology, as conceived by the author. No input rectification stage is required and the output is still a DC level. Note that since the diode drops appear in series with the (boosted) output, the impact on efficiency is less than if they were positioned at the input, as in an input rectifier stage. This also saves one diode drop actually. The circuit works by having only one FET switch at a given time, and so when the next AC half-cycle starts, the other FET starts switching instead. Feedback is accomplished by a differential amplifier and the input AC must be rectified to provide a

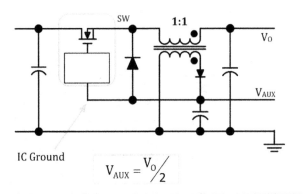

FIGURE 20.13 Floating Buck topology.

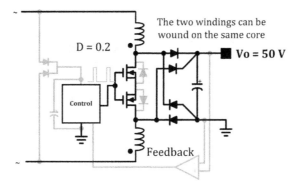

FIGURE 20.14 Symmetrical Boost topology.

DC supply to the controller. This circuit can be very useful if we don't want to require any special precautions against damage by accidental reverse polarity at the input. This circuit will just continue working normally as if nothing has happened.

A Slave Converter

Consider the equation for the output of a Buck-boost in discontinuous conduction mode

$$V_O = \frac{D^2 \cdot V_{IN}^2 \cdot 10^6}{2 \cdot I_O \cdot L \cdot f} \quad \text{V}$$

where L is in μH and f is in hertz.

This has the following proportionality

$$V_O \propto D^2 \cdot V_{IN}^2$$

But the duty cycle of a Buck converter in continuous conduction mode is

$$D \propto \frac{1}{V_{IN}}$$

So, if we use the duty cycle of a Buck in continuous mode to drive a Buck-boost in discontinuous mode, we can get the dependency on V_{IN} to cancel out as follows

$$V_O \propto \frac{1}{V_{IN}^2} \cdot V_{IN}^2 = \text{constant}$$

This was achieved in Fig. 20.15.

We have also used the fact that the output voltage of a discontinuous mode converter at a fixed duty cycle depends on its inductance. So we have "tuned" the slave to have the required output level (at its expected maximum load current) by a careful choice of inductance. Within a valid range, this technique provides completely adjustable auxiliary output voltages, something we cannot normally expect from composites based only on continuous conduction mode.

Note that the zener on the output of this slave converter is almost completely nonconducting when the slave converter is working at its designed (maximum) load. The efficiency is therefore as high as we normally expect from any conventional switching power converter. However, if the load on the slave decreases, the zener comes into play and starts automatically shunting the balance of the current away. It then behaves as a conventional shunt regulator. Therefore load regulation, which is taken for granted when dealing with single or multi-CCM stages, is not "automatic" here. It is being "enforced" by the zener, but luckily, if the inductance has been chosen correctly, this needs to happen only at less than maximum loads.

But we do have line regulation. As the input voltage increases, the feedback loop of the regulated Buck converter commands its duty cycle to decrease to maintain output regulation. It so happens that this decrease in duty cycle is exactly what was required by the discontinuous-mode Buck-boost to "regulate" its own output almost perfectly.

522 Chapter Twenty

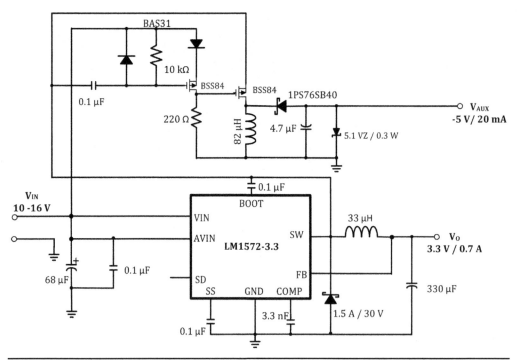

FIGURE 20.15 A slave Buck-boost in DCM riding off a master Buck in CCM.

The schematic can probably be simplified a great deal. This was rather hastily developed by the author to prove a principle for a certain *request for quotation* (RFQ) but was later granted a U.S. patent.

A Boost Preregulator with a Regulated Auxiliary Output

This is shown based around a typical Buck IC, the LM1572. The input range of the LM1572 is 8.5 to 16 V and its output is set to 5 V. As shown in Fig. 20.16, once startup has been achieved, we can make it work down to a couple of volts input, while maintaining the output at 5 V. This turns it into a step-up or step-down converter. Boost preregulators are not unknown, but here we see that no independent PWM control is required for the preregulator. This makes the solution more attractive. It is essentially a two-switch master-slave Boost-buck cascade, with the Buck stage being the master.

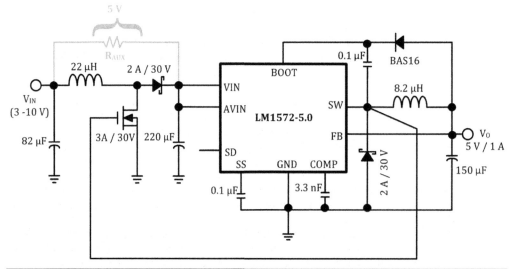

FIGURE 20.16 Boost preregulator with automatically regulated intermediate output.

But here is the interesting input-output transfer function coincidence it is based on. We have for a Boost

$$D = \frac{V_{O\_BOOST} - V_{IN\_BOOST}}{V_{O\_BOOST}}$$

where V_{IN} is the input to this stage. The output of the Boost forms the input to the Buck, so,

$$V_{O\_BUCK} = D \times V_{IN\_BUCK}$$

that is,

$$V_{O\_BUCK} = D \times V_{IN\_BOOST}$$

Eliminating D, we get

$$\frac{V_{O\_BUCK}}{V_{O\_BOOST}} = \frac{V_{O\_BOOST} - V_{IN\_BOOST}}{V_{O\_BOOST}}$$

that is,

$$V_{O\_BUCK} = V_{O\_BOOST} - V_{IN\_BOOST}$$

We see that the *strapped output* across R_{AUX} is being automatically regulated. Though it is not ground-referenced it can provide power to a stand-alone circuit block like a light emitting diode (LED) display.

APPENDIX
Design Tables and Aids, and Component FAQs

Efficiency, η	$\dfrac{P_O}{P_{IN}}$		
	In terms of P_{IN}	In terms of P_O	In terms of P_{LOSS}
Input Power, P_{IN}	P_{IN}	$P_O \times \dfrac{1}{\eta}$	$\dfrac{P_{LOSS}}{(1-\eta)}$
Output Power, P_O	$P_{IN} \times \eta$	P_O	$\dfrac{P_{LOSS} \times \eta}{(1-\eta)}$
Power Loss, P_{LOSS}	$P_{IN} \times (1-\eta)$	$P_O \times \left(\dfrac{1}{\eta} - 1\right)$	P_{LOSS}

TABLE A.1 Power Relationships

CCM Assumed	Buck	Boost	Buck-boost
Voltage across Inductor during ON-time, V_{ON}	$\approx V_{IN} - V_O$	$\approx V_{IN}$	$\approx V_{IN}$
Voltage across Inductor during OFF-time, V_{OFF}	$\approx V_O$	$\approx V_O - V_{IN}$	$\approx V_O$
Duty cycle, D	$= \dfrac{V_{OFF}}{V_{ON} + V_{OFF}}$	$= \dfrac{V_{OFF}}{V_{ON} + V_{OFF}}$	$= \dfrac{V_{OFF}}{V_{ON} + V_{OFF}}$
Duty cycle, D	$= \dfrac{V_O/\eta}{V_{IN}}$	$= \dfrac{V_O/\eta - V_{IN}}{V_O/\eta}$	$= \dfrac{V_O/\eta}{V_{IN} + V_O/\eta}$
Duty cycle, D	$= \dfrac{V_O}{\eta V_{IN}}$	$= \dfrac{V_O - \eta V_{IN}}{V_O}$	$= \dfrac{V_O}{\eta V_{IN} + V_O}$
Duty cycle, D	$\approx \dfrac{V_O}{V_{IN}}$	$\approx \dfrac{V_O - V_{IN}}{V_O}$	$\approx \dfrac{V_O}{V_{IN} + V_O}$
Duty cycle, D	$\approx \dfrac{V_O + V_D}{V_{IN} - V_{SW} + V_D}$	$\approx \dfrac{V_O - V_{IN} + V_D}{V_O - V_{SW} + V_D}$	$\approx \dfrac{V_O + V_D}{V_{IN} + V_O + V_D - V_{SW}}$
Ideal duty cycle, D_{IDEAL}	$= \dfrac{V_O}{V_{IN}}$	$= \dfrac{V_O - V_{IN}}{V_O}$	$= \dfrac{V_O}{V_{IN} + V_O}$
DC transfer function, V_O/V_{IN}	$= D_{IDEAL}$	$= \dfrac{1}{1 - D_{IDEAL}}$	$= \dfrac{D_{IDEAL}}{1 - D_{IDEAL}}$
DC transfer function, V_O/V_{IN}	$\approx D$	$\approx \dfrac{1}{1 - D}$	$\approx \dfrac{D}{1 - D}$

TABLE A.2 Design Chart for Nonisolated Topologies

CCM Assumed	Buck	Boost	Buck-boost
DC transfer function, V_O/V_{IN}	$= \eta D$	$= \dfrac{\eta}{1-D}$	$= \dfrac{\eta D}{1-D}$
Output voltage, V_O	$\approx V_{IN}D - V_{SW}D - V_D(1-D)$	$\approx \dfrac{V_{IN} - V_{SW}D - V_D(1-D)}{1-D}$	$\approx \dfrac{V_{IN}D - V_{SW}D - V_D(1-D)}{1-D}$
	Note: V_{SW} is forward drop across switch, V_D is forward drop across diode (or synch FET).		
Output voltage, V_O	$\approx V_{IN}D$	$\approx \dfrac{V_{IN}}{1-D}$	$\approx \dfrac{V_{IN}D}{1-D}$
Input voltage @ $D = 50\%$, V_{IN_50}	$\approx (2V_O) + V_{SW} + V_D \approx 2V_O$	$\approx \dfrac{1}{2} \times [V_O + V_{SW} + V_D] \approx \dfrac{V_O}{2}$	$\approx V_O + V_{SW} + V_D \approx V_O$
Center of current ramp, I_{COR} (same as average inductor current, I_L)	$= I_O$	$= \dfrac{I_O}{1-D}$	$= \dfrac{I_O}{1-D}$
Peak-to-Peak current in inductor, ΔI_L	$= 2 \times I_{AC} = r \times I_L$	$= 2 \times I_{AC} = r \times I_L$	$= 2 \times I_{AC} = r \times I_L$
Current ripple ratio, r	$= \dfrac{\Delta I_L}{I_L} = \dfrac{2 \times I_{AC}}{I_{DC}}$		
Current ripple ratio, r	$\approx \dfrac{V_O + V_D}{I_O \times L \times f} \times (1-D) \times 10^6$	$\approx \dfrac{V_O - V_{SW} + V_D}{I_O \times L \times f} \times D(1-D)^2 \times 10^6$	$\approx \dfrac{V_O + V_D}{I_O \times L \times f} \times (1-D)^2 \times 10^6$
	Note: f in Hz.		
Peak current in switch & diode & inductor, I_{PEAK}	$= I_{COR} \times \left[1 + \dfrac{r}{2}\right]$	$= I_{COR} \times \left[1 + \dfrac{r}{2}\right]$	$= I_{COR} \times \left[1 + \dfrac{r}{2}\right]$
Peak current in switch & diode & inductor, I_{PEAK}	$= I_O \times \left[1 + \dfrac{r}{2}\right]$	$= \dfrac{I_O}{1-D} \times \left[1 + \dfrac{r}{2}\right]$	$= \dfrac{I_O}{1-D} \times \left[1 + \dfrac{r}{2}\right]$
Valley (trough) current in inductor, I_{TROUGH}	$= I_{COR} \times \left[1 - \dfrac{r}{2}\right]$	$= I_{COR} \times \left[1 - \dfrac{r}{2}\right]$	$= I_{COR} \times \left[1 - \dfrac{r}{2}\right]$
Valley (trough) current in inductor, I_{TROUGH}	$= I_O \times \left[1 - \dfrac{r}{2}\right]$	$= \dfrac{I_O}{1-D} \times \left[1 - \dfrac{r}{2}\right]$	$= \dfrac{I_O}{1-D} \times \left[1 - \dfrac{r}{2}\right]$
Inductance, L (µH)	$\approx \dfrac{V_O + V_D}{I_O \times r \times f} \times (1-D) \times 10^6$	$\approx \dfrac{V_O - V_{SW} + V_D}{I_O \times r \times f} \times D(1-D)^2 \times 10^6$	$\approx \dfrac{V_O + V_D}{I_O \times r \times f} \times (1-D)^2 \times 10^6$
	Note: f in Hz.		
Peak-to-Peak current in output capacitor	$= I_O \times r$	$= \dfrac{I_O}{1-D} \times \left[1 + \dfrac{r}{2}\right]$	$= \dfrac{I_O}{1-D} \times \left[1 + \dfrac{r}{2}\right]$
Output voltage ripple (p-p) component (ESR-related)	$= I_O \times r \times ESR_{C_O}$	$= \dfrac{I_O}{1-D} \times \left[1 + \dfrac{r}{2}\right] \times ESR_{C_O}$	$= \dfrac{I_O}{1-D} \times \left[1 + \dfrac{r}{2}\right] \times ESR_{C_O}$
Output voltage ripple (p-p) component (capacitance-related)	$= \dfrac{I_O \times r}{8 \times f \times C_O}$	$= \dfrac{I_O \times (1-D)}{f \times C_O}$	$= \dfrac{I_O \times (1-D)}{f \times C_O}$

TABLE A.2 Design Chart for Nonisolated Topologies (*Continued*)

Design Tables and Aids, and Component FAQs

CCM Assumed	Buck	Boost	Buck-boost
RMS current in output capacitor	$= I_0 \times \dfrac{r}{\sqrt{12}}$	$= I_0 \times \sqrt{\dfrac{D + \dfrac{r^2}{12}}{1-D}}$	$= I_0 \times \sqrt{\dfrac{D + \dfrac{r^2}{12}}{1-D}}$
RMS current in output capacitor	≈ 0	$\approx I_0$	$\approx I_0$
Peak-to-Peak current in input capacitor	$= I_0 \left[1 + \dfrac{r}{2}\right]$	$= \dfrac{I_0}{1-D} \times r$	$= \dfrac{I_0}{1-D} \times \left[1 + \dfrac{r}{2}\right]$
Input voltage ripple (p-p) component (ESR-related)	$= I_0 \left[1 + \dfrac{r}{2}\right] \times \text{ESR}_{C_{IN}}$	$= \dfrac{I_0}{1-D} \times r \times \text{ESR}_{C_{IN}}$	$= \dfrac{I_0}{1-D} \times \left[1 + \dfrac{r}{2}\right] \times \text{ESR}_{C_{IN}}$
Input voltage ripple (p-p) component (capacitance-related)	$= \dfrac{I_0 \times D(1-D)}{f \times C_{IN}}$	$= \dfrac{I_0 \times r}{8 \times f \times C_{IN} \times (1-D)}$	$= \dfrac{I_0 \times D}{f \times C_{IN}}$
RMS current in input capacitor	$= I_0 \sqrt{D\left[1 - D + \dfrac{r^2}{12}\right]}$	$= \dfrac{I_0}{1-D} \times \dfrac{r}{\sqrt{12}}$	$= \dfrac{I_0}{1-D} \sqrt{D\left[1 - D + \dfrac{r^2}{12}\right]}$
RMS current in input capacitor	$\approx \dfrac{I_0}{2}$	≈ 0	$\approx I_0$
RMS current in inductor	$= I_0 \times \sqrt{1 + \dfrac{r^2}{12}}$	$= \dfrac{I_0}{1-D} \times \sqrt{1 + \dfrac{r^2}{12}}$	$= \dfrac{I_0}{1-D} \times \sqrt{1 + \dfrac{r^2}{12}}$
RMS current in switch	$= I_0 \times \sqrt{D \times \left[1 + \dfrac{r^2}{12}\right]}$	$= \dfrac{I_0}{1-D} \times \sqrt{D \times \left[1 + \dfrac{r^2}{12}\right]}$	$= \dfrac{I_0}{1-D} \times \sqrt{D \times \left[1 + \dfrac{r^2}{12}\right]}$
RMS current in diode (or sync FET)	$= I_0 \times \sqrt{(1-D) \times \left[1 + \dfrac{r^2}{12}\right]}$	$= I_0 \times \sqrt{\dfrac{\left[1 + \dfrac{r^2}{12}\right]}{(1-D)}}$	$= I_0 \times \sqrt{\dfrac{\left[1 + \dfrac{r^2}{12}\right]}{(1-D)}}$
Average current in diode (or sync FET)	$= I_0 (1-D)$	$= I_0$	$= I_0$
Average current in switch	$= I_0 \times D$	$= I_0 \times \dfrac{D}{1-D}$	$= I_0 \times \dfrac{D}{1-D}$
Average current in inductor, I_L	$= I_0$	$= \dfrac{I_0}{1-D}$	$= \dfrac{I_0}{1-D}$
Average input current, I_{IN}	$= I_0 \times D$	$= \dfrac{I_0}{1-D}$	$= I_0 \times \dfrac{D}{1-D}$
Peak energy–handling capability of core, ε (μJ)	$= \dfrac{1}{2} \times L \times I_{PEAK}^2 = \dfrac{\Delta\varepsilon}{8} \times \left[r \times \left(\dfrac{2}{r} + 1\right)^2\right]$		
Peak energy–handling capability of core, ε (μJ)	$= \dfrac{I_0 \times V\mu s}{8} \times \left[r \times \left(\dfrac{2}{r} + 1\right)^2\right]$	$= \dfrac{I_0 \times V\mu s}{8 \times (1-D)} \times \left[r \times \left(\dfrac{2}{r} + 1\right)^2\right]$	$= \dfrac{I_0 \times V\mu s}{8 \times (1-D)} \times \left[r \times \left(\dfrac{2}{r} + 1\right)^2\right]$
Note: See stress spiders in Figs. 2.23, 2.24, and 2.25.			

TABLE A.2 Design Chart for Nonisolated Topologies (*Continued*)

CCM Assumed	Single-Ended Forward (Like a Buck)	Flyback (Like a Buck-boost)
Transformer turns ratio, n	\multicolumn{2}{c}{$= \dfrac{N_P}{N_S}$}	
Reflected output voltage, V_{OR}	\multicolumn{2}{c}{$\approx n \times V_O$}	
	\multicolumn{2}{c}{$\approx n \times (V_O + V_D)$}	
	\multicolumn{2}{c}{$= n \times V_O / \eta$}	
Reflected input voltage, V_{INR}	\multicolumn{2}{c}{$\approx \dfrac{V_{IN}}{n}$}	
	\multicolumn{2}{c}{$\approx \dfrac{V_{IN} - V_{SW}}{n}$}	
	\multicolumn{2}{c}{$= \dfrac{\eta V_{IN}}{n}$}	
Reflected output current, I_{OR}	\multicolumn{2}{c}{$= \dfrac{I_O}{n}$}	
Reflected input current, I_{INR}	\multicolumn{2}{c}{$= n \times I_{IN}$}	
Duty cycle	$= \dfrac{V_O}{V_{INR}}$	$= \dfrac{V_O}{V_{INR} + V_O}$
	$= \dfrac{V_{OR}}{V_{IN}}$	$= \dfrac{V_{OR}}{V_{IN} + V_{OR}}$
	$= \dfrac{V_O}{\eta/n \times V_{IN}}$	$= \dfrac{V_O}{(\eta/n \times V_{IN}) + V_O}$
	$= \dfrac{V_O \times n/\eta}{V_{IN}}$	$= \dfrac{V_O \times n/\eta}{V_{IN} + (V_O \times n/\eta)}$
Ideal duty cycle, D_{IDEAL}	$= \dfrac{nV_O}{V_{IN}}$	$= \dfrac{nV_O}{V_{IN} + nV_O}$
DC transfer function, V_O/V_{IN}	$= D_{IDEAL}/n$	$= \dfrac{D_{IDEAL}/n}{1 - D_{IDEAL}}$
	$= D \times \eta/n \approx \dfrac{D}{n}$	$= \dfrac{D \times \eta/n}{1-D} \approx \dfrac{1}{n} \times \dfrac{D}{1-D}$
	\multicolumn{2}{l}{*Note:* η is the efficiency of the converter $= P_O/P_{IN}$, D is the actual/measured duty cycle, and n is the turns ratio.}	
Inductance, L (μH)	$\approx \dfrac{V_O}{I_O \times r \times f} \times (1-D) \times 10^6$	$\approx \dfrac{V_{OR}}{I_{OR} \times r \times f} \times (1-D)^2 \times 10^6$
		$\approx \dfrac{n^2 \times V_O}{I_O \times r \times f} \times (1-D)^2 \times 10^6$
	\multicolumn{2}{l}{*Note:* This refers to the inductance of the output choke of a Forward converter and the inductance of the primary side of the transformer in a Flyback.}	
	\multicolumn{2}{l}{f is the switching frequency in Hz and r is the current ripple ratio; see below.}	
	\multicolumn{2}{l}{Typically choose L such that $r = 0.4$ (i.e., inductor current swing is ± 20% of its DC or center of ramp value I_L); also, set r to this value at the highest input voltage for Forward, and at the lowest input voltage for Flyback, but also check that the max duty cycle does not exceed limit of controller at minimum input for both topologies (adjust V_{OR} and turns ratio accordingly).}	

Table A.3 Design Chart for Isolated Topologies

CCM Assumed	Single-Ended Forward (Like a Buck)	Flyback (Like a Buck-boost)
Average current in inductor, I_L (center of ramp, I_{COR})	$I_L = I_O$	Primary side: $I_{COR\_PRI} = \dfrac{I_{OR}}{1-D} \equiv \dfrac{I_O/n}{1-D}$ Secondary Side: $I_{COR\_SEC} = \dfrac{I_O}{1-D}$
	Note: See row above and Fig. A.1.	
Current ripple ratio, r of transformer	$= \dfrac{\Delta I_{COR\_PRI}}{I_{COR\_PRI}} = \dfrac{\Delta I_{COR\_SEC}}{I_{COR\_SEC}}$	
	$\approx \dfrac{V_O}{I_O \times L_{SEC} \times f} \times (1-D) \times 10^6$	$\approx \dfrac{V_O}{I_O \times L_{SEC} \times f} \times (1-D)^2 \times 10^6$
	$\approx \dfrac{V_{OR}}{I_{OR} \times L_{PRI} \times f} \times (1-D) \times 10^6$	$\approx \dfrac{V_{OR}}{I_{OR} \times L_{PRI} \times f} \times (1-D) \times 10^6$
	Note: $L_{SEC} = L_{PRI}/n^2$.	
Peak current on primary side, (I_{PRI_PK})	$\approx I_{OR}\left[1+\dfrac{r}{2}\right]$	$\approx \dfrac{I_{OR}}{1-D} \times \left[1+\dfrac{r}{2}\right]$
	Note: Ignoring transformer magnetization current.	
Peak-to-Peak current in input capacitor	$\approx I_{OR}\left[1+\dfrac{r}{2}\right]$	$\approx \dfrac{I_{OR}}{1-D} \times \left[1+\dfrac{r}{2}\right]$
	Note: Ignoring transformer magnetization current.	
Peak-to-Peak current in output capacitor	$= I_O \times r$	$= \dfrac{I_O}{1-D} \times \left[1+\dfrac{r}{2}\right]$
Input voltage ripple (p-p) component (ESR-related)	$\approx I_{OR}\left[1+\dfrac{r}{2}\right] \times ESR_{C_{IN}}$	$= \dfrac{I_{OR}}{1-D} \times \left[1+\dfrac{r}{2}\right] \times ESR_{C_{IN}}$
	Note: Ignoring transformer magnetization current.	
Output voltage ripple (p-p) component (ESR-related)	$= I_O \times r \times ESR_{C_O}$	$= \dfrac{I_O}{1-D} \times \left[1+\dfrac{r}{2}\right] \times ESR_{C_O}$
Input voltage ripple (p-p) component (capacitance-related)	$\approx \dfrac{I_{OR} \times D(1-D)}{f \times C_{IN}}$	$= \dfrac{I_{OR} \times D}{f \times C_{IN}}$
	Note: Ignoring transformer magnetization current.	

TABLE A.3 Design Chart for Isolated Topologies (*Continued*)

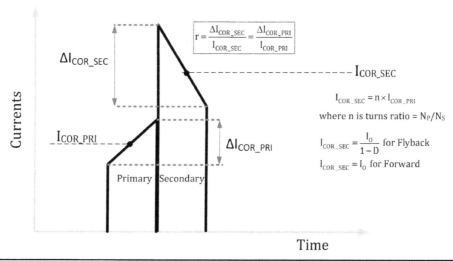

FIGURE A.1 Connecting currents and current ripple ratio for a transformer (Flyback or Forward).

CCM Assumed	Single-Ended Forward (Like a Buck)	Flyback (Like a Buck-boost)
Output voltage ripple (p-p) component (capacitance-related)	$= \dfrac{I_O \times r}{8 \times f \times C_O}$	$= \dfrac{I_O \times (1-D)}{f \times C_O}$
RMS current in input capacitor	$\approx I_{OR} \sqrt{D\left[1-D+\dfrac{r^2}{12}\right]} \approx \dfrac{I_{OR}}{2}$	$= \dfrac{I_{OR}}{1-D} \sqrt{D\left[1-D+\dfrac{r^2}{12}\right]}$
	Note: Ignoring transformer magnetization current.	
RMS current in output capacitor	$= I_O \times \dfrac{r}{\sqrt{12}} \approx 0$	$= I_O \times \sqrt{\dfrac{D+\dfrac{r^2}{12}}{1-D}}$
RMS current in inductor/transformer and windings	Primary side winding: $\approx I_{OR} \times \sqrt{D \times \left[1+\dfrac{r^2}{12}\right]}$	Primary side winding: $= \dfrac{I_{OR}}{1-D} \times \sqrt{D \times \left[1+\dfrac{r^2}{12}\right]}$
	Note: Ignoring transformer magnetization current.	
	Secondary side winding: $= I_O \times \sqrt{D \times \left[1+\dfrac{r^2}{12}\right]}$ Output choke: $= I_O \times \sqrt{1+\dfrac{r^2}{12}}$	Secondary side winding: $= I_O \times \sqrt{\dfrac{\left[1+\dfrac{r^2}{12}\right]}{1-D}}$
RMS current in switch	$\approx I_{OR} \times \sqrt{D \times \left[1+\dfrac{r^2}{12}\right]}$	$= \dfrac{I_{OR}}{1-D} \times \sqrt{D \times \left[1+\dfrac{r^2}{12}\right]}$
	Note: Ignoring transformer magnetization current.	
RMS current in diode (or sync fet)	Output diode (to transformer): $= I_O \times \sqrt{D \times \left[1+\dfrac{r^2}{12}\right]}$ Freewheeling diode (to ground): $= I_O \times \sqrt{(1-D) \times \left[1+\dfrac{r^2}{12}\right]}$	$= I_O \times \sqrt{\dfrac{\left[1+\dfrac{r^2}{12}\right]}{(1-D)}}$
Average current in switch	$\approx I_{OR} \times D$	$= I_{OR} \times \dfrac{D}{1-D}$
	Note: Ignoring transformer magnetization current.	
Average current in diode	Output diode (to transformer): $= I_O \times D$ Freewheeling diode (to ground): $= I_O \times (1-D)$	$= I_O$
Average input current, I_{IN}	Same as average switch current	
	$= I_{OR} \times D$	$= I_{OR} \times \dfrac{D}{1-D}$
Peak energy–handling capability of core, ε (µJ)	$= \dfrac{I_O \times V\mu s}{8} \times \left[r \times \left(\dfrac{2}{r}+1\right)\right]^2$	$= \dfrac{I_{OR} \times V\mu s}{8 \times (1-D)} \times \left[r \times \left(\dfrac{2}{r}+1\right)\right]^2$
	Note: This peak energy refers to the output choke of a Forward converter and to the transformer of a Flyback. For Flyback, use the $V\mu s$ appearing across the primary winding, i.e., $V_{IN} \times D/f \times 10^6$ or equivalently $V_{OR} \times (1-D)/f \times 10^6$. For a Forward converter use the $V\mu s$ across the choke.	

TABLE A.3 Design Chart for Isolated Topologies (*Continued*)

Design Tables and Aids, and Component FAQs

$n = N_P/N_S$ $V_{INR} = V_{IN}/n$ $V_{OR} = nV_O$		Switch	Catch Diode	Output Diode	Coupling/ Clamp Cap	Ideal Transfer Function
Buck		V_{INMAX}	V_{INMAX}		NA	$\dfrac{V_O}{V_{IN}} = D$
Boost		V_O	V_O		NA	$\dfrac{V_O}{V_{IN}} = \dfrac{1}{1-D}$
Buck-boost		$V_{INMAX} + V_O$	$V_{INMAX} + V_O$		NA	$\dfrac{V_O}{V_{IN}} = \dfrac{D}{1-D}$
Flyback		$V_{INMAX} + V_Z$	$V_{INRMAX} + V_O$		NA	$\dfrac{V_O}{V_{INR}} = \dfrac{D}{1-D}$
Forward		$2 \times V_{INMAX}$	V_{INRMAX}	$V_{INRMAX} + V_O$	NA	$\dfrac{V_O}{V_{INR}} = D$
2-switch Forward		V_{INMAX}	V_{INRMAX}	$V_{INRMAX} + V_O$	NA	$\dfrac{V_O}{V_{INR}} = D$
Active Clamp		$\dfrac{V_{INMAX}}{1-D_{MAX}}$	V_{INRMAX}	$V_{INRMAX} \times \dfrac{D_{MAX}}{1-D_{MAX}} + V_O$	$\dfrac{V_{IN}D_{MAX}}{1-D_{MAX}}$	$\dfrac{V_O}{V_{INR}} = D$
Half Bridge		V_{INMAX}	V_{INRMAX}	V_{INRMAX}	NA	$\dfrac{V_O}{V_{INR}} = D$
Full Bridge		V_{INMAX}	$2 \times V_{INRMAX}$	$2 \times V_{INRMAX}$	NA	$\dfrac{V_O}{V_{INR}} = 2D$
Push-Pull		$2 \times V_{INMAX}$	$2 \times V_{INRMAX}$	$2 \times V_{INRMAX}$	NA	$\dfrac{V_O}{V_{INR}} = 2D$
Cuk		$V_{INMAX} + V_O$	$V_{INMAX} + V_O$		$V_{INMAX} + V_O$	$\dfrac{V_O}{V_{IN}} = \dfrac{D}{1-D}$
Sepic		$V_{INMAX} + V_O$	$V_{INMAX} + V_O$		V_{INMAX}	$\dfrac{V_O}{V_{IN}} = \dfrac{D}{1-D}$
Zeta		$V_{INMAX} + V_O$	$V_{INMAX} + V_O$		V_O	$\dfrac{V_O}{V_{IN}} = \dfrac{D}{1-D}$

TABLE A.4 Voltage Stress Design Chart for Various Topologies

Component FAQs

1. Why are tantalum (Ta) capacitors not recommended for low-impedance applications?

   Answer: Tantalums exhibit a self-healing effect similar to that of metallized film capacitors, but they require a slow buildup of heat in fault sites within the capacitor. Low-impedance circuits may allow too much current to flow through reduced resistivity zones, thereby accelerating internal heating and turning a small material imperfection into a catastrophic component failure. Tantalum capacitors therefore have inherent limitations on the surge current they can safely pass. This usually necessitates that we limit the applied voltage to 50% of the rated voltage, especially if the capacitor is to be used as a front-end (input) capacitor. We should also be aware that manufacturers usually have special families of more robust, surge current tested tantalum capacitors, and we should prefer these in front-end applications. Note that in a Boost topology, the output capacitor also sees a high inrush surge current at power-on.

2. What are the pros and cons of tantalum capacitors versus aluminum electrolytic capacitors?

   Answer:

   a) The aluminum electrolytic has the advantage of recoverability of the oxide film, so catastrophic failure is avoided if there is an internal flaw. Thus the aluminum electrolytic fails open-circuited in most cases, whereas the Ta capacitor, because of its solid electrolyte, does not self-heal easily and can fail either with high leakage or even short-circuited.

   b) The temperature and frequency characteristics are advantageous because the leakage of Ta is also usually much better than aluminum.

   c) Solid capacitors like Ta have almost a "semipermanent" life (no wearout) and a decreasing failure rate with time. Lifetime issues have to be considered for aluminum because of the evaporation of electrolyte. Their failure rate can also climb steeply above 70°C.

   d) The ripple current an aluminum electrolytic can withstand is usually far in excess of Ta. Aluminum electrolytics also have higher reverse resistance (ability to stand momentary reverse voltages). They can also typically withstand 1.2 times their rated voltage for a second or so (but verify with vendor).

3. I have bought a 1-µF capacitor. What could be the capacitance it actually presents in my circuit?

   Answer: Capacitance as specified in a datasheet is usually measured at an applied voltage of 1 VRMS, at 1 kHz, and 25°C. In an actual circuit, we could see the following worst-case spread (after 100 kHz, with the possibility of both AC and DC voltages being considered):

   a) For C0G/NP0. Consider initial tolerance (±5%), TCC (±0.15%), voltage stability (0%), frequency stability (0%), aging (0%). Combining, we get $C = 1$ µF, +5.16%, −5.14%. (Check: $1.05 \times 1.0015 = 1.0516$, i.e., 5.16% higher, and $0.95 \times (1 - 0.0015) = 0.9486$, i.e., $1 - 0.9486 = 0.0514$, or 5.14% less)

   b) For X7R. Consider initial tolerance (±10%), TCC (+2, −10%), voltage stability (+15, −10%), frequency stability (+5, −15%), aging (−3%). Combining, we get $C = 1$ µF +35%, −40%.

   c) For Z5U. Consider initial tolerance (±20%), TCC (+2, −54%), voltage stability (+22, −56%), frequency stability (+5, −15%), aging (−25%). Combining, we get $C = 1$ µF +57%, −90%.

   Note that, in general, the tolerance reading for C0G/NP0 capacitors may already include its temperature drift.

4. I need 470 pF. Should I use a capacitor with the marking "470" on it?

   Answer: According to the standard electronics industries alliance (EIA) codes, for ceramic, film, and Ta capacitors, the significant digits are only the first two digits, whereas the third digit is the number of zeros (capacitance expressed in pF). Therefore, 470 stands for 47 pF not 470 pF. A capacitor marked 471 is 470 pF.

Note that by the same system, 4R7 is 4.7 pF. For aluminum electrolytic, the same rule applies but the capacitance is usually expressed directly in μF.

5. How could the voltage rating of a capacitor be indicated on the case?

   Answer: There is a single-digit code system in which we have A (10 V), B (16 V), C (25 V), D (50 V), E (100 V), G (200 V). There is also a two-digit code in which we have 0G (4 V), 0J (6.3 V), 1A (10 V), 1C (16 V), 1E (25 V), 1V (35 V), 1H (50 V), 1J (63 V), 2A (100 V), 2D (200 V), 2E (250 V), 2V (350 V), 2G (400 V), 2W (450 V). But, high-voltage disc capacitors may simply be marked 1kV (i.e., 1000 V), for example.

6. How is the polarity indicated on polarized capacitors?

   Answer: On a plastic packaged Ta (or rectangular-shaped/solid-electrolytic SMD polymer) capacitor, the solid band is the positive terminal (anode), not the cathode. On conformal coated Ta capacitors, the wire nib (protrusion) is the anode. On the radial aluminum electrolytic (the wet electrolyte type or the modern aluminum polymer version, both through-hole and SMD), the solid band (or shaded part of the can) is the negative terminal (cathode). Sometimes there may be an arrow along the side of the cap with nothing on it or a "−"sign indicating the cathode. There may be an arrow with a "+" sign, and that is the anode, not the cathode. In a through-hole electrolytic cap, the longer lead is usually the anode. On a printed circuit board it is customary to indicate the correct orientation by using a square through-hole pad for the positive lead and a round pad for the negative. In an SMT radial electrolytic cap, the pointy edge of its base plate is the positive terminal. Remember that on a diode, the solid band is the cathode (negative terminal), just as for the SMD aluminum cap, but unlike the Ta cap.

7. What are polymer capacitors?

   Answer: There are a very wide variety of caps here. In general, they are based on conductive polymers, which offer 1000 to 10,000 times better conductivity than the electrolyte found in conventional wet aluminum technology, and 100 to 1000 times more than MNO_2 used in conventional tantalum technology. This conductivity leads to much lower ESR. These caps are also called "dry" (solid electrolyte), and so they offer no wearout.

   As described on Wikipedia (http://en.wikipedia.org/wiki/Polythiophene): "Polythiophenes (PTs) are polymerized thiophenes, a sulfur heterocycle. They can become conducting when electrons are added or removed via doping. The study of polythiophenes has intensified over the last three decades. The most notable property of these materials, electrical conductivity, results from the delocalization of electrons along the polymer backbone, hence the term "synthetic metals."

   Historically, the first "competitor" to the conventional aluminum capacitor in terms of volumetric efficiency (capacitance per volume) came from The Sprague Electric Company, which patented the first commercially viable process for manufacturing tantalum capacitors in 1960. But tantalum acquired a subsequent reputation of worldwide supply issues and a tendency to conflagrate (which has since been curbed by using polymers instead of MnO_2 in Ta-polymer versions such as POSCAP from Sanyo/Panasonic and the KO-cap from Kemet, thus also improving ESR significantly). Later, OS-CON from Sanyo/Panasonic, appeared in 1983. In terms of conductivity it was only 100 times better than conventional aluminum electroytics, because originally it contained the electrolyte TCNQ (tetracyanoquinodimethane). Yet, it was viewed favorably as it offered very stable ESR below 0°C, in comparison to the steeply escalating ESR of wet aluminum electrolytic caps. It also offered very high life as it was solid (no outgassing). Today OS-CON also incorporates conductive polymer technology, which has blurred its boundary with modern aluminum polymer caps.

   In modern applications, where caps are often selected not for capacitance, but for low ESR, low heating, effective decoupling, low voltage ripple, and high-RMS capability, it is no longer fair to compare caps based on crude volumetric efficiency alone. So, for example, in a typical Flyback power supply, just to be able to handle the high-output RMS current, we may go in for a 1000-μF wet aluminum cap, but would be able to use a 56-μF polymer cap instead, which is clearly much smaller. In cases of holdup time (AC-DC converters), in principle, we are looking for a certain capacitance, so polymer caps won't help, unless the final selection is limited by RMS. But unfortunately, to date that is impossible to even consider, since polymer caps are available only up to a max of 25V. They were considered fundamentally limited to 28V, but in 2008, using an alternative polymerization process, Kemet announced the T521 series which could handle 35V.

At the moment, the technology choicefor power supplies' input/output caps is the AO-cap (aluminum organic polymer capacitors) from Kemet or the equivalent SP (specialty polymer) cap from Panasonic. It offers very low-ESR, long-life, high-ripple current capability, steady ESR and capacitance with respect to temperature and applied voltage, and a wide operating temperature range. Voltages are limited as mentioned.

8. What are some standard SMD component sizes?

    Answer: What we call "805" is actually 80 mils length and 50 mils wide. In millimeter,this is about 2 mm and 1.25 mm (since 1 mm is about 40 mil). In Table A.5, we present the common sizes in mm. Note that we usually talk in "imperial" sizes, whereas there are equivalent "metric" sizes too. So "805" package is also "2012" in metric sizes. In the same table we have therefore listed the metric equivalents.

    Tantalum/polymer capacitor sizes: EIA codes (and vendor size codes: length × width, with different heights possible)

    - Size 2012 (Kemet R; AVX R): 2 mm × 1.3 mm
    - Size 3216 (Kemet I, S, A; AVX K, S, A): 3.2 mm × 1.6 mm
    - Size 3528 (Kemet T, B; AVX T, B): 3.5 mm × 2.8 mm
    - Size 6032 (Kemet U, C; AVX W, C): 6.0 mm × 3.2 mm
    - Size 7260 (Kemet E; AVX V): 7.3 mm × 6.0 mm × 3.8 mm
    - Size 7343 (Kemet V, D, X; AVX Y, D, E): 7.3 mm × 4.3 mm

9. How does the air gap affect the A_L (nH/turns^2) tolerance?

    Answer: The initial permeability μ_i as defined by vendors, is only for magnetically enclosed objects, for example, toroids. For many reasons, such as the accuracy of weighed quantities or the resolution of the X-ray fluorescence analysis used to check composition, the initial permeability referred to a toroidal core can only be narrowed down to a tolerance range of ±20%.

Imperial Sizes	Length (mm)	Width (mm)	Metric Sizes
01005	≈ 0.4	≈ 0.2	0402
0201	≈ 0.6	≈ 0.3	0603
0402	40/40 = 1.00	20/40 = 0.5	1005
0603	1.6	0.8	1608
0805	2.00	1.25	2012
1206	3.20	1.60	3216
1210	3.20	2.50	3225
1806	4.50	1.60	4516
1808	4.50	2.00	4520
1812	4.50	3.20	4532
1825	4.50	6.40	4564
2010	5.00	2.50	5025
2220	5.70	5.00	5750
2225	5.60	6.35	5664
2412	6.0	3.20	6432
2512	6.35	3.20	6432

Sizes in millimeters (mm) are approximate. Exact values should be calculated by converting from mils. For example, 2225 is actually 220 mils × 250 mils, and 0402 is 40 mils × 20 mils.

TABLE A.5 Standard Sizes for SMD Component

Since μ_i is given only for a closed toroid, the effective permeability μ_e for non-toroidal (cut) cores is a function of various factors such as quality of polishing, mechanical tolerances, number of surfaces in contact, positioning of core halves, dirt on the surfaces to be mated, and the like. So, cores, supposedly "with no air gap," also have a permeability μ_e, which will be less than μ_i. In theory, for materials with $\mu_i \approx 2000$, a μ_e/μ_i of about 96% is possible. In practice this number is only 75%. For highly permeable materials ($\mu_i \approx 10{,}000$), the highest ratio of 65% by calculation is obtained for mirror surfaces with (residual) air gap of 1 μm to 2 μm. Note that *"no air gap" for cut cores should always be taken to mean at least 1 μm of air gap (default value)*. If we introduce a deliberate air gap, then as the air gap becomes large, the initial permeability plays a smaller and smaller effect on the μ_e. For example, if the air gap is 20 μm, the difference in μ_e for the above two classes of materials may be only about 20%. If the gap is greater than 100 μm, the μ_e is virtually the same. Clearly, in deliberately gapped cores, the surface texture is of little consequence. But for large cores with very small air gaps, the tolerances can be even worse than calculated, because the relationship between the grinder disk width and core width (edge effects) also comes into play. To sum up, the tolerance of the A_L value can be in general expressed as follows:

$$\frac{\Delta A_L}{A_L} = \frac{\Delta \mu_i}{\mu_i} + \frac{\Delta l_g}{l_g} + \frac{\Delta \{X\}}{\{X\}}$$

where l_g is the air gap, and $\{X\}$ stands for all the geometric influences previously indicated. However, these influences are difficult to grasp and quantify exactly. Typically, the air gap is toleranced by vendors (e.g., Epcos), as follows:

a) $l_g \leq 0.1$ mm, $\Delta l_g = 0.01$ mm, i.e., $\Delta l_g / l_g$ starting at 10%

b) 0.1 mm $< l_g \leq 0.5$ mm, $\Delta l_g = 0.02$ mm, i.e., $\Delta l_g / l_g$ is as high as 20%

c) $l_g > 0.5$ mm, $\Delta l_g = 0.05$ mm, i.e., $\Delta l_g / l_g$ is less than 10%

We, therefore, should be very cautious when dealing with small air gaps, as we can have large tolerances for the inductance.

10. What are the pros and cons of specifying A_L versus air gap?

Answer: The designer has the following choices:

a) Specify air gap tolerances.

b) Specify A_L tolerances.

c) Choose between symmetrical/unsymmetrical grinding for the mating halves.

The basic question is whether to specify the ferrite in terms of the A_L value, or the dimensions of the air gap. The two are related. If the air gap increases, A_L decreases. In power applications such as storage chokes and Flyback converters, the object is to store energy in the air gap, so the air gap (which is basically where we store the energy) must be defined fairly precisely. The A_L value by itself is not critical here. Nevertheless, we probably do have an allowed (or recommended) range for inductance for various other reasons. For example, if the inductance is too low and we are operating close to maximum power, we may hit the peak current limit and thus may not be able to deliver the required power. If the inductance is too low, we will also enter discontinuous conduction mode earlier as we decrease load and may end up with undesirable early pulse skipping eventually. Too low an inductance in current mode converters with slope compensation may also pose problems. So, a minimum A_L may need to be specified. Too large an inductance (A_L) may lead to a nonoptimum design, and cause some strange problems such as alternate cycle pulsing (which appears as traditional subharmonic instability, but is not). Therefore a maximum A_L may also need to be specified. However, we should always strive to make our design itself, relatively intolerant to variations in A_L. But we also note that the magnetics design for a switching application is further complicated by the fact that the actual operating values may not be close to those at which the A_L value is specified in the datasheet (usually 10 kHz, 0.25 mT, and 23 ± 3°C). In contrast, typical power transformer circuits work at much higher frequencies, at flux densities around 300 mT and temperatures up to 120°C. So, the correlation between the material's datasheet and our actual application may not be so trivial. As indicated, A_L and air gap are related. It can be shown by calculation that the air gap for a given

A_L value can only fluctuate within certain limits, and that a worst-case estimate is possible for the span of this fluctuation. If the A_L values are very low, and the acceptable tolerance range for both air gap and A_L sufficiently wide, it is conceivable that both A_L value and air gap could be simultaneously guaranteed. But from the manufacturers'point of view, they feel that it is safer and more precise to specify either A_L or air gap tolerances. Actually, prescribing an air gap is simple, as only geometrical tolerances are involved (though there is a problem with larger core forms). Specification of the A_L value is more complex since there are specified test conditions like a specified coil, specified temperature, and specified mating pressure (however only pressures distinctly below 10 N/m² or unrealistically high pressures can falsify measurements), all of which may also be very different from the conditions that the designer is to use in his or her application. Hence, ultimately the user's A_L value needs to be correlated and effectively converted to an equivalent manufacturer's A_L value. The magnetic material vendor should be approached for help in this matter. Historically, the early P and RM cores were specified with only an air gap, and were also supplied in pairs for maintaining tight tolerance limits. In the late 1970s, a change was made with E-cores, and they were sold only in unit delivery (and specified air gap). The idea was to give customers the choice of mating different air gaps for different applications. Nowadays, manufacturers have started specifying only A_L tolerances, while maintaining unit delivery. Now, for most applications, it is sufficient to select symmetricalcores (i.e., identical halves) *from the same pallet*, so that the end A_L value has the same tolerance. If we randomly select unsymmetrical core halves, then since the μ_i value of the two halves may be different, a statistical dual distribution can occur (as in overlapping unbalanced/unsymmetrical distributions). If unsymmetrical halves are required for some reason, the manufacturer can be approached to provide an asymmetrical form of packing, with appropriate packing codes to ensure that the asymmetrical halves are from the same sinter batch. Then the final A_L tolerances will be as expected and quality problems will not occur. Ultimately, it is advisable to create a user's test setup to correlate with standard test parameters with the help of the manufacturer, and in critical applications, to resort to 100% testing for relevant parameters.

11. Should we measure the L on an LCR meter set to series mode or parallel mode?

    Answer: Theoretically, we can model an inductor either as an ideal inductor in series or in parallel with a resistor. Here, as with capacitance, the equivalency between series- and parallel-mode values holds true only at the conversion frequency, because the quality factor used in the conversion is frequency dependent. However, as with capacitance, the Q-factor is the same regardless of the measurement technique, that is,

    $$Qp = Qs$$

    If the inductance is large, parallel mode is more suitable, since the reactance at a given frequency is large and the indicated inductance is closer to the effective inductance. Also, the parallel resistance becomes more significant than the series component. Conversely, for a low inductance, the series mode is preferred. For midvalues, a more exact comparison of reactance to resistance is called for to determine which equivalent circuit mode to use. The rules given above are obviously dependent on test frequency too. So, if the test frequency is 1 kHz, the thumb rule is that below 1 mH, we should use series mode, whereas above 1 H we should use parallel mode. In between these values, we can use the manufacturer's recommendations and/or our judgment in relation to expected resistances. Note that Q is inherently lower at lower frequencies. Distributed capacitance, if any, decreases the effective inductance, and more importantly, increases the frequency dependence of that effective inductance. A large number of turns will increase the effective series resistance and lower the Q, and a distinct difference in series-mode measurement and parallel-mode measurement will be obvious (especially at lower frequencies where Q is even lower).

Index

Note: Page numbers followed by *f* denote figures; page numbers followed by *t* denote tables.

A

AC-AC converter. *See also* LLC converters
   derating maximum power for increased input range, 488, 490
   described, 486–490
   diodes and transformer (no output capacitor), 490–491
   diodes and transformer (with output capacitor), 491–495
   step 1 of design validation process, 486–490
   step 2 of design validation process, 490–491
   step 3 of design validation process, 491–495
AC-DC converters. *See* Flyback converters; Forward converters; Front-end design
AC resistance, 222, 279, 282, 283*f*, 286, 299, 304, 317, 319*f*, 325, 332
Accuracy, bit, 19, 22–23
Active clamp reset, 417–424
Active inrush protection circuit, 519, 519*f*
AD580, Brokaw cell, 7
Adapter with battery charge function, 515, 515*f*
Adaptive deadtime, 90
Air gap:
   B-field inside, 212*t*, 213*f*
   BH curve and, 218–219, 219*f*
   effect, 208–210
   energy amount in, 217–218
   H-field inside, 212*t*, 213*f*
   varying, energy storage considerations, 214–217
Air-gap factor *z*:
   described, 210–211
   optimum design target values, 218
   origin and significance, 211–212
A_L. *See* Inductance index
Altitude, natural convection at, 355–356
Aluminum electrolytic capacitors, 2, 141, 154, 325, 326, 331, 332, 388, 404, 425, 434, 435
American Wire Gauge. *See* AWG
Ampere-seconds law, 55*f*
Ampere-turns, 36, 200, 205, 205*t*, 215, 218, 222, 227, 232, 233, 234, 262, 266, 280, 297, 413
Ampere's circuital law, 195, 197, 205
Anti-synchronization techniques, 149–155
   capacitor current reduction, 150*f*
   over wide input range, 151
   RMC cap current improvement, 152*f*

Area products, core:
   definition, 295*f*
   finer divisions, 295–296, 295*f*
   Forward converters:
      Basso/On-Semi recommended equations, 303
      Billings and Pressman recommendation, 303–304
      Fairchild Semi recommended equations, 301–302
      plotting industry recommendations, 304–306, 305*f*, 306*f*
      ST Micro recommended equations, 303
      TI/Unitrode recommended equations, 302
   margin tape sizes, 310*t*, 311*t*, 312*t*, 313*t*
   numerical example, 296*f*
   power/area product relation, 296–298
   for symmetric converters, 307–308
Area × Turns rule, 257
ASTM B-258, 271
Autotransformers, 231. *See also* Tapped-inductor converters
Auxiliary/standby self-oscillating low-cost power supply, 513–515, 514*f*
Average current problem, 370*f*
Average inductor current, 107
Average values:
   common waveforms, 55, 56*f*, 57
   piecewise linear method, 53–55, 54*f*
   sine wave-related waveforms (lookup table), 57, 57*f*
   square wave-related waveforms (lookup table), 58, 58*f*, 59*f*
AWG (American Wire Gauge):
   ASTM B-258 specification, 271
   copper wire diameters, 267*t*
   current carrying capacity, 265–266
   diameter of coated wire, 271–272, 272*f*
   different cmil/A, 268*t*
   magnet wire (bundled/Litz), 266, 282, 283, 285, 286, 316
   selection issues, 266
   skin depth, 267–269

B

B-field:
   defined, 195

B-field (*Cont.*):
  inductor selection, 240–241
  inside core and gap, 212*t*, 213*f*
Bandgap references:
  BJTs:
    CTAT, 7–13
    current mirror, 13–18
    diode-connected, 13
    Ebers-Moll model, 9, 9*f*, 14, 18
    forward active region, 7–9
    I-V curves, 7, 8*f*, 11, 13
    NPN, 7, 9, 10, 69, 70, 73, 75
    PTAT, 7–13
    understanding, 7–13
  Brokaw bandgap reference cell, 7, 13, 15, 16*f*, 17, 17*f*, 18, 21*f*
  CMOS bandgap reference cell, 17–18, 17*f*
  described, 6–7
  LT1790A, 6–7
  principle of, 13–23
  simulation and, 18
  sub-, 17
  sub-volt, 17
  Widlar bandgap reference cell, 13, 15*f*, 16*f*
  zener diodes compared to, 5–6, 19–20, 20*f*
Basic switching topologies. *See also* Boost; Buck; Buck-boost; Composite switching topologies
  composite topologies compared to, 35
  configurations, 71–74
  energy relationships of, 42–45
  explanation of, 34*f*, 35
  nuances, 83–86
  stress computations, 52–68
  wide input voltage, 59–61
Basso/On-Semi recommended equations, area product, 303
Battery charge function, adapter with, 515, 515*f*
BCM (boundary conduction mode), 62, 66–67, 66*f*
BH curve, 218–219, 219*f*
Bifilar, 165, 414
Bill of Materials, 27
Billings, Keith, 303–304
Bipolar junction transistors. *See* BJTs
Bit accuracy, 19, 22–23
BJTs (bipolar junction transistors):
  CTAT, 7–13
  current mirror, 13–18
  diode-connected, 13
  Ebers-Moll model, 9, 9*f*, 14, 18
  forward active region, 7–9
  I-V curves, 7, 8*f*, 11, 13
  NPN, 7, 9, 10, 69, 70, 73, 75
  PTAT, 7–13
  understanding, 7–13
Blanking time, 110, 191, 194, 363, 366, 378, 387
Bode plot, 84, 363, 364*f*, 371, 396, 436
Boost (step-up). *See also* DC-DC converters
  basic relations, 43*f*
  basic topologies:
    comparison, 34*f*, 35
    energy relationships, 42–45
    nuances, 83–86
  Buck to Boost, 91–95, 92*f*

Boost (step-up) (*Cont.*):
  capacitor currents:
    equations, 61*t*
    line/load changes, 61–62
    shape, 58–59
  configurations, 71, 73*f*
  current flow and energy transfer timings, 44*f*
  current ripple ratio and, 37–42, 40*f*, 41*f*
  duty cycle:
    DCM, 107, 108*f*, 116*f*
    ideal equations, 42*f*, 43*f*, 49
    real-world equations, 49–51, 50*f*
  efficiency relationships, 45–48, 46*f*
  energy relationships, 42–45
  explanation of, 34*f*, 35, 38*f*
  IC selection table, 77–79, 78*t*, 79*t*
  inductor selection criteria, 76–77
  input-to-output transfer function for, 36*t*
  key stresses, 115, 118, 120*f*
  loss relationships, 45–48, 46*f*
  negative-to-negative, 71, 82*f*, 84*f*
  parasitic synchronous, 417, 420
  PFC Boost stage, 145–146
    currents, 147*f*, 148*f*
    holdup time considerations, 148–149, 149*f*, 149*t*
    instantaneous duty cycle, 146, 148
    stress equations, 147*f*
  plant transfer functions:
    CMC, 383*f*
    VMC, 375, 376*f*
  positive-to-positive, 71, 75, 77, 80*f*
  power scaling guidelines, 38, 51–52
  ratings, 43*f*
  RHP zero, 83–85, 86*f*
  stresses:
    closed-form equations, 126*t*, 127
    key stresses, 115, 118, 120*f*
    stress spider, 62, 64*f*
  symmetrical, 520–521, 521*f*
  tapped-inductor, 236, 236*f*, 237*f*
  trace analysis, 337*f*
  type 1 ICs:
    applications, 77–79, 78*t*, 79*t*
    compared, 75–76
    described, 74–75, 74*f*
  waveforms, 43*f*
  wide input voltage, 59–61
Boost-buck topology/configuration, 102*f*. *See also* Buck-boost
Boost preregulator with regulated auxiliary output, 522–523, 522*f*
Bootstrap circuit, 70, 74, 92, 231
Boundary conduction mode. *See* BCM
Bridge rectifiers, 129, 157, 159
  paralleling, 515*f*, 516
British wire gauge. *See* SWG
Brokaw, Paul, 7, 13
Brokaw bandgap reference cell, 7, 14, 15, 16*f*, 17, 17*f*, 18, 21*f*
Brown & Sharpe. *See* AWG
Buck (step-down). *See also* Buck efficiency; DC-DC converters
  basic relations, 42*f*

Buck (step-down) (*Cont.*):
  basic topologies:
    comparison, 34*f*, 35
    energy relationships, 42–45
    nuances, 83–86
  Buck to Boost, 91–95, 92*f*
  capacitor currents:
    equations, 61*t*
    line/load changes, 61–62
    shape, 58–59
  configurations, 71, 72*f*
  current flow and energy transfer timings, 44*f*
  current ripple ratio and, 37–42, 40*f*, 41*f*
  duty cycle:
    DCM, 107, 108*f*, 115*f*
    ideal equations, 42*f*, 43*f*, 49
    real-world equations, 49–51, 50*f*
  efficiency relationships, 45–48, 46*f*
  explanation of, 34*f*, 35, 37*f*
  floating, 520, 520*f*
  IC selection table, 77–79, 78*t*, 79*t*
  inductor selection criteria, 76–77
  input-to-output transfer function for, 36*t*
  interleaving of, 95–100, 96*f*
  key stresses, 115–117, 118*f*, 119*f*, 120*f*
  loss relationships, 45–48, 46*f*
  negative-to-negative, 81*f*
  paralleling of, 95–100, 96*f*
  peak and switch RMS currents, 118*f*
  plant transfer functions:
    CMC, 378, 379*f*
    VMC, 373, 374*f*, 375
  positive-to-positive, 74, 77, 83*f*
  power scaling guidelines, 38, 51–52
  ratings, 42*f*
  stresses:
    closed-form equations, 124*t*–125*t*, 126*t*, 127
    key stresses, 115–117, 118*f*, 119*f*, 120*f*
    stress spider, 62, 63*f*
  synchronous Buck:
    modes of operation, 88–90, 89*f*
    switching losses, 465–467, 466*f*
  tapped-inductor, 232–236, 233*f*
  trace analysis, 336*f*, 337*f*
  type 2 ICs:
    applications, 77–79, 78*t*, 79*t*
    compared, 75–76
    described, 74–75, 74*f*
  waveforms, 42*f*
  wide input voltage, 59–61
Buck-boost:
  parasitic synchronous, 417, 419
  plant transfer function:
    CMC, 384*f*
    VMC, 375, 377*f*
  slave, 521–522, 522*f*
Buck-boost (step-up/step-down). *See also* DC-DC converters
  basic relations, 43*f*
  basic topologies:
    comparison, 34*f*, 35
    energy relationships, 42–45
    nuances, 83–86

Buck-boost (step-up/step-down) (*Cont.*):
  capacitor currents:
    equations, 61*t*
    line/load changes, 61–62
    shape, 58–59
  configurations, 71, 73*f*
  current flow and energy transfer timings, 44*f*
  current ripple ratio and, 37–42, 40*f*, 41*f*
  duty cycle:
    DCM, 107, 108*f*, 117*f*
    ideal equations, 42*f*, 43*f*, 49
    real-world equations, 49–51, 50*f*
  efficiency relationships, 45–48, 46*f*
  equivalent models, in Flyback converters, 169–175, 172*f*, 173*f*
  explanation of, 35–37, 39*f*
  Flyback converter compared to, 119, 121
  IC selection table, 77–79, 78*t*, 79*t*
  inductor selection criteria, 76–77
  input-to-output transfer function for, 36*t*
  input voltage variance, 118–119, 122*f*
  key stresses, 115, 118, 121*f*
  loss relationships, 45–48, 46*f*
  negative-to-negative-to-positive, 76
  negative-to-positive, 74, 76, 79*t*, 81*f*
  positive-to-negative, 74, 76, 79*t*, 81, 82*f*, 84*f*, 86
  power scaling guidelines, 38, 51–52
  ratings, 43*f*
  RHP zero, 83–85, 86*f*
  stresses:
    closed-form equations, 126*t*, 127
    key stresses, 115, 118, 121*f*
    stress spider, 62, 65*f*
  trace analysis, 337*f*
  type 1 ICs:
    applications, 77–79, 78*t*, 79*t*
    compared, 75–76
    described, 74–75, 74*f*
  waveforms, 43*f*
  wide input voltage, 59–61
Buck efficiency, 443–456
  Buck spreadsheet, 451, 454*f*
  introduction, 443–444
  losses:
    conduction, 446
    controller IC, 447
    crossover, 444*f*, 445, 445*t*
    cumulating, 447–451
    deadtime, 445, 446*f*, 447*t*
    input capacitor ESR, 445–446
    understanding, 444
  predicting, 451
  reverse-engineering tricks, 451
Bulk capacitor:
  RMS of high-frequency component, 154
  RMS of low-frequency component, 153
Bundled/Litz wire, 266, 282, 283, 285, 286, 316
Buried zener references, 6, 19, 20, 21. *See also* Zener diodes
Burn-in tests, 430
Bypass capacitor, 86, 336, 337

C

C dV/dt induced turn-on, 91

Capacitance:
  droop curves for, 131, 131f
  front-end design (in AC-DC power conversion, without PFC), 137, 140t
  input, Flyback converters, 129, 130f
  net Y-capacitance, 406
  total Y-capacitance, 440
  X-capacitance, 405, 406, 407, 440
Capacitor selection:
  Flyback converters:
    input capacitor selection, 331
    output capacitor selection, 331–332
  Forward converters:
    input capacitor selection, 325–326, 327f
    output capacitor selection, 326, 328, 328f, 329f
Capacitors:
  AC-AC converter, LLC converter:
    diodes and transformer (no output capacitor), 490–491
    diodes and transformer (with output capacitor), 491–495
  aluminum electrolytic, 2, 141, 154, 325, 326, 331, 332, 388, 404, 425, 434, 435
  bulk capacitor:
    RMS of high-frequency component, 154
    RMS of low-frequency component, 153
  bypass, 86, 336, 337
  currents, 58–59, 61–62, 61t
  input capacitor selection:
    Flyback converters, 331
    Forward converters, 325–326, 327f
  life, estimation, 434–440
  output capacitor selection:
    Flyback converters, 331–332
    Forward converters, 326, 328, 328f, 329f
Carsten, Bruce, 284
CCM (continuous conduction mode). *See also* DCM
  DCM compared to, 109–111
  equations, FCCM and, 87
  Forward converter choke, 275
  subharmonic instability, 370
CEM-1 board, 343, 344
Center of ramp (COR), 38, 40f
Chaotic pulsing, 110
Chargeable failures, 427–428
Cheap power good signal, 517, 517f
Chi-square distribution, 426–427
Choke:
  Forward converters:
    choke inductance and rating, 316
    transformer/choke comparison, 275–279
  procedure design, PFC, 155, 155f, 156f
  specifications, inductor selection, 249, 249t
Circular mils (cmil), 264–265, 265f, 266t
CISPR 22 standard, 401, 402t, 404, 405
Clearance distance, 431–432, 431f
Closed-form equations:
  Flyback converters and, 129–131
  stresses:
    Boost DCM, 126t, 127
    Buck-boost DCM, 126t, 127
    Buck DCM, 124t–125t, 126t, 127
CM. *See* Common-mode (CM) noise

CMC. *See* Current-mode control
Cmil. *See* Circular mils
CMOS bandgap reference cell, 17–18, 17f
Coated wire, diameter of, 271–272, 272f
Coefficient of thermal expansion. *See* CTE
Commercial resistor values, 26–30
Common-mode (CM) noise, 401, 405–407, 440, 518
Comparative tracking index (CTI), 344, 431
Compensation schemes:
  type 3, summary, 382f
  types, 381f
Complementary to absolute temperature. *See* CTAT
Composite switching topologies:
  basic topologies compared to, 35
  building blocks of, 100, 101f
  stress computations, 101, 105
Conducted emission limits, 401, 402t
Conduction losses, Buck efficiency, 446
Confidence levels, 426, 427
Continuous conduction mode. *See* CCM
Control-loops. *See* Loop stability
Control techniques:
  current-mode control:
    Buck plant transfer function, 378, 379f
    current-mode converters, 363
    defined, 373
    design examples, 388–391
    hysteretic control and, 386f
    summary of pros and cons, 388, 388t
    synchronous Buck converters, 92–93
    VMC versus, 378
  hysteretic control:
    changes to achieve, 386f
    defined, 378
    functional blocks, 385f
    simplified explanation, 387f
    summary of pros and cons, 388, 388t
  voltage-mode control:
    CMC versus, 378
    current-mode converters stability, 363
    design examples, 388–391
    feedforward in, 366, 368f
    hysteretic control and, 386f
    PWM in, 368f
    summary of pros and cons, 388, 388t
    synchronous Buck converters, 91–92
Controller IC losses, Buck efficiency, 447
Convection. *See also* Thermal management
  at altitude, 355–356
  equations, 347–348
    available, 348–349
    comparing two standard equations, 350–351
    historical definitions, 348
    manipulation, 349–350
    tables of standard equations, 351–353, 352t–353t
  forced air cooling, 356–357
  introduction, 345
Convection coefficient, 348
Converters. *See also* Boost; Buck; Buck-boost; Flyback converters; Forward converters; LLC converters

Converters (*Cont.*):
    AC-AC converters:
        derating maximum power for increased input range, 488, 490
        described, 486–490
        diodes and transformer (no output capacitor), 490–491
        diodes and transformer (with output capacitor), 491–495
        step 1 of design validation process, 486–490
        step 2 of design validation process, 490–491
        step 3 of design validation process, 491–495
    Cuk converters:
        building blocks, 101*f*
        capacitor currents, shape, 59
        configurations, 102*f*
        current waveforms, 104*f*, 105
        history, 100
        topologies, 102*f*
    DC-DC converters:
        duty cycle equations, ideal, 42*f*, 43*f*, 49
        duty cycle equations, real-world, 49–51, 50*f*
        efficiency relationships, 45–48, 46*f*
        ground in, 33–34
        loss relationships, 45–48, 46*f*
        power scaling guidelines, 38, 51–52
        principles, 33–52
        stress computations, 52–86
        synchronous, 87–95
        waveform analysis, 52–86
    half-bridge converters:
        diagram, 166*f*
        flux staircasing, 167–168
        LLC converter, 503–507, 506*f*
    Push-Pull converter:
        diagram, 166*f*
        flux balancing, 167
    SEPIC (single-ended primary inductance converter):
        building blocks, 101*f*
        configurations, 103*f*
        current waveforms, 104*f*, 105
        topologies, 103*f*
    tapped-inductor converters, 231–237
        Boost, 236, 236*f*, 237*f*
        Buck, 232–236, 233*f*
    Zeta converters:
        building blocks, 101*f*
        configurations, 103*f*
        current waveforms, 104*f*, 105
        topologies, 103*f*
Copper traces, PCBs, 340–341
Copper wire. *See also* Wire gauges
    diameters, AWG and SWG, 267*t*
    stacking, Forward converters, 288–289
    winding arrangements:
        Flyback converters, 332
        Forward converters, 319–325
COR. *See* Center of ramp
Cores. *See also* Cores, area products
    B-field inside, 212*t*, 213*f*
    core-loss optimization:
        tweaking geometry as frequency varies, 246–248

Cores, core-loss optimization (*Cont.*):
        tweaking geometry at fixed frequency, 244–246
    E-cores:
        effective area and effective length, 207–208
        gapped, 213–214, 214*f*
    Flyback converters:
        current limit's impact, 262–264, 262*f*
        current ripple ratio's impact, 264, 264*f*
        quick selection rules, 260–261, 293, 330
    Forward converters:
        core-loss calculations, 289–291, 290*t*, 317–318, 317*t*, 318*t*
        quick selection, 287–288, 294, 315
    gapped core:
        E-cores, 213–214, 214*f*
        general energy relationships, 256
    H-field inside, 212*t*, 213*f*
    thermal resistance of, 358*t*
    toroidal core:
        air gap's effect, 208–210
        E-core mapped in toroid, 208*f*
        effective area and effective length, 206–207, 207*f*
        mmf equation, 205–206, 205*f*
    ungapped core, energy stored, as related to core volume, 256
    utilization factor, 310*t*, 311*t*, 312*t*, 313*t*
    window areas:
        finer divisions, 295–296, 295*f*
        margin tape sizes, 310*t*, 311*t*, 312*t*, 313*t*
        numerical example, 296*f*
Cores, area products:
    definition, 295*f*
    finer divisions, 295–296, 295*f*
    Forward converters:
        Basso/On-Semi recommended equations, 303
        Billings and Pressman recommendation, 303–304
        Fairchild Semi recommended equations, 301–302
        plotting industry recommendations, 304–306, 305*f*, 306*f*
        ST Micro recommended equations, 303
        TI/Unitrode recommended equations, 302
    numerical example, 296*f*
    power/area product relation, 296–298
    for symmetric converters, 307–308
Counterintuition, in magnetics, 217, 244
Coupled diode model. *See* Ebers-Moll model
Coupled inductors, 36, 98, 165, 167
Creativity, switchers and, 3
Creepage distance, 431–432, 431*f*
Critical conduction mode. *See* BCM
Cross-conduction. *See* Shoot-through
Crossover, 362
Crossover losses, Buck efficiency, 444*f*, 445, 445*t*
CTAT (complementary to absolute temperature), 7–13
CTE (coefficient of thermal expansion), 344
CTI. *See* Comparative tracking index
Cuk converters:
    building blocks, 101*f*
    capacitor currents, shape, 59
    configurations, 102*f*

Cuk converters (*Cont.*):
  current waveforms, 104f, 105
  history, 100
  topologies, 102f
Cumulating losses, Buck efficiency, 447–451
Current carrying capacity:
  AWG, 265–266
  different cmil/A, 267t
Current density:
  conversions based on D and, 298–299
  Flyback converters, 332–333
  optimum, 299–301
  skin depth and, 280f
Current limit, core size and, 262–264, 262f
Current mirror, BJT, 13–18
Current-mode control (CMC). *See also* Hysteretic control; Voltage-mode control
  Buck plant transfer function, 378, 379f
  current-mode converters, 363
  defined, 373
  design examples, 388–391
  hysteretic control and, 386f
  summary of pros and cons, 388, 388t
  synchronous Buck converters, 92–93
  VMC *versus*, 378
Current-mode converters, 363–372
  bode-plot setup, 363, 364f
  subharmonic instability, 363, 365f
Current ripple ratio (r):
  DCM and, 107, 114
  described, 37–42, 40f, 41f
  equations, 36t, 61
  Flyback converter core, 264, 264f
  inductor selection:
    choosing r, 242–243, 243f
    mapping procedure, 249t, 250–251
    specifying r, 240–241
  key advantage of, 38–39
  power scaling guidelines, 38, 51–52
  wide input voltage and, 59–61
Current sense signal, 194, 364, 370
Current-sense trace, 341
Current-voltage (*I-V*) curves, 7, 8f, 11, 13
Currents, Forward converters, 277–279, 278f
Custom resistors, 26–27

D

DC-DC converters. *See also* Basic switching topologies; Boost; Buck; Buck-boost; Inductor selection
  duty cycle equations:
    ideal, 42f, 43f, 49
    real-world, 49–51, 50f
  efficiency relationships, 45–48, 46f
  ground in, 33–34
  loss relationships, 45–48, 46f
  power scaling guidelines, 38, 51–52
  principles, 33–52
  stress computations, 52–86
  synchronous, 87–95
  waveform analysis, 52–86
DCM (discontinuous conduction mode), 107–127. *See also* CCM
  CCM compared to, 109–111
  current ripple ratio and, 107

DCM (discontinuous conduction mode) (*Cont.*):
  defined, 87
  diode emulation mode, 87, 88f, 443
  duty cycle, 108f, 111–112, 111t–112t
  Forward converters, 164, 275
  idle time interval of, 107
  introduction, 107–111
  simplified treatment of, 114
  slave Buck-boost converter, 521–522, 522f
  stress computations, 114–115
    closed-form equations, Buck, 124t–125t, 126t, 127
    top-to-bottom calculation order, 111t–112t, 115
  terminology usage, 112–114
  treatment, in related literature, 112–114
DCR (DC resistance), 48, 49, 159, 279, 286, 289, 294, 299, 317, 517
DCR sensing, 92, 94–95, 94f
Deadtime, synchronous Buck, 88–90
Deadtime losses, Buck efficiency, 445, 446f, 447t
Dearborn, Scott, 113
Decibels:
  decibels to ratios, 361t
  ratios to decibels, 360t
Demonstrated reliability test (DRT), 425, 427, 428, 429
Design tightrope, switchers, 1–3
Differential-mode (DM) noise, 401, 404–405
Differential voltage sensing, 82–83, 85f, 85t
Diode emulation mode, 87, 88f, 443. *See also* DCM
Diodes:
  AC-AC converter, LLC converter:
    diodes and transformer (no output capacitor), 490–491
    diodes and transformer (with output capacitor), 491–495
  diode-connected BJT, 13
  Ebers-Moll model, 9, 9f, 14, 18
  Schottky diode:
    basic switching topologies and, 35
    Buck efficiency, 443
    FAN5340 lesson and, 90
  Shockley's equation for, 9
  tertiary winding diode, 165, 166f
  zener diodes:
    bandgap references compared to, 5–6, 19–20, 20f
    buried, 6, 19, 20, 21
    described, 5–6
    safety issues, 440–442
Discontinuous conduction mode. *See* DCM
Dissipation estimates, Flyback converters, 187–190, 188t
DM. *See* Differential-mode (DM) noise
Double-switch Forward converter, diagram, 166f
Dowell's equations, 280–286, 299, 304, 319, 320, 322, 323. *See also* Proximity effect
Droop curves:
  for different capacitances per watt, 131, 131f
  Forward converter, 136f
Droop positioning, 32
DRT. *See* Demonstrated reliability test

Duty cycles:
  DCM, 111–112, 111t
    Boost, 107, 108f, 116f
    Buck, 107, 108f, 115f
    Buck-boost, 107, 108f, 117f
  Flyback converter:
    100 V, 257
    optimization, 181
  Forward converter design, 316
  ideal equations, 42f, 43f, 49
  real-world equations, 49–51, 50f
  subharmonic instability, 370
  tapped-inductor converters, 236, 236f, 237f
Dynamic load response, 3–5, 5f

E

E-cores:
  effective area and effective length, 207–208
  gapped, 213–214, 214f
E series, resistors, 27–30, 27f, 30f
E24 series, 28, 29, 30
E48 series, 28, 29
E96 + E24, 29, 30f
E96 series, 29, 30f
Early, James, 7
Early effect, 7, 8f
Ebers-Moll model, 9, 9f, 14, 18
Effective area and effective length:
  E-cores, 207–208
  toroidal core, 206–207, 207f
Efficiency:
  Buck efficiency, 443–456
    Buck spreadsheet, 451, 454f
    conduction loss, 446
    controller IC loss, 447
    crossover loss, 444f, 445, 445t
    cumulating losses, 447–451
    deadtime loss, 445, 446f, 447t
    input capacitor ESR loss, 445–446
    introduction, 443–444
    predicting, 451
    reverse-engineering tricks, 451
    understanding losses, 444
  estimates, thermal management, 346–347
  onion, 447, 450
  relationships, DC-DC converters, 45–48, 46f
Electrical-magnetic analogy, 197–198, 197f
Electrolytic capacitors, 2, 141, 154, 325, 326, 331, 332, 388, 404, 425, 434, 435
EMI (electromagnetic interference) filters, 401–408
  CISPR 22 standard, 401, 402t, 404, 405
  CM filter design, 405–407
  common-mode noise, 401, 405, 406, 407
  conducted emission limits, 401, 402t
  differential-mode noise, 401, 404
  DM filter design, 404–405
  Fourier series, 402
  front-end design in AC-DC power conversion, 143
  LISNs, 401
  trapezoid, 403–404
Emitter current density, 14
EN 550022, 401. *See also* CISPR 22 standard
Energy recovery winding. *See* Tertiary winding
Energy relationships, for gapped core, 256

Energy storage requirements, Forward and Flyback, 333
Energy stored, in ungapped core, 256
Equivalent foil transformation, 286
Equivalent series resistance. *See* ESR
Error amplifiers, transfer functions, 380f
ESR (equivalent series resistance), 242, 404, 435
ESS, 430
External voltage divider, 6f, 23
Extrusion heatsinks, 358

F

Failure:
  rate, 425, 426, 428, 429, 437
  MTBF, 425–429
  MTTF, 425
  real world, simulation and, 33
Fairchild Semi recommended equations, area product, 301–302
FAN5340 lesson, 90
Faraday shield, 227, 227f
Faraday's law. *See* Voltage-dependent equation
FCCM (forced continuous conduction mode), 87
Fear, of resonant topologies, 461–462
Feedback. *See also* Control techniques
  feedback pin input bias current, 23–25
  phase margin, 4, 363, 436
  stability:
    current-mode controllers, 363–372
    loop stability, 359, 360f, 362f
Feedforward, in voltage-mode control, 366, 368f
Feucht, Dennis, 38, 67
FGA technology, 20
5-cent zener, 440–442. *See also* Zener diodes
Fixed ramp, to sensed current, 365f
Flattop approximation, 68, 69f, 99, 190
Floating Buck topology, 520, 520f
Floating drive from 384x controller, 519, 520f
Flux, 195–196
Flux linkages, 196, 197, 221
Flux staircasing, 164, 167–168, 198
Flyback converters:
  area product:
    definition, 295f
    finer divisions, 295–296, 295f
    numerical example, 296f
    power/area product relation, 296–298
  Area × Turns rule, 257
  AWG:
    ASTM B-258 specification, 271
    copper wire diameters, 267t
    current carrying capacity, 265–266
    diameter of coated wire, 271–272, 272f
    different cmil/A, 268t
    magnet wire (bundled/Litz), 266, 282, 283, 285, 286, 316
    selection issues, 266
    skin depth, 267–269
  Buck-boost compared to, 119, 121
  circular mils, 264–265, 265f, 266t
  closed-form equations and, 129–131
  copper windings, 332
  cores:
    current limit's impact, 262–264, 262f
    current ripple ratio's impact, 264, 264f
    quick selection rules, 260–261, 293, 330

Flyback converters (*Cont.*):
  current density:
    conversions based on D and, 298–299
    optimum, 299–301
    skin depth and, 280f
  current density targets, industrywide, 332–333
  design:
    basics, 255–273
    step-by-step, 328–332
  design curves, 123f
  design table:
    at 85 V, 137, 138t
    at 170 Vac, 137, 139t
  dissipation estimates, 187–190, 188t
  duty cycle:
    100 V, 257
    optimization, 181
  energy storage requirements, 333
  equivalent Buck-boost models, 169–175, 172f, 173f
  Forward-Flyback comparison, 293–333
  inductance index, 257
  inductors compared to, 221–222
  input capacitances per watt (at 85 V), 129, 130f
  input capacitor selection, 331
  introduction, 168–194
  leakage inductance:
    primary-side, 176–177, 176f
    secondary-side, 176f, 178–180
  multi-output, 175
  number of primary turns, 310–314, 314f
  optimization:
    finer points, 260
    steps, 180–187
  output capacitor selection, 331–332
  overload margin, 121–122, 123f, 124
  overload protection at high-line, 193–194
  primary inductance, 330
  primary turns, 330
  pulse-skipping, 191–193
  required preload, 191–193
  reset techniques:
    lossless snubbers, 412–413, 412f
    RCD clamps, 183–184, 410f, 411–412, 411f
    zener clamp, 330, 409, 410f
  secondary turns, 330
  600-V switches, 190–191
  skin depth, 267–269
  SWG:
    AWG compared to, 272–273, 273f
    copper wire diameters, 267t
    skin depth, 267–269
  transformer model, 222–226
  turns ratio calculation, 330
  voltage-dependent equation, practical form, 255–256
  voltage ratings, 330–331
  worked example:
    alternative design paths, 314
    part 1, 257–258
    part 2, 261–262
  zener clamp, 330
Foil:
  equivalent foil transformation, 286
  optimum foil thickness, 282–285
Forced air cooling, 356–357

Forced continuous conduction mode. *See* FCCM
Forward active region, BJT, 7–9
Forward converters:
  alternative design path, worked example, 314
  area product:
    Basso/On-Semi recommended equations, 303
    Billings and Pressman recommendation, 303–304
    definition, 295f
    Fairchild Semi recommended equations, 301–302
    finer divisions, 295–296, 295f
    numerical example, 296f
    power/area product relation, 296–298
    ST Micro recommended equations, 303
    TI/Unitrode recommended equations, 302
  Buck-derived, 166f
  choke:
    inductance and rating, 316
    transformer and choke comparison, 275–279
  configuration, 167
  cores:
    with area product, window area, utilization factor, 310t, 311t, 312t, 313t
    with basic characteristics, 309t
    core-loss calculations, 289–291, 290t, 317–318, 317t, 318t
    quick selection, 287–288, 294, 315
  current density:
    conversions based on D and, 298–299
    optimum, 299–301
    skin depth and, 280f
  currents, 277–279, 278f
  DCM, 164, 275
  design, step-by-step, 315–328
  droop curves for, 136f
  energy storage requirements, 333
  Flyback-Forward comparison, 293–333
  front-end design and, 143–144
  input capacitor selection, 325–326, 327f
  introduction, 161–168
  loss estimation:
    core loss and total estimated loss, 317–318, 317t, 318t
    overall, in transformer, 316–317
  magnetics design, 275–291
  magnetization inductance, 315
  output capacitor selection, 326, 328, 328f, 329f
  peak magnetization current, 315
  primary turns:
    design step, 315
    number of, 310–314, 314f
  primary winding, 161, 163
  proximity effect:
    Dowell's equations, 280–286
    introduction, 279
  reset techniques:
    active clamp reset, 417–424
    introduction, 413–414
    tertiary winding, 161, 414–417, 415f, 417t
  secondary turns, 316

Forward converters (Cont.):
  single-ended:
    diagram, 162f
    without PFC, 135–137
  skin depth, 279–286
  turns ratio calculation, 277, 315
  voltage ratings, 316
  winding arrangements:
    P-S, 319–323
    S-P-S, 323–325
Fourier series, EMI filter design, 402
FPS system, 204
FR-4 boards, 343–344, 345, 353, 355
Frequency foldback, 194
Fringing flux correction, 227–230
Froeschle, Thomas A., 373
Front-end design (in AC-DC power conversion), 129–159
  Boost PFC stage, 145–146
    currents, 147f, 148f
    holdup time considerations, 148–149, 149f, 149t
    instantaneous duty cycle, 146, 148
    stress equations, 147f
  Forward converters, 143–144
  PFC, 144–159
    anti-synchronization techniques, 149–155, 150f, 152f
    choke procedure design, 155, 155f, 156f
    defined, 145
    design table, 156t–157t
    nuances, 155–159
  without PFC, 129–144
    cap RMS requirements, 137, 140t
    high-frequency RMS currents, 137, 141, 141t
    lookup curves, 144, 144f
    minimum capacitances for meeting holdup time requirements, 137, 140t
    single-ended Forward converters, 135–137
    total input capacitor RMS current, 141, 142t
Fundamental topologies. See Basic switching topologies
Fundamentals of Power Electronics, 2d ed., 112

G

Gain (conversion ratio), LLC tank circuit, 478f
Gain-phase relationships, LLC tank circuit, 479–483, 479f, 481f
Gapped core:
  E-cores, 213–214, 214f
  general energy relationships, 256
Gerber file format, 345
Grashof number, 351
Green boards (FR-4), 343–344, 345, 353, 355
Ground (system ground), 33–34
Ground plane, PCBs, 341–343, 342f

H

h, thermodynamic theory, 351
H-field:
  defined, 195
  inside core and gap, 212t, 213f

Half-bridge converters:
  diagram, 166f
  flux staircasing, 167–168
  LLC converter, 503–507, 506f
Half-turns, 179
HALT, 430
Hard switching, soft and, 462–464, 463f. *See also* Resonant topologies
HASS, 430
HAST, 430
Heat transfer coefficient, 348
Heatsinking:
  extrusion, 358
  PCBs for, 353, 355
High-side active clamp, 417, 418, 419, 420–421
Hilbiber, David, 7
Holdup time:
  Flyback converter optimization, 184–185
  PFC Boost stage, 148–149, 149f, 149t
Holland, Brian, 373
Hot-air solder level finishing stage, 343
Hysteretic control. *See also* Control techniques
  changes to achieve, 386f
  defined, 378
  functional blocks, 385f
  simplified explanation, 387f
  summary of pros and cons, 388, 388t

I

I-V curves. See Current-voltage curves
I_C versus V_{BE} exponential equation, 9–12, 10f, 12f
ICs:
  selection table, 77–79, 78t, 79t
  type 1 and type 2:
    applications, 77–79, 78t, 79t
    compared, 75–76
    described, 74–75, 74f
Idle time interval, of DCM, 107
IEC61000-3-2 standard, 145
Imperial wire gauge. See SWG
Inductance:
  equations, trace analysis, 339, 339f
  large, 67–68
  leakage inductance:
    primary-side, 176–177, 176f
    secondary-side, 176f, 178–180
    tapped-inductor Buck, 235
  primary, Flyback converters, 330
  primary-side leakage inductance term, 176–177, 176f
  ratio of inductances, LLC converters, 498, 502–503
  secondary-side leakage inductance term, 176f, 178–180
  SEPIC:
    building blocks, 101f
    configurations, 103f
    current waveforms, 104f, 105
    topologies, 103f
Inductance index (A_L), 257
  BH curve, 218
  defined, 197
  Flyback converter design, 257
  self-induction and, 221
Inductor energy, Flyback optimization, 185–187
Inductor equation, 198, 241

Inductor selection (DC-DC converters), 239–253
  B in terms of current, 243–244
  choke specifications, 249, 249t
  core-loss optimization:
    tweaking geometry as frequency varies, 246–248
    tweaking geometry at fixed frequency, 244–246
  counterintuition in magnetics, 244
  criteria, 76–77
  current ripple ratio (r):
    choosing r, 242–243, 243f
    mapping, 250–251
    specifying r, 240–241
  evaluation of inductor:
    current ripple ratio, 249t, 250–251
    design conditions, 249t
    peak current, 249t, 251–252
    peak flux density, 249t, 251
    temperature rise, 249t, 252–253
  mapping procedure:
    implementation, 249–253, 249t
    volt-seconds, 241–242
  self induction, 242–243
Inductors:
  coupled, 36, 98, 165, 167
  Flyback converters compared to, 221–222
  mapping, 241–242
  multiwinding, 221, 222, 232, 275
Input bias current, voltage divider, 23–25
Input capacitor ESR losses, Buck efficiency, 445–446
Input capacitor selection:
  Flyback converters, 331
  Forward converters, 325–326, 327f
Input current waveshape, 131, 131f
Input-to-output transfer function, 36f
Input voltage:
  variance, Buck-boost, 118–119, 122f
  wide, 59–61
Inrush protection circuit:
  active, 519, 519f
  self-contained, 516, 516f
Instantaneous tempco, 22
Integrated switchers, 76, 77, 110, 169, 240, 337
Intellectual property protection, LLC converters, 461
Interleaving of Buck converters, 95–100, 96f
Intermediate (virtual) primary-side output rail, LLC converter, 486
Intersil, FGA technology, 20
Intusoft, 18
Inversely proportional to absolute temperature (iPTAT), 13
IP switch (hypothetical product), 169
IPC-9592 standard, 2
iPTAT (inversely proportional to absolute temperature), 13

K

K treatment, 114
Kirchhoff's voltage law, 465, 473

L

L. See Self-induction
Large inductance, 67–68
LDO. See Linear dropout
Leading-edge spike, 67, 67f
Leakage inductance:
  primary-side, 176–177, 176f
  secondary-side, 176f, 178–180
  tapped-inductor Buck, 235
Least significant bit (LSB), 20, 22–23
Lee, Fred C., 474, 505
Lenz's law. See Voltage-dependent equation
Lessons learned, synchronous Buck switching losses, 465–467, 466f
Liang, Yan, 474, 505
Line impedance stabilizing networks. See LISNs
Line regulation, 4
Line voltage overstress test, 430
Linear dropout (LDO), 70, 515
Linear regulator, 5
Linear Technology, LT1790A bandgap reference, 6–7, 75
LISNs (line impedance stabilizing networks), 401
Litz wire, 266, 282, 283, 285, 286, 316
Liu, Wenduo, 474, 505
LLC converters. See also Resonant topologies
  AC-AC converter:
    derating maximum power for increased input range, 488, 490
    described, 486–490
    diodes and transformer (no output capacitor), 490–491
    diodes and transformer (with output capacitor), 491–495
    step 1 of design validation process, 486–490
    step 2 of design validation process, 490–491
    step 3 of design validation process, 491–495
  half-bridge implementation, 503–507, 506f
  historical background, 474
  intellectual property protections, 461
  intermediate (virtual) primary-side output rail, 486
  introduction, 461–462
  Mathcad, 461–462, 476, 482, 487, 488, 490, 496, 498, 510
  misunderstanding, 474
  "Optimal Design Methodology for LLC Resonant Converter," 474, 505
  PD design, theoretical:
    design steps, 495–497
    validation, 497–498, 499f, 500f, 501f
  ratio of inductances (pros and cons), 498, 502–503
  solved example:
    LLC selection with maximum 1:2 frequency spread, 510–511
    LLC selection with maximum 1:3.16 frequency spread, 507–510
  trade secret protections, 461
  trial and error, 461
  unique design methodology, 461
LLC tank circuit:
  gain (conversion ratio) of, 478f
  gain-phase relationships, 479–483, 479f, 481f

LLC tank circuit (*Cont.*):
   introduction, 474, 476, 478–479
   parallel resonance, 483–484, 483*f*
   regulated voltage, 477*f*
   series resonance, 483–484, 483*f*
LM431, 24, 515, 518
LM2592, 444, 451, 456*f*
LM2593HV datasheet, 33
LM2651, 81–82
LM2676-5.0, 23
LM2676T-3.3, 23
LM2676T-ADJ, 23
Load regulation, 4
Load stress overstress test, 430
Long-term drifts, 19, 388*t*
Loop stability, 359, 360*f*, 362*f*
Losses:
   Buck efficiency:
      conduction, 446
      controller IC, 447
      crossover, 444*f*, 445, 445*t*
      cumulating, 447–451
      deadtime, 445, 446*f*, 447*t*
      input capacitor ESR, 445–446
      understanding, 444
   core-loss optimization:
      tweaking geometry as frequency varies, 246–248
      tweaking geometry at fixed frequency, 244–246
   estimation, Forward converter design, 316–318
   relationships, DC-DC converters, 45–48, 46*f*
   switching, synchronous Buck, 465–467, 466*f*
Low-cost self-oscillating standby/auxiliary power supply, 513–515, 514*f*
Low-side active clamp, 417, 418, 420–421
LSD cells, 70–71, 71*f*
LT1790A bandgap reference, 6–7
Lu, Bing, 474, 505

M

Magnet wire, AWG, 266, 282, 283, 285, 286, 316
Magnetics, 195–230
   basic concepts, 195–197
   counterintuition in, 217, 244
   definitions, 195–197
   electrical-magnetic analogy, 197–198, 197*f*
   Forward converter magnetics design, 275–291
   inductor equation, 198, 241
   misconceptions, 195
   self-induction and, 196–197, 221–222
   units conversion, 204–205, 205*t*
   voltage-dependent equation, 200–204, 204*t*
   voltage-independent equation, 198–200, 204*t*, 228–229
Magnetization inductance, Forward converter design, 315
Magnetomotive force equation (mmf), 205–206, 205*f*
Mammano, Bob, 373, 378, 463, 476
Manufacturing issues, PCBs, 343–344
Mapping inductors, 241–242
Margin tape sizes, Cores, 310*t*, 311*t*, 312*t*, 313*t*

Mathcad:
   Buck efficiency, 443–459
   droop curves for different capacitances per watt, 131, 131*f*
   input capacitances per watt, Flyback at 85 Vac, 129, 130*f*
   LLC converters, 461–462, 476, 482, 487, 488, 490, 496, 498, 510
   real-world failure and, 33
   working voltage, 433
MATLAB, 2
McLyman, W. T., 303
Mean time between failure. *See* MTBF
Mean time to failure. *See* MTTF
Mean values:
   sine wave-related waveforms (lookup table), 57, 57*f*
   square wave-related waveforms (lookup table), 58, 58*f*, 59*f*
Measurement, thermal management and, 346–347
Median values:
   sine wave-related waveforms (lookup table), 57, 57*f*
   square wave-related waveforms (lookup table), 58, 58*f*, 59*f*
Metal-oxide semiconductor field-effect transistor (MOSFET), 7
   I-V curves, 8*f*
Microlinear, 150, 151
Mil-Hdbk 217F, 429
Mil-Hdbk 781A, 426
Miller effect, 462
Minimum duty cycle, Forward converter design, 316
MKS system, 195, 204
ML4826, 150, 151
mmf. *See* Magnetomotive force equation
Morey, Taylor, 303
MOSFET. *See* Metal-oxide semiconductor field-effect transistor
MTBF (mean time between failure), 425–429
MTTF (mean time to failure), 425
Multi-output Flyback converters, 175
Multiwinding inductors, 221, 222, 232, 275

N

N-switches, 69–70, 70*f*
National Semiconductor, 2, 7, 17, 37, 62
Natural convection. *See also* Thermal management
   at altitude, 355–356
   equations, 347–348
      available, 348–349
      comparing two standard equations, 350–351
      historical definitions, 348
      manipulation, 349–350
      tables of standard equations, 351–353, 352*t*–353*t*
   forced air cooling, 356–357
   introduction, 345
Negative-to-negative:
   Boost, 71, 82*f*, 84*f*
   Buck, 81*f*

Negative-to-negative-to-positive Buck-boost, 76
Negative-to-positive Buck-boost, 74, 76, 79t, 81f
Net Y-capacitance, 406
Nomogram, 269, 270f, 332
Nonsymmetric excitation, 202, 203, 204t
NPN bipolar junction transistor, 7, 9, 10, 69, 70, 73, 75
Number of primary turns, 310–314, 314f
Nusselt number, 351, 356
Nyleze, 272

O

Onion:
    efficiency onion, 447, 450
    regulation onion, 3, 4, 5, 6f, 23, 31
Optimal Design Methodology for LLC Resonant Converter, 474, 505
Optimum current density, 299–301
Optimum foil thickness, 282–285
Optocouplers, 76, 78, 391, 432, 441, 513
Output capacitor selection:
    Flyback converters, 331–332
    Forward converters, 326, 328, 328f, 329f
Output error, voltage divider, 23–25, 24f, 26f
Overcurrent protection circuits, 517–518, 517f
Overload, Flyback converters:
    margin, 121–122, 123f, 124
    protection at high-line, 193–194
Overstress tests, 430
Overtemperature protection, 518, 518f
OVP (overvoltage protection), 22

P

P-S winding arrangements, Forward converters, 319–323
P-switches, 69–70, 70f
Parallel resonance, LLC tank circuit, 483–484, 483f
Paralleling bridge rectifiers, 515f, 516
Paralleling of Buck converters, 95–100, 96f
Parasitic synchronous Boost, 417, 420
Parasitic synchronous Buck-boost, 417, 419
Pass regulator, 5
PCBs (printed-circuit boards), 335–358
    comparative tracking index, 344
    critical components, 337t
    FR-4, 343–344, 345, 353, 355
    ground plane, 341–343, 342f
    for heatsinking, 353, 355
    introduction, 335
    manufacturing issues, 343–344
    thermal management, 345–358
        efficiency estimates, 346–347
        measurement and, 346–347
        miscellaneous issues, 357–358
        radiative heat transfer, 357
        thermal resistance of cores, 358t
    trace analysis, 335–343
        Boost, 337f
        Buck-boost, 337f
        Buck converter, 336f, 337f
        inductance equations, 339, 339f
        miscellaneous points, 337–340
        routing current-sense trace, 341
        routing feedback trace, 341
        sizing copper traces, 340–341

PCBs (printed-circuit boards) (*Cont.*):
    transition temperature, 344
    vendors, 345
    voltage divider and, 31–32, 31f
PCDs (power conversion devices), 2. *See also* Switch-mode power supplies
PD Wire and Cable, 272
Peak current:
    Flyback optimization, 181–183, 182f
    inductor evaluation, 249t, 251–252
Peak flux density, inductor evaluation, 249t, 251
Peak magnetization current, Forward converters, 315
Pease, Bob, 2, 18
PFC (power factor correction) design, 144–159. *See also* Front-end design
    anti-synchronization techniques, 149–155
        capacitor current reduction, 150f
        over wide input range, 151
        RMC cap current improvement, 152f
    choke procedure design, 155, 155f, 156f
    defined, 145
    design table, 156t–157t
    nuances, 155–159
    PFC Boost stage, 145–146
        currents, 147f, 148f
        holdup time considerations, 148–149, 149f, 149t
        instantaneous duty cycle, 146, 148
        stress equations, 147f
    turn-on snubber, 518–519, 518f
Phase margin, 4, 363, 436
Phelps Dodge Corporation, 272
Piecewise linear method, 53–55, 54f
Plant transfer functions:
    Boost:
        CMC, 383f
        VMC, 375, 376f
    Buck:
        CMC, 378, 379f
        VMC, 373, 374f, 375
    Buck-boost:
        CMC, 384f
        VMC, 375, 377f
    defined, 373
Plated through-holes, 343, 344
Pollution degree, 431, 432, 435
Positive-to-negative Buck-boost, 74, 76, 79t, 81, 82f, 84f, 86
Positive-to-positive:
    Boost, 71, 75, 77, 80f
    Buck, 74, 77, 83f
Power/area product relation, 296–298
Power conversion devices. *See* PCDs
Power Delivery for Platforms with Embedded Intel® Atom™ Processor, 1
Power factor, 145
Power factor correction. *See* PFC
Power good signal, 517, 517f
Power Integrations, 2, 40, 42, 51, 333
Power scaling guidelines, 38, 51–52
Power supplies:
    power supply (noise) rejection ratio, 17
    qualifying, 429–431
    self-oscillating low-cost standby/auxiliary, 513–515, 514f

Power supplies (*Cont.*):
  SMPS:
    creativity and, 3
    design tightrope, 1–3
    integrated, 76, 77, 110, 169, 240, 337
    IPC-9592 standard, 2
    names for, 2
    as PCDs, 2
    taking for granted, 2
    testing, 429–431
    units, 2
Pressman, Abraham, 303–304, 373
Primary inductance, Flyback converters, 330
Primary-side leakage inductance term, 176–177, 176*f*
Primary turns:
  Flyback converter design, 330
  Forward converter design, 315
  number of, 310–314, 314*f*
Primary winding, Forward converters, 161, 163
Printed-circuit boards. *See* PCBs
Proportional to absolute temperature. *See* PTAT
Proprietary voltage references, 20
Protection circuits, overcurrent, 517–518, 517*f*
Proximity effect:
  Dowell's equations, 280–286, 299, 304, 319, 320, 322, 323
  introduction, 279
PSPICE, 18, 195, 461, 509, 511
PTAT (proportional to absolute temperature), 7–13
Pulse-skipping, 114, 191–193, 443
Pulse width modulation. *See* PWM
Push-Pull converter:
  diagram, 166*f*
  flux balancing, 167
PWM (pulse width modulation):
  PWM-thinking transition to resonant topologies, 465–473
  in voltage-mode control, 368*f*

Q

Qualifying power supplies, 429–431

R

r. *See* Current ripple ratio
Radiative heat transfer, 357
Ratio of inductances, LLC converters, 498, 502–503
Ratios:
  decibels to ratios, 361*t*
  ratios to decibels, 360*t*
RCD clamps, 183–184, 410*f*, 411–412, 411*f*
Reference voltage. *See* Voltage references
Regulation onion, 3, 4, 5, 6*f*, 23, 31
Regulators:
  aim of, 3
  dynamic load response, 3–5, 5*f*
  expectations for, 1
  output voltage rail requirements, 1
  regulation onion, 3, 4, 5, 6*f*, 23, 31
  shunt, 5–6, 521
  static load response, 3–5, 5*f*

Reliability:
  calculating, 429
  chargeable failures, 427–428
  chi-square distribution, 426–427
  DRT, 425, 427, 428, 429
  MTBF, 425–429
  MTTF, 425
  warranty costs, 428–429
Reset techniques:
  Flyback converters:
    lossless snubbers, 412–413, 412*f*
    RCD clamps, 183–184, 410*f*, 411–412, 411*f*
    zener clamp, 330, 409, 410*f*
  Forward converters:
    active clamp reset, 417–424
    introduction, 413–414
    tertiary winding, 161, 414–417, 415*f*, 417*t*
Resistive voltage divider, 16*f*
Resistors:
  commercial values, 26–30
  comparison of technologies, 27*f*
  custom, 26–27
  E series, 27–30, 27*f*, 30*f*
  voltage divider resistor tolerance, 25–26, 26*f*
Resonant tank circuits:
  LLC:
    gain (conversion ratio) of, 478*f*
    gain-phase relationships, 479–483, 479*f*, 481*f*
    introduction, 474, 476, 478–479
    parallel resonance, 483–484, 483*f*
    regulated voltage, 477*f*
    series resonance, 483–484, 483*f*
  series, 473–474, 473*f*, 475*f*, 476*f*
Resonant topologies (soft switching). *See also* LLC converters
  early attempts, 461
  fear of, 461–462
  hard switching compared to, 462–464, 463*f*
  introduction, 461–462
  key concerns (guiding criteria), 464–465
  Mammano on, 463, 476
  PWM-thinking transition to, 465–473
  resonant circuits, building, 465, 468–472
  square-wave switching power conversion, 461, 462
Reverse engineering tricks, Buck efficiency, 451
Reverse saturation current, 9
RHP zero, 83–85, 86*f*
Ridley, Ray, 38, 67, 394
Ripple. *See* Current ripple ratio
RMS values (root-mean-square):
  capacitor RMS currents, 56*f*, 61*t*
  common waveforms, 55, 56*f*, 57
  high-frequency component in bulk capacitor, 154
  low-frequency component in bulk capacitor, 153
  piecewise linear method, 53–55, 54*f*
  sine wave-related waveforms (lookup table), 57, 57*f*
  square wave-related waveforms (lookup table), 58, 58*f*, 59*f*

S

S-P-S winding arrangements, Forward converters, 323–325
Safety issues:
    capacitor life estimation, 434–440
    clearance distance, 431–432, 431f
    comparative tracking index, 344, 431
    creepage distance, 431–432, 431f
    pollution degree, 431, 432, 435
    total Y-capacitance, 440
    working voltage, 431, 432–434, 435
Saturation current, 9
Schottky diode:
    basic switching topologies and, 35
    Buck efficiency, 443
    FAN5340 lesson and, 90
Secondary-side leakage inductance term, 176f, 178–180
Secondary turns:
    Flyback converter design, 330
    Forward converter design, 316
Selecting inductors. *See* Inductor selection
Self-contained inrush protection circuit, 516, 516f
Self-induction (L):
    choosing, DC-DC converters, 242–243
    magnetics and, 196–197, 221–222
Self-oscillating low-cost standby/auxiliary power supply, 513–515, 514f
SELV-EL output, 175
Sensed current, 167, 365f, 378
SEPIC (single-ended primary inductance converter):
    building blocks, 101f
    configurations, 103f
    current waveforms, 104f, 105
    topologies, 103f
Series regulator, 5
Series resonance, LLC tank circuit, 483–484, 483f
Series resonant tank circuit, 473–474, 473f, 475f, 476f. *See also* LLC tank circuit
SG1524, 372, 373
SG3524, 372
Shape factors, 38, 67, 216–217, 259, 264, 485
Shockley's equation, 9
Shoot-through:
    C dV/dt induced turn-on, 91
    current spike, 90, 109
    defined, 2, 89
Shunt regulators, 5–6, 521
Simulation. *See also* Mathcad
    AC-AC converter design validation:
        step 1, 486–490
        step 2, 490–491
        step 3, 491–495
    bandgap references and, 18
    Pease and, 2, 18
    real-world failures and, 33
    SPICE models, 10, 11, 18
    theoretical PD design validation, 497–498, 499f, 500f, 501f
    Widlar and, 2
Simulation Program with Integrated Circuit Emphasis. *See* SPICE

Sine wave-related waveforms (lookup table), 57, 57f
Single-ended Forward converters:
    diagram, 162f, 166f
    without PFC, 135–137
Single-ended primary inductance converter. *See* SEPIC
600-V switches, Flyback converters, 190–191
Skin depth. *See also* Proximity effect
    Flyback converters, 267–269
    Forward converters, 279–286
Slave Buck-boost converter, 521–522, 522f
Slope compensation. *See also* Subharmonic instability
    average current problem, 370f
    required amount, 366, 369f
    ways of expressing, 367f
SMD, 338, 343, 344
SMPS. *See* Switch-mode power supplies
Snubber, turn-on, 518–519, 518f
Soft switching. *See* Resonant topologies
SPICE (Simulation Program with Integrated Circuit Emphasis), 10, 11
    Intusoft, 18
    PSPICE, 18, 195, 461, 509, 511
Square-related waveforms (lookup table), 58, 58f, 59f
Square-wave switching power conversion, 461, 462
ST Micro recommended equations, area product, 303
Stability. *See also* Control techniques
    current-mode controllers, 363–372
    loop stability, 359, 360f, 362f
Standard topologies. *See* Basic switching topologies
Standard wire gauge. *See* SWG
Standby/auxiliary self-oscillating low-cost power supply, 513–515, 514f
Static load response, 3–5, 5f
Steady-state ratio, 359. *See also* Transfer functions
Step-down. *See* Buck
Step-up. *See* Boost
Step-up/step-down. *See* Buck-boost
Stress computations:
    basic topologies, 52–68
    Boost, 115, 118, 120f
    Buck, 115–117, 118f, 119f, 120f
    Buck-boost, 115, 118, 121f
    composite topologies, 101, 105
    DCM, 114–115
        closed-form equations, Buck, 124t–125t, 126t, 127
        top-to-bottom calculation order, 111t–112t, 115
    PFC Boost stage, 147f
Stress spiders, 62, 63f, 64f, 65f
Sub-bandgap references, 17
Sub-volt bandgap references, 17
Subharmonic instability. *See also* Slope compensation
    defined, 363
    flattop approximation, 68
    generalized rule for avoiding, 371–372, 372f
    high inductances, 67

Subharmonic instability (*Cont.*):
  mimicking, large inductances, 365*f*
  terms for, 363
SWG (standard wire gauge):
  AWG compared to, 272–273, 273*f*
  copper wire diameters, 267*t*
  skin depth, 267–269
Swinging node. *See* Switching node
Switch-mode power supplies (SMPS, switchers):
  creativity and, 3
  design tightrope, 1–3
  integrated, 76, 77, 110, 169, 240, 337
  IPC-9592 standard, 2
  names for, 2
  as PCDs, 2
  taking for granted, 2
Switchers. *See* Switch-mode power supplies
Switching losses, synchronous Buck, 465–467, 466*f*
Switching node, 35
Switching Power Supplies A-Z, 2d ed.:
  coupled inductors, 98, 167
  droop positioning, 32
  input current waveshape, 131, 131*f*
Switching Power Supply Design (Pressman, et al.), 303
Switching topologies:
  basic:
    composite topologies compared to, 35
    configurations, 71–74
    energy relationships of, 42–45
    explanation of, 34*f*, 35
    nuances, 83–86
    stress computations, 52–68
    wide input voltage, 59–61
  composite:
    basic topologies compared to, 35
    building blocks of, 100, 101*f*
    stress computations, 101, 105
Symmetric converters, area product, 307–308
Symmetric excitation, 202, 202*f*, 203, 204*t*, 205, 305, 308, 503
Symmetrical Boost topology, 520–521, 521*f*
Synchronization techniques, 149–151. *See also* Anti-synchronization techniques
Synchronizing two 3844 ICs, 513
Synchronous Buck:
  modes of operation, 88–90, 89*f*
  switching losses, 465–467, 466*f*
Synchronous DC-DC converters, 87–95
System ground. *See* Ground (system ground)

T

Tapped-inductor converters, 231–237
  Boost, 236, 236*f*, 237*f*
  Buck, 232–236, 233*f*
Tempcos (temperature coefficients):
  defined, 3, 20
  instantaneous, 22
  specifying and interpreting, 20–22
Temperature coefficients. *See* Tempcos
Temperature rise, inductor evaluation, 249*t*, 252–253
Tertiary winding (energy recovery winding), 161, 163, 171, 414–417, 415*f*, 417*t*

Tertiary winding diode, 165, 166*f*
Testing:
  burn-in, 430
  DRT, 425, 427, 428, 429
  ESS, 430
  HALT, 430
  HASS, 430
  HAST, 430
  overstress, 430
  power supplies, 429–431
  thermal ESS, 430
  vibration, 344, 430, 431, 432
Texas Instruments:
  Forward converter design, 40-W, 299
  National Semiconductor, 2, 7, 17, 37, 62
  Unitrode Corp., 167, 299, 373
Thermal ESS test, 430
Thermal expansion, coefficient of, 344
Thermal management, 345–358. *See also* Natural convection; PCBs
  efficiency estimates, 346–347
  measurement and, 346–347
  miscellaneous issues, 357–358
  radiative heat transfer, 357
  thermal resistance of cores, 358*t*
Thermal overstress test, 430
Thermaleze, 272
Thermocouples, 270, 346, 356, 429, 438
Thermodynamic theory, h and, 351
Things to try:
  active inrush protection circuit, 519, 519*f*
  adapter with battery charge function, 515, 515*f*
  Boost preregulator with regulated auxiliary output, 522–523, 522*f*
  cheap power good signal, 517, 517*f*
  floating Buck, 520, 520*f*
  floating drive from 384x controller, 519, 520*f*
  overcurrent protection circuits, 517–518, 517*f*
  overtemperature protection, 518, 518*f*
  paralleling bridge rectifiers, 515*f*, 516
  self-contained inrush protection circuit, 516, 516*f*
  self-oscillating low-cost standby/auxiliary, 513–515, 514*f*
  slave Buck-boost converter, 521–522, 522*f*
  symmetrical Boost topology, 520–521, 521*f*
  synchronizing two 3844 ICs, 513
  turn-on snubber for PFC, 518–519, 518*f*
384 controller, floating drive from, 519, 520*f*
384x series, overtemperature protection, 518, 518*f*
TI/Unitrode recommended equations, area product, 302
TL431 with Opto, design examples, 391, 394–397, 394*f*
Toroidal core:
  air gap's effect, 208–210
  E-core mapped in toroid, 208*f*
  effective area and effective length, 206–207, 207*f*
  mmf equation, 205–206, 205*f*
Total Y-capacitance, 440
Trace analysis, 335–343
  Boost, 337*f*
  Buck-boost, 337*f*
  Buck converter, 336*f*, 337*f*

Trace analysis (*Cont.*):
  inductance equations, 339, 339f
  miscellaneous points, 337–340
  routing current-sense trace, 341
  routing feedback trace, 341
  sizing copper traces, 340–341
Trade secret protections, LLC converters, 461
Transfer functions:
  error amplifiers, 380f
  plant transfer functions:
    Boost CMC, 383f
    Boost VMC, 375, 376f
    Buck-boost CMC, 384f
    Buck-boost VMC, 375, 377f
    Buck CMC, 378, 379f
    Buck VMC, 373, 374f, 375
    defined, 373
  steady-state ratio, 359
Transfer resistance, 367f
Transformer and Inductor Design Handbook (McLyman), 303
Transformer model, Flyback converters, 222–226
Transformer reset, 163, 164, 168, 414
Transformer reset techniques, 409–424
  Flyback converters:
    lossless snubbers, 412–413, 412f
    RCD clamps, 183–184, 410f, 411–412, 411f
    zener clamp, 330, 409, 410f
  Forward converters:
    active clamp reset, 417–424
    introduction, 413–414
    tertiary winding, 161, 414–417, 415f, 417t
Transition temperature, PCBs, 344
Trapezoid, EMI filter design, 403–404
Trial and error, LLC converters, 461
Trimpot, 346
Tuinenga, Paul, 18
Turn-on snubber, for PFC, 518–519, 518f
Turns ratio calculation:
  Flyback converters, 330
  Forward converters, 277, 315
Type 1/type 2 ICs:
  applications, 77–79, 78t, 79t
  compared, 75–76
  described, 74–75, 74f
Type A cell, configurations, 79, 82f, 83f, 84f
Type B cell, configurations, 79, 80f, 81f

U

UC3842/3844 series, 193, 194, 363, 364, 370, 441, 513, 518
UC3854, 151, 157, 158
UC3854A/B, 157
Undervoltage lockout. *See* UVLO
Ungapped core, 256
Units conversion, magnetics, 204–205, 205t
Units conversions, power supplies, 2
Universal-input Flybacks. *See* Flyback converters
Utilization factor, cores, 310t, 311t, 312t, 313t
UVLO (undervoltage lockout), 22
UVP, 513

V

Validation. *See also* LLC converters

Validation (*Cont.*):
  AC-AC converter:
    step 1, 486–490
    step 2, 490–491
    step 3, 491–495
  PD design, theoretical, 497–498, 499f, 500f, 501f
van Wyk, Jacobus D., 474, 505
Vendors, PCB, 345
Via, 163, 338, 339
Vibration testing, 344, 430, 431, 432
VMC. *See* Voltage-mode control
Voltage:
  input voltage:
    variance, Buck-boost, 118–119, 122f
    wide, 59–61
  Kirchhoff's voltage law, 465, 473
  ratings:
    Flyback converter design, 330–331
    Forward converter design, 316
  VRE3050A voltage reference, 6, 7, 20f, 23
  wide input voltage:
    anti-synchronization techniques, 151
    basic converters, 59–61
  working, 431, 432–434, 435
Voltage-dependent equation (Faraday's law):
  defined, 196
  Flyback converter design, 255–256
  forms, 203, 255
  inductor equation, 198, 241
  magnetics, 200–204, 204t
  practical form, 255–256
Voltage divider:
  constraints from error amplifier type, 30–31, 30f
  external, 6f, 23
  input bias current, 23–25
  output error, 23–25, 24f, 26f
  PCB placement, 31–32, 31f
  resistive, 16f
  resistor tolerance, 25–26, 26f
  useful equations, 24f
Voltage-doubler circuit, single-ended Forward converter, 135, 135f
Voltage-independent equation, 198–200, 204t, 228–229. *See also* Voltage-dependent equation
Voltage-mode control (VMC). *See also* Current-mode control; Hysteretic control
  CMC *versus*, 378
  current-mode converters stability, 363
  design examples, 388–391
  feedforward in, 366, 368f
  hysteretic control and, 386f
  PWM in, 368f
  summary of pros and cons, 388, 388t
  synchronous Buck converters, 91–92
Voltage references. *See also* Bandgap references; Zener diodes
  bit accuracy, 19, 22–23
  comparing, 19–20, 20f
  proprietary, 20
  understanding, 5–6
  VRE3050A, 6, 7, 20f, 23
  XFET, 20
Voltage regulators. *See* Regulators

Volt-seconds:
  mapping the inductor, 241–242
  volt-seconds law, 35, 36, 50, 55f, 94, 95, 163, 234
VRE3050A voltage reference, 6, 7, 20f, 23

W

Warranty costs, 428–429
Waveform analysis, 52–68
  capacitor currents, shape, 58–59
  common waveforms, 55, 56f, 57
  piecewise linear method, 53–55, 54f
  sine wave-related waveforms (lookup table), 57, 57f
  square-related waveforms (lookup table), 58, 58f, 59f
Wide input voltage:
  anti-synchronization techniques, 151
  basic converters, 59–61
Widlar, Bob, 2, 7, 13, 15, 16, 17, 18
Widlar bandgap reference cell, 13, 15f, 16f
Widlar's Leap, 7
Window areas, core:
  finer divisions, 295–296, 295f
  margin tape sizes, 310t, 311t, 312t, 313t
  numerical example, 296f
Wire gauges. *See also* Copper wire; Proximity effect
  AWG:
    ASTM B-258 specification, 271
    copper wire diameters, 267t
    current carrying capacity, 265–266
    diameter of coated wire, 271–272, 272f
    different cmil/A, 268t
    magnet wire (bundled/Litz), 266, 282, 283, 285, 286, 316

Wire gauges, AWG (*Cont.*):
    selection issues, 266
    skin depth, 267–269
  SWG:
    AWG compared to, 272–273, 273f
    copper wire diameters, 267t
    skin depth, 267–269
Working voltage, 431, 432–434, 435

X

X-capacitance, 405, 406, 407, 440
XFET voltage reference, 20
Xicor, FGA technology, 20

Y

Y-capacitance, 406, 440

Z

z. *See* Air-gap factor z
Zener clamp, 330, 409, 410f
Zener diodes. *See also* Bandgap references
  bandgap references compared to, 5–6, 19–20, 20f
  buried, 6, 19, 20, 21
  described, 5–6
  safety issues, 440–442
Zener dissipation, Flyback optimization, 180–181, 180f
Zeta converters:
  building blocks, 101f
  configurations, 103f
  current waveforms, 104f, 105
  topologies, 103f

Adding value to your research

The IET is Europe's largest professional body of engineers with over 150,000 members in 127 countries and is a source of essential engineering intelligence.

We facilitate the exchange of ideas and promote the positive role of science, engineering and technology in the world. Discover the IET online to access:

- 400 eBooks
- 26 internationally renowned research journals
- 1,300 conference publications
- over 70,000 archive articles dating back to 1872
- Inspec database containing over 13 million searchable records

To find out more please visit: **www.theiet.org/books**

The Institution of Engineering and Technology is registered as a Charity in England & Wales (no 211014) and Scotland (no SC038698). The IET, Michael Faraday House, Six Hills Way, Stevenage, SG1 2AY, UK.

CPSIA information can be obtained
at www.ICGtesting.com
Printed in the USA
LVOW04*2022190318
570377LV00018B/155/P

9 780071 798143